2nd International School and Conference "Saint-Petersburg OPEN 2015" on Optoelectronics, Photonics, Engineering and Nanostructures

Journal of Physics: Conference Series Volume 643

St. Petersburg, Russia
6 - 8 April 2015

ISBN: 978-1-5108-1470-7
ISSN: 1742-6588

Printed from e-media with permission by:

Curran Associates, Inc.
57 Morehouse Lane
Red Hook, NY 12571

Some format issues inherent in the e-media version may also appear in this print version.

Printed by Curran Associates, Inc. (2016)

For permission requests, please contact the Institute of Physics at the address below.

Institute of Physics
Dirac House, Temple Back
Bristol BS1 6BE UK

Phone: 44 1 17 929 7481
Fax: 44 1 17 920 0979

techtracking@iop.org

Additional copies of this publication are available from:

Curran Associates, Inc.
57 Morehouse Lane
Red Hook, NY 12571 USA
Phone: 845-758-0400
Fax: 845-758-2634
Email: curran@proceedings.com
Web: www.proceedings.com

Table of contents

Volume 643

2nd International School and Conference Saint-Petersburg OPEN on Optoelectronics, Photonics, Engineering and Nanostructures (SPbOPEN2015)
6–8 April 2015, St Petersburg, Russia

Papers

Invited Speakers

Crystal Growth

Optoelectronics and Lasers

Photonics

Quantum Dots and Wells

Dear Colleagues,

The 2nd International School and Conference "Saint Petersburg OPEN 2015" on Optoelectronics, Photonics, Engineering and Nanostructures was held on April 6 – 8, 2015 at St. Petersburg Academic University. The School and Conference included a series of invited talks given by leading professors with the aim to introduce young scientists with actual problems and major advances in physics and technology. The keynote speakers were

Mikhail V. Maximov (Ioffe Physico–Technical Institute RAS, Russia)

Vladimir G. Dubrovskii (St. Petersburg Academic University and St. Petersburg State University, Russia)

Anton Yu. Egorov (JSC Connector Optics, Russia)

Victor V. Luchinin (St. Petersburg State Electrotechnical University, Russia)

Vladislav E. Bugrov (St. Petersburg University of Internet Technologies, Mechanics and Optics, Russia)

Vitali A. Schukin (VI Systems, Germany)

Yuri P. Svirko (University of Eastern Finland, Finland)

During the poster session all undergraduate and graduate students attending the conference presented their works. A sufficiently large number of participants, with more than 170 student attendees from all over the world, allowed the Conference to provide a fertile ground for fruitful discussions between the young scientists as well as to become a perfect platform for valuable discussions between student authors and highly experienced scientists. The best student papers, which were selected by the Program Committee and by the invited speakers basing on the theses and their poster presentation, were awarded with diplomas of the conference – see the photos.

This year "Saint Petersburg OPEN 2015" is organized by St. Petersburg Academic University in cooperation with Peter the Great St. Petersburg Polytechnic University. The School and Conference is supported by Russian Science Foundation, SPIE (The International Society for Optics and Photonics), OSA (The Optical Society) and by Skolkovo Foundation. It is a continuation of the annual schools and seminars for youth on topical problems of physics and technology that are organized by the Academic University since 2009. We invite all the students and young scientists to attend **"Saint Petersburg OPEN" in 2016**!

Please, find details at *http://spbopen.spbau.com/*

With best wishes, Editorial Board,

Program and Organizing Committees

Organizers

St. Petersburg Academic University

Peter the Great St. Petersburg Polytechnic University

St. Petersburg Academic University RAS SPIE Student Chapter

Foundation for support of education and science (Alferov's foundation)

Acknowledgments

The Optical Society

The International Society for Optics and Photonics

Skolkovo Foundation

Opening remarks from the head of Program Committee, rector of St. Petersburg Academic University, Laureate of Nobel Prize in Physics (2000), academician *Zhores I. Alferov*

The head of Organizing Committee, Professor *Alexey Zhukov*

Poster Session

Journal of Physics: Conference Series **643** (2015) 011001 doi:10.1088/1742-6596/643/1/011001

Award ceremony

Prof. Alexey Zhukov and Prof. Vladimir Dubrovskii with students

Closing remarks from Prof. Dr. Vladimir G. Dubrovskii and

Prof. Dr. Alexey E. Zhukov

Peer review statement

All papers published in this volume of *Journal of Physics: Conference Series* have been peer reviewed through processes administered by the proceedings Editors. Reviews were conducted by expert referees to the professional and scientific standards expected of a proceedings journal published by IOP Publishing.

Novel concepts for designing semiconductor lasers

V A Shchukin[1,2], N N Ledentsov[1,2], M V Maximov[2,3], N Yu Gordeev[3,2], Yu M Shernyakov[2,3], A S Payusov[2,3] and A E Zhukov[2]

[1] VI Systems GmbH, Hardenbergstr. 7, Berlin 10623, Germany
[2] St Petersburg Academic University, 8/3 Khlopina st., 194021, St Petersburg, Russia
[3] Ioffe Institute, 26 Polytechnicheskaya, St.Petersburg, 194021, Russia

E-mail: vitaly.shchukin@v-i-systems.com

Abstract. We review novel concepts and demonstrate recent experimental data for edge-emitting semiconductor lasers with broad vertical waveguide. The ultimate case for waveguide extension in the vertical direction can be implemented by using the Tilted Wave Laser (TWL) approach. A TWL is composed of a thin active waveguide (typically 0.3-2 μm) optically coupled to a thick passive waveguide (10-150 μm). A TWL with a 26 μm-thick passive waveguide demonstrated low internal loss of 1.4 cm^{-1}, maximum pulsed power 18 W and maximum CW power 4.7 W. Vertical far field of the TWL consists of two tilted narrow lobs of 2 degrees full width at half maximum each.

1. Introduction

Development of single mode edge-emitting lasers with ultra thick and ultra broad waveguide is very important to achieve higher powers from one chip before reaching catastrophic optical mirror damage (COMD) level, reduce thermal effects and optical nonlinearities, facilitate coupling into optical fibers and simplify designs of various external-cavity systems [1, 2]. A number of different approaches for waveguide engineering have been proposed to extend the lasing mode and reduce beam divergence in edge emitting lasers [3-8]. Figure 1 represents schematically the main concepts of the laser diodes with ultra-thick vertical waveguides.

Large optical cavity laser in the frame of conventional approach is shown in Fig. 1(a). Small refractive index step between the cavity and cladding layers allows selection of the fundamental vertical optical mode. A moderately asymmetric positioning of the active region off–center in the cavity may improve the vertical mode discrimination by suppressing lasing of the high order optical modes with n=3, 5,….. Optical losses less than 0.4 cm^{-1} were reported for thick undoped waveguide laser diodes with vertical beam divergence 30° FWHM [3]. Once significantly lower beam divergence is targeted (8° FWHM was achieved in the off–centre gain approach [5]), a large part of the optical cavity above the active region must be p–doped. Then a significant part of the optical wave is located in the p–doped region which implies higher modal losses due to free carrier absorption.

An alternative concept [5, 6] employs lasers based on vertical photonic band crystals – a periodic sequence of high and low refraction index layers (Fig. 1(b)). The active region is placed asymmetrically, and only a small part of the waveguiding structure is p–doped. All vertical optical modes have different profiles in the vertical direction and the n–doped photonic band crystal provides the selection of these modes. The proper configuring of the photonic band crystal (PBC) gives

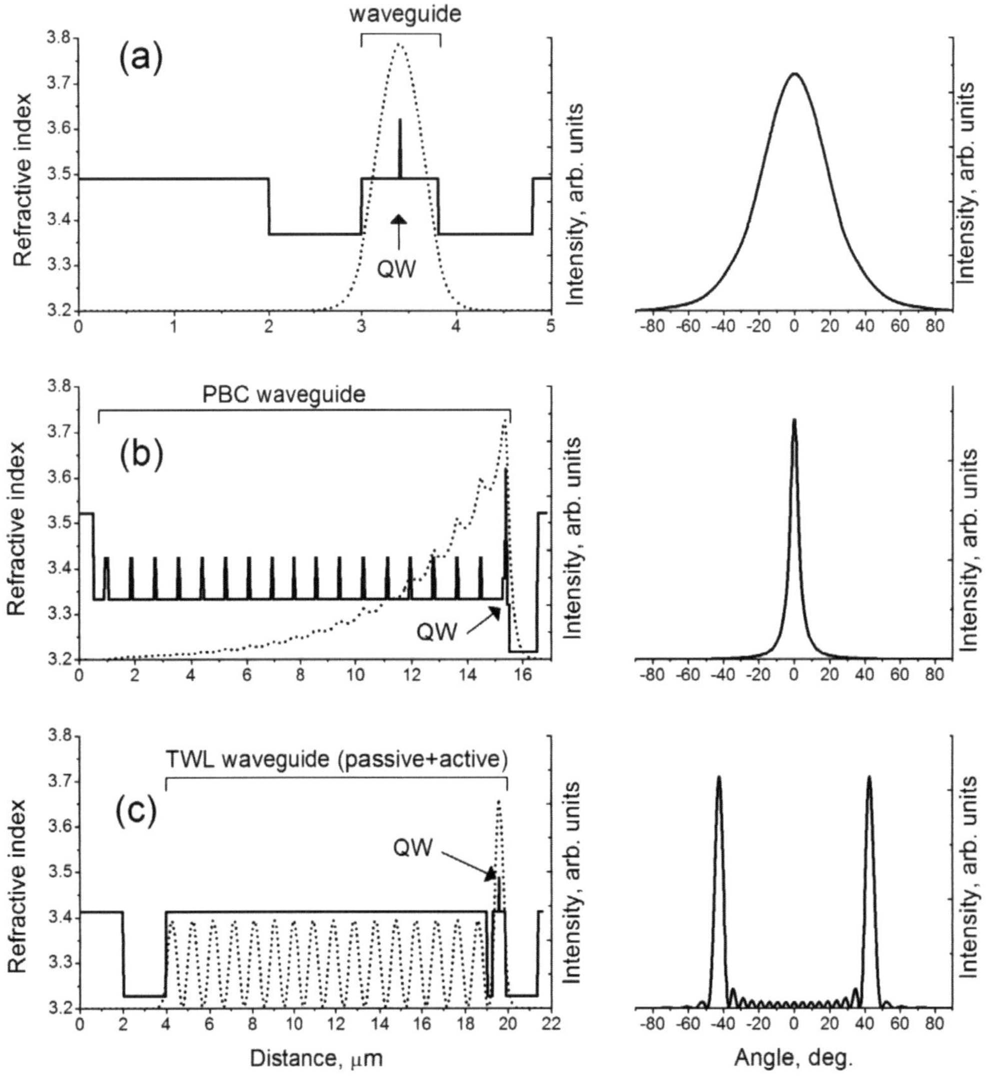

Figure 1. Schematic view of edge–emitting diode lasers with thick vertical waveguides and corresponding far field patterns: (a) large optical cavity laser, (b) laser based on a vertical photonic band crystal, (c) tilted wave laser.

preference for the fundamental vertical optical mode and discriminates high–order vertical optical modes by decreasing their optical confinement factor and/or increasing the leakage losses to the substrate. The narrow vertical beam of 5° FWHM was reached in this approach [6] while the field intensity was by far concentrated in the n–doped all–epitaxial PBC waveguide.

The ultimate case for waveguide extension in the vertical direction can be implemented by using the Tilted Wave Laser (TWL) approach [7] exploited in the present work. TWLs are based on a thin active waveguide containing quantum wells or quantum dots and an optically coupled thick passive waveguide. The lasing optical modes in the coupled waveguides are formed due to the leakage of a part of light from the active waveguide, its propagation through the thick passive waveguide in a form of a tilted wave, its return back to the active waveguide, and interference with the light amplified in the active medium. If a transparent substrate is employed as a passive waveguide, its thickness may be as high as 130 μm. The 130 μm–thick substrate–based TWLs [8] demonstrated diffraction limited vertical lobes as narrow as 0.65 degree full width at half maximum (FWHM). The disadvantage of this approach, however, is the need to use wafer backside polishing in combination with two–sided processing to fabricate low loss dielectric mirrors on the substrate side. In this paper we present recent results on high power TWL laser with 26 μm -thick epitaxial waveguide.

2. Experiment

The laser epi-structure was grown by MOCVD on GaAs substrate at low pressure by using the laboratory installation with horizontal reactor. The active region contained four compressively strained InGaAs QWs with the photoluminescence maximum at room temperature around 1040 nm. No strain compensation layers were used. Separation between neighbouring QWs was 70 nm to prevent defect formation. The thickness of the GaAs active waveguide was 700 nm. The thickness of the passive waveguide was 26 μm and its doping level was $1\cdot10^{17}$ cm^{-3}. The thickness of the intermediate cladding layer between the thin active and the thick passive waveguides was chosen as 110 nm.

The grown wafer was processed into 50 μm–wide broad area lasers, cleaved to bars, and then separated to single chips. The chips were mounted p-side down on copper heat-sinks using indium solder. The devices were characterized in pulsed (300 ns, 1 kHz) and continuous-wave (CW) modes.

3. Results

Figure 2a shows the dependence of threshold current density and lasing wavelength on the cavity length. For a 2 mm long cavity, the threshold current density is about 500 A/cm^2 (125 A/cm^2 per one QW layer). The lasing wavelength is 1039 nm. For the shortest cavity studied (400 μm), the threshold current density is 1300 A/cm^2 and the lasing wavelength is 1030 nm.

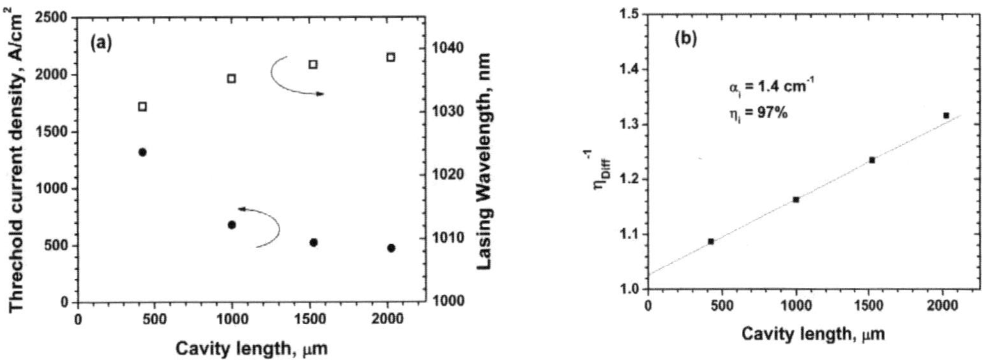

Fig. 2 Threshold current density, lasing wavelength (a) and reciprocal differential quantum efficiency (b) vs cavity length for 50 μm wide TWLs.

Fig. 2b shows the dependence of the reciprocal differential quantum efficiency ($1/\eta_{Diff}$) on the cavity length. Using the well-known equation $\eta_{Diff} = \eta_i\, \alpha_m / (\alpha_m + \alpha_i)$ we derived the internal differential quantum efficiency η_i and internal loss α_i which are equal to 97% and 1.4 cm^{-1}, respectively. Taking into account the large thickness of the doped passive waveguide the internal loss is sufficiently small. As a result, the differential efficiency for a 1.5 mm long device is as high as 81%.

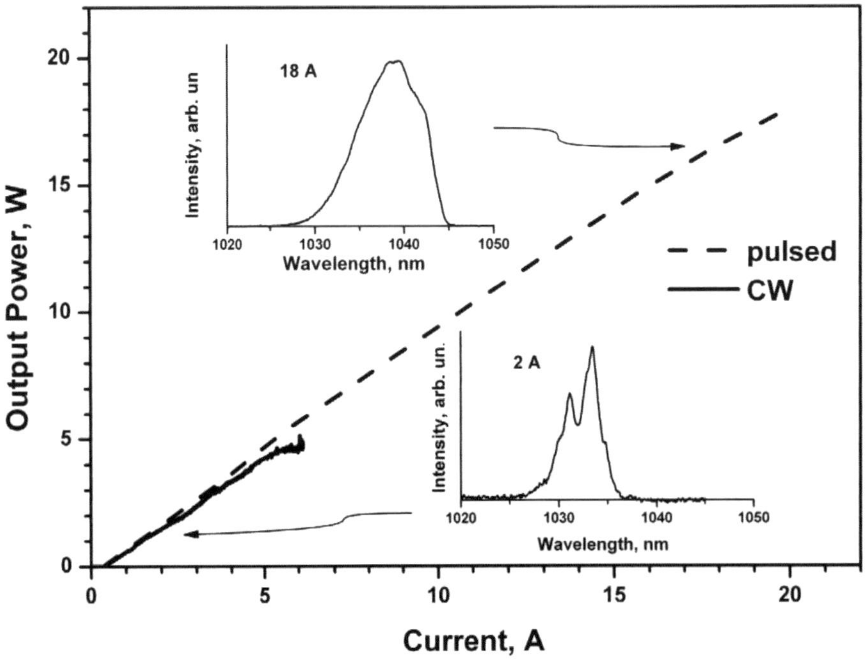

Fig. 3. Light-current characteristics in pulsed (a) and CW (b) modes measured for a 1.5-mm long and 50 μm wide TWL laser at room temperature. Insert: lasing spectra taken at various injection current.

The light-current characteristics were studied in pulsed and CW modes at room temperature. For 1.5 mm long and 50 μm–wide broad area devices, the maximum pulsed power is about 18 W limited by available current source. CW power exceeds 4.7 W without noticeable thermal roll-over (Fig. 3). This value is likely limited by our laboratory technology of laser mounting on heat sinks. The lasing wavelength is about 1035 nm.

The vertical far field pattern of TWLs consists of two tilted narrow beams and angularly broad emission between them (Fig. 4), which originates from a fraction of the light confined in the narrow waveguide. The FWHM of the lobes is 2°, which corresponds very well to the results of our simulations. No shift and broadening of the vertical lobes were observed up to the high injection current densities. The double–lobed emission pattern may be advantageous for certain applications. For example, one lobe can be used for wavelength stabilization or tuning, whereas the other one for light outcoupling. Such external cavity design will reduce parasitic interference effects characteristic to single-lobe based external resonators. In another approach, the tilted beams may be used for the vertical coupling of different laser bars into a coherent laser stack.

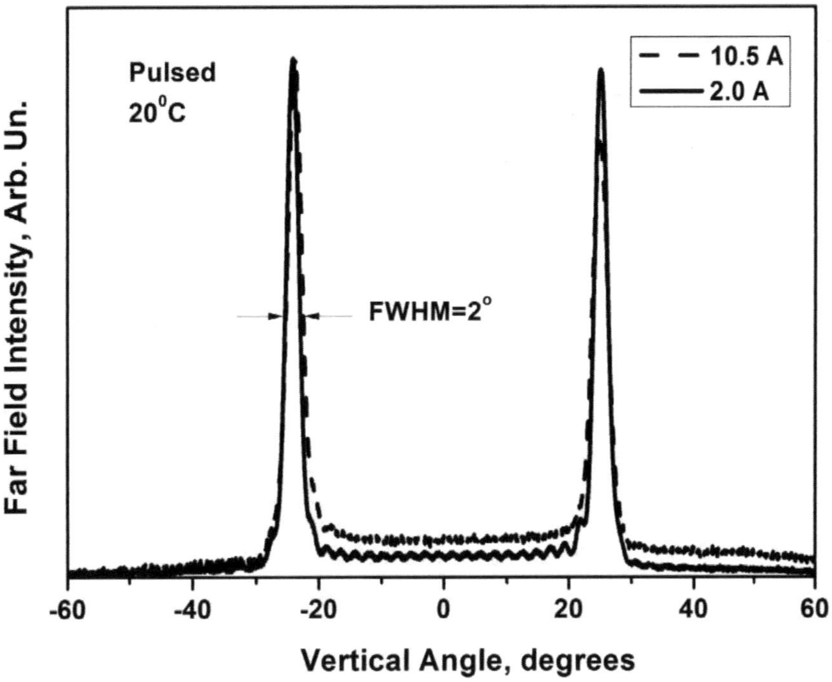

Fig. 4. Vertical far field intensity profiles at two different drive currents in the pulsed mode.

4. Conclusions

Novel device designs to achieve high performance lasing applying ultra–thick waveguides are described. The thickest vertical waveguides (up to 130 μ) can be implemented by using Tilted Wave Laser concept. Broad-area Tilted Wave Lasers with very thick waveguide (26 μm) show low internal loss of 1.4 cm^{-1} and high differential efficiency of 81% for a 1.5 mm long cavity with as-cleaved facets. The maximum pulsed power is about 18 W and maximal CW power exceeds 4.7 W. The lasers operate on a high-order single vertical mode and most of the output power is concentrated in two narrow vertical beams, each 2° FWHM in a good agreement with the theory. One of the lobes can be used for wavelength stabilization/tuning or for coherent coupling of individual devices into arrays. A very large spot size and single vertical mode operation makes TWLs particular promising candidate for applications in external cavity systems.

Acknowledgements

The work was supported by the Russian Science Foundation under Agreement 14-42-00006.

References

[1] Bachmann F, Loosen P and Poprawe R 2007 *High Power Diode Lasers. Technology and Applications* (New York: Springer)

[2] Hecht J 2011 DARPA's photonics projects – at the edge of possibility *Laser Focus World* **47** 41–44

[3] Vinokurov D A, Zorina S A, Kapitonov V A, Murashova A V, Nikolaev D N, Stankevich A L, Khomylev M A, Shamakhov V V, Leshko A Yu, Lyutetskii A V, Nalyot T A, Pikhtin N A, Slipchenko S O, Sokolova Z N, Fetisova N V and Tarasov I S 2005 *Semiconductors* **39** (3) 388-392
[4] Pietrzak A, Crump P, Wenzel H, Erbert G, Bugge F and Tränkle G 2011 IEEE J. *Select. Topics in Quant. Electron.* **17** (6) 1715–22
[5] Ledentsov N N and Shchukin VA 2002 SPIE *Optical Engineering* **41** 3193–3203
[6] Novikov I I, Gordeev N Yu, Shernyakov Yu M, Kiselev Yu Yu, Maximov M V, Kop'ev P S, Sharon A, Duboc R, Arbiv D B, Ben-Ami U, Shchukin V A and Ledentsov N N 2008 *Appl. Phys. Lett.* **92** (10) 103515
[7] Ledentsov N N, Shchukin V A and Bimberg D 2006 Merging Nanoepitaxy and Nanophotonics *Future Trends in Microelectronics: Up the Nano Creek* ed S Luryi, J Xu and A Zaslavsky (New York: Wiley & Sons) 401–432
[8] Ledentsov N N, Shchukin V A, Novikov I I, Gordeev N Yu, Maximov M V, Shernyakov Yu M, Pausov A S, Posilovic K, Kettler T, Bimberg D, Bugge F and Weyers M 2011 *Electron. Lett.* **47** 1339–40

Room temperature continuous wave operation of injection quantum dot microdisk lasers

N V Kryzhanovskaya[1], M V Maximov[1,2], A M Nadtochiy[1], S A Blokhin[2], M A Bobrov[2,3], M M Kulagina[2], S I Troshkov[2], Yu M Zadiranov[2], A A Lipovskii[1,3], E I Moiseev[1], Yu V Kudashova[1], V M Ustinov[2], A E Zhukov[1,3]

[1] Saint Petersburg National Research Academic University of RAS, St Petersburg, Russia
[2] A F Ioffe Physical-Technical Institute of RAS, St Petersburg, Russia
[3] Peter the Great St Petersburg Polytechnic University, St Petersburg, Russia

E-mail: kryj@mail.ioffe.ru

Abstract. We present results of characterization of injection microdisk lasers with InAs/InGaAs self-organized quantum dot active region grown on a GaAs substrate. The minimal microdisk diameter is 15 μm. The microlasers operate in continuous wave regime at room temperature without external cooling. Lasing wavelength is around 1.26…1.27 μm, minimal threshold current is 1.6 mA (threshold current density 900 A/cm^2). Specific thermal impedance is estimated to be $5 \cdot 10^{-3}$ °C•cm^2/W.

1. Introduction

Microdisk lasers are considered as candidates for development of planar optoelectronic circuits for on-chip optical interconnect systems (see, e.g. [1]). Owing to total internal reflection of light traveling at periphery of a microdisk resonator, very high quality factor can be achieved even in resonators several order of magnitude smaller as compared to conventional edge-emitting (macro)lasers [2]. Microdisk lasers operating at room under injection pumping in continuous wave (CW) regime are highly desired for practical usage. However, nonradiative surface recombination at microresonator sidewalls becomes an issue as the resonator size is reduced down to tens of microns. Since the first demonstration [3], most of microdisk diode lasers rely upon InP-based material system, which is characterized by low surface loss, and a quantum well active region.

Meanwhile, self-organized quantum dots (QDs) deposited on a GaAs substrate may provide characteristics very promising for their application in microlasers. In particular, edge-emitting QD lasers have reached the record-low threshold current density less than 10 A/cm^2 [4]. One property that makes QDs especially attractive for microlaser applications is lateral carrier transport suppression in QD arrays as compared to a two-dimensional quantum well structures [5]. Another advantage of InAs/InGaAs/GaAs QDs is their ability of emitting in the 1.3 μm spectral band [6] that corresponds to a transparency window of silicon- or silicon-germanium-based planar waveguides as well as standard silica optical fibers. Lasing operation of a QD microlaser as small as 2 μm has been recently reported under optical pumping at 100°C with a lasing wavelength of 1.3 μm [7]. At the same time, characteristics of injection QD microdisks reported so far are less optimistic. Room temperature operation has been reported for a 6.5 μm microdisk emitting at 1.06 μm [8], whereas the laser emitting at longer wave of 1.23 μm has much larger diameter of 80 μm [9].

In this work we report on successful realization of QD microdisk lasers emitting at 1.26…1.27 μm in CW regime at room temperature.

2. Experimental

Epitaxial structure was grown by molecular beam epitaxy on an *n*+ GaAs(100) substrate. Ten layers of $InAs/In_{0.15}Ga_{0.85}As$ QDs were deposited in the middle of a 0.44 μm thick GaAs waveguiding layer confined with $Al_{0.25}Ga_{0.75}As$ claddings. Microdisk resonators were formed by means of chemical plasma etching to have a diameter of 15…31 μm. Etch depth was about 7 μm. AgMn/NiAu (AuGe/Ni/Au) metallization was used to form ohmic contacts to *p*+ GaAs cap layer (*n*+ substrate, respectively). A micrograph of an array of microdisk lasers of different diameter is shown in Fig. 1. Lasers were mounted on a coper heatsink and tested at room temperature without external cooling. Light was collected with a microlens-tipped fiber and analysed with an optical spectrum analyser (spectral resolution 0.1 nm).

Figure 1. Scanning electron microscope image of an array of microdisk lasers of different diameters

3. Results and discussion

All the microdisk lasers demonstrate well pronounced turn-on behaviour of current-voltage dependence (example is shown in Fig. 2). Series resistance is scaled as D^{-2}; specific series resistance is estimated to be about 10^{-4} Ohm•cm^2 (see insert of Fig. 2).

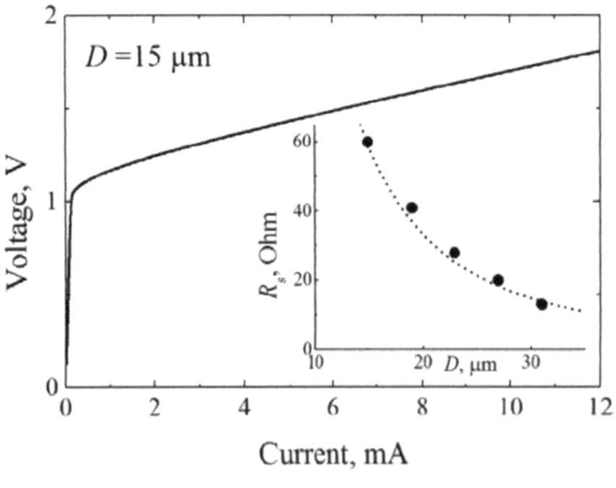

Figure 2. Current-voltage curve taken for a D = 15 μm microdisk. Insert: Series resistance as a function of diameter. Fit: ρ_S=1.05•10^{-4} Ohm•cm^2.

Integrated light intensity against bias current reveals a knee on the curve which is a characteristic of a lasing threshold (Fig. 3). The smallest threshold current measured at room temperature is 1.6 mA in a microlasers with diameter of 15 μm. Threshold current *vs* diameter dependence can be fitted well using a constant threshold current density of 900 A/cm^2.

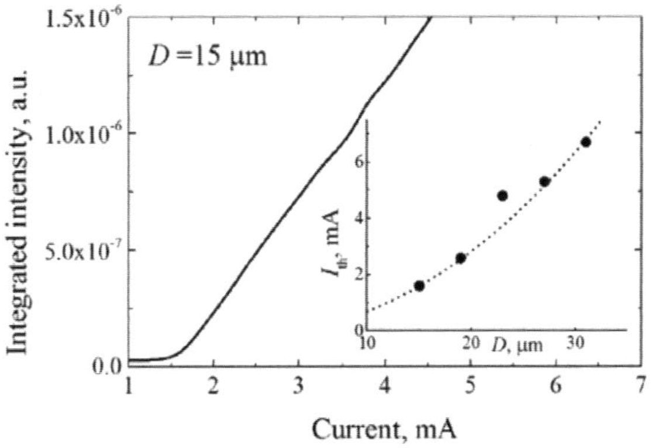

Figure 3. Light-current curve taken for a $D = 15$ μm microdisk. Insert: Threshold current as a function of diameter. Fit: $J_{th} = 900$ A/cm^2.

Emission spectra contains a dominant lasing line, which corresponds to a resonant whispering gallery mode (WGM), and one or few less intensive side lines of WGMs having other azimuthal orders (Fig. 4). In lasing regime the dominant line is typically two orders of magnitude more intensive as compared to other WGMs. In different microdisks studied the dominant line wavelength is varied in the 1.26…1.27 μm spectral interval. To the best of our knowledge, this is the longest lasing wavelength ever reported for injection QD microdisk lasers on GaAs substrates.

Mode spacing Δλ between nearest WGMs is plotted against the microdisk diameter in insert of Fig. 4. The best fit of the dependence reveals an effective group mode index of 3.64.

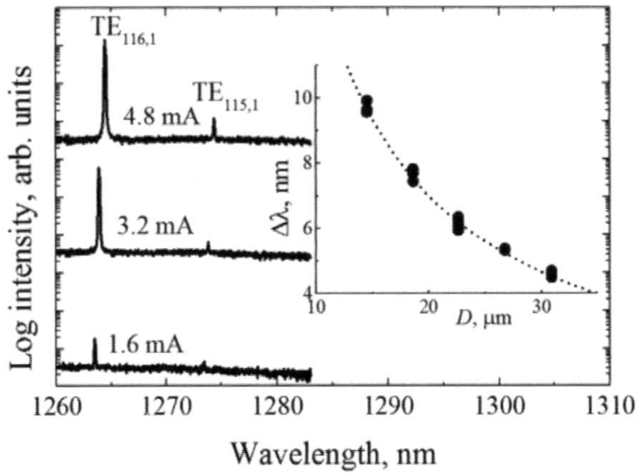

Figure 4. Series of emission spectra taken at various currents for a $D = 15$ μm microdisk. Insert: Mode spacing as a function of diameter. Fit: $N_{eff} = 3.64$.

Lasing wavelengths shifts to longer wave as bias current increases because of the laser self-heating effect. Fig. 5 displays the lasing wavelength as a function of electric power dissipated inside the device. Slope of this dependence, which is 0.207 nm/mW for the example shown, can be used to calculate a thermal impedance of the device. The results are shown in the insert of Fig. 5 for the microlasers of different diameters. We took into account a temperature induced shift of a WGM line, which is as small as 0.075 nm/°C. The thermal impedance is scaled with the diameter as D^{-2}; a coefficient of this dependence provides an estimation of a specific thermal resistance, which is $5 \cdot 10^{-3}$ °C•cm^2/W.

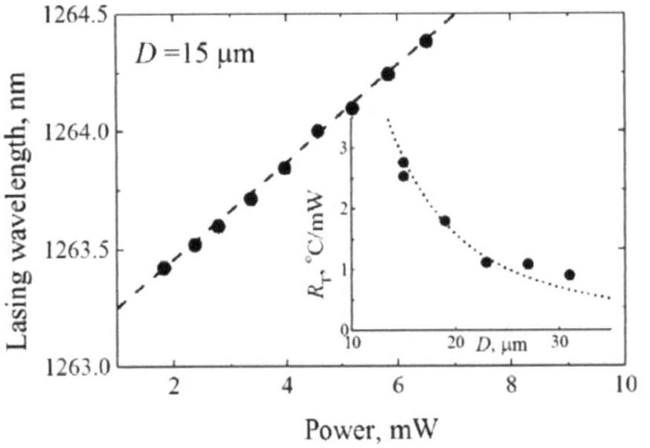

Figure 5. Lasing wavelength as a function of electric power for a $D = 15$ μm microdisk. Insert: Thermal impedance as a function of diameter. Fit: $\rho_T = 5 \cdot 10^{-3}$ °C•cm^2/W.

Data on current induced shift of the WGM spectral position (Fig. 6) were used to evaluate an increment of the active region temperature with respect to the ambient temperature. It is seen that the microlasers under study still lase even if the temperature increment is as high as 75°C. On the grounds of this finding one can suggest that the maximal operation temperature of the QD microlasers of this sort can be as high as 100 °C.

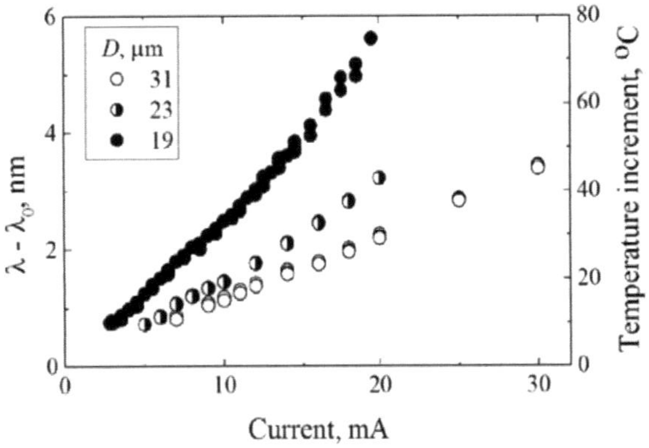

Figure 6. Left axis: Red shift of emission line as a function of current. Right axis: Temperature increment of the laser active region suggesting 0.075 nm/°C temperature coefficient.

4. Conclusions

Quantum dot based uncooled injection microdisk lasers were tested at room temperature in CW regime. A combination of device characteristics achieved (low threshold, long wavelength, operation at elevated temperatures) makes them suitable for application in future optoelectronic circuits for optical interconnect systems.

Acknowledgments
This work was supported by the Russian Foundation for Basic Research (grants ## 15-32-20238, 15-02-03624 and 13-02-12032) and the Fundamental Research Program of ONIT RAS.

References

[1] Liang D, Fiorentino M, Srinivasan S, Bowers J E and Beausoleil R G 2011 *IEEE J. Selected Topics Quantum Electron.* **17** 1528.
[2] McCall S L, Levi A F J, Slusher R E, Pearton S J and Logan R A 1992 *Appl. Phys. Lett.* **60** 289.
[3] Fujita M, Inoshita K and Baba T 1998 *Electron. Lett.* **34** 278.
[4] Deppe D G, Shavritranuruk K, Ozgur G, Chen H and Freisem S 2009 *Electron. Lett.* **45** 54.

[5] Ouyang D, Ledentsov N N, Bimberg D, Kovsh A R, Zhukov A E, Mikhrin S S and Ustinov V M 2003 *Semicond. Sci. Technol.* **18** L53.

[6] Zhukov A E, Kovsh A R, Maleev N A, Mikhrin S S, Ustinov V M, Tsatsul'nikov A F, Maximov M V, Volovik B V, Bedarev D A, Shernyakov Yu M, Kop'ev P S, Alferov Zh I, Ledentsov N N and Bimberg D 1999 *Appl. Phys. Lett.* **75** 1926.

[7] Maximov M V, Kryzhanovskay N V, Nadtochiy A M, Moiseev E I, Shostak I I, Bogdanov A A, Sadrieva Z F, Zhukov A E, Lipovskii A A, Karpov D V, Laukkanen J and Tommila J 2014 *Nanoscale Research Lett.* **9** 657.

[8] Mao M-H, Chien H-C, Hong J-Z and Cheng C-Y 2011 *Opt. Exp.* **19** 14145.

[9] Munsch M, Claudon J, Malik N S, Gilbert K, Grosse P, Gerard J-M, Albert F, Langer F, Schlereth T, Pieczarka M M, Hofling S, Kamp M, Forchel A and Reitzenstein S 2012 *Appl. Phys. Lett.* **100** 031111.

MBE growth and optical properties of GaAs nanowires grown on Si(111) substrate using two-temperature steps regime

R R Reznik[1,2], I V Shtrom[1,4,5], Yu B Samsonenko[1,3,4], A I Khrebtov[1], A D Bourauvlev[1,3,4,5], I.P.Soshnikov[1,3,4], N. V. Kryzhanovskaya[1,2] and G E Cirlin[1,2,3,4,5]

[1]St-Petersburg Academic University – Nanotechnology Research and Education Centre RAS, Khlopina 8/3, 194021, St-Petersburg, Russia
[2]St-Petersburg State Polytechnical University – Nanotechnology, Polytechnicheskaya 29, 195251, St-Petersburg, Russia
[3]Ioffe Physical Technical Institute RAS, Politekhnicheskaya 26, 194021, St-Petersburg, Russia
[4]Institute for Analytical Instrumentation RAS, Rizhsky 26, 190103, St-Petersburg, Russia
[5]St. Petersburg State University, 199034 St. Petersburg, Russia

E-mail: moment92@mail.ru

Abstract. We report on the growth and optical properties of pencil-like GaAs nanowires (NWs) grown using an original two-temperature growth mode on Si substrate. Optically, the NWs array is active up to the room temperature. The low-temperature PL spectra is composed by the narrow bands corresponding to the direct, bound to surface states and bound to impurity excitons. After toluene treatment, the fine structure is disappeared in the spectra.

1. Introduction

Si is a key material of today's semiconductor electronics. However, the realization of light-emitting devices based on Si, which are necessary for optoelectronic device integration, is still an open problem due to the indirect nature of the Si band structure. Freestanding III-V NWs have recently attracted a rapidly increasing interest due to their unique growth, structural, transport and optical properties as well as a variety of promising applications as building blocks for future nanoelectronic [1] and nanophotonic [2] devices. Due to the ability to accumulate for strain in two dimensions [3], NW geometry is ideal for monolithic integration of semiconductor materials with different lattice constants. In particular, coherent, low temperature epitaxial growth of III-V compounds on silicon is of paramount importance for applications. It has been recently shown experimentally that coherent III-V NWs can be grown epitaxially on lattice-mismatched substrates by the vapor-liquid-solid mechanism. In this paper we present MBE growth and optical examination of GaAs nanowires grown on Si(111) substrate using two-temperature steps regime allows one to obtain pencil-like NWs.

2. Experimental.

Growth experiments are carried out using Riber Compact21 MBE setup equipped with the effusion Au cell, on Si(111) substrates. Detailed procedure of NW growth is described in Ref. [4]. In brief, the desorption of an oxide layer from the substrate in a separate, vacuum-combined with the growth reactor, chamber is done. Then, to promote the NW formation by the growth catalyst, the deposition of ~ 0.5-1 nm thick Au layer is performed. The samples are then annealed in order to form liquid drops of alloy of Au with the semiconductor material of the substrate and transferred to the growth chamber. The MBE growth of GaAs NWs is carried out by the two-step temperature MBE technique. Firstly, the substrate temperature is set at 550°C and the growth of NWs is initiated. At the second stage, after 20 min. of the growth the substrate temperature is lowered to 530°C and next 40 min. of growth is performed. As a result, non-trivial, a pencil-like shape of the NWs is obtained. The process of NW nucleation and growth is monitored *in situ* by the reflection high energy electron diffraction (RHEED) technique. After several minutes of the growth, initially linear RHEED patterns are converted into the spots corresponding to the formation of wurzite-like crystallographic phase in the NWs. At the end of the process, the RHEED spots are partially converted into mixed cubic/hexagonal patterns. After the growth, the samples are studied by applying the scanning electron microscopy (SEM) and low-temperature photoluminescence (PL) techniques. Low temperature measurements is performed using neodymium laser (532 nm, exciting power density is 1.5 W/cm^2) in a He cryostat.

3. Results.

In Figs. 1 we present typical SEM image of GaAs NWs grown on the Si(111) substrate. The nanowires exhibit pencil-like shape having ~ 150 nm diameter at the base whereas ~ 25 nm at the top. This unconventional geometry is attributed to the two-step growth regime. At the first step, the NW diameter is dictated by the Au droplet size. At the second stage, at lower substrate temperature, the diffusion of the adatoms along the NWs sidewalls is suppressed. As a result, the formation of the nuclei at the sidewalls become sufficient and a lateral growth of the NWs is clearly observed.

Surprisingly, this NWs array is optically active up to the room temperature despite of the high density of the surface states typical for the Au-associated growth of GaAs NWs. The low temperature PL spectrum for as grown samples consists of a number of the narrow bands corresponding, to our opinion, to the direct (and indirect cubic/wurzite), bound to surface states and bound to impurity excitons. Indeed, PL spectra taken at 15 K consist of 1.52 eV PL band which is assisted to the direct exciton transition in GaAs NWs having cubic crystallographic phase shifted by 3 meV due to the quantization in a thinnest part of the NWs (25 nm in diameter) [5]. We note that we did not observe this shift for the NW arrays grown using conventional, one temperature step regime where cylindrical shape of the NWs were formed with a diameter ranging within 50 – 80 nm. Other bands (at 1.5 eV and 1.507 eV) are typically attributed to the cubic/wurzite transitions and unintentional doping [6]. In addition, sharp PL lines are observed which, most probably, correspond to the excitons bounded to surface states or excitons bounded to the impurity (Fig. 2). The authors[7] explain attributed them to the *bound* exciton emission from hexagonal/cubic boundaries. . In our case, we believe that these PL features correspond to surface-related exitons. In [5] it was established that the *bound* exciton related emission is quenched at ~ 40K This is clearly demonstrated in Fig.2 case where PL temperature dependence is presented. At 30K most of the fine lines are disappeared whereas PL bands corresponding to 1.5 eV, 1.507 eV and 1.52 eV remain. To prove our assumption about presence of sharp lines in PL spectra, the NWs are subjected to the toluene treatment with consequent drying during 15 min at 150^{0}C (which has to eliminate the surface-related exitons). After this procedure, the fine structure is disappeared in the spectrum (Fig.3, except of the lines corresponding to the cubic and cubic/wurzite interface excitons or the lines corresponding to the impurity states [6]) which supports the proposed origin of the narrow lines in the low-energy part of the PL spectra.

Figure 1. SEM image GaAs NWs grown on Si(111) substrate using two-step temperature mode.

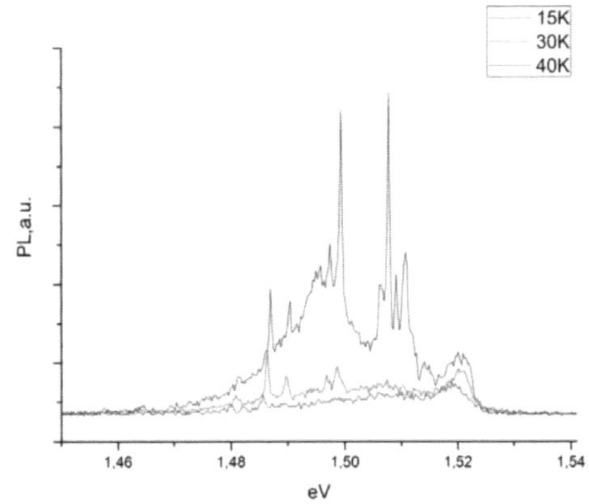

Figure 2. PL temperature dependence of GaAs NWs array grown on Si(111) substrate using two-step temperature mode.

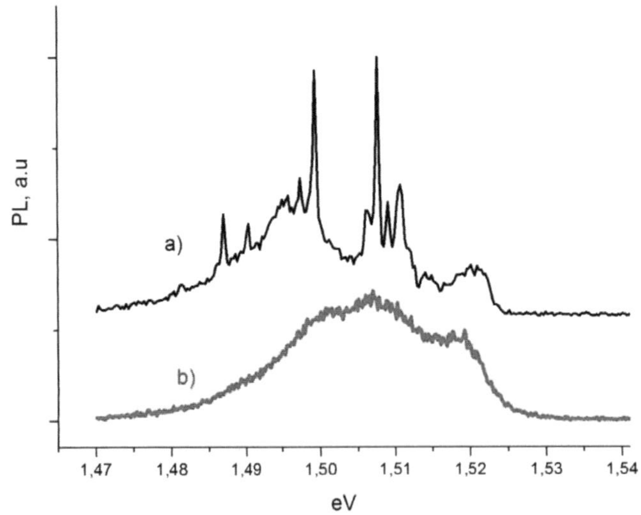

Figure 3. Low temperature PL spectra of GaAs NWs array taken at 10 K before (a) and after (b)

toluene treatment.

This work has been supported in part by the RFBR and St.Petersburg State University. The samples are grown under support from Skolkovo foundation (agreement №4 from 25.12.2015 with Academic University RAS).

References

[1] Agarwal R and Lieber C M Semiconductors nanowiers: optics and optoelectronics *Appl. Phys. A* **85** no 3 209-215

[2] Chuang L C, Moewe M, Chase C, Kobayashi N P, Chang-Hasnain C and Crankshaw S Critical diameter for III-V nanowires grown on lattice-mismatched substratest *Appl. Phys. Lett.s* **90** no 7 043115

[3] Cirlin G E, Dubrovskii V G, Soshnikov I P, Sibirev N V, Samsonenko Yu B, Buravlev A D, Harmand J C and Glas F Critical diameters and temperature domains for MBE growth of III–V nanowires on lattice mismatched substrates *Status Solidi RRL* **3** no 4 112-114

[4] Yu.B.Samsonenko, G.E.Cirlin, V.A.Egorov, N.K.Polyakov, V.P.Ulin, V.G.Dubrovskii. Semiconductors 42, 1445 (2008).

[5] B.V. Novikov, S. Yu. Serov, N. G. Filosofov, I. V. Shtrom, V. G. Talalaev, O. F. Vyvenko,E. V. Ubyivovk, Yu. B. Samsonenko, A. D. Bouravleuv, I. P. Soshnikov, N. V. Sibirev, G. E. Cirlin, and V. G. Dubrovskii "Photoluminescence properties of GaAs nanowire ensembles with zincblende and wurtzite crystal structure" , Phys. Status Solidi RRL 4, No. 7, 175–177 (2010)

[6] Alexander V. Senichev, Vadim G. Talalaev, Igor V. Shtrom, Horst Blumtritt, George E. Cirlin, Jörg Schilling, Christoph Lienau, Peter Werner."Nanospectroscopic Imaging of Twinning Superlattices in anIndividual GaAs-AlGaAs Core–Shell Nanowire", ACS Photonics, 2014, 1 (11), pp 1099–1106, DOI: 10.1021/ph5002022

[7] P. Corfdir, B. Van Hattem, E. Uccelli, A. Fontcuberta i Morral, and R. T. Phillips, «Charge carrier generation, relaxation, and recombination in polytypic GaAs nanowires studied by photoluminescence excitation spectroscopy» Appl.Phys.Lett, 103, 133109 (2013)

Acceptor states and carrier lifetime in heteroepitaxial HgCdTe-on-Si for mid-infrared photodetectors

S V Zablotsky[1,2], **N L Bazhenov**[1], **K D Mynbaev**[1,3], **M V Yakushev**[4], **D V Marin**[4], **V S Varavin**[4] **and S A Dvoretsky**[4]

[1] Ioffe Institute, 194021 Saint-Petersburg, Russia
[2] Saint Petersburg Electrotechnical University 'LETI', 197376 Saint-Petersburg, Russia
[3] ITMO University, 197101 Saint-Petersburg, Russia
[4] Rzhanov Institute of Semiconductor Physics, Siberian Branch of Russian Academy of Sciences, 630090 Novosibirsk, Russia

E-mail: zablotcky@yandex.ru

Abstract. Temperature dependences of the carrier lifetime and photoluminescence (PL) spectra in $Hg_{1-x}Cd_xTe$ (0.3< x <0.4) epilayers grown on Si substrates (HgCdTe-on-Si) and intended for fabrication of p^+-n photodiodes have been studied. It is shown that the as-grown material has a high concentration of recombination centres with energy 30 ÷ 40 meV, which reduces the lifetime. The post-growth annealings, both that converting the material to the p-type and that imitating the activation of an implanted impurity, reduce the concentration of the centres. This manifests itself in an increase in the lifetime, which is especially noticeable after 'activating' annealing and is confirmed by results of the PL studies.

1. Introduction

Recently, the increasing attention in technology of photodiodes based on $Hg_{1-x}Cd_xTe$ (MCT), one of the basic materials of infrared photo-electronics has been given to a method of fabrication of p^+-n junctions in which a p^+-layer is formed in the initially n-type material [1]. Typically, in such structures, as compared to n^+-p structures, dark currents are considerably smaller which provides a higher working temperature of photodiodes. It is believed that this is due to the fact that n-type MCT has a greater minority carrier lifetime τ than that in the p-type material, because the n-type material has a smaller concentration of deep recombination centres.

Determination of the density and energy position of such centres, as well as reduction of their concentration, are among major tasks in the technology of high-quality MCT [2]. As a rule, in order to elucidate the effect of recombination centres on parameters of the material, the experimentally measured τ values are compared with those limited by the fundamental band-to band recombination mechanisms. If the experimental τ values are much smaller than those calculated, the slope of the temperature dependence $\tau(T)$ allows evaluation of the energy position of the centres.

Fabrication of MCT-based devices includes, in addition to the growth, one or several annealings performed in order to obtain the desirable type of conductivity, improve the homogeneity of the material and reduce the number of defects. It is of interest to monitor the concentration of recombination centres at each technological step.

This paper reports on a study of the nonequilibrium charge carrier lifetime in MCT epilayers, intended for fabrication of p^+-n junctions and grown by molecular beam epitaxy (MBE) on Si substrates (HgCdTe-on-Si), one of the most promising substrate material for coupling of photodetectors with readout systems [3]. The data obtained are compared with the results of photoluminescence (PL) experiments carried out on similar epilayers.

2. Experiment

Experimental structures were grown at Rzhanov Institute of Semiconductor Physics with ZnTe and CdTe buffer layers. The $Hg_{1-x}Cd_xTe$ chemical composition x in the photosensitive layers of the structure (with thickness of 5 to 6 μm) was 0.3 to 0.4, which corresponds to cut-off wavelengths of the photo detectors of 4.9 to 2.9 μm at T=77 K. The carrier lifetime τ was studied using photoconductivity decay technique.

As-grown samples demonstrated n-type conductivity with a carrier concentration of $4 \cdot 10^{14} \div 2 \cdot 10^{15}$ cm^{-3} at $T = 77$ K. Parts of the samples were subjected to a two-stage annealing in a saturated Hg vapour (2 hours at $T = 360$ °C and 24 hours at $T = 225$ °C). This annealing was used as an imitation of the 'activating' annealing performed after ion implantation in order to create p^+-areas and fabricate p–n junctions (the samples studied were not implanted). After such an annealing, the samples remained n-type with a carrier concentration of $5 \cdot 10^{14} \div 5 \cdot 10^{15}$ cm^{-3}. Parts of some samples were annealed to convert the as-grown material to p-type conductivity in the He atmosphere and low-pressure Hg-vapour (15 hours at T=260 °C): this procedure is common in fabrication of n^+-p photodiodes. After this annealing, the hole concentration in the material was $8 \cdot 10^{15} \div 2 \cdot 10^{16}$ cm^{-3}.

3. Results and Discussion

3.1. Lifetime

The results of lifetime measurements are shown in Fig. 1, 2 and 3.

Figure 1 shows temperature dependences of the lifetime for as-grown samples. For samples with the composition $x = 0.31$ at high temperatures τ typically decreases with increasing temperature, which is due to the contribution of the intrinsic conductivity and the domination of the band-to-band recombination. Therefore, the curves have a maximum; however the curve for the sample with $x = 0.4$ has no maximum of this kind up to 300 K. As the temperature decreases below 200 K, τ becomes smaller, with an exponential (linear in the chosen coordinate axes) section being clearly visible; then the $\tau(T)$ dependence for samples with $x = 0.31$ becomes nonmonotonic. Figure 2 shows temperature dependences of the lifetime for samples after their annealing into p-type material. The $\tau(T)$ dependences are qualitatively similar to those shown in Fig. 1, but τ becomes about ten times smaller after the annealing. Figure 3 shows $\tau(T)$ for samples after the 'activating' annealing, when the material remains n-type. As before, the $\tau(T)$ dependence is similar to that in Fig. 1, but τ becomes about ten times greater.

On the whole, Figs. 1, 2 and 3 indicate that τ in the samples studied is governed by the Shockley-Read recombination via local centres and the annealings significantly affect the concentration of these centres. To evaluate the effect of recombination centres, let us compare experimental data with the calculated lifetimes, governed by the band-to-band recombination mechanisms: radiative and non-radiative. In MCT, the main non-radiative mechanisms are the follows: in an n-type material, Auger recombination involving two electrons and a heavy hole, with the excitation of an electron into a higher energy state (CHCC or A1 process); and for a p-type material, Auger recombination involving two heavy holes and an electron, with the conversion of a heavy hole into a light hole (CHHL or A7 process). The most rigorous calculations of τ with consideration for these mechanisms can be carried out using expressions from [4]. The radiative recombination in MCT of the compositions studied is of importance at

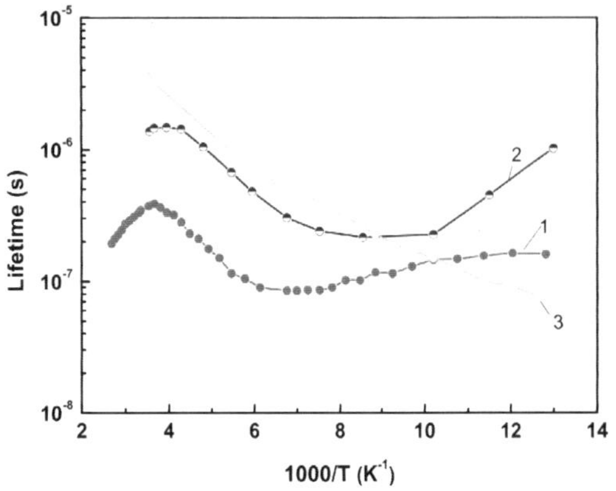

Figure 1. Temperature dependences of the lifetime for as-grown samples. The composition x is equal to 0.31 (*1*, *2*) and 0.4 (*3*).

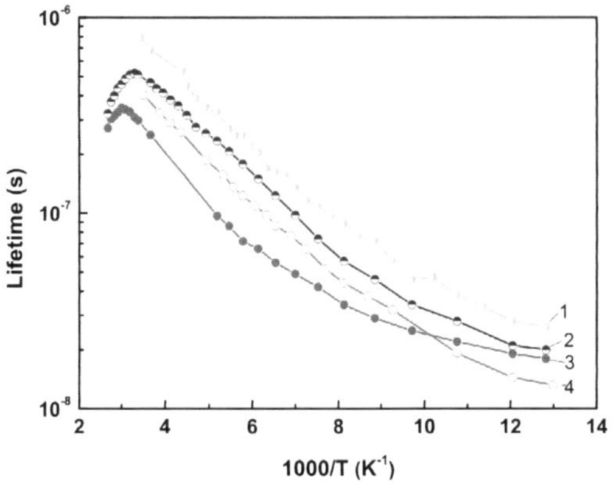

Figure 2. Temperature dependences of the lifetime for samples after annealing into p-type. The composition x is equal to 0.31 for all the samples.

temperatures below 200 K. To calculate τ governed by the radiative recombination, we used expressions from [5].

Figure 4 combines experimental and calculated τ for a sample with $x = 0.31$. The carrier concentration at $T = 77$ K was $n = 4.1{\cdot}10^{14}$ cm^{-3} in the as-grown sample, $p = 1{\cdot}10^{16}$ cm^{-3} after the annealing with conversion into p-type and $n = 1 \cdot 10^{15}$ cm^{-3} after the 'activating' annealing. It can be seen that in the as-grown sample τ determined by the band-to-band mechanisms is about an order of magnitude greater than the experimental data. After the annealing with conversion into p-type, τ decreases due to the increase in the equilibrium carrier concentration; however, experimental τ values become closer to those calculated. This relative closeness of the experimental and calculated τ values holds after the 'activating' annealing, which unambiguously indicates that the concentration of recombination centres decreases.

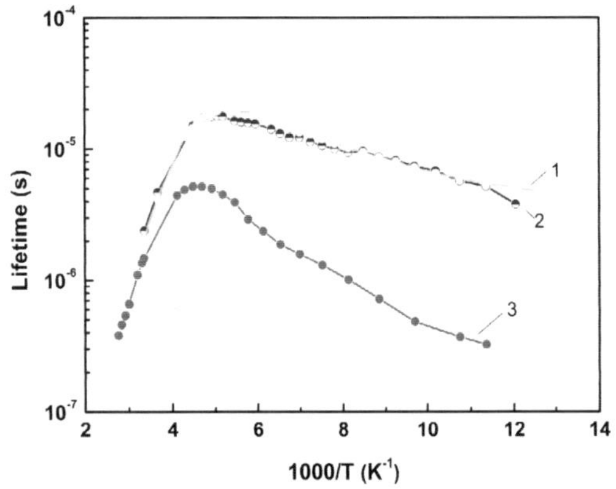

Figure 3. Temperature dependences of the lifetime for samples after the activation annealing. The composition x is equal to 0.31 for all the samples.

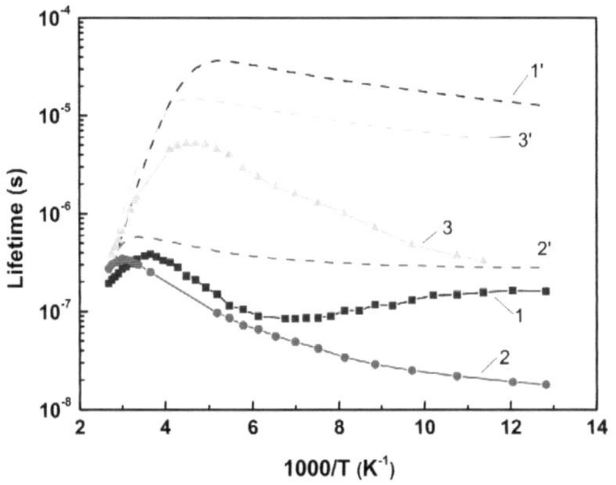

Figure 4. $\tau(T)$ dependences for a sample with x=0.31: as-grown (*1*), p-type (*2*) and after activation annealing (*3*). Dashed curves were calculated for band-to-band recombination.

The energy position E_t of the centres was estimated using the linear section of the $\tau(T)$ curves at temperatures below 200 K. The values obtained were within the range $30 \div 40$ meV.

3.2. Photoluminescence

$\tau(T)$ data were compared with those of photoluminescence (PL) measurements. The PL signal was excited by a pulsed semiconductor laser with wavelength $\lambda = 1.03$ μm and detected by a cooled InSb photodiode and a lock-in amplifier. Figure 5 shows PL spectra for a sample with $x = 0.4$ at $T = 4.2$ K (as-grown and after the 'activating' annealing). The discrete band corresponds to band-to-band recombination (band A), while an additional structure in the long-wavelength part of the spectrum can be attributed to transitions to local centres (B, C and D). The annealing slightly modified the composition x and shifted the spectrum to higher energies. Note, that after the annealing the long-wavelength structure in the PL spectrum disappeared (except for line D'), which indicated that concentration of the recombination centres involved in the radiative recombination decreased. Thus, PL data confirmed the decrease of the density of states in the energy gap after the annealing, which was observed in the lifetime study.

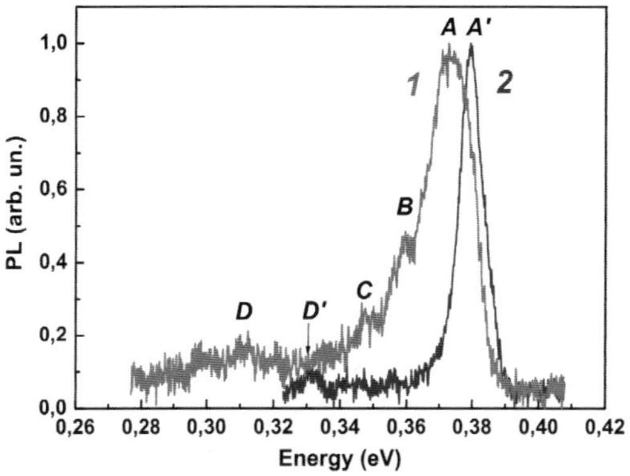

Figure 5. PL spectra of as-grown (1) and annealed (2) HgCdTe-on-Si samples.

4. Conclusion

Studies of temperature dependences of the carrier lifetime and PL spectra in HgCdTe-on-Si grown by MBE have shown that the as-grown material has a considerable concentration of recombination centres, which reduces the lifetime in the material. The conversion of the initial n-type material into that of p-type through an appropriate annealing leads to a decrease in the recombination centre concentration. An even greater decrease was provided by the 'activating' annealing, which imitates that performed to fabricate implanted p^+-n photodiode structures.

Acknowledgments

The study at Rzhanov Institute of Semiconductor Physics was supported by the Ministry of Education of the Russian Federation (Grant RFMEFI60414X0134).

References

[1] Mollard L, Bourgeois G, Lobre C, Gout S, Viollet-Bosson S, Baier N, Destefanis G, Gravrand O, Barnes J, Milesi F, Kerlain A, Rubaldo L and Manissadjian A 2014 *J. Electron. Mater.* **43** 802–807
[2] Edwall D D, DeWames R E, McLevige W V, Pasko J G and Arias J M 1998 *J. Electron. Mater.* **27** 698–702
[3] Simingalam S, Brill G, Wijewarnasuriya P and Rao M 2015 *J. Electron. Mater.* **44** 1321–1326
[4] Bazhenov N L, Mynbaev K D and Zegrya G G 2015 *Semiconductors* **49** 432–436
[5] Gel'mont B L and Zegrya G G 1991 *Sov. Phys. Semicond.* **25** 1216–1218

Surface defects formation on strained thin films growing via chemical reaction: a model

A V Redkov[1,2], S A Kukushkin[1,2,3] and A V Osipov[1,2,3]

[1] Institute of Problems of Mechanical Engineering, 8/3 Khlopina str., St.Petersburg, 194021, Russia
[2] St.Petersburg Academic University, 8/3 Khlopina str., St.Petersburg, 194021, Russia
[3] ITMO University, Kronverkskiy pr. 49, St.Petersburg, 197101, Russia

E-mail: avredkov@gmail.com

Abstract. We investigated in a general way evolution of the elastic flat surface of the crystal during its growth from multicomponent media. The morphological instability caused by mechanical stress has been demonstrated and conditions under which a multicomponent thin film retains its shape have been found. As an example, we have analyzed growth of GaN on different substrates and discussed the impact of III/V ratio on the stability of the GaN epitaxial film growing by HVPE.

1. Introduction

The study of growth processes of epitaxial thin films is of great interest of science and industry, since almost all electronic components are being made using thin-film technology. One of the main goals is to find the reasons of formation of different defects in the film during growth and to eliminate them. For example, common type of defects are so-called V-defects [1], which affect dramatically the quality of the heterostructures. In this paper, we assume that one of the possible causes of the development of surface defects can be local fluctuations of stress and concentrations of the components in strained areas of the film, leading to the further development of the instability of the growth front by mechanism of Mullins and Sekerka [2]. The choice of the ratio of III / V components may affect the growth rate of the instability. Generally, there are a lot of studies, devoted to stability of the growing crystals, in which authors considered various systems, differing in geometry, the cause of instability, and the factors affecting the development of the instability [3,4]. For example, in [5] authors discussed the stability of one-component strained epitaxial film commending the contribution of vacancies. Most studies, however, are devoted to one-component systems, whereas many of the modern technologies for growing GaN epitaxial layers often include chemical reaction in a multicomponent medium, for example, MOCVD [6] or HVPE [7].

2. The model

Let us consider a flat surface of the crystal of phase S growing from multicomponent medium in the reactor. Near the surface there is always boundary hydrodynamic layer of thickness l because of viscosity, where components move due to diffusion. The flow of the reactants from the atmosphere into the hydrodynamic layer is constant (see. Fig. 1a). Components can diffuse along the surface and react chemically (1) forming substance S, which incorporates directly into the surface and increase the height at that point.

$$\nu_1 A_1 + \nu_2 A_2 + \cdots + \nu_N A_N \leftrightarrow S^{(s)} + \nu_{N+1} B_1 + \cdots + \nu_{N+M} B_M \quad (1)$$

Hereinafter stoichiometric coefficients of reaction products will be redenoted as negative for convinience.

Initial reactants as well as reaction products can also leave the hydrodynamic layer. Let's imagine some defects below the surface like dislocations, pores, or substrate with a lattice parameter different from the growing phase S, or other defects that create stress field σ, which in general may change with height of the film ($\partial\sigma_0/\partial z \neq 0$[8]).

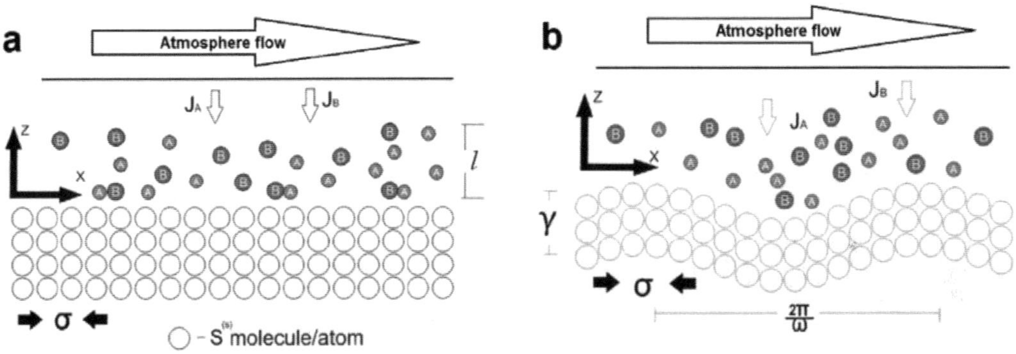

Figure 1.Illustration of the unperturbed (a) and perturbated (b) surfaces of thin film, growing via reaction of the type A + B↔S. In general, the chemical reaction in the valleys and on the tops can occur with different rates.

During steady growth the system of equations describing the concentrations of all components in hydrodynamic layer can be written as (2):

$$\begin{cases} 0 = D_i \frac{\partial^2 C_i}{\partial x^2} + J_i - K_p \nu_i \prod_{k=1}^{N} C_k^{\nu_k} + K_o \nu_i \prod_{k=N+1}^{M} C_k^{-\nu_k} - \frac{C_i}{\tau_i}, \quad i = A_1..B_M \\ \rho_S \frac{dz}{dt} = K_p \prod_{k=1}^{N} C_k^{\nu_k} - K_o \prod_{k=N+1}^{M} C_k^{-\nu_k} \end{cases} \quad (2)$$

Here C_i - the concentration of component i in the hydrodynamic layer, which changes due to incident flux J_i, diffusion, chemical reaction and flying back to the reactor. D_i - diffusion coefficient of substance i in gas, K_p - the rate of direct chemical reaction of S formation, K_o - the rate of reverse chemical reaction. τ_i - the lifetime of the component i in hydrodynamic layer, dz/dt - growth rate of the surface, ρ_s - concentration of molecules in solid phase S. We can estimate the time τ_i for each component as $\tau_i \sim l^2/D_i$, where l can be an order of 0.5-1mm [9].

Now we should made a small sinusoidal perturbation of surface Z with amplitude γ:

$$Z = z + \gamma \sin(\omega x)$$

Due to the change of surface geometry local stresses will also change according to [5,10]:

$$\sigma|_{z+\gamma\sin(\omega x)} = \sigma_0 + (\frac{\partial\sigma_0}{\partial z} - 2\sigma_0\omega)\gamma\sin(\omega x) \quad (3)$$

This will affect the rate of a chemical reaction, since the stress variation contributes to the energy of the reaction product S. Additional elastic energy density ΔU, which is caused by the perturbation of the surface can be expressed by modified equation from [10]:

$$\Delta U(x) = U\gamma\sin(\omega x) = \frac{{\sigma_0}^2}{\bar{M}}\left[\frac{1}{\sigma_0}\frac{\partial\sigma_0}{\partial z} - 2(1+\nu)\omega\right]\gamma\sin(\omega x)$$

Where $\bar{M} = E/(1-\nu)$ is biaxial elastic modulus, E is Young's modulus of the film, ν - Poisson ratio [10]. Therefore the local chemical potential and thus equilibrium constant will change and reaction rates can be described by the formula:

$$\left.\frac{K_o}{K_p}\right|_{z+\gamma\sin(\omega x)} = \frac{K_o}{K_p}(1 + \left[\frac{\theta V_s}{kT}\omega^2 + \frac{UV_s}{kT}\right]\gamma\sin(\omega x)) \quad (4)$$

Here $V_s = \sum_{i=1}^{M} \nu_i V_i / \sum_{i=1}^{M}|\nu_i|$, V_i- volume of i component, and θ - surface tension coefficient. Fluctuations of reaction rate (4) in its turn will lead to a corresponding change in the concentrations and growth rates of the surface.

$$C_k|_{z+\gamma\sin(\omega x)} = C_k^0(1 + \delta_k\sin(\omega x)),\ k = A_1..B_M \quad (5)$$

$$\dot{Z} = \frac{dz}{dt} + \frac{d\gamma}{dt}\sin(\omega x)$$

Taking into account expressions (4) and (5), substitute them into the system of equations (2). Then we should apply a similar chain of mathematics as in [11], taking into account the first order in γ, and we can obtain the formula for the rate of growth of perturbations in case of small multicomponent oversaturation.

$$\frac{\rho_s}{\gamma}\frac{d\gamma}{dt} = -\frac{{\mu_o}^{os}{\mu_p}^{os}}{{\mu_o}^{os}+{\mu_p}^{os}}\left(\frac{UV_s}{kT} + \frac{\theta V_s}{kT}\omega^2\right) \quad (6)$$

Where ${\mu_o}^{os}$, ${\mu_p}^{os}$ are generalized coefficients:

$${\mu_p}^{os} = \left(\sum_{k=1}^{N}\left(\frac{{\nu_k}^2{\tau^0}_k}{C_k^0({\tau^0}_k D_k\omega^2+1)}\right)\right)^{-1},\ {\mu_o}^{os} = \left(\sum_{k=N+1}^{M}\frac{{\nu_k}^2{\tau^0}_k}{C_k^0({\tau^0}_k D_k\omega^2+1)}\right)^{-1} \quad (7)$$

It can be seen that the perturbation will not grow if the right side of (6) is zero or negative, i.e., the inequality is satisfied

$$\frac{\rho_s}{\gamma}\frac{d\gamma}{dt} \sim \frac{{\sigma_0}^2}{M}\left[\left(2(1+\nu)\omega - \frac{1}{\sigma_0}\frac{\partial\sigma_0}{\partial z}\right)\right] - \theta\omega^2 \le 0$$

The stability criterion fully corresponds to the formula obtained in [10] for one-component film, but the growth rate of perturbation depends on properties of all components. Also, from (6) it can be concluded that for any harmonic there is a critical stress at which the perturbation will increase with time. If the stress gradient is absent or small $\sigma_0\omega \gg \frac{1}{2}\frac{\partial\sigma_0}{\partial z}$, then the surface is stable at frequencies satisfying the inequality:

$$\frac{2{\sigma_0}^2}{\theta\bar{E}} < \omega \quad (8)$$

Where $\bar{E}$ is the plain strain modulus. If the main contribution comes from the stress gradient, the critical frequency can be defined as

$$\sqrt{\left|\frac{\sigma_0}{2\bar{E}\theta}\frac{\partial\sigma_0}{\partial z}\right|} \le \omega$$

One should note, that stability criterion (8) doesn't itself contain individual properties of components, because in such formulation of the problem stress affects each component indirectly, only through the chemical reaction rate. If we considered growth of perturbation via surface diffusion of adatoms, stress could affect differently adsorption-desorption processes of each component.

3. Stability of GaN film

Let us consider as an example the growth of GaN by methods involving chemical reaction e.g. HVPE (9) and make some estimates of film stability on different substrates.

$$\mathrm{GaCl} + \mathrm{NH_3} \leftrightarrow \mathrm{GaN^{(s)}} + \mathrm{HCl} + \mathrm{H_2} \tag{9}$$

We suppose that the gradient of the stress along z axis is absent ($\partial\sigma_0/\partial z = 0$). Thus the dependence of the critical size of perturbation $\lambda_{cr} = 2\pi / \omega$, at which the surface becomes unstable, can be found from the formula (8). Knowing the type of substrate on which GaN grows, we can estimate the characteristic dimensions of the perturbations and defects that may appear on the surface of the film via considered mechanism (see Fig.2).

The surface tension of GaN $\theta \approx 2\ \mathrm{J/m^2}$ [12], Young's modulus E ≈ 301 GPa[12]. We also should estimate stresses in the film during growth process (see Table 1).

Table 1. Stress (tensile if negative) in ~1 μm-thick GaN film on different substrates.

Substrate	Stress at growth temperature (estimation), GPa
Sapphire(Al_2O_3)	-0.3 [14,15]
Si with AlN-buffer	-0.25 [16]
SiC	-0.1 [17] (in this work GaN was Si-doped)

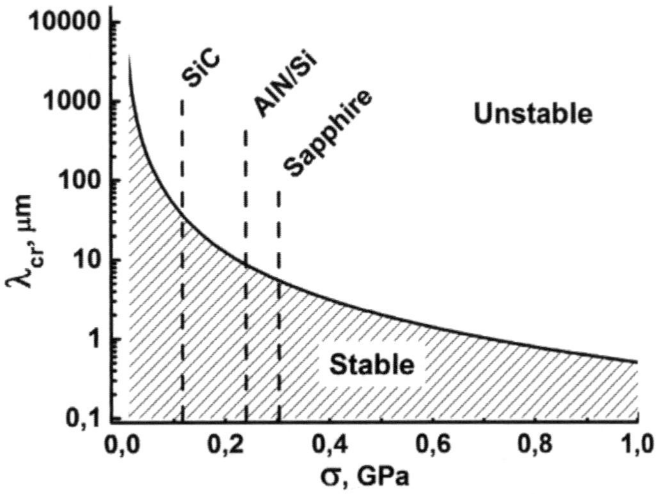

Figure 2. Dependence of the critical wavelength on stress for GaN film. Estimates for common substrates are shown on the graph. The shaded area corresponds to stable growth.

4. Impact of III/V ratio on surface stability

We have shown, that growth rate of perturbation via abovementioned mechanism depends dramatically on the generalized coefficients $\mu_p{}^{os}$, $\mu_o{}^{os}$, which are determined by diffusion coefficients and concentrations of components. We should analyze how the ratio of III/V affects HVPE (9) growth process. Let's fix chemical reaction rate:

$$C_{GaCl}C_{NH3} \approx \text{const}$$

One can see that coefficient before parenthesis in (6) is maximal when concentrations of III/V components obey the rule $C_{NH3} = (D_{GaCl}/D_{NH3})\, C_{GaCl}$ (for case $D_i\omega^2 \gg 1$). Thus, for stable frequencies maximal "stabilizing force" via considered mechanism will be achieved if $\frac{C_{NH3}}{C_{GaCl}} \sim 0.4$ (see Table 2). We should note, that unstable harmonics will also grow with maximal rate in such conditions. We emphasize that this condition involve concentrations in hydrodynamic layer, not in reactor atmosphere. For calculation of them one should take into account growth rate and lifetimes of components in particular growth process and solve (2).

Table 2. Diffusion coefficients of components involved in HVPE reaction in gaseous phase [18].

	GaCl	NH_3	HCl	H_2
D, 10^{-4} m^2/s (at 1000K)	1.08	2.68	1.83	7.82

5. Conclusions

We have studied the morphological stability of epitaxial strained film, growing via chemical reaction in multicomponent system. A criterion of stability connecting the stress in the film and the critical wavelength of the perturbation from which the surface becomes unstable is derived and applied to growth of GaN films via HVPE method which involves chemical reaction. Under standard growth stresses of 0.1-0.5 GPa on different substrates the critical wavelength for this mechanism of instability corresponds to tens or hundreds of microns.

Acknowledgements
This work was supported by Russian Foundation for Basic Research (projects #15-03-06155 and #13-02-12040-ofi-m)

References

[1] Wu X H, Elsass C R, Abare A, Mack M, Keller S, Petroff P M, Den Baars S P, and Specka J S 1998 *Appl. Phys. Lett* **72** 692
[2] Mullins W W and Sekerka R F 1963 *J. Appl. Phys.* **34** 323
[3] Cahn J W 1967 Crystal Growth (Suppl. J. Phys. Chem. Solids. Oxford) ed H.S. Peiser. p 681
[4] Кукушкин S A and Kuz'michev S V 2008 Phys. Solid State+ **50** 1390
[5] Panat R, Jimmy Hsia K and Cahill D G 2005 *J. Appl. Phys.* **97** 013521
[6] Fujieda S, Mizuta M and Matsumoto Y 1987 *Jpn. J. Appl. Phys.* **26**, 2067
[7] Richter E, *et al* 2005 *J. Cryst. Growth* **277** 6
[8] Paskova T *et al* 2001 *J. Cryst. Growth* **230** 381
[9] Arendarenko A A, Vigdorovich E N and Sveshnikov Yu N 2003 *Phys. Stat. Sol. (c)*, **3**, 827–830

[10] Freund L B and Suresh S 2008 *Thin Film Materials Stress: Defect Formation and Surface Evolution* (Cambridge University Press, Cambridge)
[11] Kukushkin S A, Osipov A V and Redkov A V 2014 *Phys. Solid State+* **56** 2530–2536
[12] Telyatnik R S, Osipov A V and Kukushkin S A 2015 Phys. Solid State+, **57** 162–172.
[13] Barghout K and Chaudhuri J 2004 *J. Mat. Sci.* **39** 5817 – 5823
[14] Hearne S *et al* 1999 *Appl. Phys. Lett.* **74** 356
[15] Krost A, Dadgar A, Strassburger G and Clos R 2003 *Phys. Stat. Sol. (a)* **200** 26–35
[16] Raghavan S and Redwing J M 2005 *J. Appl. Phys.* **98** 023514
[17] Ager III J W *et al* 1996 *MRS Proceedings*, **449** 775
[18] Kempisty P *et al* 2006 J. Cryst. Growth **296** 31–42

InAsSb on GaAs (001): influence of the arsenic molecules form on composition and crystalline properties of MBE layers

E A Emel'yanov, A V Vasev, B R Semyagin, A P Vasilenko, A A Komanov, A K Gutakovskii, M A Putyato, V V Preobrazhenskii
ISP SB RAS, pr. Lavrentieva 13, Novosibirsk, 630090, Russia

E-mail: e2a@isp.nsc.ru

Abstract. The influence of As molecular form on the composition and crystalline properties of $InAs_xSb_{1-x}$ solid solutions with MBE has been experimentally investigated. A series of samples has been grown at different growth temperatures. The grown samples were studied with the HRXRD and TEM methods. The incorporation coefficient of As_4 and As_2 molecules were determined at different growth temperatures. It has been found that the incorporation coefficient of As_4 much more dependent on growth temperature compared to As_2. It has been found that at a low growth temperature a step-like increase of Sb fraction in an $InAs_xSb_{1-x}$ film leads to a decrease of threading dislocations density in a layer with a smaller x.

1. Introduction

The $InAs_xSb_{1-x}$ solid solutions get much attention because of a possibility of creating IR-sensitive optoelectron devices on their base with a wavelength to 12.4 μm [1-5]. However, there are a number of problems concerned with obtaining high-quality epitaxial $InAs_xSb_{1-x}$ layers.

One of them is the absence of a substrate material for the whole range of $InAs_xSb_{1-x}$ compositions [6]. This leads to a degradation of the grown structures quality caused by the introduction of defects conditioned by mismatch film and substrate lattice constants.

Moreover, the formation of layers with a specific composition in the group V sublattice is a nontrivial problem too. The composition of the solid solution depends on both the value of Sb and As molecule fluxes (J_{Sb} and J_{As}) and their incorporation coefficient S_{Sb} и S_{As}. Incorporation coefficient is understood as a ratio of the atoms built in a crystal to their number arrived with the corresponding molecular flux for a time unit per surface unit. According to the definition of incorporation coefficients, arsenic composition x in a $InAs_xSb_{1-x}$ solid solution is connected with the density of group V molecular fluxes by the ratio:

$$x = \frac{1}{1 + \frac{J_{Sb}}{J_{As}} \times \frac{S_{Sb}}{S_{As}}} \quad (1)$$

Relation J_{As}/J_{Sb} is easily set and controlled during MBE, whereas relation S_{Sb}/S_{As} is a complex function of growth conditions. The ratio of incorporation coefficients depends on substrate temperature (T_S), value and ration of group III and V elements molecular fluxes, molecular form of group V elements in a flux [7]. These parameters allow immediate control of the surface state and the growing $A^{III}B^{V}$ film properties. Therefore, studying their influence on the S_{Sb}/S_{As} ratio at MBE of solid $InAs_xSb_{1-x}$ solutions is of interest.

The purpose of this work is experimental investigation of the influence of As molecular form and growth temperature on the composition and crystalline properties of $InAs_xSb_{1-x}$ solid solutions grown on GaAs (001) substrate with the MBE method.

2. Experimental methods

The work has been carried out on the handmade MBE-machine "Shtat"-type equipped by a valve-type As source with a cracking zone to obtain As_2 and As_4 fluxes. The In and Sb_4 fluxes were formed by crucible sources with shutters. The density of atom and molecular fluxes coming up to the surface was determined by the indications of vacuum controller Granville Phillips 307 whose sensor was placed onto the substrate position [8] during the measurement. Substrate temperature control was realized by the indications of the thermocouple placed in the heating element. The flux sensor and the thermocouple indications were calibrated using the phase diagrams of GaAs (001) [9] surfaces on the methods formulated in [10].

A series of samples was grown at 320°C, 330°C, 355°C and 370°C. Each sample consisted of two layers. The first layer was grown using the As_2 molecular flux. After the end of the first layer growth, the growth manipulator with the substrate was turned to the molecular fluxes measurement position. The cracking zone temperature of the arsenic source was decreased from 950°C to 400°C for As_4 flux generation. The second solid solution layer was grown using the As_4 molecular flux equivalent to the As_2 in the atomic expression.

The In flux was equal to $5.45 \cdot 10^{14}$ atom/cm$^2 \cdot$s, which corresponds to the growth rate of one monolayer per second. The fluxes of group V, in the atom expression, was ~ $8.6 \cdot 10^{14}$ atom/cm$^2 \cdot$s for Sb_4, for As_2 and As_4 ~ $1.0 \cdot 10^{15}$ atom/cm$^2 \cdot$s. The thickness of epitaxial layers was 500 nm when using the As_2 and 350 nm when using the As_4.

3. Investigation methods

The grown samples were investigated by HRXRD and TEM.

3.1. Investigated by HRXRD

The grown samples were investigated by HRXRD. Registration of the X-Ray rocking curves was made on a two-crystal diffractometer using the crystal Ge (004) monochromator for line Cu $K_{\alpha 1}$.

The data on the dependence of the composition of $InAs_xSb_{1-x}$ solid solution on T_s are presented in Figure 1.

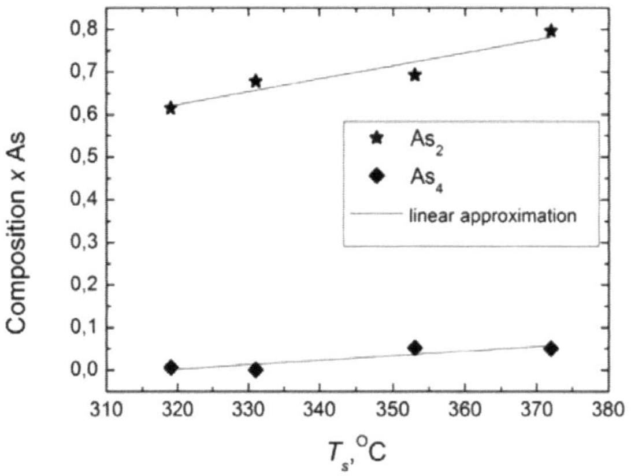

Figure 1. The arsenic fraction in the $InAs_xSb_{1-x}$ solid solution as a function of T_s. The data for the films grown with As_2 are designated with asterisks, the data for the films grown with As_4 are designated with a diamond.

The data of FWHM for X-Ray rocking curve (W), which are given in Table 1, were obtained depending on the $InAs_xSb_{1-x}$ film growth temperature. Here h_1 and h_2 are the first and second film thicknesses, respectively, R – relaxation degree, %.

Table 1. HRXRD data.

	T_s °C	h_1 nm (As_2)	h_2 nm (As_4)	W h_1	W h_2	R h_1(%)	R h_2(%)
st 1468	320	500	350	2486	1690	96.65	99.75
st 1469	330	500	350	2313	1678	95	100
st 1470	355	500	350	2237	1629	95	97
st 1471	370	500	350	2486	1690	91	100

It follows from the data of Table 1 that, in case of using As_2, a decrease of $InAs_xSb_{1-x}$ film relaxation degree is observed with an increase of growth temperature.

Assisted with expression 1, the values of S_{Sb4}/S_{As2} and S_{Sb4}/S_{As4} ratios were calculated for arsenic and antimony molecules. The data about the composition of solid solution layers obtained with the HRXR method and the J_{As2}, J_{As4} и J_{Sb4} values measured during the samples growth (see black squares in Figure 2 and Figure 3) were used for the calculation. The linear approximation of experimental data was carried out for dependence of the As fraction in the solid solution on the growth temperature (see Figure 1). This procedure allowed building the model dependences of S_{Sb4}/S_{As2} and S_{Sb4}/S_{As4} ratios on T_S.

The Figure 2 and 3 show that both S_{Sb4}/S_{As2}, and S_{Sb4}/S_{As4} values decrease with increasing of T_S in temperature range 320 ÷ 370°C, but decreasing of S_{Sb4}/S_{As4} about 5 times more. Such a difference in the behavior of these ratios is explained by the thing that As_2 molecules interact with a surface predominantly by the exchange mechanism, and As_4 molecules – by the vacancy mechanism [11].

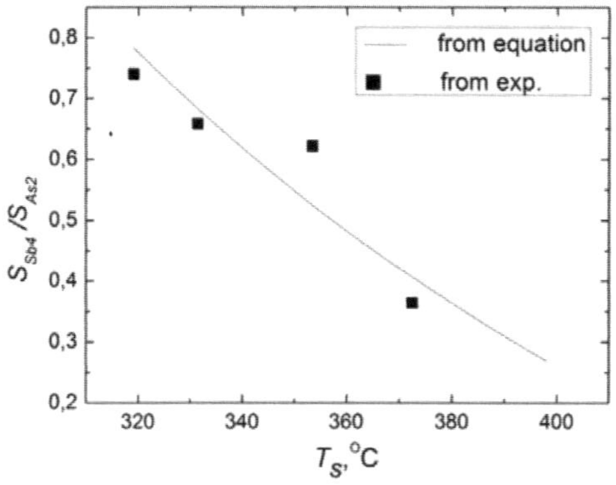

Figure 2. S_{Sb4}/S_{As2} as a function of T_s.

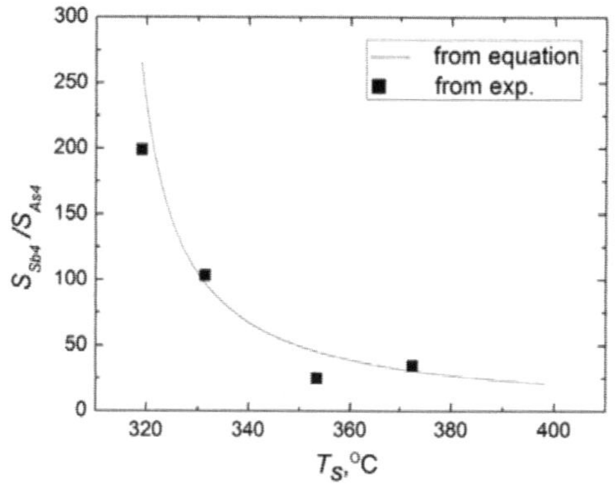

Figure 3. S_{Sb4}/S_{As4} as a function of T_s.

The data of Figure 3 are in good qualitatively agreement with the results of Lee G S *et al.* [12]. It is important to note that although temperature ranges of our and G. S. Lee *et al.* investigations are analogous, the beam equivalent pressure ratios for group V elements *BEP*(Sb_4)/*BEP*(As_4) is essentially different (from 0.08 to 0.38 – for ref. [12] and 0.72 – for our work).

3.2. Investigated by TEM

The structure of grown samples investigations were performed by a JEOL-4000EX electron microscope operated at 400 kV. The (110) cross sections for TEM studies were prepared in a standard manner by etching with argon ions. Bright-field / dark-field TEM and HRTEM were used.

It was found that a transition layer, that forms at the interface of $InAs_xSb_{1-x}$ films with different compositions at low T_s, decrease the threading dislocations density to the 5-6 times (see Figure 4).

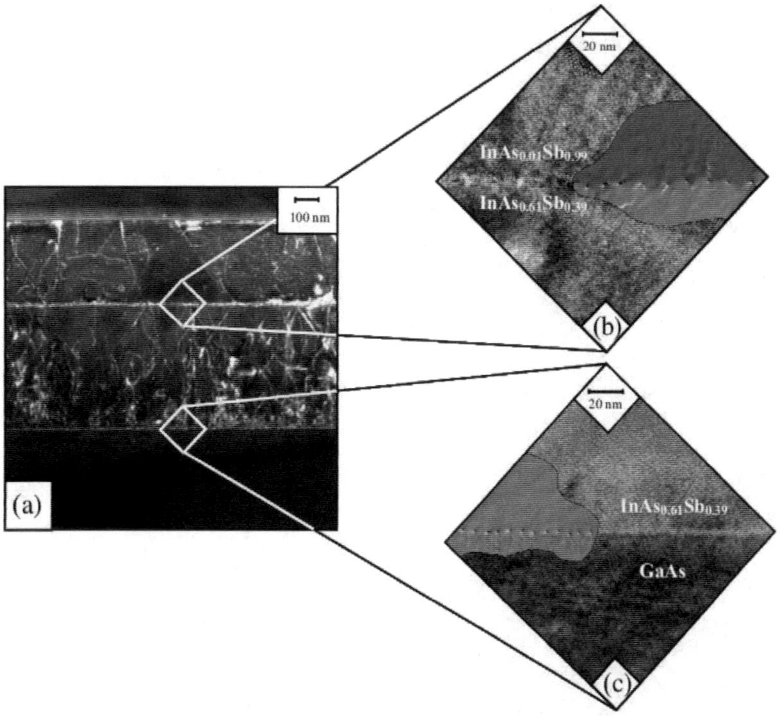

Figure 4. (a) – The (110) cross-sectional TEM image of the sample with two $InAs_xSb_{1-x}$ epilayers. The lower layer was grown with As_2, top – with As_4 (T_s = 320°C); (b) – HRTEM image of the *film / film* interface; (c) – HRTEM image of the *substrate / film* interface. The arrays of misfit dislocations are shown on insets (derived by Fast Fourier Transform analysis of HRTEM images).

Such effect was not observed at high growth temperatures (see Figure 5).

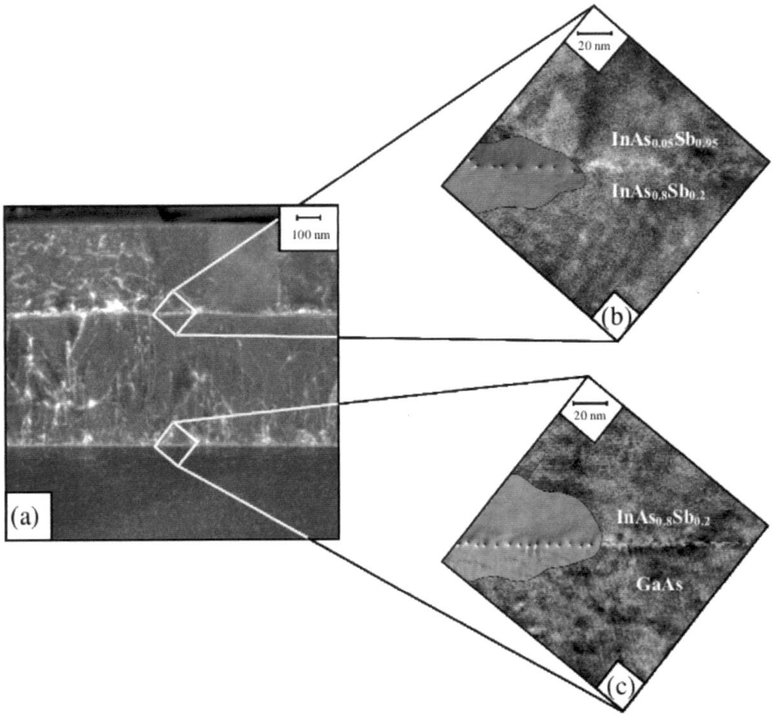

Figure 5. The same as Figure 4, but for the case of T_s = 370°C.

4. Conclusion

It was clarified that, during $InAs_xSb_{1-x}$ MBE, the efficiency of As_2 molecules incorporation, at other equal conditions, is considerably higher than that of As_4 molecules. It was found that the As_4 efficiency incorporation is more temperature sensitive compared to As_2. It was found that a decrease of threading dislocations density occurs in the layer with a smaller x value at a low growth temperature in case of a step-like increase of the Sb fraction in a $InAs_xSb_{1-x}$ film. This phenomenon can be used in the creation of dislocation filters which will provide growing high-perfection $InAs_xSb_{1-x}$ solid solution layers on various $A^{III}B^{V}$ semiconductor substrates.

The work was carried out supported by grants RFBR – 14-02-31054, 13-02-12114, 13-02-12124 and RSCF (project No 14-22-00143)

References

[1] Rogalski A 1989 *Prog. Quant. Electr.* **13** 191
[2] Bethea C G, Yen M Y, Levine B F, Choi K K and Cho A Y 1987 *Appl. Phys. Lett.* **51** 1431
[3] Dobbelaere W, De Boeck J, Heremans P, Mertens R, Borghs G, Luyten W and Van Landuyt J 1992 *Appl. Phys. Lett.* **60** 3256
[4] Besikci C, Ozer S, Van Hoof C, Zimmermann L, John J and Merken P 2001 *Semicond. Sci. Tech.* **16** 992
[5] Gao Y Z, Gong X Y, Fang W Z, Wu G H and Feng Y B 2009 *Japn. J. Appl. Phys.* **48** 080202

[6] Gao H, Wang W, Jiang Z, Liu L, Zhou J and Chen H 2007 *J. Cryst. Growth.* **308** 406
[7] Semenov A N, Sorokin V S, Solovyev V A, Meltser B Ya and Ivanov S V 2004 *Semiconductors* **38** 266
[8] Preobrazhenskii V V, Putyato M A, Pchelyakov O P and Semyagin B R 1999 *J. Cryst. Growth.* **201/202** 170
[9] Preobrazhenskii V V, Putyato M A, Pchelyakov O P and Semyagin B R 1999 *J. Cryst. Growth.* **201/202** 166
[10] Preobrazhenskii V V, Putyato M A and Semyagin B R 2002 *Semiconductors* **36** 837
[11] Emelyanov E A, Feklin D F, Vasev A V, Putyato M A, Semyagin B R, Vasilenko A P, Pchelyakov O P and Preobrazhenskii V V 2011 *Optoelectronics, Instrumentation and Data Processing* (*Avtometriya*) **47** 452.
[12] Lee G S, Lo Y, Lin Y F, Bedair S M and Laidig W D 1986 *Appl. Phys. Lett.* **47** 1219

Modeling of axial heterostructure formation in ternary III-V nanowires

A A Koryakin[1,2], N V Sibirev[1,2], D A Zeze[2,3] and V G Dubrovskii[1,2]

[1] St. Petersburg Academic University, Khlopina 8/3, St.Petersburg, 194021, Russia
[2] ITMO University, Kronverkskiy pr. 49, St. Petersburg, 197101, Russia
[3] School of Engineering and Computing Sciences, South Road, Durham, DH1 3LE, UK

E-mail: alexkorya@gmail.com

Abstract. A model is proposed to depict the formation of axial heterostructure in ternary III-V nanowires (NW) grown by the catalytic vapor-liquid-solid (VLS) method. Our approach is based on the determination of chemical potential of a four-component liquid using the regular solution model and Stringfellow's scheme for the computation of the interaction coefficients of species present in the droplet. The model allows the estimation of the heterojunction width dependence on the growth temperature. This dependence has not been reported before by any previous theoretical studies. The AlGaAs/GaAs heterojunction formation in the Au-catalyzed AlGaAs NWs was considered as an example of ternary system. The heterojunction width was found to increase with the growth temperature with a second-order polynomial dependence.

1. Introduction

Heterostructure nanowires (NWs) have a wide range of potential applications in opto- and nanoelectronics where the presence of an abrupt heterojunction is highly desirable for designing devices with high-performance characteristics [1, 2]. Vapor-liquid-solid growth (VLS) is probably the method most used to synthesize NWs [3] because it facilitates the formation of relatively sharp interfaces for several III-V semiconductors, elements from group V may be easily interchanged (e.g. InAsP/InAs, GaAsP/GaP [4]). In turn, the formation of abrupt interfaces is much more difficult (e.g. for AlAs/GaAs, GaInAs/InAs, InGaAs/GaAs [4]) when elements from group III are interchanged formation. This difference of interface sharpness has been explained by the fact that solubility of group III atoms in the catalyst droplet (10-40%) is usually higher than that of group V atoms, e.g. arsenic and phosphor (1-6%) [5]. Sophisticated growth techniques have been proposed to decrease heterojunction widths. For instance, a growth method with pulsed switching of material source was recently developed [6] resulting in a decrease of the heterojunction width down to 5-10 nm. Despite the variety of experimental data, theoretical studies to describe heterojunction formation in NWs are very limited [7-10]. To the best of our knowledge, none of the existing model deals with heterostructures in ternary III-V NWs. Moreover, only a small set of materials can be considered within the assumptions made. In reports describing the Si/Ge heterostructure formation in Au-catalyzed NWs [7-9], the material balance equations were considered for a fixed droplet volume, i.e. the amount of material coming into the droplet from the vapor is equal to the amount of material coming through the droplet-NW interface, albeit the droplet volume is not a constant in the general case. The equilibrium concentrations of dissolved materials were determined by means of pseudo-binary [7] and pseudo-ternary [8] phase diagrams. This simplification was direct consequence of the difficulty to calculate accurately multicomponent system phase diagrams. Nevertheless, for a proper description of III-V NW growth, it is important to consider a concentration change for all materials [5, 11, 12].

In this paper, we propose a model describing heterostructure formation in ternary III-V NWs grown by catalytic VLS method. Our approach is based on the determination of chemical potentials of dissolved materials using the regular solution model and the Stringfellow formula for the computation of the interaction coefficients of species present in the droplet [5, 11]. In contrast to the existing works where only a three-component liquid droplet was considered, the model developed here utilizes a four-component liquid droplet, i.e. three NW growth materials and one catalyst. Furthermore, the droplet volume is not a fixed constant but can vary during the heterojunction formation. As a result, the model allows the estimation of the heterojunction width dependence on the growth temperature that was not demonstrated in previous theoretical reports. In what follows, the Au-catalyzed AlGaAs/GaAs NWs grown by the molecular beam epitaxy method are considered as an example of ternary system. The data presented in the literature [13] are used to compare our calculations with experimental results.

2. Model

In the model proposed, we assume that NW growth proceeds in a mononuclear layer-by-layer growth regime, in line with the case for NWs with small radii and under the usual conditions for molecular beam epitaxy NW growth [14]. The material fluxes from the vapor phase dissolve in the catalyst droplet and reach the droplet-NW interface. If the solution is saturated, two-dimensional islands (nuclei of solid phase) form at the interface. In the case of mononuclear nucleation, the lateral growth rate of islands is high and a complete monolayer forms before the next sequence of nucleation [14, 11, 12]. The driving force behind nucleus formation is a deviation from the chemical equilibrium condition [15]: $\Delta\mu \equiv \sum \nu_j \mu_j = 0$, where μ_j is the chemical potential of the material involved in the chemical reaction and ν_j, the stoichiometric coefficient. Our model accounts for AlAs and GaAs nuclei formation. Therefore, two reactions were considered: $Al + As \rightarrow AlAs$ and $Ga + As \rightarrow GaAs$, respectively. Two variations of the chemical potentials were also introduced: $\Delta\mu_{AlAs} = \mu_{Al}^{L} + \mu_{As}^{L} - \mu_{AlAs}^{S}$, $\Delta\mu_{GaAs} = \mu_{Ga}^{L} + \mu_{As}^{L} - \mu_{GaAs}^{S}$, where $\mu_{Al}^{L}, \mu_{Ga}^{L}$ and μ_{As}^{L} are the chemical potentials of Al, Ga and As in the droplet, μ_{AlAs}^{S} and μ_{GaAs}^{S} are the AlAs and GaAs chemical potentials in solid state. The Gibbs-Thomson effect was neglected since the correction to the chemical potentials is small for the radii to be considered (about 30 nm). For the calculation of the chemical potentials in the droplet, we considered the liquid as a regular solution. For Al atoms, for example, the chemical potential of is written as: $\mu_{Al}^{L} = \mu_{Al}^{0L} + k_B T \ln a_{Al}$ where μ_{Al}^{0L} is the chemical potential of pure Al, k_B - the Boltzmann constant, T – the growth temperature, a_{Al} - the Al activity in the four-component solution such that

$$k_B T \ln a_{Al} = k_B T \ln c_{Al} + \omega_{AlGa} c_{Ga}^2 + \omega_{AlAs} c_{As}^2 + \omega_{AlAu} c_{Au}^2 + c_{Ga} c_{As} \left(\omega_{AlGa} + \omega_{AlAs} - \omega_{GaAs} \right)$$
$$+ c_{Ga} c_{Au} \left(\omega_{AlGa} + \omega_{AlAu} - \omega_{GaAu} \right) + c_{Au} c_{As} \left(\omega_{AlAs} + \omega_{AlAu} - \omega_{AuAs} \right), \quad (1)$$

c_{Al}, c_{Ga} and c_{As} are the Al, Ga, As molar fractions, $\omega_{AlGa}, \omega_{AlAs}, \ldots$ - the energy coefficients that allows for pair interactions between species. The energy coefficients are calculated using the Stringfellow formula [16]

$$\omega_{XY} = \frac{1}{N_A} \frac{V_X V_Y}{c_X V_X + c_Y V_Y} \left[\left(\delta_X - \delta_Y \right)^2 - \frac{C \left(\chi_X - \chi_Y \right)^2}{\left(V_X V_Y \right)^{1/2}} \right], \quad (2)$$

where V_X is the molar volume, δ_X, the Hildebrand solubility parameter, χ_X; the Pauling electronegativity constant, N_A, the Avogadro constant, $C = 1.256 \times 10^5$ if all quantities are in SI units.

To find the chemical potentials in solid state, μ^S_{AlAs} and μ^S_{GaAs}, tabular values from the computations were used [17,18].

The Gibbs free energies of AlAs and GaAs island formation on the droplet-NW interface have the form of $G_k = a_k i^{1/2} - \Delta\mu_k i$, $k=$ AlAs, GaAs, where i is the number of AlAs or GaAs pairs in the nucleus, $a_k = 2^{3/2}(\pi h_{ML}\Omega)^{1/2}\gamma_k$ (for disk-shaped nuclei), γ_k ,the surface energy of nucleus side walls, h_{ML}, a monolayer height , Ω, volume per atom. For the mononuclear growth, the NW growth rate in monolayers equals to the number of AlAs and GaAs nuclei that appear on the NW top facet per second [14] $dL/dt = \pi R^2(I_{AlAs} + I_{GaAs})$, where $I_k = A_k \exp(-a_k^2/(4k_B T\Delta\mu_k))$ is the intensity of AlAs and GaAs island formation. Two intensities are summed since the AlAs and GaAs nucleation events are considered as independent. The factor A_k is a slow function (compared with the exponent) of atomic concentrations of species in the droplet and temperature. For simplicity, it was assumed that $A_{AlAs} = A_{GaAs} = A$, i.e. in the model the island formation intensity of AlAs and GaAs differ in strong exponential dependence only. The NW radius R is a constant since it does not change during the heterojunction formation in the experiment [13].

The lateral growth of the nucleus proceeds by attaching Al, Ga, As atoms from the liquid. The incorporation rate of Al and Ga atoms into the growing monolayer is proportional to the differences of corresponding chemical potentials: $v_{Al} = b_{Al}\Delta\mu_{AlAs}$, $v_{Ga} = b_{Ga}\Delta\mu_{GaAs}$ [19, 9], b_{Al} and b_{Ga} are kinetic coefficients. Therefore, the Al and Ga molar fraction in the growing monolayer can be presented in the following form:

$$X_{Al} \equiv \frac{v_{Al}}{v_{Al}+v_{Ga}} = \frac{\Delta\mu_{AlAs}}{\Delta\mu_{AlAs} + b_{Ga}/b_{Al}\Delta\mu_{GaAs}}, \quad X_{Ga} \equiv \frac{v_{Ga}}{v_{Al}+v_{Ga}} = 1 - X_{Al}. \tag{3}$$

Note that a monolayer composition is not influenced by the critical nucleus composition because its lateral size is much less than the NW radius. The material balance equations for material fluxes in and out of the droplet are derived on the basis that the chemical potential differences of species depend strongly on Al, Ga and As concentrations [5], leading to:

$$\begin{aligned} \frac{dN_{Al}}{dt} &= J_{Al}s(\alpha,\beta) - \frac{1}{2}\frac{dL}{dt}X_{Al}, \\ \frac{dN_{Ga}}{dt} &= J_{Ga}s(\alpha,\beta) - \frac{1}{2}\frac{dL}{dt}(1-X_{Al}), \\ \frac{dN_{As}}{dt} &= J_{As}s(\alpha,\beta) - \frac{1}{2}\frac{dL}{dt} - s_t(\beta)\frac{c_{As}}{\tau_{As}}, \end{aligned} \tag{4}$$

Where N_{Al}, N_{Ga}, N_{As} are the number of Al, Ga and As atoms in the droplet divided by the number of atoms in one monolayer ($N_{ML} = h_{ML}\pi R^2/\Omega$),respectively. J_{Al}, J_{Ga}, J_{As} are the molecular fluxes of Al, Ga and As species (in monolayer/s). τ_{As} is the desorption time for As atoms, $s(\alpha,\beta)$, part of the surface area of the droplet which the molecular fluxes intersect with [20] divided by πR^2 (the droplet has a spherical cap shape). α is the incidence angle of molecular beams (0 in this model), β, the droplet contact angle and $s_t(\beta) = 2/(1+\cos\beta)$, the total surface area of the droplet divided by πR^2. To obtaining the formulae in equation (4), the differences in atomic volumes in liquid and solid state between Al, Ga and As atoms were neglected ($\Omega_{Al} \approx \Omega_{Ga} \approx \Omega_{As}$). Evaporation of group III atoms are negligible [21, 22] and thus, the desorption time for As atoms only is introduced. The growth rate dL/dt and the monolayer composition X_{Al} are a function of the molar fractions c_{Al}, c_{Ga} and c_{As} which equal to $c_l = N_l/(N_{Au} + N_{Al} + N_{Ga} + N_{As})$, $l = Al, Ga, As$ by definition. Therefore, equation (4) is a system of ordinary differential equations with three unknown functions N_{Al}, N_{Ga}, N_{As} (N_{Au} is fixed). As initial conditions, N^0_{Al}, N^0_{Ga} and N^0_{As}, a steady state regime system solution is used, i.e. $dN_{Al}/dt = dN_{Ga}/dt = dN_{As}/dt = 0$.

3. Results and discussion

The model enables us to find the dependence of the heterojunction width on the growth temperature. For this purpose, it is essential to estimate the values of the material constants and how they vary with the temperature. In the general case, the following constants: A, τ_{As}, b_{Al}, b_{Ga}, Ω, are dependent on the temperature. A has an exponential temperature dependence where it is believed that the argument of the exponent is small enough to neglect this dependence [22]. The desorption time of As equals to $\tau_{As} = \tau_{As0}\exp\left(E_{des}/k_B T\right)$ [3]. $E_{des} \approx 0.5\ eV$ was estimated from the saturated vapor pressure dependence of As on the temperature [24]. The term b_{Al}/b_{Ga} reflects the ratio of two Avogadro dependences with $(E^{(Al)}_{des} - E^{(Ga)}_{des})/k_B T$ as an argument [19], where E_{des} is the energy barrier for atoms incorporating to the growing monolayer from the droplet. We suppose that this ratio has a weak dependence on temperature since the difference, $E^{(Al)}_{des} - E^{(Ga)}_{des}$, should be small in $k_B T$ units. Likewise for a simplified consideration, $b_{Al}/b_{Ga} = 1$. The surface energy of the nucleus side facets should be a slow linear function of the temperature and this dependence can be neglected in the fraction $\gamma_k^2/\Delta\mu_k$, where $\Delta\mu_k$ has a stronger polynomial dependence on T [5]. Given the exact values of the surface energies of nanoislands are unknown, the following simplification was made to account for the difference in AlAs and GaAs surface energies: $\gamma_{AlAs} = 1.2\gamma_0$ and $\gamma_{GaAs} = \gamma_0$, where the factor 1.2 is a typical ratio between the AlAs and GaAs surface energies [3,23] and γ_0 is left as a fitting parameter. Change in Ω is negligible for the considered temperature range (500-600 ^oC). As a result, there are three fitting parameters in the model: A, τ_{As0} and γ_0 that are determined from a comparison with the experimental data [13]. In the experiments, AlGaAs NWs with embedded GaAs quantum well were grown. The substrate temperature was fixed and equal to 550 or 580°C. The $As_4/(Ga+Al)$ flux ratio was set at 1. The nominal speed of growth (i.e. growth of a pure surface) was chosen equal to 1 monolayer per second for GaAs and 0.4 of monolayers per second for AlAs. The Al content along the NW was monitored by Raman spectroscopy. It was found that Al concentration was in the range of 0.24–0.26 while the NW radius was 20-30 nm.

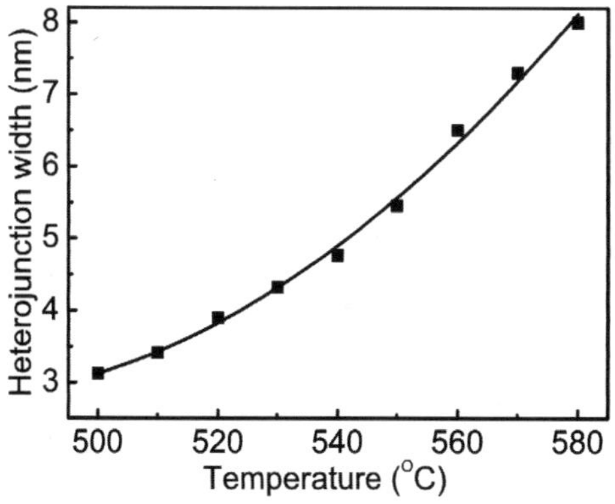

Figure 1. AlGaAs/GaAs heterojunction width as a function of the growth temperature. The solid line is the best-fit function $h = 103 - 0.4T + 4.5\cdot10^{-4}T^2$, where h is the heterojunction width.

To compute the AlGaAs/GaAs heterojunction width at the temperature *T*, equation (4) is solved when J_{Al} is turned off ($J_{Al} = 0, J_{Ga} \neq 0, J_{As} \neq 0$) and the numbers $N^0_{Al}, N^0_{Ga}, N^0_{As}$ from the steady state regime are used as initial conditions. To estimate the number of Au atoms N_{Au}, it is assumed that if the contact angle β equals 90°, the percentage of Au atoms is in the range of 50-80% [5], leading to $N_{Au} = 50N_{ML}$ in the computations. The results of the modelling are presented in figure 1, where the condition $X_{Al}(t) < X_{Al}(t=0)/e$ are used as a criterion that the heterojunction has formed (i.e. the Al flux is turned off at $t=0$). The following values of the fitting parameters ($A = 11.3\ nm^{-2}s^{-1}$, $\tau_{As0} = 3.2\times10^{-6}$ c, $\gamma_0 = 0.2\ J\cdot m^{-2}$) enabled us to obtain the heterojunction width and the NW growth rate similar to those in the experiment (about 3 nm and 1 nm/s, respectively). For comparison, $A \approx 0.5\ nm^{-2}s^{-1}$, $\gamma_0 \approx 0.1\ J\cdot m^{-2}$ were reported in the literature [22] to describe the formation of GaAs self-catalyzed NWs.

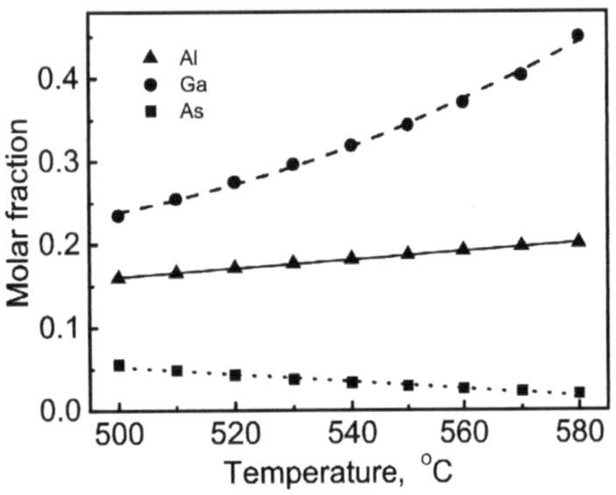

Figure 2. Al, Ga and As molar fractions versus the growth temperature. The lines are the best-fit functions $c_{Al} = -0.094 + 5.1\cdot10^{-4}T$, $c_{Ga} = 3.1 - 0.013T + 1.4\cdot10^{-5}T^2$ and $c_{As} = 0.27 - 4.4\cdot10^{-4}T$.

An increase of the heterojunction width from 4.1 nm to 7.7 nm (88% increase) when the temperature was varied from 500^oC to 580^oC is attributable to the increase of the number of Al atom N_{Al} in the droplet (108%). Although the change of the Al molar fraction (figure 2) is only 25%, the variation of the Ga molar fraction is much larger (91%) and results in a significant change of the droplet volume (66% increase that corresponds to the contact angle β change from 99° to 115° (figure 3) that leads finally to the increase of N_{Al}. The NW growth rate does not change much (increases by 18%, see figure 3) and equals to 1.2-1.4 nm/s. Thus, the values of heterojunction width and growth rate obtained in the model are similar to those in the experiment.

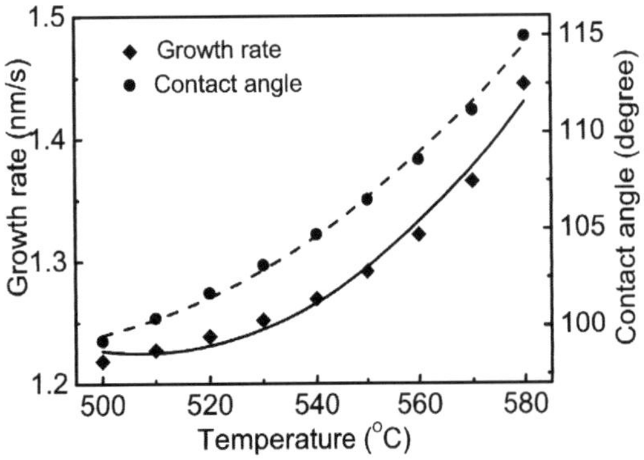

Figure 3. The dependence of NW growth rate and contact angle on the growth temperature. The lines are the best-fit functions $h_{ML}dL/dt = 11 - 0.04T + 3.9\cdot10^{-5}T^2$ and $\beta = 438 - 1.4T + 1.5\cdot10^{-3}T^2$.

It was found that the heterojunction width, contact angle, growth rate and Ga molar fraction increase with the temperature as second-order polynomial functions (figure 1, 2 and 3) but the Al and Ga molar fraction show a close to linear dependence on the temperature (figure 2).

4. Conclusion

A model of heterostructure formation in ternary III-V NWs grown by catalytic VLS method is presented. The approach proposed is based on the determination of chemical potential of a four-component liquid using the regular solution model and Stringfellow's scheme for the computation of the interaction coefficients of species. The AlGaAs/GaAs heterojunction formation in AlGaAs NWs was described using the model. Although the calculations were performed for the particular case of ternary material system, the model may be extended to other ternary III-V systems. In contrast to the previously developed models, the droplet volume can vary during the heterojunction formation. The heterojunction width was found to increase with the growth temperature with a second-order polynomial dependence. Another advantage of the model is an ability to describe both the steady-state and transient regime of NW growth. These results can be used for further improving of the NW growth techniques, e.g. for the optimization of the growth method with pulsed switching of material sources.

Acknowledgments

This work was partially supported by several grants of the Russian Foundation for Basic Research and the FP7 project FUNPROB.

References

[1] Vallett A L, Minassian S, Kaszuba P, Datta S, Redwing J M, Mayer T S 2010 *Nano Lett.* **10** 4813
[2] Borg B M, Dick K A, Ganjipour B, Pistol M E, Wernersson L E, Thelander C 2010 *Nano Lett.* **10** 4080
[3] Dubrovskii V 2014 *Nucleation Theory and Growth of Nanostructures* (Heidelberg: Springer)
[4] Messing M *et al.* 2011 *Nano Lett.* **11** 3899
[5] Glas F 2010 *Appl. Phys.* **108** 73506
[6] Dick A *et al.* 2012 *Nano Lett.* **12** 3200
[7] Li N, Tan T, Gosele U 2008 *Appl. Phys.* A **90** 591
[8] Perea D, Li N, Dickerson R, Misra A, Picraux S 2011 *Nano Lett.* **11** 3117
[9] Periwal P *et al.* 2014 *Nano Lett.* **14** 5140
[10] Sibirev N V 2015 *Technical Physics Letters* **41** 209
[11] Dubrovskii V G 2013 *Physical Review* B **87** 195426
[12] Sibirev N V, Nazarenko M V, Zeze D A and Dubrovskii V G 2014 *Journal of Crystal Growth* **401** 51
[13] Kochereshko V *et al.* 2012 *Journal of Surface Investigation. X-ray, Synchrotron and Neutron Techniques* **6** *722*
[14] Dubrovskii V G and Sibirev N V 2004 *Phys. Rev.* E **70** 031604
[15] Landau L D and Lifshitz E M 1980 *Statistical Physics* vol 5 (Oxford: Butterworth-Heinemann)
[16] Stringfellow J C 1972 *J. Phys. Chem. Solids* **33** 665
[17] Dinsdale A T 1991 *Calphad* **15** 317
[18] Ansara I *et al.* 1994 *Calphad* **18** 177
[19] Chernov A A *et al.* 1984 *Modern Crystallography* vol 3 (Berlin: Springer)
[20] Glas F 2010 *Phys. Status Solidi* B **247** 254
[21] Dubrovskii V G 2014 *Appl. Phys. Lett.* **104** 053110
[22] Glas F *et al.* 2013 *Phys. Rev.* B **88** 195304
[23] Sibirev N V *et al.* 2010 *Physics of the Solid State* **52** 1531
[24] Hultgren R *et al.* 1973 *Selected Values of Thermodynamic Properties of the Elements* (Ohio: American Society for Metals)

Modelling the growth process of porous aluminum oxide film during anodization

E M Aryslanova[1], A V Alfimov[1], S A Chivilikhin[1]

[1]ITMO University, Kronverkskiy Ave, 49, St. Petersburg, 197101, Russia

E-mail: elizabeth.aryslanova@gmail.com

Abstract. Currently it has become important for the development of metamaterials and nanotechnology to obtain regular self-assembled structures. One such structure is porous anodic alumina film that consists of hexagonally packed cylindrical pores. In this work we consider the anodization process, our model takes into account the influence of layers of aluminum and electrolyte on the rate of growth of aluminum oxide, as well as the effect of surface diffusion. In present work we consider those effects. And as a result of our model we obtain the minimum distance between centers of alumina pores in the beginning of anodizing process.

1. Introduction

Currently, the development of nanotechnology and metamaterials requires the ability to obtain regular self-assembled structures with different parameters [1-2]. One such structure is porous alumina film, which are self-organizing structures that consist of hexagonally packed cylindrical pores (Figure 1). Artificially on the surface of the aluminum may be built a thick layer of porous aluminum oxide film. Highly ordered pores may be obtained using two-stage anodization process, proposed in 1995 by Masuda et al. [3]. Pore size and the distance between them can be varied depending on the anodization voltage, the electrolyte and the anodization time (pore diameter - from 2 to 350 nm, the distance between the pores - from 5 to 50 nm).

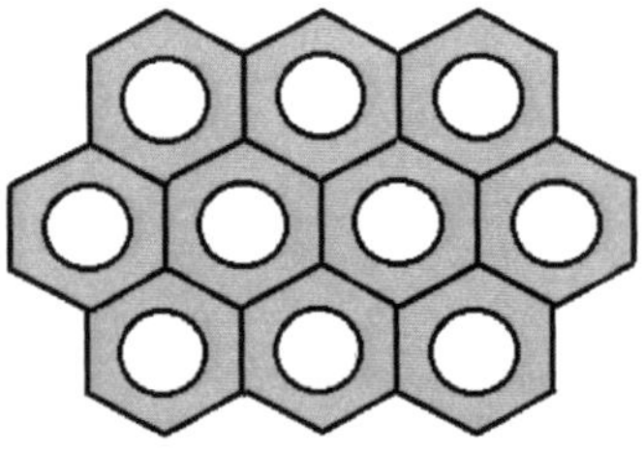

Figure 1. Hexagonal packing of the porous AAO

Anodizing is carried out in a vessel with electrolyte (aqueous solution of acid: oxalic, phosphoric acid; chromium, etc.), where are placed anode (aluminum) and cathode (inert conductive material), which are respectively connected to the positive and negative power supply output (Figure 2).

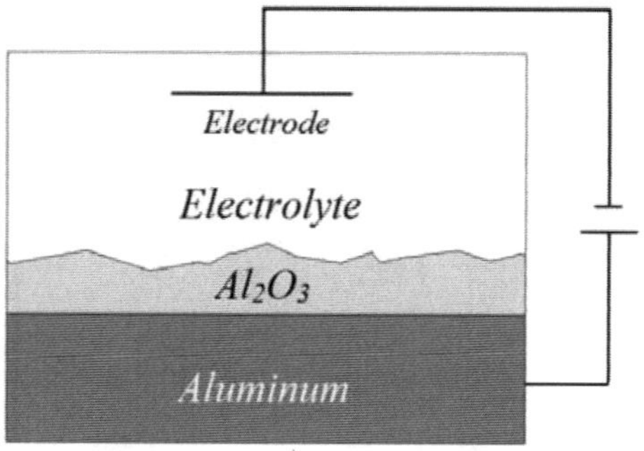

Figure 2. Anodizing process

On the surface of the aluminum grows thin aluminum oxide film. As an electric current flows at the aluminum - electrolyte border there grows a thin dense electrolyte film – barrier layer. Barrier layer grows due to the migration of aluminum ions towards oxygen ions. The thickness of this barrier layer (0.01 – 0.1 nm) does not change throughout the process as it dissolves at the outer side exposed to the electrolyte [4].

Electrochemical field localizes on inhomogeneities of the surface of oxide, therefore the oxide dissolves more intense the higher the inhomogeneity of the field. Growth then occurs in the area of inhomogeneity. After some time the competition between the pores leads to a stabilization process. However, the thicker the structure, the more uneven grows the pore walls. Alumina layer is then dissolved, leaving a regular array of porous aluminum. The whole anodizing process is repeated a second time, which leads to growth of regular porous alumina.

2. Model

In this paper we describe an analytical model of the growth of anodic alumina. We consider the motion of the interfaces between the electrolyte-Al_2O_3 (dissolution) and between the Al_2O_3-aluminum (oxidation) (Figure 2).

We consider the dynamics of moving boundaries and the change of small perturbations forms of these boundaries. In each of the areas under consideration Laplace equation for the electric potential is solved. The process of growth of porous alumina is described by the theory of small perturbations. In zero approximation boundaries are considered flat and the speed of their movement proportional to the current density at these boundaries. The first approximation takes into account the small perturbations of the interface, which lead to small changes in the potentials and currents at these boundaries. The evolution of small perturbations of the interface is defined as a perturbation of the current density at the borders, and the process of surface diffusion.

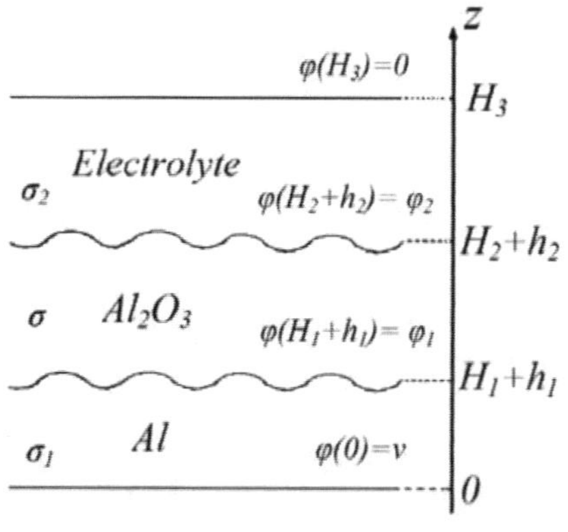

Figure 3. Inhomogeneous film with rough boundaries.

Using conditions of continuity of the current density at both interfaces we obtain the system of equations relating the potential disturbance on the interfaces:

$$\begin{cases} \left(-\sigma_2 sh(kH) - \sigma ch(kH)\right)\hat{\varphi}_{2\bar{k}} + \sigma\hat{\varphi}_{1\bar{k}} = \beta e^{kH} - \alpha \\ -\sigma\hat{\varphi}_{2\bar{k}} + \left(\sigma_1 sh(kH) + \sigma ch(kH)\right)\hat{\varphi}_{1\bar{k}} = \beta - \alpha e^{kH} \end{cases} \quad (1)$$

where $\alpha = (v/\vartheta)h_{1\bar{k}}$, $\beta = (v/\vartheta)h_{2\bar{k}}$. $\hat{\varphi}_{\bar{k}}$ - the Fourier transform of the potential disturbance to the coordinates x, y, $\varphi = \varphi^0_{(z)} + \hat{\varphi}$, $\varphi^0_{(z)}$ - stationary potential for the homogeneous problem, $\hat{\varphi}$ - potential oscillations, $_1$ – electrostatic potential of the aluminum layer on Al-Al_2O_3 interface, $_2$ – electrostatic potential of the aluminum layer on electrolyte-Al_2O_3 interface, $_1$ – conductivity of the aluminum, – conductivity of aluminum oxide, $_2$ – conductivity of the electrolyte, $\vartheta = \left(H/\sigma + H_1/\sigma_1 + \left(H_3 - H_2\right)/\sigma_2\right)$, v – anodization voltage, H– thickness of aluminum oxide, H_1+h_1 – thickness of the aluminum layer, $H_3-(H_2+h_2)$ - thickness of electrolyte, $h_{1\bar{k}} = \int_{-\infty}^{\infty} \exp\left[-ik_y y\right] \int_{-\infty}^{\infty} \exp\left[-ik_x x\right] h_1 dxdy$, $h_{2\bar{k}} = \int_{-\infty}^{\infty} \exp\left[-ik_y y\right] \int_{-\infty}^{\infty} \exp\left[-ik_x x\right] h_2 dxdy$.

We considered the solutions of the system (1) for different values of kH.

The rate of change of small perturbations of the boundaries Al-Al_2O_3 and Al_2O_3-electrolyte without the influence of surface diffusion is proportional to the perturbation of the current density at these interfaces $dh/dt = a\sigma\, d\varphi/dz$. Since the conductivity of the electrolyte and alumina is small compared with aluminum we see that the perturbation on the interface of Al_2O_3-electrolyte increases indefinitely with time.

The rate of change of small perturbations of the interfaces Al-Al_2O_3 and Al_2O_3-electrolyte with the influence of surface diffusion is described by the relation $\frac{\partial h}{\partial t} = V + D\Delta^2 h$, where D – is the surface diffusion coefficient.

For Al-Al_2O_3 interface we get the wavelength corresponding to the limit of stability:

$$\lambda_1 = 2\pi\sqrt[3]{a_1 v/\vartheta D_1}. \quad (2)$$

All perturbations of this interface with wavelength $\lambda > \lambda_1$ are unstable.

For Al_2O_3-electrolyte interface we get the wavelength corresponding to the limit of stability:

$$\lambda_2 = 2\pi\sqrt[3]{a_1 v/\vartheta D_2}. \quad (3)$$

All perturbations of this interface with wavelength $\lambda > \lambda_2$ are unstable.

We assume that surface diffusion coefficient D_2 on the Al_2O_3-electrolyte interface is larger than the same coefficient on D_1 the Al-Al_2O_3 interface. According to (2), (3), the correspondent critical wavelength satisfies of inequality $\lambda_2 < \lambda_1$. Therefore, for perturbations with wavelength λ in interval $\lambda_2 < \lambda < \lambda_1$ the upper boundary of alumina layer is unstable and bottom boundary is stable. The perturbations of upper boundary in this interval of wavelength are the source of porous formation. For wavelength $\lambda > \lambda_1$the bottom boundary of alumina is unstable. The evolution of perturbations in this wavelength region determines the irregularities of porous structure.

3. Conclusion

As a result of the developed model we obtained the minimum distance $2\pi\sqrt[3]{a_1 kv / \vartheta D_2}$ between centres of aluminum oxide pores in the beginning of anodizing process. The irregularities of porous structure have wavelength $\lambda > 2\pi\sqrt[3]{a_1 kv / \vartheta D_1}$. The dependence of distance between centres of inhomogeneities on anodizing voltage is shown in Figure 4. It follows that the bigger the anodizing voltage, the less smooth the surface should be at the beginning of anodization process.

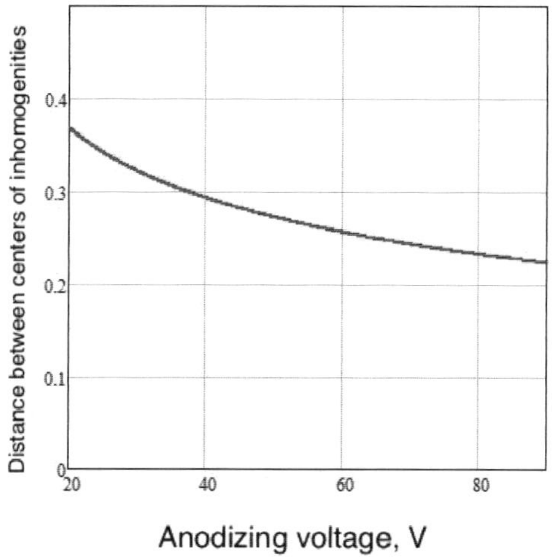

Figure 4. Dependence of distance between centres of inhomogeneities on anodizing voltage

Acknowledgments
This work was financially supported by the Government of Russian Federation, Grant 074-U01.

References

[1] Limonov A G 2010 *Matem. Mod.* **22** 97–108
[2] Petukhov D I, Napolskii K S and Eliseev A A 2012 *Nanotechnology* **23** 335601
[3] Masuda H, Fukuda K 1995 *Science*, **268(9)** 1466-8
[4] Patermarakis G 2009 *Journal of Electroanalytical Chemistry* **635** 39–50

Modelling of field desorption of monocrystal nanotip

K A Nikiforov and A A Krasnova

Saint-Petersburg State University, 7-9, Universitetskaya nab., St. Petersburg, 199034, Russia

E-mail: knikiforov@cc.spbu.ru

Abstract. Mathematical and computer model of field desorption process from metal nanocrystal tip is proposed. The radius of curvature on the top of the emitter is about 50 lattice parameters. The model includes initial calculation of intersection between the crystal lattice and emitter shape for bcc and fcc crystal structures. Arbitrary axisymmetric shapes (figures of rotation) can be used for the emitter model. The algorithm for allocation of atoms being desorbed at given time step is based on an analysis of geometric environment with specified local electric field. Polyhedron nanostructured shape of emitter is obtained as result of evaporation. Computer program realization (Matlab stand alone application) is presented.

1. Introduction

The object of this study is a tip-shaped metal field emitter with a monocrystal on its top. The main goal is constructing a model of structure of emitter surface under influence of an electric field at given temperature. Field desorption can lead to a particular surface structure that does not depend on the starting shape of the emitter but characterizes the lattice type and size of the monocrystal. This fact can probably be used for further determining of emission experiment conditions and corresponding models of both electron [1-12] and ion [13] field emission. As such, this study is of interest for such fields as field desorption microscopy and atomic probe tomography, that are among the modern methods of studying of 3d structure in solid bodies [14].

In accordance with this goal the following problems have to be solved: detection of the outline of the experimental sample emitter tip on a microphoto and finding the coordinates of its cross-section points using the Sobel operator; choosing a second-order rotary surface that would be the best approximation for the emitter shape using the method of the least squares; finding an equidistant line for the forming surface for modeling thin subsurface layer using the method of bisection; calculation of atomic structure of emitter tip surface based on the chosen approximating shape and crystal lattice type; choosing a model criterion of field desorption from the emitter surface and construction of corresponding model algorithm that would satisfy the desorption criteria; implementation of this algorithm and computation of emitter surface structure under the influence of temperature and electric field. This paper focuses on the latter two of those problems. The preceding ones are in greater detail considered in [2].

2. Simulation algorithm for field desorption process

For a given value of the electric field strength on the surface of the monocrystal nanotip the most protruding atoms evaporate firstly, as having the highest number of broken bonds with neighboring atoms of the crystal lattice.

In [1, 2, 13] it was been shown that without a substantial loss of accuracy, we can confine ourselves to the fifth order of nearest neighbors in the analysis of crystallographic faces on the surface of the emitter tip. Therefore, analysis of the number of broken bonds in this paper was carried out with the same view of the nearest neighbors up to the 5-th order.

We used the method described in [15] to take into the atomic environment to simulate the field evaporation of the crystals.

For the atoms with the fewest nearest neighbors following characteristics have to be calculated: attributed coordination number N_{attr}, characterizing the binding energy of the atom with a given environment on the sample surface; the binding energy *E*; the average evaporation time Δt_m by the following expressions (1)−(4).

$$N_{\mathrm{attr}} = \sum_{i=1}^{5} N_i k_i, \tag{1}$$

where N_i is the number of i-th order neighbors of an given atom, k_i is factor describing the dependence of the energy Ψ of pair interaction of atoms on the distance between them. In the simplified assumption that the interaction energy (binding energy) of an atom with its neighbors decreases in proportion to the sixth power of the distance [15]

$$k_i = g_i^6 g_1^{-6}, \tag{2}$$

where g_i is the distance to the i-th order neighbor.

Full binding energy of atom on the surface or in the volume of the crystal

$$E = \sum_{i=1}^{5} N_i \Psi_i = k_1 N_1 \Psi_1 + k_2 N_2 \Psi_1 + k_3 N_3 \Psi_1 + k_4 N_4 \Psi_1 + k_5 N_5 \Psi_1 = \Psi_1 \sum_{i=1}^{5} N_i k_i = N_{\mathrm{attr}} \Psi_1, \tag{3}$$

where Ψ_i is the energy of pair interaction with i-th order neighbouring atom.

For selected atoms with the lowest binding energy *E* the evaporation time (can be interpreted as the mean value of the time during which an atom with a given environment stays on the surface in a bound state) was calculated

$$\Delta t_m = c \exp\left[10^4 \eta T^{-1}\left(c_1 F^{1/2} + c_2 N_{\mathrm{attr}}\right)\right], \tag{4}$$

where η, c, c_1, c_2 are constants [15], *T* is the temperature; *F* is the electric field strength.

It specifies the time Δt of stable existence of corresponding configuration of the surface as the least evaporation time for all atoms with the lowest binding energy.

Simulation of the evaporation process at this stage is completed by removal of atoms, which have the evaporation time equal to Δt, and algorithm is repeated again for the new configuration of the nanotip by simulating the next stage of the process.

It is necessary to note that although in this approach the strength of field *F* can be calculated for each step of the algorithm taking into account the changing surface cross-section, in this research we used a constant value for *F*, which may be seen as a drawback. On the other hand, this algorithm takes in account the local neighborhood (i.e. number of neighbours) for each of the surface atoms and its changes during the desorption process. This is a positive side of the model.

3. Modeling results

Simulation algorithm was implemented in Matlab. Table 1 presents calculated data for tungsten monocrystal nanotip (bcc crystal structure, radius of apex curvature is 50 lattice parameters) characterizing the process of field desorption (field strength is 48 GV·m^{-1}, temperature 78 K).

Figure 1 shows the dependence of number of desorbed atoms on time of evaporation. The interesting result from figure 1 is that during desorption process an increasing number of atoms leaving the surface, it shows that evaporated emitter decreases with time taking the form of a polyhedron. One can suppose that the sides would be more smooth if local increases in electric field strength on particular areas of emitter surface were to be taken in account.

Figure 2 shows a model of a segment of a hemispherical apex of the emitter, formed as result of the process of field evaporation. Area of some crystal faces increased on the surface of the emitter comparing to the original form.

Table 1. Atomic environment (type of bond), values of N_{attr} and evaporation times Δt_m of atoms in several different configurations on the tungsten nanotip surface.

Coordination numbers $N_1 - N_2 - N_3 - N_4 - N_5$	N_{attr}	Δt_m (s)
$4-3-6-11-3$	5.91	$2.93 \cdot 10^{-8}$
$4-3-6-12-3$	5.93	$2.43 \cdot 10^{-7}$
$4-4-6-12-4$	6.38	$7.58 \cdot 10^{-3}$
$6-3-7-12-4$	8.00	$1.92 \cdot 10^{9}$
$6-4-12-4-6$	8.21	$3.96 \cdot 10^{11}$
$6-4-7-14-6$	8.51	$2.30 \cdot 10^{16}$

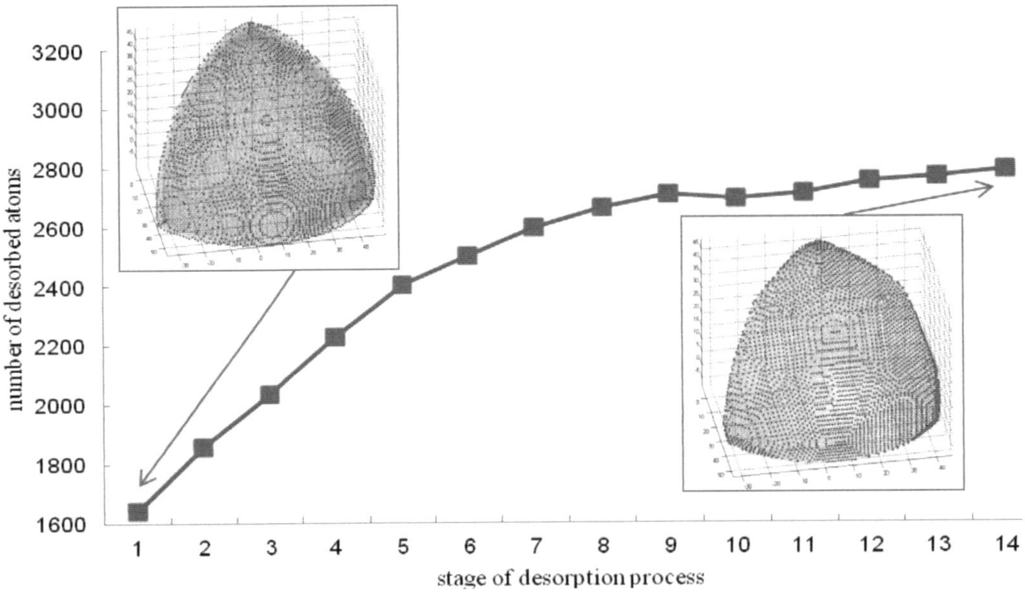

Figure 1. Number of evaporated atoms at several stages of field desorption process. Insets show surface configuration at the first and the last stages (surface atoms are shown in blue, desorbed atoms are marked red).

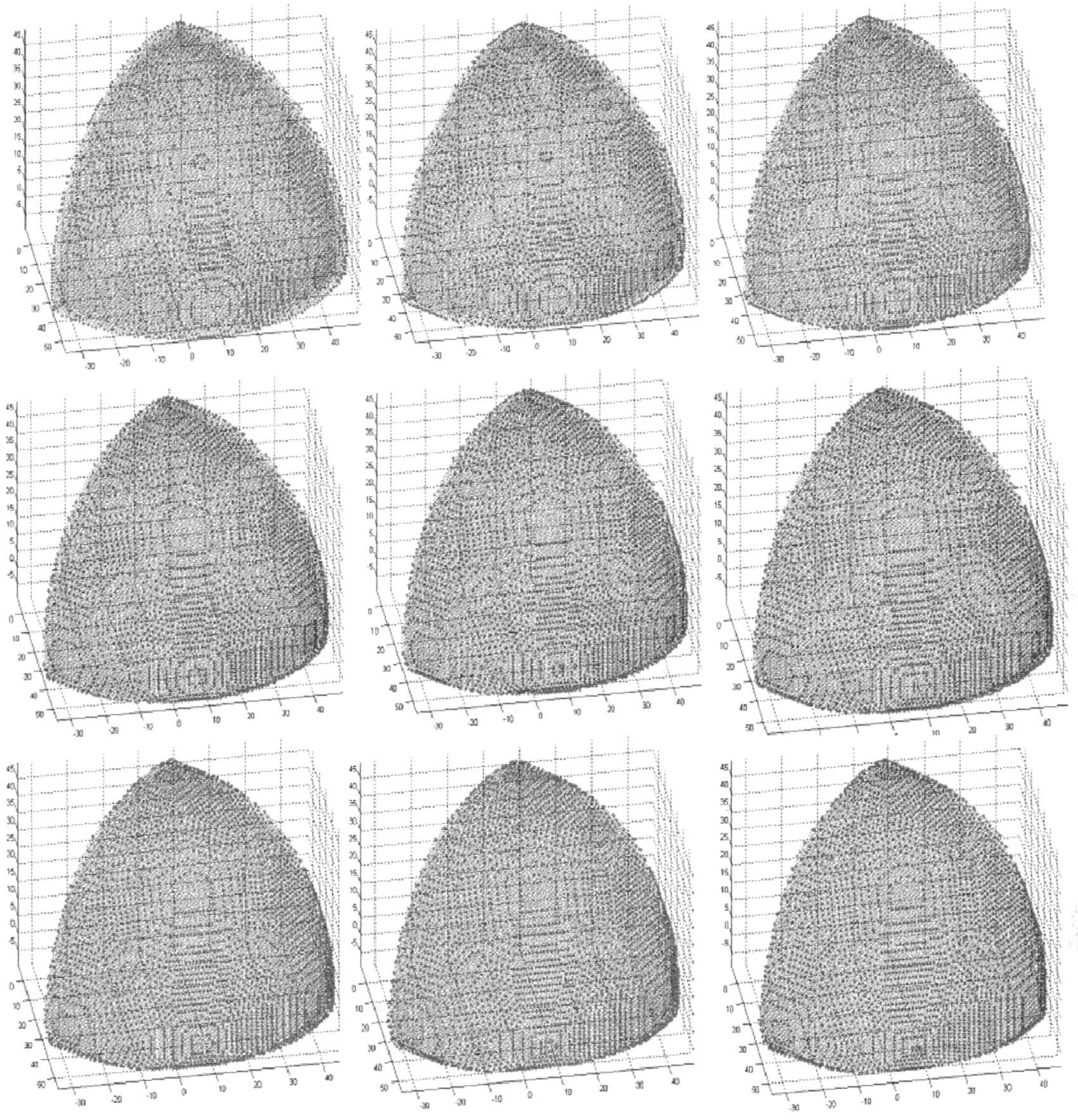

Figure 2. Several stages of field evaporation in the model of emitter tip surface segment.

4. Conclusion

In this paper we've constructed a model of emitter surface formed by the process of field-temperature evaporation. Some of the crystallographic faces of the emitter surface grew in this process relative to the initial shape (described in [2]). Thus, we can see how in field-temperature evaporation the structure of the emitter surface becomes more regular, which can have practical application for making experimental conditions for study of adsorption more precisely defined and predictable [3]. In this paper we have assumed the electric field constant, although it is known that field distribution is influenced by the surface structure that changes due to desorption. On the other hand, we do take into account those changes while calculating the binding energies. Even this simplified approach requires a considerable amount of computational resources, which makes studying of its computational complexity and looking for ways to increase its efficiency an important area of future research [16-18].

Acknowledgments

Financial support has been obtained from RFBR (13-01-00150) and partially from Saint-Petersburg State University (9.38.673.2013). Research was carried out using computational resources provided by Resource Center "Computer Center of SPbU" (http://cc.spbu.ru/en).

[1] Nikiforov K A, Egorov N V and Shen C-C 2009 Surface reconstruction of a field electron emitter *Journal of Surface Investigation: X-Ray, Synchrotron and Neutron Techniques* **3** 833-9

[2] Nikiforov K and Krasnova A 2014 Model of field electron emitter surface structure *2014 Tenth International Vacuum Electron Sources Conference, IVESC (Saint Petersburg)* (Piscataway: IEEE) 6892018

[3] Plšek, J, Hrubý P, Nikiforov K and Knor Z 2005 Properties of physisorbed water layers on gold revealed in a FEM study *Applied Surface Science* **252** 1553–60

[4] Nikiforov K A, Antonova L I, Egorov N V, Trofimov V V, Makeev V V and Ogurtsov O F 2012 Non-gated field emission array as low-energy electron source: Experiment and simulation *RuPAC 2012 Contributions to the Proc. of 23 Russian Particle Accelerator Conf. (St. Petersburg)*, (Cern: JACoW) 218–20

[5] Nikiforov K A and Egorov N V 2012 Program complex for vacuum nanoelectronics finite element simulations *RuPAC 2012 Contributions to the Proc. of 23 Russian Particle Accelerator Conf. (St. Petersburg)*, (Cern: JACoW) 409–11

[6] Sayfullin M and Nikiforov K 2014 Fringe field effect study of a field emitter array *2014 Tenth International Vacuum Electron Sources Conference, IVESC (Saint Petersburg)* (Piscataway: IEEE) 6892072

[7] Nikiforov K and Andrievskiy D 2014 Comparison of injection algorithms for electron beam simulation by particle-in-cell method *2014 20th International Workshop on Beam Dynamics and Optimization, BDO (Saint Petersburg)* (Piscataway: IEEE) 6890057

[8] Nikiforov K 2014 Modelling of emission processes in Matlab *International Conference on Computer Technologies in Physical and Engineering Applications, ICCTPEA (Saint Petersburg)* (Piscataway: IEEE) 6893317

[9] Sayfullin M F and Nikiforov K A 2014 Studying fringe field effect of a field emitter array *Journal of Physics: Conference Series* **541** 012020

[10] Gallyamov Z R and Nikiforov K A 2014 Model of electron transport in cell of a thin-film vacuum nanotriode *Journal of Physics: Conference Series* **541** 012035

[11] Nikiforov K A and Zartdinov A N 2014 Studying field emission characteristics of point and wedge–shaped surface defects *Journal of Physics: Conference Series* **541** 012009

[12] Nikiforov K 2015 Mathematical model and software complex for computer simulation of field emission electron sources *AIP Conf. Proc.* **1648** 450014

[13] Nikiforov K A and Egorov N V 2015 Simulation of specimen structure in atom probe tomography and field electron microscopy *2015 International Conference on Mechanics – Seventh Polyakhov's Reading (Saint Petersburg)* (Piscataway: IEEE) 7106762

[14] Gault B, Moody M P, Cairney J M and Ringer S P 2012 *Atom Probe Microscopy* (Berlin: Springer)

[15] Suvorov A L, Razinkova T L, Kukavadze G M, Bobkov A F, Kuznetsov B Ya and Kuznetsov V 1975 Kinetics of field evaporation (continuous regime) *Zh. Eksp. Teor. Fiz.* **68** 1460-70

[16] Kiktenko A A, Lunkovskiy M N and Nikiforov K A 2014 Confidence complexity of computer algorithms *2014 2nd International Conference on Emission Electronics, ICEE (Saint Petersburg)* (Piscataway: IEEE) 6893971

[17] Kiktenko A, Lunkovskiy M and Nikiforov K 2014 Studying Informational Sensitivity of Computer *Algorithms Advances in Intelligent Systems and Computing* **285** 421–31

[18] Kiktenko A, Lunkovskiy M and Nikiforov K 2015 Informational sensitivity as algorithm's complexity function characteristic *AIP Conf. Proc.* **1648** 450015

Journal of Physics: Conference Series **643** (2015) 012010 doi:10.1088/1742-6596/643/1/012010

Modelling of field emitter surface structure

K A Nikiforov, N V Egorov and M N Lunkovskiy

Saint-Petersburg State University, 7-9, Universitetskaya nab., St. Petersburg, 199034, Russia

E-mail: knikiforov@cc.spbu.ru

Abstract. The paper presents a mathematical model of the crystal structure of the emitter, which is the subject of research in atom probe tomography and in the field electron microscopy. Electronic, mechanical and other properties of the sample having the shape of the tip with a radius of curvature at the top of the order of hundred nanometers, these methods are considered on the basis of field desorption and emission (electron or ion). Strong electric field simultaneously performs generating, focusing, accelerating and transport function. Therefore, multi-scale modeling of the properties of the sample, including an approximation of its shape on the micro-, meso- and nanolevel, calculation of atomic packing density and size of crystal faces is a topical applications and is used to interpret the results of field emission experiment.

1. Introduction

Field emission properties of crystals are depended on surface structure. Information about the processes that occur on the metal surface in conditions of influence of strong electric fields, are important from the practical point of view for technologies of manufacture and use of electron and ion emitters, atom probes, field emission and scanning electron microscopy. Studying those phenomena is also useful for theoretical science [1−17] as it allows to obtain the data of mechanisms and properties of diffusion processes, crystal growth, self-organization and phase transitions, to measure such values as surface tension (free surface energy) in solids, energies of interaction of atoms and molecules with the surface and between each other.

The object of study is a metal single crystal emitter tip, which can be made out of metal wires of diameters about 0.1−0.15 mm by the use of anode electric etching.

Emission properties of crystal substances are in a great extent defined by the electron work function. It is usually assumed that this value has both surface and volume parts. Most probably, both of them have a non-negligible influence. The impact of the surface structure can be observed in the fact that faces with different Miller indexes have different work function values. But it's the volume that supplies the electrons to the surface and it's unacceptable to neglect this.

The goal of this research is development of model for surface structure of field emitter, based on its meso-scale shape approximation and atomistic calculation of crystal lattice inside this shape. Subproblem of this modeling is work function map calculation for the surface of emitter.

2. Surface structure calculation

Calculation of the atomic structure of the surface of monocrystal emitter tip is a basic step in many models of both electron and ion field emission. As the emission properties are mostly defined by relatively thin layer below the surface, it is only necessary to take in account the coordinates of atoms that lie up to certain depth and store them in computer memory. The surface atoms are defined as ones having an incomplete set of neighbors of a certain order, which distinguish them from the atoms in volume.

For example Figure 1 shows possible shapes approximating emitter as equipotential surfaces of the electric field [1] generated by charged cone with a sphere at the top (dotted line). As can be seen , the emitter apex shape is close to hemisphere. The model structure of the apex surface is shown in Figure 1c and 1d, where the surface layers of the atoms of different depths (0.2 and 0.4 lattice parameters) and Miller indices for major crystal faces are represented. The model structure of the apex surface is shown in Figure 2, where the surface layers of the atoms of different depths (0.2 and 0.4 lattice parameters) and Miller indices for major crystal faces are represented. Computational algorithm is based on [2, 3] and implemented in the Matlab environment.

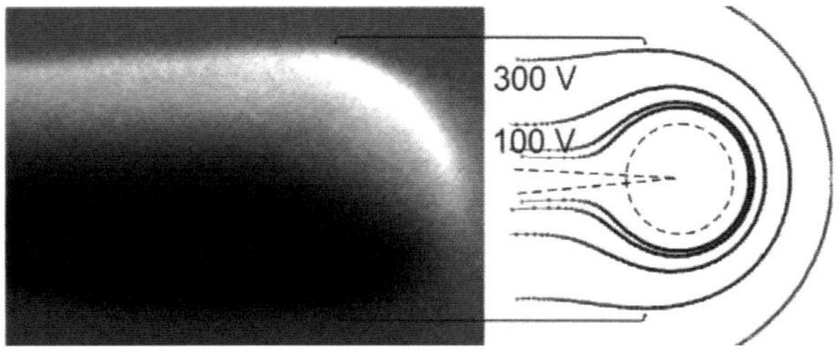

Figure 1. Emitter profile and shape approximation.

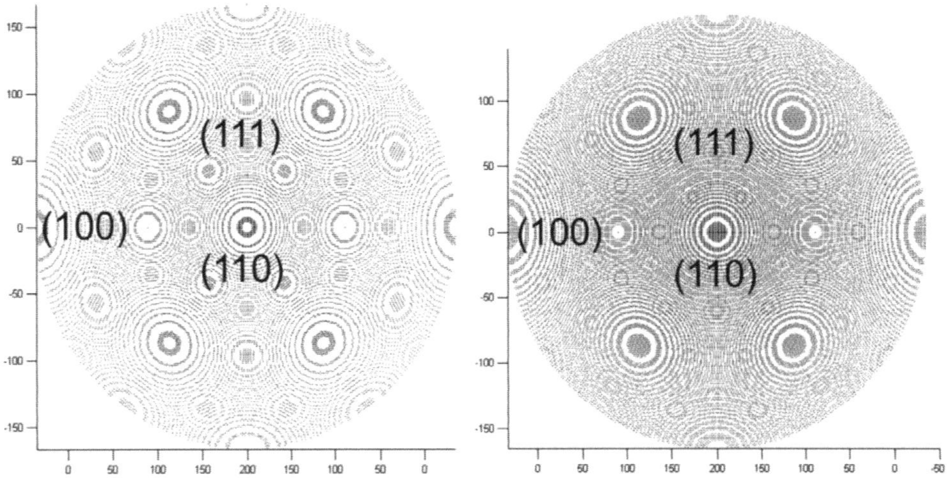

Figure 2. Apex surface structure models with depth 0.2 and 0.4 lattice parameters.

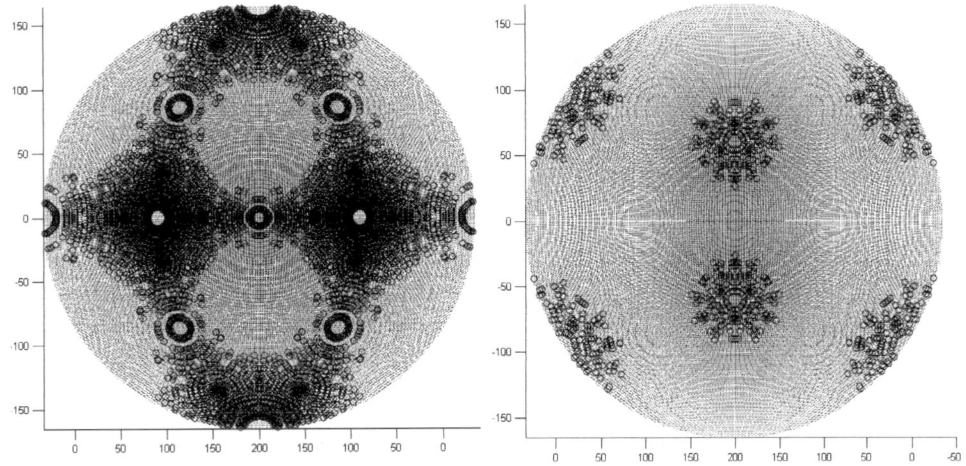

Figure 3. Apex surface structure models with depth 0.6 and 0.8 lattice parameters.

Figure 3 shows surface layer with depth 0.6 lattice parameters, where 5160 atoms (highlighted) have full set of first order neighbours and 0.8 lattice parameters, where only 838 atoms (highlighted) have broken bounds with nearest neighbours of first order.

3. Work function surface map

Although, in principle, work function can be calculated directly (by the DFT method and the cluster model of the surface, but it is practically difficult to implement for whole surface, because of the large numbers of atoms on the surface. Another approach is more practical, although it does not take into account effects such as relaxation and reconstruction of the surface, but also requires an analysis of the coordination numbers that characterize the types of atoms (step, kink, with different binding energy and different local values of the work function

The results of paper [18] show that the semiempirical model proposed by Surma can be used successfully to predict metal work functions near the crystal faces, experimental data on which are lacking, for example, in the simulation of the phenomenon of field electron emission. The error of such a prediction can be at a level of 0.2 eV. It is important to mention that, in tested cases of tungsten, molybdenum, platinum and iridium the Surma model was assumed to be statistically significant [18].

In this paper work function simulation is based on Surma model and this aspect is new for field emitter surface structure model, applicable to field electron/ion microscopy simulation, as well as for field desorption microscopy and atom probe tomography [4].

The simulation of the work function values nonuniform distribution (due to crystallographic anisotropy) over the surface of the emitter is carried out within the framework of the broken-bounds approach. Figure 4 shows work function map of the emitter surface.

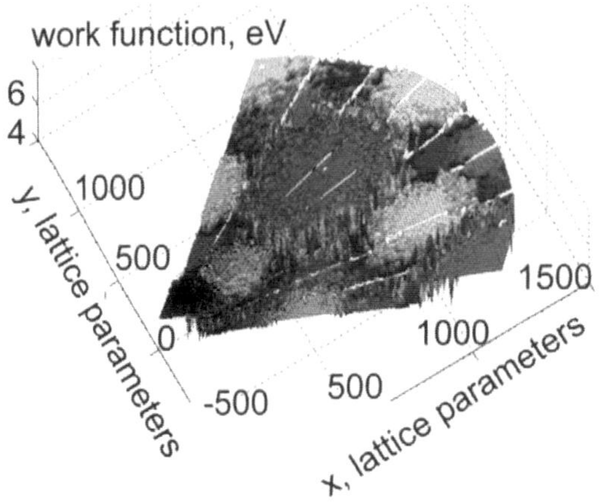

Figure 4. Model map of work function distribution over the tungsten emitter surface (one quarter of the hemispherical apex of emitter).

4. Conclusion

The developed mathematical model describes the surface structure of the field emitter metal tip taking into account the arrangement of different crystallographic faces at the top of the emitter. This study can be associated with the model of patch field effect due to contact potential difference established between areas having different work function values.

Acknowledgments

Financial support has been obtained from RFBR (13-01-00150) and partially from Saint-Petersburg State University (9.38.673.2013). Research was carried out using computational resources provided by Resource Center "Computer Center of SPbU" (http://cc.spbu.ru/en) and using experimental equipment of the Interdisciplinary Resource Center for Nanotechnology of St. Petersburg State University.

[1] Nikiforov K A, Egorov N V, Shen C-C 2009 Surface reconstruction of a field electron emitter *Journal of Surface Investigation: X-Ray, Synchrotron and Neutron Techniques* **3** 833-9

[2] Nikiforov K, Krasnova A 2014 Model of field electron emitter surface structure *2014 Tenth Int. Vacuum Electron Sources Conf. (Saint Petersburg)* (Piscataway: IEEE)

[3] Eaton H C, Lee L 1982 The simulation of the images of the field ion microscope: Specimen of arbitrary crystal structure and orientation *Journal of Applied Physics* **2** 988-94

[4] Gault B, Moody M P, Cairney J M, Ringer S P 2012 *Atom Probe Microscopy* (Berlin: Springer)

[5] Nikiforov K A, Egorov N V and Shen C-C 2009 Surface reconstruction of a field electron emitter *Journal of Surface Investigation: X-Ray, Synchrotron and Neutron Techniques* **3** 833-9

[6] Nikiforov K and Krasnova A 2014 Model of field electron emitter surface structure *2014 Tenth International Vacuum Electron Sources Conference, IVESC (Saint Petersburg)* (Piscataway: IEEE) 6892018

[7] Plšek, J, Hrubý P, Nikiforov K and Knor Z 2005 Properties of physisorbed water layers on gold revealed in a FEM study *Applied Surface Science* **252** 1553–60

[8] Nikiforov K A, Antonova L I, Egorov N V, Trofimov V V, Makeev V V and Ogurtsov O F 2012 Non-gated field emission array as low-energy electron source: Experiment and simulation *RuPAC 2012 Contributions to the Proc. of 23 Russian Particle Accelerator Conf. (St. Petersburg)*, (Cern: JACoW) 218–20

[9] Nikiforov K A and Egorov N V 2012 Program complex for vacuum nanoelectronics finite element simulations *RuPAC 2012 Contributions to the Proc. of 23 Russian Particle Accelerator Conf. (St. Petersburg)*, (Cern: JACoW) 409–11

[10] Sayfullin M and Nikiforov K 2014 Fringe field effect study of a field emitter array *2014 Tenth International Vacuum Electron Sources Conference, IVESC (Saint Petersburg)* (Piscataway: IEEE) 6892072

[11] Nikiforov K and Andrievskiy D 2014 Comparison of injection algorithms for electron beam simulation by particle-in-cell method *2014 20th International Workshop on Beam Dynamics and Optimization, BDO (Saint Petersburg)* (Piscataway: IEEE) 6890057

[12] Nikiforov K 2014 Modelling of emission processes in Matlab *International Conference on Computer Technologies in Physical and Engineering Applications, ICCTPEA (Saint Petersburg)* (Piscataway: IEEE) 6893317

[13] Sayfullin M F and Nikiforov K A 2014 Studying fringe field effect of a field emitter array *Journal of Physics: Conference Series* **541** 012020

[14] Gallyamov Z R and Nikiforov K A 2014 Model of electron transport in cell of a thin-film vacuum nanotriode *Journal of Physics: Conference Series* **541** 012035

[15] Nikiforov K A and Zartdinov A N 2014 Studying field emission characteristics of point and wedge–shaped surface defects *Journal of Physics: Conference Series* **541** 012009

[16] Nikiforov K 2015 Mathematical model and software complex for computer simulation of field emission electron sources *AIP Conf. Proc.* **1648** 450014

[17] Nikiforov K A and Egorov N V 2015 Simulation of specimen structure in atom probe tomography and field electron microscopy *2015 International Conference on Mechanics – Seventh Polyakhov's Reading (Saint Petersburg)* (Piscataway: IEEE)

[18] Egorov N V, Antonov A Y, Gribkova I M Statistical test of a single semiempirical work function model 2014 *Journal of Surface Investigation: X-Ray, Synchrotron and Neutron Techniques* **8** 138–43

Synthesis of silicon carbide at room temperature from colloidal suspensions of silicon dioxide and carbon nanotubes

D A Zhukalin, A V Tuchin, T V Kulikova, L A Bityutskaya

Voronezh State University, 394006, Universitetskaya pl.1, Voronezh, Russia

E-mail: d.zhukalin@mail.ru

Abstract. Experimental and theoretical approaches were used for the investigation of mechanisms and conditions of self-organized nanostructures formation in the drying drop of the mixture of colloidal suspensions of nanoscale amorphous silicon dioxide and carbon nanotubes. The formation of rodlike structures with diameter 250-300nm and length ~4μm was revealed. The diffraction analysis of the obtained nanostructures showed the formation of the silicon carbide phase at room temperature.

1. Introduction

The use of carbon nanotubes (CNT) is commonly assumed while producing composite and hybrid nanomaterials as well as elements of nanoelectronics, sensing, energetic, microsystems technology [1–10]. There is a great interest to Si/CNT composites as nanomaterials for structural stabile anodes of lithium-ion batteries with the high energy density [11–13]. The synthesis of Si/CNT composites often includes a high-temperature anneal [6, 13]. Bolotov et al. [1] showed that during high-temperature treatment of the composite SnO_{2-x}/multi-walled CNT in vacuum, there was a transformation of its electronic structure along with local synthesis of tin oxycarbide. In addition to external high-energy CNT activation methods, it is important to consider the internal reserve of nanotubes themselves related to their qualities, surface curvature and defects [5]. The uniqueness of carbon nanotubes is based on the dependence of the reactivity of carbon atoms on their topological positions: the increasing curvature is the reason of the decreasing activation energy of chemical reactions [14–16]. Caps of carbon nanotubes have the greatest curvature, which allows to consider caps as active centers. Defects of the hexagonal structure define the local charge redistribution near caps [5, 16]. The charged active center is the core of the interaction between carbon nanotubes and second nanocomponent of the similar size. Zhukalin et al. [5] using this approach investigated the formation of nanostructures from colloidal suspensions of nanoscale amorphous silicon dioxide particles and carbon nanotubes. The indirect methods indicated the formation of the silicon oxycarbide phase from the drying drop at the room temperature. In the present work we carried out a further investigation of mechanisms and conditions of aggregation and synthesis of self-organized structures in the nanosystem SiO_2-CNT using the methods of numerical and natural experiments.

2. Methodology

The investigation of the interaction of amorphous silicon dioxide (aerosil brand 300, the average size of particles 7nm) (figure 1a) with the array of short carbon nanotubes (average diameter and length are 20–40 nm and ~500 nm respectively) in homogeneous environment was conducted by mixing colloidal solutions of SiO_2 and CNTs using the drop method [17–19], with the subsequent evaporation of a dispersion medium – water. A drying drop is viewed as nanoreactor with dynamic walls and inernal parameters, such as, the concentration of components, pressure, temperature [20–21].

Colloidal suspensions of initial components were dispersed by the ultrasound disperser for 15 minutes at 60 W. The morphology of obtained structures was investigated on the scanning electron microscope JSM-6390LV (JEOL, Japan). The elemental analysis was conducted with the help of the analytical console Bruker AXS Microanalysis GmbH (Germany). The investigation of powder-like samples of pure aerosil and nanostructures SiO_2/CNT was controlled on IR-spectrometer VERTEX-70 (Germany) by the method of interrupted full internal reflection in the range from 600–4000 cm^{-1}. The transmission electron microscopy (TEM) was performed on Carl Zeiss Libra 120. The analysis was performed on PANalytical EMPYREAN with the use the Cu $K\alpha_1$ radiation. Measurements were made in the reflection mode ($\theta/2\theta$ scan with a step $0.013°/2\theta$). The measure range 2θ was 5–60 °.

Nnumerical simulations of the interaction of carbon atoms of the charged interface of the capped armchair SWCNT (5, 5) with atoms of different electronegativity (H, O, F) were carried out using hybrid DFT (density functional theory) method of Becke [22] and Lee, Yang, Parr [23] (B3LYP) and 6-31G basis set [24] with the help of the Gaussian09 program package [25] in the Supercomputing center of Voronezh State University.

3. Results and discussions

During the interaction of certain concentrations of amorphous SiO_2 and CNTs in the drying drop at room temperature 2 kinds of self-organized structures: rodlike and spherical was revealed (figure 1b). These structures differ not only by the shape, but also by the ratio of atomic concentrations of Si:O:C – 3.5:1:1.5 for rodlike structures, and 1.9:3.9:1 for spherical structures. It is known that the absorption band in the IR spectra in the region 1100 (SiO_2) – 980 – (SiO) – 943 cm^{-1} (Si) are linked to stretching vibrations of the bridging oxygen Si–O–Si (stretching mode Si–O–Si) and widely used for the determination of the composition of SiO_x layers. A displacement of the oscillation mode to lower frequencies indicates the increase of the silicon concentration in rodlike structures SiO_2/CNT. On the basis of IR spectroscopic studies and elemental analysis, we can conclude that composite SiO_2/CNT rodlike structures have a high content of the silicon.

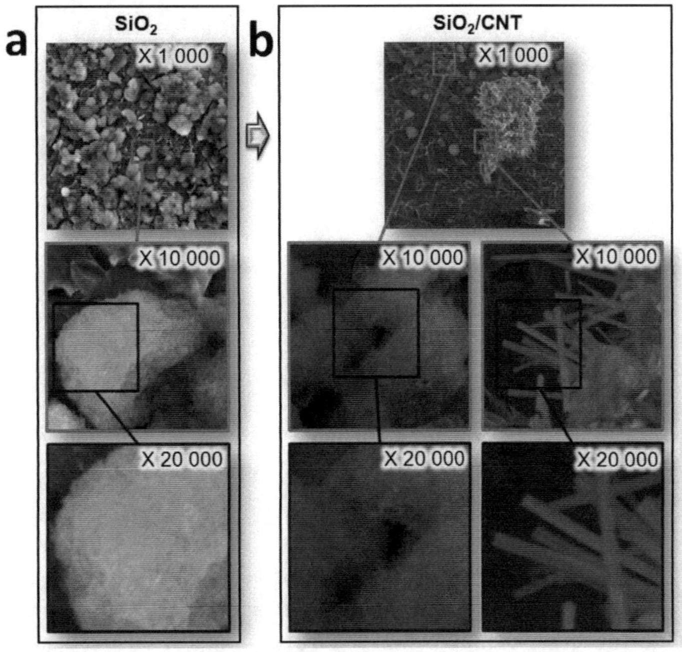

Figure 1. The electron microscopy of initial silicon dioxide (a) and nanostructures obtained after the dry drop (b).

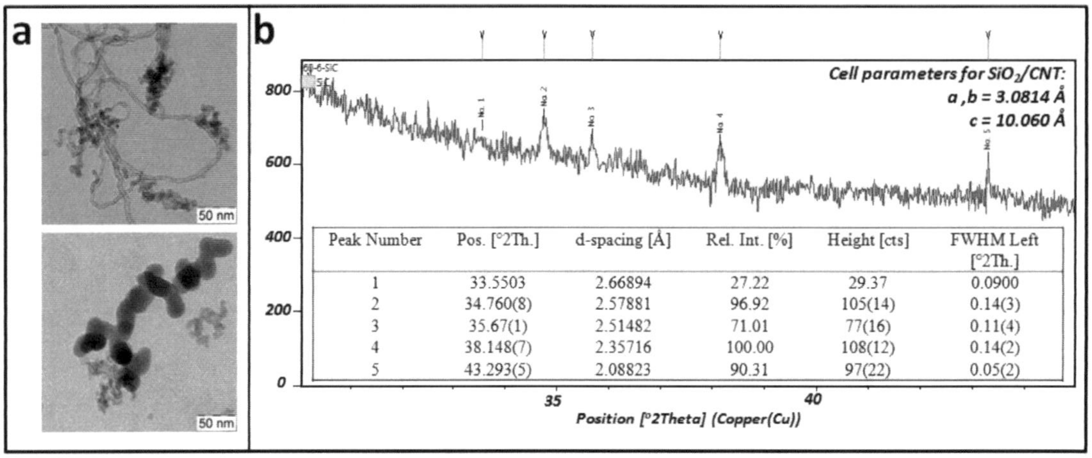

Peak Number	Pos. [°2Th.]	d-spacing [Å]	Rel. Int. [%]	Height [cts]	FWHM Left [°2Th.]
1	33.5503	2.66894	27.22	29.37	0.0900
2	34.760(8)	2.57881	96.92	105(14)	0.14(3)
3	35.67(1)	2.51482	71.01	77(16)	0.11(4)
4	38.148(7)	2.35716	100.00	108(12)	0.14(2)
5	43.293(5)	2.08823	90.31	97(22)	0.05(2)

Figure 2. TEM image (a) and diffraction pattern of the nanostructure of SiO_2/CNT (b).

The XRay diffraction analysis of obtained nanostructures revealed the formation of the silicon carbide phase with a polytype 4H-SiC (figure 2). Lattice parameters are a=b=3.081 Å, c=10.060 Å. The formation of oxycarbide phases during the high temperature treatment of amorphous SnO_{2-x} with multi-walled CNTs, was convincingly shown by Bolotov et al. [1]. The synthesis of silicon carbide commonly occurs at 2200–2500°C. Reasons of the formation of silicon carbide at the room temperature are not obvious and relate to complex dynamical processes in the drying drop and features of the CNT structure.

A capped armchair single walled carbon nanotube (SWCNT) (5, 5) was taken as a model object. Capes of SWCNT (5, 5) are two halves of the fullerene C_{60} dissected perpendicular to the five-fold axe [26, 27]. The boundary ring of carbon atoms of the cap is formed by sides of pentagons. The boundary ring of carbon atoms of the SWCNT's body is formed by sides of hexagons. That results in the shift of the electron density from the body of SWCNT to near lying ring of atom of the cap. Hence the charged interface forms.

The calculated potential curves are presented in figure 3. The energies of covalent bonds of hydrogen and fluoride with the atoms of carbon from negatively charged ring of the cap were 3 and 1.5 times higher than with positively charged ring of the tube's body. Therefore, negatively charged ring of the cap can be viewed as local activity center with increased reactivity.

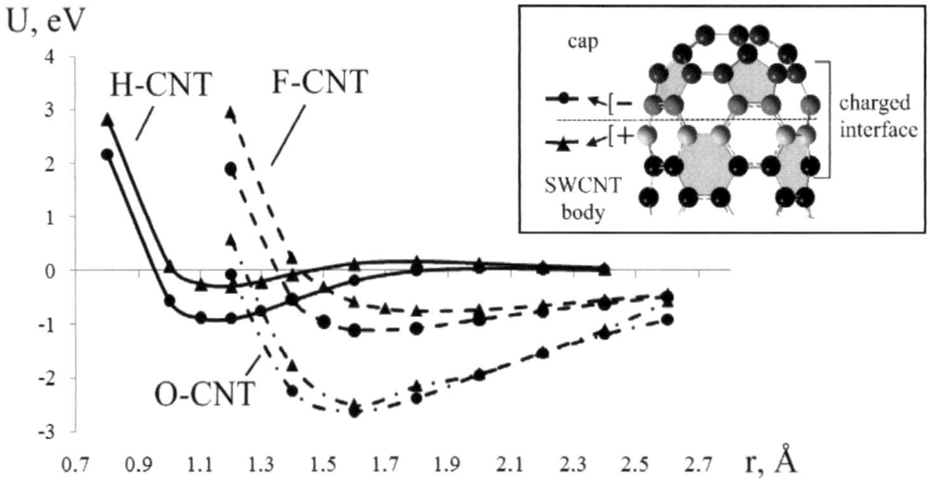

Figure 3. Potential curves of interaction of hydrogen atoms (solid lines), fluorine (dash lines) and oxygen (dot-dash lines) with negatively-charged ring of carbon atoms of the cap and positively charged ring of SWCNT's body (5, 5). Full energy of non-interacting nanotube and H, F, O was set as a zero point.

The strong interaction with the oxygen was observed (figure 3). Binding energies of the oxygen with carbon atom of negatively and positively charged ring are 2.629 and 2.502 eV which exceeds the binding energy H-C and F-C by more than two times. The structure of the amorphous silicon dioxide is tetrahedral [28]. All silicon atoms are surrounded by four oxygen atoms. Base elements of the surface of the amorphous silicon dioxide are two tetrahedrons with the common oxygen atom. In that case the strong interaction of CNT with the nanoscale silicon dioxide is quite realistic. We conducted simulation of local interaction between nanofragment of SiO_2 and carbon nanotube. If the initial distance is greater than $r^{cr}_{Si\text{-}C}$, then the calculated value is $r_{Si\text{-}C}$ ~2.54 Å, which is characteristic of Van-der-Waals interactions. The formation of covalent bonds Si-C and O-C, as well as bridges C-Si-O-C was revealed by the geometry optimization if the initial distance is less than $r^{cr}_{Si\text{-}C}$. Hence, when the silicon dioxide and CNT are enough close each other the formation of silicon carbide is going on. This condition is fulfilled during the dry drop due to the rise of pressure, active flow and increase of concentration of CNT and silicon dioxide.

4. Conclusion

The interaction of nanoscale amorphous silicon dioxide and carbon nanotubes in the drying drop of colloidal suspension with the certain concentration of components leads to the formation of self-organized rodlike structures with diameter 250-300nm and length ~4μm. The elemental analysis and IR- spectroscopy showed the high content of the silicon in rodlike structures. The diffraction analysis of the obtained nanostructures revealed the formation of the phase of silicon carbide at the room temperature. Theoretical analysis confirmed that the mechanism of formation of such structures is due to the curvature effects and presence of charged active centre near caps of CNTs, which increases their reactivity. Complex dynamical processes in the drying drop and high reactivity of short CNT with high concentration of chemically active atoms are the basis for the development of energy-saving technologies of synthesis of functional composite nanomaterials and high-temperature phases.

Acknowledgments

This work was supported by the project FP7-IRSES-295260 "ECONANOSORB" under Marie Curie Actions of the 7th Framework Program of European Union, by RFBR, research projects N 14-02-31315 mol_a.

References

[1] Bolotov V V, Nesov S N, Korusenko P M, Povoroznyuk S N 2014 *Physics of the Solid State* **56** 1834
[2] Timmermans M Y, Estrada D, Nasibulin A G, Wood J D, Behnam A, Sun D, Ohno Y, Lyding J W, Hassanien A, Pop E, Kauppinen E I 2012 *Nano Res.* **5** 307
[3] Constantopoulos K T, Shearer C J, Ellis A V, Voelcker N H, Shapter J G 2010 *Adv. Funct. Mater.* **22** 557
[4] Novikov L S, Voronina E N, Chirskaya N P 2013 *J. Adv. Mater.* **11** 12
[5] Zhukalin D A, Tuchin A V, Goloshchapov D L, Bityutskaya L A 2015 *Tech. Phys. Lett.* **41** 157
[6] Song J, Jensen D S, Hutchison D N, Turner B, Wood T, Dadson A, Vail M A, Linford M R, Vanfleet R R and Davis R C 2011 *Adv. Funct. Mater.* **21** 1132
[7] Choi N-S, Yao Y, Cui Y, Cho J 2011 *J. of Materials Chemistry* **21** 9825
[8] Liu R, Duay J, Lee S B 2011 *Chem. Commun.* **47** 1384
[9] Li L, Yang Y, Yang G, Chen X, Hsiao B S, Chu B, Spanier J E, Li C Y 2006 *Nano Letters* **6** 1007
[10] Bai J B, Allaoui 2003 *Composites: Part A* **34** 689
[11] Guo J, Sun A, Chen X, Wang C, Manivannan A 2011 *Electrohimica Acta* **56** 3981
[12] Chou S-L, Zhao Y, Wang J-Z, Chen Z-X, Liu H-K and Dou S-X 2010 *J. Phys. Chem.* C **114** 15862
[13] Gundiah G, Madhav G V, Govindaraj A, Seikh Md M, Rao C N R 2002 *The RS of Chemistry* **12**

1606
[14] Wu J, Hagelberg F 2010 *Phys. Rev. B* **81** 155407
[15] Park S, Srivastava D, Cho K 2003 *Nano Letters* **3** 1273
[16] Zhukalin D A, Tuchin A V, Goloshchapov D L, Bityutskaya L A. 2014 *Kondensirovannye sredy i mezhfaznye granitsy* **16** 431
[17] Bin Su, Shutao Wang, Yanling Song, Lei Jiang *Nano Research*, 2011, vol. 4, no. 3, pp. 266-273.
[18] Vodolazskaya I V, Tarasevich Yu Yu 2011 *Mod. Phys. Lett. B* **25** 1303
[19] Tarasevich Yu Yu, Pravoslavnova D M 2007 *The European Physical Journal E* **22** 311
[20] Su B, Wang S, Song Y, Jiang L 2011 *Nano Research* **4** 266
[21] Barash L Yu 2015 *International Journal of Heat and Mass Transfer* **84** 419
[22] Becke A D 1993 *J. Chem. Phys.* **98** 5648
[23] Lee C, Yang W, Parr R G 1988 *Phys. Rev. B* **37** 785
[24] Ditchfield R, Hehre W J, Pople J A 1971 *J. Chem. Phys.* **54** 724
[25] http://www.gaussian.com
[26] Cioslowski J, Rao N and Moncrieff D 2002 *J. Am. Chem. Soc.* **124** 8485
[27] Tuchin A V, Nestrugina A V, Bityutskaya L A, Bormontov E N 2014 *J. of Phys.: Conf. Ser.* **541** 012008
[28] Nekrashevich S S, Gritsenko V A 2014 *Phys. of the Solid State* **56** 207

Self-limiting growth and bimodal size distribution of Au nanoislands on InAs(111)B surface

Y. Berdnikov[1], N. V. Sibirev[1,4], V.G Dubrovskii[1,2,3], J.H. Kang[5]

[1] St. Petersburg Academic University, Khlopina 8/3, 194021 St. Petersburg, Russia
[2] Ioffe Physical Technical Institute RAS, 26 Polytekhnicheskaya str, 194021 St. Petersburg, Russia
[3] ITMO University, Kronverkskiy pr. 49, 197101 St. Petersburg, Russia
[4] St. Petersburg Polytechnic University, 29 Politekhnicheskaya St., St. Petersburg, 195251, Russia
[5] Braun Center for Submicron Research, Department of Condensed Matter Physics, Weizmann Institute of Science, Rehovot 76100, Israel
E-mail: yury.berdnikov@gmail.com

Abstract. Gold nanoparticles are widely used for catalytic growth of different semiconductor nanowires. It is important therefore to have a full appreciation of the factors influencing the size distribution of such particles on semiconductor substrates. Here, we describe some interesting phenomena observed during low-temperature deposition of Au onto InAs(111)B substrates where Au islands are solid. In particular, we find a bimodal character of the size distribution of these islands. We then develop a nucleation model which is based on a specific form of the island formation energy. The model yields a self-limiting growth behavior in which the growth rate of large islands is suppressed e.g. due to dislocations. We show that the regime of self-regulated nucleation with a few pulses is observed for short growth times, leading to a bimodal size distribution or even a more complex shapes. For longer growth times, the system acquires a steady-state regime with a time-independent supersaturation and hence the pulses disappear.

1. Introduction

Au-catalyzed growth of nanowires allows for fabrication of a variety of sophisticated non-planar nanostructures suitable for optical and electronic applications. These applications usually require a good control over the nanowire diameters and thus the size distribution of the catalytic particles from which the nanowire growth starts [1,2]. Usually, Au is deposited on semiconductor surfaces at low temperatures where the initial Au nanoparticles are solid (before anneal). In this case, the strain-induced effects due to the lattice mismatch between the islands and the substrate becomes very important [3-5] and may strongly influence the size distribution. The latter determines the initial condition for the annealing step and thus influences the resulting surface assembly of the nanowire seeds. If we restrict ourselves to solid Au nanoparticles, the strain-induced effects may lead to a suppression of the growth rate for large sizes and consequently to the regime of self-limiting growth with secondary nucleation. This was previously discussed, e.g., in Ref. [6] in connection with the bimodal size distribution of three-dimensional (3D) Ge islands on Si(100). However, in this work the strain-induced suppression of the growth rate was introduced in modeling without changing the island formation energy. On the other hand, several studies revealed a saturation of the size of Au [7] and some other metal nanoislands [8] at a certain steady-state value in the course of growth. Similar phenomenon was reported for dislocated 3D GaN islands on Si(111) surfaces covered with AlN mismatching layer [9] or even for dislocation-free 3D GaN islands on an amorphous SiN interlayer (without any strain) [10-12] before the island-to-wire growth transformation. The last example suggests that the self-limiting growth behavior is not necessarily associated with strain relaxation but may be caused by other reasons such as the surface energy balance and shape transformation.

Consequently, here we study the morphology of 3D Au nanoislands deposited on InAs(111)B substrates at a low temperature of 350°C where the islands are solid. We find experimentally a bimodal character of the size distribution of such islands. We then develop a nucleation model which is based on a specific form of the island formation energy with a strain-induced contribution. The model describes a self-limiting growth and enables us to explain the experimental findings. Furthermore, from theoretical point of view, we present an unusual behavior of supersaturation with damped oscillations which produce two differently shaped peaks of the size distribution which are well-separated at the beginning but merge together on a later stage of growth.

2. Experiment

The deposition of Au on InAs(111)B substrates was performed at 350 °C which is below the melting point for these materials [13,14]. We thus deal with solid strained Au nanoislands with a huge lattice mismatch ($\varepsilon_0 = 33\%$) with the substrate which is expected to induce misfit dislocations in the islands soon after their nucleation. In the growth time series, the Au deposition was stopped after 0.5, 1, 2, 2.5, 4 and 5 min. Low resolution transmission electron microscopy (TEM) images shown in Fig. 1 did not allow us to confirm precisely the presence or absence of crystalline defects or dislocations. This will be elaborated elsewhere.

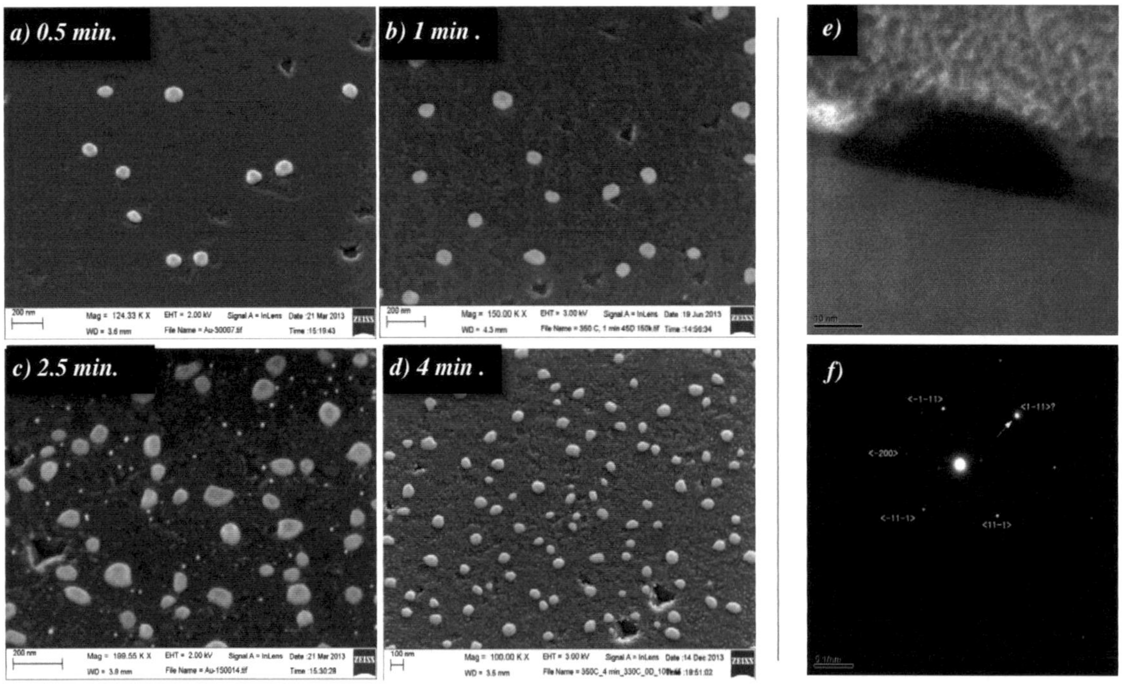

Figure 1(a)-(d) SEM images of Au nanoislands on InAs(111)B substrate for the deposition times of 0.5, 1, 2.5 and 4 min, respectively; (e) TEM image and (f) diffraction pattern of Au nanoislands on InAs(111)B substrate

The analysis of scanning electron microscopy (SEM) images (Fig. 1 a-d) reveals several interesting features of the time-dependent growth behaviour. First, within the time interval from 0.5 to 4 min, the mean diameter of the island bases remains within the interval from 75 to 100 nm but increases to 125 nm after 5 min of deposition. Thus, we observe a quick saturation of the island size with a slow increase at a later growth step. At the same time, the island density gradually increases with time from 2.2 μm^{-2} at 0.5 min to 16 μm^{-2} at 4 min. Surprisingly, in the 2.5 min sample (Fig. 1c) we can see a co-existence of large islands with much smaller islands having the mean diameter of only 10 nm. The corresponding distribution over lateral sizes has two separate peaks which are not observed for shorter or longer deposition times.

The island density and the mean size at different growth times are summarized in Table 1. Since the island density only increases with time, the evolution of the size distribution has nothing to do with the Ostwald ripening [1]. In a qualitative way, one could think of the two generations of islands whose nucleation events are separated in time. The first generation quickly emerges and grows until reaching a steady-state size which is stabilized by some inhibition of the material consumption by larger islands. Since stable islands cannot continue growing and the material influx remains on, supersaturation starts increasing again and this gives rise to nucleation of the next generation of islands, as described in Ref. [6].

Table 1 Experimental data on the island density and size at different deposition times

Deposition time	0.5 min	1 min	2 min	2.5 min	4 min
Island density (μm^{-2})	2.2	6.8	5.8	15	16
Mean size (nm)	81	78	94	75	72

3. Theoretical model

Standard macroscopic expression for the island formation energy F as a function of the number of monomers (Au adatoms in our case) i yields a function with one maximum after which the islands grow infinitely [1,15,16]. In order to describe the observed self-limiting growth, we modify the formation energy of a 3D island with the linear size r by adding the new $i^{4/3}$ term which dominates at large enough i. If we attribute the inhibition of growth to the lattice mismatch, $i^{4/3}$ is proportional to r^4 and can be considered as the energy of interaction between a linear dislocation of length r with the island volume r^3. On the other hand, the residual strain is known to modify the volume term in the formation energy.

Therefore, our model for $F(i)$ has the form

$$F(i) = Ai^{2/3} - [B + \ln(1+\zeta)]i + Ci^{4/3}. \tag{1}$$

Here A, B and C are the material-related constants and $\zeta = {}^{n_1}/_{n_{eq}} - 1$ is the supersaturation of 2D adatom sea with concentration n_1 with respect to their equilibrium concentration n_{eq}. The $Ai^{2/3}$ terms stands for the surface energy and $B\,i$ is volume energy gain due to strain relaxation. With an obvious relationship between the coefficients in Eq. (1), the formation energy has a maximum at the critical size as in the standard theory [1] and a minimum at a certain steady-state size which corresponds to the growth equilibration [5,6,17-19]. The typical shape of the formation energy versus the base size r is shown in Fig. 2.

Following the standard procedure described in detail in Refs. [1,15,16], we obtain the growth (W_i^+) and decay (W_i^-) rates in the form

$$W_i^-(i) = W_{i-1}^+ exp\left(\frac{d\,F(i)}{d\,i}\right); \quad W_i^+ = \frac{m}{\tau}(1+\zeta)i^{\frac{m-1}{m}}. \tag{2}$$

Here, the growth index m depends on the mechanism of material transport of Au adatoms into the islands and τ is the characteristic diffusion time of adatoms on the substrate. The rate equations describing the time evolution of the surface concentration of differently sized islands $n_i(t)$ is taken in the standard form [1,15,16]:

$$\frac{dn_i}{dt} = W_{i-1}^{+}(\zeta)n_{i-1} + W_{i+1}^{-}(\zeta)n_{i+1} - \left(W_i^{+}(\zeta) + W^{-}(\zeta)\right)n_i. \quad (3)$$

The time evolution of supersaturation $\zeta = \zeta(t)$ is obtained from material balance equation

$$\zeta(t) + \sum_{i=2}^{\infty} i\, n_i(t) = \Phi(t), \quad (4)$$

where $\Phi(t)$ is the so-called "ideal supersaturation" which equals the appropriately normalized total volume of the deposited material and increases linearly with time [20].

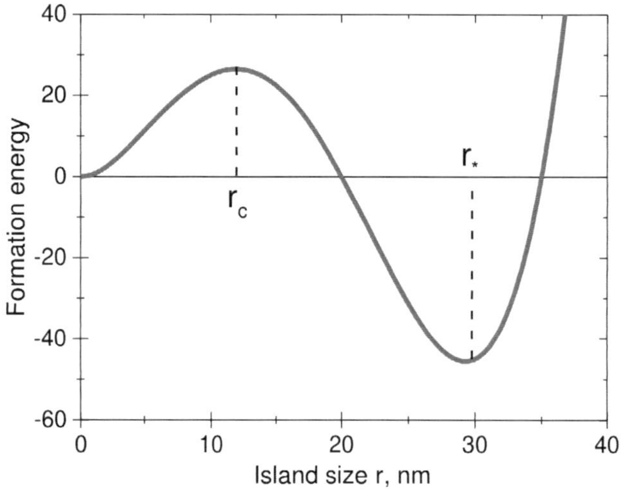

Figure 2 Formation energy $F(r)$ versus the base size of 3D nanoisland $r = i^{1/3}$ (measured in the units of lattice spacing) with the critical size R_c and the steady-state size R_*. This shape corresponds to $A = 0.7$, $B + \ln(1+\zeta) = 0.055$ and $C = 0.001$ in Eq. (1)

The nucleation rate gives the number of islands emerging per unit surface area per unit time and is obtained from the Zeldovich expression [1]:

$$I(\zeta(t)) \sim (1+\zeta)^2 \ln(1+\zeta)\, exp\left[-\frac{\alpha}{\ln^2(1+\zeta)}\right], \quad (5)$$

where α is a material-related constant corresponding to the maximum of the formation energy. The most important feature of Eq. (5) is that the nucleation rate is extremely sensitive to the value of supersaturation ζ due to the exponential dependence on the nucleation barrier with a large parameter $\alpha >> 1$.

4. Results and discussion

Numerical solution of Eqs. (3) to (5) was performed by introducing the dimensionless time $z \equiv m\, ^{t}/_{\tau}$, scaled size $s = {}^{i}/_{i_0}$ and concentrations $f_i \equiv {}^{n_i}/_{n_{eq}}$ and implementing the explicit finite difference method. The parameter values used in calculations were the following: $m = 1.5, A = 10, B = 3.6, C = 0.34$. The time dependences of supersaturation, size distribution, mean size and density of islands were obtained and compared with the experimental data.

In particular, Fig. 3 (a) shows the dependence of supersaturation ζ on the dimensionless time z. It is seen that this dependence has the form of dumped oscillations whose amplitude rapidly decreases with the number of pulses. After that, supersaturation stabilizes at a certain value corresponding to a time-independent nucleation of the new islands (at a low rate) and a slow growth of all previously formed islands.

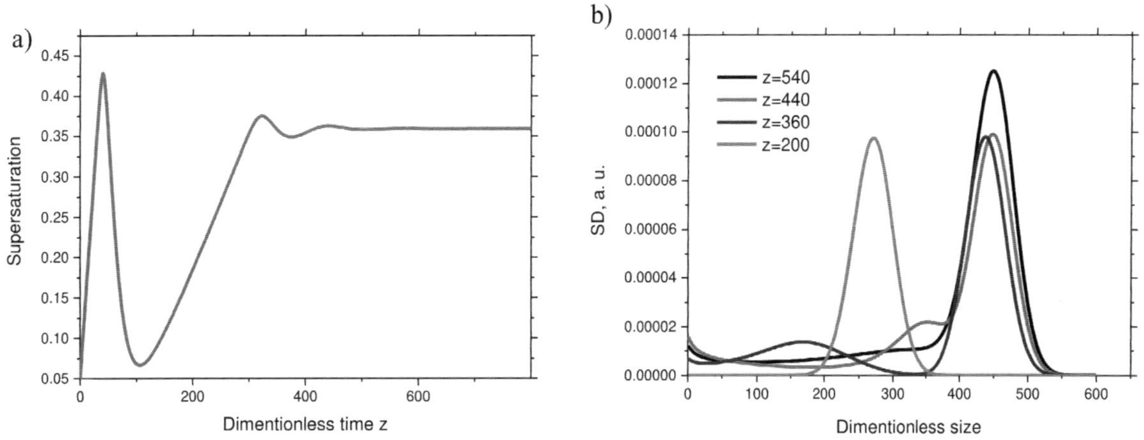

Figure 3 (a) Time dependence of supersaturation, (b) Size distribution shapes at different times ***z***

A much more informative picture of the growth process can be deduced from the analysis of the size distributions shown in Fig. 3 (b). Following the pulsed behavior of supersaturation, the initial distribution at short times ($z = 200$) is monomodal which corresponds to the time interval before the growth stabilization. After that (at $z = 360$) we observe the bimodal distribution with a lower peak at smaller sizes caused by secondary nucleation. At longer times, this second peak moves toward larger sizes and merges with the first peak, as demonstrated by the curve at $z = 440$. The curve at $z = 540$ illustrates the follow-up evolution of the system: the distribution becomes monomodal again and is taller than the others, corresponding to a steady-state process with a slowly increasing mean size.

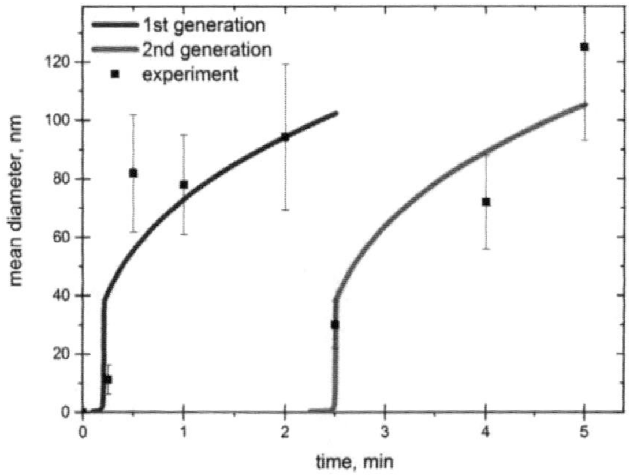

Figure 4 Time dependences of the islands mean diameter, with error bars corresponding to half-widths of the size distributions

As regards the interpretation of our experimental data, we believe that the two populations of islands seen in our SEM images correspond to the situation with two distribution peaks soon after secondary nucleation. After that, the two peaks merge as shown in Fig. 3 (b) and we do not observe any size separation on a later growth stage. The fits to the mean island size versus the deposition time are shown Fig. 4. With our model parameters, we are able to reproduce the observed behavior. The first generation of islands features growth at a decreasing rate from 0.5 to 2. Around 2.5 min, secondary nucleation of new islands occurs and then the two generations become indistinguishable in terms of their mean size.

In conclusion, we have shown experimentally an interesting growth behavior of Au nanoislands deposited onto InAs(111)B surfaces at 350 °C. We clearly observe the initial nucleation step with a monomodal size distribution, then the two populations of islands with a distinct size separation, transformed to a monomodal system again in the course of deposition. Such a behavior is well-described by a theoretical model where the island formation energy is modified to yield the growth inhibition for large islands. Our model describes a very unusual growth with self-regulated nucleation pulses producing multiple populations of differently sized islands which merge together on a later growth stage.

Acknowledgement

This work was supported by grants of the Russian Foundation for Basic Research № 13-02-00622, 13-02-12045, 15-52-53041, 15-52-78057 and the FP7 UC project "Nanoembrace" (Grant Agreement No. 316751).

References

[1] Dubrovskii V G 2014 *Nucleation Theory and Growth of Nanostructures* (Springer-Verlag, Berlin Heidelberg)
[2] Kang J H, Cohen Y, Ronen Y, Heiblum M, Buczko R, Kacman P, Popovitz-Biro R and Shtrikman H 2013 *Nano Lett.* **13** 5190
[3] Dubrovskii V G, Sibirev N V, Zhang X and Suris R A 2010 *Cryst. Growth Des.* **10** 3949
[4] Znang X, Dubrovskii V G, Sibirev N V and Ren X 2011 *Cryst. Growth Des.* 11 5441
[5] Aqua J-N, Berbezier I, Favre L, Frisch T and Ronda A 2013 *Physics Reports* **522** 59
[6] Osipov A V, Kukushkin S A, Schmitt F and Hess P 2001 *Phys. Rev. B* **64** 205421
[7] Valden M, Lai X and Goodman D W 1998 *Science* **281** 1647
[8] Liu F *Phys. Rev. Lett.* **89** 246105.
[9] Consonni V, Hanke M, Knelangen M, Geelhaar L, Trampert A, Riechert H 2011 *Phys. Rev. B* **83**, 035310
[10] Dubrovskii V G, Consonni V, Trampert A, Geelhaar L and Riechert H 2012 *Phys. Rev. B* **85** 165317
[11] Dubrovskii V G, Consonni V, Geelhaar L, Trampert A, and Riechert H 2012 *Appl. Phys. Lett.* **100** 153101
[12] Consonni V, Dubrovskii V G, Trampert A, Geelhaar L and Riechert H 2012 *Phys. Rev. B* **85** 155313
[13] Liu H S, Cui Y, Ishida K, Jin Z P 2003 *Thermodynamic Reassessment of the Au-In Binary System* CALPHAD **27** 27
[14] Tchernycheva M, Travers L, Patriarche G, Glas F, Harmand J C, Cirlin G E and Dubrovskii V G 2007 *J. Appl. Phys.* **102** 094313
[15] Dubrovskii V G and Nazarenko M V 2010 *J. Chem. Phys.*132, 114507
[16] Dubrovskii V G and Nazarenko M V 2010 *J. Chem. Phys.***132**, 114508
[17] M. Kastner et al. 1999 *Phys. Rev. Lett.* **82** 2745
[18] Ross F M, Tersoff J, and Tromp R M 1998 *Phys. Rev. Lett.* **80** 984
[19] Guyer J E, Voorhees P W 1995 *Phys. Rev. Lett.* **74** 4031
[20] Dubrovskii V G 1992 *Phys. Stat.*

Photoluminescence observation from zinc oxide formed by magnetron sputtering at room temperature

D Kudryashov[1], A Babichev[1,2], E Nikitina[1], A Gudovskikh[1], P. Kladko[3]

[1] St. Petersburg Academic University RAS 194021 St. Petersburg, Russia
kudryashovda@spbau.ru

[2] ITMO University, 197101 St.Petersburg, Russia

[3] Connector Optics LLC, 194292 St. Petersburg, Russia

Abstract: The photoluminescence (PL) of ZnO thin films grown by magnetron sputtering at room temperature has been observed. The PL spectra were measured using an instrument from Accent Optical Technologies with a solid state UV laser (λ = 266 nm) as the pumping source and at the temperature of 300 K. Samples grown at sputtering power of 100-200 W show a strong photoluminescence (PL) at wavelength of 377 nm and its intensity shows non-linear dependence with magnetron power. At values of sputtering power less then 100 W PL signal was not observed. A correlation between PL, XRD intensity and ZnO grain size was shown.

1. Introduction

Recently, there has been a growing interest in oxide photovoltaics. ZnO is a one of promising materials for wide-gap emitter in oxide heterojunction solar cells, for example AZO/ZnO/Cu_2O structure. Such structure has showed an efficiency value of 5.38% at AM1.5 [1]. Moreover ZnO thin films have attracted a great deal of a base for visible-blinde photodetectors [2]. Zinc oxide is a direct- and wide bandgap semiconductor with Eg of 3.3 eV [3]. Due to numerous of native point defects it has n-type doping. Defects have also influence on its optical and electrical properties [4]. Solar cell efficiency is determined primarily by a quality of structure layers. The last can be directly determined by PL measurements. The aim of this work is to determine a ZnO thin films quality in dependence of its growth parameters.

2. Experimental

Deposition of Zinc Oxides thin films was carried out using Radio frequency (RF) magnetron sputtering setup (BOC EDWARDS Auto 500RF) using 3" ZnO-target supplied by LTS Chemicals. Distance between target and substrate was 100 mm. The sputtering chamber was evacuated to less than $5\cdot10^{-6}$ mbar prior to deposition. Deposition of thin films performed at room temperature on Si(100)-wafers and glass slides. Pure argon (99.999 %) was used as working gas. Control of the oxide layers thickness was carried out using a profilometer AMBiOS XP-1. Refractive and absorption indexes of ITO film were measured using HORIBA JOBIN YVON and Horiba UVISEL2 Spectroscopic ellipsometers.. Optical transmittance was measured by a spectrometer based on a monochromator Solar Laser Systems M266. SEM images were analyzed using a scanning electron microscope SEM (Zeiss SUPRA 25). The PL spectra were measured using an instrument from Accent

Optical Technologies with a solid state UV laser (λ=266 nm) as the pumping source and at the temperature of 300 K. XRD spectra were measured using PANalytical X'Pert PRO equipment.

3. Results and discussion

SEM image of 500 nm ZnO thin films deposited on Si-wafer by magnetron sputtering is shown in figure 1. A small growth defects are observed on smooth surface. Zinc oxide grown on glass shows good transmittance in visible region (Fig. 2) with absorption region below 400 nm.

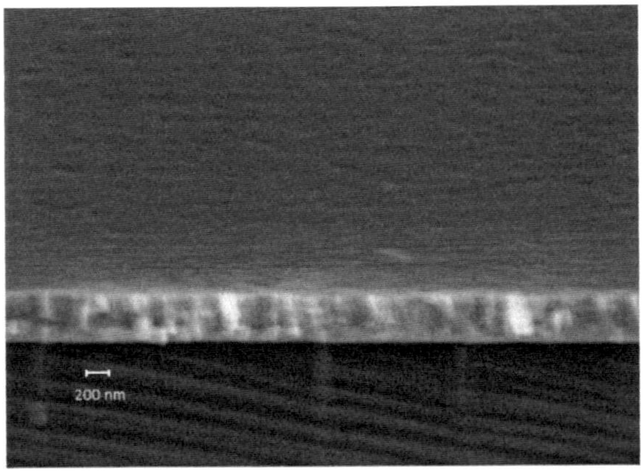

Figure 1. SEM image of rf-magnetron sputtered ZnO thin film deposited at 150 W on Si-wafer.

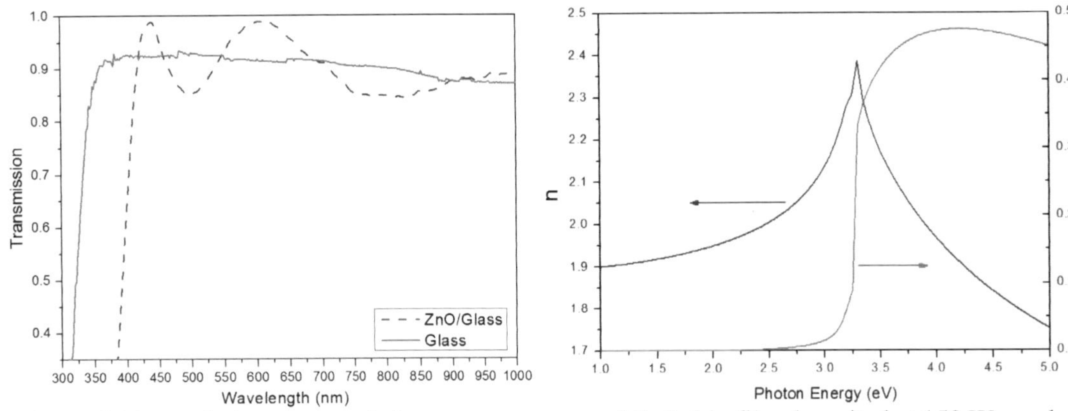

Figure 2. Transmission spectra of rf-magnetron sputtered ZnO thin film deposited at 150 W on glass (left) and extinction coefficient for ZnO on silicon substrate (right)

Figure 3 shows the PL spectra for ZnO thin films deposited at different values of magnetron power at room temperature. Band-edge PL peak (at λ=377 nm) corresponds to direct band-to-band radiative recombination. The PL intensity measurements show nonlinear RF power dependent behavior in the range of 100-200 W. It has maximum value at the magnetron RF power of 150 W. At power values of 50W and less no PL signal was observed. XRD measurements show similar behavior (Fig. 4). Diffraction peaks observed in (0002) via x-ray diffraction analysis show that the zinc oxide thin film grown on glass is oriented following the c-axis.

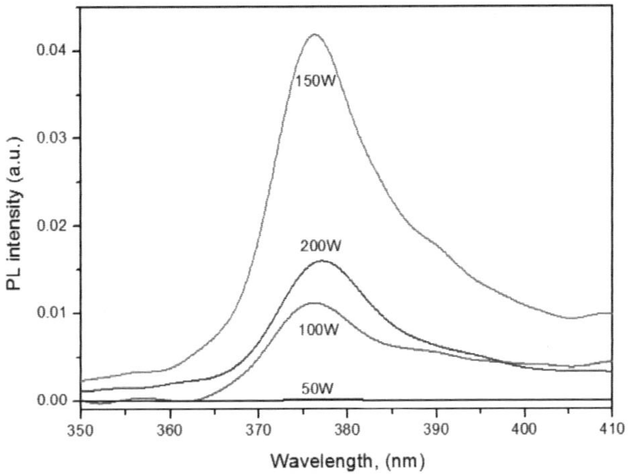

Figure 3. Room temperature PL spectra of ZnO thin films (500 nm) as a function of the RF power.

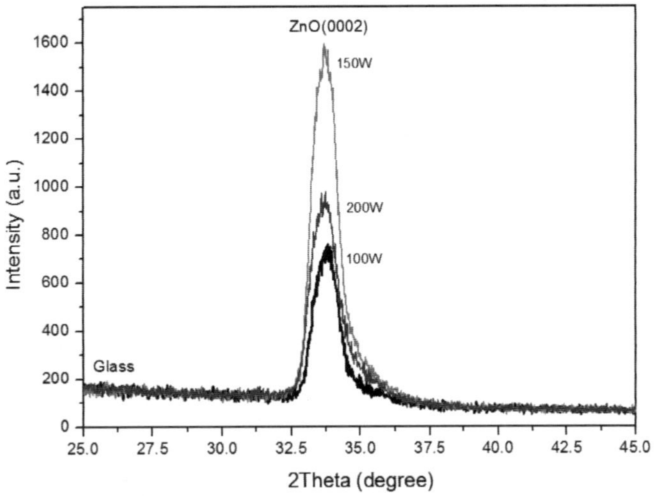

Figure 4. XRD patterns of ZnO films (500 nm) deposited on glass as a function of the RF power.

The vertical grain size estimated from the full width at half maximum of XRD diffraction peak by using Sherrer formula [5] shows that ZnO grown at rf-power of 150W has the higher value of grain size. Increasing or decreasing the value of magnetron rf-power leads to the grain size and PL intensity decreasing. Similar relations were obtained in [6] where authors shown changes in ZnO structure (grain size and Zn/O ratio) with increasing of rf-power. It was shown that at rf-power of 175W the Zn/O ratio has a maximum value of 1.11 and it decreases with further increasing of rf-power. But the authors found that the ZnO grain size is the lowest in case of magnetron power of 175W. It should be noted that properties of magnetron sputtered ZnO are sensitive toward least changes in magnetron power value.

4. Conclusions

The photoluminescence of ZnO thin films grown by magnetron sputtering at room temperature has been observed. Samples grown at sputtering power of 100-200 W show a strong photoluminescence at wavelength of 377 nm and its intensity shows non-linear dependence with magnetron power caused by changes in Zn/O ratio and zinc oxide grain size. An optimal ZnO growth parameters were determined based on measured PL and XRD data. It will provide a basis for further of ZnO/Cu_2O based solar cells and visible-blind ZnO-based photodetectors manufacturing.

5. Acknowledgment

The reported study was partially supported by RFBR, research project № 15-08-06645A, № 14-02-31485 and № 15-02-08282 A, and by Scholarship of the President of the Russian Federation for young scientists and graduate students (Grant number SP-4716.2015.1) (2015-2017)

References

[1] T. Minami, T. Miyata, Y. Nishi // Solar Energy, V. 105, 2014, Pages 206–217
[2] H.Zhang, A. Babichev et al. // Journal of Applied Physics, 114(23), 2013 Page 234505
[3] H. Morkoç and Ü. Özgür, General Properties of ZnO, in Zinc Oxide: Fundamentals, Materials and Device Technology, Wiley-VCH Verlag GmbH & Co. KGaA, (2009) Weinheim, Germany
[4] J. Anderson and G. Chris Van de Walle Rep. Prog. Phys. V72, 2009, Page 126501
[5] J.I. Langford and A.J.C. Wilson, "Scherrer after Sixty Years: A Survey and Some New Results in the Determination of Crystallite Size" // J. Appl. Cryst. V 11, 1978. Pages 102-113
[6] A. Ismail, M. J. Abdullah // Journal of King Saud University - Science V25, 2009, Pages 209–215

Investigation of condensation of small portions of Ag at thermal evaporation in vacuum

D.G. Gromov[1], E.A. Lebedev[1], A.I. Savitskiy[1], A.Yu. Trifonov[2], V.V. Rubcov[1], N.I. Borgardt[1], Ya.S. Grishina[1]
[1]National Research University of Electronic Technology, 124498, Russia, Moscow, Zelenograd, Bld. 5, Pas. 4806
[2]Scientific Research Institute of Physical Problems named after F.V. Lukin, 124460, Russia, Moscow, Zelenograd, Bld. 6, Pas. 4806

E-mail: gromadima@gmail.com

Abstract. The formation of arrays of Ag nanoparticles from vapor phase on the unheated surface of the amorphous carbon by sequential evaporation of the two portions with decreasing masses has been studied using scanning electron microscopy. The dependence of particle size and their surface density on the amount of evaporated Ag was detected. The precipitate after condensation is in an unstable state: thermal treatment at low temperatures (200-250°C) leads to significant changes and the array goes into a metastable state. This is manifested in the absence of further changes after repeated heat treatment at the same temperature or 100°C higher. It is shown that sequential evaporation of the two portions of the silver with a decrease in mass leads to bimodal distribution of particle sizes in the array.

1. Introduction

The objects whose dimensions are in the nanometer range are of particular interest recently. Such systems are attracting the attention of researchers due to their unexpected properties which are different from bulk materials. One of the representatives of such objects are metal nanoparticles. For example, the surface-enhanced Raman scattering on the array of Ag and Au nanoparticles was observed [1,2]. The enhancement factor can be as much as 10^{10} to 10^{11}, which means the technique may detect single molecules.

The method of vacuum thermal evaporation is one of the techniques of forming an array of nanoparticles. Its attractiveness lies in the fact that, as has been shown, it allows to control the average particle size of Ag and Au on the substrate surface [3-12].

Nevertheless, there is still a number of phenomena that occur when small physical amounts are condensing in vacuum, which is not fully clear.

In the present study, we investigated the formation of arrays of silver nanoclusters by sequential vacuum thermal evaporation of two small amounts of the substance. The purpose was to enhance the understanding of this phenomenon in order to form arrays of nanoparticles of a given size.

2. Experimental

The investigated arrays of silver clusters were formed by vacuum thermal evaporation and condensation of small weight portions of the material from molybdenum evaporator at a distance from the evaporator to the substrate 20 cm. Preliminary experiments were conducted to determine the thickness of the thicker layers of silver deposited from weight portions 100-500 mg. The thickness of the layers was measured by atomic force microscopy. Using the results of measurements the calibration curves of the film thickness as a function of the silver weight portions were constructed. To determine the weight portions required to form the metal film with a thickness of a few nanometers, the calibration curves were extrapolated to the corresponding region of small thicknesses. According to this data, the silver weight portion of 0.6 mg was necessary to form 1 nm thick metal film. It should be noted that during evaporation of such small amounts of silver the condensate immediately turns into an array of clusters, so the specified film thickness in the range of 1-10 nm are hypothetical i.e. obtained as a result of extrapolation.

Thermally oxidized Si (100) was used as a substrate. The substrates were pre-washed in a standard way in Caro solution H_2SO_4:H_2O_2 (1:1) and then washed with deionized water and dried in isopropyl alcohol vapor. Then 20 nm thick layer of amorphous carbon was sputtered on the substrate by magnetron sputtering. The number of samples corresponds to the number of operations: one sample was withdrawn for investigation after each operation.

To produce the samples two weight portions of silver were subsequently vaporized: 4.5 mg and 2.7 mg. After each deposition vacuum annealing at 230°C for 30 minutes was performed. It should be noted that the heavier weight portion was evaporated first.

The resulting samples were examined using dual beam electron-ion microscope FEI Helios NanoLab 650. Images of nanoscale silver clusters were obtained in the immersion mode in secondary electrons at an accelerating voltage of 2 keV and a current of 100 pA. Then, images were analyzed and for each of the obtained samples the distribution histograms of the number of particles according to their size were constructed. This allowed to define the average particle size of the resulting array, as well as to determine the number of particles per area unit

Using this microscope by in-situ lift-out method [13] the cross-section samples of silver clusters deposited on the ~20 nm thick amorphous carbon film at various stages of formation, were prepared.

The resulting samples less than 50 nm thick were investigated in the transmission electron microscope Philips CM 30 with lanthanum hexaboride cathode at an accelerating voltage of 200 kV.

3. Results and discussion

Fig. 1 shows micrographs of the obtained arrays of clusters and cluster size distribution histogram. Condensate resulting from the first evaporation has "insular" structure where the "islands" have an elongated serpentine shape, but are completely separated and are tightly arranged relative to each other (Fig. 1). The length of some of them is 80 nm and the width generally ranges from 10 to 20 nm. In this structure, on an area of 1 μm^2 there are ~3000 of such particles, and the average spacing is about 4 nm. Measurements have shown that the electrical conductivity in such a layer is missing.

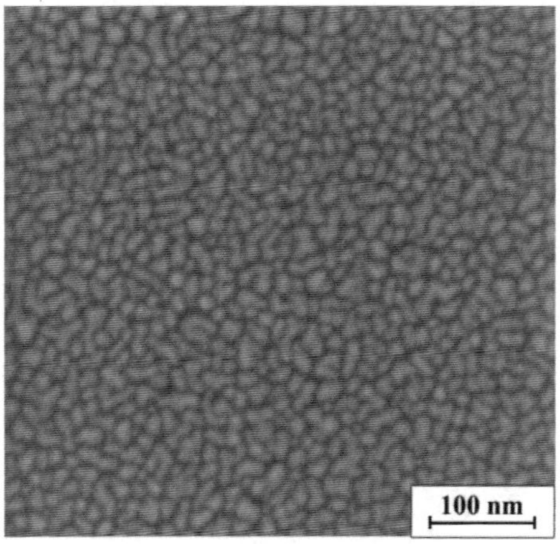

Figure 1. SEM image of the sample obtained by evaporation of 4.5 mg weight portions

Subsequent annealing at quite low temperature of 230°C leads to significant changes in the array: particles of various shapes form an array of clusters, the shape of which is close to the spherical (Fig.2). Moreover the surface density of the particles markedly reduces. In this case this value is about 1500 pcs/μm2, and the average distance between particles is ~8 nm. The diameter of the largest one of the clusters is ~40 nm whereas the smallest are in the range from 4 to 8 nm. Although somewhere one can find the smaller clusters whose diameter is estimated to be about 2 nm. The average size of the particles in the formed array is ~21 nm.

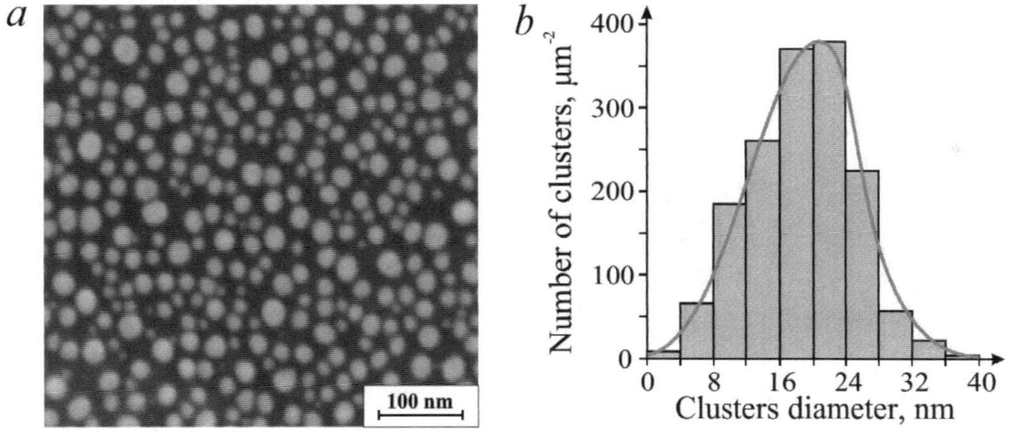

Figure 2. SEM image (a) and the cluster size distribution histogram of silver particles of the sample obtained by evaporation of 4.5 mg weight portions after annealing for 30 min at 230 °C (b)

Subsequent evaporation of the finer silver weight portion leads to the fact that there is a large amount of small particles on the surface of amorphous carbon. These particles are located between the clusters which were already formed at the previous step (Fig. 3). Thus, the density of particles on μm2 now is ~3000 pcs. Furthermore, bimodal size distribution of the particles begins to emerge. Preferred diameters are 5 and 21 nm.

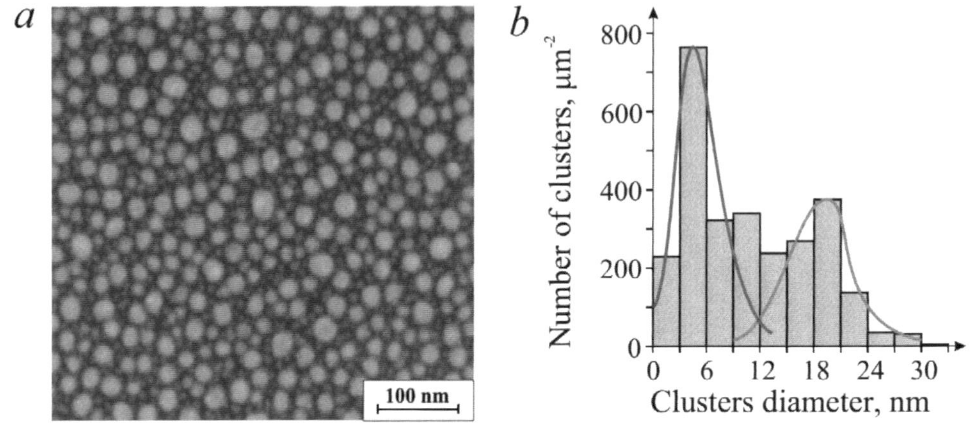

Figure 3. SEM image (a) and the cluster size distribution histogram of silver particles of the sample with consecutive evaporation of 4.5 and 2.7 mg weight portions (b)

Despite the fact that in general, after the annealing the total number of particles decreased (now 2000 pcs/μm2), which is directly related to their enlargement, the clusters size distribution still is bimodal (Fig. 4). Predominant sizes shift to the area of larger values: 11 and 24 nm after annealing versus 5 and 21 nm before this heat treatment, respectively.

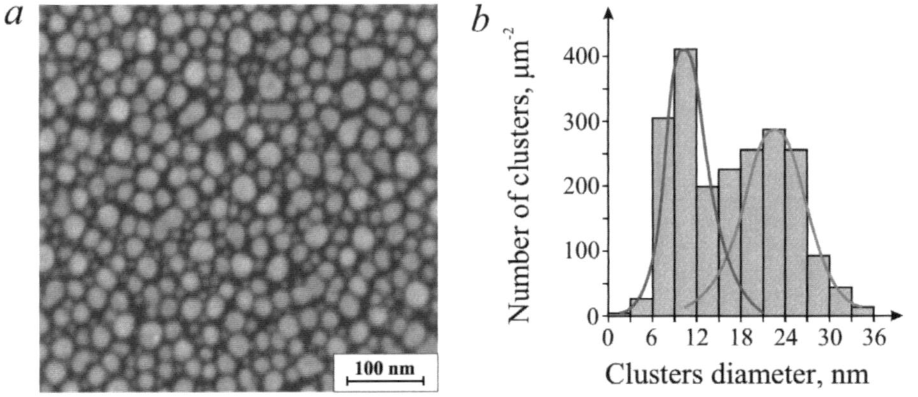

Figure 4. SEM image (a) and the cluster size distribution histogram of silver particles of the sample with consecutive evaporation of 4.5 and 2.7 mg weight portions after annealing at 230 °C during 30 min (b)

Repeated heat treatment of already annealed samples was carried out at a higher temperature of 300°C. Additional annealing has not led to any noticeable changes: for the samples with both unimodal and bimodal distributions, the average particle size and surface density of the particles remained unchanged.

Thus, investigation of the samples after sequential evaporation of the two small portions of silver when the second weight portion is smaller than the first, demonstrate the following features:

1. The initial state of silver condensate is extremely unstable. Insignificant thermal impact causes a change in it and leads the system to a state where clusters are more separated from each other;
2. At repeated evaporation, the formation of smaller particles occurs both between and over previously formed clusters;
3. After the second evaporation bimodal size distribution of the particles is observed;
4. Repeated thermal impact onto previously annealed array of clusters at even higher temperatures does not lead to any noticeable change in it.

Upon substance evaporation from a point source in vacuum, evaporated atoms retain high temperature, because their trajectories diverge, and the atomic concentration is small. This makes the exchange of energy between them unlikely. In case of silver evaporation, we expect that the temperature of the vaporized atoms corresponds to the evaporation temperature of silver ~1300 K. Therefore, during the condensation we actually deal with the impact of hot steam onto the deposition surface. For this reason, the dependence of the size of nanoparticles on quantity of evaporated substance is observed: the larger silver weight portion, the more intensive steam action on the surface, and as a consequence, larger the average size of the nanoparticles in the array, as it has been previously demonstrated in a number of papers [3,5,11,12].

When we first vaporize a larger weight portion and larger clusters are formed, and then re-evaporate a smaller number of substance, this smaller amount of substance is not able to heat larger clusters to a high enough temperature. Therefore, we observe features 2) and 3): smaller clusters are formed between the previously formed large clusters and over them, and as a result, there are two predominant particle sizes.

The cross-sectional images obtained by transmission electron microscope allowed to determine the height of the nanoparticles (Fig. 5).

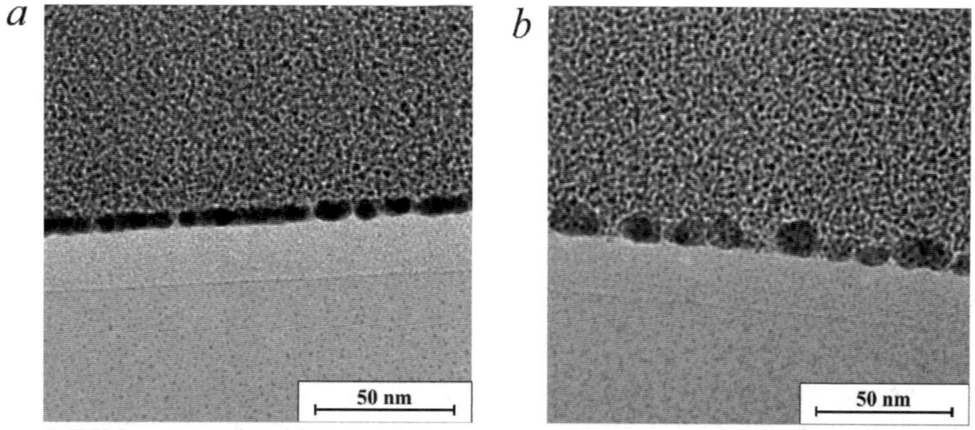

Figure 5. TEM cross-sectional images of the samples obtained by evaporation of 4.5 mg weight portions before annealing (a) and after annealing for 30 min at 230 °C (b)

Using electron-microscopic images of silver clusters shown in Fig. 1 and 2, a comparative evaluation of the amount of substance in the array of nanoparticles after deposition and after heat treatment was carried out. Because the shape of the particles obtained after deposition is complex and not spherical, the volume of silver was calculated as the product of the surface area occupied by silver particles to their height (Fig. 5a). For the annealed sample, the particle shape is close to spherical, but nevertheless there are some deviations (Fig. 5b). Given these nuances, the estimate calculations showed that the mass of silver was ~ 3.2×10^{-14} g/μm^2 and ~ 4.6×10^{-14} g/μm^2 for the initial and annealed samples, respectively. This evaluation showed that after annealing the total amount of matter in an array of nanoparticles is larger than before annealing. This, of course, in no way means that there is a violation of the mass conservation law, or large calculation error, because the calculations were carried out repeatedly and in various samples. We are more inclined to think that this is due to the resolving power of the electron-ion microscope we used, and we just do not see on the original sample any part of the deposited silver, which is not in the form of clusters. Obviously, this part of the material is the cause of feature 1): when the initial condensate is in an unstable state and a small thermal impact leads system into a more stable state. This explains why the re-annealing does not lead to any change (feature 4).

4. Conclusions

Thus, our investigation of condensation of small amounts of Ag during thermal evaporation in vacuum showed that Ag condensate after the deposition is in an unstable state, and a relatively small thermal impact leads to the formation of an array of separated from each other clusters. It is important to note that, despite the small size of the clusters in the array, and consequently, the high surface energy of the system, this state is sufficiently stable. One would expect that repeated annealing of such a system would lead to coalescence into larger particles, but after repeated annealing no changes were observed. Moreover, second deposition of smaller weight portion of silver allows to form new particles with another characteristic size without noticeable enlargement of existing particles.

Acknowledgments

This work was financially supported by the Ministry of Education and Science of the Russian Federation within the state program "Organization of research" (code 486). The equipment of the center for collective use "Diagnosis and modification of microstructures and nano-objects" was used in this work.

References

[1] Sanles-Sobrido M *et al* 2009 *Journal of the American Chemical Society* **131** 7 pp 2699-2705
[2] Shen C *et al* 2008 *Chemistry of Materials* **20** 22 pp 6939-6944
[3] Hamilton J F *et al* 1973 *Thin Solid Films* **16** pp 49-63
[4] Goldby I M *et al* 1996 *Appl. Phys.* **69** 19 pp 2819-2821
[5] Schmidt A A *et al* 1996 *Surf. Sci.* **349** 3 pp 301-316
[6] Carrey J *et al* 2002 *Surf. Sci.* **504** pp 75-82
[7] Gatel C *et al* 2006 *Surf. Sci.* **600** pp 2650-2662
[8] Sui M *et al* 2013 *Nanoscale Res. Lett.* **8** pp 525
[9] Gromov D G *et al* 2013 *Physics of the Solid State* **55** 3 pp 619-623
[10] Belov A N *et al* 2014 *Calphad* **44** pp 138-141
[11] Gromov D G *et al* 2015 *Appl. Phys* **118** 4 pp 1297-1303
[12] Gromov D G *et al* 2015 *Physics of the Solid State* **57** pp 173-180
[13] Giannuzzi L A *et al* 1998 *Microscopy Research and Technique* **41** 4 pp 285-290

Modeling of the behavior and statistical analysis of compressibility in the process of Langmuir monolayer structurization

H.V. Grushevskaya, G.G. Krylov, N.G. Krylova, I.V. Lipnevich

Belarusian State University, 4 Nezavisimosti Ave., Minsk 220030, Belarus
E-mail: grushevskaja@bsu.by

Abstract. An approach has been developed to perform data processing for experiments on two-dimensional phase transition of first order. The approach represents itself the two stage filtration with subsequent merging of processed data and comparison with simulation results based on geometro-thermodynamical methods. The approach has been successfully applied to description of the phase transition in compressed monolayers of steric acid.

1. Introduction

Development of formation methods for two-dimensional (2D) solid-state perfect structures, which can operate at room temperature, is an important problem of flexible electronics. Langmuir-Blodgett (LB) technique can be used to produce almost perfect two-dimensional structures, provided that the conditions on a phase boundary are under the control [1].

To utilize LB monolayers as a material for nanoelectronics it is necessary to correlate structural changes in a metastable compressed monolayer with changes of phenomenological parameters of the monolayer (e.g., compressibility coefficient). This requires a statistical analysis of experimental dependences of the surface tension $\tilde{\pi}$ upon the area A per molecule, followed by compressibility calculation.

When modeling the Langmuir monolayer compression by Monte Carlo methods, a metastable state is excluded from consideration that does not allow for such a control [2]. A compressibility κ, for $\tilde{\pi}$–A isotherms being dependencies of surface tension $\tilde{\pi}$ on an area A per one molecule, presents itself a derivative from the dependence $\ln A(\tilde{\pi})$ over $\tilde{\pi}$. But, instead of it, the "apparent" compressibility $\kappa_{\text{app}} = 1/K'$ being an inverse elasticity of a Langmuir monolayer is calculated [3–5].

To construct the dependence of compressibility coefficient in the whole range of the surface pressure $\tilde{\pi}$, it is necessary to use the procedure of the statistical data processing on a huge experimental data set due to large fluctuations in critical regions for 2D phase transitions. Nowadays such data processing of $\tilde{\pi}$–A compression isotherms have not been proposed.

In [6–8], a geometric approach to the thermodynamics of the first order phase transitions in compressed monolayers on the interphase boundary of air/water subphase has been developed.

The goal of this work is to propose a method of compression isotherms statistical analysis for metastable Langmuir monolayers. It allows to evaluate the compressibility coefficient based on experimental data. We model the 2D phase transitions of first order using geometrodynamic methods and demonstrate that theoretical predictions and experimentally obtained compressibility of monolayers of stearic acid are close to each other.

2. Materials and methods

Langmuir monolayers are fabricated on an automated home-built Langmuir trough with controlled deposition on a substrate [9]. Hexan solution of stearic acid was dripped with the help of a micropipette on a liquid subphase surface. Langmuir monolayers were obtained by compressing of stearic acid molecules on air / aqueous subphase interface. Deionized water was used as a subphase for stearic acid monolayer formation.

3. Data processing

In this paper, we propose a method of statistical analysis when at first stage we smoothen 2D fluctuations of the vector $\vec{r}_i \equiv (\tilde{\pi}_i, A_i)$ over an interval of n_1 values of the surface tension $\tilde{\pi}_{i+j}$ and corresponding to them values of the area A_{i+j} per one molecule: $\vec{r}_i \equiv (\tilde{\pi}_i, A_i) = \frac{1}{n_1}\sum_{j=-n_1/2}^{n_1/2} \vec{r}_{i+j}$. Then, a filtering procedure based on the sliding mean with given window width m_1 is applied to the smoothed set, when an experimental value is to replace by a mean value obtained from the m_1 nearest points as: $\tilde{\pi}_i \to \langle \tilde{\pi}_i \rangle = \frac{1}{m_1}\sum_{j=-m_1/2}^{m_1/2} \tilde{\pi}_{i+j}$. The filtering on the sliding mean was applied only to the readings of the surface tension sensor, as its noise prevails over the noise of "inertial" position sensor. To find the compressibility $\kappa(\tilde{\pi})$ of monolayer, the logarithmic derivative of the dependence $A(\tilde{\pi})$ obtained after the data processing with filtering parameters $\{n_1, m_1\}$ was calculated numerically. The existence of scale invariance property of phase transitions was accounted by merging several filtrations of original data.

4. Compressibility of amphiphilic two-dimensional membranes with Finsler–Lagrange space structure

The amphiphilic molecules of fat acids form a monolayer on the surface of the water subphase. Let us consider the motion of a particle in the monolayer under the action of electrocapillary forces that take place on the interface and determine the magnitude of the surface tension σ of such a 2D membrane [10]. The first order phase transition occurs for long compression time $\Delta t \gg 1$ (long observation time). The experimentally measured quantity, interpreted as the surface pressure $\tilde{\pi}$ of the monolayer, is defined by the expression:

$$\tilde{\pi} = \sigma_{H_2O} - \sigma_{hydrocomp} \tag{1}$$

where σ_{H_2O} is the surface tension of the subphase (e.g., water), $\sigma_{hydrocomp}$ is the surface tension of the subphase with hydrate complexes. According to (1), molecules which leave the hydrate complexes do not contribute to the experimentally measurable value of $\tilde{\pi}$. The surface pressure of the hydrated monolayer changes, but as a consequence of the diminishing of the Langmuir monolayer density ρ due to the complexes decay, the increment of surface pressure $\Delta\tilde{\pi}$, observed during the phase transition of Δt, is equal to zero: $\Delta\tilde{\pi} = 0$. In this sense, at the first order phase transition, the parameters of molecules which leave hydrated complexes are hidden from the observer. This allows us to introduce an evolution parameter τ as a parameter of the current state of the process of the first order phase transition in a hydrated complexes monolayer only. Then, the derivative $\frac{d\Delta\tilde{\pi}}{d\tau}$ is the rate of surface pressure change for the hydrated complexes monolayer. Since $\frac{\partial(S\tilde{\pi})}{\partial\tau} \propto -\frac{\partial(V\Delta t_{pht}\tilde{\pi})}{\partial\tau} \propto k_B T\frac{\partial\rho}{\partial\tau} \propto k_B T\tilde{C}$ and $\frac{\partial r}{\partial\tau} \propto C'V$, C'; $\tilde{C}$ and the temperature $k_B T$ are constants, and the change of the area S per one molecule is proportional to the square of variation $\delta r(\tau)^2$, then we find a space–time derivative of the compressibility coefficient $\kappa = \frac{1}{S}\frac{\partial S}{\partial \tilde{\pi}}$:

$$\frac{\partial \kappa'_\tau}{\partial r} = -\frac{1}{S\Delta t_{pht}}\frac{\partial^2 S}{\partial r \partial \tilde{\pi}} = \frac{2}{k_B T C'' \delta r}\frac{\partial^2 \delta r}{\partial \tau^2} \tag{2}$$

where $\kappa'_\tau = \frac{\partial \kappa}{\partial \tau} \propto \frac{\kappa}{\Delta t_{pht}}$, $C'' = C'\tilde{C}$, $C', \tilde{C} > 0$; $S \square \delta r^2$, Δt_{pht} is time of the phase transition, V is a compression rate.

We apply a geometrodynamics of the first order phase transitions in two-dimensional systems to examine metastable compressed monolayers [6–8]. δr is determined by the flag curvature K as $\frac{\partial^2 \delta r(\tau)}{\partial \tau^2} = -K \cdot \delta r$ [11]. We choose the trace B_c of Berwald curvature B^k_{ikj} as K: $B_c = g^{ij} B^k_{ikj} = g^{ij}\frac{\partial G^k}{\partial y^i \partial y^k \partial y^j}$, $\vec{y} = (\dot{\xi}, \dot{r}, \dot{\varphi})$, $\dot{\xi} = \frac{dt}{d\tau}, \dot{r} = \frac{dr}{d\tau}, \dot{\varphi} = \frac{d\varphi}{d\tau}$. Here, a vector G^k is determined by equating the variation of action $\int dl_F = \int F(x^j, dx^j)$ to zero. One uses a following metric function $F(x^j, dx^j)$ [6–8]:

$$F^2(x^j, \Delta x^j) = A\frac{\Delta \xi^3}{\Delta r} + B\Delta\xi - C\frac{(\Delta r^2 + r^2 \Delta\varphi^2)}{2c^2} \tag{3}$$

where A, B and C are defined as follows:

$$\begin{aligned}
A &= p|V|r^5 e^{\frac{2|V|t}{r}},\\
B &= mc^2 - p\left(\left(-\frac{4}{3}r^5 + \frac{16}{15}(|V|t)r^4 + \frac{1}{30}(|V|t)^2 r^3 + \frac{1}{45}(|V|t)^3 r^2 + \frac{1}{45}(|V|t)^4 r \right.\right.\\
&\left.\left. + \frac{2}{45}(|V|t)^5\right)e^{\frac{2|V|t}{r}} - \frac{4}{45}\frac{(|V|t)^6}{r}\mathrm{Ei}\left[\frac{2|V|t}{r}\right]\right),\\
C &= mc^2
\end{aligned}$$

where m is a mass of a particle, $p = \frac{\pi^2 q^2}{\varepsilon\varepsilon_0}\frac{\rho_0^2}{R_0^2}$ is a parameter dependent on a charge q of monolayer molecule, density ρ_0 of molecules at initial moment, dielectric constant ε_0, dielectric permittivity ε of the subphase, and radius R_0 of a radial trough, c is speed of light; r, φ determine a displacement of the particle in the polar coordinate system; a special function Ei is an exponential integral function. The metric tensor is determined through the metric function as follows:

$$g_{ij}=\frac{1}{2}\frac{\partial^2 F^2}{\partial y_i \partial y_j}=\begin{pmatrix} 3A\dfrac{\dot{\xi}}{\dot{r}} & -\dfrac{3}{2}A\left(\dfrac{\dot{\xi}}{\dot{r}}\right)^2 & 0 \\ -\dfrac{3}{2}A\left(\dfrac{\dot{\xi}}{\dot{r}}\right)^2 & A\left(\dfrac{\dot{\xi}}{\dot{r}}\right)^3-\dfrac{C}{2c^2} & 0 \\ 0 & 0 & -\dfrac{C}{2c^2}r^2 \end{pmatrix}. \tag{4}$$

And the components of the vector G^k :

$$G^i(y)=\frac{1}{4}g^{il}(y)\left\{2\frac{\partial g_{jl}}{\partial x^k}(y)-\frac{\partial g_{ik}}{\partial x^l}(y)\right\}y^j y^k \tag{5}$$

are determined as

$$\begin{aligned} G^1 &= \frac{A}{4\Delta}\left[\frac{1}{2}A_t\frac{\dot{\xi}^6}{\dot{r}^4}+B_t\frac{\dot{\xi}^5}{\dot{r}^3}+\frac{1}{2}B_r\frac{\dot{\xi}^4}{\dot{r}^2}+\frac{3}{2}\frac{C}{c^2}r\frac{\dot{\xi}^2\dot{\varphi}^2}{\dot{r}^2}\right]-\frac{C}{8c^2\Delta}\left[2A_t\frac{\dot{\xi}^3}{\dot{r}}+B_t\dot{\xi}^2+3A_r\dot{\xi}^2+3B_r\dot{\xi}\dot{r}\right], \\ G^2 &= \frac{3A}{4\Delta}\left[-\frac{1}{2}A_r\frac{\dot{\xi}^4}{\dot{r}^2}+\frac{1}{2}B_t\frac{\dot{\xi}^4}{\dot{r}^2}+\frac{C}{c^2}r\frac{\dot{\xi}\dot{\varphi}^2}{\dot{r}}\right]-\frac{B}{4\Delta}\left[A_t\frac{\dot{\xi}^4}{\dot{r}^2}+2A_r\frac{\dot{\xi}^3}{\dot{r}}+B_r\dot{\xi}^2-\frac{C}{c^2}r\dot{\varphi}^2\right], \\ G^3 &= \frac{1}{r}\dot{r}\dot{\varphi} \end{aligned} \tag{6}$$

where $\Delta=\frac{3}{4}A^2\left(\frac{\dot{\xi}}{\dot{r}}\right)^4+AB\left(\frac{\dot{\xi}}{\dot{r}}\right)^3-\frac{3}{2}A\frac{C}{c^2}\frac{\dot{\xi}}{\dot{r}}-\frac{1}{2}B\frac{C}{c^2}$ and $A_t=\frac{\partial A}{\partial t}$, $A_r=\frac{\partial A}{\partial r}$, $B_t=\frac{\partial B}{\partial t}$, $B_r=\frac{\partial B}{\partial r}$.

Assuming $\frac{\partial^2\delta r(\tau)}{\partial\tau^2}=-B_c\delta r$ one gets the compressibility after integration (2):

$$\kappa=\kappa'_\tau\,\Delta t_{pht}\approx-\frac{1}{S}\int_{r(t)}^{r(t+\Delta t_{pht})}B_c d^2 r. \tag{7}$$

Here one takes into account that $\frac{\partial\tilde{\pi}}{\partial\tau}=\frac{\partial\tilde{\pi}}{\partial t}\frac{\partial t}{\partial\tau}\approx\Delta t_{pht}$. The Berwald curvature B_c is equal to zero everywhere except of the isotherm plateau. Therefore the expression (7) can get the form

$$\kappa\approx-\frac{1}{S}\int_{r(t)}^{r(t+\Delta t_{pht})}B_c d^2 r=-\frac{1}{S}\int_{r(0)}^{r(\Delta t_{pht})}\frac{\partial B_c}{\partial\vec{r}}d^2 r=-\left(B_c(r(\Delta t_{pht}))-B_c(0)\right). \tag{8}$$

Here $B_c(0)\equiv B_c(r(0))$ is a constant of the phase transition.

The theoretical dependences B_c on displacement r at $\varphi=0$ are shown for compression rates $V=2$ and $V=10$ in Fig. 1. One can see that $B_c(r)$ has two singularities, position of which depends on the compression rate. At increase of V, the singular points shift to each other, and, respectively, a region $\Delta\tilde{\pi}_{pht}$ of phase transition becomes narrower according to Fig. 1.

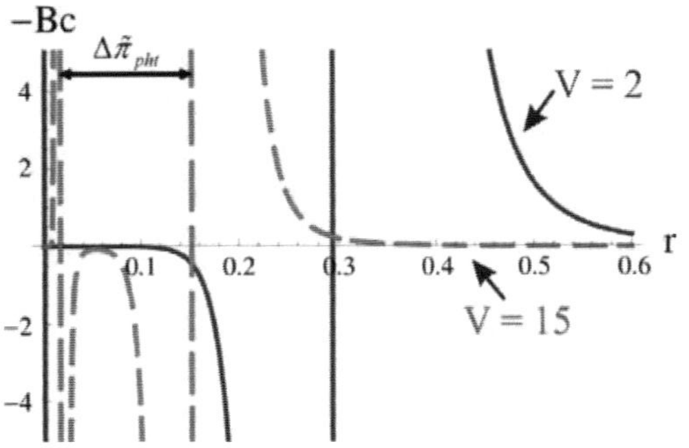

Figure 1. The dependences of Berwald curvature B_c on displacement r for compression rates $V = 2$ (blue solid line) and $V = 10$ (magenta dashed line).

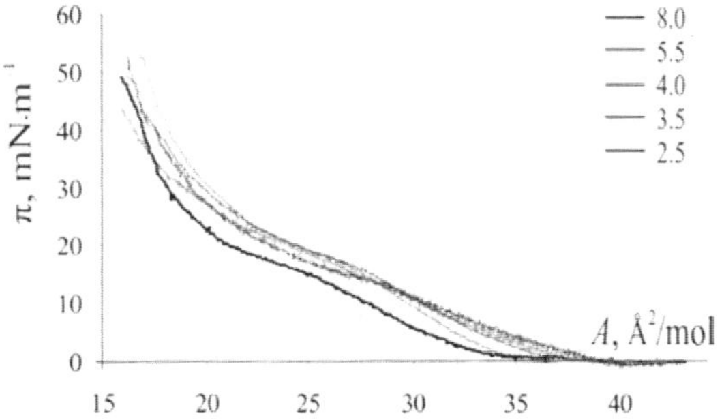

Figure 2. $\tilde{\pi} - A$-isotherms for a stearic acid monolayer formed at compression rates V equal to 8.0 (blue), 5.5 (green), 4.0 (red), 3.5 (magenta), 2.5 mm/min (black).

5. Experimental results and discussion

Typical compression $\tilde{\pi} - A$ isotherms at different compression rates V are shown in Fig. 2. One can see that depending on V, the phase transitions can vary in area per molecule and start at different molecular area A. At low V the surface pressure increases shaper in range of high value A than at high V: the first order phase transition begins at lower A and is smoother (compare isotherm for rates 2.5 and 8.0 mm/min). These isotherms have been used to get the dependences of compressibility on surface pressure according to data processing, described above.

The dependences of compressibility $\kappa(\tilde{\pi})$ on $\tilde{\pi}$ at different compression rates V are represented in Fig. 3. Each of them represents itself three sets of data which correspond to three filtration parameters: $\{n_1 = 6, m_1=6\}$, $\{n_1 = 6, m_1=7\}$, $\{n_1 = 7, m_1=6\}$ for $V = 3.5$, 4.0, and 5.5 mm/min (Fig. 3 (b – d)) and

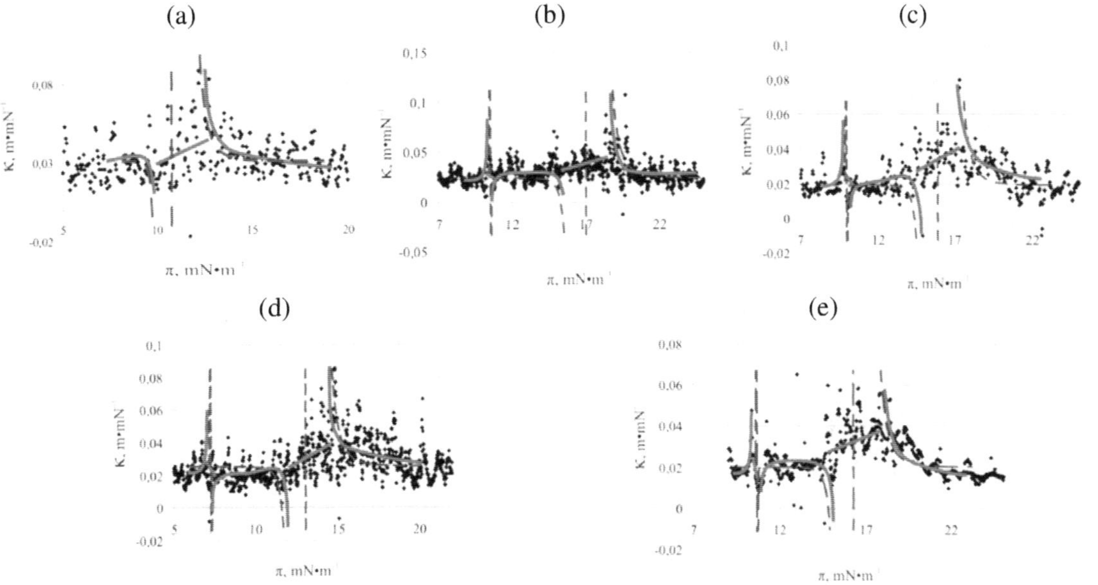

Figure 3. The dependences of the compressibility κ on $\tilde{\pi}$ at the 2D-phase transition of first order in the monolayer at different compression rates: $V = 2.5$ (a), 3.5 (b), 4.0 (c), 5.5 (d), 8.0 mm/min (e). Symbols denote experimental data processing results, solid lines are fitting, dashed lines are the theoretical predictions.

$\{n_1 = 3, m_1=3\}$, $\{n_1 = 4, m_1=5\}$, $\{n_1 = 5, m_1=3\}$ for $V = 8.0$ mm/min (Fig. 3 (e)). One can see that the dependences $\kappa(\tilde{\pi})$ have a complex non-linear character in the region of the first order phase transition.

The experimental dependences of compressibility κ on surface pressure $\tilde{\pi}$ for the stearic acid monolayer at different compression rates V have been approximated by the theoretical curves (8), red dashed lines in Fig. 3. As one can see the theoretical curves elucidate the origin of dispersion of compressibility. The phase transition width $\Delta\tilde{\pi}_{pht}$ was defined as a difference between two singular points on a theoretical curve in Fig. 1. Table 1 demonstrates that $\Delta\tilde{\pi}_{pht}$ is a function of the compression rate V. The phase transition width was found out to decrease with the increasing of the compression rate that is in accordance with our theoretical results shown in Fig. 1.

Table 1. The value of phase transition width $\Delta\tilde{\pi}_{pht}$ at different compression rates V.

V, mm/min	$\Delta\tilde{\pi}_{pht}$, mN/m
2.5	-
3.5	6.5
4.0	6.0
5.5	5.7
8.0	5.5

6. Conclusion

Thus, the approach has been proposed that allows to examine the anomalous behavior of compressibility at 2D phase transition of first order. The proposed approach is used to describe the formation of Langmuir stearic acid monolayers.

References

[1] Motschmann H, Möhwald H 2001 *Langmuir–Blodgett Films.* In: *Handbook of Applied Surface and Colloid Chemistry.* Ed. Holmberg K (John Wiley & Sons, Ltd)
[2] Fiore C E, Carneiro C E I 2008 *ArXiv [cond-mat.stat-mech]* 0809.4023v1
[3] Modlińska A, Bauman D 2011 *Int. J. Mol. Sci.* **12** 4923
[4] Toma L M, Gengler R Y N, Prinsen E B, Gournis D, Rudolf P 2010 *Phys. Chem. Chem. Phys.* **12** 12188
[5] Bhande R S, Landge Y A, Giri P 2012 *J. Chem. Pharm. Res.* **4(6)** 3297
[6] Balan V, Grushevskaya G, Krylova N, Neagu M, Oana A 2013 *Applied Sciences* **15** 30
[7] Grushevskaja G V, Krylova N G 2011 *Hypercomplex Numbers in Geometry and Physics* **15** 128 (In Russian)
[8] Neagu M, Krylova N G, Grushevskaya H V 2013 *J. Math. Phys.* **54** 031508
[9] Grushevskaya H V, Lipnevich I V, Orekhovskaya T I 2013 *J. Mod. Phys.* **4** 7
[10] Landau L D, Lifshits E M 1982 *Electrodynamics of continuous media.* In: *Theoretical Physics.* Vol. 8 (Moscow: Science)
[11] Bao D, Chern S, Shen Z 2000 *An Introduction to Riemann-Finsler Geometry* (Berlin: Springer)

The flow reactor system for in-line synthesis of semiconductor nanoparticle

O A Ryzhov[1], L B Matyushkin[2]

[1]Student, St. Petersburg State Electrotechnical University "LETI", Pr. Popova 5, St.Petersburg, 197376 Russia

[2]Postgraduate, St. Petersburg State Electrotechnical University "LETI", Pr. Popova 5, St.Petersburg, 197376 Russia

Abstract. A flow reactor nanoparticle synthesis technique is proposed as replacement for «hot injection» synthesis of semiconductor nanocrystals in a glass flask. The main advantages are possibility of continuous nanoparticles production, technology flexibility and lower cost of the final products in comparison with currently applied methods.

1. Introduction

The traditional synthesis in a chemical glass flask allows to obtain a variety of nanoparticle types with different functional properties: luminescent quantum dots [1], magnetic structures [2], plasmonic active elements [3] etc. In some of these methods materials are synthesized at room temperature, while others require a heating step. For instance, in the case of "hot-injection" synthesis of colloidal quantum dots temperatures about 200–300°C are used. Nucleation and growth stages dividing allows to obtain monodisperse nanoparticles. However, in the case of chemical flask it is impossible to separate these stages completely, because the reaction takes place in a restricted volume. This technique also limits the product yield – a single synthesis gives only a limited amount of particles, and for large quantities it must be repeated many times. This factor limits the creation of industrial systems for production of colloidal quantum dots and hinders development of their applications.

1.1. Types of flows

Flow reactors are divided into two basic types: continuous and segmented flow reactors. The latter type is conventionally divided into two groups: the "gas/liquid" and "liquid/liquid".

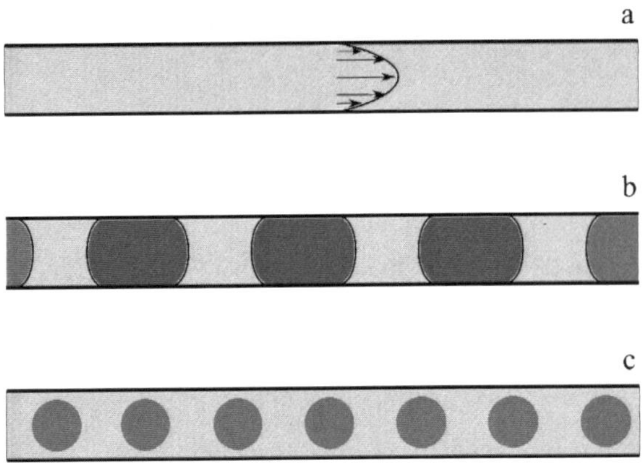

Figure 1 – A diagram illustrating the continuous (a) and segmented (b, c) flow regimes in channels with a circular cross section

The reactors of the continuous flow type (Figure 1a) are the first system which has been implemented in the synthesis of nanoparticles in a stream mode. Along with the manufacturing ease of such reactors they have some disadvantages: firstly, they are soiled over time and secondly, the liquid front has a parabolic velocity profile. The reason for such profile known as Hagen-Poiseuille is the friction between flow and wall which results in a corresponding decrease in velocity near the walls. That leads to a fatal issue of relatively large nanoparticles size dispersion. In low viscosity solvents the velocity dispersion effect is partly compensated by lateral diffusion through the channel – growing nanoparticles and reagents cross channel numerous times thereby experiencing all speeds equally. But in case of high-temperature synthesis we must use solvents with high boiling point, which usually expect a high viscosity solvent. Particle size dispersion reducing can be achieved by the transition from a continuous flow to a segmented one.

A segmented stream is created by introducing to it an additional component which is immiscible with the reactive basic solutions (liquid or gas) so that discrete droplets stream of a liquid is separated by layers of another liquid or gas. Using the segmented flow is becoming increasingly popular for the synthesis of nanoparticles. The method has been applied to a wide range of materials [4 – 7].

When it is used the "gas/liquid" mode viscous friction of the channel walls causes convective mixing within each segment, providing chemical homogeneity, important to obtain monodisperse nanoparticles product. The particles pass through the reactor at the same speed, therefore dispersion of particles size is reduced in comparison with a continuous stream. Not only inert gases can be used for segmentation, but also, for example, volatile amines or just air.

However, a disadvantage of using "gas/liquid" mode is a contact of the reaction mixture with the walls, resulting in gradual clogging of the channel. This problem can be solved by using the next type of flow regime "liquid/liquid", where the reactants move in a series of discrete droplets which are separated not only from each other but also from the channel walls by an immiscible liquid (Figure 1c). This occurs when the separating phase is a liquid, preferably wetting the channel walls and being in excess in relation to the reactive components.

This type of flow has an important advantage over the above described: the reaction mixture isolated from the channel walls, any reaction products reliably enclosed in the droplet volume, thus minimizing the risk of reactor fouling. Therefore, this type of synthesis is the most interesting for the industrial implementation of the entrained flow reactor.

2. Experiment

In comparison with traditional chemical synthesis in a flask, a flow chemistry nanoparticle synthesis is performed in a stream of fluid flowing in a narrow capillary. Precursors can be solutions of water soluble salts or organic compounds in a nonpolar medium. There are two organic immiscible fluids in our system: one fluid serves as the liquid carrier with good wetting of capillary wall, while the other consists of precursor solution drops which are distributed in the flow of medium. Syringe pumps are used for setting transfer rates of components (carrier liquid and precurors). Number of pumps equals to the number of independent components that are placed in corresponding syringes.

Our system works as follows: 1) the precursors A and B are moved by syringe pumps to the mixing cell, 2) the mixed solution gets into a coil with a temperature T_n corresponding to the nucleation of nanocrystals, 3) the core particles solution flows into the coil with a temperature T_g corresponding to the particles growth stage, 4) in the last step the solutions are selected and cooled to room temperature. Time control of the particle growth is determined by the flow rate of carrier fluid. Due to the immiscible nature of the carrier and the solvent, the liquids separates into two layers of different density, so nanoparticles are isolated from the solvent by standard flocculation technique by polar liquid such as acetone or methanol.

At the last stage and each of the intermediate steps the solutions can be tested by optical spectroscopy. For this purpose we have developed a compact optical spectrophotometer Eltech-M-14 allowing to obtain in situ luminescence spectra. The syringe pumps flow rate and the heating devices temperatures T_n and T_g are controlled by computer program, microcontroller board and several additional printed circuit boards created ourselves.

2.1. Implementation of precursors flow in the capillary

An important part of the reactor is precursors mixing module. The scheme of such a module is shown in Figure 2.

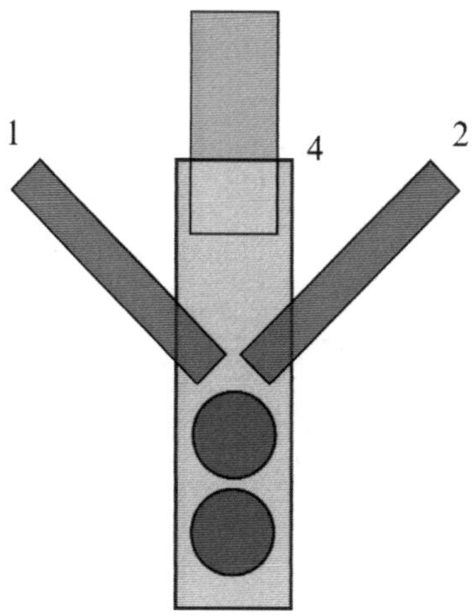

Figure 2 – Scheme module for mixing: 1, 2 – needle precursors; 3 – needle with a liquid carrier; 4 – capillary

Two precursors flow through the needles *1* and *2* (angled 90°), mixes together and forms droplets in a stream of carrier liquid from needle *3* and go in the polytetrafluoroethylene capillary *4*. Since the capillary has a diameter less than 1 millimeter, it is the best to use needle from the insulin syringes. The feed rate of the carrier fluid should be higher than the rate of precursors feed to control the distance between drops.

2.2. Software implementation management syringe pump

Syringe pump system is based on a stepper motor and transmission system of movement to the syringe, which is printed using the 3D-printer UP Plus 2. The appearance is shown in Figure 3. Inject syringes can have different volumes: 5 ml, 10 ml, 20 ml.

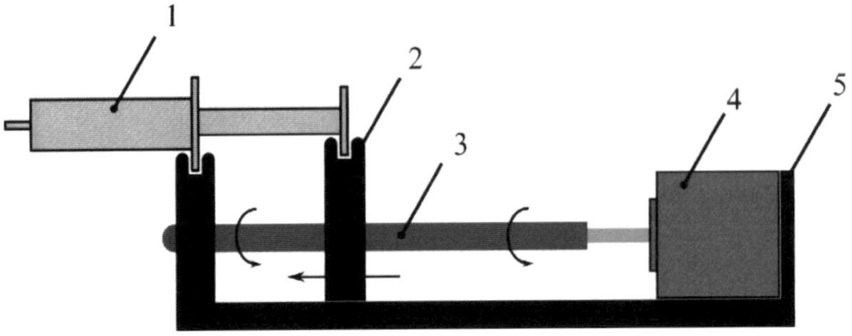

Figure 3 – The syringe pump model: 1 – a syringe, 2 – the movable unit with an internal thread, 3 – the threaded rod, 4 – the stepping motor, 5 – module housing

Pump control is done by a single-board microcontroller Arduino Uno and software developed in NI LabVIEW with LabVIEW Interface for Arduino. Figure 3 shows a flowchart of this program. Work can be divided into three stages of the pump: 1) "initialization" – attaching the movable unit with an internal thread (element 2 in Figure 3) to its initial position, which depends on the filling of the syringe being used; 2) a syringe pump setup; 3) "start operation" – extruding the solution from a syringe at a determined speed.

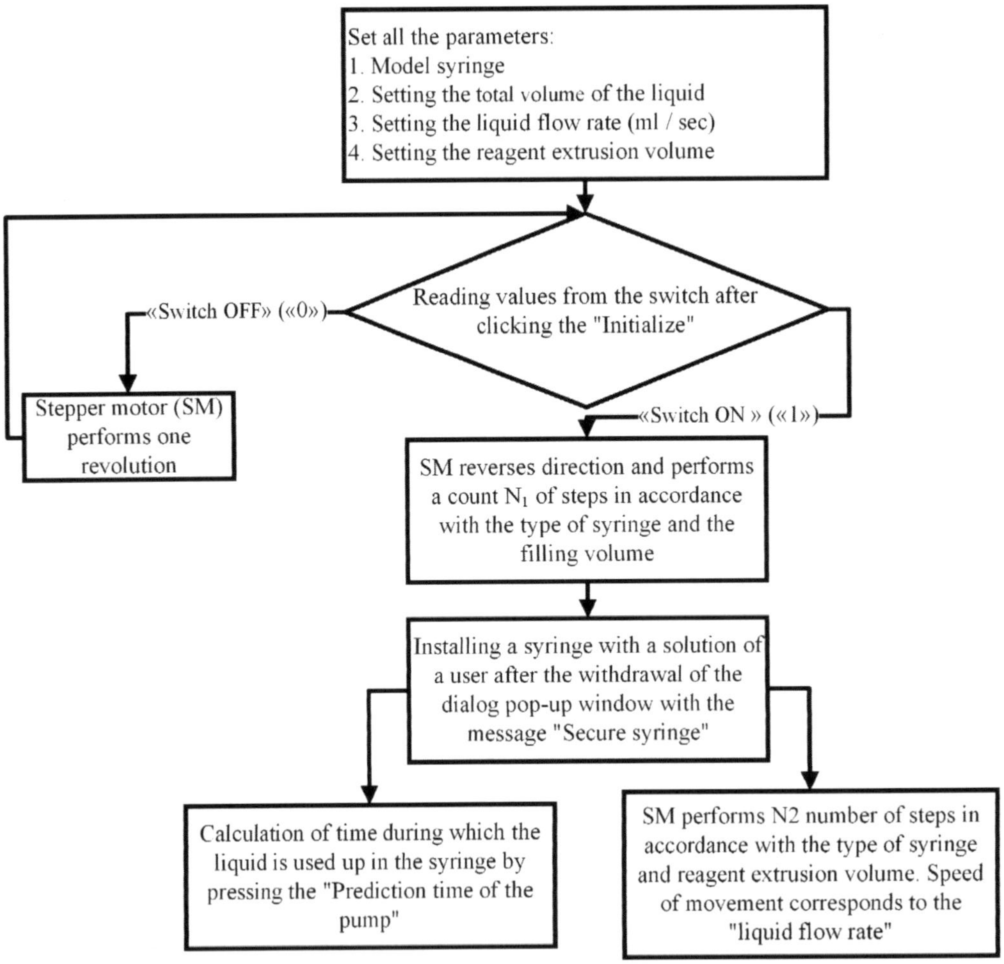

Figure 4 – Schematic diagram of the software is running for pump control

2.3. Diagnosis of the optical properties of nanoparticles

The new technology of synthesis is needed to measure the optical characteristics of produced quantum dots *in situ*. At the same time photometric device on the output of the reactor should be very compact and economically competitive with the total cost of the reactor. Spectrophotometric instruments for the diagnostics of such materials are traditionally used. So there was a need to create a compact device for the rapid analysis of the luminescence of colloidal quantum dots.

When measuring the luminescence spectra of the exciting radiation (excitation by LED) passes through the cell and causes the sample photoluminescence. Secondary radiation passes through a narrow slit and divides by diffraction grating to different wavelengths with intensities registered by CMOS webcam. CMOS sensor allows to simultaneously record the entire wavelength range of the visible spectrum, which reduces analysis time and allows the measurement of the optical spectrum in a continuous mode. In addition, webcam is fully prepared to host that does not require additional electronic components. Adjacent rows of pixels can increase the number of experimental points and accordingly improve the measurement accuracy.

As a result of analysis on the computer receives a radiation image of the diffraction pattern, the next stage of the work was to develop a program that transforms the spectrogram (two-dimensional array of pixels) in the spectrum. During the analysis program in the form of two-dimensional array of picture, data transmitted diffraction pattern. The position of a pixel corresponds to a particular wavelength value, while the brightness of the pixel is proportional to the intensity of the light coming on cell CMOS sensor. Signal is formed of three colors (red, green, blue), brightness is due to the averaging of the intensity values on the three channels:

$$I = \frac{R+G+B}{3} \quad (1)$$

where I – brightness (intensity) of the pixel; R, G, B – the intensity of red, green, and blue channel of the pixel, respectively. Three modes of the device are implemented:

1. In the "Calibration" mode conversion factors are being identified diffraction. As a gauge used emission spectrum compact fluorescent lamp having a fixed characteristic intensity peaks at known wavelengths. To determine the scaling factors in the resulting emission spectrum of the fluorescent lamp two selected intensity maxima are assigned their own wavelengths and calculated ratios of corresponding pixels of the wavelength of the diffraction pattern. In fact, it is necessary to know the left border of the spectrum and the coefficient corresponding to step (a kind of dividing the price of one pixel). The last factor is calculated as follows:

$$h = \frac{P_2 - P_1}{N_2 - N_1} \quad (2)$$

where P_i and N_i – respectively the position of the i-th maximum of the selected pixel relative to the left boundary and the position of that peak, expressed in nanometers.

Calculation of the left end of the spectrum in nanometers is made, taking into account the calculated ratio:

$$L = \frac{N_1 - hP_1 + N_2 - hP_2}{2} \quad (3)$$

The calculated data are used as calibration data in other modes of operation of the program.

2. In the "Emission spectrum" mode – measured photoluminescence spectrum of a solution of nanoparticles. When measuring the luminescence spectra of the exciting radiation (LED $\lambda = 365$ nm) passes through the sample and causes the photoluminescence. Software converts the captured image in dependence of the emission intensity of the wavelength, assigning each pixel column wavelength based calibration data.

3. In the "Time response" mode luminescence intensity is measured on the same wavelength with respect to time. This mode is particularly interesting for the *in situ* measurement and research questions degradation of quantum dots under the influence of high-energy excitation.

The software also realized the possibility of averaging analysis results for several consecutive spectra measurements. The body of device is also made by 3D-printer. The cell may be of arbitrary shape, its role in the synthesis reactor stream may carry functional nozzles allowing fast move without any difficulties spectrofluorimeter at different reactor units. Thus, the developed device can measure the optical characteristics of the nanoparticles at a nucleation step and during the analysis of growth of quantum dots.

3. Results

The reactor uses the principle of separation of nucleation and growth of nanoparticles. The offered technique increases the production rate, makes synthesis more reproducible and reduces the reagents consumption. Such system allows to obtain nanoparticles of metal chalcogenides A^2B^6, A^4B^6.

Acknowledgements

The authors acknowledge financial support from the Russian Research Foundation (project 14-15-00324).

References

[1] Mikhailov I I et al 2014 Functional Materials 21, №2, pp 186-189
[2] Bogachev Yu V et al 2014 Applied Magnetic Resonance, 45, № 4, pp. 329-337
[3] Viazmitinov D V, Matyushkin L B and Maximov A I 2014 J. Phys.: Conf. Ser 541 p. 012015
[4] Taton T A, Lu G, Mirkin C A 2001 Journal of the American Chemical Society, № 21, pp. 5164-5165.
[5] Panyam J, Labhasetwar V 2003 Ad-vanced Drug Delivery Reviews №55, pp. 329-347.
[6] El-Sayed l H, Huang X H, El-Sayed M A 2005 Nano Letters №5, pp. 829-834.
[7] Veiseh, O et al 2005 Nano Letters №5, pp. 1003-1008.

Modelling polytypism in III-V nanowires: role of group V and nucleation patterns during the growth

J. Grecenkov[1], **V.G. Dubrovskii**[1,2,3],

[1] St.Petersburg Academic University, Khlopina 8/3, 194021 St. Petersbur, Russia
[2] Ioffe Physical Technical Institute RAS, Politekhnicheskaya 26, 194021, St. Petersburg, Russia
[3] ITMO University, Kronverkskiy pr. 49, 197101 St. Petersburg, Russia

Abstract. This theoretical work deals with polytypism in nanowires by utilizing nucleation theory and by modifying standard expression for nucleation rate. Polycentric and monocentric nucleation cases are also considered. Results show good agreement with experimental data.

1. Introduction

This is a theoretical work on describing wurtzie/zincblend polytyism that is present in III-V nanowires. Nanowires are usually grown by VLS (vapour-liquid-solid method), which means that after material is introduced into a MBE chamber it is absorbed by liquid catalyst droplet on top of the nanowire and only then incorporated into the solid body of the nanowire [1].

Nanowires are promising objects that may help to integrate silicon electronics with III-V optolectronic devices in future.

2. The model

Material balance equations and nucleation theory are used to calculate concentrations inside the droplet. Comparing nanowire growth rate measured in experiment with the growth rate calculated with the help of Zeldovich formula nucleation rate yields the group III concentration inside the droplet and group V concentration is obtained through solution of material balance equations and careful calculation of chemical potential. Knowing the concentrations one can calculate wurtzite content in nanowire ensemble by comparing nucleation probabilities for islands of different crystal phases. [2]

It is speculated that the classical formula for nucleation rate must be modified due to the fact the calculation of nucleation barrier uses surface energies of nucleation islands, which is essentially macroscopic variable. Hence, for small nucleation islands consisting of just a couple of adatoms one has to make sure that the nucleation barrier is introduced correctly – nucleation probability should be zero for a nucleation island with the size of just one adatom. [3]

3. Results

Calculation results show good agreement with experimental data.[4] Fig. 1 depicts how wurtzite content is calculated. Scales in graph are set in such a way that the intercrossing of doted and solid lines corresponds to group III concentration in catalyst droplet. Knowing this concentration one can refer to the solid line to see the wurtzite content.

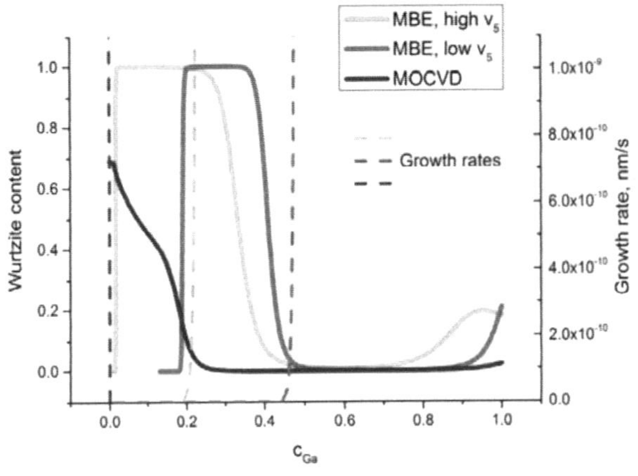

Fig.1 Growth rates and wurtzite percentage in GaAs nanowires as function of Gallium concentration in catalyst droplet. Intersections of same color corresponds to model prediction on Ga concentration inside the droplet and wurtzite percentage

The source of inconsistency of results of the model can be attributed to the fact that for higher group V flows polycentric nucleation may occur. Hence, for higher group V flows the model should be remastered to account also for polycentric nucleation case.

4. Conclusion

To conclude: the presented model uses modification of nucleation rate to describe zincblend/wurtzite polytypism in III-V nanowires. The correctness of this approach is backed by experimental results.

References

[1] V. G. Dubrovskii, *Nucleation Theory and Growth of Nanostructures*, Springer (2014)
[2] V. G. Dubrovskii, *Appl. Phys. Lett.* **104**, 053110 (2014)
[3] V. G. Dubrovskii and J. Grecenkov, *Cryst. Growth Des.* 2015, *15* (1)
[4] E. Gil et al, *Nanoletters* (2014)

Combination of laser correlation and dielectric spectroscopy in albumin investigations

E. Nepomnyashchaya[1], A. Cheremiskina[1], E. Velichko[1,2], E. Aksenov[1], T. Bogomaz[1]

[1]Saint Petersburg State Polytechnic University, Saint Petersburg, Russia
[2]Saint Petersburg National Research University of Information Technologies, Mechanics and Optics, St. Petersburg, Russia

E-mail: elina.nep@gmail.com

Abstract. Joint use of laser correlation and dielectric spectroscopies for studies of biomolecular properties of albumin in water solution is considered. The conditions and parameters of the experiments are discussed. Similar behaviours of albumin molecular sizes and maximum frequency of peak of dielectric dissipation factor with increasing acidity were revealed. Using the suggested approach, biomolecular aggregation dynamics and changes in electrophysical properties on transition from one molecular structure to another may be investigated.

1. Introduction

Biomolecular objects attract an ever increasing attention of researches in different areas [1,2]. The interest is caused by a wide range of possible application of biomolecular technologies — from diagnostics and drug manufacturing in medicine to development of new elements and devices in molecular electronics [3–6].

Biomolecular technologies are invaluable in modern chemical biology and medicine, helping to advance fundamental studies of biophysical interactions and recognition, drug discovery, and the translation of new insights into clinical application [7]. For goals of molecular electronics biomolecules have the advantages of functionality, high-density, low-consumption and specificity [3]. Biomolecular electronics is considered to be an alternative or supplement to microelectronics in continuing miniaturization. A number of researches believe that biomolecular materials have great promise for potential applications in electronics, optoelectronics, nonlinear optics; and conductive materials for electronics [3, 8]. Amino acids and proteins are considered to be the most promising materials and investigation of their electrophysical properties is the actual task.

One of the objects of research in our study is human serum albumin — the globular protein Albumin is the most abundant protein in human blood plasma which transports hormones, fatty acids, and other compounds, buffers pH, and maintains osmotic pressure, among other functions.

It is known that shift in pH cause conformational changes in protein molecules and their dysfunction. So, investigation of changes in molecular structure and dielectric features under the influence of different factors including pH can give valuable information. There are a lot of different methods of investigation of molecular structure and properties but electrophysical properties of proteins are understudied.

The goal of our study was to estimate molecular aggregation dynamics and to analyze the changes of electrophysical properties of albumin depending on solution acidity by joint use of laser correlation and dielectric spectroscopies.

2. Methods and objects

We used combination of two methods — laser correlation spectroscopy (LCS) and dielectric spectroscopy in frequency range 20 Hz – 1MHz.

2.1. Samples

In our experiments we used the solutions of human serum albumin in distilled water. The concentration of albumin was 0.01. The acidity of the solution was changed by addition of acid and alkaline buffers in the range pH = 1.4 – 8.2.

2.2. Laser correlation spectroscopy

The measure procedure of laser correlation spectroscopy was described in [9]. This method was used for evaluation of molecular size and dynamics of molecular aggregation in solution.

The experimental setup is presented in figure 1.

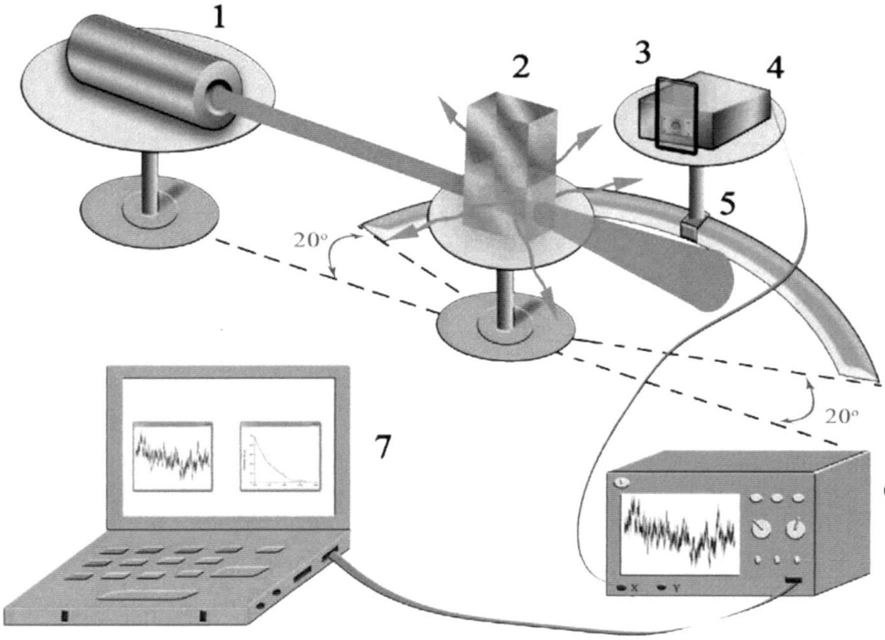

Figure 1. LCS setup. 1 — junction laser, 2 — optical cell with water suspension of biomolecules, 3 — filter, 4 — photomultiplier, 5 — rotating table, 6 — oscilloscope, 7 — computer with software for correlation and statistical analysis.

In the setup, a coherent light beam from a junction laser (1) (KLM-G650-13-5) was passed through the optical cell with a water suspension of particles to be studied (2). The scattered light at a scattering angle from 20 to 160 degrees (in figure 1, the angle is 90 degrees) was detected by a photomultiplier (4) (H11706-01). The signal from the photomultiplier was transmitted to the oscilloscope (6) and computer (7) for further processing.

2.3. Dielectric spectroscopy method

The dielectric spectroscopy (DS) method occupies a special place among the numerous modern methods used for physical and chemical analysis of material, because it enables investigation of dielectric relaxation processes. Intermolecular interactions and cooperative processes may be monitored by help of dielectric spectroscopy.

The experimental setup consisted of a special liquid cell with two pole plates and LCR meter "MNIPI E7-20" (figure 2).

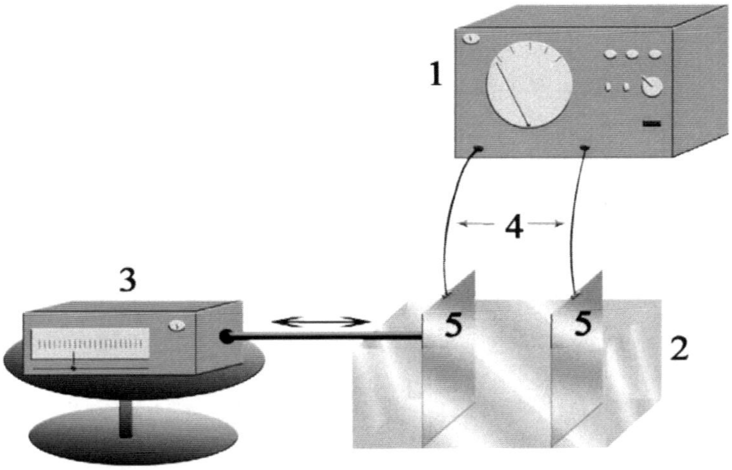

Figure 2. Experimental setup for DS measurements.
1 — LCR-meter, 2 — optical cell with water suspension of biomolecules, 3 — micrometer, 4 — coaxial wires, 5 — moving and motionless electrodes.

The albumin solution was placed in the optical cell (2). The distance between two flat electrodes was fixed and controlled by micrometer (3). The impedance Z and phase angle φ were detected by LCR-meter. For further analysis the dielectric dissipation factor was calculated:

$$tg\delta = \frac{1}{\omega \cdot R \cdot C},$$

where

$$R = \frac{Z}{\cos(\varphi)}, \quad C = \frac{\sin(\varphi)}{\omega \cdot Z}.$$

The dielectric dissipation factor dependences on frequency in different conditions were studied. It was revealed that albumin solution has peaks of dielectric dissipation factor, and the peak is corresponds to different frequencies depending on pH of solution.

3. Results and discussion

The results of our study revealed similar behaviors of two characteristics: the size of the molecules in the solution (measured by LCS) and frequencies of maximum dielectric dissipation factor peaks of albumin (measured by DS) depending on acidity of solution. The obtained results are presented in figures 3 and 4.

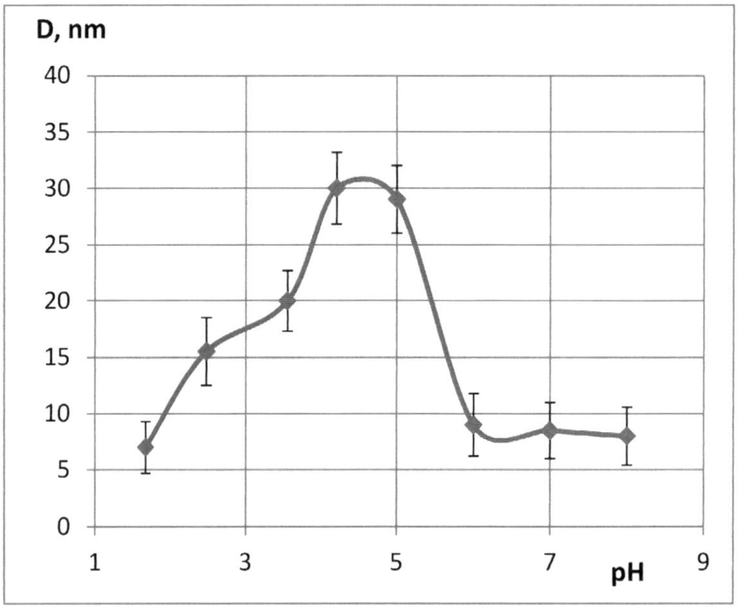

Figure 3. Dependence of molecular sizes of albumin on solution acidity

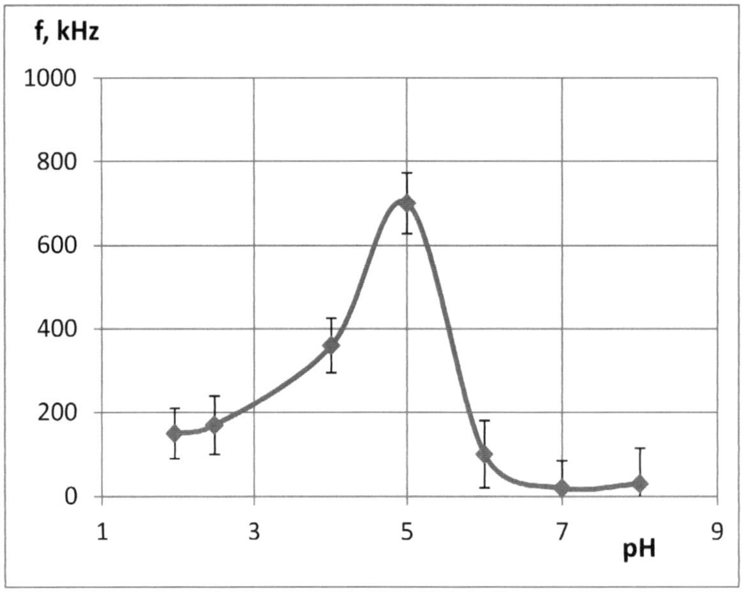

Figure 4. Dependence of frequencies of maximal dielectric loss angle tangent on acidity of albumin solution.

The maximal sizes of albumin molecules and maximal frequencies of dielectric loss angle tangent peaks were observed near the isoelectric point of albumin (pH = 4.8). This may be concerned with forming of special molecular structures in isoelectric solution.

4. Conclusion

Joint use of laser correlation and dielectric spectroscopies for studies of biomolecular properties is presented. Using the suggested approach, biomolecular aggregation dynamics and changes in electrophysical properties on transition from one molecular structure to another may be investigated.

References

[1] www.scopus.com/results/results.url?sort=plf-f&src=s&st1=biomolecular+electronics 2015
[2] Liao C.-D., Tsai T.C. et al. 2015 *Device Architecture and Biosensing Applications for Attractive One- and Two-Dimensional Nanostructures*, ed. M.C. Vestergaard *Nanobiosensors and nanobioanalyses* (Japan: Springer) p.41
[3] Phadke R.S. 2001 *Applied Biochemistry and Biotechnology* **96** 279
[4] Cingolani R, Rinaldi R., Maruccio G., Biasco A. 2002 Phisica E **13** 1229
[5] Maruccio G. et al 2004 Electroanalysis **16** (22)
[6] Jiang W. et al 2011 J. Am. Chem. Soc. **133** 1
[7] Kindt J.T, Bailey R.C 2013 *Current Opinion in Chemical Biology* **17** (5) 818
[8] Velichko E., Tsybin O. 2011 *Biomolecular electronics* (SPb, SPbSPU) 256 p.
[9] Nepomnyashchaya E, Velichko E, Aksenov E and Bogomaz T 2014 *IOP Journal of Physics: Conference Series* **541** 012039

Theoretical study of the amphoteric oxide nanoparticle surface charge during multi-particle interactions in aqueous solutions

A V Alfimov, E M Aryslanova, S A Chivilikhin

ITMO University, Kronverkskiy, 49, St. Petersburg, 197101, Russian Federation

E-mail: alfimov.anton@gmail.com

Abstract. Nanoparticle surface charge plays an important role in many biological applications. In this study, an analytical surface charging model for the amphoteric oxide nanoparticles has been presented. The model accounts for the particle's electric double layer self-action on the charging process and the charge regulation during multi-particle interactions in aqueous solutions. The employment of the model allows to explicitly describe the nanoparticle agglomeration process and the accompanying agglomerate surface charge variation.

1. Introduction

Recent studies in the fields of nanotoxicology, nanobiotechnology and drug delivery suggest that the surface charge of a nanoparticle can significantly influence it's interactions with biological matter. It has been shown that the cellular uptake, cytotoxicity and overall reactivity of amphoteric oxide nanoparticles are highly dependent on their charge [1, 2]. In the presence of agglomeration, the interparticle interaction and morphology of the agglomerates are also governed by the nanoparticle surface charge [3].

The present study proposes an analytical model for the amphoteric oxide nanoparticle surface charging mechanism which accounts for the charge regulation during the multi-particle interaction in aqueous solutions. This allows for the explicit description of the nanoparticle agglomeration process and the accompanying agglomerate surface charge variation.

2. Methods

The suggested colloid nanoparticle interaction and agglomeration model has been developed in the framework of the DLVO (Derjaguin, Landau, Verwey, Overbeak) theory. This theory assumes that the interaction between colloid particles can be represented using the attractive van der Waals term and the repulsive electroosmotic term. The latter arises due to the overlap of particles' electric double layers (EDLs).

We use a system of Debye-Huckel equations for the electrostatic potential φ to describe the ion atmosphere of a single spherical nanoparticle suspended in electrolyte solution:

$$\begin{gathered} \Delta\varphi_{\mathrm{I}} = 0, \quad a < r < b \\ \Delta\varphi_{\mathrm{II}} = \kappa^2\varphi_{\mathrm{II}}, \ r > b \end{gathered}, \tag{1}$$

with boundary conditions:

$$\begin{array}{ll} \left.\dfrac{d}{dr}\varphi_{\mathrm{I}}\right|_{r=a} = -\dfrac{\sigma}{\varepsilon_h\varepsilon_0} & \left.\varepsilon_h\dfrac{d}{dr}\varphi_{\mathrm{I}}\right|_{r=b-} = \left.\varepsilon\dfrac{d}{dr}\varphi_{\mathrm{II}}\right|_{r=b+} \\ \varphi_{\mathrm{I}}(b) = \varphi_{\mathrm{II}}(b) = \varphi_0 & \left.\varphi_{\mathrm{II}}\right|_{r\to\infty} = 0 \end{array}. \tag{2}$$

where r - radial coordinate of the spherical coordinate system; a - particle radius; b - effective radius of the hydrated particle; φ_{I}, φ_{II} - potential in the hydration layer and the diffuse ion layer correspondingly; κ - reciprocal of the Debye screening length; σ - nanoparticle surface charge density ε_h, ε - effective dielectric permittivity in the hydration layer and the diffuse ion layer correspondingly; ε_0 - dielectric constant; φ_0 - zeta potential.

Solving (1) with (2) we can show that the zeta-potential expression in given by:

$$\varphi_0 = \frac{1}{4\pi\varepsilon\varepsilon_0} \cdot \frac{4\pi a^2 \cdot \sigma}{b(1+\kappa b)} \tag{3}$$

We also introduce the original surface charge model for amphoteric oxide nanoparticles that accounts for a number of charging mechanisms including competitive protonation and deprotonation of the [–OH] surface groups, chemical adsorption of H^+ and OH^- ions and van der Waals adsorption onto the particle surface and operates in the limit of non-saturated adsorption and desorption. The model also accounts for the external electric field effect and the nanoparticle's electric double layer self-action on the surface charging process.

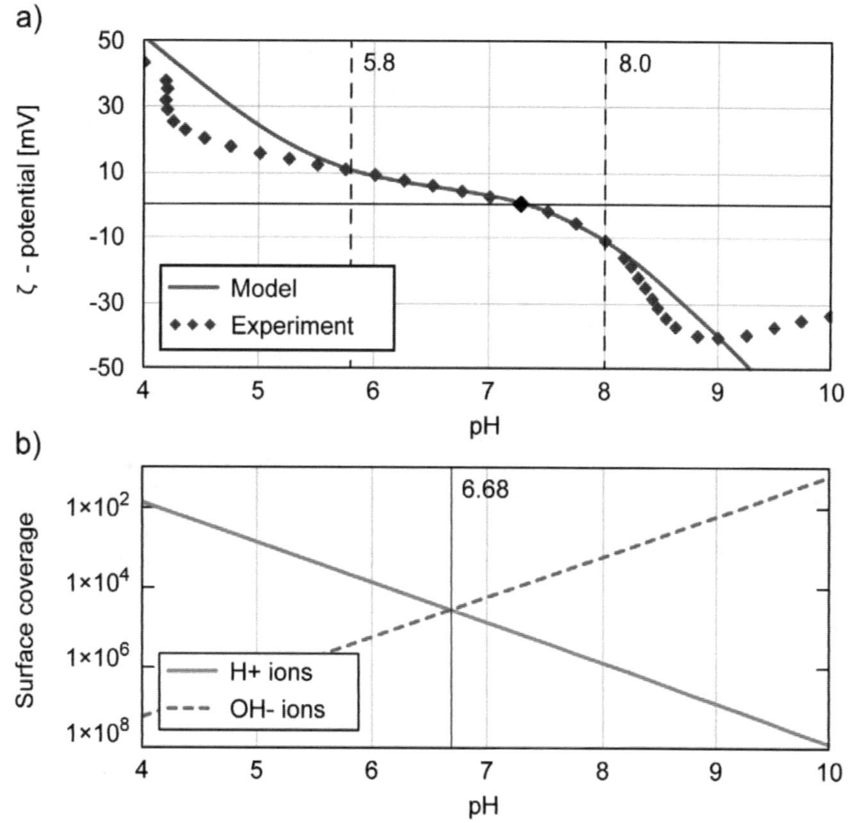

Figure 1. Nanoparticle surface charge model. a) – fitting of the experimental zeta potential dependency on pH [2] with the suggested model; dashed vertical lines indicate the limits of applicability of the model. b) – fraction of the nanoparticle surface covered with H^+ and OH^- ions as a function of pH for the suggested model.

Linearization of the potential-dependent H^+ and OH^- concentrations at the particle surface yields a simple expression for the nanoparticle surface charge density σ:

$$\sigma=\left\{\sigma_0+e\left[\beta_+ n_{\mathrm{H}}\left(1-\frac{e\varphi_\Sigma}{kT}\right)-\beta_- n_{\mathrm{OH}}\left(1+\frac{e\varphi_\Sigma}{kT}\right)\right]\right\}\cdot\left\{1+\frac{e^2\left[\beta_+ n_{\mathrm{H}}+\beta_- n_{\mathrm{OH}}\right]}{4\pi\varepsilon\varepsilon_0\cdot kT}\left[\frac{1}{b(1+\kappa b)}+\frac{b-a}{ab}\right]\right\}^{-1}, \quad (4)$$

where σ_0 - constant portion of charge; $n_{\mathrm{H}}, n_{\mathrm{OH}}$ - bulk H^+ and OH^- concentrations; a, b - physical and hydrodynamic particle radii; β_+, β_- - coefficients proportional to the sums of the rate constants for the adsorption processes involved; φ_Σ - potential of the external field; e - elementary charge.

By substituting (4) into (3) we obtain the model expression for the nanoparticle zeta-potential as a function of pH (H^+ and OH^- concentrations). By fitting the experimental zeta potential versus pH dependency for ZnO 30 nm colloid particles taken from [2] with the obtained theoretical expression we can find the model parameters σ_0 and β_+, β_-. The results of such data fitting are represented in figure 1. It can be seen that the model provides good agreement with the experiment except for the adsorption saturation effects at low and high pH-values.

The surface charge model can be implemented into the many-body nanoparticle interaction model, first introduced by the authors in [4]. Using the monopole in the framework of the classic the monopole approximation, we substitute each particle with its equivalent screened point-charge [5]:

$$Q=4\pi\varepsilon\varepsilon_0\cdot\varphi_0\cdot b\mathrm{e}^{\kappa b}. \quad (5)$$

As it has been shown in [4], the equivalent screened point-charges of multiple interacting particles can be than found by solving the following linear system of algebraic equations:

$$Q_\alpha+\sum_{\beta\neq\alpha}\theta_{\alpha,\beta}\cdot Q_\beta=Q_0 \quad (6)$$

with the coefficient given by:

$$\theta_{\alpha,\beta}=\frac{1}{4\pi\varepsilon\varepsilon_0}\cdot\frac{e^2}{kT}\cdot\frac{1}{R_{\alpha,\beta}(1+\kappa b)}\cdot\frac{\left(\beta_+ n_{\mathrm{H}}+\beta_- n_{\mathrm{OH}}\right)\cdot\exp\left[-\kappa\left(R_{\alpha,\beta}-b\right)\right]}{1+\frac{e^2\left[\beta_+ n_{\mathrm{H}}+\beta_- n_{\mathrm{OH}}\right]}{4\pi\varepsilon\varepsilon_0\cdot kT}\left[\frac{1}{b(1+\kappa b)}+\frac{b-a}{ab}\right]} \quad (7)$$

Q_0 - the equivalent charge of an isolated particle; α, β - particle numbers; $R_{\alpha,\beta}$ - centre-to-centre separation between particles α and β.

3. Results and discussion

The presented colloid nanoparticle interaction and charge regulation model allows us to calculate the equivalent effective charges of multiple interacting particles using (6) and (7). The model also allows to study the multi-particle interaction potential energy, which was previously discussed in [4]. It has been shown that the electroosmotic repulsion between nanoparticles can give rise to orientation effects during their agglomeration. Namely, depending on the primary particle concentration and pH the most energetically favourable agglomerate configuration might be either compact, or dendritic (fractal-like).

In the present work the proposed interaction and charge regulation model was used to study the effective charge and surface charge density transformations during nanoparticle agglomeration process. To account for the agglomerate morphology differences mentioned above we have considered both compact and linear (as a limiting case of dendritic) agglomerates of varied sizes.

The limits of applicability for the presented model are illustrated on figure 2 and figure 1a. Figure 1a illustrates the limits of applicability of the non-saturated adsorption and desorption assumption for pH. Figure 2a demonstrates the limitations on the primary particle size and pH arising from the monopole approximation. Finally, figure 2b shows the maximum electrostatic potential in the agglomerate's ion atmosphere as a function of the number of it's constituent particles and pH. The Debye-Huckel equations (1) are only applicable when the potential in the system less or equal 10 mV at 283 K. Therefore, figure 2a provides another set of limitations for the possible agglomerate sizes and pH values.

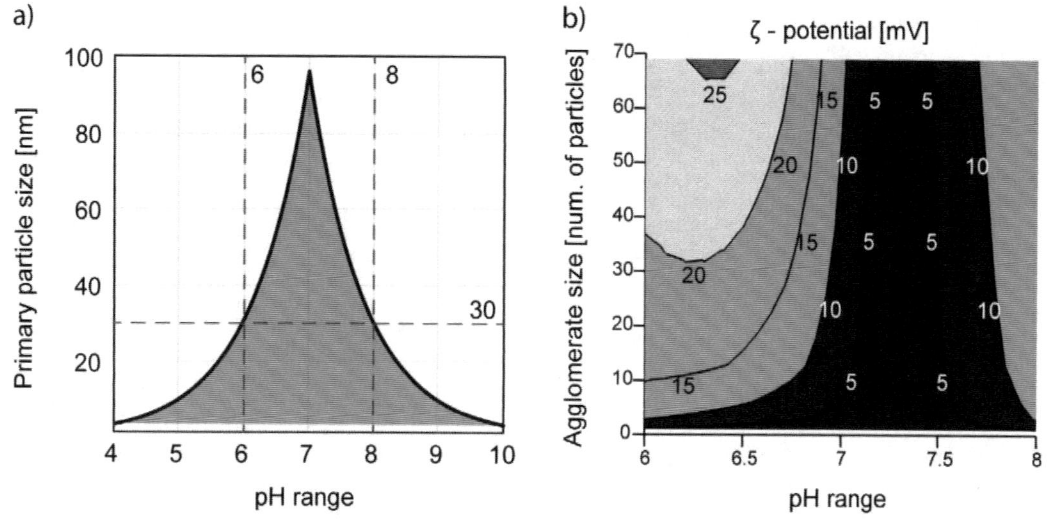

Figure 2. Limits of applicability for the presented model. a) – primary particle diameter and pH limitations; filled area represents the limits of applicability of the model based on the diffuse ion layer thickness. b) – contour plot of the agglomerate's zeta potential as a function pH and number of constituent primary particles; the suggested model is applicable when the agglomerate zeta potential is less than *10 mV* at room temperature.

Figure 3 illustrates the effective nano-agglomerate charge as a function of the number of it's constituent primary particles for the case of compact (a) and linear (b) agglomerates. The figure shows that the increase of the agglomerate size results in a much faster effective charge growth for the linear agglomerates rather than for the compact ones. This is due to the fact that particles in compact

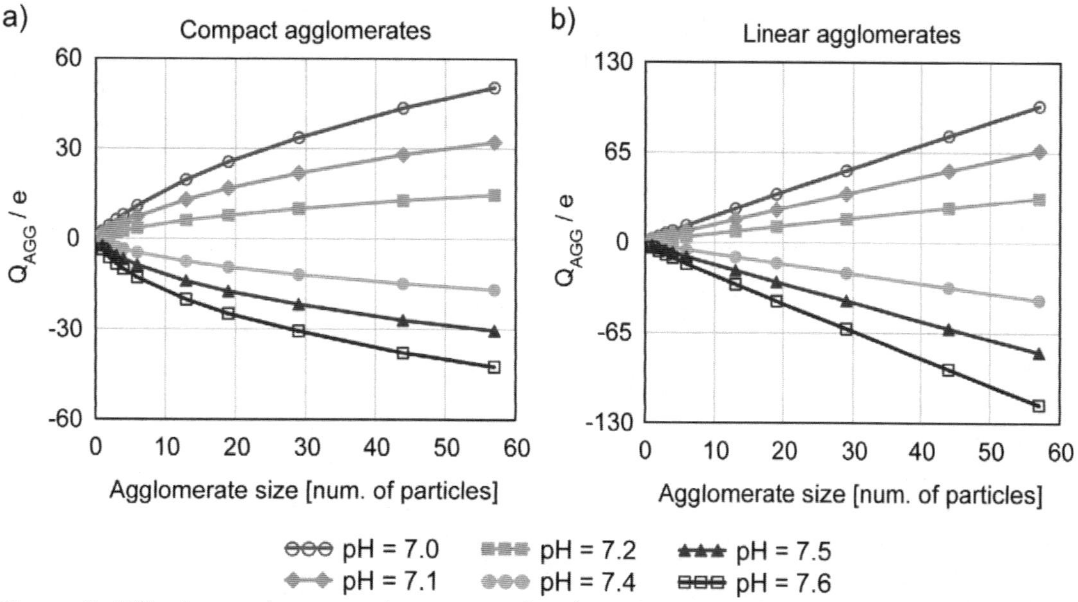

Figure 3. Effective agglomerate charge as a function of the number of its constituent primary particles for a) compact agglomerates and b) linear agglomerates at different pH values. The effective agglomerate charges are normalized to the elementary charge.

agglomerates have more neighbours than in linear agglomerates. As all particles have a charge of the same sign, particles in compact agglomerates tend to decrease their charges though the charge regulation process more, than in linear agglomerates.

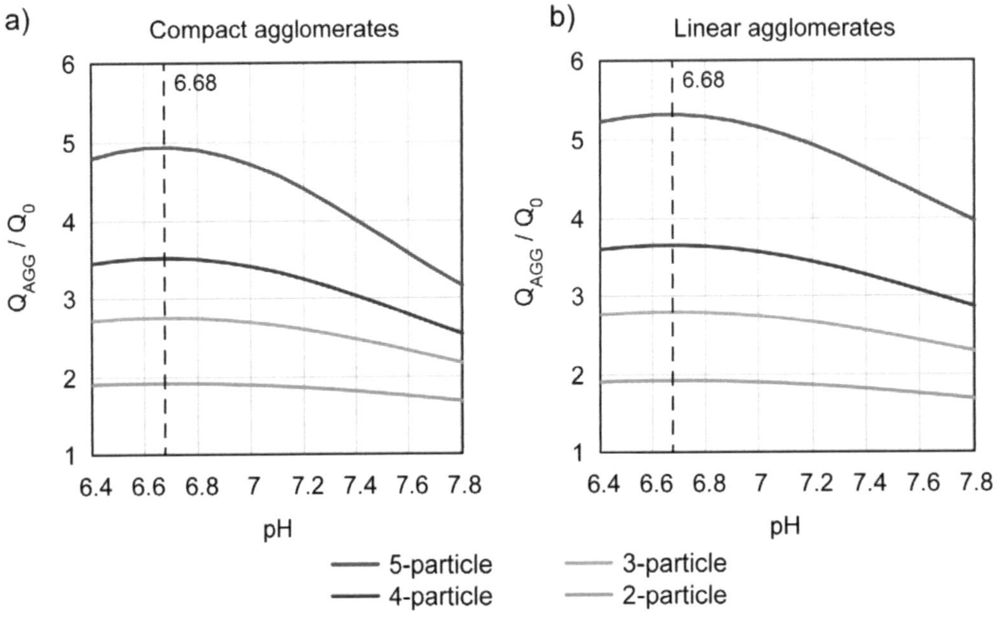

Figure 4. Relative effective agglomerate charge as a function of pH for a) compact agglomerates and b) linear agglomerates with different number of constituent particles. The effective agglomerate charges are normalized to the effective charge of the isolated primary particle at given pH.

Figure 4 depicts the effective agglomerate charge normalized the charge of an isolated primary particle as a function of pH. This dependency demonstrates a maximum at pH = 6.68. According to figure 1, said value corresponds to equal amounts of H^+ and OH^- ions adsorbed on the surface. But, more importantly, at this pH value the nanoparticle's surface has the least amount of adsorbed ions that are actively involved in chemical reactions. This means that the charge regulation process is least active at pH = 6.68, hence the maximum in relative charge. It should also be noted that the point of zero charge pH = 7.3 (see figure 1a) does not significantly impact the relative curves shown in figure 4, while causing the minimum for the pH dependency of the absolute charge values (see figure 3).

Finally, figure 5 shows the dependency of nano-agglomerate surface charge density on the number of it's constituent primary particles. The agglomerate surface charge densities are normalized to those of the individual isolated primary particles. Presented are the charge densities for compact and linear agglomerates at 4 different pH values. The size-dependent surface charge density of a compact agglomerate is generally demonstrates a maximum. The location of this maximum depends on pH and shifts into bigger agglomerate sizes when moving further away from the pH = 6.68 value. Similarly to the curves on figure 4, the point of zero charge pH = 7.3 does not impact the pH-dependent surface charge maximum location for compact agglomerates.

Figure 5 also shoes that the mean surface charge density is generally higher for compact agglomerates than for linear ones. However, the surface charge of linear agglomerates varies significantly based on the location in the agglomerate and reaches its maximum value on the tip of the particle chain. This is explained by the least amount of neighbours for the particles sitting on the tips of the linear or dendritic agglomerates. As shown in figure 5d, the maximum surface charge density of a linear agglomerate can overtake the mean surface charge of a compact agglomerate given the significantly high pH or constituent particle count.

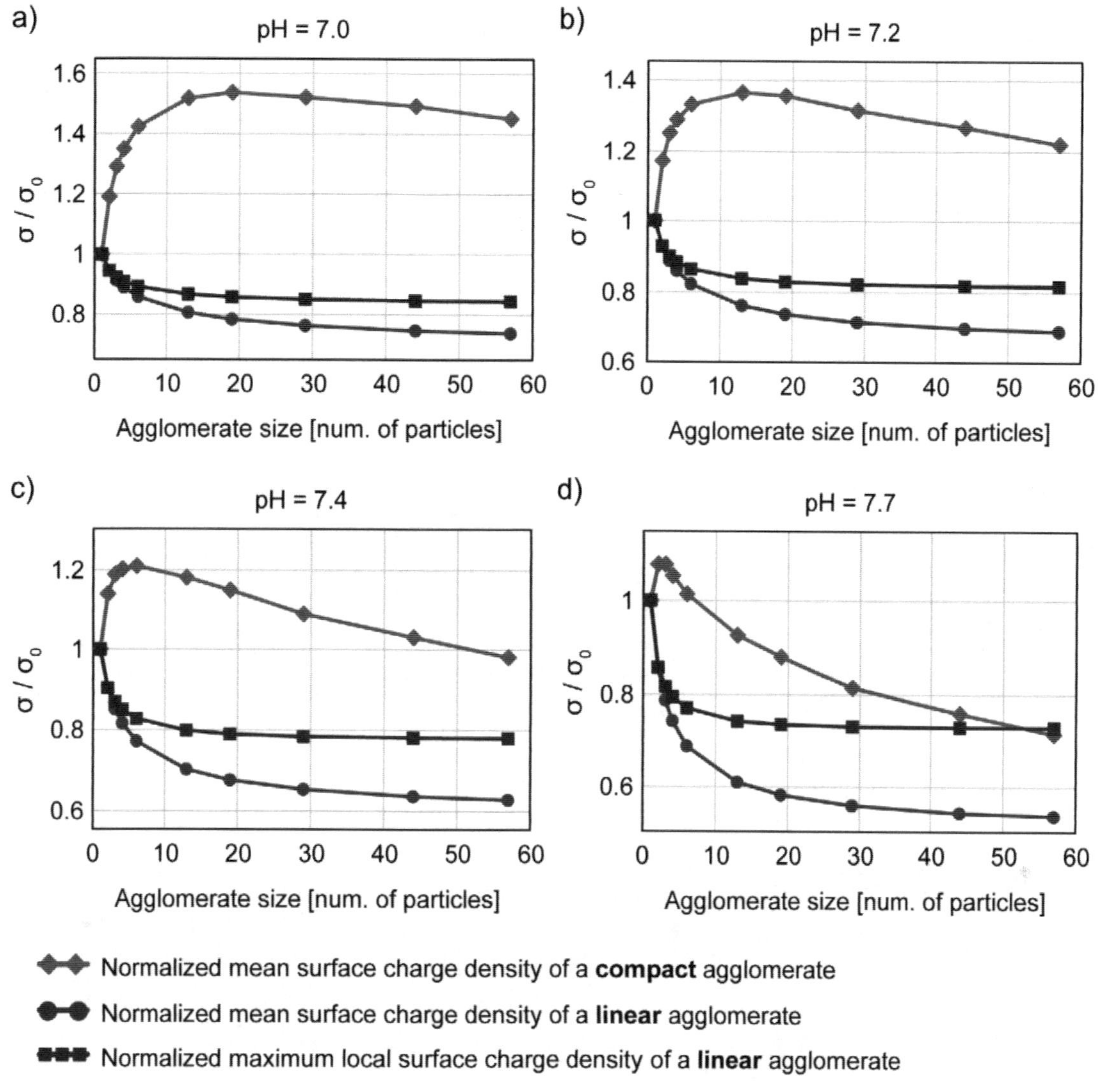

Figure 5. Normalized effective agglomerate surface charge density as a function of the number of its constituent primary particles for a) pH = 7.0; b) pH = 7.2; c) pH = 7.4; d pH = 7.7. The effective agglomerate surface charge density is normalized to the effective surface charge density of the isolated primary particle at given pH.

Acknowledgments
This work was financially supported by the Government of Russian Federation, Grant 074-U01.

References

[1] Barisik M *et al* 2014 *J. Phys. Chem. C* **118** (4) 1836
[2] Berg J M *et al* 2009 *Nanotoxicology* **3** (4) 276
[3] Chowdhury I *et al* 2013 *Environ. Sci.: Processes Impacts* **15** 275
[4] Alfimov A V *et al* 2014 *J. Phys.: Conf. Ser.* **541** 012063
[5] Ohshima H 1995 *Journal of Colloid and Interface Science* **174** (1) 45

Journal of Physics: Conference Series **643** (2015) 012020 doi:10.1088/1742-6596/643/1/012020

Multi-color localization microscopy of fixed cells as a promising tool to study organization of bacterial cytoskeleton

A D Vedyaykin[1], V V Gorbunov[1], A V Sabantsev[1, a], V S Polinovskaya[1], I E Vishnyakov[2], A S Melnikov[1], P Yu Serdobintsev[1] and M A Khodorkovskii[1]

[1] Research Institute of Nanobiotechnologies, Peter the Great St.Petersburg Polytechnic University, Polytechnicheskaya 29, Saint-Petersburg, Russia

[2] Institute of Cytology, Russian Academy of Sciences, Tikhoretsky av. 4, Saint Petersburg 194064, Russia

E-mail: sabantsev.a.v@gmail.com

Abstract. Localization microscopy allows visualization of biological structures with resolution well below the diffraction limit. Localization microscopy was used to study FtsZ organization in *Escherichia coli* previously in combination with fluorescent protein labeling, but the fact that fluorescent chimeric protein was unable to rescue temperature-sensitive *ftsZ* mutants suggests that obtained images may not represent native FtsZ structures faithfully. Indirect immunolabeling of FtsZ not only overcomes this problem, but also allows the use of the powerful visualization methods arsenal available for different structures in fixed cells. In this work we simultaneously obtained super-resolution images of FtsZ structures and diffraction-limited or super-resolution images of DNA and cell surface in *E. coli*, which allows for the study of the spatial arrangement of FtsZ structures with respect to the nucleoid positions and septum formation.

1. Introduction.

Localization microscopy is a new method of fluorescence microscopy that increases its resolution about ten-fold and well below the diffraction limit. The main idea behind this method is the temporal separation of individual molecules fluorescence, their separate detection and subsequent localization [1, 2]. Increased resolution is especially valuable for the study of prokaryotic organisms as the sizes of most of them are comparable to the diffraction limit of optical microscopy which makes localization microscopy a method of choice for the investigation of bacteria and archaea internal organization [3].

Bacterial division is a robust and complex process with a number of cytoskeleton-like proteins involved in it. They play crucial roles in determination of division site and management of the following division stages. FtsZ, one of prokaryotic homologs of eukaryotic tubulin, plays central role in the division of most bacteria, being the first actor in the formation of the Z-ring at the middle of dividing cells (Figure 1,A) [4, 5]. Previously structures formed by FtsZ in *Escherichia coli* were visualized with the use of FtsZ fluorescent chimeras [6, 7], but chimeric proteins were unable to substitute wild-type FtsZ in these cells. It suggests that fluorescent labelling may perturb native FtsZ structures. It is therefore important to compare these results with the ones obtained using an alternative immunolabelling method, which lacks some of the drawbacks of fluorescent protein labelling.

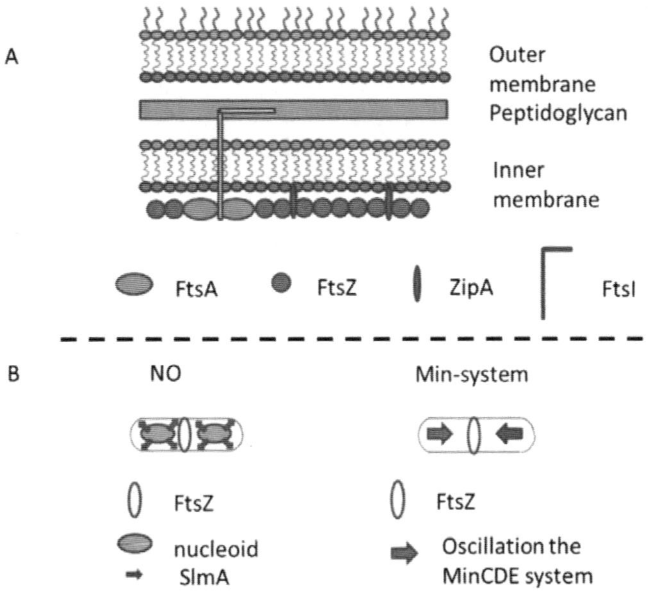

Figure 1. *E. coli* division machinery. A) Cross-sectional scheme of the division machinery, depicting major proteins that form the Z-ring. FtsZ is a central division protein that polymerizes to form filaments that further assemble to form the Z-ring. ZipA anchors the Z-ring to the inner membrane. FtsA is an ATP-binding protein that stabilizes the Z-ring and links it to the transmembrane protein FtsI. **B)** Nucleoid occlusion (NO) and Min-system are the two inhibitory systems that assure proper Z-ring positioning by preventing the Z-ring assembly at random positions along the cell. NO prevents Z-ring formation over nucleoid by the means of DNA-binding protein SlmA. Min-system, comprised of MinC, MinD and MinE proteins, prevents Z-ring formation at cell poles by oscillating along the cell axis.

Proper FtsZ positioning in the cell is controlled by two major systems: Min-system and nucleoid occlusion (Figure 1,B). Oscillating MinCDE system inhibits FtsZ polymerization to prevent septum formation at the cell poles [8]. Nucleoid occlusion protein SlmA binds to DNA and inhibits FtsZ polymerization over nucleoid therefore preventing improper septum positioning that may lead to DNA breakdown [9]. It is therefore of great interest to investigate spatial relations between structures formed by FtsZ, nucleoid and cell surface. Immunofluorescent labeling of fixed cells provides numerous possibilities of DNA, cell surface and protein visualization using wide selection of dyes, allowing for dual-color and possibly three-color super-resolution microscopy of FtsZ, DNA and cell surface, which shall provide deeper understanding of FtsZ functioning and regulation in *E.coli*.

2. Materials and methods.

In all experiments Top10 *E. coli* strain transformed with pGEX-4T-2 plasmid was used. The use of a plasmid encoding an ampicillin resistance protein was aimed to simplify bacteria handling and prevent possible contamination by the means of ampicillin resistance selection. Cells were inoculated from the frozen stock into fresh LB medium with added ampicillin (100 µg/ml), grown for 5 hours at 37°C to OD≈0.3 and fixed directly in culture medium by the addition of sodium-phosphate buffer (pH 7.4, final concentration 30 mM), formaldehyde (final concentration 2.6%) and glutaraldehyde (final concentration 0.04%) for 10 minutes at room temperature followed by 50 minutes on ice. To quench autofluorescence cells were incubated for 10 minutes in 30 mM sodium borohydride in PBS. After that cells were washed 3 times with PBS and immobilized on poly-L-lysyne coated coverslips with nonspecifically adsorbed polystyrene microspheres (2.35 µm, Bangs Laboratories). Then cells were permeabilized with 0.1% Triton X-100 in PBS and 10 μg/ml lysozyme in GTE buffer (glucose 50 mM, Tris 32.5 mM pH=7.5, EDTA 10 mM), each for 5 minutes. After blocking with 2% w/v BSA in PBS (PBS-BSA) for 20 minutes, cells were incubated with anti-FtsZ polyclonal rabbit antibodies (Agrisera) diluted 1 to 200 in PBS-BSA overnight at +4°C. Then cells were washed 5 times with 0.01% Tween-20 in PBS, incubated with secondary goat anti-rabbit antibodies diluted in PBS-BSA for 1 hour at room temperature and washed another 5 times with 0.01% Tween-20 in PBS. In the case of DNA and FtsZ localization microscopy Alexa 555 conjugated secondary antibodies from Bioss were used in 1 to 100 dilution and cells were then stained with DNA dye TOTO-3 (250 nM in PBS) for 10 minutes. In case of FtsZ and cell surface (cell wall and cytoplasmic membrane stained by wheat germ agglutinin(WGA) [10]) localization microscopy Alexa 647 conjugated $F(ab')_2$ fragments of goat anti-rabbit antibodies (LifeTechnologies) were used in 1 to 1000 dilution to visualize FtsZ and cells were additionally incubated with 5 μg/ml WGA conjugated with Alexa 488 dye for 10 minutes and 100 nM propidium iodide (LifeTechnologies) for another 10 minutes to visualize DNA. The dye combinations were chosen according to the compatibility of the dye spectra and conditions required to perform localization microscopy using these dyes. After that cells were post-fixed with 2.6% formaldehyde for 10 minutes.

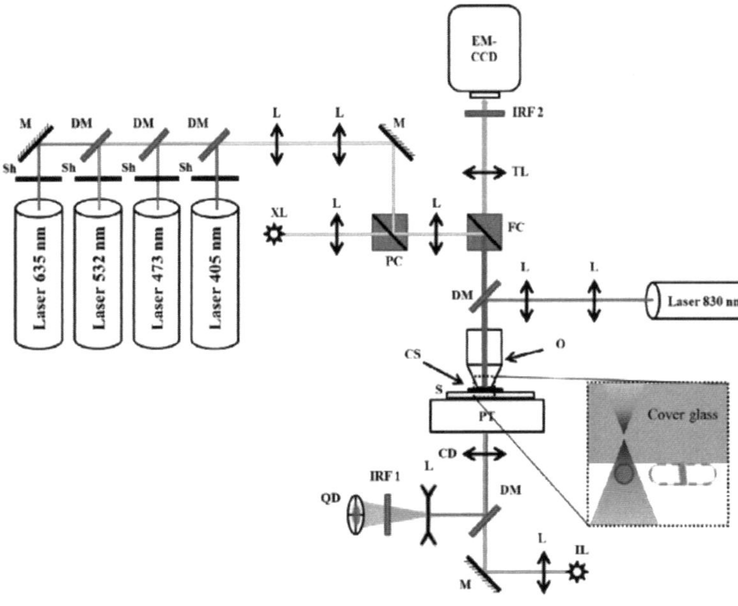

Figure 2. Experimental setup. CD – microscope condenser, CL – cylindrical lens, CS – coverslip, DM – dichroic mirror, EM-CCD – microscope camera, FC – fluorescence filter cube, IL – transmitted light lamp, IRF 1 – infrared filter, passing 830 nm, IRF 2 – infrared filter, blocking 830 nm, L- lens, M – mirror, ML – mercury lamp for fluorescence excitation, O – microscope objective, PC - polarizing cube, PS – piezo stage, QD – quadrant detector, S – sample, Sh – shutter, TL – microscope tube lens.

Super-resolution images were obtained using custom experimental set-up (Figure 2) based on Axioimager.Z1 (Carl Zeiss) upright microscope, equipped with custom laser excitation block and feedback sample position stabilization system based on microsphere position detection in 3D using a quadrant detector on which the image of the bead is created by a near-infrared diode laser [11, 12]. DNA-FtsZ super-resolution images were obtained in the buffer consisting of 10% glucose, 0.5 mg/ml glucose oxidase (Sigma), 40 μg/ml catalase (Sigma), 100 mM 2-mercaptoethylamine, 2 mM cyclooctatetraen in PBS. In the case of FtsZ, DNA and cell surface imaging GOC oxygen scavenging system was replaced with PCA-PCD oxygen scavenging system (25% glycerin, 100mM TRIS-HCl pH 7.4, 100 mM 2-mercaptoethulamine, 5 mM protocatechuic acid, 60 nM protocatechuate 3,4-

dioxygenase and 2 mM cyclooctatetraene) [13]. Super-resolution images were reconstructed using ImageJ [14, 15] plug-in QuickPALM [16].

3. Results and discussion.

Dual-color localization microscopy of FtsZ and DNA in *E.coli* (Figure 3) allowed relative distribution of FtsZ and DNA to be analyzed in fine details. These images showed that FtsZ distribution in dividing cells mirrors that of DNA throughout the cell (Figure 4), with DNA localizing primarily in two bands halfway from the cell center to the poles, while FtsZ forms single- and multi-band structures at the cell center and also concentrates at DNA-free regions near the cell poles.

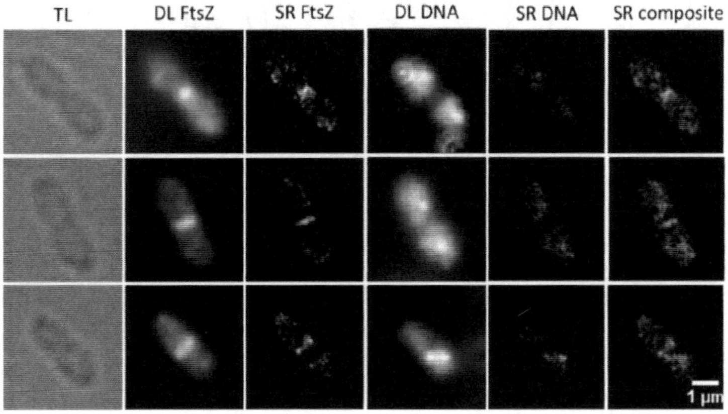

Figure 3. DNA and FtsZ super-resolution images. Columns: 1 - transmitted light, 2 – diffraction-limited (DL) images of the FtsZ structures, 3 - super-resolution (SR) images of the FtsZ structures, 4 – diffraction-limited images of the DNA (nucleoid), 5 – super-resolution images of the DNA (nucleoid), 6 – composite super-resolution image, DNA pseudo colored red and FtsZ pseudo colored green. Scale bar corresponds to 1 μm.

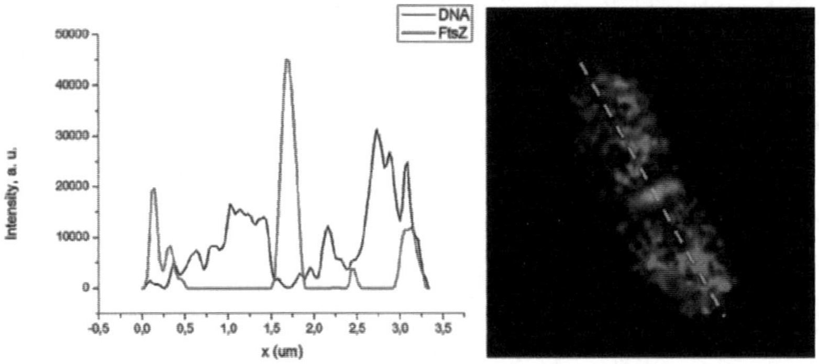

Figure 4. FtsZ and DNA are distributed inversely in the *E. coli* cells. Characteristic distributions of DNA (pseudo colored red) and FtsZ (pseudo colored green) along the cell.

Simultaneous cell surface and FtsZ localization microscopy combined with diffraction-limited DNA visualization (Figure 5) allows us to reveal the spatial arrangement of FtsZ structures relative to the nucleoid and the place of the septum formation. These images also clearly show that FtsZ forms not only the Z-ring, but also a variety of spiral-like and other off-center structures running along the cell axis (Figure 3, 5), which was not observed with the use of fluorescent protein fusions. This emphasizes the need to compare data obtained by different staining methods to create an accurate model of FtsZ functioning.

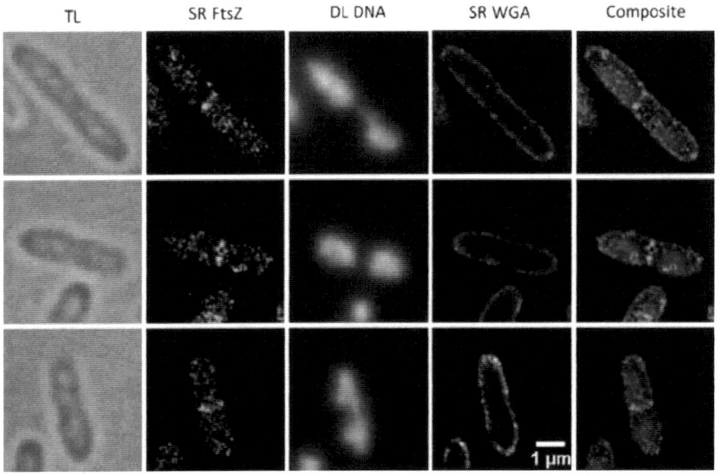

Figure 5. Cell surface and FtsZ super-resolution and diffraction-limited DNA images. Columns: 1 –transmitted light, 2 – super-resolution (SR) images of the FtsZ structures, 3 – diffraction-limited (DL) images of the DNA (nucleoid), 4 – super-resolution images of the cell surface stained with WGA, 5 – composite images; DNA pseudo colored blue, FtsZ - green and cell surface - red. Scale bar corresponds to 1 μm.

4. Conclusions.

Multicolor super-resolution localization microscopy combined with immunofluorescent labeling proves to be a powerful method for the study of bacteria division machinery. Immunolabeling allows the visualization of FtsZ structures outside the Z-ring, which was not achieved with fluorescent protein fusion super-resolution microscopy. Simultaneous DNA and FtsZ super-resolution images enable deeper study of nucleoid occlusion system, whereas cell surface super-resolution imaging allows to visualize sub-membrane localization of the FtsZ structures in the cells and the place of the septum. Developed methods can be applied to the in-depth study of the FtsZ assembly regulation and its rearrangements during stress response. Three-dimensional localization microscopy should allow whole-cell visualization of FtsZ structures.

The reported study was supported by Russian Science Foundation, research project No. 14-34-00023. The work was carried out using scientific equipment of the Center of Shared Usage «The analytical center of nano- and biotechnologies of SPbPU.

5. References.

1. Hess, S.T., T.P. Girirajan, and M.D. Mason, *Ultra-high resolution imaging by fluorescence photoactivation localization microscopy.* Biophys J, 2006. **91**(11): p. 4258-72.
2. Rust, M.J., M. Bates, and X. Zhuang, *Sub-diffraction-limit imaging by stochastic optical reconstruction microscopy (STORM).* Nat Methods, 2006. **3**(10): p. 793-5.
3. Coltharp, C. and J. Xiao, *Superresolution microscopy for microbiology.* Cell Microbiol, 2012. **14**(12): p. 1808-18.
4. Addinall, S.G., E. Bi, and J. Lutkenhaus, *FtsZ ring formation in fts mutants.* J Bacteriol, 1996. **178**(13): p. 3877-84.
5. de Boer, P., R. Crossley, and L. Rothfield, *The essential bacterial cell-division protein FtsZ is a GTPase.* Nature, 1992. **359**(6392): p. 254-6.

6. Fu, G., et al., *In vivo structure of the E. coli FtsZ-ring revealed by photoactivated localization microscopy (PALM)*. PLoS One, 2010. **5**(9): p. e12682.
7. Buss, J., et al., *In vivo organization of the FtsZ-ring by ZapA and ZapB revealed by quantitative super-resolution microscopy*. Mol Microbiol, 2013. **89**(6): p. 1099-120.
8. Kretschmer, S. and P. Schwille, *Toward Spatially Regulated Division of Protocells: Insights into the E. coli Min System from in Vitro Studies*. Life (Basel), 2014. **4**(4): p. 915-28.
9. Monahan, L.G., et al., *Division site positioning in bacteria: one size does not fit all*. Front Microbiol, 2014. **5**: p. 19.
10. Morioka, H., M. Tachibana, and A. Suganuma, *Ultrastructural localization of wheat germ agglutinin-binding sites in Escherichia coli with wheat germ agglutinin-gold complexes*. J Electron Microsc (Tokyo), 1987. **36**(1-2): p. 36-9.
11. Sabantsev, A., et al., *«Study of structures formed by FtsZ in Escherichia coli and Mycoplasma hominis cells»*. FEBS Journal, Special Issue: 38th FEBS Congress, Saint Petersburg, Russia, 2013. **280**(Sup. s1): p. 469.
12. Vedyaykin, A.D., et al., *Localization microscopy study of FtsZ structures in E. coli cells during SOS-response*. Journal of Physics: Conference Series, 2014. **541**(1).
13. Olivier, N., et al., *Resolution Doubling in 3D-STORM Imaging through Improved Buffers*. PLoS ONE, 2013. **8**(7): p. e69004.
14. Schindelin, J., et al., *Fiji: an open-source platform for biological-image analysis*. Nature methods, 2012. **9**(7): p. 676-682.
15. Schneider, C.A., W.S. Rasband, and K.W. Eliceiri, *NIH Image to ImageJ: 25 years of image analysis*. Nature methods, 2012. **9**(7): p. 671-675.
16. Henriques, R., et al., *QuickPALM: 3D real-time photoactivation nanoscopy image processing in ImageJ*. Nat Methods. 2010 May;7(5):339-40. doi: 10.1038/nmeth0510-339.

Numerical modeling of the liquid front propagation in inhomogeneous nanoporous media

D S Golovina and S A Chivilikhin

ITMO University, Kronverkskiy, 49, St. Petersburg, 197101, Russia

E-mail: dheaddd@gmail.com

Abstract. In the last decade nanoporous materials applications in biology and medicine have been widely researched. One of the problems in this field are the transport processes in nanoporous media. The paper presents the numerical model, which simulates the motion of front of the liquid, two-dimensionally permeating inhomogeneous nanoporous medium with altering porosity and pore size. This model allows us to analyze the correlation between the inhomogeneities of the media and the shape and transition of the front. The evolutions of the liquid front profile for nanoporous media with various parameters have been obtained using the program written in C++.

1. Introduction

In recent years nanoporous materials have become a subject of various scientific researches. Considerable attention is given to their application in biology and medicine. For example, nanoporous materials can be used for drug delivery [1, 2], immunoisolation [3], tissue engineering [4], as biological sensors [5], etc.

One of the topical problems in the theoretical studying of the nanoporous media are the transport processes [6]. It should be noted that transport processes in nanoporous systems are considerably more complex, than in homogeneous phases or even in macroporous systems [7]. The purpose of our work is to numerically model the motion of the front of the liquid, two-dimensionally permeating inhomogeneous nanoporous medium with altering porosity and pore size. Such model allows us to analyze how the inhomogeneity of a medium influences the shape and the transition of the liquid front.

2. Description of the numerical model

The capillary forces in the liquid front create the pressure $P_H = -\frac{2\sigma}{R}$, where σ is the surface tension coefficient of the liquid, R is the pore radius. The flowing through the medium creates the pressure in the liquid.

The presented model is based on the continuity equation for the incompressible flow $\nabla\vec{v} = 0$ and the approximation of Darcy's law $\vec{v} = -\gamma\nabla P$, where $\vec{v}$ is the velocity of the flow and P is the pressure in the fluid. Their combination results in the linear differential equation (LDE):

$$\nabla(\gamma\nabla P) = 0 \tag{1}$$

with boundary conditions:

$$P_0\big|_{y=0} = 0$$
$$P_H\big|_{y=H} = -\frac{2\sigma}{R}$$
$$\left.\frac{\partial p}{\partial n}\right|_{x=0} = \left.\frac{\partial p}{\partial n}\right|_{x=L} = 0 \qquad (2)$$

Here the coefficient γ describes the permeability of the porous medium [8]:

$$\gamma = \frac{\Omega R^2}{8\mu} \qquad (3)$$

where Ω is the porosity of medium, R is the average pore radius and μ is the dynamic viscosity coefficient of the fluid.

The γ coefficient describes the structural properties of the nanoporous medium. By introducing the set of γ values as a matrix of the same size as the numerical scheme grid we are able to specify the γ value in every grid node. This allows us to take into consideration the inhomogeneity of the medium.

The numerical calculation procedure includes two steps:

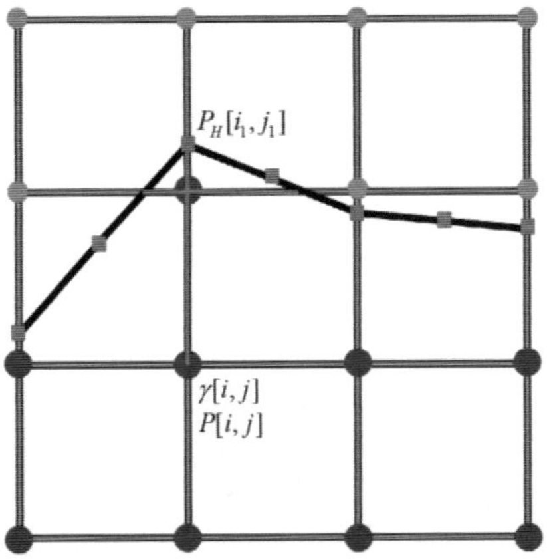

Figure 1. The numerical calculation procedure, step 1: solving LDE (1) by finite-difference method.

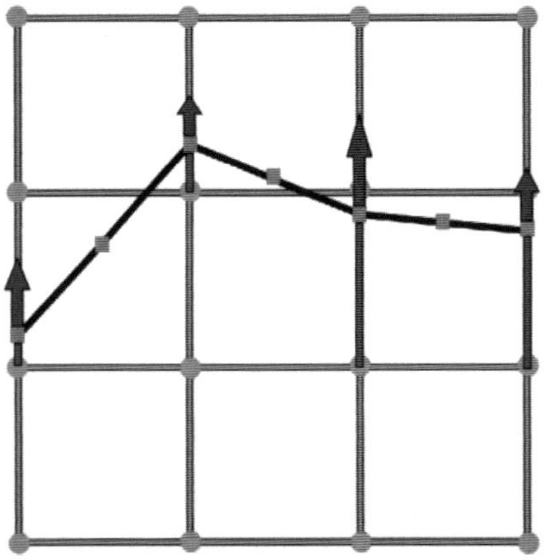

Figure 2. The numerical calculation procedure, step 2: translocation of the front and interpolation.

Firstly, by solving LDE (1) we obtain the pressure distribution in the medium's part which has been permeated by liquid. Secondly, on the basis of the derivative of the pressure we translocate the front points, which lie on the grid, and interpolate the results. Since a medium is inhomogeneous, the capillary pressure P_H changes with fronts motion and must be recalculated after each iteration using a matrix, which comprises the values of possible capillary pressures and also is of the same size as the grid.

3. Results

The computer simulation program, which numerically models two-dimensional motion of the front of the liquid, permeating inhomogeneous nanoporous medium, has been written in C++. The evolutions of the liquid front profiles for different inhomogeneous nanoporous media are demonstrated in the figures 3, 4, 5 and 6.

The following parameters were used for time evaluation: average pore size $\bar{R} = 10$ nm and parameters of the water $\sigma = 0.07$ N/m, $\mu = 0.8$ mPa s. In most cases porosity has been taken as constant $\Omega = 0.8$.

Darker areas in the figures' 3, 4, 5 and 6 backgrounds indicate smaller pore size and lighter, accordingly, larger.

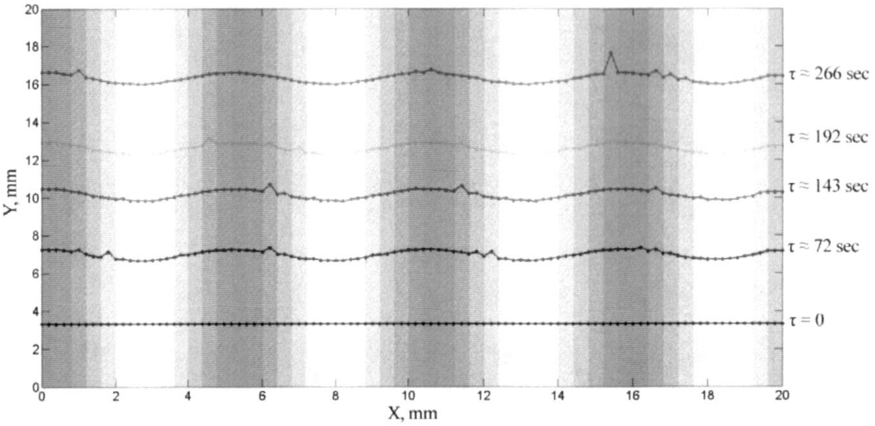

Figure 3. The evolution of the water front profile in case of pore size changing periodically in one direction: $R = \bar{R} + 3.5\sin^2(kx)$ nm. The porosity of the medium is constant: $\Omega = 0.8$. Darker areas in the background correspond to smaller pore size and lighter ones, respectively, to larger.

Figure 3 shows the modification of the fronts shape in accordance with the alteration of the pore size in the medium, assuming that the porosity is constant. The evolution of the fronts shape in the medium in which both pore size and porosity vary can be seen at the figure 4:

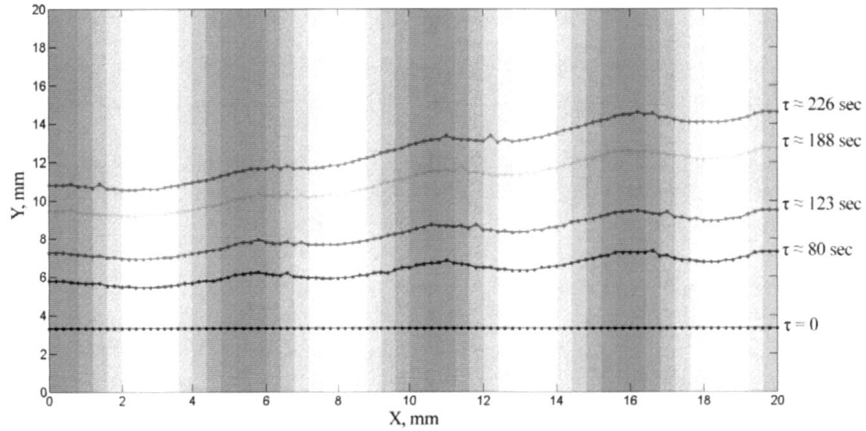

Figure 4. The evolution of the water front profile in case of pore size changing periodically in one direction: $R = \bar{R} + 3.5\sin^2(kx)$ nm. The porosity of the medium Ω changes stepwise from 0.4 to 0.8 from left to right. Darker areas on the background correspond to smaller pore size and lighter ones, respectively, to larger.

If medium's porosity changes stepwise, the main characteristic of the modification of the fronts shape remains the same. However, the front moves faster in the parts of the medium with larger porosity, which leads to additional linear deviation of the front from its original shape. Also, in the areas with larger porosity the impact of the pore size variation is more considerable, which results in more notable bending of the front in the mediums parts with larger porosity, if the pore size alters in the same way.

Figure 5 shows the evolution of the liquid front profile in the medium in which pore size vary in both directions:

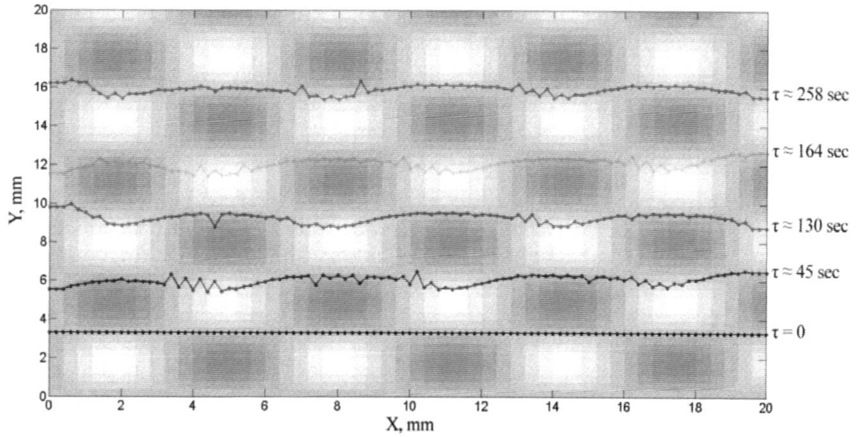

Figure 5. The evolution of the water front profile in case of pore size changing periodically in two directions: $R = \bar{R} + 3.5\sin(kx)\sin(ky)$ nm. The porosity of the medium is constant: $\Omega = 0.8$. Darker areas in the background correspond to smaller pore size and lighter ones, respectively, to larger.

Both figure 3 and figure 5 show that if mediums porosity is constant, liquid flows faster through the areas with smaller pores, which results in bending of the front in accordance with the difference between the sizes of the nearby pores.

Figure 6 shows the evolution of the water front profile in the medium with random distribution of the pore size. As expected, fronts shape remains generally linear, with comparatively small deviations:

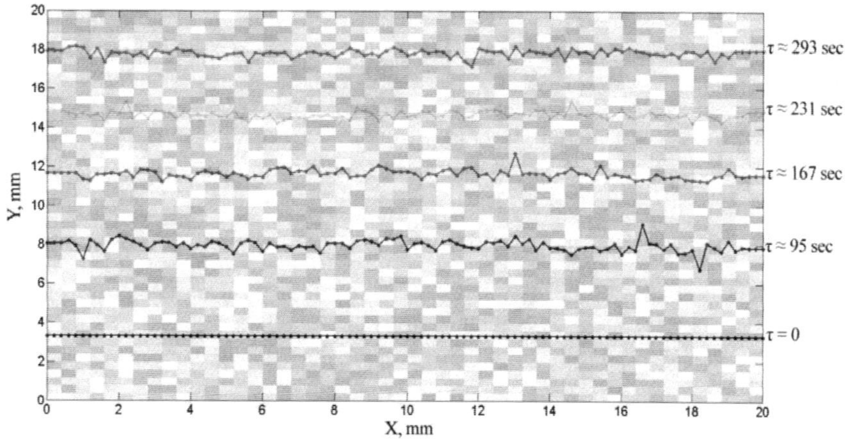

Figure 6. The evolution of the water front profile in case of pore size changing randomly from 10 to 13.5 nm. The porosity of the medium is constant: $\Omega = 0.8$. Darker areas in the background correspond to smaller pore size and lighter ones, respectively, to larger.

4. Conclusion.

The numerical model, which simulates the motion of front of the liquid, two-dimensionally permeating an inhomogeneous nanoporous medium, has been proposed. The evolutions of the water front profiles for different inhomogeneities of the media (such as pore size changing periodically in one or two directions and randomly) have been obtained and the time of water propagation has been evaluated.

In case of equally altering pore size the bending of the front is more notable in the parts of the medium with larger porosity. The stepwise changing of porosity also adds the additional linear deviation of the front from its original shape. If the porosity is constant, the front bends in accordance with the relation between pore sizes.

Acknowledgments
This work was financially supported by the Government of Russian Federation, Grant 074-U01.

References

[1] Sun CY, Qin C, Wang CG, Su ZM, Wang S, Wang XL, Yang GS, Shao KZ, Lan YQ and Wang EB 2011 *Adv. Mat* **23** 5629–32
[2] Gultepe E, Nagesha D, Sridhar S and Amiji M 2010 *Adv. Drug Deliv. Rev.* **62** 305–15
[3] La Flamme K E, Popat K C, Leoni L, Markiewicz E, La Tempa T J, Roman B B, Grimes C A and Desai T A 2007 *Biomaterials* **28** 2638–45
[4] Cardea S, Pisanti P and Reverchon E 2010 *J. of Supercritical Fluids* **54** 290–95
[5] Hu K, Lan D, Li X and Zhang S 2008 *Anal. Chem.* **80** 9124–30
[6] Beckstein O and Sansom M S P 2004 *Phys. Biol.* **1** 42–52
[7] Conner WC and Fraissard JP 2006 *Fluid Transport in Nanoporous Materials* (Dordrecht: Springer)
[8] Scheidegger A E 1958 *The Physics of Flow Through Porous Media* (Toronto: University of Toronto Press)

Porous silicon nanoparticles for target drag delivery: structure and morphology

Yu M Spivak[1], A O Belorus[1], P A Somov[1], S S Tulenin[2], K A Bespalova[1], V A Moshnikov[1,3]

[1]Saint Petersburg Electrotechnical University "LETI",
5[th]Prof.Popova Str., Saint-Petersburg, 197376, Russia
[2]Ural Federal University named after the first President of Russia B.N. Yeltsin, 19th Mira street, Ekaterinburg, 620002, Russia
[3]Peter the Great St. Petersburg Polytechnic University, 29, Polytechnicheskaya st., 195251, St. Petersburg, Russia

mop_92@mail.ru; ymkanageeva@yandex.ru

Abstract. Nanoparticles of porous silicon were obtained by electrochemical anodic etching. Morphology and structure of the particles was investigated by means dynamic light scattering and scanning electron microscopy. The influence of technological conditions of preparation on geometrical parameters of the porous silicon particles (particle size distribution, pore shape and size, the specific surface area of the porous silicon) is discussed.

1. Introduction

Porous silicon (por-Si) and nanocomposites based on it offers many properties useful for various fields as micro- and nanoelectronics, solar and hydrogen energies as well as in biomedical technology [1-8]. For example, ultra-thin layers consisting of monodisperse por-Si nanoparticles due to photoluminescence properties of por-Si could dramatically improve the efficiency of photovoltaic solar cells, including cells obtained by hydro-chemical deposition [9-11]. At the moment application of por-Si in biomedicine is the most actively developing area [12-22], namely:

- support material for targeted drug delivery (both single and multiple drugs simultaneously);
- photosensitizers in photodynamic therapy;
- tissue engineering;
- biosensors of various types;
- biomedical imaging, including tumor imaging;
- treatment of eye diseases and etc.;

The great attention to the por-Si for applications in biomedicine caused by following important properties:

- biocompatibility i.e. ability to be incorporated into the body, without causing side effects, and the ability to induce a cellular or tissue response necessary to achieve optimum therapeutic effect;
- biodegradability (i.e., the ability to dissolve and be absorbed);
- processability;

- possibility to control the parameters of the porous structure (porosity, pore diameter, surface development, hydrophobicity / hydrophilicity) and properties (surface charge state, the local distribution of charges and m. p.) in a rather wide range;
- multifunctional (not only the carrier matrix, and simultaneously the sensor, marker and so on).

Advantages of por-Si as carriers of drugs due are also the high surface area and large pore volume. Nanosystems based on por-Si exhibit improved parameters needed for drug delivery: increase the solubility of the test drug, high stability in vitro, low cytotoxicity, and others. So, nanoparticles of porous silicon can effectively deliver antiviral drugs to infected cells. In [23] investigated the possibility of increasing the solubility of saliphenylhalamide (SaliPhe) by loading the compound into nanoparticles thermohydrocarbonized porous silicon (THCPSi). Nanoparticles loaded SaliPhe, were further tested for ability to inhibit influenza infection. It was shown that after release from THCPSiSaliPhe inhibit influenza infection in vitro and reduces the number of progeny in influenza virus-infected cells. In [24] described the preparation of biodegradable porous silicon nanoparticles functionalized with antibodies that target cancer cells and the loaded hydrophobic anticancer drug camptothecin. Successful targeting demonstrated by flow cytometry and immunocytochemistry, both cell lines and primary cells. The analysis of cells after incubation for viability confirmed the selective destruction of cancer cells.

An important task is the creation of drug delivery systems capable of carrying multiple loads simultaneously, for example, the simultaneous delivery of small molecule drugs and relatively large peptides [18]. In [18] investigated the possibility of simultaneous delivery of multiple drugs with the joint loading of hydrophobic and hydrophilic peptide indomethacin human YY3-36 (PYY3 -36). The authors argue that the startup sequence of these two drugs in porous silicon nanoparticles increases the rate of release of each drug. Loading PYY3-36 also significantly improved compatibility cytological nanoparticles spanking. Conformational analysis showed that after the release of por-Si nanoparticles, PYY3-36 drug exhibits bioactivity and the ability to penetrate the intestinal cells.

Currently, porous silicon and porous SiO_2were used to demonstrate the release of the steroid dexamethasone, ibuprofen, Peyronie chloride, doxorubicin and many other drugs. Note that the nanoparticles themselves spanking can have bioactivity [25].

In the preparation of nanoscale powders of porous silicon on the one hand, there is a need for the development of processing methods for the preparation of porous silicon nanopowder, separating them by size and choice of storage conditions. On the other hand, need information about the properties, composition and the evolution of the time properties of individual particles of porous nanosized powders, as well as the impact on their properties or environmental conditions (temperature, light, composition, medium). It is also important to investigate the effect of nanoparticles Spanking (especially modified other substances) from the point of view of safety of living organisms.

This work is devoted to characterization of the structure and morphology of por-Si nanoparticles as well as the study of the influence of various dispersion media.

2. Experiment

Preparation of porous silicon nanoparticles was performed in several stages [8, 12, 26]. First, porous silicon layers very obtained by electrochemical etching using the single electrochemical cell (Figure 1). Such configuration of the electrochemical cell allows the formation of por-Si on both sides of the original silicon wafer (rather than only in one side as in the conventional cells). Thus, it leads to increasing productivity of the method. The electrolyte is an aqueous solution of hydrofluoric acid with the addition of isopropanol and hydrogen peroxide. There were studied three series of por-Si powders which were differ anodizing current density, resistivity silicon substrate. The anodization conditions are shown in Table 1.In a second step the samples were placed in a vessel containing isopropanol and sonicated for 30 min. Thus it was obtained the porous silicon powder dispersion in isopropanol media.

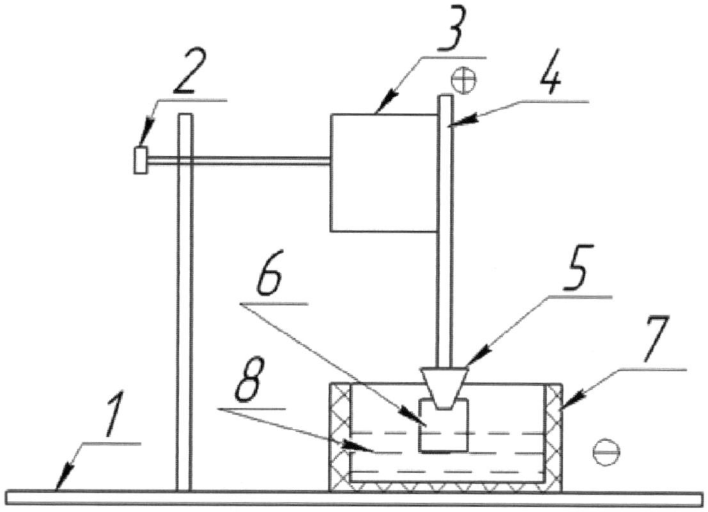

Figure 1. Schematic representation of laboratory stand for porous silicon powder preparation: 1-tripod with stand; 2- knob position of the sample; 3-holder; 4- metal rod (anode); 5 - copper holder; 6 - silicon sample; 7 - glassy carbon crucible (cathode); 8- electrolyte

Table 1. The anodizing process parameters for the obtained por-Si nanoparticles

Series	Si parameters	Si resistivity, Ω·cm	Anodizing time, min.	Current density J_A, mA/cm^2
I	n-Si (111)	1,0	20	30
II		1,0		70
III		0,3		70

Subsequently, the por-Si dispersions were used to study the distribution of particle size by dynamic light scattering using Photocor mini (Scientific Laboratory UrFU named after the first President of Russia Boris Yeltsin). The device is designed according to the traditional scheme of dynamic light scattering spectrometer. It allows multi-angle measurements both for dynamic and static light scattering. The dynamic light scattering allows determining the diffusion coefficient of the particle in the fluid by analysis fluctuations of the correlation function scattered light intensity. Further, from the diffusion coefficient is calculated the radius of the nanoparticles. Directly before measuring the dispersion of porous silicon powders in isopropyl alcohol were additionally sonicated.
The structure and morphology of porous silicon particles were studied using Tescan MIRA II scanning electron microscope. For scanning electron microscopy the samples were prepared as follows: dispersion droplets were deposited on the prewashed monocrystalline silicon substrate by dispenser, and then were dried.

3. Results and discussion

Figure 2 shows typical SEM micrographs of porous silicon particles for series I-III. As expected, the por-Si powders are consisting of particles of different size, which is obviously due to the grinding method of por-Si layers. The lowest number of large particles observed in the sample №1, the largest number in the sample №2.Por-Si particle of series II and III are sharp, but in I series particles more smooth and round, and of smaller sizes generally. Por-Si Series I and II are of mixed type and contain meso- and micropores as was expected by previous studies [27].Thus, by the capillary condensation it has been found that por-Si particles are present in the pores with the diameter less than 5 nm (the proportion of which was 42.4% of the total pore fraction) and rather larger mesopores with sizes of 10-

20 nm and 50 nm [27]. Mesopores for Series II might be distinguished on Fig. 2, e. Observation of micropores by conventional SEM is extremely difficult. For por-Si particles of Series I, which are obtained at a lower current density, are expected even smaller pores [28].

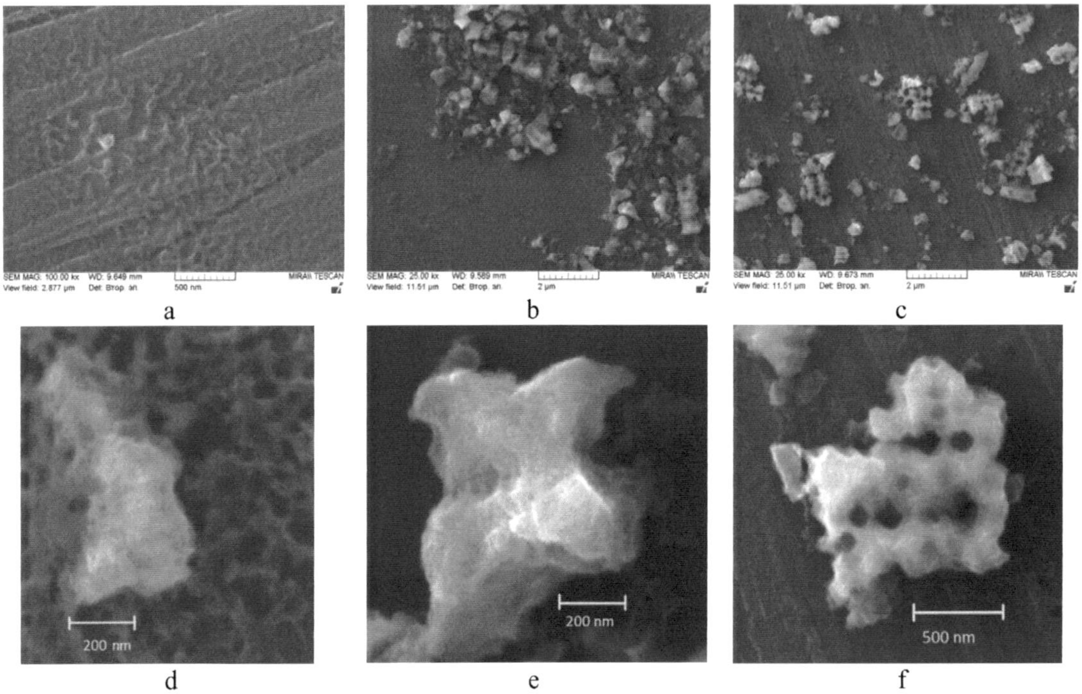

Fig. 2. Morphology of por-Si particles fixed on Si substrate for series: I – a, d; II – b, e; III – c, f

According to Fig. 2,a and 2,d, it may be assumed that in the series I contains a significant proportion of smaller particles of porous silica, which aggregate to form a porous layer consisting of porous particles on the surface of the silicon substrate. It has been found by means of SEM, so that a porous layer made of por-Si particles series I covers almost the entire silicon substrate. Thus a "secondary" porous layer is a macro- and mesoporous with typical pore diameter of about 20-150 nm. A typical thickness of the "secondary" porous skeleton is of the order 20-40 nm. It can be assumed that a vivid manifestation of aggregation effects in dispersions series I due to the fact that the particles of porous silicon in this series have a large specific surface area and tend to minimize the surface energy through the formation of aggregates. Also, there are large particle with a size of 200-400 nm.

Samples Series III contains preferably macropores of 50-100 nm and less sizes with a square shape of the pore channels; the skeleton of porous silicon perhaps retains monocrystalline (Fig. 2, c, f). This form and the geometry of the pore channels corresponds to the propagation channel along the crystallographic directions <100>, characterized maximal etch rate for single crystal silicon [29].

The estimates of distribution for por-Si dispersions obtained by the method of dynamic light scattering showed the following (Table 2). In general, the data for estimating the size of particles and its size distribution obtained according to the dynamic light scattering are consistent with SEM data for the same samples: in all studied samples present por-Si particle with size of tens and hundreds of nanometers in varying proportions. Also, common to all studied sample series is that the main contribution –more than 50%, was given by particles of 100-200 nm sizes. In contrast to the series II and III, a series I possesses quite large contribution of particles of about ≈14-20 nm sizes – about 40%. Evidently, the porous layer observed in Fig. 2, a and 2, d, is composed of such particles.

Table 2. Estimation of por-Si particle size and percentage of particles in the dispersions according to the dynamic light scattering

Series	Media	The particle size and its accuracy, nm	The percentage of particles in dispersion
I	Isopropanol	17.8±3.3	40.6
		194.1±42.4	59.4
II	Isopropanol	18.4±1.7	11.8
		197.2±29.1	73.9
		691.9±81	14.3
III	Isopropanol	9.3±0.6	7.4
		94.7±12.2	53.1
		609.4±67.6	39.5

4. Conclusions

Particles of porous silicon were obtained by electrochemical anodic etching of n-Si (111) under different technological conditions: 30 and 70 mA/cm2 of current density, 1.0 and 3.0 Ω·cm of specific resistivity of initial n-Si. Por-Si particles of series II and III are sharp, but in I series particles more smooth and round, and of smaller sizes generally. ion current density, 0.3 and 1.0 Ω·cm of specific resistivity of Si. The morphology, texture, sizes and size distribution of porous particles and pore diameter in it were studied by SEM and dynamic light scattering. SEM data revealed pores in all sample series.

It was found that por-Si nanoparticles(Series I and II) are of mixed type and contain mesopores, series III – meso- and macroporous silicon particles (≈50-100 nm). Por-Si particles of series II and III are sharp, but I series particles more smooth and round, and of smaller sizes generally. As could be noticed, the pore diameter increases with increasing doping level of the substrate and with increasing current density anodizing, which is in good agreement with the experimental data of other authors [30]. It was found that por-Si particles of series I unlike the other two series covered the surface of the monocrystalline Si substrate with porous layer in which is difficult to distinguish individual particles (with the exception of large ones order of 100-200 nm). The pore diameter in a "secondary" layer on the order of 20-150 nm, and the cross section of skeletal branches - the order of 20-40 nm. It is assumed that the formation of such layers involves particles aggregation processes in the origin dispersion to minimize surface energy.

Both methods (SEM and dynamic light scattering method) have shown that porous silicon powders are characterized by polydispersity: the main contribution – more than 50%, was given by particles of 100-200 nm sizes. In contrast to the series II and III, a series I possesses quite large contribution of particles of about ≈14-20 nm sizes – about 40%. Apparently this is due to the fact that the particles obtained in one series of lower current density and nonporous be expected with a high degree of porosity. Such layers are typically more mechanically strained, more fragile, so are ground by ultrasound into smaller parts.

From the viewpoint of practical use obtained in this paper por-Si particles are perspective as containers for targeted drug delivery. Thus, the creation of a composed container consisting of porous particles with different morphology and pore sizes are very promising for the target delivery of several types of drugs with different molecular sizes, different properties and so on. Also, porous layers composed of porous silicon particles of one or more types may be of interest for sensors, considering advantages and simplicity of the technology for producing porous silicon.

Acknowledgments
The reported study was this work was supported by project "Preparation and study of porous systems, functionalized nanomaterials, applications in photonics, sensor technology and medicine" (under the Russian Ministry of Education state task № 16.2112.2014/K.

References
[1] Punzon-Quijoma E, Sanchez-Vanquero V, Munoz-Noval A 2012 *Research Lett* **7** 396
[2] Martin-Palma R J, Manso-Silvan M, Torres-Costa V 2010 *Nanophoton* **4** 042502
[3] Dhanekar S, Jain S, Islamia J M, etc. 2013 *Biosensors and Bioelectronics* **41** 54-64
[4] Astrova E V,Li G V, Parfen'eva A V, etc. 2015 *Technical Physics* **60** 531-540
[5] Moshnikov V A, Gracheva I E, Lenshin A S, etc. 2012 *J. of Non-Cryst. Sol.* **358** 590-595
[6] Nalimova S.S., Spivak Yu. M., Moshnikov V.A., etc. 2014 *Smart Nanocompos* **5** 5
[7] Patel P N, Mishra V, Panchal A K *Journal of Optoelectronics and Biomedical Materials* **4** 19-28
[8] Lenshin A S, Kashkarov V M, Spivak Yu M, Moshnikov V A 2012 *Materials Chemistry and Physics* **135** 293-297
[9] Chufarov A Yu, Ermakov A N, Samigulina R F, Zainulin Yu G, Forostyanaya N A, Maskaeva L N and Markov V F 2013 *Russian J. of Inorganic Chem.* **58** 1362-1369
[10] Markov V F, Tulenin S S, Maskaeva L N and Kuznetsov M V 2014 *J. of Surface Investigation. X-ray, Synchrotron and Neutron Techniques.* **8** 659-665
[11] Fedorova E A, Tulenin S S, Maskaeva L N, Kuznetsov M V, Chufarov A Yu, Markov V F 2012 *Butlerov Communications* **30** 144-149
[12] Spivak Yu M, Maraeva E V, Belorus A O, Molchanova A V, Nigmadzyanova N R 2013 *Smart Nanocomsite* **4** 115-118
[13] Syshchyk O, Skryshevsky V A, Soldatkin O O, Soldatkin A P 2015 *Biosensors and Bioelectronics* **66** 89-94
[14] Belorus A O 2013 *Molodoi Uchenyi, in Russian* **8** 20720297
[15] Liu D, Mäkilä E, Zhang H, etc. 2013 *Adv. Func. Mat.* **23** 1893-1902
[16] Santos H A, Bimbo L M, Herranz B, etc. 2013 *Journal of Materials Research* **28** 152-164
[17] Barnes T J, Karyn L J, Prestidge C A 2013 *Therapeutic Delivery* **4** 811-823
[18] Liu D, Bimbo L M, Mäkilä E, etc. 2013 *J.Cont.Releas.* **170** 2268-278
[19] Xu W, Riikonen J, Lehto V P 2013 *International Journal of Pharmaceutics* **453** 181-197
[20] Anglin E L, Cheng L C, Freeman W R, etc. 2008 *Adv. Drug Deliv. Rev.* **60** 1-32.
[21] Guo D-J, Zhang H, Li J-B, etc. 2013 *J. Mater. Chem. B* **1** 379-386
[22] Shahbazi M-A, Hamidi M, Mäkilä Er M, etc. 2013 *Biomaterials* **34** 7776-7789.
[23] Bimbo L. M, Denisova O. V., Mäkilä E, etc. 2013 *ACS Nano* **7** 6884-6892
[24] Secret E., Smith K., Dubljevic V., etc. 2013 *Advanced Healthcare Materials* **2** 718-727
[25] Iskusnykh I Yu, Popov A L, Popova T N, etc. 2012 *Vestnik VGU. Chemistry, Biology, Pharmatsia, in Russian* **1** 96-102
[26] Belorus A O, Spivak Yu M, Moshnikov 2014 *The 9th International Conf. on European Science and Technology* **II** 268-277
[27] Belorus A O, Maraeva E V, Spivak Yu M, Moshnikov V A 2015 *Journal of Physics: Conference Series* **586** 012017.
[28] Zhang X G *J. Electrochem. Soc.* 2004 **151** 69-80
[29] Seidel H, Csepregi L, Heuberger A, Baumgartel H. 1990 *J. Electrochem. Soc.* **137** 3612-3626
[30] Aslanov L A, Fetisov G V, Ishchenko A A 2012 *Phismatli, in Russian* 648

Molecular Beacon CNT-based Detection of SNPs

V P Egorova[1], H V Krylova[2], I V Lipnevich[2], A A Veligura[1], B G Shulitsky[3], L.Y. Fedotenkova[1]

[1]Belarusian State Pedagogical University, 18 Sovetskaya Str, 200005 Minsk, Belarus
[2]Physics Department, Belarusan State University, 4 Nezavisimosti Ave., 200030 Minsk, Belarus
[3]Belarusian State University of Informatics and Radioelectronics, 6 P. Brovki Str, 200013 Minsk, Belarus

E-mail: alla.veligura@mail.ru

Abstract. An fluorescence quenching effect due to few-walled carbon nanotubes chemically modified by carboxyl groups has been utilized to discriminate Single Nucleotide Polymorphism (SNP). It was shown that the complex obtained from these nanotube and single-stranded primer DNA is formed due to stacking interactions between the hexagons of the nanotubes and aromatic rings of nucleotide bases as well as due to establishing of hydrogen bonds between acceptor amine groups of nucleotide bases and donor carboxyl groups of the nanotubes. It has been demonstrated that these complexes may be used to make highly effective DNA biosensors detecting SNPs which operate as molecular beacons.

1. Introduction

SNP is the most prevailing mutation of genome. A large number of SNPs makes them unique genetic markers for preventive medicine and biomedical diagnostics. The molecular beacons (MB) technique is one of the most common used methods to detect SNPs. Using of the MB to discriminate SNPs has substantial disadvantages such as the presence of residual fluorescence and exposure to intracellular nucleases. This problem can be solved by using nanostructured materials to modify MBs. Single stranded deoxyribonucleic acid (ssDNA) can form a stable complex with individual single-walled nanotubes (SWCNTs) or multi-walled nanotubes (MWCNTs) by the aromatic interactions between nucleotide bases and SWCNT or MWCNT sidewalls. It has been shown that the carbon nanotubes (CNTs) are effective quenchers of fluorescent oligonucleotides which are labelled by dye molecules [1]. The fluorescence is quenched by touching the rings of the dye molecule bound to the FAM-oligonucleotide, with the surface of CNTs [1–3]. If the hybridization of the FAM-oligonucleotide labelled by fluorophore FAM (fluorescein phosphoramidites) with complementary ssDNA target in solution occurs then the fluorescence is recovered. Stacking $\pi-\pi$-interaction reduces the distance between the surfaces of FAM-rings and CNTs to such an extend, that the charge transfer occurs to acceptor CNTs. Complexes – non-modified CNT/linear probe DNA are less effective then MB based on Fluorescence Resonance Energy Transfer with chemically linked fluorescent and quenching [1].

DNA selectively disperse different CNTs, with unfilled coating CNTs. In order to uniformly distribute the DNA on CNTs, the last are functionalized by polymers. Probe DNA chemically linked to the functionalized by polymer MWCNT is placed transversely to the CNT surface. It makes it possible discrimination of SNP [4]. Unfortunately, it does not prevent the folding of DNA with subsequent compactification into globules and inefficient charge transfer, that sets a nanomole limit for concentration detection [3, 4]. Preferential adhesion of oligonucleotide sequence $(GT)_n$ to SWCNT bundle in comparison with other oligonucleotides allowed to use $(GT)_n$ as reporter sequences chemically bound to ssDNA target. Since the bundle are not selected by the ssDNA targets, then the last are located transversally to SWCNT surface and are homogeneously distributed. It makes possible to perform hybridization. In paper [5] it has been suggested that because of the dependence of the light absorption coefficient on the density of the hybrids, the intensity of Raman scattering in SWCNTs will be reduced linearly with density of the hybrids. However ssDNA complexes with unmodified and non-functionalized SWCNTs do not allow to discriminate SNP. It was shown that DNA single strands self-assemble on carbon nanotubes (CNTs) with the formation of electron-dense layers of the oligonucleotide [6].

In this study, we have utilized the fluorescence quenching effect due to few-walled CNTs (FWCNTs) chemically modified by carboxyl groups to develop a new approach which allows us to identify SNP. We show in this paper that the complex ssDNA/FWCNT is formed due to stacking interactions between the hexagons of CNTs and aromatic rings of nucleotide bases as well as due to establishing of hydrogen bonds between acceptor amine groups NH of nucleotide bases and donor carboxyl groups of the nanotubes. Hydrogen bonds, orienting aromatic rings in a plane transversal to the surface of CNT, prevent $\pi-\pi$-interactions to minimize the distance between the aromatic rings and the surface of CNTs in the complex. The inclination of planes of nucleotide bases allows not only self-assembling of the oligonucleotide shell of complex, but an epitaxial growth of oligonucleotide molecular crystallites on electron-dense self-assembling oligonucleotide. DNA strands of epitaxial layers in the complex are vertically oriented, contrary to horizontally oriented DNA strands of self-assembling oligonucleotide. Therefore, as we will demonstrate the hybridization of probe DNA with DNA target can take place on complex ssDNA/FWCNT.

2. Materials and methods

ssDNA primer dye-labeled sequence 5'-FAM-GCCATATACTCTCCTTGGTGACA-3' (fluorescent oligonucleotide FAM-ON1) is used as DNA probe. For the heteroanalysing we utilize oligonucleotide targets 5′-TGTCACCAAGGAGAGTATATGGC-3′ (ON2) and 5′-GTAGGCTTGGGAATAGTACGA-3′ (ON4), complementary and non-complementary to the ON1, as well as single–base mismatched target 5′-TGTCACCAAGG<u>C</u>GAGTATATGGC-3′ (ON5) were synthesized at «Praymteh» (Minsk, Belarus).

We will utilize FWCNTs with diameters ranging from 2.0 to 5 nm and length of ~ 50–100 μm obtained by chemical vapor deposition method [7]. Then, FWCNTs were covalently modified by carboxyl groups.We used the suspension of carbon nanotubes obtained by dispersing of FWCNTs (from 400 to 600 mg) in 400 ml of TE buffer or deionized water with resistivity 18.2 MΩ·cm. A complexification is performed by ultrasound treatment of a mixture from fluorescent FAN-ON1 with FWCNTs, suspended in TE buffer at various concentrations. Preliminary, it was added 100 μL of suspension of various concentrations of nanotubes in TE buffer to 13.5 μL of oligonucleotide FAM-ON1 (c_{ssDNA} = 0.5 pmol/μL) in TE buffer. The reaction mixture was kept in a cuvette at room temperature or sonicated for 1 to 10 minutes (22 kHz). The structural and functional properties of the resulting FWCNT/FAM-oligonucleotide complexes was characterized by using transmission electron microscopy (TEM), PAGE electrophoresis, and the method of heterostructure analysis (HA).

Electron microscopy images are obtained by a transmission electron microscope JEM-100CX TEM (JEOL, Japan) at an accelerating voltage of 100 kV.

The electrophoresis was performed in polyacrylamide gel in 0.5-x 0.09 M TBE buffer (pH 8.0) at room temperature during 45 min. Control of the electrophoretic system electrical parameters which are reset by power, was carried out on the equipment Electrophoresis Power Supply EPS 601 (GE Healthcare Bio-Sciences AB, Sweden). PAGE electrophoresis was repeated at least 3 times.

3. Results and Discussion

3.1. *Structural analysis* The original carbon nanotubes are shown in figure 1a. TEM images of complexes FWCNT/ssDNA in figures 1b and 2 demonstrate that FWCNTs were covered by electron-dense layers of self-assembled oligonucleotide. As one can see in figure 1b, molecular crystallites can grow epitaxially on the surface of electron-dense layers of complexes FWCNT/ssDNA fabricated in deionized water. The TEM images of complexes FWCNT/ssDNA fabricated in TE buffer was used to estimate the dependence of thickness l_{DNA} of DNA layer included in the complexes, upon FWCNT concentration c_{FWCNT}, used at fabrication of these complexes. Measured values of thickness of the oligonucleotide shell in the complex are presented in table 1.

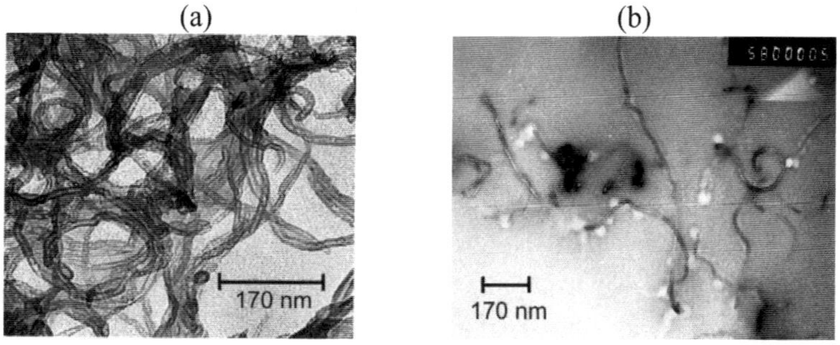

Figure 1. (a) TEM-image of original FWCNTs. (b) TEM-image of complexes ssDNA/FWCNT in deionized water.

According to table 1, l_{DNA} takes on the highest values at low c_{FWCNT} equals to 0.05 mg/ml. Due to the remoteness, stacking interaction between the upper monolayers of the self-assembled oligonucleotide and CNT is a weak. Therefore, at these concentrations the epitaxial growth of molecular ssDNA crystallite does not occur, and, respectively, there always is the ssDNA probe which is not linked with CNTs (figure 2a). At increasing c_{FWCNT} (0.1 mg/ml and more) the thickness l_{DNA} of FWCNT/DNA is decreases, since the weakly bound oligonucleotide of top monolayers of the complexes is redistributed to other CNTs. For l_{DNA} equal to 2.5 nm and less the increased stacking interaction of self-assembled ssDNA with CNT is able to hold the oligonucleotide molecules involved in the process of epitaxial growth of monocrystals as one can see in figures 1b and 2b. At high value c_{FWCNT} equals to 0.15 mg/ml due to the decrease l_{DNA} to value 1.2 nm the stacking interaction increases in such an extend DNA and holds the epitaxial DNA polycrystallites (figure 2c).

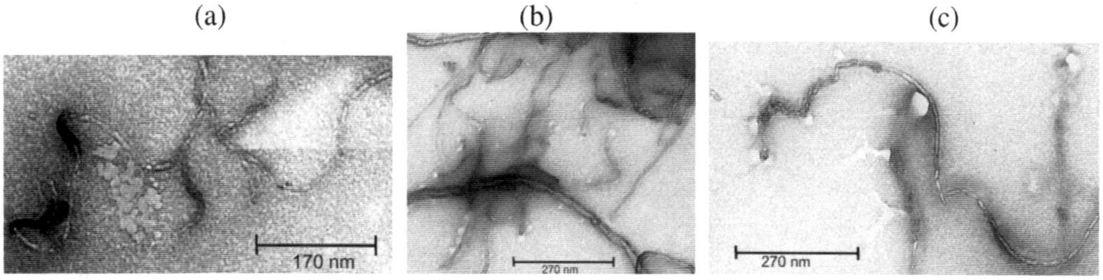

Figure 2. TEM-image of complexes ssDNA/FWCNT which are fabricated at FWCNT concentrations 0.05 (a), 0.1 (b), and 0.15 (c) mg/ml in TE buffer.

3.2. *Fluorescence-Quenching.* The influence of temperature on complexation was estimated assuming inversely proportional dependence of ionic conductivity of the solution on the temperature T, by analogy with the electrical conductivity of solids [8]:

$$\sigma = q^2 Nk/(DT) \tag{1}$$

where q is the electric charge of an ion, k is the reaction rate of ion with counterion, D is a diffusion coefficient, N is the ion density. Using (1), relative temperature of electrophoresis was calculated by the formula

$$\frac{T_m}{T_1} = \frac{W_m I_1^2}{W_1 I_m^2} \tag{2}$$

where T_m, I_m, W_m, m = 1, 2, 3 are temperature, electric current and power of m-th electrophoresis, respectively. Three electrophoresis with parameters listed in table 2 are represented by the curves with symbols «•», «■» и «♦» in figure 3 for relative temperatures $\{T_n/T_1\} = \{0.59, 1, 1.26\}$ respectively. As one can see from figure 3, up to concentrations c_{FWCNT} ~ 0.08 mg/ml inclusive, the increase of T leads to increase in luminescence intensity I and, respectively, the number of free molecules of FAM-ON1. It is caused by probability decrease of π–π-bonding of a selected oligonucleotide with FWCNT at increase of T due to the increase of fluctuations intensity.

Table 1. Thicknesses l_{DNA} of ssDNA layer in complexes FWCNT/ssDNA in depend on FWCNT concentration c_{FWCNT} utilized at fabrication of these complexes.

c_{FWCNT}, mg/ml	l_{DNA}, nm
0.05	5.3, 3.7
0.1	3.7, 2.5
0.15	3.7, 2.5, 1.25

Table 2. Electrophoresis parameters in depend on the relative temperature at voltage 100 V.

No.	I_n, mA	W_n, W	T_n/T_1
1	19.1	2	1
2	30.5	3	0.59
3	17.0	2	1.26

At high concentrations c_{FWCNT} ~ 0.1 and 0.15 mg/ml, when FAM-ON1 molecules start to interact each other, the decrease of relative temperature from 1 to 0.59, contrary, leads to the increase of I, because low diffusion mobility of FAM-ON1 prevent fast relaxation of FWCNT/ FAM-ON1 complex to an equilibrium state with a minimum of the free energy by minimizing interparticle orientational interaction in the oligonucleotide coating, as a result of which part of FAM-ON1 molecules do not have time to link with FWCNTs. However, very high temperatures are also undesirable at high c_{FWCNT}, because they disturb the equilibrium state with the minimum of the free energy.

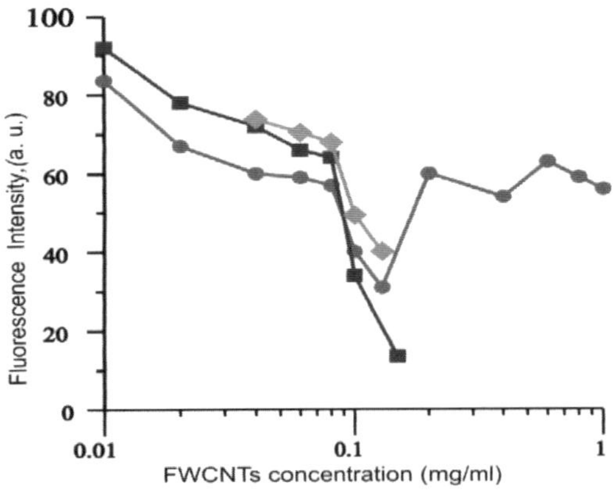

Figure 3. The intensity dependence of fluorescence of FAM-ON1 on FWCNT concentration c. Lines with symbols «•», «■» и «♦» denote dependencies for relative temperatures 0.59, 1, and 1.26 respectively.

At concentration c_{FWCNT} ~ 0.15 mg/ml and fixed value of c_{FAM} (FAM-ON1 concentration) the mean number of free oligonucleotide molecules, which are not linked with carbon nanotubes, decreases practically to zero with formation of FWCNT/ON1 complex (see curve with symbols «■» in figure 3). However, in accord with figure 3 (curve with symbols «•»), effective quenching of FAM-fluorescence in mixture of FWCNTs and ON1 occurs not at all concentrations the oligonucleotide ($c_{FWCNT} \geq 0.15$ mg/ml), but only for some optimal thicknesses of oligonucleotide coating $l_{st(1)}$, $l_{st(2)}$, $l_{st(3)}$ of carbon nanotube, corresponding to $c_{FWCNT} \approx 0.15$, 0.4, 0.9 mg/ml. This testifies the existence of a discrete set of FWCNT/ON1 complex sizes with thermodynamically stable structure.

3.3. *Hybridization Assays.* The method of HA have demonstrated that the presence of complementary DNA target in the solution leads to the dissociation of the FAM-oligonucleotide/FWCNT complexes due to the hybridization with homoduplexes formation. As consequence, the recovering of the fluorophore fluorescence occurs. The intensity of the fluorescence is much more than the fluorescence intensity for free FAM-oligonucleotides. The presence of the target SNP oligonucleotide sequence drops both the fluorescence intensity of the fluorophore and the electrophoretic mobility of heteroduplexes containing single mismatched base.

Figure 4 demonstrates a fluorescence intensity diminishing for FAM-oligonucleotide due to the formation of oligonucleotide hybridization duplex (c_{FWCNT} = 0.04 mg/ml). Forming homoduplexes ON2/FAM-ON1 (lane 6 in figure 4a,b) leaves the surface of ON1/FWCNT complexes, because on DNA-electrophoregram this homoduplexes emits in the field of molecular weights of freely formed homoduplexes (lane 5 in figure 4a,b).

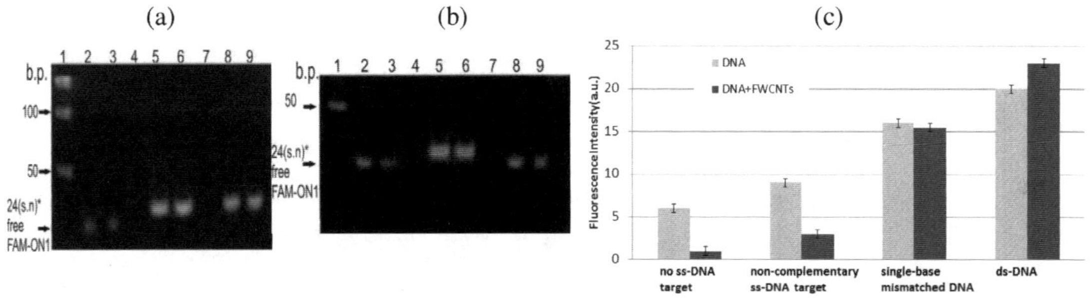

Figure 4. (a) Electrophoregram of DNA hybridization with nucleotide, containing single mismatch. Lanes: 1 – DNA-marker; 2 – DNA probe (fluorescent-oligonucleotide, FAM-ON1); 3 – FAM-ON1/FWCNT complex; 4 – complementary oligonucleotide (ON2) as DNA target, 5 and 6 – FAM-ON1/ON2 homoduplex without and with FWCNT, respectively; 7 – nucleotide with SNP (ON5) as DNA target; 8 and 9 – FAM-ON1/ON5 heteroduplex without and with FWCNT, respectively. (b) Electrophoregram of DNA hybridization with complementary and noncomplementary oligonucleotides. Lanes: 1 – DNA-marker; 2 – fluorescent-oligonucleotide (FAM-ON1) as DNA probe; 3 – FAM-ON1/FWCNT complex; 4 – complementary DNA target (oligonucleotide ON2), 5 and 6 – FAM-ON1/ON2 homoduplex without and with FWCNT, respectively; 7 – noncomplementary DNA target (oligonucleotide ON4); 8 and 9 – FAM-ON1/ON4 heteroduplex without and with FWCNT, respectively. (c) Intensity of FAM-ssDNA (light bars) and complexes FAM-ssDNA/FWCNT (dark bars) fluorescence in the absence of DNA target (left), at the presence non-complementary ssDNA target ON4 and single–base mismatched target ON5 (in center), and at the presence of perfect complementary ssDNA (right). Molecular weight FAM-ON1 of order 24 is given in single-nucleotides (s.n.).

According to figure 4c the FAM-oligonucleotide fluorescence quenching by few-walled CNTs stipulates the significant luminescence intensity differences of homoduplexes, heteroduplex, and free fluorescent oligonucleotide FAM-ON1 in comparison with their fluorescence without quenching.

Since the mobility of homoduplexes with and without FWCNTs are practically undistinguishable, the cooperative binding causes a detachment of epitaxial oligonucleotide along the crystal planes. Remaining in the monocrystal oligonucleotide remains bound to self-assembled FAM-ON1 of FAM-ON1/FWCNT complexes.

The method of HA have demonstrated that the presence of complementary target-DNA in the solution leads to the dissociation of the FWCNT/FAM-oligonucleotide complexes due to the hybridization with homoduplexes formation. As consequence, the recovering of the fluorophore fluorescence occurs. The intensity of the fluorescence is much more than the fluorescence intensity for free FAM-oligonucleotides. The presence of the target SNP oligonucleotide sequence reduces both the fluorescence intensity of the fluorophore and the electrophoretic mobility of heteroduplexes containing a single mismatch.

4. Conclusion

Thus, FAM-oligonucleotide/FWCNT complexes may be used to make highly effective DNA biosensors detecting SNPs which operate as molecular beacons.

References

[1] Yang R, Jin J, Chen Ya, Shao Na, Kang H, Xiao Z, Tang Z, Wu Ya, Zhu Zhi, and Tan W *J. Am. Chem. Soc.* **130** 8351

[2] Veligura A A, Egorov A S, Egorova V P, Krylova H V, Lipnevich I V and Shulitsky B G 2013 *Scientific Bulletin of the Uzhgorod University. Series Physics.* **32** 94

[3] Aparcin E K, Novopashina D S, Nastaushev Y. V and Ven'jaminova A G 2012 *Russian Nanotechnologies.* **7** 38

[4] Liu S and Guo X 2012 *NPG Asia Materials* . **4** e23

[5] Bansal J, Singh I, Bhatnagar P K and Mathur P Ch 2013 *J. Biosci. Bioeng.* **115**(4) 438

[6] Egorov A S, Egorova V P, Krylova H V, Lipnevich I V, Orekhovskaya T I, Veligura A A, Govorov M I and Shulitsky B G 2014 *J.Phys. CS.* **541**(1) 012037

[7] Labunov V, Shulitski B, Prudnikava A, Shaman Y P and Basaev A S 2010 *Semiconductor Physics, Quantum Electronics & Optoelectronics.* **13**, 137

[8] Krylova H, Hursky L 2013 *Spin polarization in strong-correlated nanosystems.* (Saarbrucken: LAP LAMBERT Academic Publishing GmbH & Co.)

Journal of Physics: Conference Series **643** (2015) 012024 doi:10.1088/1742-6596/643/1/012024

The FNS-based analyzing the EEG to diagnose the bipolar affective disorder

O Yu Panischev[1,2], S N Panischeva[1,2], S A Demin[1,2]
[1] Kazan Federal University, Institute of Physics, Kazan, 420008 Russia
[2] Kazan Federal University, Engelhardt Astronomical observatory, Zelenodolski region, st. Observatory, 422526 Russia

E-mail: opanischev@gmail.com

Abstract. Here we demonstrate a capability of method based on the Flicker-Noise Spectroscopy (FNS) in analyzing the manifestation bipolar affective disorder (BAD) in EEG. Generally EEG from BAD patient does not show the visual differences from healthy EEG. Analyzing the behavior of FNS-parameters and the structure of 3D-cross correlators allows to discover the differential characteristics of BAD. The cerebral cortex electric activity of BAD patients have a specific collective dynamics and configuration of the FNS-characteristics in comparison with healthy subjects.

1. Introduction

Modern medical science has the powerful capabilities in solving many tasks in diagnosing and monitoring of the treatment quality. However, there are many difficulties connected with correct diagnosing in specific fields. First of all, it can be illustrated on example of the psychiatric disorders. The most disorders haven't the objective diagnostic criteria. At this rate, schizophrenia or a bipolar affective disorder are diagnosed based on criteria in either the American Psychiatric Association's fifth edition of the Diagnostic and Statistical Manual of Mental Disorders (DSM 5), or the World Health Organization's International Statistical Classification of Diseases and Related Health Problems (ICD-10). However in various papers [1, 2] it is shown that BAD and epilepsy are manifested in the specific behavior of synchronization effects and in spectral properties of alpha rhythm. In [3] on the basis of the flicker-noise spectroscopy (FNS) authors have formulated the criteria of degree of the susceptibility to schizophrenia in children/adolescents. In [4] it is demonstrated that the method based on the memory function formalism (MFF) can discover the susceptibility to schizophrenia by analyzing the collective phenomena in the cerebral cortex bioelectric activity.

Bipolar affective disorder (earlier called manic-depressive illness) is a mental disorder characterized by periods of elevated mood and periods of depression. The elevated mood is significant and is known as mania or hypomania depending on the severity or whether there is psychosis. During periods of depression there may be crying, poor eye contact with others, and a negative outlook on life. In this paper within the framework of the flicker-noise spectroscopy [5, 6] we demonstrate a capability to formulate the objective diagnosing criteria for BAD using the analysis of EEG signals. The FNS method separates the analyzed signal into three components: low-frequency regular component corresponding to system-specific "resonances" and their interferential contributions, stochastic random-walk component at larger frequencies corresponding to dissipation effects, and highest-frequency inertial "spike" component corresponding to flicker noise. Also the method contains equations for describing the frequency-phase synchronization.

We show here that the BAD existence leads to decreasing the resonance components and to increasing the chaotic components of the EEG-signal from the certain electrodes. Also we have discovered the crucial cerebral cortex areas for the BAD diagnosing. According to the hypothesis about dependence between the pathological abnormalities and the synchronization properties of bioelectrical activity we have revealed the characteristic features of the EEG collective dynamics in healthy people and BAD patients. It is shown that the cerebral cortex signals from healthy people are characterized by high degree of long-range synchronization, whereas EEG of BAD patients has a strongly pronounced short-range synchronization.

2. Basic relations of the Flicker-Noise Spectroscopy

The essence of the FNS [3, 5, 6] method consists in assigning the information sense to distinct types of irregularities in considering time series: spike-like irregularities ("chaotic" components) and jump-like irregularities ("resonant" components). It allows parameterizing the studied dynamics by means of short set of variables describing the evolution in different frequency ranges.

Within the framework of FNS six parameters are introduced T_{01}, $S_S(T_{01}^{-1})$, n, σ, T_1, H_1: T_{01}, the correlation time for jump- and spike-like irregularities after which the self-similarity observed in power spectrum estimate breaks down; $S_S(T_{01}^{-1})$, the "spikiness" factor – power spectrum estimate at frequency T_{01}^{-1}, which accounts for the "intensity" of jump- and spike-like irregularities in the highest-frequency interval; n, the flicker-noise parameter, which characterizes the rate of loss of correlations in the series of high-frequency irregularities in time intervals T_{01}; σ, the standard deviation of the value of the measured dynamic variable from the slowly varying resonant (regular) component, which is based solely on jump-like irregularities; T_1, the correlation time for jump-like irregularities in stochastically varying signal $V(t)$; H_1, the Hurst exponent (this estimate of the Hurst component is also referred to in literature as the Hausdorff exponent), which describes the rate at which the dynamic variable "forgets" its values on the time intervals that are less than T_1.

3. Experimental Data

The background EEG signals were recorded from electrode locations Fp1, Fp2, F7, F3, F4, F8, T3, C3, C4, T4, T5, P3, P4, T6, O1, O2 on the scalp according to the standard 10-20 International electrode placement system. The electrodes are numbered (figure 1) from 1 to 16. An average reference was used. The subjects were placed in a sound proof, light attenuated air-conditioned (20^0 C) room and instructed to relax and close their eyes for some time during the data acquisition period. The sampling frequency was 200 Hz and the signal was filtered between 0.1-70 Hz. A notch filter of 50 Hz was also used. The data were stored on an optical disk drive and then later transported to a hard disk for further analysis. Sixteen subjects were chosen belonging to two broad groups (each group consisted of eight subjects): control subjects with no reported psychiatric or neurological disorders, and subjects with bipolar affective disorder. The subjects were within the age range 18-65 years with the mean ages of 32.50, and 30.44 years for the two groups respectively. All the subjects gave written consent prior to the recording. An epoch length of ~ 10 s of uninterrupted EEG data, which were free from any visual complexes, were chosen for analysis. Baseline drift was removed by subtracting a polynomial of 2nd order [1, 2].

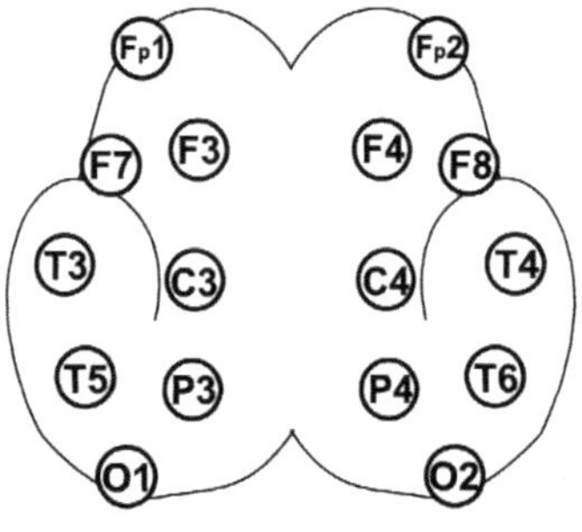

Figure 1. Positions of the 16 electrodes including their number and their designations. The schemata are based on the internationally established 10-20 system. Midline electrodes are exempted from this study.

4. Discussion

Figure 2 (a – c) demonstrates the FNS parameters $S_S(T_{01}^{-1})$, n, σ, :which are not demonstrate the differences for EEG dynamics of healthy and patients. But the following parameters (Figure 2 d – f) T_{01}, T_1, H_1: allow to determine the differences in correlation characteristics of EEG from healthy and patients. The main differences are in frontal and top cerebral cortex areas. Also we observe the differences in other areas.

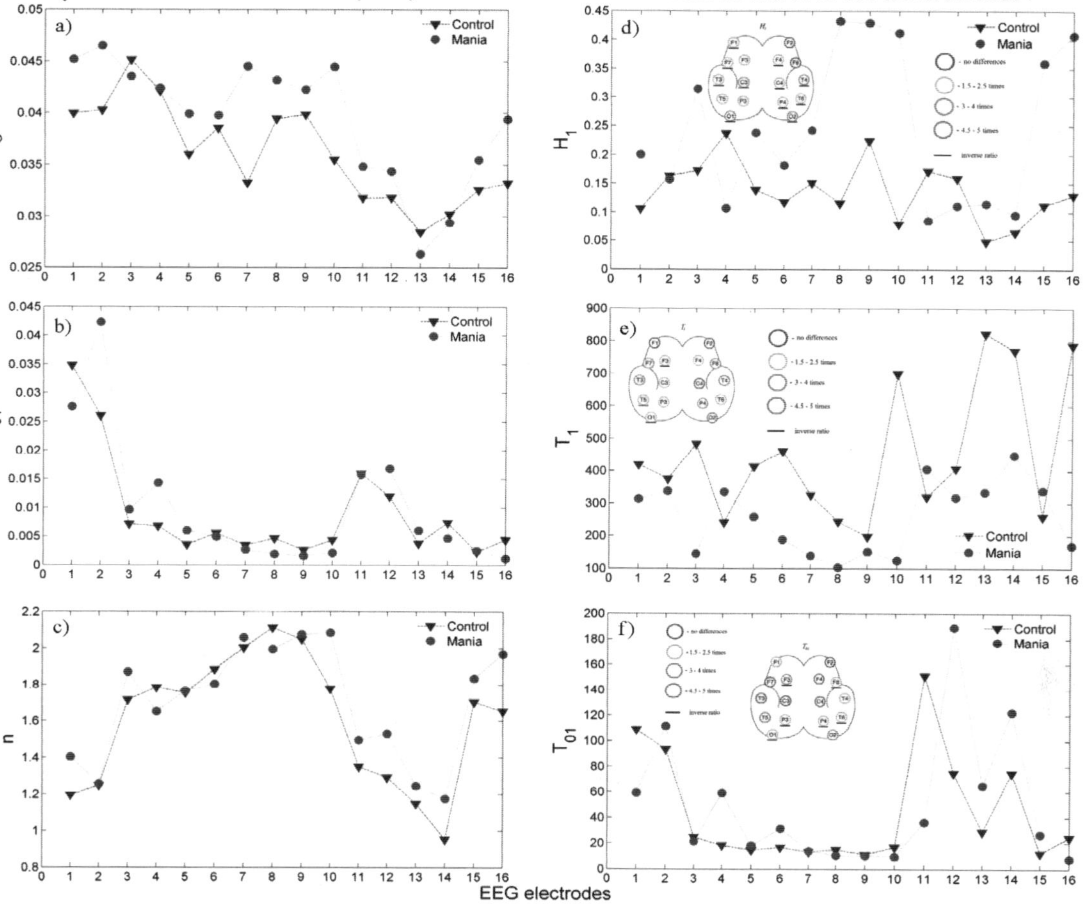

Figure 2. The ratios of mean values of FNS parameters for healthy and patients.

Cross correlator, introduced within the frameworks of FNS (figure 4) shows in healthy people the strong phase synchronization between the long range areas and more low its degree for short range areas. Another picture is observed for BAR patients: close areas have a significant synchronization at just long range areas demonstrate low synchrony.

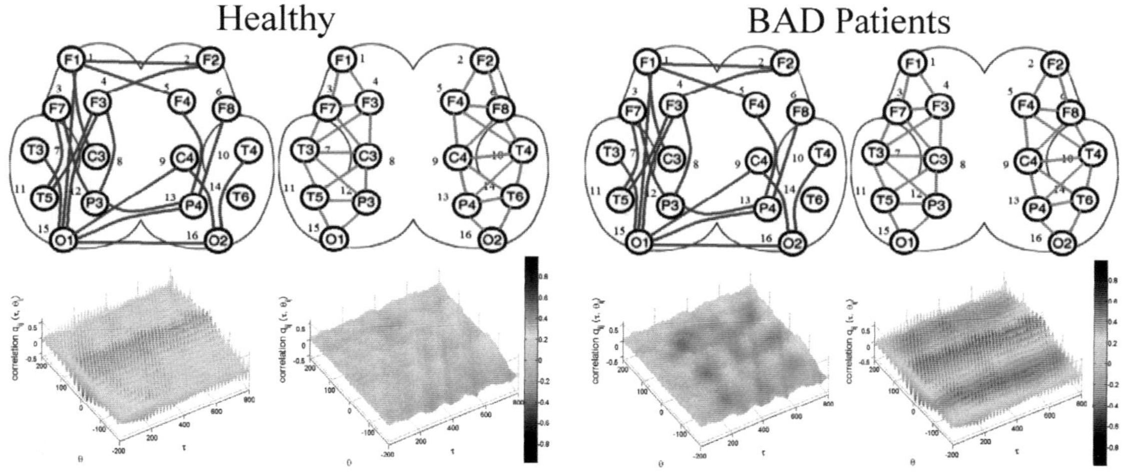

Figure 3. Coupling the remote and neighboring cortical areas in BAD patients and healthy people.

Journal of Physics: Conference Series **643** (2015) 012024 doi:10.1088/1742-6596/643/1/012024

5. Conclusions

In this paper we have presented a diagnosing method of psychiatric disorders based on the FNS analysis. To find the diagnosing criteria for bipolar affective disorder (BAD) we use the FNS method to analyze the behavior of EEG signals and collective phenomena in the cerebral cortex electric activity. We reveal the differences in "chaotic" and "resonant" characteristics of EEGs in healthy and BAD patients. Also we identify the differences incoupling of the neighboring and remote cortical areas in the BAD patents and healthy subjects. At the pathology the high degree of synchronization is observed in close proximity cortical areas, at that in healthy people have a high synchrony in remote ones. Thus we have demonstrated a capability to use the EEGs to develop a new diagnostic methods of the psychiatric disorders.

Acknowledgments

The work is performed according to the Russian Government Program of Competitive Growth of Kazan Federal University. This work was supported by the Russian Foundation for Basic Research, grant No 14-02-31385_mol_a.

References

[1] Bhattacharya J, Kanjilal P P and Nizamie S H 2000 *IEEE Trans. Biomed. Engin.*, **46** 738
[2] Bhattacharya J 2001 *Acta Neurobiol. Exp.*, **61** 309
[3] Timashev S F, Panischev O Yu, Polyakov Yu S, Demin S A and Kaplan A Ya 2012 *Phys. A*, **391** 1179
[4] Panischev O Yu, Demin S A, Kaplan A Ya and Varaksina N Yu 2013 *Biomed. Engin.*, **47** 153
[5] Timashev S F 2007 *Flicker-Noise Spectroscopy: Information in Chaotic Signals* (Moscow: Fizmatlit) (in Russian)
[6] Timashev S F, Polyakov Yu S, Yulmetyev R M, Demin S A, Panischev O Yu, Shimojo S and Bhattachatya J 2010 *Laser Phys.* **20** 604

Microfluidic device for unidirectional axon growth

E.Malishev[1*], A.Pimashkin[2], A.Gladkov[2,3], Y.Pigareva[2] A.Bukatin[1], V.Kazantsev[2], I.Mukhina[2,3], M.Dubina[1]

[1] Saint-Petersburg Academic University - Nanotechnology Research and Education Centre of the RAS, Saint-Petersburg, Russia
[2] N.I. Lobachevsky State University of Nizhny Novgorod - Neurocenter, Nizhny Novgorod, Russia
[3] Nizhny Novgorod State Medical Academy, Nizhny Novgorod, Russia
[*] corresponding author. E-mail: malishevg@gmail.com

Abstract. In order to better understand the communication and connectivity development of neuron networks, we designed microfluidic devices with several chambers for growing dissociated neuronal cultures from mice fetal hippocampus (E18). The chambers were connected with microchannels providing unidirectional axonal growth between “Source” and “Target” neural sub-networks. Experiments were performed in a hippocampal cultures plated in a poly-dimethylsiloxane (PDMS) microfluidic chip, aligned with a 60 microelectrode array (MEA). Axonal growth through microchannels was observed with brightfield, phase-contrast and fluorescence microscopy, and after 7 days *in vitro* electrical activity was recorded. Visual inspection and spike propagation analysis showed the predominant axonal growth in microchannels in a direction from “Source” to “Target”.

1. Introduction

Microfluidic chips combined with microelectrode arrays are used in a wide range of electrophysiological studies of neural networks at subcellular, cellular and network-wide scale. Such devices provide a method to control morphology of the grown culture of dissociated neurons in order to model a brain regions, study interaction or coupling of different cell types and formation of functional signalling pathways in the brain. Dissociated neuronal cultures can be grown in two separated chambers connected with microchannels which are long enough to provide a growth only for axons of the neurons [1]. Specific design of the microchannels can define axon outgrowth direction and organize functional connectivity in the neural culture. In such circuits presynaptic sub-population is connected with post-synaptic sub-population of the neurons [2]. Such dual compartment *in vitro* system, coupled with a microelectrode array and can provide unique approach to study network-wide synaptic plasticity as a basics of information coding, memory and learning. In this study we proposed a microfluidic device with different asymmetric microchannel designs to investigate unidirectional axonal growth during hippocampal culture development in order to localize presynaptic and postsynaptic subnetworks.

2. Materials and Methods

2.1 Chip design

Proposed design of microfluidic chip is based on the probabilistic nature of the direction of axon outgrowth. In order to maximize the probability of axon growth in desired area it is sufficient to place some "focusing" side walls, limiting available directions of axon growth (fig. 1.a). For solving the reverse problem of axon backward outgrowth minimization it is necessary to reduce the area from which axons may grow in an undesired direction. This can be done, for example, by creating a narrow "bottleneck" and false directions for growth (fig. 1.b). In our chip we have combined both these methods and proposed microchannels consisting of the sequence of triangular segments (fig. 1.c,d)

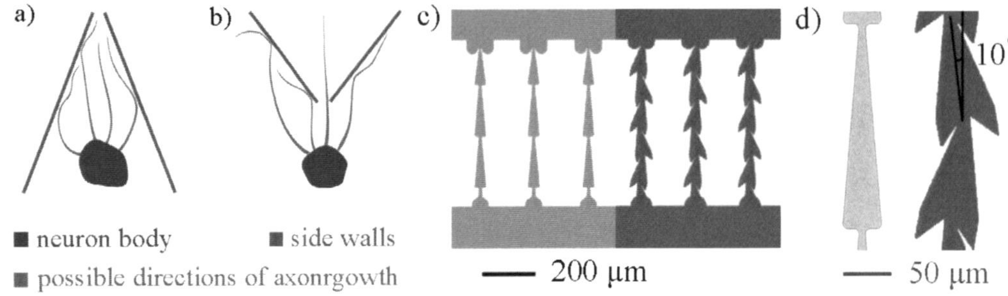

Figure 1. Microchannels' design: **a)** model of axon "focusing", **b)** model of a "bottleneck" for backward growing axons, **c)** proposed designs of microchannels: "Straight" type (left) and "Zig-zag" type (right), **d)** triangular segments' design, blue bar – 200 μm.

2.2 Microfluidic device fabrication

PDMS microfluidic chips containing an array of microchannels between two chambers were fabricated by two layer lithography and PDMS molding techniques (see fig. 2.a). Mold design contained: first 5 μm-thick layer, which formed microchannel structure and second 50 μm-thick layer, which formed chambers. Microchannels' structure was based on two types of segments, shown in figure 1.d. and their length varied from 200 μm to 1000 μm (2 - 10 segments).

For mold fabrication, the negative photoresist SU8 2025 and SU8 Thinner (MicroChem, USA) were used. Some portion of SU8 2025 was diluted by SU8 Thinner in a way to decrease the solids concentration from 68.55% (original concentration) to 45% (desired concentration, similar to SU8 2005) in order to achieve 5 μm layer thickness.

Silicon wafers were spin-coated with SU8 using a headway spinner at speeds of 4000 rpm. Softbake was done on a hotplate at 95 °C for 4 and 7 minutes for the first and second layer respectively. The lithography was performed with a MJB4 (SUSS Microtec, Germany) and an UV-filter PL-360-LP (Omega Optics, USA) with a cutoff wavelength 350 nm. Then wafers were post exposure-baked on a hotplate at 95 °C for 5 and 7 minutes respectively. Patterns were developed by using SU8 Developer. Finally, molds were hard baked at 200 °C for 7 minutes.

In order to create PDMS chip, PDMS (Dow Corning, USA) was mixed in a 10:1 ratio. Then, uncured PDMS was degassed in a desiccators and poured onto the master molds and cured in an oven at 70°C at least for two hours. After that PDMS chips were removed from molds and chamber structures were punched.

In order to investigate spike propagation inside the microchannels, each PDMS chip was positioned and mounted onto the surface of a planar microelectrode array (MEA), so as to locate several electrodes in the microchannels (fig 2.b). PDMS chips were adhered to MEA plate by heating them after mounting to 120°C for 20 minutes. Before cell plating the device was pre-treated with the adhesion promoting molecule polyethyleneimine (Sigma P3143).

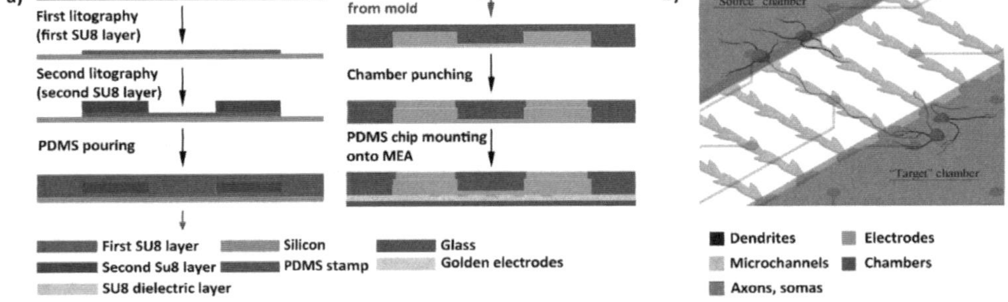

Figure 2. **a)** microfluidic device fabrication process **b)** Scheme of the microfluidic device with plated cells and branches.

2.3 Culture preparation

Hippocampal cells were dissociated from embryonic mice (E18) and plated into separate subcompartments on MEAs at a final density of approximately 15,000–20,000 cells/mm^2. C57Bl/6 mice were euthanized via cervical dislocation, according to the protocols approved by the Russian National Ministry of Public Health for the care and use of laboratory animals. The cultures were stored in incubator at 36-37°C and 5% CO_2 for 3-4 weeks.

2.4 Microscopy

Axon growth and culture development monitoring was performed using brightfield and phase-contrast mode microscopy under a fluorescent inverted microscope DMIL HC (Leica, Germany), with a 20x/1.0 objective. Also fluorescence micrographs were done. For that immunostained cultures were fixed for 15 min in 4% formaldehyde containing phosphate-buffered saline (PBS) (pH=7.4), washed in PBS and permeabilized for 30 min with 0.1% Triton X-100 (Sigma 93443-100ML) and 2% bovine serum albumin (BSA). Subsequently, the cells were incubated for 2 hours at room temperature in PBS containing 1% BSA and the primary antibodies: rabbit polyclonal Anti-Tau (314002, SySy, Germany) to stain axons. After washing in PBS, the cell cultures were incubated for 2 hours at room temperature with the following secondary antibodies: goat anti- rabbit conjugated Alexa Fluor 647 (A21245, Molecular Probes®, US).

2.5 Electrical recording and stimulation

Electrical activity was recorded after 7 days in vitro by USB-MEA system (Multichannel systems, Germany). Signals were recorded simultaneously from microelectrode array with 60 planar electrodes with 30 μm diameter and 200 μm interelectrode space. Electrical stimulation was applied using a STG-4004 stimulator (Multichannel Systems, Germany). Spikes were detected using threshold detection estimated from the noise statistics in each channel separately. The amplitudes of detected spikes were in the range of 10–40 μV. All signal analysis and statistics were performed using custom made software in Matlab® (see [3] for details).

3. Results and Conclusions

3.1 results

First we produced the microfluidic devices with different microchannel designs (see Materials and Methods) and plated hippocampal cells into interconnected chambers (fig. 3). In order to find optimal design of the microchannels for unidirectional connectivity between neuronal sub-populations we studied axon growth dynamics varying microchannels' length and shape. We examined a microsope images taken each day after plating and found that in most cases axons, growing in "forward" direction of microchannels (from Source to Target chamber) reached their ends, while only a few number of the axons, growing "backward" passed more, than 2 segments (see fig. 4.a,b). In summary we analyzed axon growth in 26 cultures with two shapes of microchannel segments and 5 types of microchannel length and found that in most cases efficiency of "Zig-zag"-shaped microchannels was less than in "Straight"-shaped. We confirmed that the branches in microchannels were axons by staining axon with fluorescent markers using immunostaining methods (fig. 4 b).

Figure 3. General view of a microfluidic device.

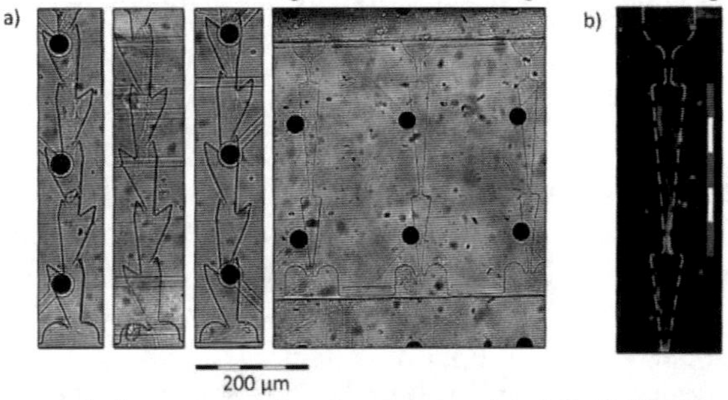

Figure 4. Axon growth microscopy **a)** brightfield micrographs of axons in different types of microchannels, **b)** fluorescence micrographs of Tau immunostained neurons.

We tested functional connections between neurons in microchannels using microelectrode array. Bioelectrical activity of the hippocampal culture starting form 7 DIV consists of network-wide synchronous spiking activity which referred as a bursts. Such bursts last for several hundreds of milliseconds with a few seconds of interburst interval. At the initiation phase of the burst spikes propagate through the network recruiting most of the neurons which induces reverberating activity for the rest of the burst.

Such bursts from 7 DIV to 10 DIV emerged spontaneously and independently in each of two chambers. We found that after 20 DIV the spontaneously emerged bursts in the Source chamber consequently evoked the bursts in the Target chamber through axons in microchannels (fig.5. a,b,c,d). A probability of such burst propagation was higher than the probability of evoking the bursts in opposite direction. The results for different 3 cultures are represented in fig. 5.e as a percentage of "Source to Target" bursts (blue bar) and "Target to Source" (green bar). We summarized the results and found that in average the probability of burst propagation through grown pathways was by 4 ± 1.5 times higher than in opposite direction.

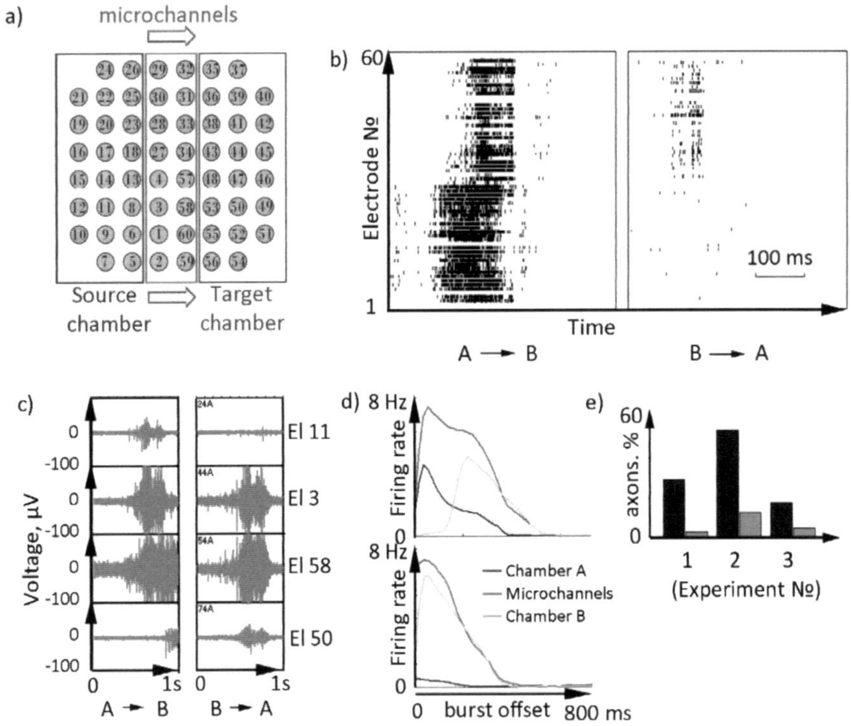

Figure 5. **a)** electrode numeration map, blue area depicts electrodes coupled with Source chamber, red - with microchannels, green - with Target chamber. **b)** raster plots of the bursts from Source chamber (A) evoked the burst in the Target (B) chamber (left). Burst in the Target chamber didn't propagate backward to Source chamber (right). **c)** Signals from 4 selected electrodes in Source chamber (top), two middle electrodes - coupled with microchannels, bottom electrode - electrode in Target chamber. Left column – burst spontaneously emerged in Source chamber (A), Right - in Target chamber (B). **d)** Top panel - Average spontaneous burst profile emerged in Source chamber. Each line represents spiking firing rate in each 5 ms time bin in Source chamber (A) (blue line), axons (red line) and Target chamber (B) (green line). Bottom panel – Average spontaneous burst profile emerged in Target chamber in opposite to axon grow direction. **e)** burst evoke probability from "Source to Target" chamber (blue bar) and "Target to Source" chamber (green bar). Each pair of bars represents on culture 20 min. recording on 20-25 DIV.

We also observed spike propagation through a grown synaptic pathway by single short pulse of electrical voltage stimulation of the neural network. We have randomly chosen the electrode for stimulation and applied a bipolar voltage pulse (500 μs, ±600 mV) in order to evoke bursting activity in both chambers. Stimulus responses propagated between chambers only from Source chamber into Target, but not vice versa (fig. 6).

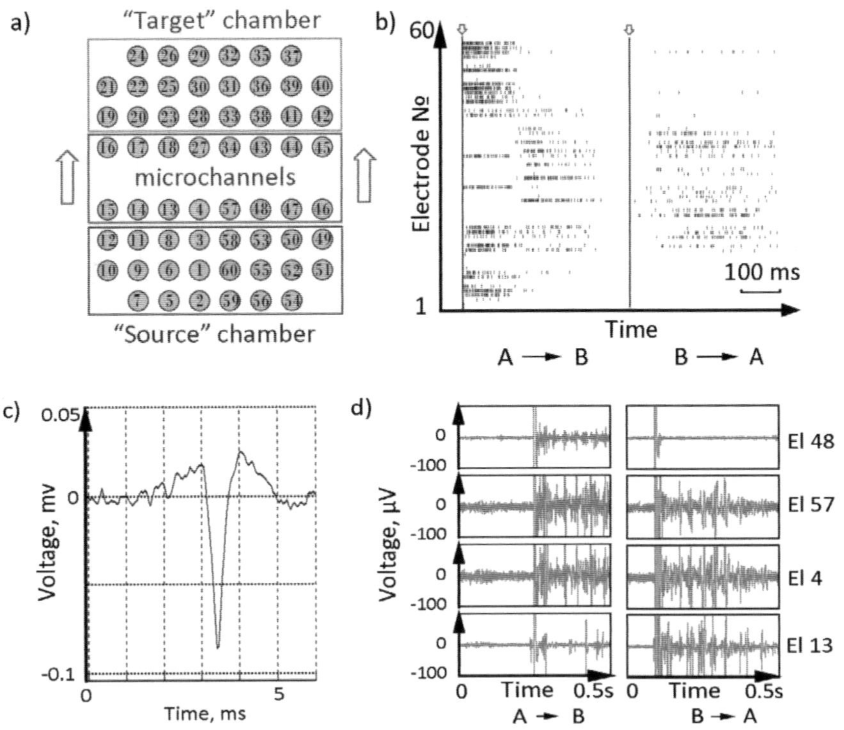

Figure 6. Electrical activity of hippocampal culture: **a)** electrode numeration map. **b)** raster plots of spiking activity in response to electrical stimulus (red arrow) applied to the electrode in Source chamber (A) (left) and Target chamber (B) (right). Each vertical line represents a spike timestamp, **c)** example of a spike timestamp, **d)** electrophysiological signals in response to electrical stimulus recorded from 4 selected electrodes: top - electrode from Source chamber (A), two middle electrodes - coupled with microchannels, bottom electrode - Target chamber (B). Left column (A->B) - burst emerged in Source chamber, Right (B->A) - burst emerged in Target chamber.

3.2 conclusions

In this study we developed microfluidic chips with dual chambers coupled by specifically designed microchannels for unidirectional neural axon outgrowth between sub-populations of dissociated hippocampal networks. We tested several designs varying length and inner design to find optimal design of microfluidic device. We found that the "Straight" type of the microchannel was optimal for unidirectional synaptic spike propagation during spontaneous and stimulus evoked bursts. The minimal length for microchannel to provide unidirectional coupling was found to be 400 μm, indicating that dendrites cannot overgrow such length from Source chamber to Target and axons are not able to grow backward from Target chamber to Source Source chamber in order to provide backward synaptic connections in Target chamber or Source chamber respectively. Such microfluidic chip coupled with MEA can be widely used in the research of network dynamics in a realistic morphology, network-wide synaptic plasticity and interaction between cell types which is the basics of the higher cognitive functions in the brain.

Acknowledgments

This research was supported by Russian Science Foundation (№ 14-19-01381)

4. References

1 - Brewer, G. J., Boehler, M. D., Leondopulos, S., Pan, L., Alagapan, S., DeMarse, T. B., & Wheeler, B. C. (2013). Toward a self-wired active reconstruction of the hippocampal trisynaptic loop: DG-CA3. Frontiers in Neural Circuits, 7(October), 165.

2 - Le Feber J., Postma W., Rutten W., De Weerd E.,Weusthof M.. Two-Chamber MEA to Unidirectionally Connect Neuronal Cultures. Proceeding of 9th Int. Meeting on Substrate-Integrated Microelectrode Arrays, 2014. pp. 259-262

3 - Pimashkin A, Kastalskiy I, Simonov A, Koryagina E, Mukhina I and Kazantsev V (2011) Spiking signatures of spontaneous activity bursts in hippocampal cultures. Front. Comput. Neurosci. 5:46. doi: 10.3389/fncom.2011.00046

Application of a LED–photodiode optocouple for the study of human respiratory function

A G Malyshev[1], N K Zhumashev[1], G N Lukyanov[1], K D Mynbaev[1,2] and A A Rassadina[1]

[1]ITMO University, Saint-Petersburg 197101, Russia
[2]Ioffe Institute, Saint-Petersburg 194021, Russia

E-mail: gen-lukjanow@yandex.ru, karim.mynbaev@niuitmo.ru

Abstract. Application of a LED-photodiode optocouple for the study of human respiratory function is considered. The optocouple operating in the middle-infrared part of the spectrum (with the peak of the LED emission at 4.1 µm at 300 K) was applied for monitoring changes in CO_2 concentration during respiration of a human. Studies were carried out simultaneously with the measurements of pressure in the left and right halves of the nose cavity. The results of the study open possibilities for using infrared optocouples as sensitive elements in compact and cheap sensors detecting breath abnormalities associated with respiratory and other diseases.

1. Introduction

Early diagnostics of respiratory diseases and development of its methods are among the important trends in modern otorhynolaryngology. According to the data by The Ministry of Healthcare of Russian Federation (2013), nearly 39% of population in Russia suffer from respiratory diseases. Also, diagnosed are asthma (1%), allergic rhinitis (0.2%) and chronic obstructive pulmonary disease (0.6%) [1]. All these diseases are accompanied with nasal breath anomalies. At the present time, anterior active rhinomanometry is considered to be a clinically approved method of unbiased investigation of nasal breath (Fig. 1(a,b)) [2]. Measured datum in this method is the nasal airway resistance, which is the ratio of pressure difference to flux difference in the tube supplied to the organs of the respiratory system. Non-invasiveness of the measurements and the compactness of the equipment are the advantages of the method. However, this method distorts natural breathing because of using a tube, thus reducing the chances of accurate early diagnostics. Working with such diagnostic equipment demands a high professional level of staff. The equipment is quite expensive ($ 10 000 per piece at average) and only specialized clinics can afford it. Using that kind of equipment in Russian Federation for prevention remains rather difficult.

For early diagnosis of respiratory diseases, it is relevant to study natural breathing without additional tube. The results of the studies carried out earlier [3] showed a non-linear dynamic character of such diagnostic parameters of natural breathing as airflow velocity, temperature, and pressure. Therefore, the basic requirements for the sensor element are high sensitivity and fast response. The method described in [3] had a significant drawback: the quick response was achieved at the cost of the miniature size of the sensor (a thermistor). Such miniature elements are brittle and not suitable for wide use. Replacement of the sensors described in [3] by optical sensors should be more beneficial. For example, Russian company "LED Microsensor NT", Ltd., produces infrared (IR) light-emitting diodes (LEDs) and photodiodes with fast response (dozens of nanoseconds), miniature size of the LED chip (0.3×0.3 mm), small angular divergence of LEDs ($\pm 3^0$), low power consumption (about 1 mW in a pulsed mode), long service life (80 000-100 000 hours), and reasonably low cost. These LEDs and photodiodes can be easily integrated into optocouples (Fig. 1(c)) [4].

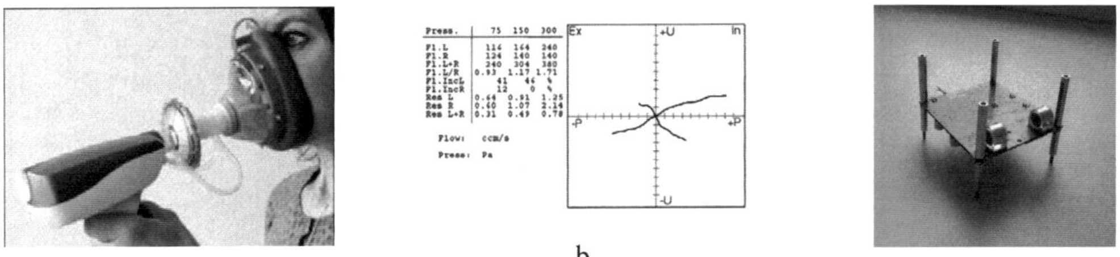

a b c

Figure 1. Anterior active rhinomanometry equipment (a) and diagnostic result (b), IR optocouple (c).

2. Materials and method

This work presents the results of the study of the possibility of using a LED–photodiode optocouple ("LED Microsensor NT", Ltd.) as a sensor device for contactless measurements of the dynamic fluctuations of CO_2 concentration in the human breath. Carbon dioxide plays a leading role in the humoral mechanism of the regulation of respiration. It is known that the concentration of CO_2 changes during respiration: the proportion of CO_2 at inhalation is 0.03% and that at exhalation is 4.00%. This parameter is important in the breath diagnostics and is very informative.

The basic absorption band of CO_2 corresponds to the spectral region between 4200 and 4320 nm. Weaker absorption bands are located at 2700 nm and 2000 nm. In our study, we have used LEDs with dominating wavelength at 3.4, 3.8 and 4.1 μm at the temperature 300 K. First, we have studied their spectral characteristics. For this purpose, pulse excitation with a pulse width of 1 μs and pulse repetition frequency of 1 kHz was applied. Electroluminescence (EL) spectra were recorded with the use of cooled InSb photodiode. During the measurements, the LEDs were placed in a liquid nitrogen cryostat with a sapphire window. A lock-in amplifier was used for signal detection.

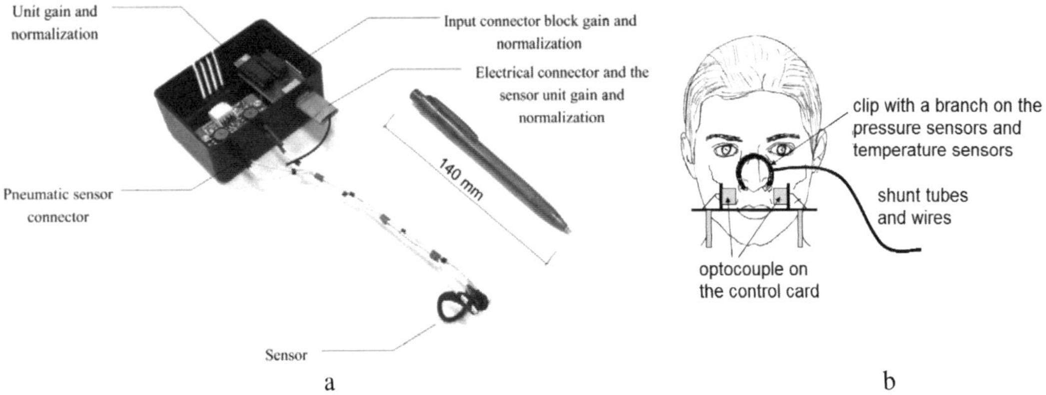

a b

Figure 2. Sensing device based on the pressure sensors (a) and the experimental scheme used (b).

One of the studied IR LEDs was used as a part of LED-photodiode optocouple for the measurements of CO_2 during the breathing of a patient. The photodetector used in the optocouple was based on the same semiconductor material system as the LED, namely narrow–gap semiconductor InAs(Sb,P) system. To test the sensitivity of the method, the measurements of CO_2 concentration oscillations during the breathing were performed synchronically with those of pressure changes. Sensing device made on the basis of pressure sensors and gain block are shown in Fig. 2(a) [5]. The

Journal of Physics: Conference Series **643** (2015) 012026 doi:10.1088/1742-6596/643/1/012026

experimental scheme of synchronous measurements is presented in Fig. 2(b). Clip with pressure and temperature sensors tap was attached to the nasal septum. The board has been placed near to patient's nose, at a distance of ~5 cm. The signal from the pressure sensors and the optocouple was analyzed with the LeCroy WaveSurfer 3000 oscilloscope.

3. Results

The results of the study of the spectral characteristics of IR LEDs with dominating wavelengths at 3.4, 3.8 and 4.1 μm are shown in Figs. 3 and 4. As can be seen, at 300 K (Fig. 3), which should be the working temperature of the sensing device, only emission of the LED with dominating wavelength at 4.1 μm (Fig. 3(c)) corresponded to the strong absorption band of a gas present in the air, namely, CO_2. While the other LEDs could have been used for detecting other gases in the human exhalation, for our purposes it seemed reasonable to use the LED emitting at 4.1 μm, as low power of IR LEDs (typically just tens of μW) makes it difficult to obtain a strong signal at 300 K. For comparison, spectra of the same LEDs recorded at 77 K are shown in Fig. 4. The emission intensity of the LEDs at this temperature was much higher than at 300 K (the driving currents used at 77 K were much smaller, which is the reason for noise seen in the spectra), the spectra were narrowed and red–shifted. Comparison of the results obtained at 300 K and 77 K suggested that the reason for low power of the studied IR LEDs at the room temperature was the dominating contribution of non–radiative (Auger) recombination. Another problem with this type of LEDs is related to carrier leakage over the barriers [6]. To increase the quantum efficiency of these LEDs, one needs to improve their design so the Auger recombination and carrier leakage would be suppressed.

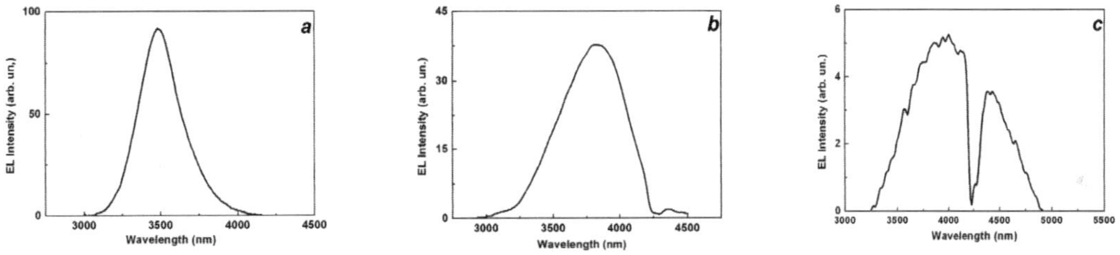

Figure 3. Electroluminescence spectra of three LEDs at T=300 K.

Note, that while the LEDs emitting at 3.4 and 3.8 μm were based on p-n junction structures with the active layer made of InAs (3.4 μm) or InAsSb (3.8 μm), the studied LEDs emitting at 4.1 μm were based on a multiple quantum well InAs/InAsSb structure with size quantization for holes in the well. This type of IR LEDs structure is considered to be very promising in future applications [7].

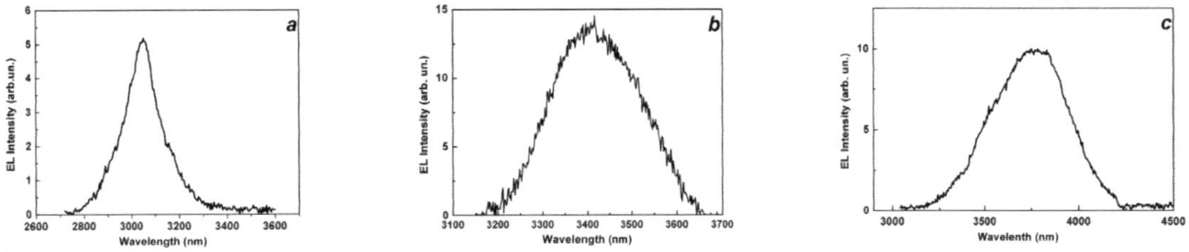

Figure 4. Electroluminescence spectra of three LEDs at T=77 K.

The example of the application of the LED-photodiode optocouple for the study of human respiratory function is shown in Fig. 5. As mentioned above, the sensitivity of the optocouple in the dynamic measurements of CO_2 concentration was studied simultaneously with the measurements of pressure in the left and right halves of the nose cavity using the sensor developed earlier [5]. The results of the present study showed the high sensitivity of the optocouple in respect to the changes of CO_2 concentration during respiration (Fig. 5). The graphs of power spectral density revealed a coincidence of the main frequency peaks registered with the two methods (Fig. 5). Thus, the results of our study open possibilities for using LED-photodiode optocouple as a sensitive element in the breath diagnostics sensor.

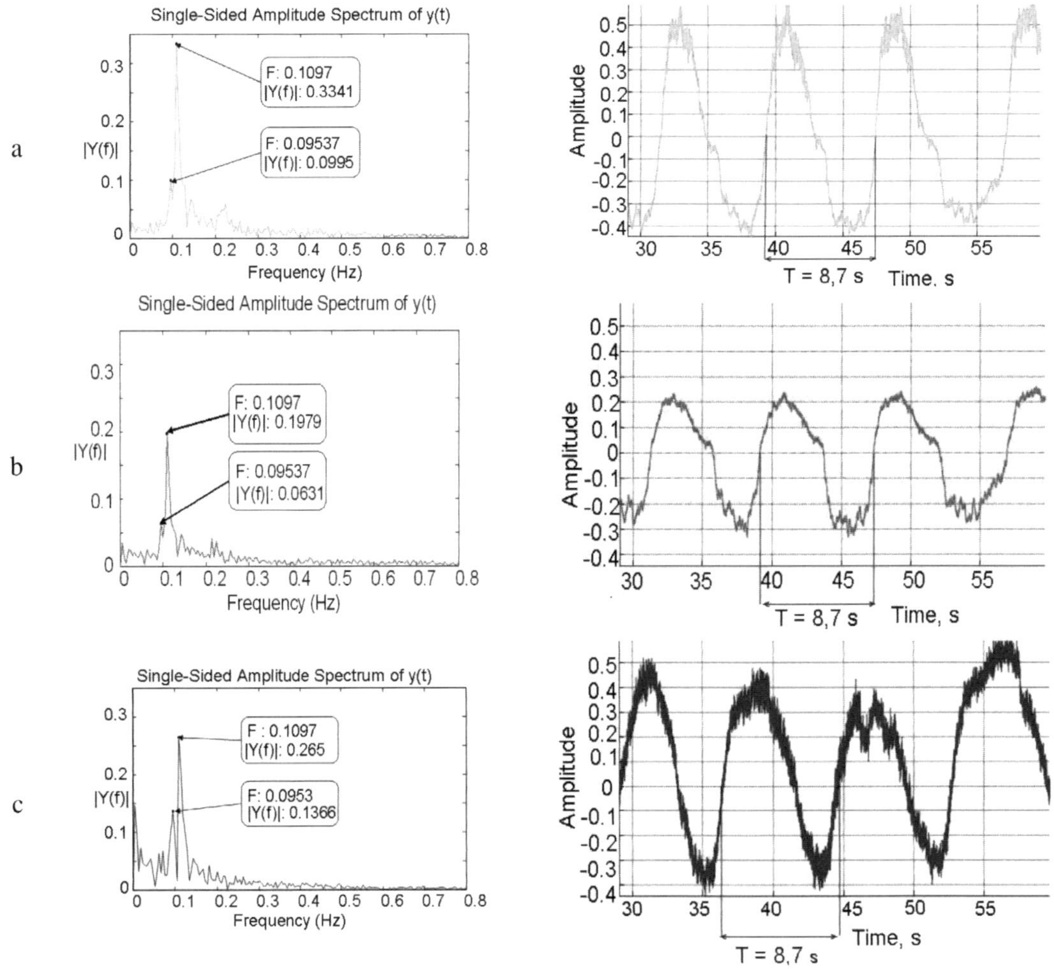

Figure 5. The results of the measurements of respiration fluctuations and the graphs of the power spectral density: (a) for measured pressure in the right nasal cavity; (b) for measured pressure in the left nasal cavity; (c) for optocouple.

4. Conclusion

Application of a LED-photodiode optocouple for the study of human respiratory function was considered. The optocouple based on InAs(Sb,P) heterostructures and operating in the middle-infrared part of the spectrum (with the peak of the LED emission at 4.1 μm at 300 K) was applied for monitoring changes in carbon dioxide concentration during respiration of a human. Studies were carried out simultaneously with the measurements of pressure in the left and right halves of the nose cavity according to the method developed earlier and based on a thermistor. The results of the study open possibilities for using infrared optocouples as sensitive elements in compact and cheap sensors detecting breath abnormalities associated with respiratory and other diseases. Parameters of the sensor need to be improved, and in particular, quantum efficiency of the LED should be increased. For this purpose, Auger recombination processes and carrier leakage over the barriers should be suppressed.

Acknowledgments
The authors are grateful to their collaborators from "LED Microsensor NT", Ltd. (N Stoyanov, S Kizhaev, S Molchanov and T Gurina) for supplying them with the LEDs and optocouples and for useful advice and assistance in carrying out this research.

References:

1. The Ministry of Healthcare. The incidence of the entire population of Russia in 2013 URL: http://www.rosminzdrav.ru/open/usefull/stat *(in Russian)*
2. Clement P A 1984 *Rhinology* **22**(3) 151-5
3. Lukyanov G, Rassadina A and Usachev V 2005 Comparison and the analysis of processes of the movement of air through the human breathing system and its natural model *2005 International* Conference on *Physics and Control (Saint-Petersburg, Russia, 24-26 August 2005)* IEEE Catalog Number: 05EX1099C ed A L Fradkov and A N Churilov 872-5
4. Matveev B A 2006 LED-Photodiode Opto-pairs *Mid-infrared Semiconductor Optoelectronics* ed A Krier *Springer Series in Optical Sciences v 118* (London: Springer–Verlag) 395
5. Lukyanov G, Voronin A, Rassadina A, Neronov R and Polish S 2015 Method and Device for Researching and Diagnostics of Respiratory Diseases *Advances in Biotechnology and Bioscience, Proceedings of the 6th International Conference on Bioscience and Bioinformatics (ICBB '15) (Dubai, United Arab Emirates, 22-24 February 2015)* 17–21
6. Petuhov A A, Il'inskaya N D, Kizhaev S S, Stoyanov N D and Yakovlev Y P 2011 *Semiconductors* **45** 1501–4
7. Kroemer H 2004 *Physica E* **20** 196–203

The influence of mobile impurities on photoelectric properties of PbSe-based multiphase photosensitive structures

A N Belenkov[1], V V Davydov[1], S B Boyko[2], Z N Petrovskaya[2] and A Yu Karseev[1]

[1] Saint Petersburg State Polytechnical University, Saint Petersburg, Russia
[2] JSC "NII Giricond", Saint Petersburg, Russia

E-mail: belenalexban@gmail.com, 21@giricond.ru

Abstract. The PbSe-based photoelectric sensor technology is considered. The polycrystalline PbSe films are used for the production. The influence of mobile impurities in a material on the mechanism of the photovoltaic effect is revealed and investigated. The experimental results of investigation of the photovoltaic effect are discussed.

1. Introduction.

Fire-fighting measures are very important for fire safety of all technological processes. Despite the significant progress in fire-fighting systems, fires stay a serious problem of modern society because they continue to claim lives and cause immense damage. So timely fire detection is essential for efficient fire fighting. The number of factories which use highly flammable and explosive materials increases. Thereby the requirements to a technological process monitoring were strengthened.

Nowadays there are several types of flame detectors which are widely used in fire monitoring systems. Among them are thermal, smoke, gas and optical flame detectors. Each of them responds to a special fire signature (heat, smoke, gas or optical radiation). Operating experience has demonstrated that optical systems were the most efficient devices for fire safety control of technological processes. Optical flame detectors respond to flame optical radiation, allowing detection of fire in its early stage. Optical flame detectors also can be used in large halls and outdoors. The main disadvantage of such systems is probability of false alarms caused by natural and industrial interferences. The interference power can be several orders of magnitude greater than the flame radiation power [1]. So the optical flame detectors must respond to some specific radiation properties of flame. In modern optical systems two fire detection methods are used: frequency and spectral selection of registered signal [1, 2].

The frequency selection means the detection of low-frequency oscillations of radiation intensity (from 1 to 20 Hz), caused by the turbulent nature of fire. This method has a number of disadvantages. Among them are low interference protection and high response rate of the optical detector (1-10 s). Moreover the frequency selection is useless for smoldering fire detection [1, 2].

The experimental data shows that the spectral selection is more efficient for flame detection. This method allows detecting the definite spectral bands in registered signal. There are the spectral band of hot water vapor with spectral characteristic peak at 2.9 μm and the spectral band of hot carbon dioxide CO_2 with peak at 4.4 μm in the flame spectrum (figure 1). So, several photoelectric sensors should be used for the spectral selection. Some of them detect the spectral bands of water vapor and CO_2. And the others are used for monitoring of optical interference radiation in the range 0.9-1.1 μm. This method allows us to eliminate the disadvantages of the frequency selection and to solve the problem more efficiently.

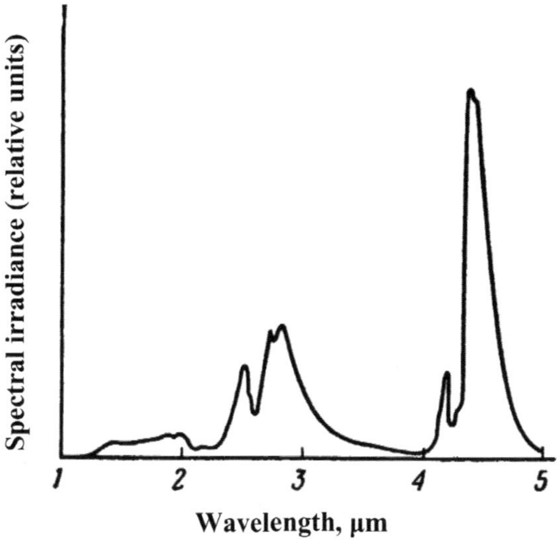

Figure 1. The radiation spectra of a flame.

To improve the fire-fighting optical systems based on the spectral selection it is necessary to increase the spectral sensitivity, response rate and operational reliability of integrated photodetectors. The use of the photovoltaic infrared sensors based on polycrystalline PbSe films and PbSe–CdSe solid solutions is a promising approach. The spectral sensitivity of these photoelectric sensors in the range 1.5-4.7 μm is considered to be satisfactory. These photovoltaic infrared sensors can operate in temperature range from -60°C to +60°C without cooling [1]. The compositional change of semiconductor material allows us to control the spectral characteristics of the photoelectric sensors. So they can detect radiation from one or several types of fires or from interference sources [2].

In this paper we present the experimental results of investigation of mechanism of the photovoltaic effect in photoelectric sensors based on PbSe. This data enables to improve properties of developed photoelectric sensors.

2. The photovoltaic PbSe based sensor technology.

A new photovoltaic sensor technology is developed in research laboratory of JSC "NII Giricond". We have used the thin-film polycrystalline structures based on A^4B^6 materials and A^4B^6-A^2B^6 solid solutions. The main materials for polycrystalline structures are PbSe doping with group V impurities (e. g. Bi) and solid solutions of PbSe and CdSe. Conductivity of these materials is near to intrinsic (this is achieved by the doping with group V atoms). We can change the band gap of the material directionally between 0.2 and 0.6 eV by addition of CdSe to PbSe that allows us to control the spectral characteristics of the photoelectric sensors. The manufacturing technology of different configurations photoresistors and photovoltaic cells with the use of these polycrystalline structures is successfully used in JSC "NII Giricond" [4-6].

The polycrystalline films are created by vacuum thermal evaporation of basic p-type semiconductor material followed by its deposition onto a dielectric glass substrate [1, 5, 7]. High temperature heat treatment (T>300°C) causes the collective recrystallizing of crystallites, their dynamic oxidation and interaction of these oxides with the substrate material. As result the photosensitive grid-type structure covered by oxide-glassy substance arises [7]. Next the Au and In contacts are printed on the surface of photosensitive layer by photolithography. An electric field arises in the clearance between In and PbSe because of electronegativity difference between them. The electric field causes the enhancement of this area by electrons. The energy bands bend downward. N-type area arises. The p-n-junction is formed at the boundary between areas with different types of conductivity (base area). The p-n-junction is necessary for design of self-powered photodetectors operating in photovoltaic mode. The additional heat treatment (200-300°C) boosts the process described above.

3. The experimental technique and the results.

The mechanism of the photovoltaic effect in developed photoelectric sensors has not been fully investigated. It has been previously assumed that the barrier structure formation was caused by the diffusion of indium atoms from the indium contacts to the photosensitive layer during thermal treatment. Based on the received data it can be expected that the photovoltaic effect was induced by the migration of the mobile impurities from the glass substance surrounding PbSe crystallites (mostly Na^+ ions) to the photoelectric layer. The ion mobility increases with temperature. There is a positive charge accumulation on the surface of the crystallites. The Fermi level of PbSe grain is gradually shifted toward conduction band when the crystallites are sufficiently small (dimensions are much smaller than the shielding length $l_S=10^{-4}$-10^{-5} sm). As a result there is conductivity type conversion from p to n [8].

A scheme of the experimental setup for the investigation of the impurity migration is presented in figure 2.

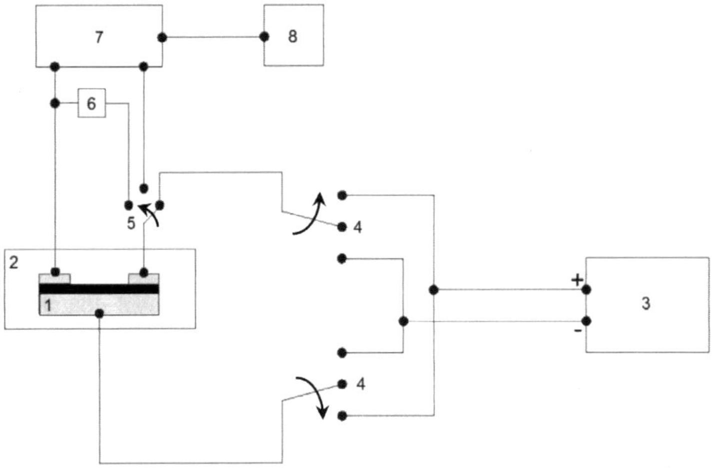

Figure 2. A scheme of the experimental setup: 1 – photoelectric sensor (photoresistor or photovoltaic cell); 2 – thermostat; 3 – voltage power supply; 4 – switches for changing direction of the electric field; 5 – the research mode switch; 6 – multimeter GDM 354 A; 7 – the device for automatic measurement of volt-ampere characteristics; 8 – PC.

Na^+ ions have a positive charge. So we can control its migration by the electric field. The direction of the electric field can be changed by switches 4. The voltage U=250 V from power supply 3 was applied to the investigated photoelectric sensor. The photoelectric sensor was also affected by the temperatures T=120-150°C (the sensor was placed into the thermostat 2).

The investigated photoelectric sensor based on $Pb_{0.96}Cd_{0.04}Se$ (photoresistor) is presented in figure 3. The p-n-junction doesn't arise in the photosensitive layer of this sensor because there is no In electrode and this sensor was not affected by additional heat treatment. But there are mobile impurities in the substrate. To investigate the impurity migration we affected the photoresistor by electric field and temperature and measured the resistance of it (see a scheme of the experimental setup in figure 2).

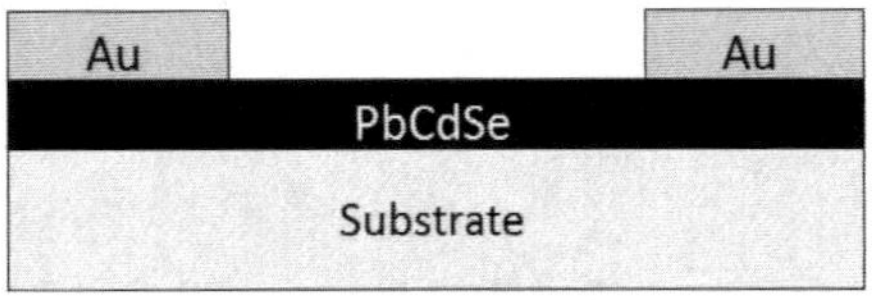

Figure 3. The photoresistor based on $Pb_{0.96}Cd_{0.04}Se$.

The photoresistor was affected by electric field in the following way: a positive voltage from power supply 3 was applied to substrate, a negative voltage to golden (Au) electrode. Then the direction of the electric field was reversed by the switches 4. The dependencies of the photosensitive layer resistance upon time are presented in figure 4.

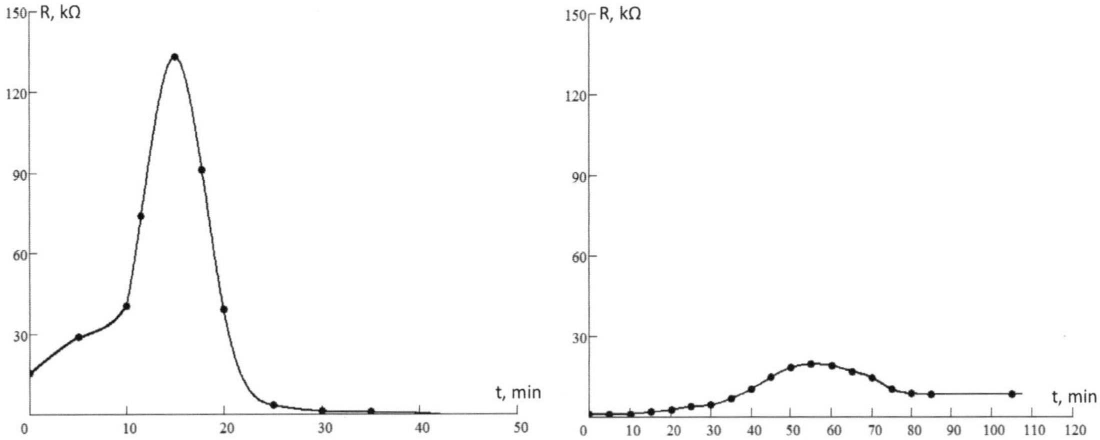

Figure 4. Time dependence of the photosensitive layer resistance: a) a positive voltage is applied to substrate, a negative voltage to golden electrode; b) a negative voltage on substrate, a positive voltage to golden electrode.

Figure 4a shows the increase of the photosensitive layer resistance until it reaches a maximum. Then resistance decreases and reaches values lower than initial. We detected a conductivity type conversion of the photosensitive layer from p to n with use of hot-probe method. Then the direction of the electric field was reversed (switches 4 in figure 2). In figure 4b we can also see the increase of the resistance to a maximum. But its value is less than in figure 4a. Then the resistance decreases to a fixed value. A conductivity type changes from n to p after reaching a maximum. The resistance curve in figure 4b is shallower than in figure 4a.

We explain this process as a result of the migration of the positive charge Na^+ ions moved with the electric field from the substrate to the photosensitive layer. This causes the increase of the photosensitive layer resistance. A conductivity type of the material changes when the ion concentration in the photosensitive layer reaches some value. Then the photosensitive layer decreases with enhancement of the Na^+ concentration because of the increase of the electron concentration. We observe an opposite effect when the direction of the electric field was reversed. There is outflow of positive ions Na^+ from the photosensitive layer, a conductivity type of the layer changes again. The final state of the photoresistor is similar to initial. The obtained results show high mobility of impurities in the developed photoelectric sensor.

The investigated photoelectric cell based on $Pb_{0.96}Cd_{0.04}Se$ is presented in figure 5 (top view). The p-n-junction was formed in the photosensitive layer as a result of an additional heat treatment. The comb structure of the electrodes was developed to expand the junction plane.

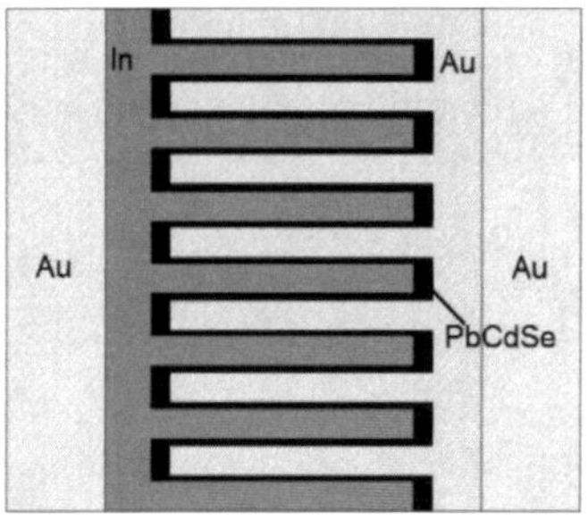

Figure 5. Top view of the investigated photoelectric cell.

The designed photoelectric cell was placed in the thermostat (see figure 2) where it was affected by the electric field for 40 minutes. The electric field was created by a voltage applied to the photoelectric cell (U=250 V). The switch 5 (figure 2) of the experimental setup was switched to a developed device for automatic measurement of volt-ampere characteristics based on a 8-channel ADC. The data from the device was processed by PC. The results of measurement are presented in figure 6.

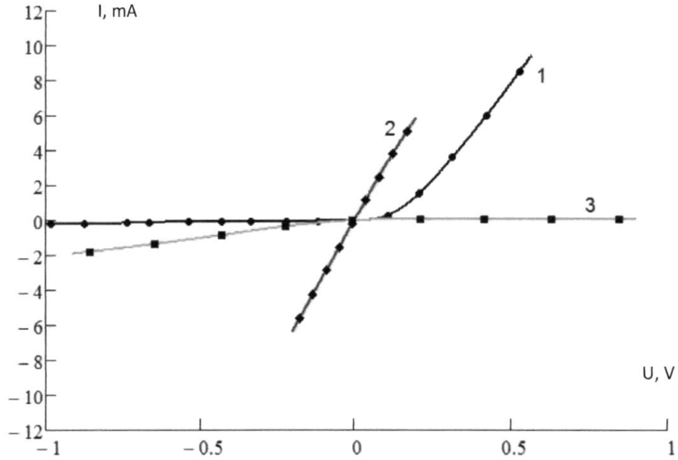

Figure 6. The volt-ampere characteristic of the photovoltaic cell (T=150°C). Graphs 1, 2, 3 correspond to the following voltage values and electric field polarity: 0, +250, -250 V.

The results showed that there is outflow of positive ions Na^+ both from base area and electrode (In) sheath after applying a positive voltage to the photosensitive layer. Thereby the base area resistance decreases. The volt-ampere characteristic becomes practically linear (graph 2). It means that the p-n-junction disappears. There is enhancement of base area and electrode sheath by positive ions after applying a negative voltage to the photosensitive layer (graph 3). The base area resistance increases; the slope of the volt-ampere characteristic decreases. The probability of the reverse conductivity type conversion indicates that diffusion of indium atoms has no dominant role in process of conductivity type conversion of electrode sheath of the photosensitive layer.

Experimental data also indicates that the Fermi level of PbSe grain is shifted by the positive ions Na^+ accumulated near the boundary between grain and dielectric glass substance. Thereby the photosensitive layer resistance increases. Positive charge at the grain boundary increases and this causes a conductivity type conversion. The p-n-junction is formed at the boundary between areas with different types of conductivity.

4. Conclusion

The results obtained in our study indicate that the mechanism of the photovoltaic effect is a result of the migration of the mobile impurities. This enables to improve properties of developed photoelectric sensors by control of the migration of the mobile impurities. But it is also important to determine the configuration of p-n-junction arising in the photosensitive layer and its localization near the In electrode. The solution of the problems will be the goal of our further studies.

References

[1] Medvedev F K, Smirnov A E, Sak A V and Kulagov V B 2014 The Present Nomenclature and Prospects for Future Development and Production of Fire Safety Systems Incorporating Optical IR Sensors *Elektronnaya Promyshlennost'* **2** 67-78

[2] Gorbunov N I, Varpholomeev S P, Diykov L K and Medvedev F K 2005 *Electronics: Science, Technology, Business* **2** 30-33

[3] Hudson R 1969 *Infrared System Engineering* (New York: Wiley–Interscience) p 29

[4] Diykov L K, Varpholomeev S P and Medvedev F K 2004 *Components & Technologies* **6**

[5] Tropina N E *Scientific and Technical Journal of Information Technologies, Mechanics and Optics* **6(64)** pp 47-53

[6] Diykov L K, Medvedev F K, Shelehin Yu L et al 2000 *Electronics: Science, Technology, Business* **6** 26-29

[7] Sak A V, Boyko D B, Tropin A N, Tropina N E 2014 Current State and Future Developments Concerning Photoelectric and Optoelectronic PbSe Based Components in JSC "NII Giricond" *Elektronnaya Promyshlennost'* **2** 99-111

[8] Volkenshtein F F 1960 *The electronic theory of catalysis on semiconductors* (Moscow: Gos. izd-vo fiz.-mat. lit.) p 146

Increasing the noise immunity of optical-electronic systems based on video cameras with an optical converter

M.P. Boronenko, P.Yu. Gulyaev, A.E. Seregin, K.G. Poluhina

Ugra State University, Chehov st. 16, Khanty-Mansiysk, 628012 Russian Federation

e-mail: m_boronenko@ugrasu.ru

Abstract. The luminophor coating of an electro- optical converter afterglow introduces an additional error to the measurement. The ratio that allows to calculate the intensity of spurious illumination at each subsequent frame have been determinate according to experimental data of luminescence kinetics. The proposed method increases the noise immunity of the electro-optical converter by eliminating luminophor afterglow.

1. Introduction

Application of optoelectronic systems based on video cameras with an electro-optical converters in scientific experiments [2], related to the problem of low sensitivity at high speed shooting and low exposures. The afterglow time of luminophor, covering the screen of OEC, presence is another problem. For some fields fast processes research the screen residual glow is the main limiting factor [1].In different with low speed experiment where afterglow can be cleaned by several miniature lamp brief (5-10 seconds) screen backlight [4, 5] the proposals for solving this problem for high speed experiment without changing the hardware of the OEC were not found.

2. Experimental techniques

Experimental stand, which is composed of following equipment (fig. 1a): 1 - optical bench; 2 - sparkler; 3 – high speed camera VideoSprint based on 4 electron-optical converter (EOC); 5 – PC, was build for accomplish the task. Self-luminous sparkling fire was used as source of luminescence particles provoking glow luminophor. Flying glowing particles are shifted by a distance depending on their velocities during the accumulation of charge in the light-sensitive sensor area. Stroboscopic effect achieving by additional EOC shutter firing on the interval photocathode – micro channel substrate. At the same time on the EOC screen the number of images equal to the number of additional shutter releases N is forming. This track of moving particle is divided into N parts. The average brightness of each track part is forming with contribution of particles radiation which was absorbed by photosensitive element during the exposure time. It is obvious that in the filmed sequence of combustion particles frames, the brightness of the left track will fade gradually from frame to frame, if the time of luminophor afterglow is longer than frame interval. According to that it becomes possible to study the kinetics of luminophor afterglow.

3. Luminophor afterglow time measurement

To determine the decay time [3] the search for a sequence of frames, which was clearly evident gradual extinction of the moving object luminous trail registered in the image, was carried out. The study track must be chosen to avoid crossing with other particles tracks precipitating into the viewing area of camera. Afterglow luminophor observed in a frame sequence and the brightness profile line is shown in Figure 1. b). The luminescence intensity of the luminophor corresponds to the brightness of the pixels constituting the tracks. The measurements were carried out by the ImageJ program. The brightness measurement of photosensitive cells, which registering radiation one of the particle track part, is held in each frame where the track is still visible. The measured brightness was normalized by the maximum value. The measured data of time depended changing the track brightness shown in Figure 2. The coefficient, characterizing residual luminophor luminescence from frame to frame decreasing, is possible to determine by the plot.

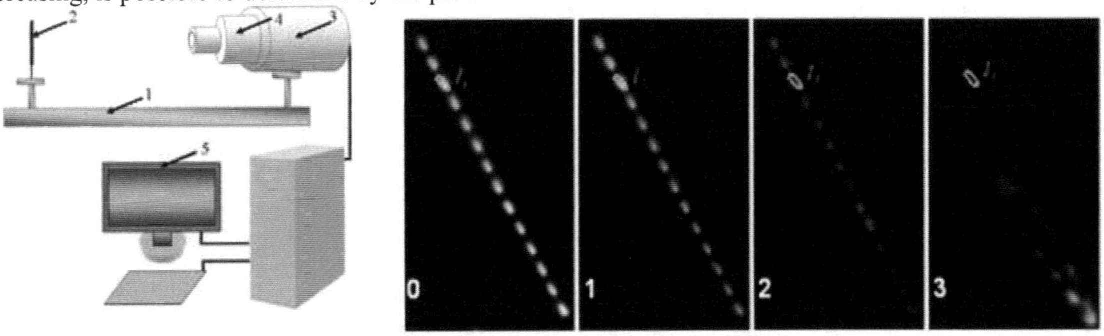

Figure 1. The experimental setup (a), observation of luminophor afterglow

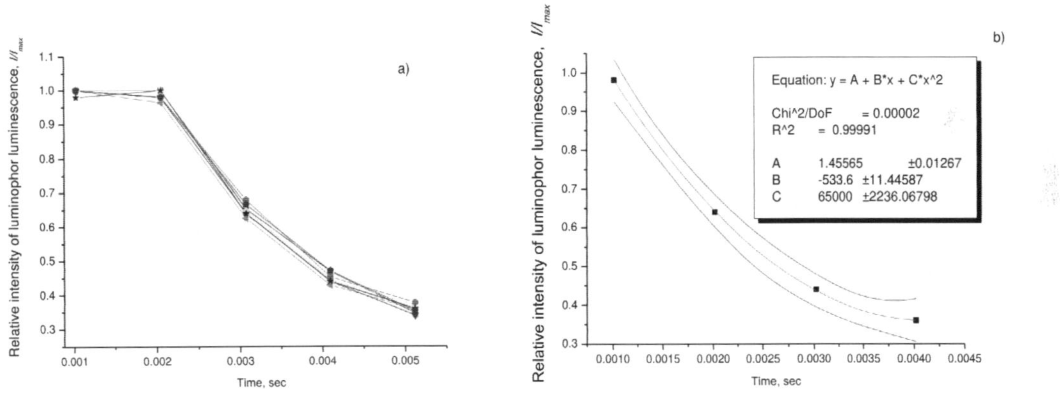

Figure 2. The time dependence of the residual luminescence brightness (a), the experimental data approximation (b).

4. Electro-optical converter noise immunity increasing method

In order to reduce the error introduced by the luminophor afterglow, it is necessary to subtract remaining afterflow background of previous frame from the study frame. Drawings explaining the process of removing the residual luminescence OEC screen shown in Figure 3.

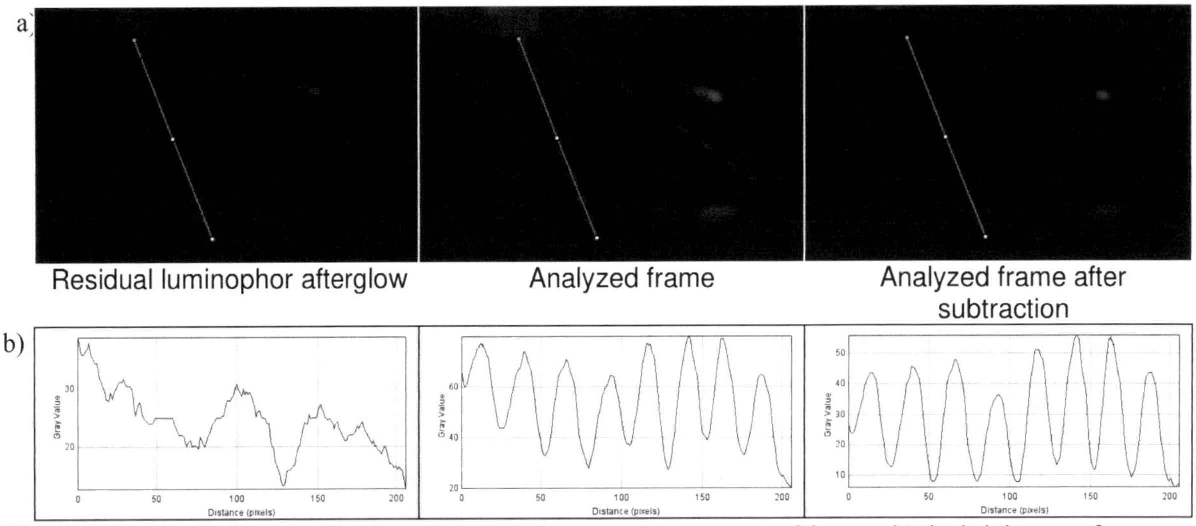

Figure 3. Removal of residual luminescence OEC screen: a) a sequence of frames; b) the brightness of pixels along the profile line.

Let registered by each detector brightness on the first frame I_1 has noise I_{noise} components and useful signal I_{01}:

$$I_1 = I_{noise} + I_{01},$$

Then the next frame is due to I_{noise}, I_{02} and afterglow of luminophor I_{1res} :

$$I_2 = I_{noise} + I_{02} + I_{1res},$$

And I_{1res} may be calculated from the calculated law of luminophor luminescence decay. Using inter frame subtraction, the accuracy of the tracks brightness measuring on the 30% can be achievable:

$$I_{1res} = \mu I_1,$$

$$I_2 = I_{noise} + I_{02} + I_{1res} - \mu I_1$$

where μ is the determined from the graph and depending on the type of luminophor afterglow decay coefficient.

The brightness measurement accuracy increasing method performed in ImageJ program consists four items:

1. Creation a duplicate of processed frame. This step creates a new window containing a copy of the active frame or rectangular selection on it. For stack, specify a range of channels (c) fragments (z) and shots (t) is possible.

2. Deleting the last frame in original file and first frame of duplicate file.

3. In accordance with the luminescence decay graph the coefficient, which is necessary to multiply the brightness value of each pixel in subtracted file, is selected. Changing the brightness of each pixel is performed by using command "Macro". This command performs arithmetic operations on each pixel of the image, using the user specified equation. This tool can be used to create a completely new image, or perform precise pixel manipulation of existing images.

4. Inter frame subtraction (substacting the original file from duplicate) is carried out in the image calculator.

5. Conclusion

The proposed method of increasing the noise immunity of the OEC allows to consider remaining afterglow in the study frame, subtract it, thereby increasing the accuracy of brightness temperature measurements up to 30% without changing the hardware. The method is limited to the specific type of the luminophor and the detected objects radiation intensity.

References

[1] Antonenkov D. A. Sredstva poluchenija informacii o parametrah vzvesi, osnovannye na obrabotke fotoizobrazhenij //Vostochno-Evropejskij zhurnal peredovyh tehnologij. – 2014. – №. 4 (9). – P. 15-20.

[2] Boronenko M.P., Guljaev P.Ju. Metrologicheskie harakteristiki televizionnoj izmeritel'noj sistemy nanosekundnogo razreshenija//Vestnik Jugorskogo gosudarstvennogo universiteta, Informacionno-izdatel'skij centr Jugorskogo gosudarstvennogo universiteta, Chanty-Mansiysk, — 2014. — №2(33). — P. 68-73

[3] Poluhina K.K. Exploration of time of afterglow of luminescent screen electro - optical of converter / K.K Poluhina, M.P. Boronenko. — Proceedings of the 3rd International conference on Eurasian scientific development. «East West» Association for Advanced Studies and Higher Education GmbH. Vienna. — 2014. — P. 112-117.

[4] Hirvonen L. M. et al. Sub-μs time resolution in wide-field time-correlated single photon counting microscopy obtained from the photon event phosphor decay //New Journal of Physics. – 2015. – T. 17. – №. 2. – P. 023032.

[5] Marchenko S. V. Spectral variability of Wolf-Rayet stars. – AN Ukrainskoj SSR, Kiev. Inst. Teoreticheskoj Fiziki, 1987.A reference

Journal of Physics: Conference Series **643** (2015) 012029 doi:10.1088/1742-6596/643/1/012029

The new ultra high-speed all-optical coherent streak-camera

R.M.Arkhipov[1,2], M.V.Arkhipov[1], V.S.Egorov[1], I.A.Chekhonin[1], M.A.Chekhonin[1], S.N.Bagayev[3]

[1]St.Petersburg State University, St.Petersburg, Russia
[2]Weierstrass Institute for Applied Analysis and Stochastics, Leibniz Institute in Forschungsverbund Berlin e.V, Berlin, Germany
[3]Institute of Laser Physics, Novosibirsk, Russia

E-mail: arkhipovrostislav@gmail.com

Abstract. In the present paper a new type of ultra high-speed all-optical coherent streak-camera was developed. It was shown that a thin resonant film (quantum dots or molecules) could radiate the angular sequence of delayed ultra-short pulses if a transverse spatial periodic distribution of the laser pump field amplitude has a triangle shape.

1. Introduction

For solving various problems of data transfer and processing, it is necessary to change at high speed the direction of propagation of laser radiation in space. The problem of rapid angular laser beam deflection is one of the most difficult problems of controlling the parameters of laser radiation. Nowadays mirror galvanometers and acousto-optic deflectors are widely used. The ratio between the angle of deviation α and setting time τ for mirror galvanometers and acousto-optic deflectors has order of magnitude $90^{o}/10^{-4}$ sec and $2^{o}/10^{-6}$ sec, respectively.

In this paper we address the following question: how to increase angular deflection rate? In the present work, we consider all-optical angular deflectors of laser radiation. To date it has been suggested several types of deflectors with different principles of action. In Refs. [1-8] the optical deflector using the effects of light-induced transverse gradient index of refraction $n(x)$ of the medium, leading to a deflection of the laser beam propagating through an optically inhomogeneous medium was studied. In Refs. [9,10] an effect of optically steering phase matching in the multiwave parametric excitation in photonic crystals has been analyzed. Effect of all-optical steering of light was considered for gazes of three-level and four-level atoms. This effect is based on the phenomenon of electromagnetically induced transparency (EIT) under inhomogeneous optical pumping of the medium leading to the changes of the index of refraction [11-13].

In [14] we have proposed for the first time a new principle of the ultra high-speed all-optical coherent streak-camera and deflectors, which uses the effect of self-diffraction of ultrashort coherent pulse on the grating of polarization $P(t,x)$ of resonant two-level medium with an optically controlled grating pitch. In [15], this principle was generalized for the pump-probe method using the diffraction effect of the probe pulse on the grating of the population difference $N(t,x)$ induced by the pump pulse in the resonant medium.

The using of the resonant media seems to be promising. One can use resonant media based on atoms, molecules, excitons and quantum dots. Coherent laser radiation can rapidly change the polarization of the medium $P(t,x)$ and the level populations $N(t,x)$ in these media. The ratio of the deflection angle α to the setting time τ can reach the value $10^{o}/10^{-11}$ sec, which is much greater than angular deflection rate of mirror galvanometer and acousto-optic deflectors.

2. All-optical coherent streak-camera: a new principle of operation

The principle of operation of a coherent deflector is based on the sequence in time of the excitation of nonlinear spatial harmonics of the polarization $P(t,x)$ and the population difference $N(t,x)$ in a thin layer of resonant medium. A diffraction of the incident pulse on the grating leads to the appearance of the sequence of n pulses separated in time and each of them is emitted at angle α_n.

In the first case, laser pumping pulse is divided into the series of pulses with variable direction of the propagation in space due to self-diffraction, see Fig. 1. In the second case, illustrated in Fig.2 pumping pulse creates in the resonant layer periodic grating of population $N(t,x)$ which is used for the diffraction of the probe pulse.

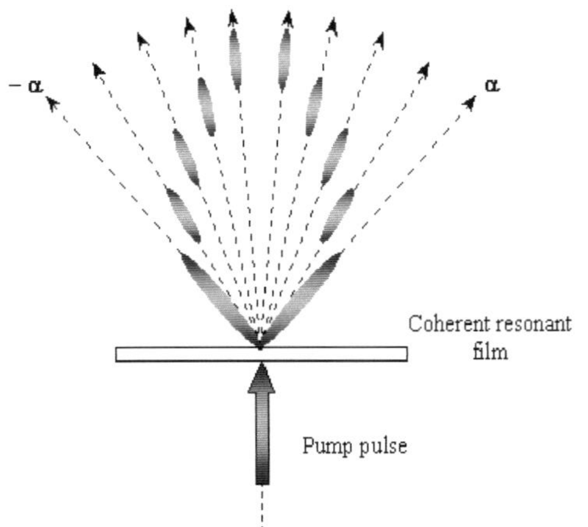

Figure 1. Dynamics of the angular distribution of the film radiation under the self-diffraction of the pumping pulse

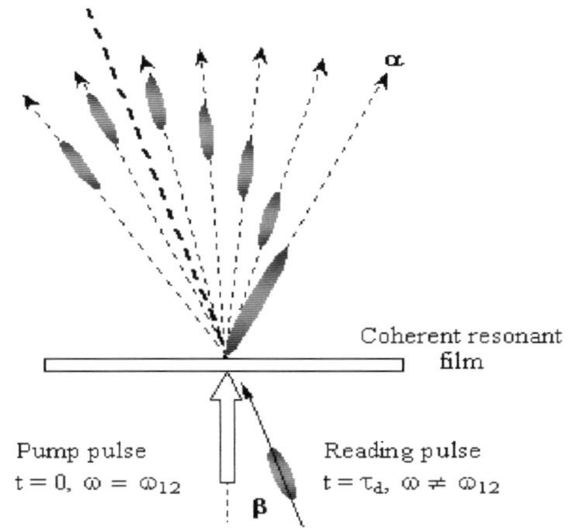

Figure 2. Dynamics of the angular distribution of the film radiation under the diffraction of the probe pulse

The angular deflection can be achieved by laser excitation of a resonant layer with a periodic "sawtooth" transverse spatial profile of the field, $E(t,x) = \varepsilon(t)saw(x)$ with a period Δ. Graphic of the unit function $saw(x)$ ($|saw(x)| \leq 1$) is illustrated in Fig.3.

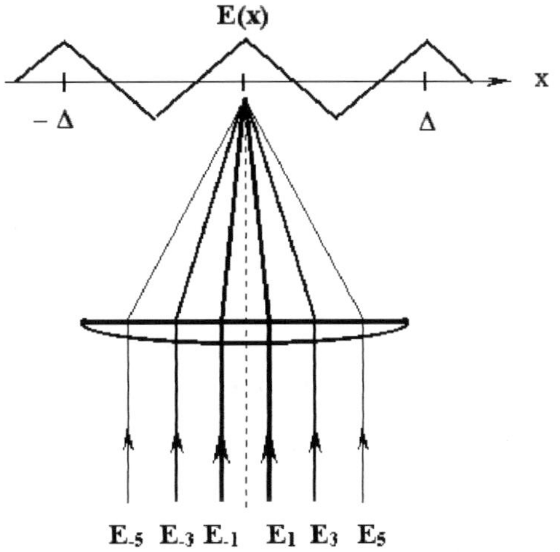

Figure 3. Fourier synthesis of the spatial distribution of the electric field $E(t,x) = \varepsilon(t)saw(x)$
The ratio of the wave amplitudes - $E_{\pm1}$: $E_{\pm3}$: $E_{\pm5}$: $E_{\pm7}$… = 0.8106: 0.0901: 0.0324: 0.0165…

The pumping field for obtaining "sawtooth" distribution $E(t,x) = \varepsilon(t)saw(x)$ can be simply realized as illustrated in Fig.3. Due to the interference of the wave pairs electric field profile $E(t,x)$ formed in the focal plane of the lens can be described by the Fourier series (1):

$$\mathrm{saw}(x) = \sum_{m=1}^{\infty} a_m \cdot \cos(Q \cdot x \cdot m) \tag{1}$$

$$a_m = 2 \cdot \mathrm{sinc}\left(\pi \cdot \frac{m}{2}\right)^2, \tag{2}$$

where $Q=2\pi/\Delta$ и $sinc(Z)=sin(Z)/Z$. This Fourier synthesis is effective due to the fast decreasing of the Fourier series components (2) - a_1:a_3:a_5:a_7 ... = 0.8106: 0.0901: 0.0324: 0.0165 ...

3. Dynamics of radiation emitted by a grating of resonant medium polarization

To calculate the dynamics of the medium polarization $P(t,x) = d_{12}N_o(u(t,x)+iv(t,x))$ and population difference of the layer interacting with the field $E(t,x) = \varepsilon(t)saw(x)$ we used a numerical solution of the system of optical Bloch equations [16]:

$$\frac{d}{dt}u(t,x) = -\Delta\omega \cdot v(t,x) - \frac{1}{T_2} \cdot u(t,x) \tag{3}$$

$$\frac{d}{dt}v(t,x) = \Delta\omega \cdot u(t,x) - \frac{1}{T_2} \cdot v(t,x) + \Omega_R(t,x) \cdot w(t,x) \tag{4}$$

$$\frac{d}{dt}w(t,x) = -\frac{1}{T_1} \cdot (w+1) - \Omega_R(t,x) \cdot v(t,x). \tag{5}$$

Here N_0 - particle density in the layer, $u(t,x)$, $v(t,x)$ - components of the medium polarization, $w(t,x)$ - population difference per unit atom, $\Delta\omega$ - frequency detuning of the field frequency from transition frequency ω_{12} of the two-level particles d_{12} - transition dipole moment, T_1 - population difference relaxation time, T_2 - polarization relaxation time, $\Omega_R(t,x) = d_{12}E(t,x)/\hbar$ – Rabi frequency of the pumping pulse.

In particular case when detuning $\Delta\omega = 0$ one can neglect medium relaxation times, $T_1 = T_2 = \infty$, the system of optical Bloch equations (3)–(5) can be solved analytically using local "area" of the pumping pulse $\Theta(t,x)$ [17, 18]:

$$\Theta(t,x) \equiv \frac{d_{12}}{\hbar} \cdot \int_{-\infty}^{t} E(t',x)dt'. \tag{6}$$

The analytical solution of (3)-(5) allows one to obtain the following expressions for the medium polarization $P(t,x)$ and population difference $N(t,x)$ in the form [17, 18]:

$$P(t,x) = d_{12}N_0 \sin(\Theta(t,x)) = d_{12}N_0 \sin(\Theta(t)saw(x)) \tag{7}$$

$$N(t,x) = N_0 \cos(\Theta(t,x)) = N_0 \cos(\Theta(t)saw(x)). \tag{8}$$

From (7)-(8) one can immediately see that current pulse area $\Theta(t)$ has a significancy of current spatial frequency of the polarization grating and population difference grating $N(t,x)$. Spatial period of the both grating depends on time. Therefore, one can expect an occurrence of the radiation of different diffraction orders emitted by polarization $P(t,x)$ grating with a time delay, see Fig.1.

Spatial Fourier spectrum of the resonant polarization $P(t,x)$ contains only odd harmonics of grating wave vector $Q = 2\pi/\Delta$: $Q_n = (2n+1)Q$, $n = 0,1,2\ldots$ - integer number. The basic idea of the streak camera (or deflector) can be described in the following way. Each of the spatial harmonics of the polarization $P_n(t)$ radiates in series in the far-field zone a short pulses, propagating at the angle which is satisfying to the Vavilov-Cherenkov condition, $\alpha_n = arcsin(Q_n/k)$, where $k = 2\pi/\lambda$ - is the wave number.

Therefore, self-diffraction of the pumping pulse looks as the separation of it on the series of n short pulses each of them is propagating at his own angle α_n (Fig. 1).

Fig.4 illustrates an evolution of the spatial Fourier spectrum of the polarization $P(t,x)$ for the unipolar pumping pulse having Gaussian shape $\varepsilon(t) = \varepsilon_0 \, exp(-(t/\tau_p)^2)$ with duration $\tau_p = 25$ psec and total pulse area $\Theta_\infty = 19.5\pi$. It is clearly seen the appearance of the harmonics of the polarization $P_n(t)$ successive in time with wave vectors $Q_n = (2n+1)Q$.

A supreme value of the deflection angle α_{max} for the unipolar pumping pulse is determined only by the total pulse area, $\Theta_\infty = \Theta\,(t{=}\infty)$: $\alpha_{max} = arcsin(2\Theta_\infty Q/\pi k)$. Similar situation for the bipolar pumping pulse (breather) with total pulse area $\Theta_\infty = 9.75\,\pi$ is illustrated in Fig. 5.

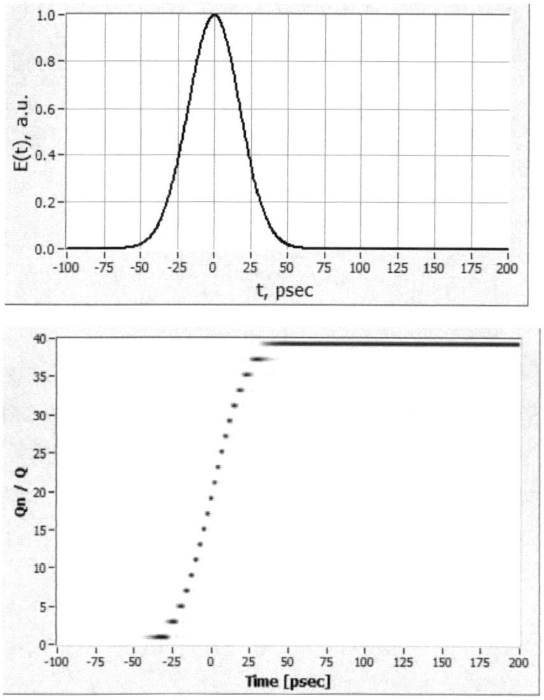

Figure 4. Dynamics of the Fourier spectrum of the polarization $P(t,x)$ (low figure) under pumping by unipolar pulse (upper panel) $E(t,x) = \varepsilon(t)saw(x)$, $\Delta\omega = 0$, $T_1 = T_2 = \infty$, $\Theta_\infty = 19.5\,\pi$

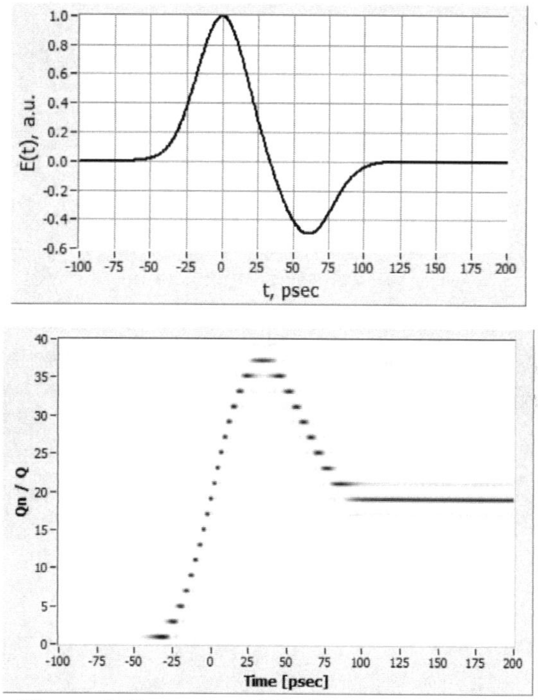

Figure 5. Dynamics of the Fourier spectrum of the polarization $P(t,x)$ (low figure) under pumping by bipolar pulse (upper panel) . $E(t,x) = \varepsilon(t)saw(x)$, $\Delta\omega = 0$, $T_1 = T_2 = \infty$, $\Theta_\infty = 9.75\,\pi$

4. Dynamics of the radiation emitted by resonant polarization grating under realistic conditions

It should be noted that sawtooth profile of the pumping field $E(t,x) = \varepsilon(t)saw(x)$ is not practical because of the using of the infinite numbers of interfering harmonics in Fourier series (1). However, as it was mentioned above, the components of the series (2) rapidly decrease with the increase of the index m. Therefore, to create the sawtooth profile of the pump field one can take a finite number of interfering waves.

Fig.6 illustrates the dynamics of the resonant polarization radiation in the case when three pairs of the waves are used - $E_{\pm 1}$, $E_{\pm 3}$ and $E_{\pm 5}$. It is clear that dynamics of emission from resonant layer is almost ideal for the sawtooth pumping $E(t,x) = \varepsilon(t) saw(x)$.

Note, that in the case of pumping by one pair of waves $E_{\pm 1}$ the distribution of the pumping field has a form $E(t,x) = \varepsilon(t)\cos(Qx)$ and angular deflection does not occur. In this case, we have a huge amount of diffraction orders emitted by the polarization grating. It was observed experimentally with sodium vapors [19].

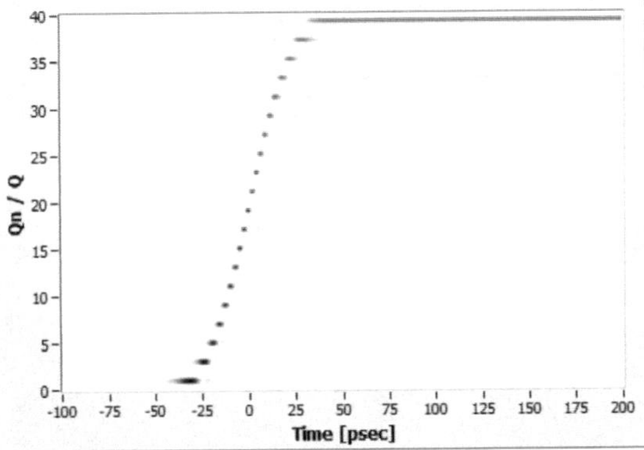

Figure 6. Dynamics of Fourier spectrum of polarization $P(t,x)$. $\Delta\omega = 0$, $T_1 = T_2 = \infty$, $\Theta_\infty = 19.5\,\pi$,
$E_{\pm 1}$: $E_{\pm 3}$: $E_{\pm 5}$ = 0.8106: 0.0901: 0.0324

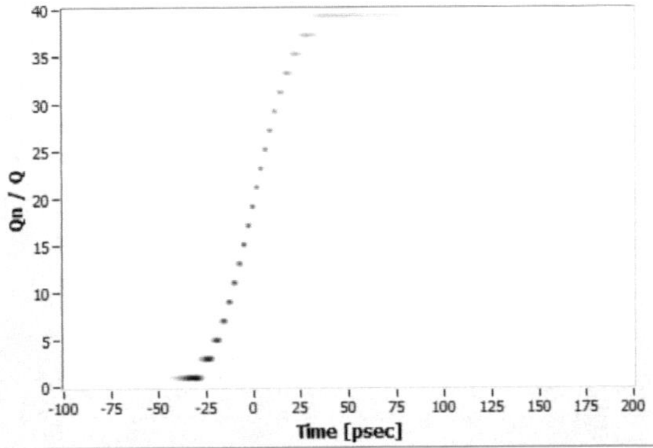

Figure 7. Dynamics of Fourier spectrum of polarization $P(t,x)$. $E(t,x) = \varepsilon(t) saw(x)$,
$\Delta\omega = 0$, $T_1 = 1$ nsec, $T_2 = 100$ psec,
$\Theta_\infty = 19.5\,\pi$

Effect of coherent angular deflection was also investigated in the general case when relaxation times T_1 and T_2 have finite values, see Fig.7. It is seen from this figure that relaxation leads only to monotonic decrease of the amplitude of high orders of diffraction and effect of angular deflection is preserved.

We have found that greatest negative impact on the decrease of the emission field amplitude with deflection angle α_{max} provides the frequency detuning $\Delta\omega$, see Fig. 8. In this case it is necessary to use pump-probe method [15] (Fig. 2). Linear and nonlinear diffraction of a probe pulse on the resonant harmonic grating of population difference $N(x)$ was considered in detail in [20].

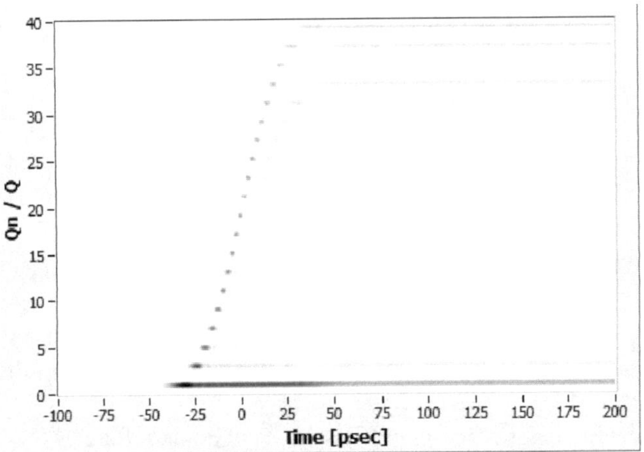

Figure 8. Dynamics of Fourier spectrum of polarization $P(t,x)$. $E(t,x) = \varepsilon(t) saw(x)$,
$\Delta\omega = 0.05\,\Omega_R$, $T_1 = T_2 = \infty$, $\Theta_\infty = 19.5\,\pi$

Situations considered above are valid for optically thin layer of resonant medium and the propagation of the pumping pulse in the layer was not taken into account. In general case it is necessary to include a finite length of the medium and diffraction of the optical field during its propagation in the extended medium. Numerical simulations for this case were performed by us using general system of Maxwell-Bloch equations in the slowly varying envelope approximation of the electric field, polarization and population difference in a two-level extended resonant medium. Numerical simulations showed that the including of the diffraction leads to the spreading of spatially periodic structures of polarization and population difference. Therefore, it is necessary to use thin optical layers with large values of relaxation times.

5. Acknowledgements

The authors acknowledge Saint-Petersburg State University for a research grant 11.0.41.2010. This work was partially supported by EU FP7 ITN PROPHET, Grant No. 264687.

References

[1] C.H.Sarantos, J.E.Heebner 2010 Opt. Lett. **35** (9) 1389-1391.
[2] Y.Li, D.Y.Chen, L.Yang, R.R.Alfano 1991 Opt. Lett. **16** (6) 438-440.
[3] J.Hubner, H.M. van Driel, J.S.Aitchison 2005 Opt. Lett. **30** (23) 3168-3170.
[4] S.Shwartz, M.Segev, U.El-Hanany 2004 Opt. Lett. 2004 **29**, (7) 760-762.
[5] G.A.Swartzlander, Jr., H.Yin, A.E.Kaplan J. Opt. Soc. Am. B. 1989 **6** (7) 1317-1325.
[6] H.-S.Albrecht, P.Heist, J.Kleinschmidt, D.V.Lap Appl. Phys. B 1993 **57**, 193-197.
[7] S.Blair, K.Wagner Appl. Opt. 1999 **38** (32) 6749- 6772.
[8] A.Dreischuh, E.Eugenieva, S.Dinev IEEE Journal of Quant. Electron. 1994 **30** (7) 1656-1661.
[9] T.Ellenbogen, A.Ganany-Padowicz, A.Arie Opt. Express 2008 **16** (5) 3077-3082.
[10] S.M.Saltiel, Y.S.Kivshar Opt. Lett. 2002 **27** (11) 921-923.
[11] C.Hang, V.V.Konotop Phys. Rev. A 2010 **81** 053849.
[12] O.N.Verma, T.N.Dey Phys. Rev. A 2015 **91** 013820.
[13] Q.Sun, Y.V.Rostovtsev, M.S.Zubairy Phys. Rev. A 2006 **74** 033819.
[14] R.M.Arkhipov, S.N.Bagayev, V.S.Egorov, I.A.Chekhonin, M.A.Chekhonin. Proceedings of the VI Intern. conf. HOLOEXPO-2009 81-86 Kiev Ukraine 01-02 July 2009.
[15] R.M.Arkhipov, M.V.Arkhipov, S.N.Bagayev, V.S.Egorov, I.A.Chekhonin, M.A.Chekhonin Scientific notes of the Physics Department of Moscow State University 2013 **5** 135034 (1-4).
[16] L.Allen, J.H.Eberly *Optical Resonance and Two-Level Atoms.* John Wiley, New York, 1975.
[17] V.S.Egorov, I.A.Chekhonin Zh. Tech. Phys. 1986 **56** (3) 572-574. [Sov. Phys.-Tech. Phys. 1986 **31** 344].
[18] V.S.Egorov, I.A.Chekhonin, N.N.Shubin Opt. Spektrosk. 1987 **62** (4) 853-859. [Opt. Spectr. 1987 **62** (4) 509-513].
[19] N.Tan-no, K.Ohkawara, H.Inaba Phys. Rev. Lett.. 1981 **46** (19) 1282-1285.
[20] R.M.Arkhipov, I.Babushkin, M.K.Lebedev, Yu.A.Tolmachev, M.V.Arkhipov Phys. Rev. A. 2014 **89** 043811.

Silicon photodiode as the two-color detector

D B Ponomarev[1] and V A Zakharenko[2]

[1] Senior lecturer, OmSTU, Omsk, Russia

[2] D., Professor, OmSTU, Omsk, Russia

E-mail: dimi34@gmail.com

Abstract. This paper describes a silicon photodiode as the two-color photodetector. The work of one photodiode in two spectral ranges is achieved due to the changes of the spectral sensitivity of the photodiodes in the transition from photodiode mode for photovoltaic in the short circuit mode. On the basis of silicon photodiode FD-256 the layout of the spectral ratio pyrometer was assembled and the results of theoretical calculations was confirmed experimentally. The calculated dependences of the coefficient of error of the spectral ratio pyrometer from temperature reverse voltage 10 and 100 V was presented. The calculated dependence of the instrumental error and the assessment of methodological errors of the proposed photodetector spectral ratio was done. According to the results of the presented research was set the task of development photodiode detectors which change the spectral sensitivity depending on the applied voltage.

1. Introduction

Currently, the spectral relations pyrometry is used "tandem" and mosaic receivers radiation with a broad band of spectral sensitivity. "Tandem" photodiodes have a two-layer structure, wherein the allocated plots of spectral sensitivity are negligible spectral overlap of the intervals, which determines the methodical error of measurement associated, primarily, with the changes of spectral and integral emissivity $\varepsilon(T,\lambda)$ of the measurement object. When mosaic detectors using spectral selection is performed at the expense of the photosensitive elements with different spectral sensitivities, secured or optical filters or the use of radiation detectors with different spectral sensitivities. The photodetectors are arranged in one plane. Reduction of truncation error in the latter case is possible due to the convergence and spectral overlapping working ranges [1].

In this paper we propose to implement a radiation sensitive element for spectral pyrometer relationship on one photodiode without the use of emit spectral plots of optical filters [2].

The work of one photodiode in two spectral ranges is achieved due to the changes of the spectral sensitivity of the photodiodes in the transition from photodiode mode with applied reverse voltage to the photovoltaic in the short circuit mode. According to the physics of the operation of p-n junction photodiode [3] with the increase of the reverse voltage applied to the p-n transition in photodiode mode, grow the width of the space charge region, the height of the potential barrier and increases the width of the p-n junction. This leads to an increase of the coefficient of collecting minority charge carriers and, consequently, to increase the maximum spectral sensitivity and its shift in the long wavelength region. The change in the characteristic of the spectral sensitivity of the silicon photodiode on the applied voltage, according to [3], is presented in Fig. 1.

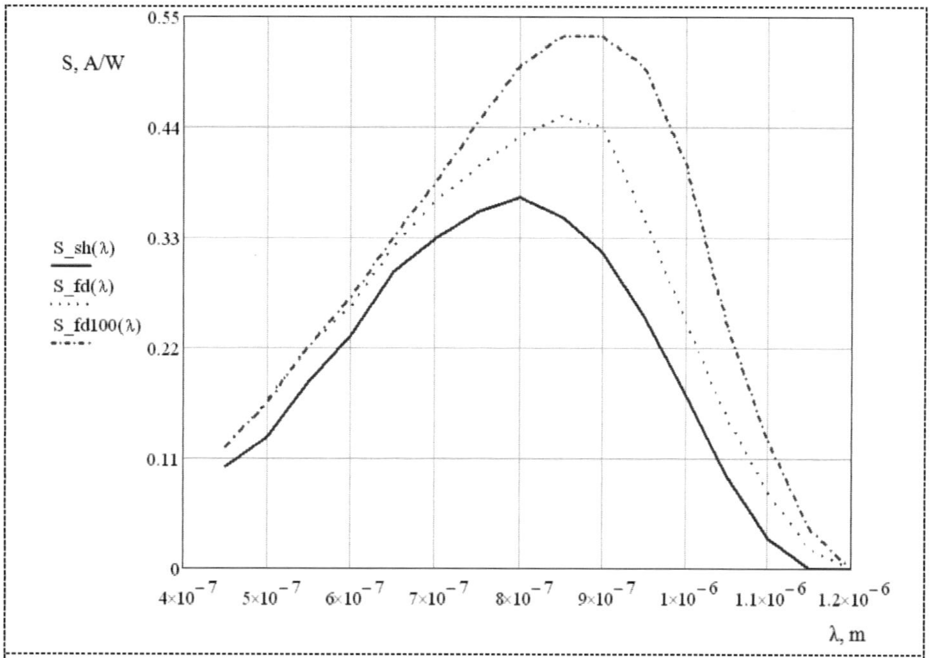

Figure 1. Changes of the spectral sensitivity of the silicon photodiode FD-256: S_sh - when the voltage on the photodiode Upn=0; S_fd - when Upn= 10 V, S_fd100 - when Upn= 100 V.

Therefore, it is proposed in the pyrometric device during one measurement consistently use the inclusion of a photodiode with a maximum reverse voltage and the voltage is zero (short circuit). Thus the output signal of the pyrometer to form as the ratio of the output signals obtained at different modes of inclusion.

2. Calculation justification

Theoretical calculations of the flow characteristics of the pyrometer for each of the modes of the photodiode based on the Planck equation and experimental functions changes the spectral sensitivity of the photodetector shown in Fig.1.

For two spectral intervals λ1 - λ2 and λ3 - λ4 functions of the spectral sensitivities of the photodetectors output signal of the pyrometer in accordance with the Planck function is represented as

$$K = \frac{I_{f1}(T)}{I_{f2}(T)} = \frac{\int_{\lambda 1}^{\lambda 2} S_1(\lambda) r(\lambda, T) d\lambda}{\int_{\lambda 3}^{\lambda 4} S_2(\lambda) r(\lambda, T) d\lambda}, \qquad (1)$$

where $r(\lambda, T)$ is the Planck function; $I_{f1}(T)$ and $I_{f2}(T)$ is the photocurrent for different inclusions photodiode; $S_1(\lambda)$ and $S_2(\lambda)$ is a function of the spectral sensitivity of the photodetector.

The share of total energy luminosity converted by a photodetector into an electrical signal photocurrent in the photodiode modes and short-circuited, can be calculated using the product of

$$I_{fi}(T) = A \cdot \int_{0.45 \cdot 10^{-6}}^{1.2 \cdot 10^{-6}} S_i(\lambda) \cdot r(\lambda, T) d\lambda, \qquad (2)$$

where A is the aperture coefficient taking into account the ratio of the geometric parameters between the receiver and the emitter.

In accordance with the expression (2) were obtained by calculation of the dependence of the photocurrent on the temperature of the radiating blackbody cavity for silicon photodiode FD-256. The calculated dependence of the ratio of the output signals obtained as

$$K(T) = \frac{I_{sh}(T)}{I_{fd}(T)} \quad (3)$$

is shown in Fig. 2, where *Ish(T)* is the photocurrent in the short circuit mode; *Ifd(T)* is the photocurrent in the photodiode mode.

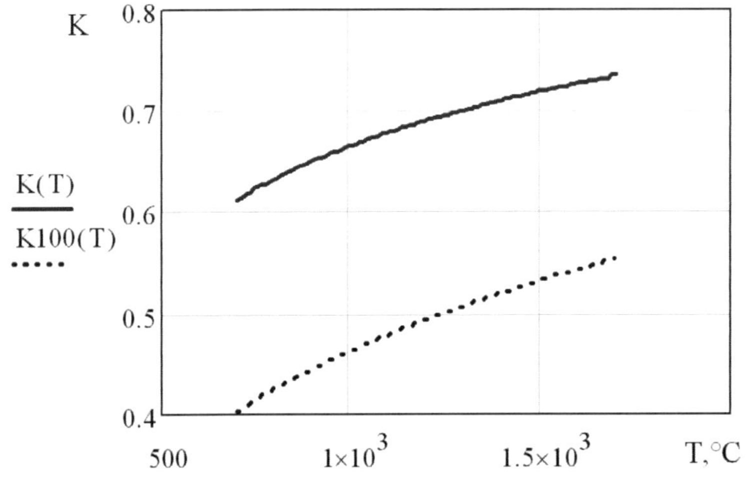

Figure 2. The calculated function according to the relationship of the output signals of the pyrometer from the temperature of the blackbody cavity: K(T) - Upn= 10 V, K100(T) - Upn= 100 V.

3. The results of experimental studies

On the basis of the photodiode FD-256 the layout of the pyrometer was assembled and produced experimental studies of the dependency of K functions on T. In Fig. 3 shows the experimental dependence of the normalized output signal of the pyrometer on the basis of silicon photodiode FD-256 from the cavity temperature emitter type blackbody model. When the experiment was used model of a black body MTP-2M-50-500 manufactured by the "Etalon" (Omsk). The output signal of the pyrometer was measured in accordance with the expression (3). Thus the experimental value of the reverse voltage in the photodiode mode Upn was 10 V.

4. Estimation of errors

According to [1] the accuracy of the pyrometer spectral ratio increases with the decrease of the slope of the conversion. The steepness of the conversion is determined by private derivative with respect to T relationship pyrometric signal $K(T) = I_{f1}(T)/I_{f2}(T)$. The increase in error can be evaluated using the coefficient kp. When measuring the temperature of grey bodies, this coefficient is calculated as:

$$kp(T) = \frac{K(T)}{T \cdot \left(\frac{d}{dT} K(T) \right)} \quad (4)$$

The coefficient kp shows how many times the measurement error in the spectral ratio pyrometer. In Fig. 4 shows the calculated dependences of the coefficient kp of the temperature of the blackbody for spectral pyrometer relations on the basis of a silicon photodiode FD-256, calculated in accordance with expression 4.

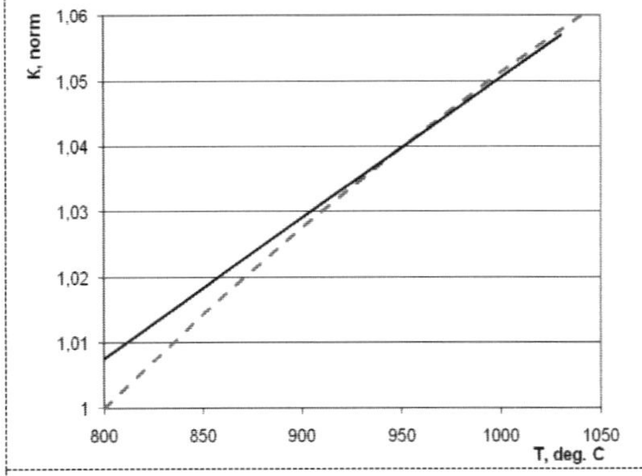

Figure 3. Dependence of the output signal of the pyrometer on the basis of silicon photodiode FD-256 temperature: calculate (- - -), experiment (—).

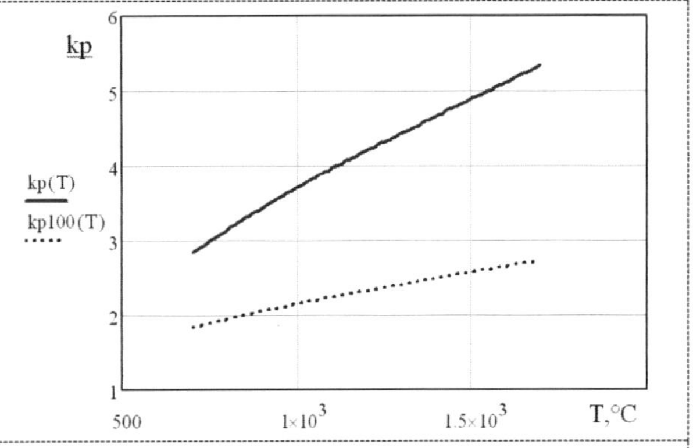

Figure 4. The calculated function of the dependency of the coefficient kp of the temperature of the blackbody cavity: kp(T) - Upn= 10 V, kp100(T) - Upn= 100 V.

Relative instrumental error δTi determining the temperature of the pyrometer spectral ratio in accordance with [1] can be found by the expression:

$$\delta Ti = kp \cdot \sqrt{(\delta Tr)^2 + \left[\frac{\Delta I_{1H}}{I_1}\right]^2 + \left[\frac{\Delta I_{2H}}{I_2}\right]^2} \qquad (5)$$

where δTr is the relative error caused by radiation noise of the received radiation, the intrinsic noise of the photodetector and the dependence of the dark current of the photodetector temperature. The calculation of this error for a silicon photodiode FD-256 gives a small amount $\delta Tr = 3 \cdot 10^{-5}$. The instability of amplification channels $\Delta I_H/I$ is determined by the circuit parameters of gain, bit ADC conversion and the error function linearization. For modern solutions in the industry $\Delta I_H/I = 10^{-4}$. While the relative instrumental error, respectively δTi = kp·0,000141. The calculated dependence of the instrumental error on the temperature is shown in Fig. 5.

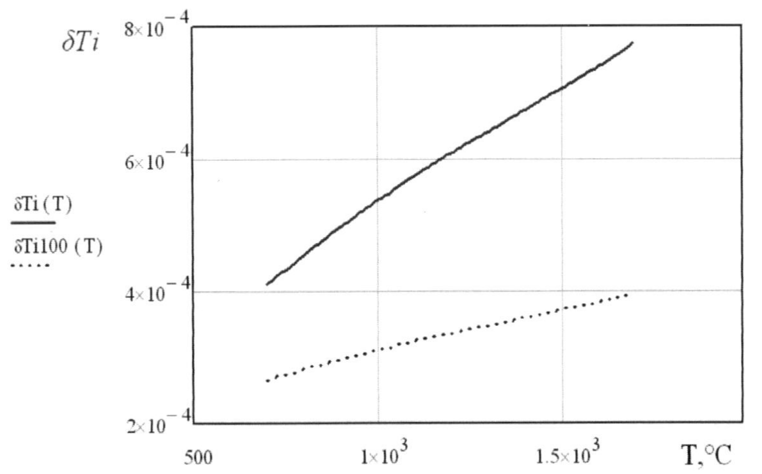

Figure 5. The calculated feature dependencies relative instrumental error δTi from the temperature of the blackbody cavity: δTi (T) - Upn= 10 V, δTi100(T) - Upn= 100 V.

For bodies with emissivity ε(λ) depend on the wavelength of the relative truncation error δTε according to [1] can be described by the expression:

$$\delta T\varepsilon = \left| kp \cdot \left(\frac{\varepsilon_1^{эф}}{\varepsilon_2^{эф}} - 1 \right) \right|$$

where $\varepsilon_1^{ef}(\lambda)$, $\varepsilon_2^{ef}(\lambda)$ is the effective emissivity [4], defined by the expression:

$$\varepsilon_i^{эф} = \frac{\int_{\lambda 1}^{\lambda 2} \varepsilon(\lambda) S_i(\lambda) r(\lambda, T) d\lambda}{\int_{\lambda 3}^{\lambda 4} S_i(\lambda) r(\lambda, T) d\lambda}$$

Figure 6 according to [5] shows an example of the spectral dependence of the emissivity ε(λ) for iron. The calculated dependence of the relative methodical error δTε(T) for iron are shown in figure 7. For both cases Upn= 10 In and Upn= 100 In it matches.

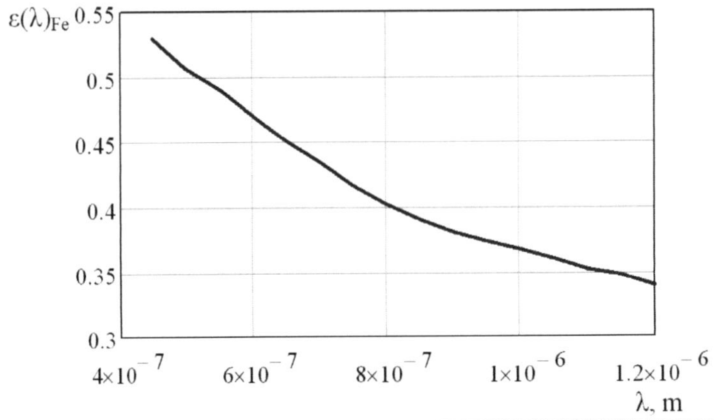

Figure 6. The spectral dependence of the emissivity ε(λ) iron.

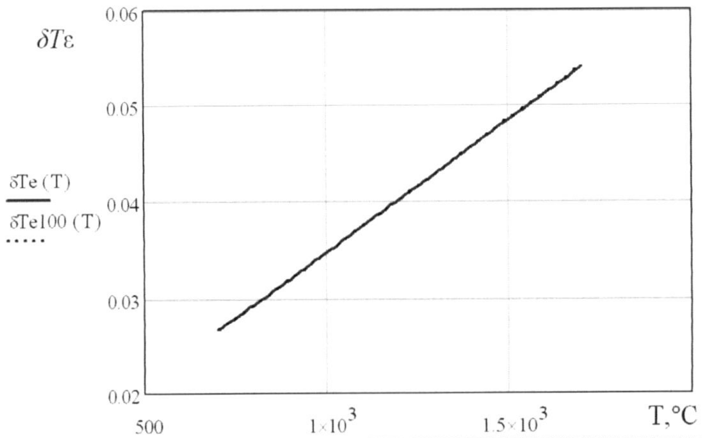

Figure 7. The calculated function according to the relative systematic error δTi from the temperature of the iron.

5. Conclusions

It should be noted that thanks to the work of silicon photodiode in a wide spectral sensitivity plots for different modes of inclusion decreases the relative systematic error of measurement. This instrument accuracy comparable with thermometers that use narrowband optical filters or detectors with different spectral sensitivity ranges "tandem" and mosaic types.

The results are a prerequisite for developers not only a new type of thermometers, but also outlines the challenges for developers of photodiodes to create specialized photodetectors providing maximum difference of the spectral sensitivity functions with changes applied to p-n - transition voltages, for tasks spectral pyrometry relations. Such detectors will allow in accordance with expressions (4) and (5) to reduce the sampling error when a small reverse voltage in the photodiode mode.

References

[1] Senkov A .G., Firago V.A. *Optimization of the characteristics of the spectral ratio pyrometers//* Vestnik St.Petersburg University. BSU. Ser. 1. 2009. No. **1**. S. 47.

[2] RF patent №2485458, MKI G 01 J 05/60. *The pyrometer of spectral relations* /V.A. Zakharenko, D.G. Lobov, D. B. Ponomarev, A.G. Skayew, applicant and paten holder of the State educational institution of higher professional education "Omsk state technical University". No. 2011149297/28; Appl. 02.12.2011; publ. 20.06.2013 g, bull. No. 17.-4C.: the sludge.

[3] Isanin G.G. Sources and detectors: a manual for students of the optical specialties/ Isanin GG [and others] - SPb.: Polytechnic, 1991 - 240 C.

[4] Senkov A .G., Firago V.A. *Reduction of methodological errors in determining the temperature of metals by two-color pyrometers* Journal of Engineering Physics and Thermophysics. July 2006, Volume 79, Issue 4, pp 768-772.

[5] Frunze A. V. *A numerical method of determining the spectral-ratio temperature* Measurement Techniques, October 2010, Volume 53, Issue 6, pp 664-667

Formation of elements of integrated acousto-optic cell based on $LiNbO_3$ films by methods of nanotechnology

O A Ageev[1], E G Zamburg[2], A S Kolomiytsev[3], D O Suchkov[4], I A Shipulin[5], and A V Shumov[6]
Southern Federal University, Institute of Nanotechnology, Electronics and Electronic Equipment Engineering, 2 Shevchenko St., 347900, Taganrog, Russia
[1] PhD, Professor, Directors of Institute of Nanotechnology, Electronics and Electronic Equipment Engineering, tel.: +7(8634) 371611, email: ageev@sfedu,ru
[2] Researcher, Department of Nanotechnology and MEMS, tel.: +7(8634) 371611, email: egzamburg@sfedu.ru
[3] PhD, Assistant Professor, Department of Nanotechnology and MEMS, tel.: +7(8634) 371611, email: askolomiytsev@sfedu.ru
[4] Student, Department of Radio-Technical Electronics, tel.: +7(8634) 371611, email: suchkov@sfedu.ru
[5] Student, Department of Nanotechnology and MEMS, tel.: +7(8634) 371611, email: shipulin@sfedu.ru
[6] Student, Department of Nanotechnology and MEMS, tel.: +7(8634) 371611, email: shumov@sfedu.ru

Abstract. In the experiments we defined modes, and developed the technology of formation of elements of input-output laser emission and microlens of integrated acousto-optic cell by Pulsed Laser Deposition and Focused Ion Beams by using nanotechnology cluster complex, allowing controlled creation of elements in a single process cycle.

1. Introduction

Acousto-optical devices are widely used in all kinds of radio-astronomy observatories, including terrestrial, airborne and space-based ones, as well as in atmospheric monitoring systems in many countries. Acousto-optic modulators and deflectors are used to control the amplitude, frequency, phase, polarization, direction of propagation of light beams. One of the budding directions is the development and study of the technology of the formation of elements of integrated acousto-optic cells, which allows reducing the size of devices, power consumption, and increasing sensitivity to vibrations, reproducibility of the parameters and provides for cost-effective mass production [1]. One of the budding materials for the production of acousto-optic cells is $LiNbO_3$ [2, 3].

At present time, one of the most promising approaches to the production of micro-, nano-, opto- and acousto-elecronic devices are those, which combine capabilities of analytical and technological methods in a single manufacturing process and controlled environment without venting. It excludes negative influence of the environment at all stages of production and research.

Besides, an important and urgent issue in the technology of manufacturing of structures for integral acousto-optics is the method of sublimated profiling of substrate surface. Traditional methods of the microelectronic technology, based on processes of optical lithography and liquid and plasma etching, have a number of basic limitations linked to complexity of achieving nanometer space resolution, the necessity of using photoresists, development of specialized masks and templates, and impossibility of forming complex profile structures in a single process cycle.

One of the promising methods to overcome the limitations of traditional production methods is the method of Focused Ion Beams (FIB). The method is based on the interaction of a beam of accelerated gallium ions, focused to ~ 10 nm in diameter, with a solid surface which results in local physical sputtering of the substrate with a resolution collated with the diameter of the beam. The possibility of controlling FIB parameters within a wide range, and using a 16 bit digital image generator allows forming micro- and nanoscale structures with high accuracy and resolution.

Cluster nanotechnological complex NANOFAB NFS-9 (NT-MDT, Russia) is one of the most budding systems as it combines several methods, including pulsed laser deposition (PLD), local nanoscale profiling by focused ion beam FIP, scanning probe microscopy (SPM), scanning electron and ion microscopy (SEM), diffraction of fast reflected electrons, and creates new opportunities for research, development of manufacturing technology, and the production of advanced micro- and nanoelectronics devices [4-7].

2. Formation of elements of integrated acousto-optic cell

During experimental part of the work, we carried research of rate of etching $LiNbO_3$ and Al_2O_3 by FIB. Also, during the experimental research on a substrate by ion-beam etching, we formed a rectangular shape array of recess with the size of each element of 2x2 μm at a constant time of etching of 30 sec, the accelerating voltage of 30 keV, and values of FIB current in the range from 1 pA to 1 nA (Figure 1). The research of topology of the formed structures were carried out by methods of SEM and AFM.

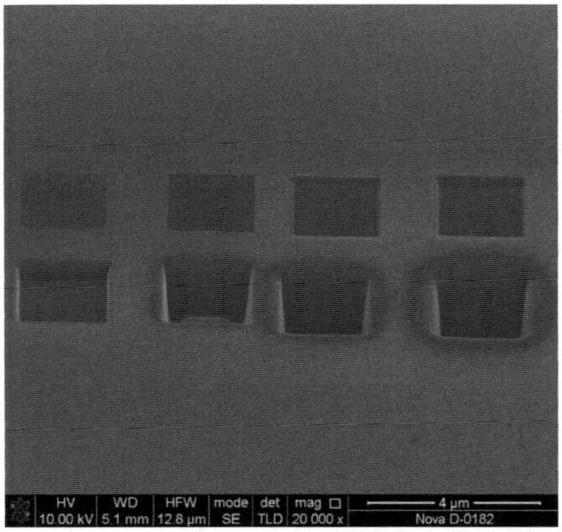

Figure 1. SEM image of an array of test structures on sapphire substrates for the investigation of the etching rate of materials by FIB

The speed of ion beam etching was calculated as the ratio of the volume of material removed per unit to etching time. During experiments it was found that in the range of FIB currents from 1 pA to 1 nA, the

Journal of Physics: Conference Series **643** (2015) 012031 doi:10.1088/1742-6596/643/1/012031

speed of ion-beam etching was as follows: from 0.20 to up to 322 nm^3/sec for Al_2O_3, and from 0.33 to 395 nm^3/sec for $LiNbO_3$. Dependence of the speed of etching on the ion beam current are shown in Figure 2.

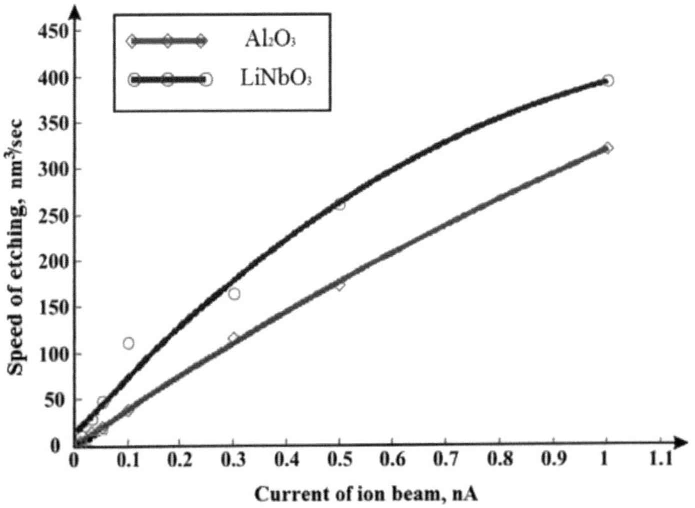

Figure 2. Relationship between the speed of the etching of Al_2O_3 и $LiNbO_3$ and the ion beam current

Analysis of the experimental curves shows that the speed of ion-beam etching increases linearly with the increase in ion beam current. It's due to the fact that with the increase in current, diameter of FIB increases due to the increase the number of ions, pulled out of the source per unit of time. With increase in diameter of FIB, the effective area of the ion beam interaction with the surface increases, hence the intensity of the substrate material spraying increases.

For forming $LiNbO_3$ films we used a Pulsed Laser Deposition module (Neocera Inc., USA) of complex NANOFAB, which provides for controlled deposition of films consisting of multicomponent oxides. The films were deposited on sapphire at the energy density of the laser emission of 2.5 J/cm^2, the laser wavelength of 248 nm (KrF), and substrate temperature of 600°C. In order to control the in-situ width and the structure of $LiNbO_3$ under formation, RHEED system kSA 400 (STAIB Instruments, Germany), as part of PLD, was used. Film thickness - 400 nm.

After the formation of $LiNbO_3$ films, the samples were passed through a vacuum transportation system to the modules of complex NANOFAB with Focused Ion Beams (Orsay Physics, France), which are used for local nanoscale operations, including local spraying, cutting, ion implantation, and local growth. Resolution of nano-profiling operations by means of ion beam is 10 nm. Etching was performed for each element at an accelerating voltage of 30 keV ion beam, the ion beam exposure time at each point of the pattern is 10 μsec. Control of geometrical parameters generated structures was performed by SEM with built-in image processing software, as well as by semicontact AFM module SPM.

Figure 3 shows the SEM images of obtained elements of input-output laser emission and microlens in integrated version formed on $LiNbO_3$ film.

To sum up, as a result of experimental studies we identified modes and developed the technology of formation of elements of acousto-optic cell using nanotechnology cluster complex, allowing controlled creation of elements in a single process cycle.

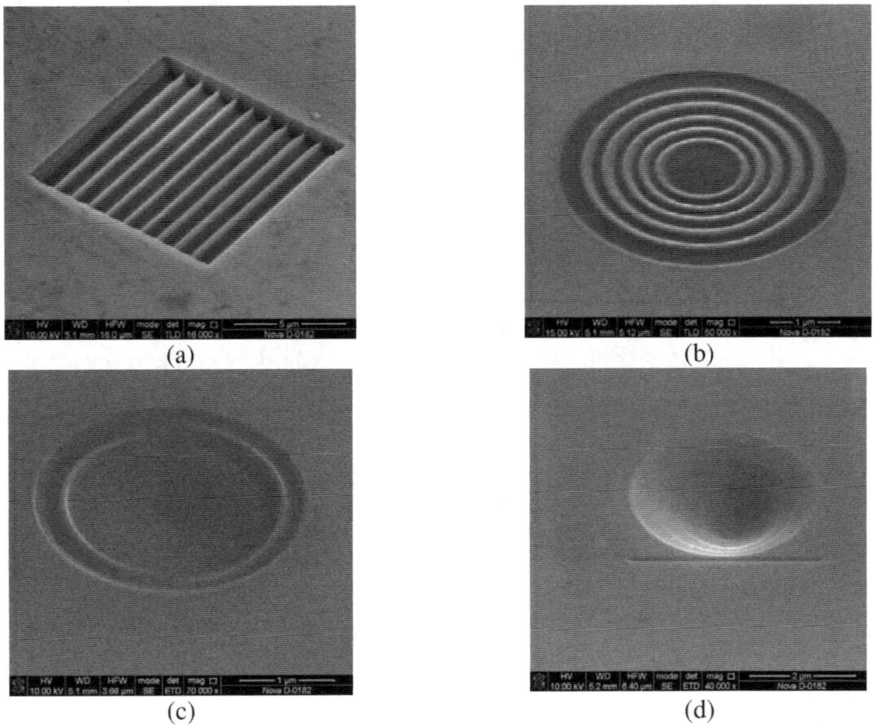

Figure 3. Element of input-output laser emission (a) and microlens (b)-(d) in integrated version.

3. Conclusion

The results were obtained by using the equipment of the Center for Collective Use "Nanotechnology" and the Research and Education Center, "Nanotechnologies" Southern Federal University.

This work was financially supported by the Ministry of Education and Science of the Russian Federation within the framework of the project stage of state contracts for scientific activities (job 16.1154.2014 / K), Federal Target Program "Scientific and Scientific-Pedagogical Personnel of Innovative Russia" for 2009-2013 (agreement №14.575.21.0045 , project's unique ID - RFMEFI57514X0045).

References

[1] Voloshinov V B, Nikitin P A, Trushin A S, Magdich L N 2011 *Latter in JTF* **37** 22-28

[2] Dostanko A P, Ageev O A, Golosov D A, Zavadski S M, Zamburg E G, Vakulov D E, Vakulov Z E 2014 *Semiconductors* **48 (9)** 1242-1247

[3] Zamburg E G ,Ageev O A, Golosov D A, Alexeev A M, Vakulov D E, Vakulov Z E, Shumov A V, Ivonin M N 2014 *Applied Mechanics and Materials* **481** 55-59

[4] Kolomiitsev A S, Konoplev B G, Ageev O A 2011 *Semiconductors* **45** 89–92

[5] Ageev O A, Smirnov V A, Kolomiitsev A S, Serbu N I, Konoplev B G 2012 *Russian Microelectronics* **41** 41-50

[6] Ageev O A, Vnukova A V, Gromov A L, Ilin O I, Kolomiytsev A S, Konoplev B G, Lisitsyn S A 2014 *Nanotechnologies in Russia* **9** 145-150

[7] Ageev O A, Gusev E Y, Zamburg E G, Vakulov D E, Vakulov Z E, Shumov A V, Ivonin M N 2014 *Applied Mechanics and Materials* **475-476** 446-450

Raman spectroscopy for the control of soil contamination by copper ions

E.V. Timchenko[1], P.E. Timchenko[1], L.A. Zherdeva[1], N.V. Tregub[1], L.A. Shamina[1]

[1]Samara State Aerospace University (SSAU), 443086, Samara, Russia

E-Mail: laser-optics.timchenko@mail.ru

Abstract. The experimental results of changes in the plant optical characteristics depending on the concentration of copper ions in the soil were presented by a method of Raman spectroscopy. Raman spectral features of plant tissue growing under the influence of copper ions in the soil that excess the MPC were obtained. The basic changes were registered at Raman bands 1325 cm^{-1}, 1527 cm^{-1} and 1547 cm^{-1} that responsible for the stretching vibrations of chlorophyll a and b, carotene and chlorophyll a. The optical coefficient based on the dependence of the copper ion concentration from the optical parameters of plants was introduced. When the concentration of copper ions in the soil of 3 MPC observed increase optical coefficient values on the first day of experiments.

1. Introduction

In today's, environmental problems in their public importance left on one of the first places. The priority issue in ecology is to develop methods to control atmospheric emissions. Especially dangerous are the industrial enterprises, which are the sources of emissions of heavy metals, which have a negative impact on the entire ecosystem. One of the most common heavy metals emitted by enterprises is copper [1]. At concentrations of copper ions in the soil, it is much higher than the one of the MPC biotoxic heavy metals. However, the localization and accumulation of copper ions in the soil is not of a permanent nature, as It depends on the "wind rose". Considering that the plant leaves are the most sensitive to the action of atmospheric pollutants, plants can be used as "markers" environmental state of the atmosphere and their optical characteristics to serve as a quantitative measure of the environment. In order to control atmospheric emissions using physical and chemical methods of environmental monitoring, including optical [2] and chromatographic techniques [3]. Because of the specific characteristics of electrochemical and chromatographic methods require time-consuming and quite laborious. Optical methods are deprived of these drawbacks are not destructive analysis methods based on the relationship and the optical properties of the system and its composition. The most common methods include optical fluorescence analysis, the method of inverse scattering, absorption analysis, the method of Raman spectroscopy (RS) [4-7]. In the paper [8] describes the use of Raman spectroscopy to elucidate structural changes in plant tissues and individual cells under the influence of pollutants. In [9] the possibility of finding TiO2 using Raman spectroscopy. Therefore, Raman spectroscopy was selected as the basic method for the control the copper ions in the soil, which has high sensitivity and provides the most complete information about the state of a living system, in contrast to other optical methods.

2. Materials and methods of research

As the test is selected bioindicator copper peas (Písum). We investigated the following groups of plants that are in the same environmental conditions (light and humidity) control samples, samples of plants grown in soil with different concentrations of copper ions in the soil 0.2MPC, 0.5MPC, 1MPC, 2MPC, 3MPC. The concentration of copper ions was entered once before the start of the experiment. Raman spectroscopy method has been implemented with the help of experimental stand shown in Figure 1.

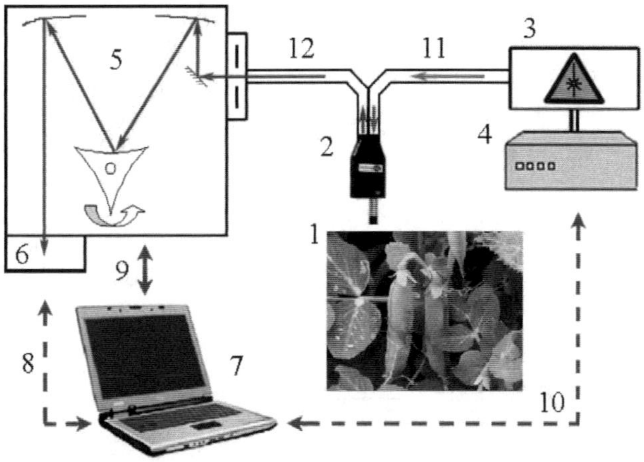

Figure 1. Experimental setup: 1 - the object; 2 - Raman Probe RPB785; 3 - Laser Module LuxxMaster Raman Boxx; 4 - power supply of the laser module; 5 - spectrometer Shamrock sr-303i; 6 - Built-in cooling chamber DV420A-OE; 7 - the computer; 8, 9, 10 - information electric cables; 11 - transmitting fiber; 12 - the receiving fiber

Processing Raman spectra was carried out in a software environment Mathematica'8. To suppress background fluorescence and Raman spectra allocation method was used polynomial approximation. The result of the test experiments, it was found that the error factor used is 7%.

3. Research results

Figure 2 shows a typical Raman spectrum of the test plants. The following Table 1 provides a breakdown of the main bands of Raman spectrum.

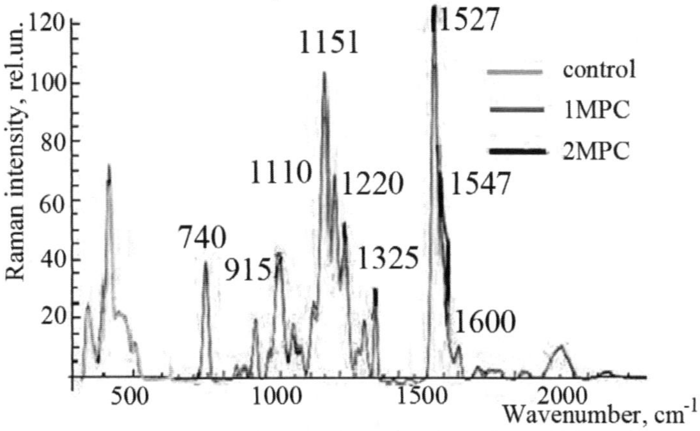

Figure 2. The intensity of the Raman wavenumbers for the selected bioindicator.

Table 1. Explanation of the main bands of Raman spectrum.

Wavenumber , cm^{-1}	**Component**
740	Chlorophyll a and b
915	D - glucose
1110	D - glucose

1151	C–C stretching vibrations bands of β - carotene
1220	C-N stretching vibrations of alkaloids
1284	Chlorophyll a and b
1325	Chlorophyll a and b
1527	C=C stretching vibrations of β - carotene
1547	Chlorophyll a
1600	Lignin

As seen in Figure 1, the main optical changes of plants grown in the presence of copper ions in the soil is 1325 cm^{-1}, 1527 cm^{-1} and 1547 cm^{-1}, responsible for the stretching vibrations in chlorophyll a and b, carotene and chlorophyll a. However, it is known that the change of chlorophyll concentration "a" in the plant leaves is informative factor of environmental influences on the plant [10, 11], which in turn is reflected in the intensity of the RS. As can be seen from Fig. 1 Raman band at 1547 cm^{-1}, corresponding to stretching vibrations of chlorophyll "a" clearly expressed for the plants grown under the influence of copper ions with a concentration 3MPC.

As it has been shown in [12] a relatively constant parameters considered lignin in plants. The deposition of lignin in the cell walls increases their strength. Thus, the ratio of the intensity of the Kyrgyz Republic, proportional to the concentration of chlorophyll "a" and carotenoids to the intensity of the Kyrgyz Republic, proportional to the concentration of lignin, reflect, including changes in the concentration of pollutants and can serve as an indicator of the effect of hydrogen on plants. Thus, it was introduced the following optical coefficient R, which allows to determine the effect of copper on the plant:

$$R = \frac{I_{1527}}{I_{1600}} \frac{I_{1547}}{I_{1600}}$$

where I_{1527} – Raman intensity on the wavenumber 1527 cm^{-1}, which is proportional to C = C stretching vibrations in the b-carotene in the leaves of plants; I_{1547} - the intensity of Raman for the wave number 1547 cm^{-1}, which is proportional to the concentration of chlorophyll in the leaves of plants; I_{1600} - the intensity of Raman for the wave number 1600 cm^{-1}, proportional to the concentration of lignin in the plant leaves.

Further, Figure 3 shows the dependence of optical coefficient values plants grown in soils with different concentration of copper ions of time.

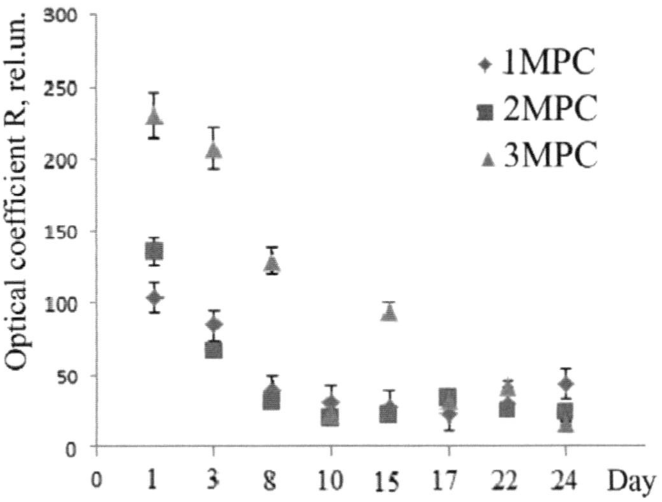

Figure 3. The dependence of the optical coefficient for soils with different concentrations of heavy metals on the time.

Figure 3 shows that the optical coefficient values (R) for plants grown in soil at a concentration of copper ions 3MPC higher than the values for R 1MPC. While in the last days of the life cycle of the plant occurs reverse pattern, which is probably due to the fact that the greatest concentration of copper ions in the soil causes the acceleration of the life cycle of plants, resulting in degradation of the major pigments in plants, which in turn affects the optical parameters change plants.

Further, Figure 4 shows the dependence of optical coefficient values plants concentration of copper ions.

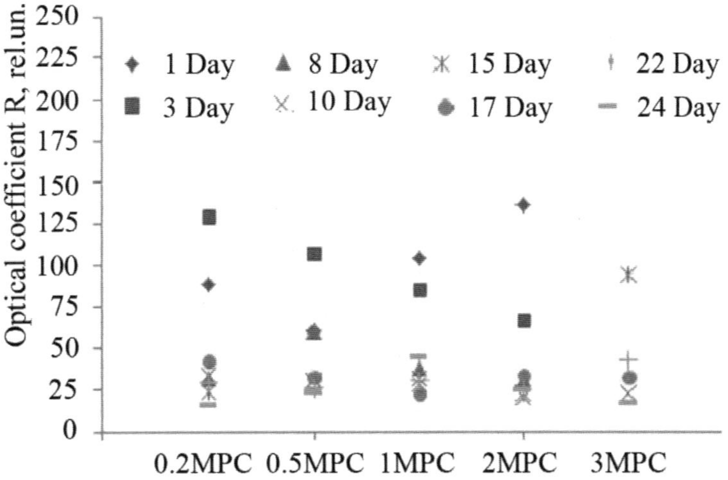

Figure 4. The dependence of the optical coefficient on the concentration of copper in the soil.

Figure 4 shows that the concentration of copper ions in the soil equal to 3 MPC, an increase in the values of the optical coefficient in the first days of the experiment, indicating that the nature of the impact volley of copper ions in the plant.

4. Conclusion

Was particularly the Raman spectrum of plants growing under the influence of copper ions in the soil exceeding the MPC. Basic changes were registered in the wave numbers 1325 cm^{-1}, 1527 cm^{-1} and 1547 cm^{-1}, responsible for the stretching vibrations in chlorophyll a and b, carotene and chlorophyll a.

Was introduced optical coefficient on the basis of which the dependence of the concentration of copper ions from the optical parameters of the plants. When the concentration of copper ions in the soil of 3 MPC observed increase optical coefficient values on the first day of experiments.

Acknowledgement

This work was supported by the Ministry of Education and Science of the Russian.

References

[1] Vetoshkin A 2007 Environmental protection from the impacts of energy Publishing Penza State University 386

[2] Timchenko E, Timchenko P, Tregub N, Taskina L and Selezneva E 2014 Optical methods for control of hydrogen influence on plants SPIE 9221-14

[3] Becker Yu 2009 Chromatography. Instrumental analytics. Methods of chromatography and capillary electrophoresis M .: Technosphere 472

[4] Allen J 2004 Probing the Mechanism of State Transitions in Oxygenic Photosynthesis by

Chlorophyll Fluorescence Spectroscopy, Kinetics and Imaging. – In: Chlorophyll fl uorescence: a signature of photosynthesis Springer The Netherlands, Dordrecht 467–461

[5] Timchenko Taskina E and L 2012 Application of optical methods for the control plants under external influence of the Samara Scientific Center of the Russian Academy of Sciences 14 (4) 221-226

[6] Arsar G 1984 Estimating absorbed photosynthetic radiation and leaf area index from spectral reflectance in wheat Agron. J 76 211-220

[7] Brandmyuller I, Moser G 1964 Introduction to Raman spectroscopy World 628

[8] Batirtze Prats Mateu, Stefke Barbara, Notburga Gierlinger 2014 Elucidating structural and compositional changes in plant tissues and single cells by Raman spectroscopic imaging, VOL. 26 NO.5

[9] Jose Antonio Carrero, Gorka Arana and Juan Manuel Madariaga, An important use of Raman spectroscopy to help understand the impact of traffic on roadside soils and plants

[10] Lysenko V, Vardun T Sawyer and The Reds in the 2013 Basic research 4112

[11] Gitelson A, Gao B, Li R, Berdnikov S and Saprygin V 2011 Environ. Res. Lett. 6 024023

[12] Timchenko E, Timchenko P, N Zherdev, Tregub H, E and Seleznev Yakovlev In 2015 Optics and Spectroscopy T.119

Improving the efficiency of emission extraction from nitride LED structures with textured interfaces

A S Evseenkov[1], I A Lamkin[1] , S A Tarasov[1], A V Solomonov[1]
[1]St. Petersburg Electrotechnical University "LETI", 5 Prof. Popova Street, St. Petersburg 197376, Russia.

E-mail: as.evseenkov@gmail.com

Abstract. The light-emitting diode (LED) structures based on gallium nitride (GaN) and grown on sapphire substrates were investigated. The obtained experimental data include the emission spectrum, optical power and current-voltage characteristics. Monte Carlo simulation based on the parameters of the structures was conducted. During the simulation optimal parameters of surface texturing were evaluated, and alternative methods of structure modification were shown.

1. Introduction

Realization of the potential advantages of blue light-emitting diodes (LEDs) based on gallium nitride (GaN) and its solid solutions, as light sources satisfying modern specifications, requires considerable development and improvement efforts. Primarily, significant increase of power and external quantum efficiency (EQE) of the LEDs is required. There are various methods to improve the parameters of the structures based on GaN [1-2].

The main physical effect, which reduces the ratio of the output optical power to the emission power generated inside the crystal, is a total internal reflection [3,4] at the interface of materials with high and low refraction indexes .

The effective way to solve the problem of limiting the light output is to obtain surfaces that scatter light [5-7, 9-13]. When creating a surface texturing forms with predetermined parameters, the impact effect of total internal reflection is reduced, and part of the light output from the device can be increased. At present, the texturing of both substrates and epitaxial layers is used widely [6]. In the LED industry two basic approaches are applied: random texturing technology and patterned sapphire substrate. Random texturing of the top layer of p-GaN is an economical method where additional efforts are not required.

At present, the transition from experimental methods to the analytical prediction and the calculation of the value of the efficiency of the emission output depending on the size and geometry of textures interfaces is of current interest [7]. The aim of the work was to find the optimum variant of texturing interfaces in the Al_2O_3/GaN/InGaN flip-chip LED structure leading to the maximum efficiency of the emission output. To reach this goal, a mathematical model describing the process of the emission output of the nitride LED structures was created, and numerical calculation was carried out by the Monte Carlo method. The results of calculation are compared with the analytical solution of this problem. The experimental studies of emitting structures have been undertaken, and a more than three-fold increase of the emission output efficiency when using texturing was demonstrated.

2. Numerical simulation Monte Carlo

Below, we call an elementary particle of light propagating according to the laws of geometrical optics a photon.

In the simulation of the emission propagation through the optical semiconductor structure by the Monte Carlo method, two main approaches are used: wave and corpuscular [9]. Since the geometrical dimensions of the roughness greater than the wavelength emission, it is possible to use the Monte Carlo method. For the purpose of increasing the calculation accuracy, absorption in the thicker layers should be taken into account. Since the calculation is carried out in geometrical optics approach, where the absorption in the medium is not considered, the Bouguer-Lambert-Beer law for the light intensity is used for numerical calculations:

$$I(l) = I_0 e^{-k_\lambda l}, k_\lambda = \frac{4\pi k}{\lambda}.$$

Since the Bouguer-Lambert-Beer law is applicable to the flux intensity consisting of a great number of photons, the corpuscular Monte-Carlo method operates with single photons, the problem in modeling using the Monte-Carlo method is the choice of criterion, according to which one determines whether a particular photon has been absorbed or continues its movement through the material. We suggest a following statistical approach to establishing a relation between a single particle and a flux. The criterion of absorbing a photon on the interface is based on comparison between a randomly chosen number with the intensity calculated according to the Bouguer-Lambert-Beer law, in which the length of the path is substituted with the actual path length of the photon. To find the transmission efficiency using the Monte-Carlo method one has to calculate the ratio of the number of photons that passed through the structure to their total number, i.e. take into account all the possible scenarios of interaction between the photons and the structure with randomly distributed parameters.

As randomly distributed variables the following ones were taken: the initial position of a photon generated in the active region, the vector of its movement (which defines the initial solid angle), and the factor of interaction with the medium (characterizes the photon absorption probability, namely to take into account the absorption of defects, free charge carriers, contacts, and other types of absorption). The angle of its incidence on the phase boundary during traversal of a photon through the structure was calculated. Depending on this angle either a rotation of the photon's velocity vector or its reflection from the interface occurs. After a series of reflections and refractions the photon can be absorbed by the multilayer structure or it can pass through it. Having a sufficient number of generated photons one can determine the transmission coefficient of the structure according to their statistics, which is what has been done within the framework of our model.

As the basis of parameters modeling the structure type "A" was taken, which is described in Experiment. The superlattice was accounted for as a "mirror" under the active region. Generation occurred only in the active region where the internal quantum efficiency was considered equal to 1. In the neighboring areas it was not considered. The peak emission was at 455 nm, the absorption coefficient was 10^3 cm^{-1}[9].

The computer program was developed in the graphical programming environment of LabVIEW. The grid step was 10 nm, the number of cross-sectional nodes was $8.25 \cdot 10^{14}$, and the number of generated photons was 10^6. The choice of these numbers is the result of finding

a compromise between the precision and the speed of calculations as well as finding optimal conditions of interaction between the photon and the structure (with a smaller grid step a photon would be sensitive to too small-scaled roughness and with a larger one the calculation would describe the interaction with the texturing inadequately).
In the course of generation of photons in the active region and their traversal through the LED structure a significant part of radiation is lost due to reflection from the interface. Making pyramids with different refraction coefficients on the surface of semiconductors leads to an observable increase of transmission, which is explained by photons finding optimal trajectories that lead out of the structure as a result of multiple reflections (see Figure 1).
In reality it is impossible to produce identical texturing over the entire area of the interface. In order to take this into account in modeling one has to introduce a statistical distribution of roughness parameters with main variable parameter being the aspect ratio (roughness ratio of base to height) on the interface. Also, the presence or absence of lenses and the width of the semiconductor layer is used as a parameter. The width of the layer was 1.5, 4.5 and 7.5 μm.

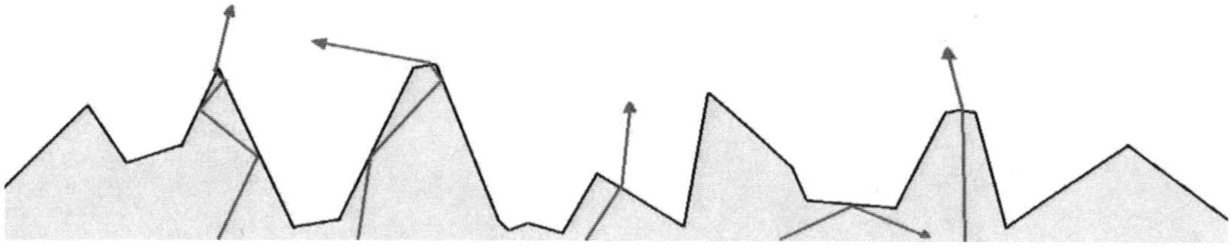

Figure 1. Ray diagram of light exiting a textured layer.
Depending on the parameters of the structure different values of shape and size parameters are used for the texturing. The shape has the strongest influence since it defines the process of photon emission. Presently, the sphere-shaped, well-shaped, and pyramid-shaped (both sharp and canted) texturing is used. In the present work we use the well-shaped texturing in the modeling.
The texturing can be created using a variety of methods: the technique of wet etching in combination with the Laser-lift-off technology, creation of microoptic elements on a sapphire substrate [10], micromachining [11], using porous surfaces [12] etc.
The modeling was performed using the parameters of real nanoheterostructures based on solid solutions (Al, Ga, In)N with multiple quantum dots [13] and superlattices [14] created with MOCVD method on a sapphire substrate. The nanoheterostructures had identical structure except for the last p-GaN epitaxial layer. The buffer and contact n-GaN layers with total thickness of 4 microns were grown on a c-Al_2O_3 substrate. The active region contained one 2.5 nm-thick $In_{0.15}Ga_{0.85}N$ quantum well. Then a low-doped i-GaN layer with 50 nm thickness was grown. The final p-GaN layer was of 4 following types: a superlattice consisting of 9 pairs of 1.5 nm-thick $Al_{0.15}Ga_{0.85}N$ and GaN layers (type A), a p-AlGaN with 15% of aluminum (type B), a p-AlGaN with 10% of aluminum (type C), and a p-GaN layer (type D).

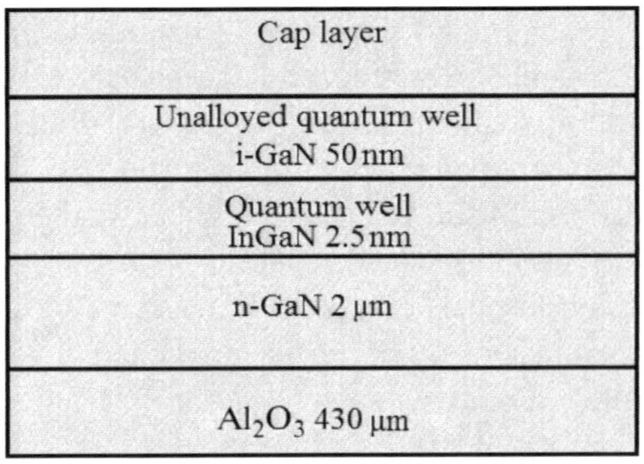

Figure 2. Design of the LED heterostructure.

A calculation of transmission coefficient for structures with different values of thickness of GaN (Figure 3) and sapphire layers was performed. It was shown that the increase of the thickness of the semiconductor layer greatly reduces the transmission coefficient. For optimal transmission this layer should be made as thin as possible. The increase of thickness of the sapphire layer does not influence the light emission. Figure 3 also demonstrates the effect of a lens that collects the emitted light, which was placed on the sapphire layer.

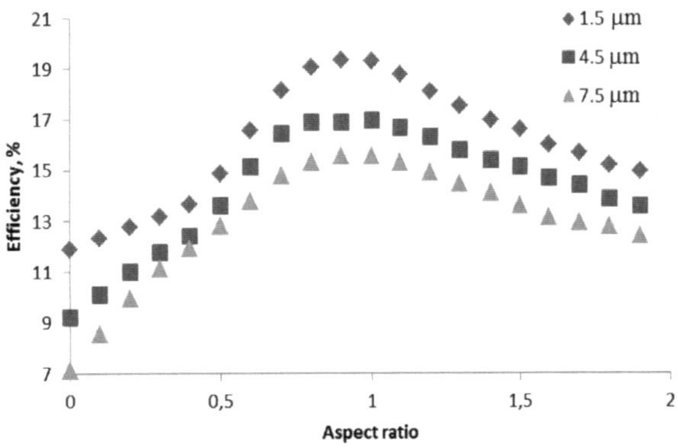

Figure 3. The dependence of the transmittance of the aspect ratio depending on the thickness of the semiconductor layer

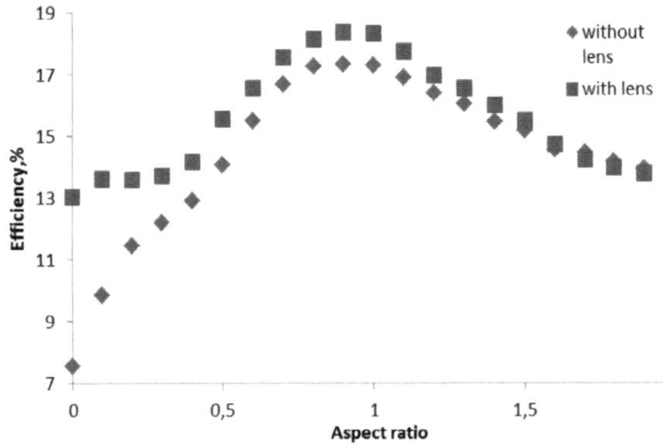

Figure 4. The dependence of the transmittance of the aspect ratio in the lens available on a sapphire layer

The obtained dependencies suggest the possibility to increase the transmission through the structure by decreasing the thickness of the semiconductor layer and adding a lens that gathers incoming light (particularly in absence of texturing).

The first effect is explained by a decrease in the average path length of photons that propagate through the layers with the highest light absorption coefficient. As a result, reduction in the thickness of the GaN layer from 7.5 to 1.5 um increases the transmittance coefficient almost by one third (figure 3). This suggests that the thickness of the semiconductor layer has a determining influence on the efficiency of the optical emission output of the LED. Improving effect is observed for all the aspect ratio texturing, since absorption occurs for all photon paths.

Adding a lens allows to achieve the greatest improvement transmission at low aspect ratio texturing (figure 4), because in this case the effect of texturing is weak. In this case, one can achieve almost doubling the transmission. For large aspect ratio effect is minimized.

3. Experiment

The investigations of nanoheterostructures were carried out using a test-diagnostic system operating parameters of light-emitting nanoheterostructures developed by the authors [2, 15]. The system allows for defining all the main characteristics of solid-state light sources and devices based on them: spectral, watt and volt-ampere characteristics, power and efficiency of emission. In addition, it provides an opportunity to determine the temperature of the active region nanoheterostructures. For this purpose a new rapid non-contact method of determining the temperature of the active region is used [16]. This method is based on a detailed analysis of the electroluminescence spectra and allows us to investigate not only the separated crystals on the plates, but also LEDs including in the LED lamps. In general, the diagnostic system allows the measurement of the most important optical and electrical characteristics of any LED and devices based on them in the following ranges: wavelengths 200 – 1100 nm, forward currents: 10^{-8} – 10 A, voltages: 0 – 250 V, temperatures: 10 – 500 K. In this contribution, the samples were a non-separated emitting chips on a sapphire substrate, so the diagnosis was carried out by using probe station. For the test samples all the major parameters were determined: the emission spectra, power, current-voltage and temperature dependence.

Figure 5 shows the dependence of the optical power of the current samples with various finishing layer. It can be seen that the greatest power and other parameters were achieved for the sample "A". Next, this sample used in the simulation as a research object

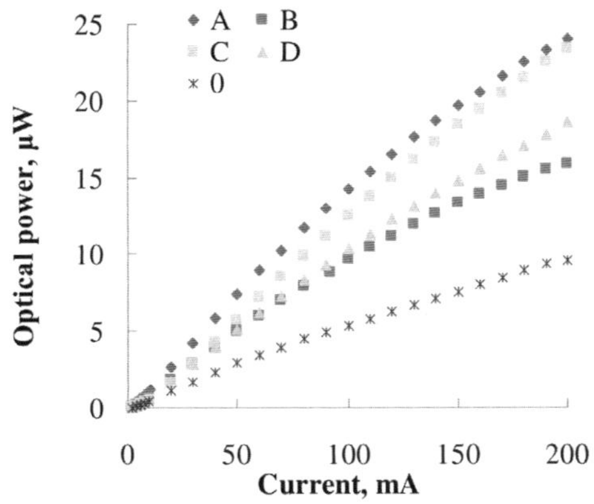

Figure 5. The optical capacity dependency on the current.

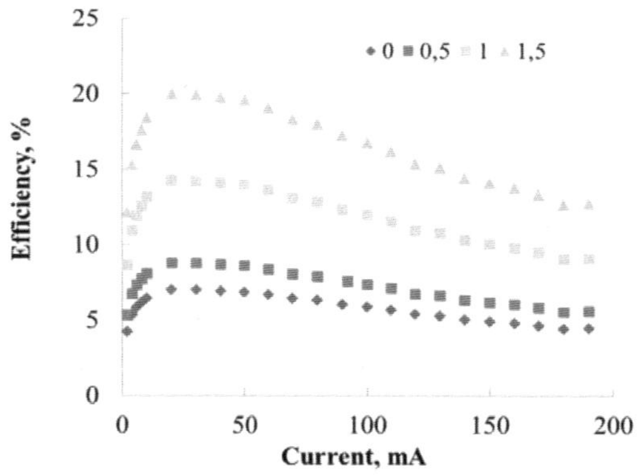

Figure 6. The dependence of the efficiency of the current for different aspect ratios (0 - absence of texturing, characterizing experimental data the sample A)

Figure 6 shows the results of the experiment (graph A) and the expected efficiency, if the surface of the LEDs texturing start. The greatest effect of the semiconductor/sapphire, since the emission passing across the first interface through the roughness comes in sapphire layer at the enough right angles and passes on without problems. Maximum efficiency occurs when the aspect ratio corresponds to the critical angle of the interface GaN/sapphire - for this structure, it is about unity, i.e. facet of the pyramid are arranged at an angle 45^0 to the interface.

The graph shows the possible efficiency at the various currents and texturing parameters. The main variable parameter is the average statistical aspect ratio obtained roughness. Maximum value on the proposed LED structures may be obtained by texturing with an aspect ratio equal to one, and a current value of 20 mA. This is due to the choice of the critical angle interface sapphire/semiconductor, in which the middle path of photons from the point of generation to the point where they exit from the structure is minimal.

4. Experiment Conclusion

In this paper LED structures based on GaN grown on sapphire substrates have been investigated (the sample with two SL of InGaN and AlGaN showed the best stability and performance). All the basic characteristics were obtained: the emission spectra, power, voltage and current characteristic. Based on the parameters of the structures numerical simulation of the emission output was made by Monte Carlo method. As a result of simulation of the LED structures, the patterns of the effect of the interface and the input roughness on the part of the emission output from the structure were obtained. Thanks the data obtained during the simulation, the change of the structures, that come as a result of surface texturing, were predicted, as well as other changes - for example the addition of lens or reducing the thickness of the semiconductor layer. It was found that the output of the structure can be increased more than 3 times by creating a pyramid with an aspect ratio equal to 1, by a decreasing the layer of gallium nitride, as well as the creation of the lens on the sapphire interface, and may be more than 30%.

The work was supported by the Ministry of Education of the Russian Federation in the framework of the project of the state task in the field of scientific activity, project number 16.1307.2014K.

References

[1] Panfeng J, Naixin L, Tongboo W 2011 Journal of Semiconductors Vol. **32** 11

[2] Menkovich E A,Tarasov S A, Lamkin I A, Kurin S Yu, Antipov A A, Roenkov A D, Barash I S, Helava H I, Makarov Yu N 2013 Journal of Physics: Conference Series **461** 012027

[3] Kim J K Alhmed N, Frank W. 2008 Appl Phys Lett **93** 221111

[4] Son J H, Jong U K, Yang Hee Song 2012 Advanced Materials **24** 2259-2262

[5] Lee K D,Sjodin R, Eriksson T 2010 LED clusters and assembly 22

[6] Rooman C, Kuijk M, Jonge S D 2005 Photonics Technology Letters 2649-2651

[7] Deinega A, Valuev I, Potepkin B, Lozovik Y 2011 J.Opt.Soc. Am. A **28** 5

[8] V Kandidatov et al 2006 Quantum Electronics **36** 1003-1008

[9] Wong W S, Sands T, Cheung N W 1999 Appl. Phys. Lett. **75** 1360

[10] Osincky V, Goncharenko T, Lyahova N 2003 FIP **1**. - № 1. – P. 94

[11] Lee K D, Szhodin R, Eriksson T 2010 Semiconductor lighting **5** 22

[12] Muratova E N, Spivak Yu M, Moshnikov V A, Petrov D V, Shemukhin A A, Shimanova V V 2013 Glass Physics and Chemistry **39** 320-328

[13] Menkovich E A, Tarasov S A, Lamkin I A 2012 Function Materials **19** 233-237

[14] Kurin S Yu, Antipov A A, Barash I, Roenkov A D, Usikov A,Helava H I, Ratnikov V, Shmidt N, Sakharov A, Tarasov S A, Menkovich E A, Lamkin L A, Papchenko B, Makarov Yu N 2014 Physica status solidi **3** 813–816

[15] Solomonov A V, Tarasov S A, Men'kovich E A, Lamkin I A, Kurin S Yu, Antipov A A, Barash I S, Roenkov A D, Helava H I, Makarov Yu N 2014 Semiconductors **48** 245–250

[16] Menkovich E A, Solomonov A V, Tarasov S A, Yurgin P A 2014 Functional materials **21** 186-189

Optical tomography in reacting flows based on Stokes Raman scattering

D Sharaborin[1,2], L Chikishev[1,2], V Dulin[1,2], D Markovich[1,2]

[1]Kutateladze Institute of Thermophysics, 1 Ak. Lavrentyeva Avenue, Novosibirsk, 630090, Russia
[2]Novosibirsk State University, 2 Pirogova Str., Novosibirsk, 630090, Russia

E-mail: sharaborin.d@gmail.com

Abstract This paper reports on development of an optical system for tomographic imaging in reacting flows. The measurement principle is based on registration of Stokes Raman scattering (rovibrational transitions for molecular nitrogen), when a transparent object is illuminated by a laser sheet. The method allows to retrieve local gas density and concentration.

1. Introduction

State-of-the-art optoelectronic devices provide detection of weak emission signals. It allows for non-intrusive measurements in reacting flows with high spatial and temporal resolution. Powerful lasers, efficient photocathodes with photo-multiplying tubes, and new optical filters can be combined into an optical system for registration of 2D images of Raman scattering, characterized by weak signal. The present paper reports on development of such system for local density and concentration measurements in flames based on detection of Stokes Raman scattering from molecular nitrogen.

2. Experimental setup

Sketch of the experimental setup for 1D and 2D measurements is shown in figure 1 The third (355 nm) or the second (532 nm) harmonic of a pulsed Nd:YAG laser (QUANTEL BrilliantB) was used for the illumination of sampling volume. The third harmonic of the laser was used only in 1D Raman system. The second harmonic of a pulsed Nd:YAG laser (Quanta-Ray) was used for 2D Raman system with combination of collimation optics. The energy of 6 ns laser pulses was monitored by an energy meter (Coherent LabMax-TOP). On average, the energy was 120 and 690 mJ for 355 and 532 nm, respectively, with 5% RMS fluctuations. Figure 1b shows the 2D image when a flame was illuminated by a non-focused laser beam. Orientation of linearly polarized laser light was controlled by half-wave plates. This was necessary to record signal for two perpendicular polarizations of the incident laser light. The images for the horizontal polarization were subtracted from those for the vertical polarization in order to minimize influence of the dark-current, reflections, and possible fluorescence (for UV).

3. 1D Raman system

1D system captured spectrum of the scattered light along the laser beam by using a Czerny-Turner spectrograph (Newport MS127i 1/8 m) equipped with a UV lens. The spectrometer alignment ensured the parallel orientation of the entrance slit and the laser beam. Grating with 1 200 lines/mm was used to provided spectral resolution of 1 nm. The spectra along the beam were registered by a 12-bit ICCD

Journal of Physics: Conference Series **643** (2015) 012034 doi:10.1088/1742-6596/643/1/012034

camera (PCO DICAM Pro, , GEN II photocathode type with quantum efficiency 20% in UV spectral region) mounted at the output of the spectrograph. To reduce noise, the scattered light from 100 laser pulses was collected during acquisition of each image. The collection time during each shot was 30 ns. For each measurement point, 10 frames were captured and averaged in order to reduce contribution of read-out noise. The image size of 1280×800 pixels corresponded to 108 nm on 105 mm. The full width at half maximum of the registered Stokes peaks was 7 pixels.

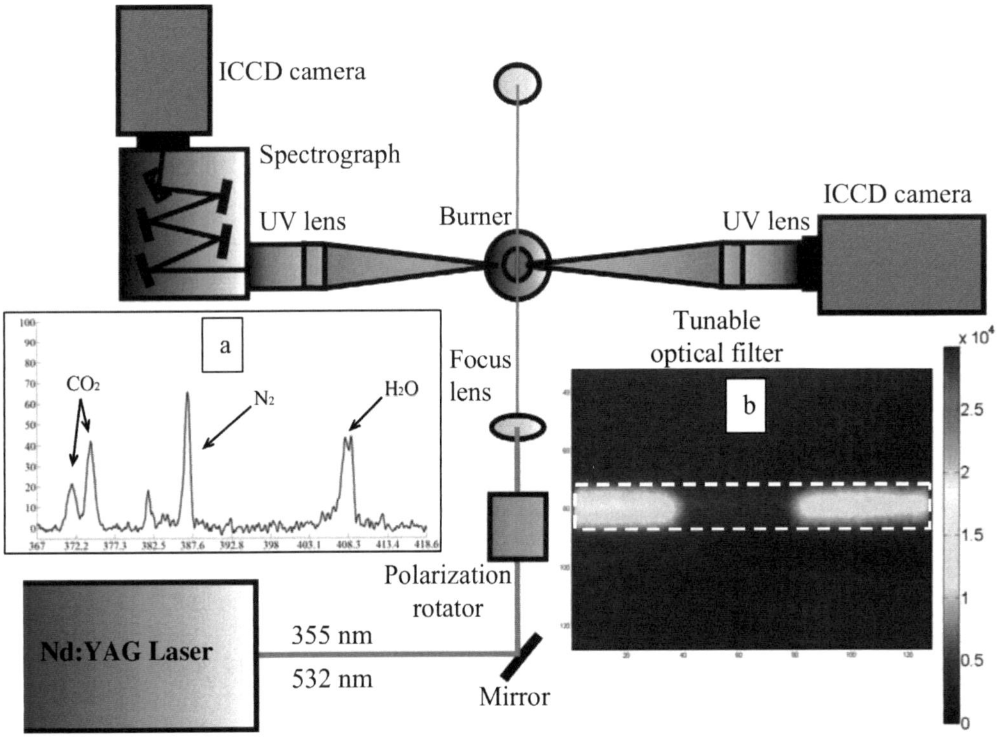

Figure 1. Scheme of the experimental setup for Raman thermometry and data example (a- Stokes band of Raman spectrum, b- 2D distribution of Raman signal for N2 vibronic transition) in flame.

The used temperature evaluation approach is based on the paper of Rabenstein and Leipertz [1]. The beam passed trough the ambient air and the flame simultaneously; thus, the Raman spectrum was imaged for both regions (see the example Fig. 1b). The air was considered as the control volume (it's temperature was monitored). Following Egermann et al. [2], the Raman spectra were recorded for two perpendicular polarizations of the laser light (vertical and horizontal). The images for the horizontal polarization were subtracted from those for the vertical polarization in order to minimize influence of the dark-current, reflections, and fluorescence. Unfortunately, the light scattered at 355 nm was not removed by a special filter, Rayleigh signal was present on the images, that limited dynamic range. The photo of the studied flame is shown in figure 2.

Figure 2. Photo of the studied flame. The black line indicates the area of study, the green line indicates the direction of the laser beam .

Intensity of Stokes lines depends on the temperature and volume fraction of the molecules. The line for molecular nitrogen was used to evaluate ratio between the flame and ambient air temperatures. The spectra in both regions was averaged over 60 pixels (i.e., for 8 mm along the beam). The background from the Rayleigh signal was approximated by an analytical fit function and was removed. According to [1], the dependence of the effective Raman scattering cross-section on temperature should be taken into account. This was done from calculations of the Raman spectra by using RAMSES code [3]. The fraction of nitrogen in the flames was evaluated from the numerical simulations. In addition, the approach was tested for the flows when the mixture was issuing from the burner, but the flame was not ignited. The results are compared with the temperature measured by the thermocouple are shown in figure 3.

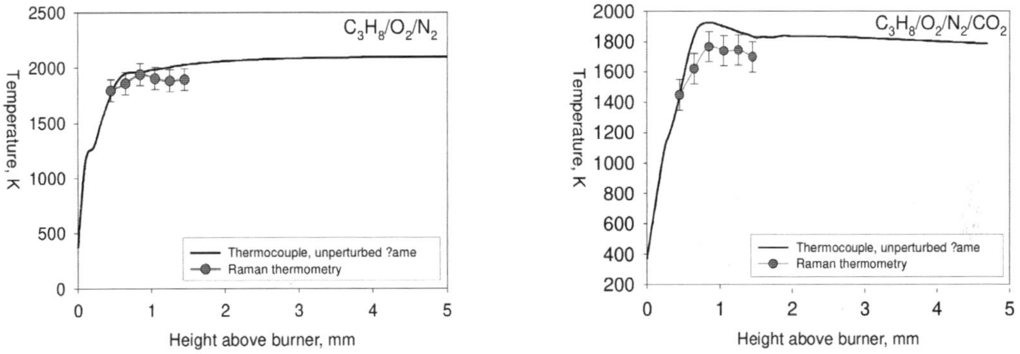

Figure 3. Temperature profiles in fuel-rich C3H8/O2/N2 and C3H8/O2/N2/CO2 flame measured by thermocouple (solid black line) and by Raman thermometry technique (solid red line and circles) [4].

4. 2D Raman system

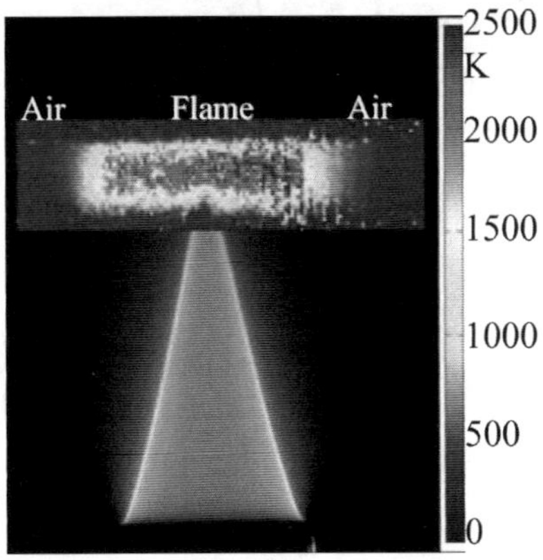

Figure 4. Temperature distribution in laminar Bunsen flame cone measured by 2D Raman technique with cylindrical laser beam and photo of Bunsen cone.

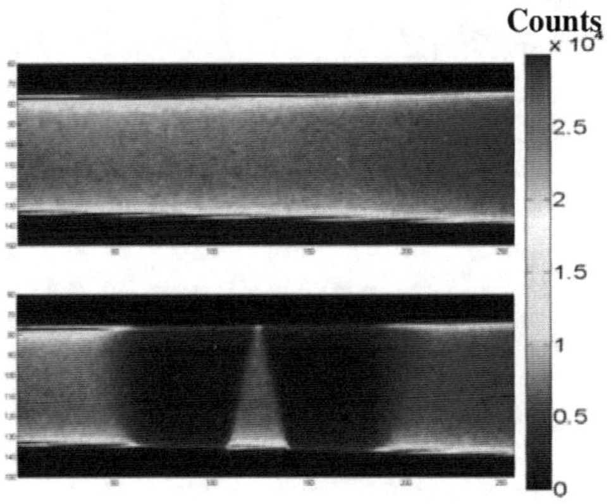

Figure 5. Example of Raman signal distributionin air and in the laminar Bunsen flame cone.

4.1. 2D Raman system (cylindrical laser beam)

For registration of the Raman signal, a 16-bit ICCD camera (Princeton instruments PI-MAX-4) equipped with a tunable optical filter (VariSpec LC), based on liquid crystals, was used to collect the emission for the desired wavelength, corresponding to the vibronic transition of nitrogen (607.3 nm when excited at 532 nm). A multi-notch holographic filter was also used to block the emission from Rayleigh scattering, which is about 104 times greater than the Raman scattering and was not blocked completely by the liquid-crystal filter. The scattered light from 200 laser pulses was collected during acquisition of each image with 8x8 pixels binning. The collection time was 200 ns. Figure 4 shows the temperature distribution in laminar Bunsen flame cone measured by 2D Raman technique with cylindrical laser beam.

4.2. 2D Raman system (collimated laser sheet)
To increase the spatial resolution and reduce an error was used a more powerful laser (Quanta-Ray). To provide a measurement in plane the cylindrical laser beam was transformed into the laser sheet by using collimating optics. Examples of Raman signal intensity distribution in air and the flame is shown in figure 5. Figure 6 demonstrates temperature distribution in a laminar Bunsen flame cone. Black line shows the borders between three temperature fields measured at different heights above the burner.

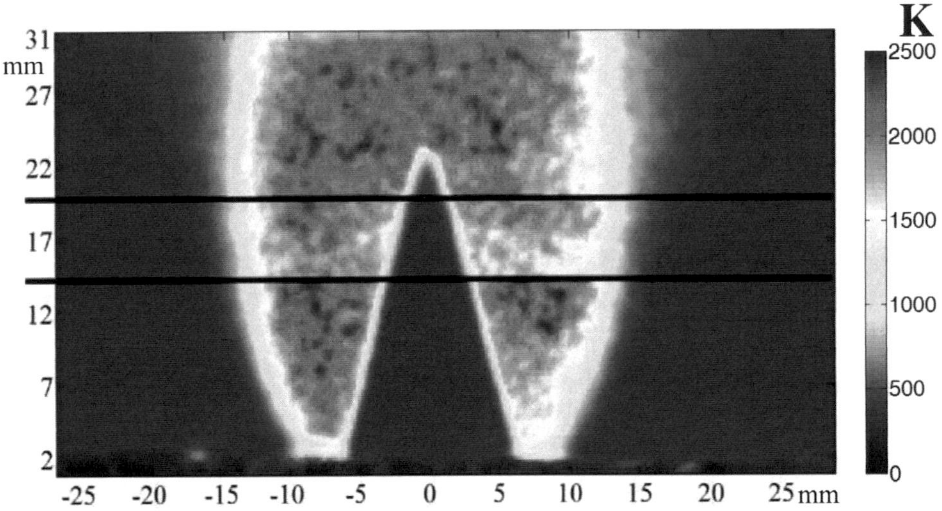

Figure 6. Temperature distribution in laminar Bunsen flame cone measured by 2D Raman technique with collimated laser sheet.

5. Conclusion

Comparison of 1D Raman temperature measurements with results of thermocouple (type R) measurements shows acceptable accuracy of the optical system for temperatures below 1800 K. The characteristic temperature of the propane flame, as can be seen from Figure 3, is about 2100 K. For temperatures above 1800 K the accuracy of the temperature measurements was reduced by, a limited dynamic range of the ICCD sensor. In case of 2D Raman system more sensitive 16-bit ICCD camera (Princeton instruments PI-MAX-4) was used to provide precision temperature measurements. Figure 5 shows that the results of temperature measurements slightly higher than expected, which is associated with low signal intensity. This work demonstrates the possibility of measuring the spatial temperature distribution in the reacting flow by method based on Raman scattering. The increase in signal-to-noise ratio for further studies by using more precisely selected conditions of the experiment will reduce the error in determining the temperature.

Acknowledgments
Financial support has been obtained from RFBR (grant № 14-08-01233 A).

References

[1] Rabenstein F and Leipertz A 1998 J. Appl. Opt. **37** 4937
[2] Egermann J, SeegerT and Leipertz A 2004 J. Appl. Opt. **43** pp 5564
[3] Hassel E P 1993 J. Appl. Opt. **32** 4058
[4] Shmakov A, Knyazkov D, Korobeinichev O, Bolshova T, Dmitriev A, Lobasov A. and Sharaborin D 2014 *Book of abstracts of the 8th International Seminar on Flame Structure* (Berlin) pp 1-6.

Development of non-contact laser vibrometer

I. Kononov[1], E. Velichko[1,2], E. Aksenov[1]

[1]Saint Petersburg State Polytechnic University, Saint Petersburg, Russia
[2]Saint-Petersburg National Research University of Information Technologies, Mechanics and Optics, Saint-Petersburg, Russia

E-mail: kononov.ivan.1993@gmail.com

Abstract. Modified schemes of noncontact laser sensors for vibrometry are considered. The results of investigation of two laboratory models of laser speckle vibrometers are presented. Advantages and disadvantages of the vibrometer sensors are discussed.

1. Introduction

Development of methods and devices for nondestructive control of physico-mechanical properties of different objects is one of the important tasks of modern technique and machinery. There are a number of methods and devices for remote control of mechanical characteristics; however, the challenges and new tasks require the development of methods and devices of vibrational monitoring of different objects.

The laser-based optical methods of non-contact remote monitoring are in fact an extremely powerful tool for the study of the behavior of components during operation. One of the contactless optic methods is laser speckle-interferometry based on rather simple and cost-effective optical sensors. At the same time the method accuracy is comparable with the precision of more complex holographic systems.

The aim of our work was to develop methods and create laboratory models of laser speckle-vibrometry sensors for non-contact remote vibrometry of large- and small-scale objects. Vibration characteristics are important for evaluation of physical and mechanical parameters of different objects, such as bridges, dams, jet engines, electric motors, biological cells and tissues, etc.

2. Experimental study

Registration of input light signal by optical fiber with core diameter which is smaller than size of single speckle is mostly used in studies of speckle field. A magnitude of signal transmitted through the fiber is small and photomultiplier is necessary. However, this scheme can be simplified by using of a diaphragm and wide-aperture photodetector.

Speckle pattern is formed in plane of aperture with average radius of speckle ρ

$$\rho = {}^{\lambda z}/_{d}, \quad (1)$$

where λ is wavelength of the laser; z is the distance between surface of the object and photodetector plane; d is the diameter of illuminated area of object's surface.

Speckle pattern is formed in plane of photodetector by changing d and z with average speckle size ρ, which fulfil the condition:

$$h < \rho < D \tag{2}$$

where h is the shift of speckle in plane of photodetector; D is the size of aperture in plane of photodetector.

In this study special attention was paid to non-contact registration of man's pulse shape of the radiocarpal artery. To solve this problem two laboratory models of speckle-vibrometers were developed: the first one was based on the point contactless sensing of skin and the second one was based on ringed light field sensing. The scheme of the speckle-vibrometer model with point sensing is presented in fig. 1.

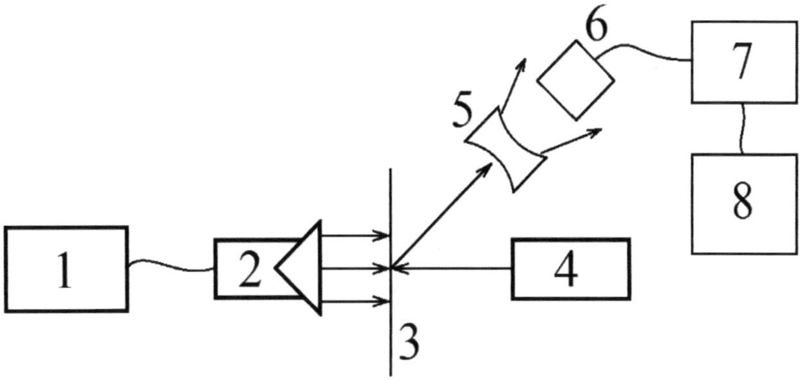

Figure 1. Speckle vibrometer scheme. 1 – sound frequency generator, 2 – speaker, 3 – target, 4 – semiconductor laser, 5 – diverging lens, 6 – photodiode, 7 – oscilloscope, 8 – computer.

We had revealed a number of limitations of the speckle-vibrometer model with point sensing:

- criticality to location of the sensing laser beam;
- detecting of signal from various light sources on photodetector and so the presence of pulse interferences;
- loss of signal in case of laser beam shifting from the artery.

To reduce the influence of sensor positioning on the signal it was proposed to use several infrared LEDs circumferentially. To create the circular illumination the optical system with axicon was developed (fig. 2). Vibrations with intended frequency were simulated by membrane which was excited by sound wave from generator of acoustic oscillations GZ-112.

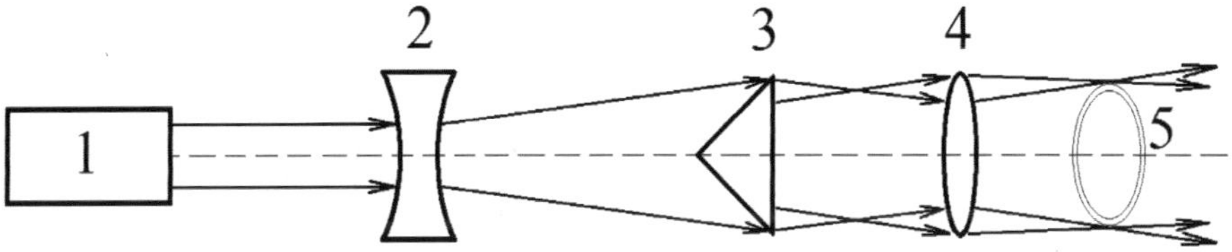

Figure 2. Optical system of ringed sensing. 1 – semiconductor laser, 2 – diverging lens, 3 – axicon, 4 – focusing lens, 5 – circular illumination.

The experimental setup is presented in fig. 3.

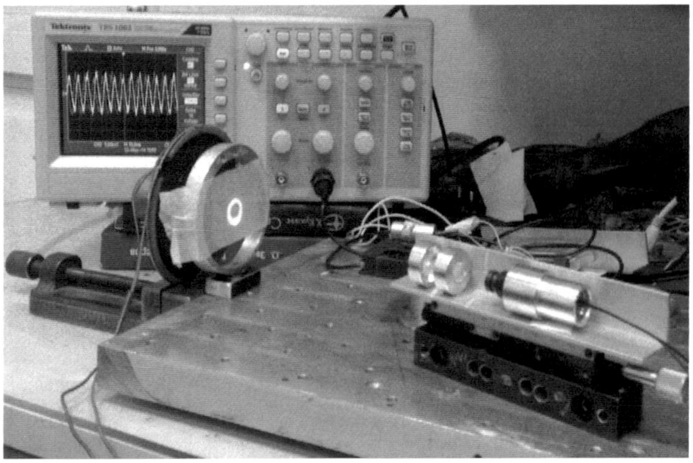

Figure 3. Experimental setup.

Diameter of ring illumination remained constant for longitudinal displacement of optical system within at least ten centimeters (Fig. 4).

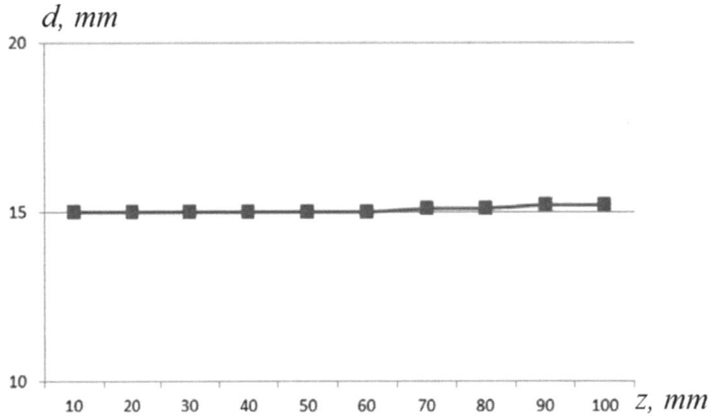

Figure 4. Ring diameter versus distance from the surface of the object.

Measured dynamic range of laboratory model of the sensor was 30 dB. Linear mode was observed up to output voltage of generator 0.3 V. After this threshold nonlinear mode of target stimulation was observed (Fig. 5).

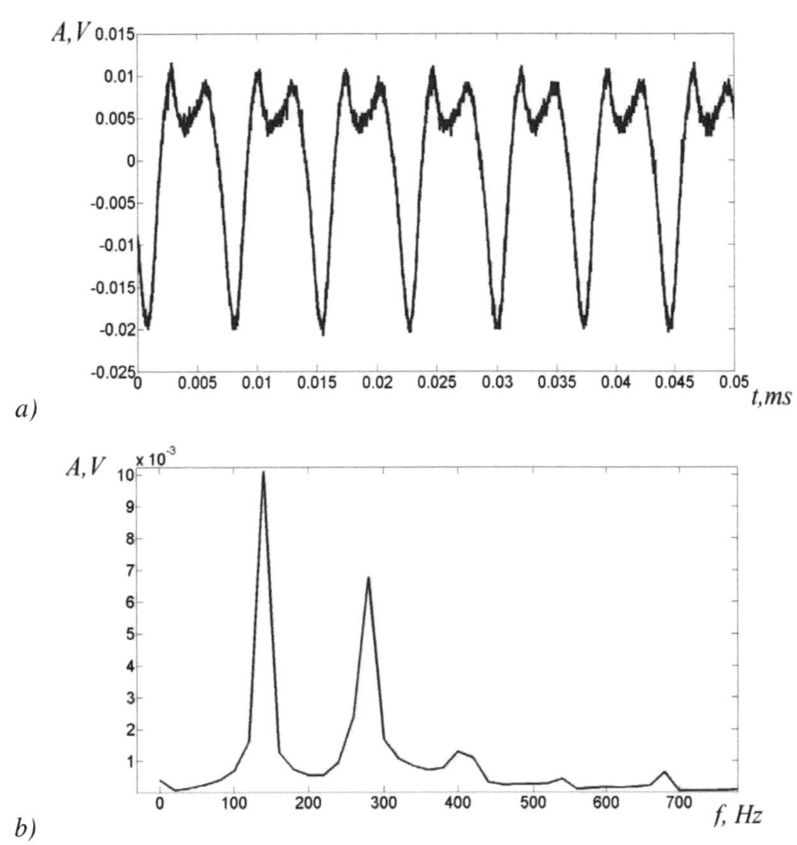

Figure 5. Shape (a) and spectrum (b) of the signal.

3. Conclusion

The results of our study are the following:

1. Technique of non-contact opto-electronic registration of pulse wave form was developed;
2. Scheme of non-contact sensor based on registration speckle field was proposed and created;
3. Axicon-based scheme of optical sensor with ring illumination was developed and investigated;
4. Advantages of scheme based on axicon were confirmed:
   - high sensitivity;
   - relative insensibility to positioning on arm;
   - independence of diameter of the formed ring on distance to study object;
   - contactless measurement of recording pulse signal.

The obtained results provide the basis for further investigation and development of laser vibrometer.

References

[1] Korotaev V, Timofeev A and Ivanov A 2000 *Optical Journal* **67**
[2] Desova A 1998 *Instruments and systems of control* **8**

Journal of Physics: Conference Series **643** (2015) 012036 doi:10.1088/1742-6596/643/1/012036

Monte-Carlo modeling of drift mobility in thin organic layers: lattice type influence

V M Sukharev and V R Nikitenko

National Research Nuclear University MEPhI, Moscow, Russia

E-mail: VMSukharev@mephi.ru, VRNikitenko@mephi.ru

Abstract. Hopping transport of charge carriers in thin (up to 100 nm) organic layers, which are suitable for organic light-emitting diodes and solar cells, is modeled in the framework of Gaussian disorder model. Drift mobility dependence within ranges of layer thickness, energy disorder and field strength is considered. Monte-Carlo simulations for low concentration case show weak dependence on type of lattice – simple cubic, body-centered and face-centered lattice. Results of this work are compared with the results obtained recently by Cottaar et al., done within master equation approach. The comparison shows the deviation at low disorder. Additionally we take into account influence of barrier-limited injection in case of cubic lattice.

1. Introduction

Organic electronics devices consist of thin active layers. Predictive modeling of operation of such devices unavoidably needs correct characterization of charge transport in thin organic layers. Charge transport, being of hopping nature, can be described in the framework of Gaussian disorder model [1]. Values of mobility, which have been obtained from experiments for thicker layers (> 1 micrometer), are often used for modeling the light-emitting diodes. The recent review [2] combined all the results that had been published in this scientific area for the last few decades. As it was shown, lattice type of an organic volume does not have big influence on dynamic characteristics of a particle. It can be explained in terms of percolation theory [3]: correlation radius R is much bigger than lattice constant a of a cell, so there is no difference what type of a lattice is. However, all main computing results, dedicated to dependence on lattice type, were obtained within master equation approach to electron transport with periodic boundary conditions [4]. So it would be of interest to check numerically, whether mobility in thin films depend on lattice type, or not, using complete Monte-Carlo (MC) modeling, which is superior respective to master equation approach.

2. Monte-Carlo procedure and scheme of modeling

In our previous work [5] we took into account the scenario of MC simulations without the influence of the electrode (dummy injecting electrode), in order to model properties of a thin organic layer itself, irrespective to conditions of injection. Here we use a scenario, where an event starts with an electron being born in the first layer and this scenario allows the particle to be reflected from left boundary, in order to compare results with more realistic case of energetic barrier-limited injection, with barrier being reduced by image-charge Coulomb potential (thermionic injection). A carrier, being born in the 1-th layer, walks through the whole volume until it reaches the last layer. Influence of image charge was obtained for simple cubic (SC) lattice only, but it is enough to estimate main dependences of the phenomenon. We consider not only SC, but two more options: body-centered (BC) and face-centered (FC) lattice (see Fig. 1, *a-c*). Injection of a charge carrier from the electrode into the layer of organic material should be presented as a drift and diffusion via manifold of localized states, [6] and the potential energy of a carrier has to be given in the form: $U(x) = H - e^2/16\pi\varepsilon\varepsilon_0 x - eF_0 x$, where H is the height of the barrier (see Fig.2), which is equal to *0,7 eV* (this barrier is reduced by one-dimensional Coulomb interaction with image charge) , and ε in our calculations is equal to *3* [6].

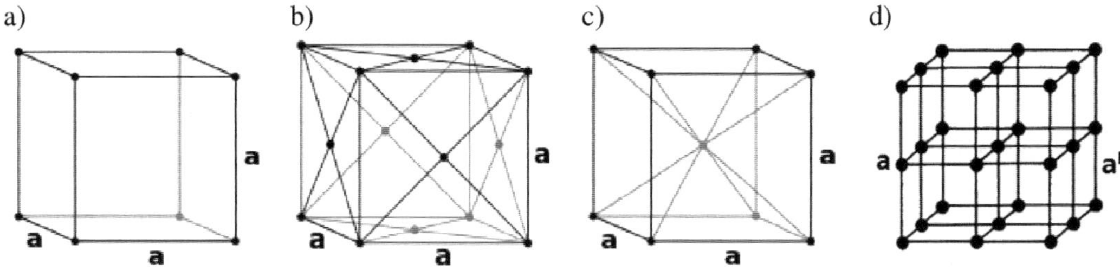

Fig.1 Lattice types involved in MC modeling. Fig. 1, *a-c*, represents cubic, face-centered and body-centered lattice types, respectively. Fig. 1, *d* shows a feature of MC modeling for being able to model different lattice types: each standard cubic cell is subdivided into eight elementary cells with a lattice constant a'.

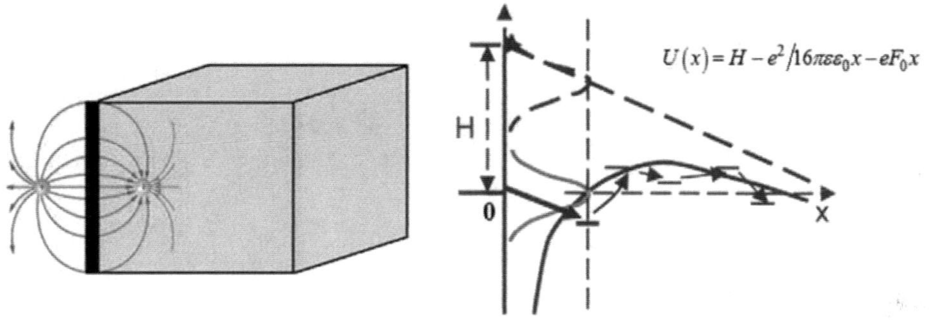

Fig.2 Scheme of electron injection in the presence of an energy barrier of height *H*.

A conventional model is used, where hopping centers are located at sites of a lattice. Number of sites along any transversal direction exceeds 10 times a number of sites along the electric field direction. Energies of sites are distributed randomly according to Gaussian distribution $g(E) = \left(1/\sqrt{2\pi\sigma^2}\right)\exp\left[-E^2/2\sigma^2\right]$ with the variance σ. If an electron appears in the nearest to right electrode layer, event is over. Drift mobility is calculated from the average drift velocity, $\mu = L/F\langle 1/t_{tr}\rangle$, where L – length of the volume, F – strength of the field applied, t_{tr} – transition time. It is also important to notice, that hopping rates that we use in our calculations, are determined by usual Miller-Abraham's expression: $\nu_{ij} = \omega_0 \exp\left[-2\gamma r - \left(\Delta E_{ij} + \left|\Delta E_{ij}\right|\right)/2kT\right]$, where ω_0 is frequency factor, γ is inverse radius of localization, r is a distance between two sites, $\Delta E_{ij} = E_i - E_j - eFr\cos\theta$, T is the absolute temperature, k_B is Boltzmann's constant and θ is the angle of directions of electric field $\boldsymbol{F}$ and jump $\boldsymbol{r}$. The value $2\gamma a$ = 10 is assumed. The main difference between modeling SC and, for example, FC lattice is that for face-centered lattice option we have additional sites that an electron can visit. To make it simple from the programing point of view, one should subdivide an original cell into eight elementary cells with a constant lattice a', which is twice smaller than the original one, a, and allow visiting only those sites, that provide a cell of concrete shape needed (see Fig. 1, d). Although this procedure is simple, but it takes much more time to get required statistics because of considering more sites for hopping.

The analytic expression for drift mobility has the form [7]:

$$\mu(T) = B\frac{e\omega_0}{N_t^{2/3}\sigma}\hat{\sigma}^{1-\lambda}\exp\left(-\frac{1}{2}\hat{\sigma}^2 - \frac{E_{crit}}{k_B T}\right), \quad (1)$$

in case of low relative concentration of charge carriers, that is true in MC-modelling,

$$c = n/N_t \,\square\, c_* = (1/2)\exp\left[-(\sigma/kT)^2/2\right], \quad (2)$$

where B is the numerical prefactor, e – the electron charge, λ – the exponent, $\hat{\sigma} \equiv \sigma/k_B T$, E_{crit} – the critical (transport) energy, N_t – the site density, ω_0 is the frequency factor. In [6] authors use both Miller-Abrahams and Marcus hopping rates [8], [8]. Parameters λ and E_{crit} are claimed to be influenced by lattice type. It has to emphasize, that in [7] Eq. (1) is taken to fit obtained mobility data, that are calculated for low field strength and relative carrier concentration, c, being in range of *~10^-5^ – 10^-1^*, see also [4]. Numerical values of B, λ and E_{crit}, that depends on σ, are given in [7] for different lattice types (except BC lattice, that is not considered there).

3. Results and discussion

The results in [7] were obtained using periodic boundary conditions and master equation approach, modeling the infinite media with applied small electric field for SC and FC lattice types. The values of σ in [7] are considered in the whole range of disorder, which is technically gives us the opportunity to compare our results for SC and FC with [7] for $L = 100\ nm$ and $F = 10^7\ V/m$, but one has to remember what conditions Eq. (1) was obtained under, plus authors in [7] do not calculate mobility in low disorder. The brief comparison is summarized in Table 1, where values of the mobility, calculated according to Eq. (1) with all the variables used for MC modeling, are divided by MC results themselves.

Table 1. The comparison of Eq. (1) relative to MC results for SC and FC lattice types

$F = 10^7\ V/m$, $L = 100\ nm$	Eq. (1) / MC, SC	Eq. (1) / MC, FC
$\sigma = 0.1\ kT$	4.766	7.963
$\sigma = 3\ kT$	0.516	1.339
$\sigma = 4\ kT$	0.189	0.651

As one can see, for low disorder the relative comparison shows the biggest deviation. One has to note, that in [7] comparison of analytic (1) and numerical results is absent in case of low disorder. Fig. 3 *a*,*b* shows that both MC and analytic results have only slight dependence on a lattice type, in accord with previous suggestions [3]. However, the difference increases along with the increase of disorder (it reaches 50% at $\sigma/kT = 4$, see Fig. 1*b*), so one has to take it into account in quantitative modeling. MC results exceed mobility from [7] at high disorder due to finite-size effect on mobility in thin films [5]. Deviations in low-disorder limit result from absence of applicability of the analytic approach [7]. One has to expect that with field strength increasing all the deviations will be negligible quantities. The truth of this statement is demonstrated on Fig.3,4, where three lattice types are compared with each other in strong field modes, $2.5\ 10^7$ and $5\ 10^7$ *V/m*. For the case of barrier-limited injection, see Fig. 5, it is shown, that this realistic effect smoothens slightly the thickness dependence of drift mobility, which has been reported in [5], but deviations are not principal (values of mobility being slightly reduced).

Journal of Physics: Conference Series **643** (2015) 012036 doi:10.1088/1742-6596/643/1/012036

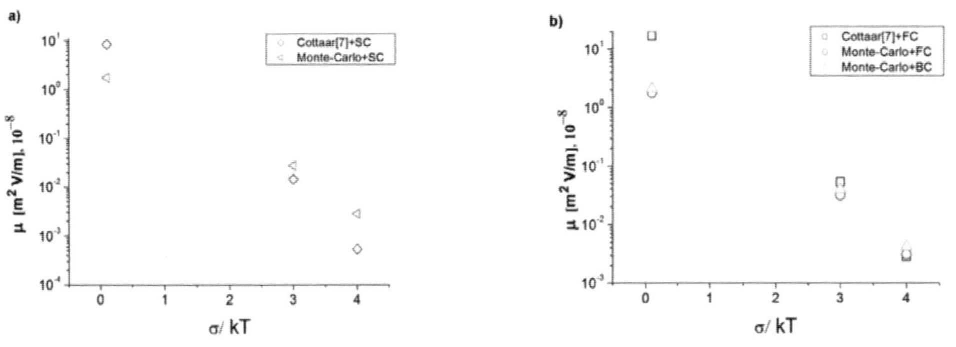

Fig. 3 Comparison MC results with [7] for two different cases: a) SC and b) FC and BC lattice. MC results were obtained for field strength 10^7 *V/m*, *σ/kT = 0.1, 3, 4* and c □ c_*.

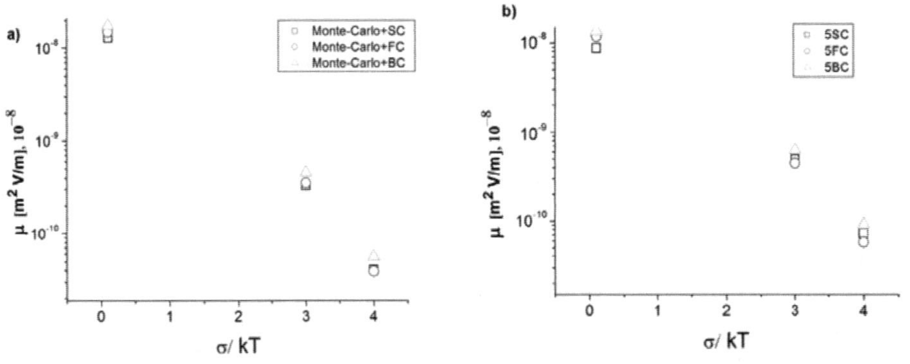

Fig. 4 Comparison MC results for different lattice types for two different cases: a) for field strength *2.5* 10^7 *V/m* and b) for field strength *5* 10^7 *V/m*; *σ/kT = 0.1, 3, 4* and c □ c_*.

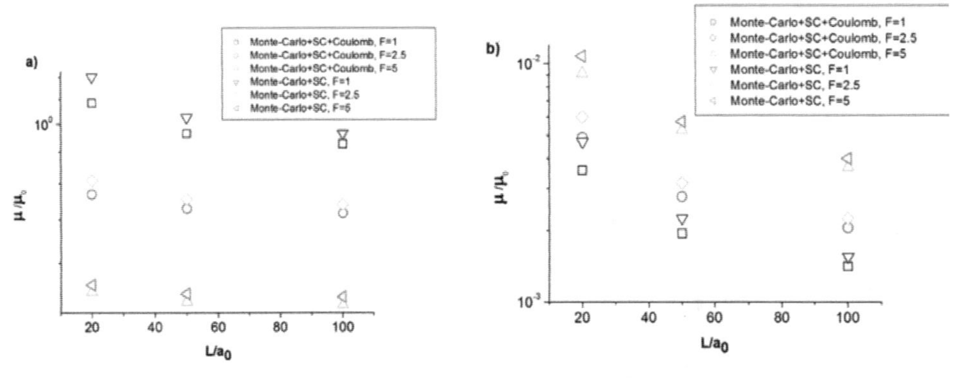

Fig. 5 Comparison MC results for SC lattice type, taking into account Coulomb interaction, for two different cases: a) for low disorder *σ/kT = 0.1*and b) for high disorder *σ/kT = 4*; *F = 1, 2.5, 5* 10^7 *V/m* and c □ c_*.

This work was supported by Ministry of Education and Science of Russian Federation, Agreement No. 14.575.21.0002, ID RFMEFI57514X0002.

References

[1] Bassler H Charge Transport in Disordered Organic Photoconductors a Monte Carlo Simulation Study 1993 *Phys. Status Solidi B* **175** 15
[2] Baranovskii S D Theoretical description of charge transport in disordered organic semiconductors 2014 *Phys. Status Solidi B* **251** 487–525
[3] Shklovskii B I and Efros A L Electronic Properties of Doped Semiconductors 1984 *Springer Berlin*
[4] Cottaar J Coehoorn R and Bobbert P A Scaling theory for percolative charge transport in molecular semiconductors: Correlated versus uncorrelated energetic disorder 2012 *Phys. Rev. B* **85** 245205
[5] Nikitenko V R Sukharev V M and Sannikova N A Drift mobility in thin organic layers: joint Monte-Carlo and analytic modeling 2014 *J. Phys.: Conf. Ser.* **541** 012028
[6] Arkhipov V I Wolf U and Bässler H Current injection from a metal to a disordered hopping system. II. Comparison between analytic theory and simulation 1999 *Phys. Rev. B* **59** 7514
[7] Cottaar J Koster L J A Coehoorn R and Bobbert P A Scaling theory for percolative charge transport in disordered molecular semiconductors 2011 *Phys. Rev. Lett.* **107** 136601
[8] Miller A and Abrahams E Impurity conduction at low concentrations 1960 *Phys. Rev.* **120** 745
[9] Marcus R A Electron transfer reactions in chemistry. Theory and experiment 1993 *Rev. Mod. Phys.* **65** 599
[10] Zvyagin I P A Percolation Approach to the Temperature and Charge Carrier Concentration Dependence of the Hopping Conductivity in Organic Materials 2008 *Phys. Stat. Sol.* **5** 725
[11] Coehoorn R Pasveer W F Bobbert P A and Michels M A J Charge-carrier concentration dependence of the hopping mobility in organic materials with Gaussian disorder 2005 *Phys. Rev. B* **72** 155206.

Multifunctional optical processor for biological micro-objects investigation

K. Domnin[1], E. Aksenov[1]

[1]Saint Petersburg State Polytechnic University, Saint Petersburg, Russia

E-mail: konstant.domnin@gmail.com

Abstract. The combination of two methods: diffractometry and polarimetry was considered. The optical processor device was realized on base of these methods. Biological objects and their phantoms investigations were carried out.

1. Introduction

Optical methods of researches time are usually applied for studying of various objects, including bio-objects at the microlevel. These methods have a huge potential for noninvasive medical express diagnostics.

Different changes at the pathological conditions of an organism courses some deviations in blood morphological and chemical composition. These deviations reflect physiological shifts in organs and their functions, showing pathological processes developing in them.

Abnormalities in morphological composition of blood are expressed in reduction or increase of erythrocytes and leukocytes quantity, in change of uniform blood elements ratios and erythrocytes size and form. At shifts of physical, chemical and biological blood properties it is usually observed viscosity change, violation of blood coagulability process, etc.

Erythrocytes (red blood cells) have a highest sensitivity to any pathologic changes in an organism, therefore they can be convinient objects for research and assessment physiological condition of an organism.

There exist some optical methods for microobjects researches. One of them is a laser diffractometry method, based on diffraction of the laser radiation on a single or multiple biological microobjects [1,2]. This method is characterized by high precision, sensitivity, speed, minimum invasivity, possibility of a huge number of small particles simultaneous registration. The diffraction picture parameters are connected with microobject parameters, so it allows to determine their sizes, form, internal structure.

The other method having a rather high potential for application in bioobjects researches is a laser polarimetry. Polarimetric diagnostics allows researchers to receive information about optical anisotropy of biological tissues.

Optical methods usually allow one to solve problems which are insoluble or difficult for other methods. They are characterized by high sensitivity but make great demands of experiment conditions because of methods high sensitivity to external influences. The potential opportunities of optical processing are obvious: two-dimension research, real time working, high speed processing of big arrangements of data, wide frequencies bandwidth, etc. The theory describing operation of devices is rather well developed. In connection with this, we determined the objectives of this work: to develop a multifunctional optical processor for biological microobject and its phantoms research.

2. Method and experimental device

The conceptual design of the experimental processor is presented in fig. 1. The device has three information channels: two measuring and one control. To make available an experiment it is necessary to provide a maximum uniformity of object lighting. Uniformity of object lighting can be accomplished due to cut only the central part of the light beam formed by the collimator.

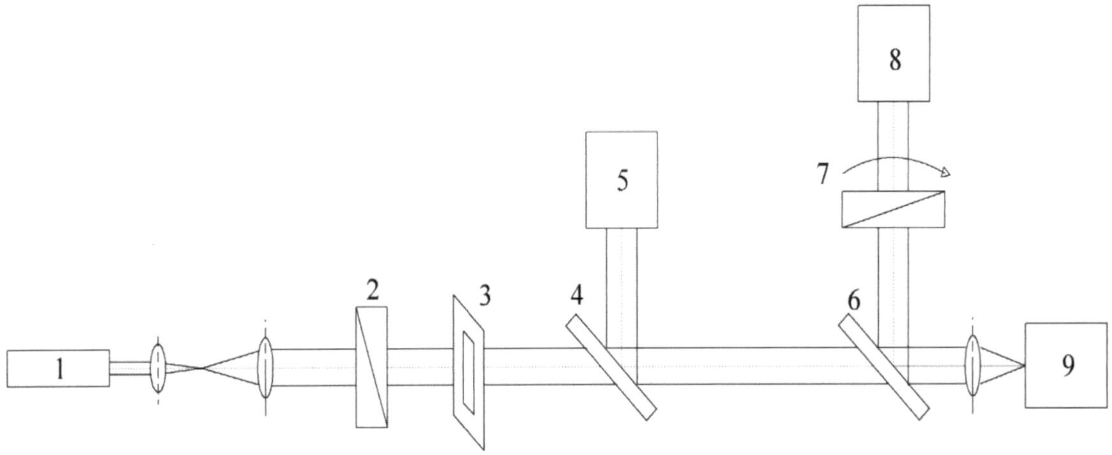

Figure 1. Structural scheme of experimental setup. 1 – laser, 2,7 – polarizers, 3 – object, 4,6 – beam splitters, 5,8,9 – CCD cameras

This is made by installing a pinhole in a focal spot of the laser beam. For focusing or expansion of a beam it is possible to use a microlens. Behind a lens the spherical front of wave is formed. For receiving the flat wave front it is necessary to add the collimating lens. Thus focus of a lens has to coincide with the central point of a diaphragm. At the exit we receive a parallel beam of light. The diaphragm allows changing diameter of a beam.

Thus, on the transparent with object of interest, placed before a lens, falls the flat wave. In the back focal plane of a lens is formed a complex amplitudes distribution which can be described as Fourier transform of sample transmission function t(x, y). It is correct irrespective of distance between a lens and the transparent. The additional phase of a spherical wave can be disposed by setting the transparent in the forward focal plane of a lens.

Further there is a division of a light beam into three channels: polarimetric, spectral and visual. In the spectral channel we registered an intensity of the diffraction picture (DP) by means of the CCD camera. Then data were saved on the personal computer and the section of a DP is analyzed. For the data analysis it was used the program "Vision assistant". It is included in the "National Instruments Vision Development Module" which is intended for engineers and scientists using technical sight divices in industrial and scientific tasks. By means of the vision assistant the received DP were examined. The DP section in various directions were taken and visualized by means of the "Mathcad" program.

In fig.2 and fig. 3 the diffractometric image of blood smear and DP section for transparent with human blood smear are presented. Pictures were taken to define how does erythrocytes transform their shapes in time after blood taken. The diffraction picture became more ellipsoidal after 10 minutes that can specify a change in erythrocytes state, it this case it is supposed that red blood cells formed structures called coins pill. We have seen this structures visually by microscope.

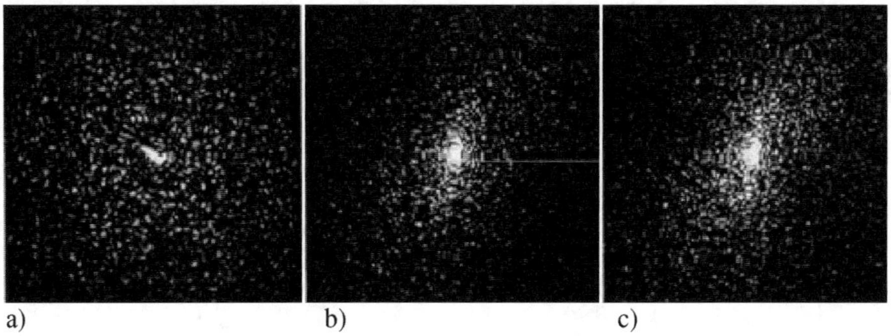

Figure 2 Diffractometric image of blood smear a) a start of experiment; b) after 10 minutes; c) after 35 minutes

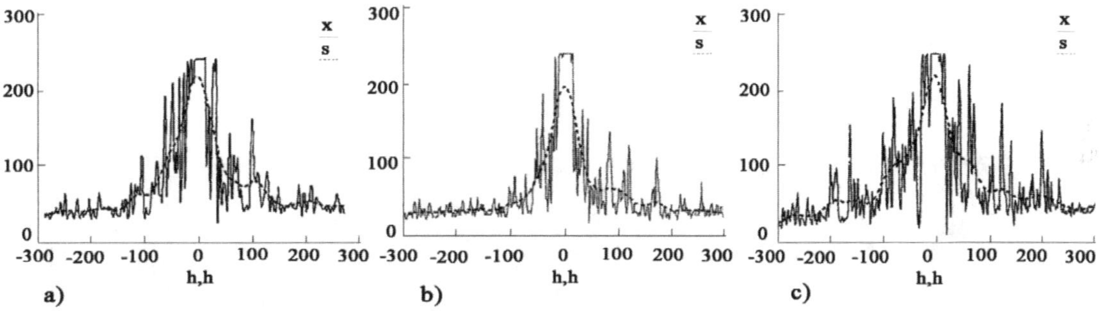

Figure 3 The DP section for transparent with human blood smear. a) a start of experiment; b) after 10 minutes; c) after 35 minutes

The polarimetric channel registered the image with consideration for polarization and used for detection of an object anisotropy. In this channel in front of the CCD camera was located the polarizer. In experiment two types of measurements were performed: then the polarizer plane of transmission was parallel to the polarization plane of the probing light beam, and then polarizer plane of transmission was perpendicular to the polarization plane of the probing light beam. In fig. 4 and 5 are presented pictures from CCD camera, taken on different transparent in two different output polarizer positions.

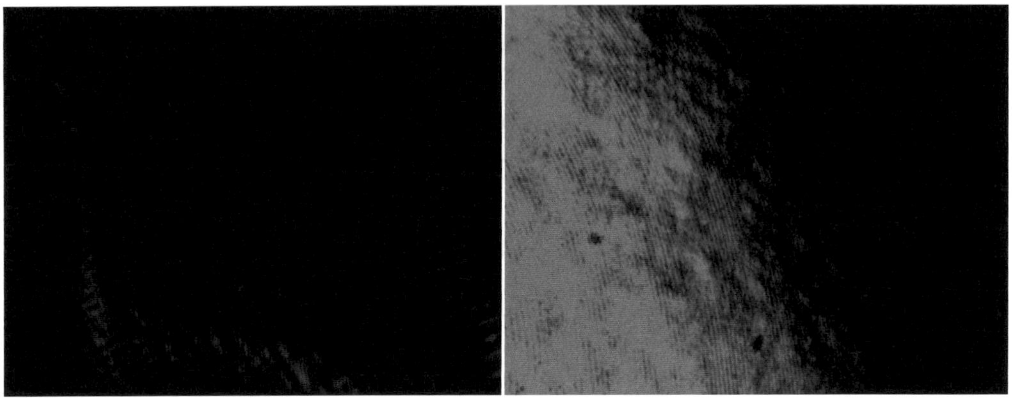

Figure 4 Results of research of the plate from glass varnished (the induced anisotropy) on the polarimetric channel. At the left – perpendicular orientation of an output polarizer, on the right – parallel orientation of an output polarizer

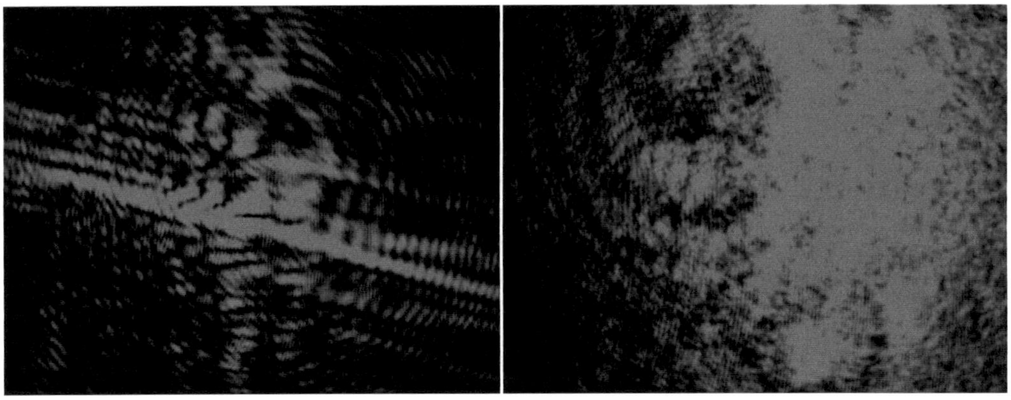

Figure 5 Results of research of the human blood placed between subject glasses on the polarimetric channel. At the left – perpendicular orientation of an output polarizer, on the right – parallel orientation of an output polarizer.

3. Conclusion
For carrying out preliminary pilot studies the measuring setup which is based on design model of the processor was mounted. Measurements on channels in the isolated and joint modes were taken.

Preliminary measurements within the polarimetric channel of the optical processor demonstrated its sensitivity to objects optical anisotropy, and measurements within the diffractomeric channel proved its sensitivity to the scatters size. Results of measurements under a microscope and the measurements received by a diffractomeric method are in good compliance.

4. References
1. Bessmeltsev S, Lendiaev A, Tarlykov V 2004 *Proc. SPIE* **5381**151-156
2. Bessmeltsev S, Tarlykov V, Hodus I 2004 *Proc. SPIE* **5381**146-150

High dynamic extinction ratio and pulse modulation of optical signals

A. Petrov[1,2], A.V. Tronev,[3] V. Lebedev[1], I.V. Il'ichev[1], E. Velichko[2,3], A.Shamray[1,2,3]

[1]Ioffe Institute, RAS, St.Petersburg, Russia
[2] Peter the Great St.Petersburg Polytechnic University, St.Petersburg, Russia
[3]Saint-Petersburg National Research University of Information Technologies, Mechanics and Optics, St.Petersburg, Russia

alexey-np@yandex.ru

Abstract. The use of a saturable absorber for increasing the extinction ratio at external modulation of optical signals is considered. An erbium doped fiber was used as the saturable absorber in the experiments. A considerable increase in the static extinction ratio (up to 50 dB) was demonstrated. A rather long erbium doped fiber relaxation time (about 10 ms) was a limiting factor in the case of pulse modulation. Ways of overcoming this drawback are discussed.

1. Introduction

Pulsed optical radiation sources are widely used in modern systems of high-precision fiber-optic sensors. The extinction ratio of optical pulses in time multiplexers, interrogators and distributed systems based on coherent reflectometry defines the noise level, dynamic range and sensitivity of the system [1]. One of approaches to optical pulse generation involves the use of a highly coherent DFB laser diode in the CW regime and an external high-frequency integrated optical modulator [2]. Typical values of the electrooptical modulator extinction ratio are in the range 20–30 dB, however, a number of applications require the ratio of about 50–60 dB.

In our work the nonlinear transmission of saturable absorber (with high losses at the lower input power and lower losses at high input power) is used to improve the extinction ratio. An erbium doped fiber was chosen as the saturable absorber because of the required wavelength range (1520 – 1565 nm), lower saturation power, and excellent compatibility with fiber optic systems.

2. Experimental setup

Figure 1 shows an optical scheme of the experimental setup. A DFB laser diode with a central wavelength of 1530 nm and controlled output laser power of up to 20 mW was used as an optical source. The external modulator was a Mach-Zehnder lithium niobate integrated optical modulator with a half-wave voltage of 4 V, modulation bandwidth of about 10 GHz and optical losses of 3 dB. The maximum extinction ratio of the modulator was about 30 dB which was checked in the static regime. To investigate the dynamic extinction ratio, the modulation signal from the waveform generator was fed to the modulator RF input. A special scheme for the flat bottom pulse generation was used [3]. The DC input was used to measure the static extinction ratio and to control the bias at pulse modulation. A section of the erbium doped fiber EDF (HE980 Lucent) of a 3-m length was used as the saturable absorber. A fiber optic coupler with a 50/50 ratio at the input of the saturable absorber was used to monitor the input power and control the extinction ratio at the modulator output. An optical power meter was connected to the 50% output of the coupler in static measurements. In dynamic measurements, a high frequency photodetector was connected to this output. A WDM fiber optic filter with the central wavelength at 1530 nm and wavelength bandwidth of about 0.1 nm was used to reduce noise and background caused by spontaneous emission at the EDF output. The final extinction ratio was measured by an optical powermeter in the static (DC) regime or by a fast photodetector in the dynamic mode. To observe a very high extinction ratio, a nonlinear signal amplifier with the gain characteristic close to the logarithmic one was designed. The output optical pulse shape was monitored at an oscilloscope (Tektronix TDS 1012C-EDU).

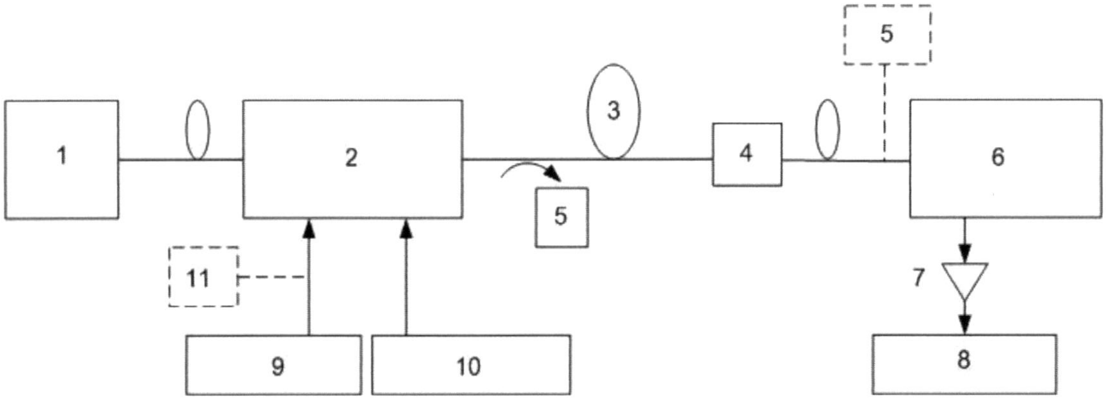

Figure 1. Experimental setup: 1 – DFB laser diode; 2 – Mach-Zehnder lithium niobate integrated optical modulator; 3 – saturable absorber (EDF); 4 – WDM fiber optic filter; 5 – optical powermeter; 6 – fast photodetector; 7 - nonlinear signal amplifier; 8 – oscilloscope; 9 – waveform generator; 10 – DC power supply for biasing; 11 – scheme for pulse bottom formation.

3. Static (DC) regime

At first the nonlinear transmission of the saturable absorber was studied in the static mode. A coarse input power control was performed by changing the DFB laser power, and fine tuning was performed by biasing the modulator. A typical saturation curve is shown in figure 2.

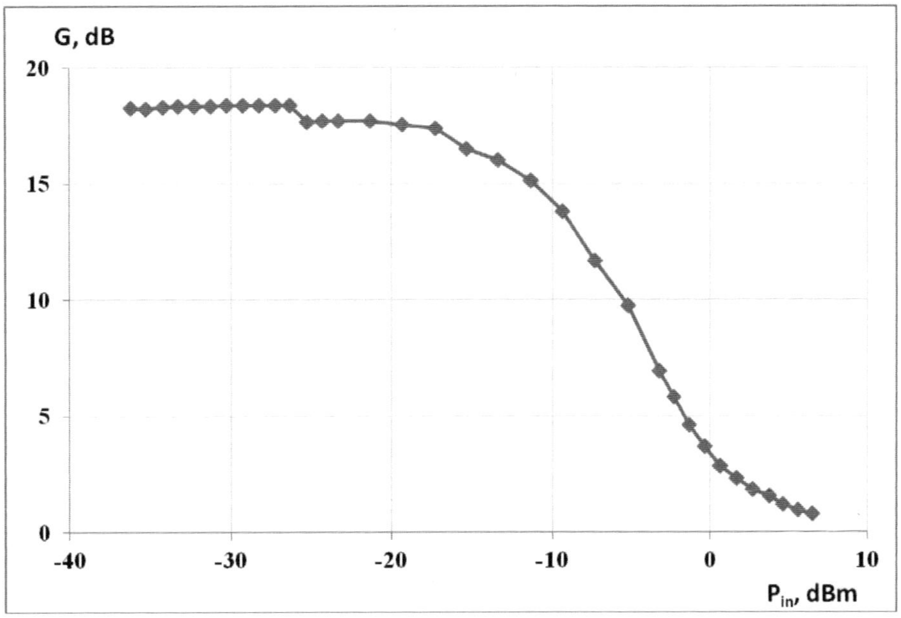

Figure 2. Optical transmission of saturable absorber as a function of input light power (saturation curve)

High losses (high absorption) of about 20 dB were observed at low input light powers. Low losses of about 2 dB were observed when the input light power exceeded the saturation power (~5 mW). The use of the saturable absorber allowed an about 20 dB increase in the extinction ratio as compared with the initial modulator extinction ratio. The extinction ratios for different input light powers are listed in table 1. The maximum static extinction ratio of 47 dB which was close to the sensitivity limit of our measurement equipment was demonstrated.

Table 1. Extinction ratio for different light powers at the modulator input

Input power	Extinction ratio
5 mW	7000
10 mW	22200
15 mW	37700
20 mW	46100
27 mW	52000

4. Dynamic mode at pulse modulation

To investigate the dynamic extinction ratio, the modulation by square shape pulses was used. The pulse duration was 10 ms, the repetition time was 10 ms. The pulse rise and fall time was < 1 μs. The oscilloscope traces of the detected optical pulse after the modulator are shown in figure 3a. A high reproducibility of the modulated signal was observed. The maximum extinction ratio calculated as the ratio between the upper and lower pulse levels with a correction for the nonlinearity of the transmission characteristic of the nonlinear signal amplifier was about 30 dB, which is in good agreement with the results of static measurements. The oscilloscope traces of the optical pulses after the saturable absorber are shown in figure 3b. The light power at the EDF input corresponding to the upper level of pulses was about 10 mW, which is enough for saturation. A rather long transition process (~10 ms) was observed at the trailing pulse edge. We attribute this transition process to the relaxation of erbium ions excited to the metastable level [4]. Because of a long relaxation time the EDF cannot be used for improvement of the extinction ratio when the modulation frequency is higher than several hundred Hertz. However, for the signal reaching the lowest level (very close to the noise bottom) estimation of the dynamic extinction ratio gives more than 40 dB, which is the evidence of applicability of the suggested approach in the case other saturable absorbers with short relaxation time are used. Semiconductor saturable absorbers can provide nanosecond and even picosecond relaxation times [5]. Acetylene-filled hollow-core photonic crystal fibers with typical relaxation times of several nanoseconds [6] are of particular interest since they provide the all-fiber implementation, like EDF.

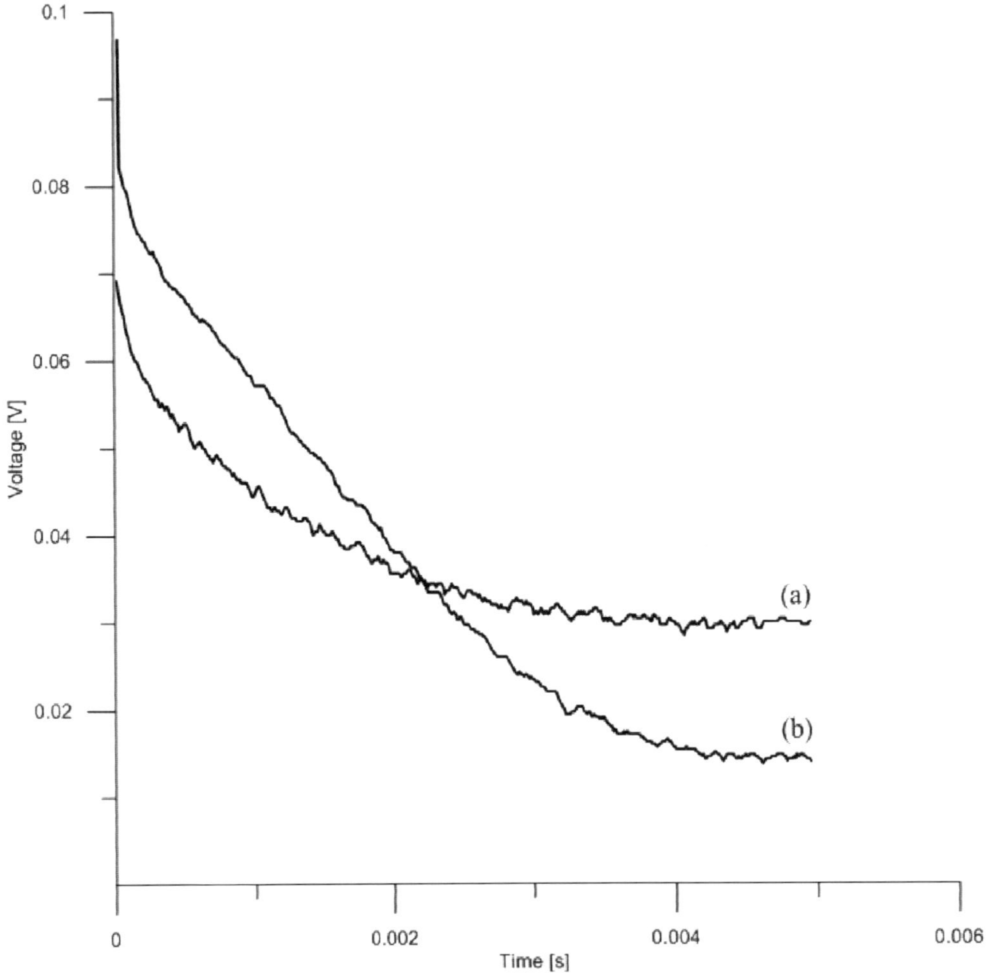

Figure 3. Oscilloscope traces: a) optical pulses at the modulator output; b) optical pulses after saturable absorber

5. Conclusions

To summarize, it is shown that the use of a saturable absorber allows one to considerably increase the extinction ratio of an optical pulse.

6. Acknowledgment

The sample of Mach-Zehnder lithium niobate integrated optical modulator was designed and fabricated in the framework of Project RFMEFI60414X0055 funded by the Ministry of Education and Science of the Russian Federation

References

[1] Brooks J L, Boslehi B, Kim B Y and Shaw H J 1987 *J. Lightwave Technol.* **5** 1014

[2] Huang S-C, Tsay J-S and Lin W-W 2000 *Opt. Eng.* **39** 2214

[3] Schoenwetter H.K. 1983 *IEEE Trans. Instrum. Meas.* 32 22

[4] Becker P C, Olsson N A, Simpson J R, 1999 *Erbium-Doped Fiber Amplifiers* (New York London, Boston: Academic press)

[5] Keller U 1999 *"Semiconductor nonlinearities for solid-state laser modelocking and Q-switching," in Semiconductors and Semimetals* vol. 59A, ed A Kost and E Garmire (Boston: Academic Press).

[6] Agruzov P, Shamray A, Miramontes M O, Hernández E H, Stepanov S 2012 *Applied Physics B* **108** 827

Journal of Physics: Conference Series **643** (2015) 012039 doi:10.1088/1742-6596/643/1/012039

Modulation doping of quantum dot laser active area and its impact on lasing performance

S S Konoplev, A V Savelyev, V V Korenev, M V Maximov, A E Zhukov

St Petersburg Academic University, St. Petersburg, 194021 Russia

E-mail: `Konoplev.Sergey@yandex.ru`

Abstract. We present a theoretical study of modulation doping of active region in the quantum dot (QD) laser and corresponding issues of QD charge neutrality violation, a band diagram of the laser and charge carriers distribution in the structure. Modulation doping is discussed as a possible technique to control laser output characteristics. It was shown that modulation doping leads to an increase of threshold current of lasing through excited QD optical transition together with power emission from QD ground state.

1. Introduction

The GaAs/AlGaAs semiconductor laser with InAs QD has several advantages compared to quantum well laser such as low threshold current and better temperature stability. The reason of such improvements is δ-like density of states in QD. However, when the excited states (ES) of quantum dots are involved QD laser advantages become less pronounced. The major goal of this work is to study possibilities of suppression of ES lasing via modulation doping of thin layers near QD planes. It was shown earlier [1] that QDs charging plays an important role in lasing though high-power lasing regime and ES emission were not discussed. Later, an impact of charge redistribution between QDs and surrounding matrix was intensively discussed (see [2, 3] and reference wherein) on a phenomenological basis. The ratio of hole to electron capture rates ('h-factor') was suggested as a phenomenological parameter that is constant with respect to injection current. We present a thorough theoretical study of a laser band diagram in high injection regime in order to clarify the origin of the non-unity value of the h-factor, its dependence on laser design and injection current. Finally, we study L-I curves in modulation doped structures and compare them with known experimental results [4].

2. Model of the QD layer

In order to discuss the interplay between ground state (GS) and excited state lasing the 3-energy level model has been chosen to describe charge carriers in QDs. There are two levels for electons (GS and ES) that are separated by $\sim$ 60 meV energy gap and one effective level for holes. The idea of an effective hole level derives from small energy gap between 1st and 2nd hole energy states. Due to that, one can assume fast exchange of holes between ES and GS levels and almost same values of filling factors $f_{h1} \approx f_{h2} \equiv f_h$. The rate equations for carriers' dynamics are:

$$\dot{f_{e2}} = g_{cn}\left(1 - f_{e2}\right) - g_{en} f_{e2} - g_{e21} f_{e2}\left(1 - f_{e1}\right) + g_{e12} f_{e1}\left(1 - f_{e2}\right) - \frac{P_2}{D_2}$$
$$\dot{f_{e1}} = \frac{D_2}{D_1}\left[g_{e21} f_{e2}\left(1 - f_{e1}\right) - g_{e12} f_{e1}\left(1 - f_{e2}\right)\right] - \frac{P_1}{D_1}$$
$$\dot{f_h} = g_{cp}\left(1 - f_h\right) - g_{ep} f_h - \frac{P_1 + P_2}{D_1 + D_2}, \quad (1)$$

where $f_{e1,2}$ stands for GS and ES electron level filling respectively, $P_{1,2}$ are the stimulated emission rates from GS and ES optical transitions measured per one QD (s^{-1}), $D_{1,2}$ are the degree of degeneracy of GS, ES state (2 and 4, respecively), g_{cn}, g_{cp} are the capture rates of electons and holes into empty QD from GaAs matrix, g_{en}, g_{ep} are escape rates of electrons and holes from QD to the matrix, g_{e12}, g_{e21} are the electron transfer rates from GS to ES and vice versa. It is assumed for electons that only ES level is connected with the matrix.

The proposed model describes a few regimes of laser operation such as:

$$(NL) \qquad P_1 = 0, P_2 = 0 \quad (2)$$
$$(GS) \qquad f_{e1} + f_h - 1 = \gamma_1, P_2 = 0 \quad (3)$$
$$(GS+ES) \qquad f_{e1} + f_h - 1 = \gamma_1, f_{e2} + f_h - 1 = \gamma_2 \quad (4)$$
$$(ES) \qquad P_1 = 0, f_{e2} + f_h - 1 = \gamma_2, \quad (5)$$

depending on injection current and model parameters. The $\gamma_{GS,ES} = \alpha/G_{GS,ES}$ are population inversion required for lasing through GS and ES respectively; α is modal losses and $G_{max,i}$ is maximum modal gain of QD layer. If the injection is low enough (NL) regime takes place and there is no stimulated emission. While injection, i.e. g_{cn} and g_{cp}, is increasing there are few possible scenarios: '(NL) → (GS) → (GS+ES)', '(NL) → (GS) → (GS+ES) → (ES)' or '(NL) → (ES)'. The first scenario at high injection results in simulataneous ES and GS lasing, the second one is called 'damping of ground state lasing' and the last one corresponds to lasing only through ES when GS modal gain is not high enough. The details are discussed in [3]. It was shown there that laser operation is sensitive to $g_{cp}/g_{cn} \equiv h$ ratio. For example, any of three scenarios could unfold depending just on the value of h. The h value itself depends on the electron and hole concentration in the matrix near QDs. The latter should be calculated by means of modeling of a band diagram of a working laser and is discussed in the next section.

3. Drift-diffusion model of semiconductor structure with δ-doped quantum dots

Previously [1] the band diagram was modeled on the assumption of negligible carrier concentration in the i-region of p-i-n laser structure. This assumption results in the absence of screening effects and constant electric field between the emitter/i-region boundary and QD layer. Moreover, it was assumed in [1] that the size of a potential curve near the AlGaAs/GaAs boundary is well below the T (temperature in energy units). As one can see below both of these assumptions could be unfulfilled in the real QD laser. Hence, the method to model a laser band diagram that takes into account electric field of charges that are located in the i-area is required. In the paper the set of drift-diffusion equations (DDE) [6, 7] is used based on Boltzman statistics of charge carriers:

$$n = n_{ir} e^{(\varphi + \psi_n - \phi_n)/T}\ ;\ p = n_{ir} e^{-(\varphi - \psi_p - \phi_p)/T}, \quad (6)$$

where n_{ir} - intrinsic concentration at reference location, φ -electrostatic potential, $\phi_{n,p}$ - quasi-Fermi levels of electrons and holes respectivly, $\psi_{n,p}$ - band parameters of modeling heterostructure.

$$\frac{d}{dx}\left(\epsilon\frac{d\varphi}{dx}\right) = -q\left(p - n + N_D^+ - N_A^-\right) - \rho_{QD} \tag{7}$$

$$\frac{dJ_n}{dx} = qR = -\frac{dJ_p}{dx} \tag{8}$$

$$J_n = -\mu_n n\frac{d}{dx}\left(\varphi + \psi_n\right) + T\mu_n\frac{dn}{dx} \tag{9}$$

$$J_p = -\mu_p p\frac{d}{dx}\left(\varphi - \psi_p\right) - T\mu_p\frac{dp}{dx} \tag{10}$$

$$q\psi_n = \left(\chi(x) - \chi_r\right) + T\ln\left(\frac{N_c(x)}{N_{cr}}\right) \tag{11}$$

$$q\psi_p = \left(\chi(x) - \chi_r\right) - \left(E_g(x) - E_{gr}\right) + T\ln\left(\frac{N_v(x)}{N_{vr}}\right) \tag{12}$$

where q - electrical charge, ρ_{QD} - QD charge, n, p - concentration of electrons and holes, N_D^+, N_A^- - impurity concentration of donors and acceptors, $J_{n,p}$ - current densities, $\mu_{n,p}$ - carrier mobility, R - recombination rate, E_g - band gap energy, χ - electron affinity, $N_{C,V}$ - conduction and valence effective densities of states, subscript r refers to value at references material.

The recombination rate includes three terms, i.e. bimolecular recombination, Shockley-Reed-Hall recombination and recombination in quantum dots:

$$R = Bnp + R_{SRH} + R_{QD}N_S\Phi(x) \tag{13}$$

$$R_{SRH} = \frac{pn - n_i^2}{\tau_n\left(p + n_i\right) + \tau_p\left(n + n_i\right)}, \tag{14}$$

where B is the constant of bimolecular recombination, R_{QD} is the recombination rate in a single quantum dot, N_S is the total quantum dot surface density, and τ_n, τ_p are electron/hole lifetimes in the GaAs matrix. The $\Phi(x)$ function is non-zero in the area where QDs are placed and should

Table 1. Table of models parameters. QD parameters [2], AlGaAs parameters[9], GaAs parameters [10], [11].

QD model		Drift-diffusion model	
Electrone capture rate, g_{en}	219 ns^{-1}	Temperature, T	300 K
Hole capture rate, g_{ep}	19 ns^{-1}	Active area width, b	300 nm
Electron down-hopping rate, g_{e21}	1250 ns^{-1}	n-emitter doping, N_D	$5 \cdot 10^{17}$ cm^{-3}
Electron up-hopping rate, g_{e12}	125 ns^{-1}	p-emitter doping, N_A	$5 \cdot 10^{17}$ cm^{-3}
Internal losses, α_{in}	1.5 cm^{-1}	Active layer doping, N_{act}	No doping
External losses, α_{out}	3.1 cm^{-1}	QD surface density, N_S	$5 \cdot 10^{10}$ cm^{-2}
GS level degeneracy, D_1	2	Al mole fraction, x	0.8
ES level degeneracy, D_2	4	Band gap, E_g ($AlGaAs$)	2.09 eV
Energy of GS optical transition, E_1	983 meV	Band gap, E_g ($GaAs$)	1.42 eV
Energy of ES optical transition, E_2	1050 meV	Conductive band density, N_C	$1.6 \cdot 10^{19}$ cm^{-3}
GS maximum gain, G_{maxGS}	6 cm^{-1}	Valence band density, N_V	$1.5 \cdot 10^{19}$ cm^{-3}
ES maximum gain, G_{maxES}	12 cm^{-1}	SRH relaxation time, $\tau_{n,p}$	10^{-8} s^{-1}
Localization energy of electron ES level, E_e	0.16 eV	Bimolecular coefficient, B	$5 \cdot 10^{-11}$ $s^{-1}cm^{-3}$
Localization energy of hole level, E_h	0.22 eV		

be normalized to unity: $\int \Phi(x)dx = 1$. At the same time the charge density formed by a QD layer and included in (7) could be described as:

$$\rho_{QD} = qQ_{QD}N_S\Phi(x) \ , \tag{15}$$

where Q_{QD} is the dimensionless charge of quantum dots.

One can see that quantum dots are characterized in the DDE by only two functions $R_{QD}(n(x_0), p(x_0))$ and $Q_{QD}(n(x_0), p(x_0))$, where $n(x_0)$,$p(x_0)$ are carriers concentrations at the point where QD layer is placed. These functions can be written down as

$$Q_{QD} = (D_1 + D_2)f_h - D_2 f_{e2} - D_1 f_{e1} \tag{16}$$
$$R_{QD} = (D_1 + D_2)\left(g_{cp}\left(1 - f_h\right) - g_{ep} f_h\right) \ , \tag{17}$$

where $f_{e1,2}$ and f_h are the solution of (1) at the given rates of captures into empty QD g_{cp} and g_{cn}. The rates themself are obviously proportional to carrier concentration near a QD:

$$g_{cn} = V_{cn}n(x_0) \ ; \ g_{cp} = V_{cp}p(x_0), \tag{18}$$

and proportionality constants can be calculated with the detailed equilibrium principle:

$$N_C V_{cn} \exp\left(-\frac{E_e}{T}\right) = \frac{1}{2}g_{en} \ ; \ N_V V_{cp} \exp\left(-\frac{E_h}{T}\right) = \frac{1}{2}g_{ep} \ , \tag{19}$$

where $E_{e,h}$ are electron ES and hole localization energies in a QD, and $N_{C,V}$ are effective state density in conductive and valence GaAs bands.

In order to numerically solve DDE system we used the method of [5] based on J_n, J_p effective approximation and Newton-Rapson iterative relaxation scheme [8].

4. Results and discussion

The band diagram near the threshold of GS lasing is shown in Figure 1.

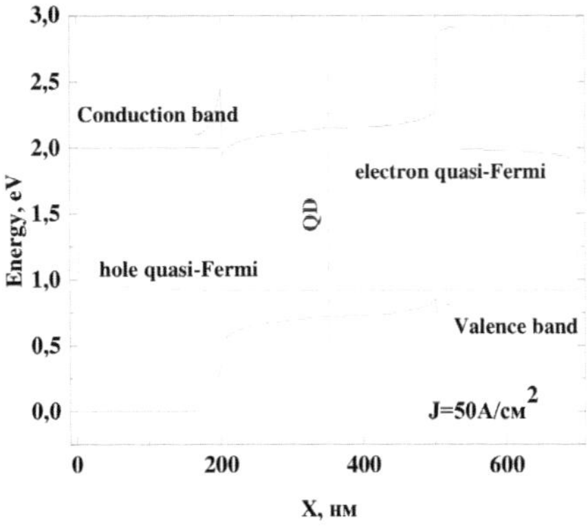

Figure 1. *Band diagramm of QD laser at the injection current just above the onset of GS lasing. The QD layer is placed in the middle of i-region. The n-emitter is at the left side of the figure. Quasi-Fermi levels for electron and holes are also shown.*

One can clearly notice a significant potential step on the emitter/i-region interface. The value of a potential rise in n-emitter is about 150 meV, which is much higher than T, and should be taken into account in any modeling approach. Also, the profile of the potential is clearly asymmetric with respect to the QD plane. This happens because of the significant drop of electric field while moving through the QD plane in positive direction. A possible explanation is following: due to higher electron mobility uncharged quantum dots attract more electrons than holes or, in other words, in the steady laser operation QDs tend to be negatively charged (see also [4]). This charging reveals itself as an electric field drop near the QD plane that results in boosted flow of holes from p-emitter to QDs that compensate higher electron mobility.

There are two widely used simplifications in modeling of light-emitting p-i-n structures: high Coulomb screening in the i-region and no–screening approach in the i-region. The first one refers to the idea that electron/hole concentration is so high in the i-region that no significant electric field could exist at the T/qb scale, i.e. the free carriers screens fields of the charges in QD and emitters. That is often called 'flat' band theory. On the other hand, the second one assumes that charge carrier concentration is low enough to neglect any screening effects and, therefore, electric field is constant between emitters and QD layer [1]. We state that in the types of laser which have been studied both assumptions are not valid. As it can be seen from Figure 1 neither bands are flat nor screening can be neglected. That is why numerical modeling seems to be required in order to achieve quantitative results.

The dependence of h-factor on injection current and δ-doping is shown in Figure 2a. The h-factor is well below unity which is in complete agreement with band diagram analysis i.e. negative charging of QDs and low hole concentration (depletion) in the QD region comparing with electrons (not shown in the current paper). One may also notice that the h-factor is growing with the injection current. Probably, this is due to increased boosting of holes current towards the QDs layer. The impact of δ-doping on h-factor is quite obvious: p-doping increases h-factor and n-doping decreases it. In simplified terms p-doping, for example, leads to additional negative charge of acceptors near the QD plane, that clearly, increases holes flow and holes concentration. The dependence of lasing power through GS/ES optical transitions on the injection current is

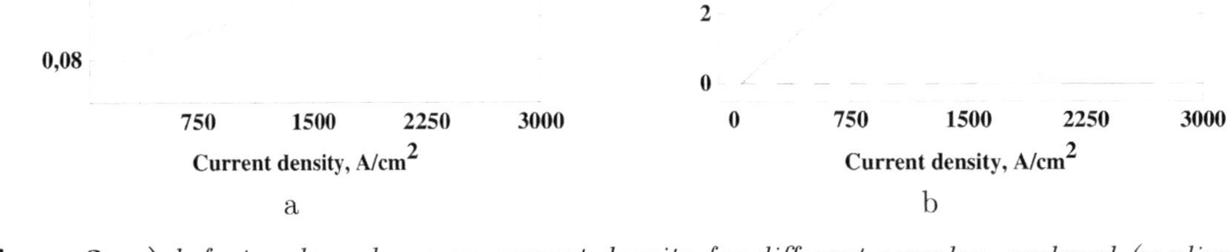

Figure 2. a) *h-factor dependence on current density for different samples: undoped (medium curve), n-doped (bottom curve) and p - doped (top curve). The onset of lasing through ES is also shown for undoped and p-doped samples.* ***b)*** *Dependence of the radiation power of the current at different modulation doping. Dash lines represents emission from ES, solid lines represents emission from GS.*

shown on Figure 2b. The major result is that p-doping increases both maximum power emitted through GS and threshold of ES lasing. These results are closely connected because after the onset of ES lasing GS power stabilizes or decreases depending on laser parameters. The detailed explanation of the way higher value of h results in GS power rising was presented earlier [2]. We mention here simple qualitative picture: let us assume that QD charge should be equal -6q in order to balance the flows of electrons and holes toward the QD region, i.e. all electron levels should be filled and hole levels should be empty. In that case only transparent condition could fulfilled (no filling inversion) and no gain is possible. Otherwise, the charge neutrality of QDs, for example, when all electron and hole levels are filled, corresponds to maximum gain. As a result, charging of QDs suppresses QD gain and since the ES maximum gain is higher than GS one favors to the ES lasing.

In the end we should notice that performed calculations are also of significant importance for multimode or comb emission of QD laser. The multimode lasing is intrinsically connected with spacial-hole burning in active area that is ruled out by non-homogeneous distribution of light intensity and carrier diffusion along the direction of light propagation. The latter is described by the solution of DDE set studied in this paper. Thus, suggested approach to QD laser modeling can be discussed as a first step to accurate calculation of comb laser spectral properties, spatial hole burning and underlying physical effects.

5. Conclusion

It has been shown that h-factor is variable and depends on the carrier concentration near QD, doping and injection current. In the non-doped laser QDs become charged, ground electron state (GS) fully filled that suppress gain and lasing through GS in favor to ES emission. Our calculations demonstrate that p-type modulation layer leads to neutralization of QD and increases the ground state gain. That, in turn, leads to an increase of threshold current for lasing via ES and increases power radiated from ground state. On the other hand, n-type doping leads to decrease of threshold current and can completely suppress GS lasing.

Acknowledgement

The authors acknowledge the nancial support of Russian Foundation for Basic Research (grant num. 13-02-12016).

References

[1] L.V. Asryan, R.A. Suris. Charge Neutrality Violation in Quantum-Dot Lasers IEEE J. Sel. Top. Quant. Electron., vol. 3, no. 2, 1997

[2] V.V. Korenev, A.V. Savelyev, A.E. Zhukov, A.V. Omelchenko, M.V. Maximov. Effect of carrier dynamics and temperature on two-state lasing in semiconductor quantum dot lasers, Semicond. 47 (10), 1397 (2013).

[3] V.V. Korenev, A.V. Savelyev, A.E. Zhukov, A.V. Omelchenko, M.V. Maximov. Appl. Phys. Lett., 102, 112101 (2013).

[4] I. C. Sandall, P. M. Smowton, J. D. Thomson, T. Badcock, D. J. Mowbray, H.-Y Liu, and M. Hopkinson, Appl. Phys. Lett. 89, 151118 2006.

[5] Scharfetter, D.L., Gummel, H.K., Large signal analysis of a silicon Read diode. IEEE Trans. Electron Dev. 16, 6477 (1969)

[6] Lundstrom M.S., Schwartz J., Gray J.L. Transport equations for the analysis of heavily doped semiconductor devices Solid-Staft Electronics. 1981. Vol. 24. Pp. 195202.

[7] Lundstrom M.S., Schuelke R.J. Modeling semiconsuctor heterojunctions in equlibrium Solid-State Electronics. 1988. Vol. 25, no. 8. P. 683691.

[8] A. M. Ostrowski. Solution of Equations and Systems of Equations, Academic Press, New York, 1960

[9] Goldberg Yu.A. Handbook Series on Semiconductor Parameters, vol.2, M. Levinshtein, S. Rumyantsev and M. Shur, ed., World Scientific, London, 1999, pp. 1-36.

[10] Levinshtein M.E., S.L. Rumyantsev Handbook Series on Semiconductor Parameters, vol.1, M. Levinshtein, S. Rumyantsev and M. Shur, ed., World Scientific, London, 1996, pp. 77-103.

[11] Dargys A. and J. Kundrotas Handbook on Physical Properties of Ge, Si, GaAs and InP, Vilnius, Science and Encyclopedia Publishers, 1994

Microdisk lasers based on GaInNAsSb/GaAsN quantum well active region

E I Moiseev[1], N V Kryzhanovskaya[1], Yu V Kudashova[1], M V Maximov[1], M M Kulagina[3], S I Troshkov[3], A A Lipovskii[2,1], V-M Korpijärvi[4], H Karjalainen[4], T Niemi[4], M Guina[4], A E Zhukov[1]

[1] St. Petersburg Academic University, St. Petersburg, 194021 Russia
[2] Peter the Great St.Petersburg Polytechnic University, St. Petersburg, 195251 Russia
[3] Ioffe Physical–Technical Institute, Russian Academy of Sciences, St. Petersburg, 194021 Russia
[4]Optoelectronics Research Centre, Tampere University of Technology, Tampere , Finland

moiseev@spbau.com

Abstract. Microdisk lasers based on novel InGaAsNSb/GaAsN quantum well active region are developed and studied under optical pumping. Room temperature lasing at 1.55 μm in 2.3 μm in diameter microdisks with InGaAsNSb/GaAsN QW is demonstrated.

1. Introduction

Semiconductors microdisk and microring lasers have been widely investigated as possible building blocks for photonic integrated circuits [1]. The circular symmetry of the cavity results in unique advantages such as low threshold, small foot-print, small mode volume, in-plane emission, control of emission wavelength by the device size, etc. [2]. The use of quantum dots (QDs) as active region in such lasers have the advantages of low threshold, high quantum efficiency, and high thermal stability of characteristics [3]. However since the ground state QDs optical gain is limited due to the finite number of the QDs the decrease of the resonator diameter to a limiting value results in lasing via excited states of QDs. Quantum well (QW) active region provides higher gain compare to QDs. Thus the use of QW active region may help to work out the problem of gain saturation when the laser`s size are scaled down. In this work we have studied microdisk lasers based on single diluted InGaAsNSb/GaAs(N) QW active region.

The anomalous reduction in both bandgap and lattice constant of GaAs, with the introduction of dilute amounts of nitrogen was reported in work [4]. In 1996, Kondow et al. [5] recognized that introducing appropriate amounts of nitrogen and indium into GaAs shifts the emission wavelength to the technologically important 1.3 μm regime. Growth of GaInNAs-based lasers emitting at 1.55 μm proved much more difficult, since it was usually observed that the incorporation of nitrogen deteriorates the optical properties due to ion damage [6, 7], nonradiative traps [8, 9], and phase separation [10, 11]. Yang and coworkers found that adding antimony as a surfactant during GaInNAs growth improved morphology and device performance, and allowed longer wavelength devices [12]. This is due to the reactive surfactant properties of antimony, which reduce the group III surface diffusion length suppressing phase segregation and roughening and thereby improving alloy homogeneity [13]. CW laser operation is demonstrated in GaInNAsSb/GaAs lasers at room temperature [14,15] and at temperatures up to 80°C with T_0 = 70K [16]. Compared with an InGaAsP/InP system, GaInNAsSb/GaAs has a larger conduction-band offset which leads to a higher characteristic temperature [17].

We investigated the microdisk lasers at different temperatures (78-300 K). The single-mode lasing at room temperature in microdisk laser based on GaInNAsSb/GaAsN quantum well active region with diameter 2.3 μm is demonstrated.

2. Experiment

The epitaxial structure with InGaAsNSb/GaAsN QW was grown by MOCVD on n-doped GaAs substrate. The active region comprises InGaAsNSb QW inserted in the GaAsN strain compensating layers. The active region was placed in the middle of the GaAs waveguide confined from both sides with 10 nm thick $Al_{0.3}Ga_{0.7}As$ barriers to prevent carrier leakage. Total thickness of the waveguide layer was 240 nm. This waveguide layer was grown on the top of the 1500-nm-thick $Al_xGa_{1-x}As$ layer that later forming a pedestal of the microdisks (x=0.7-0.8). Microdisks of different diameters from 2.3 μm to 7 μm were fabricated using photolithography and two step wet etching. The structures were investigated under optical pumping with YAG: Nd laser (λ = 532nm).

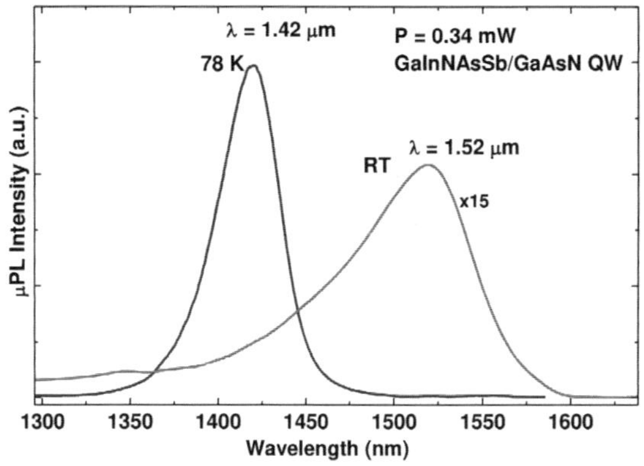

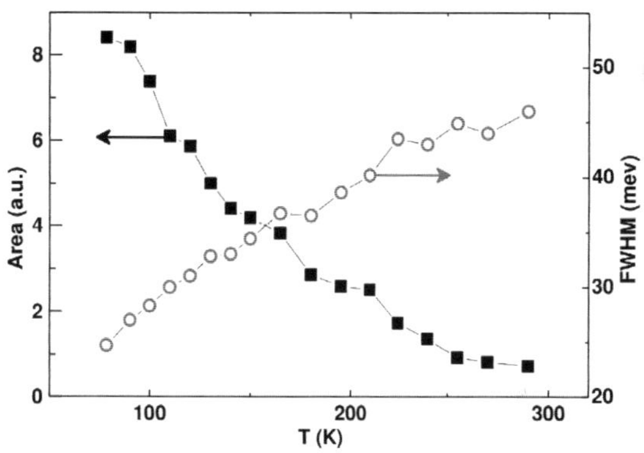

Figure 1. μPL spectra of a single GaInNAsSb/GaAs(N) QW at 78K and at room-temperature.

Figure 2. Temperature dependence of integrated PL intensity and FWHM of spectra GaInNAsSb/GaAs(N) QW

3. Results

The microphotoluminescence (μPL) spectra of the lasers were studied in a temperature range from 78 K to 300 K. The μPL spectra at different excitation power of unprocessed sample with single GaInNAsSb/GaAsN QW obtained at 78K and room temperature are shown at figure 1. The wavelength of PL spectra maximum is 1.52 μm at room temperature and 1.42 μm at 78K. Increase of the temperature result in intensity drop only in 15 times revealing low non-radiative recombination in the epitaxial structure. The dependence of integrated μPL intensity and full width at half maximum (FWHM) on the temperature are shown at figure 2. At the temperature 78K the FWHM is 25 meV that reveals good homogeneity of the GaInNAsSb/GaAsN layers composition and thickness.

The μPL spectrum of the microdisk with diameter D = 2.3 μm obtained at 78 K is shown at fig.3. Scanning electron micrograph of the microdisk laser is shown in insert in fig.3. Sharp line observed at the spectra corresponds to high quality whispering-gallery (WG) mode. Lasing was observed for dominant resonance line at 1.47 μm. In fig. 4 shows the dependence of integrated intensity and FWHM of the resonant line for 2.3 μm microdisk on the pump power. The threshold pump power is 220 μW. The FWHM of the 1.47 μm line near the threshold is 30 pm and is limited by the resolution of the measurement system. We can estimate the quality factor (obtained from λ/Δλ) to be

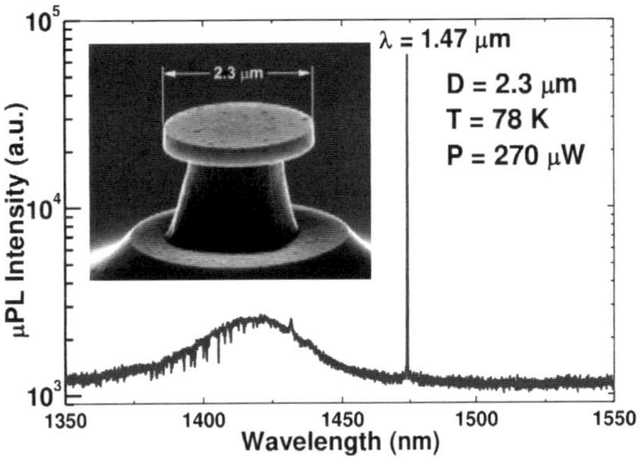

Figure 3. μPL spectrum of the microdisk laser obtained at 78 K. Inset: SEM image of a microdisk laser with D = 2.3 μm

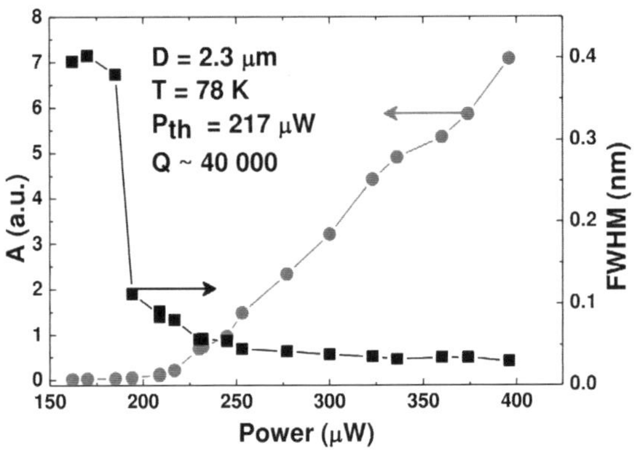

Figure 4. Spectrally integrated emission intensity and FWHM of a whispering gallery mode as a functions of incident optical pump power

more than 4 × 10^4. Free spectral range (FSR) and threshold power as a function of the microdisk diameter are shown at fig.5. Decrease of the resonator diameter from 7 μm to 2.3 μm results in increase of the resonator FSR from 26 nm to 80 nm. The obtained values agree very well with theoretically predicted dependence FSR ~ $\lambda^2/(n_{eff}(2\pi R))$. Reducing the diameter of the resonator leads to a decrease of the threshold power proportionally to the squared diameter of the microdisk.

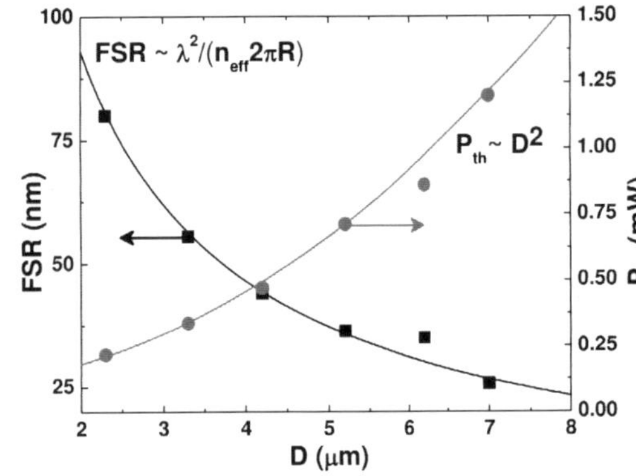

Figure 5. FSR and threshold power as a functions of the microdisk diameter

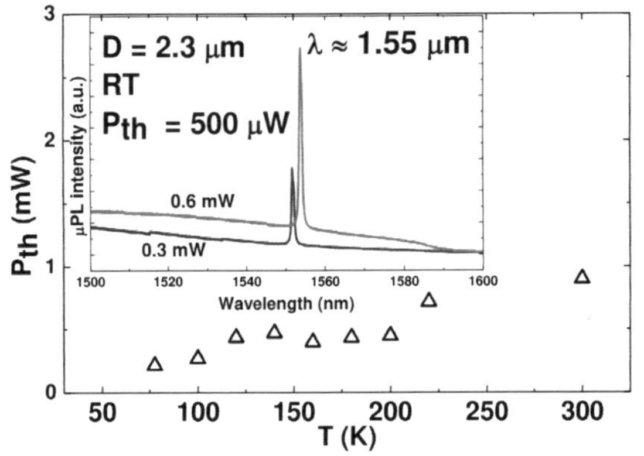

Figure 6. Threshold power as a function of the temperature. Inset shows μPL spectra of the microdisk laser obtained at room-temperature.

The lasing on WG mode of the resonator was observed in temperature range from 78K to room-temperature. The fig.6 shows temperature dependence of the threshold power of the microdisk laser. The μPL spectra of the microdisk with diameter D = 2.3 μm at different pump powers obtained at room-temperature are shown in insert in fig.6. The threshold pump power is 0.5 mW.

To conclude, microdisk lasers with novel active region based on GaInNAsSb/GaAsN QW were investigated in a temperature range 78-300K. Room temperature single-mode lasing in microdisk laser with diameter 2.3 μm is demonstrated.

4. Acknowledgments
The work is supported by the Skolkovo Foundation (grant agreement for Russian educational and scientific organization no.4 dd. 25.12.2014).

References
[1] S. Koseki et al, Applied Physics Letters, 94, 051110 (2009).
[2] K. Srinivasan et al, Optics Express, 14 (3), 1094-1105 (2006).
[3] Y. Arakawa, H. Sakaki, Appl. Phys. Lett. 40 939 (1982).
[4] M. Weyers, M. Sato, and H. Ando, "Red shift of photoluminescence and absorption in dilute GaAsN alloy layers," Jpn. J. Appl. Phys., vol.31, no. 7A, pp. L853–L855, Jul. 1992.
[5] M. Kondow, K. Uomi, A. Niwa, T. Kitatani, S. Watahiki, and Y. Yazawa, Jpn. J. Appl. Phys., Part 1 35, 1273 (1996)
[6] M. A. Wistey, S. R. Bank, H. B. Yuen, H. P. Bae, and J. S. Harris Jr., J. Cryst. Growth 278, 229 (2005).
[7] H. B. Yuen, M. A. Wistey, S. R. Bank, H. P. Bae, and J. S. Harris Jr., J. Vac. Sci. Technol. B 23, 1328 (2005).
[8] A. J. Ptak, S. W. Johnston, S. Kurtz, D. J. Friedman, and W. K. Metzger, J. Cryst. Growth 251, 392 (2003).
[9] S. R. Kurtz, A. A. Allerman, E. D. Jones, J. Gee, J. J. Banas, and B. E. Hammons, Appl. Phys. Lett. 74, 729 (1999).
[10] K. Volz, T. Torunski, and W. Stolz, J. Appl. Phys. 97, 014306 (2005).
[11] X. Kong, A. Trampert, E. Tournie, and K. H. Ploog, Appl. Phys. Lett. 87, 171901 (2005)
[12] X. Yang, M. J. Jurkovic, J. B. Heroux, and W. I. Wang, "Molecular beam epitaxial growth of InGaAsN:Sb/GaAs quantum wells for long-wave-length semiconductor lasers," Appl. Phys. Lett., vol. 75, pp. 178–180 (1999)
[13] Massies J, Grandjean N: Surfactant effect on the surface diffusion length in epitaxial growth. Phys Rev B, 48:8502–8505 (1993)
[14] IEEE JOURNAL OF QUANTUM ELECTRONICS, VOL. 40, NO. 6, JUNE (2004)
[15] Bank, S. R., et al. "Room-temperature continuous-wave 1.55 μm GaInNAsSb laser on GaAs." Electronics Letters 42(3) 156-157 (2006)
[16] Korpijarvi, V-M., et al. "1.55 μm GaInNAsSb/GaAs ridge waveguide lasers and semiconductor optical amplifiers for photonic integrated circuits." International Semiconductor Laser Conference (2014)
[17] Z. C. Niu et al, GaAs-based room-temperature continuous-wave 1.59 m GaInNAsSb single-quantum-well laser diode grown by molecular-beam epitaxy, APPLIED PHYSICS LETTERS 87, 231121 (2005)

Journal of Physics: Conference Series **643** (2015) 012041 doi:10.1088/1742-6596/643/1/012041

Injection current minimization of InAs/InGaAs quantum dot laser by optimization of its active region and reflectivity of laser cavity edges

V V Korenev[1,3], A V Savelyev[1,3], A E Zhukov[1,2,3], M V Maximov[2,1]
[1]Saint–Petersburg Academic University RAS, Saint–Petersburg 194021, Russia
[2]Ioffe Physico–Technical Institute RAS, Saint–Petersburg 194021, Russia
[3]Peter the Great St. Petersburg Polytechnic University, St. Petersburg 195251, Russia
E-mail: korenev@spbau.ru

Abstract. The ways to optimize key parameters of active region and edge reflectivity of edge-emitting semiconductor quantum dot laser are provided. It is shown that in the case of optimal cavity length and sufficiently large dispersion lasing spectrum of a given width can be obtained at injection current up to an order of magnitude lower in comparison to non-optimized sample. The influence of internal loss and edge reflection is also studied in details.

1. Introduction

Compact lasers emitting in broad wavelength range near 1.3 μm are needed for the wide range of practically important applications, from medicine to high-speed data transfer [1 – 3]. The usage of InAs/InGaAs quantum dots (QDs) having long-wavelength emission spectra allows not only to cover the required wavelength range, but also to simplify the production process and decrease costs in comparison to the distributed-feedback (DFB) lasers [4 – 5]. Therefore, the task of injection current minimization of such QD lasers at a given lasing spectrum width tends to play prominent role. In this paper we show how to tackle this issue using the active region having optimized structure and size as well as reflection coefficients of its edges.

2. The ways to optimize laser active region and reflection of the edges

In the case of MBE-grown InAs/InGaAs QD laser there is a series of parameters, which can be varied even during the growth process. These parameters are QD layers number in laser active region (Z), dispersion of inhomogeneous broadening (σ) and laser cavity length (L). Moreover, the reflectivity of laser cavity edges, which can be also varied by using coatings of different types, may also affect characteristics of such lasers. Thus, the key goal of current paper is to figure out at which parameters it is possible to obtain lasing spectrum of a given width (Ω) at minimum injection current.

To study the influence of laser cavity length (L) on the injection current (I_{inj}), we have calculated the dependence of I_{inj} on L for different numbers of QD layers in laser active region (Z) at a given dispersion (σ) and lasing spectrum width (Ω) – see Fig. 1, where σ= 20.5 meV and Ω= 18 meV, while the other parameters are in accord with [4, 6 – 9]. As it can be seen from the figure there is always an optimal laser cavity length allowing one to attain lasing spectrum of a given width at minimum injection current given σ and Z. This is because of high injection current in the case of sufficiently short or long cavity. Indeed, as laser cavity length is too small, output loss tends to increase resulting in the increase of injection current required to attain given Ω. In the case of long cavities the number of QDs in laser active region increases meaning that the injection current should be increased in order to maintain population inversion in certain number of QDs required to attain Ω. The dashed line marked with (A) in the Fig. 1 corresponds to the optimal injection currents for different numbers of QD layers *Z* in laser active regions. Fractional Z, e.g. Z= 1.5, means integer layers number, but adjusted QD density in laser active region.

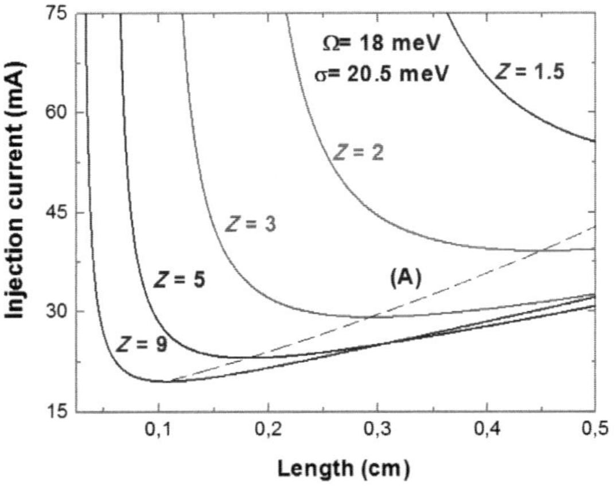

Figure 1. The dependence of the injection current (I_{inj}) on laser cavity length (L) for different values of QD layers number in laser cavity (Z) varying from 1.5 to 9. Dashed line corresponds to the optimal injection currents (I^L_{opt}) at optimal length (L_{opt}) for different Z. Here Ω= 18 meV and σ= 20.5 meV.

As it can be seen from the figure the larger is *Z*, the smaller is injection current and corresponding cavity length. This is due to the increase in maximum gain (G_{max}), which is directly proportional to *Z*, that allows to obtain lasing spectrum of the same width Ω at lower injection current and even higher output loss, i.e. at shorter laser cavities.

In order to elucidate how the variation of inhomogeneous broadening may influence injection current, which is already optimized by laser cavity length (I^L_{opt}), we have calculated the dependence of I^L_{opt} on the value of dispersion (σ) for different numbers of QD layers in laser active region at given lasing spectrum width Ω= 18 meV – see Fig. 2a.

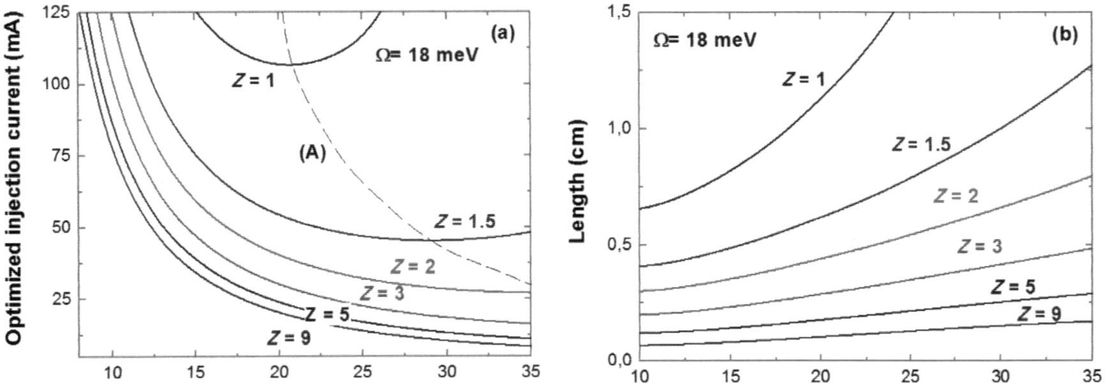

Figure 2(a, b). **(a)** The dependence of injection current (I^L_{opt}), which is optimized by laser cavity length (L), on the value of dispersion (σ) of inhomogeneous broadening varying from 10 meV to 35 meV for the different values of *Z*. Dashed line corresponds to the minimum of I^L_{opt} at different values of *Z*. **(b)** The dependence of optimal length L_{opt} corresponding to the minimum of I^L_{opt} on dispersion for different numbers of QD layers in laser active region (Z). In both cases Ω= 18 meV.

The rate equation model used to model dynamics of charge carriers in QDs as well as values of the other parameters is chosen to be in accord with [6 – 9]. As it can be seen from the figure 2a (e.g., see curve Z= 1), there is always an optimal value of dispersion (σ) minimizing I^L_{opt} for lasing spectrum of a given width. Indeed, if dispersion is sufficiently small, lasing spectrum is narrow and, therefore, the injection current (I_{inj}), as well as optimal injection current (I^L_{opt}), should both increase in order to maintain lasing spectrum of a given width Ω. At large dispersions maximum gain (G_{max}) turns out to be sufficiently small and can be even lower than the total loss (α) at certain dispersion due to its reciprocal proportionality to the value of dispersion ($G_{max} \sim 1/\sigma$). This results in higher values of I^L_{opt}, which is needed to attain given Ω, as σ increases. Therefore, there is always an optimal value of dispersion (σ_{opt}) that allows one to attain lasing spectrum of a given width at minimum injection current ($I^{L,\sigma}_{opt}$) – see figure 2a, e.g. curve Z= 1. Furthermore, as it is shown in the figure 2b, such an optimal value of injection current ($I^{L,\sigma}_{opt}$) corresponds to the attainable cavity length that does not exceed 1.5 cm.

As it can be seen from the comparison of figure 1 and figure 2a, optimization of dispersion plays the key role, while optimization of laser cavity length is only fine tuning of injection current value corresponding to the lasing spectrum of a given width. Indeed, the usage of optimal cavity length allows one to decrease injection current by 30 % from 30 mA for L= 0.5 cm to 20 mA for L= 0.1 cm – see figure 1 for Z= 9. At the same time, the increase of dispersion (σ) from 15 to 25 meV results in the decrease of injection current from 35 to 18 mA, i.e. practically by 48 %.

In order to elucidate if there is an optimal number of QD layers (Z) in laser active region, which allows one to decrease $I^{L,\sigma}_{opt}$ further at the same width of lasing spectrum (Ω), we studied the dependence of $I^{L,\sigma}_{opt}$ on Z in the case of different internal (α_{in}) and external (α_{out}) loss in laser cavity. The output loss was changed by the variation of reflectivity of laser cavity edges that, in its turn, can be achieved by the usage of different coatings – see figure 3(a, b).

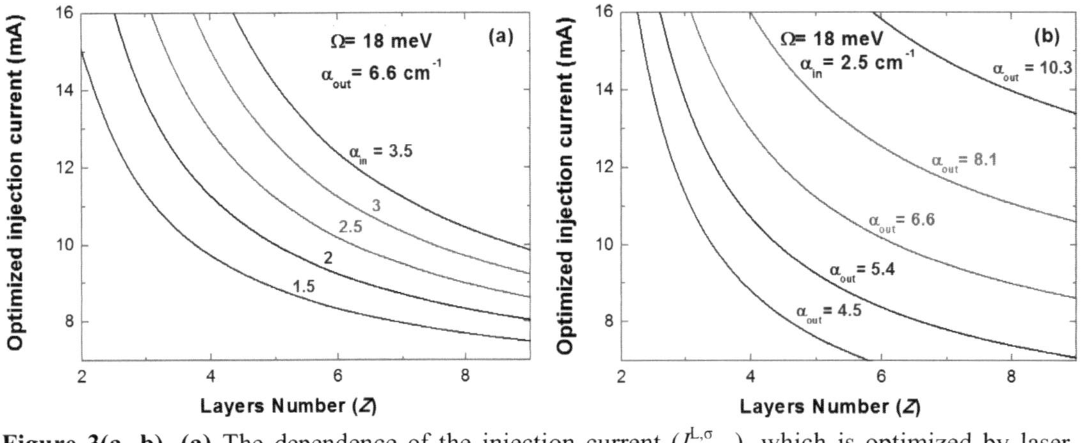

Figure 3(a, b). (a) The dependence of the injection current ($I^{L,\sigma}_{opt}$), which is optimized by laser cavity length and dispersion, on the number of QD layers in the laser cavity (Z) for different values of internal loss (α_{in}) and fixed output loss α_{out}= 6.6 cm^{-1}. **(b)** The dependence of $I^{L,\sigma}_{opt}$ on the number of QD layers in the laser cavity (Z) for different values of external loss (α_{out}), which is varied by the selection of different coatings to change reflectivity of cavity edges, and at fixed internal loss α_{in}= 2.5 cm^{-1}. The number of QD layers in laser active region (Z) and lasing spectrum width (Ω) were the same for both figures and equal to 5 and 18 meV respectively.

As it is seen from the figure 3a, the variation of α_{in} from 3.5 to 1.5 cm^{-1} results in smaller values of $I^{L,\sigma}_{opt}$ required to achieve lasing spectrum of a given width Ω. This is because of the decrease in $\alpha = \alpha_{in} + \alpha_{out}$, which results in smaller threshold population inversion in average QDs $\gamma_0 = \alpha/G_{max}$, which should be maintained in order to involve these QDs into lasing. The increase in the number of QD layers in laser active region (Z) has the same effect on the value of injection current $I^{L,\sigma}_{opt}$ and also leads to its decrease as $\gamma_0 = \alpha/G_{max}$ decreases due to the increase in $G_{max} \sim Z$.

As it is shown in the figure 3b, the decrease of α_{out} from 10.3 to 4.5 cm^{-1} due to variation of cavity edge reflectivity also results in smaller injection current $I^{L,\sigma}_{opt}$ at a given lasing spectrum width Ω due to the same reason: the decrease in α_{out} results in lower population inversion $\gamma_0 = \alpha/G_{max}$, which is required for the lasing start in average QDs.

3. Conclusion

It is shown that in the case of InAs/InGaAs quantum dot lasers, there always exists an attainable value of laser cavity length realizing lasing spectrum of a given width at minimum injection current for any given dispersion of inhomogeneous broadening. At the same time, the optimization of dispersion of inhomogeneous broadening plays the major role for the minimization of injection current, while the optimization of laser cavity length is only fine tuning of its value. In the case of multi-layered structures, the increase of dispersion as well as decrease of internal and external loss leads to further decrease of injection current corresponding to the lasing spectrum of a given width.

As a result, in order to minimize injection current realizing lasing spectrum of a given width, one should always increase the number of QD layers in laser active region and dispersion using multi-layered structures with broad distribution of QDs over the sizes, while internal and external loss should both decrease. The decrease of external loss should be realized by the usage of special coatings for laser cavity edges, while the length of laser cavity should be chosen in such a way that it has an optimal value, which minimizes the value of injection current for the given lasing spectrum width.

Acknowledgements

This work was supported by the Russian Scientific Foundation (project 14-42-00006). One of the authors (VVK) is also supported by the Committee for Science and Higher Education of the Government of Saint–Petersburg.

References

[1] Zhukov A E, Kovsh A R 2008 *Quantum Electronics* **38** 409
[2] Zhukov A E, Kovsh A R, Nikitina E V, Ustinov V M, Alferov Zh I 2007 *Semicond.* **41** 606
[3] Sugawara M, Mukai K, Nakata Y, Ishikawa H, Sakamoto A 2000 *Phys. Rev. B* **61** 7595.
[4] Savelyev A V, Novikov I I, Maximov M V, Shernaykov Yu M, Zhukov A E 2009 *Semicond.* **43** 1597.
[5] Zhukov A E, Maximov M V 2009 *Modern Injection Lasers* (St. Petersburg: Publishing House of St. Petersburg State Polytechnical University)
[6] Korenev V V, Savelyev A V, Zhukov A E, Omelchenko A V, Maximov M V 2012 *Semicond.* **46** 645
[7] Korenev V V, Savelyev A V, Zhukov A E, Omelchenko A V, Maximov M V, Shernyakov Yu M 2013 *Proc. of SPIE* **8772** 84320W
[8] Korenev V V, Savelyev A V, Zhukov A E, Omelchenko A V, Maximov M V 2014 *Appl. Phys. Lett.* **102** 112101
[9] Korenev V V, Savelyev A V, Zhukov A E, Maximov M V, Omelchenko A V 2015 *Proc. of SPIE* **9503** 950305

Diode lasers with asymmetric barriers for 850 nm spectral range: experimental studies of power characteristics

F.I. Zubov[1,2], A.E. Zhukov[1,2], Yu.M. Shernyakov[3,1], M.V. Maximov[1,3,2], E.S. Semenova[4], L.V. Asryan[5]

[1] St. Petersburg Academic University, 194021, St. Petersburg, Russia
[2] St. Petersburg State Polytechnical University, 195251, St. Petersburg, Russia
[3] Ioffe Physical-Technical Institute of RAS, 194021 St. Petersburg, Russia
[4] DTU Fotonik, Technical University of Denmark, Kgs. Lyngby, DK-2800, Denmark
[5] Virginia Polytechnic Institute and State University, Blacksburg, Virginia 24061, USA

E-mail: fedyazu@mail.ru

Abstract. It is demonstrated that the use of asymmetric barrier layers in a waveguide of a diode laser suppress non-linearity of light-current characteristic and thus improve its power characteristics under high current injection. The results are presented for 850-nm AlGaAs/GaAs broad-area lasers with GaInP and AlInGaAs asymmetric barriers.

1. Introduction

It was proposed to use so-called asymmetric barrier layers (ABLs) on both sides of active region to improve temperature stability of diode lasers [1]. As shown on Fig. 1, ABL that is on *n*-emitter side should pass only electrons and another ABL, on *p*-emitter side, should pass only holes. Such selectivity of ABLs is achieved through formation of a high energy barrier in one band (e.g., in conduction band for *p*-ABL) together with negligible barrier in another band (in valance band for *p*-ABL) at the ABL/waveguide heteroboundary. As a result, charge carriers in such laser coexist and, consequently, recombine only in the active region. Implementation of this concept is aimed at suppression of parasitic recombination in waveguide layers – the main problem of diode lasers with small localization energy of charge carriers in the active region, which in particular leads to decrease of efficiency of conversion of electrical energy into energy of laser radiation. The most pronouncedly this issue appears in structures with broad waveguide [2], at elevated temperatures [3] and under high pumping [4].

Earlier [5] we have fabricated for the first time diode lasers with asymmetric barriers (LABs). Their experimental studies revealed that the introduction of ABLs leads to reduction in threshold current, increase in external differential quantum efficiency and decrease in internal optical losses near the lasing threshold, as well as increase in temperature stability. Moreover, recent calculations [6] have shown that ABLs are capable of suppression of light-current characteristic (LCC) saturation, associated with parasitic recombination, and thus to improve diode lasers power characteristics. In fact, more recent measurements [7] demonstrated moderate enhancement of LCC linearity owing to

ABLs. However, a low threshold of catastrophic optical mirror damage (COMD) due to utilization of narrow waveguide did not allow conducting measurements at high pumping, when a significant saturation of LCC takes place.

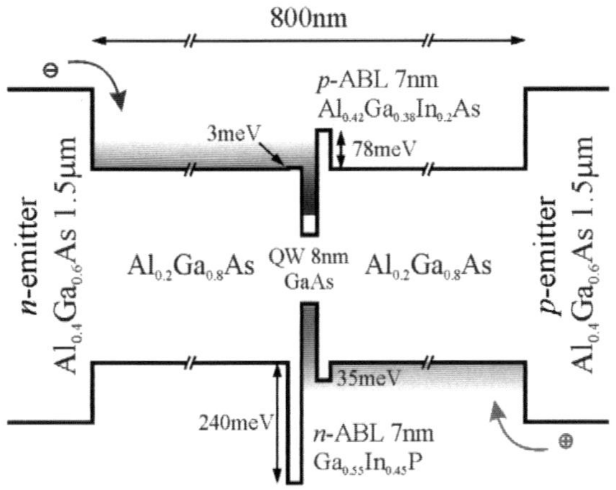

Figure 1. Schematic band diagram of studied LAB.

In the present work, we study experimentally the effect of ABLs on power characteristics of diode lasers of 850 nm spectral range with wide waveguide. We show that LAB is characterized by considerably more linear LCC as compared to the reference laser of conventional design without ABLs both at room and elevated temperatures.

2. Experiment

Studied laser diodes of 850 nm spectral range were fabricated from heterostructures of two types: with and without ABLs. Synthesis was performed by low pressure MOVPE on n^+-GaAs (100) substrates. The heterostructure without ABLs served as reference and, in fact, was a conventional heterostructure of a semiconductor quantum well (QW) laser for the given spectral region. A single GaAs QW of 8 nm width was placed in the middle of wide (800 nm) $Al_{0.2}Ga_{0.8}As$ waveguide. We used $Al_{0.4}Ga_{0.6}As$ for n- and p-type emitters with dopant concentration on the order of 10^{18} cm^{-3}. Silicon was used as the n-type impurity, while carbon was used as the p-type impurity.

The structure of the second type (Fig. 1) varied from the first structure in that its waveguide comprised two 7-nm-thick ABLs on both sides of the QW. $Ga_{0.55}In_{0.45}P$ was used as n-ABL, while $Al_{0.42}Ga_{0.38}In_{0.2}As$ – as p-ABL. The calculated energy barriers at the n-ABL/waveguide interface were 240 meV for holes and 3 meV for electrons, while those at the p-ABL/waveguide interface were 78 and 35 meV for electrons and holes, respectively. The band edge positions were calculated taking in account the effect of both the chemical composition and elastic strains. The lattice mismatch between ABLs and the substrate did not exceed 1.5%, which allowed a pseudomorphic synthesis of the structure.

Laser strips of 50 μm width were processed on the surface of epitaxial wafers by the standard postgrowth techniques including optical lithography, ion beam etching as well as deposition of metals and insulators. Laser with cavities of various lengths were fabricated by cleaving the crystals. No coatings were deposited onto the mirrors. The laser chips were then soldered with the p-side down on a copper heat sink. During measurements the heat sink temperature was stabilized with a temperature controller.

The saturation of LCC of diode laser may be due to its overheating (the so-called thermal rollover) or to the parasitic carrier recombination in the waveguide. To eliminate the influence of QW overheating on the device characteristics, the samples were pumped with short rectangular current pulses of 200 ns duration and low repetition frequency of 100 Hz. To measure the LCCs, we used a large area (1 × 1 cm^2) silicon photodiode mounted close to the sample. To eliminate saturation of the photodiode by high-power laser radiation, calibrated optical filters were mounted between the laser and the photodiode.

3. Results

Fig. 2 shows the dependence of the threshold current density on the cavity length for both types of lasers at 20°C. The presented data corresponds to ground state lasing only. It is seen that the inclusion of ABLs in the heterostructure has insignificant effect on the threshold current. For example, 2-mm long laser diodes had threshold current of about 400 mA. The inset of Fig. 2 shows spectra of lasing emission via ground state optical transition of the LAB and the reference laser with 2 mm length measured at 20°C and 2 A pumping current. It is seen that the lasing line in the LAB is blue shifted compared to the lasing line of the conventional reference laser without barriers. The lasing peaks of the LAB and the test laser are positioned at 841 and 853 nm, respectively. The reason for the lasing line blue shift is in that the additional barriers close to the active region lead to increase in ground state energy with respect to bottom of the QW and corresponding growth of the ground state optical transition energy.

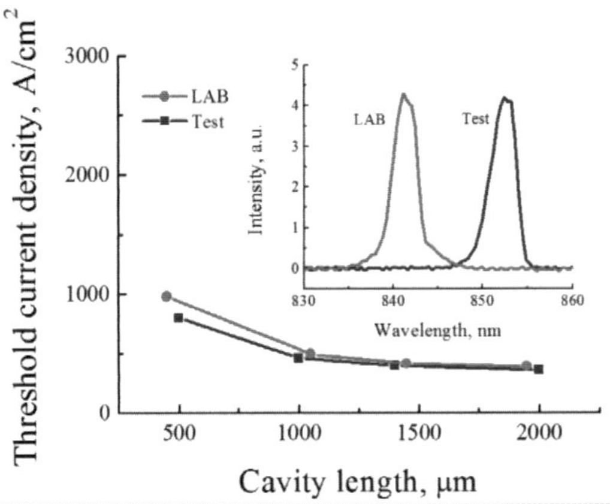

Figure 2. Threshold current density against cavity length for LAB and reference laser at 20°C. Inset shows lasing emission spectra of 2-mm long lasers measured at 20°C and 2 A current.

Fig. 3 shows the degree of deviation from linearity δ for the LCCs measured at 20 and 60°C, calculated as $\{[\eta_0(I-I_{th})-P]/\eta_0(I-I_{th})\}\times 100\%$, where η_0 is the external differential quantum efficiency determined in the initial part of the LCCs, near the threshold current I_{th}, I is the drive current and P is the measured output optical power. At room and elevated temperature the LAB demonstrates significantly better linearity of the LCC than the conventional laser without ABLs. Thus at 7(5) A the deviation from linearity δ in the test laser made 33(16)% against 9(5)% in the LAB at 20(60)°C.

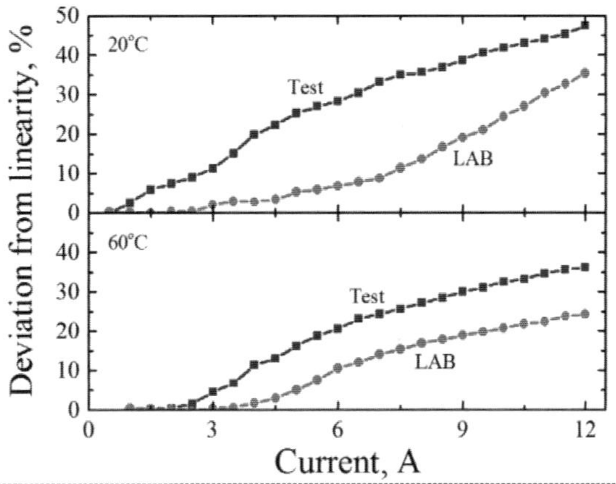

Figure 3. Current dependences of deviation from linearity for LCCs measured at 20 and 60°C for LAB and test laser of 2-mm-length.

Output optical power dependences of the pumping current for both types of lasers of 2-mm-length at room and elevated temperatures are shown in Fig. 4. Owing to suppressed saturation of the LCC the same lasing power is achieved at a considerably lower operating current in the laser with ABLs as compared to the reference laser. For example, at 20(60)°C output optical power of 8(6) W was reached with 45(17)% less pumping as a result of ABLs implementation. It should be mentioned that the maximum lasing power in both lasers is the same, being about 9 W at 20°C and is limited by catastrophic optical mirror damage.

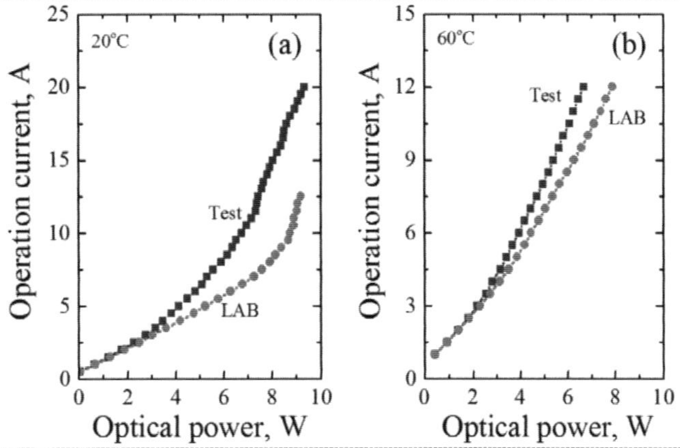

Figure 4. Operating current as a function of output optical power for LAB and the test laser at 20°C (a) and 60°C (b).

4. Conclusion

We have shown that the saturation of LCC in QW lasers at high pumping currents, which is caused by parasitic recombination in the waveguide, can be significantly suppressed by two ABLs adjoining the active region on both sides. This effect was observed at room and elevated temperatures. The deviation of LCC from linear dependence in the LAB was shown to be less dramatic as compared to that in the reference laser without ABLs. For example, at 20°C and pumping current of 7 A the deviation of LCC from linearity in the test laser was as high as 33%, whereas it was only 9% in the LAB. While the maximum output power, being limited by the catastrophic optical mirror damage, is the same in both lasers, the operation current, at which it is reached, is more than 40% lower in the LAB as compared to the reference laser.

5. Acknowledgement

This work is supported by the Russian Scientific Foundation (project 14-42-00006 «A novel type of diode lasers with characteristics improved by using asymmetric barriers»).

References

[1] Asryan L V and Luryi S 2003 Temperature-insensitive semiconductor quantum dot laser *Solid-State Electron.* **47** 205.
[2] Livshits D A, Kochnev I V, Lantratov V M, Ledentsov N N, Nalyot T A, Tarasov I S and Alferov Zh I 2000 Improved catastrophic optical mirror damage level in InGaAs/AlGaAs laser diodes *Electron. Lett.* **36** 1848.
[3] Kazarinov R F and Shtengel G E 1997 Effect of carrier heating on static linearity of MQW InGaAsP/InP lasers *J. Lightwave Technol.* **15** 2284.
[4] Lee J J, Mawst L J and Botez D 2003 MOCVD growth of asymmetric 980 nm InGaAs/InGaAsP broad-waveguide diode lasers for high power applications *J. Cryst. Growth* **249** 100.
[5] Zhukov A E, Kryzhanovskaya N V, Zubov F I, Shernyakov Yu M, Maximov M V, Semenova E S, Yvind K and Asryan L V 2012 Improvement of temperature-stability in a quantum well laser with asymmetric barrier layers *Appl. Phys. Lett.* **100** 021107.
[6] Asryan L V, Kryzhanovskaya N V, Maximov M V, Zubov F I and Zhukov A E 2013 Light-current characteristic of a quantum well laser with asymmetric barrier layers *J. Appl. Phys.* **114** 143103.
[7] Zubov F I, Zhukov A E, Shernyakov Yu M, Maximov M V, Kryzhanovskaya N V, Yvind K, Semenova E S and Asryan L V 2015 The effect of asymmetric barrier layers in the waveguide region on power characteristics of QW lasers *Tech. Phys. Lett.* **41** 61.

The effect of the sulfide passivation on the luminescence of microdisk mesas with quantum wells and quantum dots

N V Kryzhanovskaya[1], M V Lebedev[2], T V Lvova[2], Yu V Kudashova[1], I I Shostak[1], E I Moiseev[1], A E Zhukov[1,3], M V Maximov[1,2], M M Kulagina[2], A M Nadtochiy[1] and S I Troshkov[2]

[1] St Petersburg Academic University, 8/3 Khlopina st., 194021, St Petersburg, Russia,
[2] Ioffe Institute, 26 Polytekhnicheskaya, St.Petersburg, 194021, Russia
[3] St Petersburg Scientific Center of RAS, 5 Universitetskaya Nab., 199034 St Petersburg, Russia

Yu.v.kudashova@mail.ru

Abstract. Sulphide passivation was used for reducing the surface recombination velocity of microdisk mesas based on (AlGaIn)As/GaAs heterostructures with active region formed either by ten GaAs/AlAs quantum wells or by 1 layer of InAs/$In_{0.15}Ga_{0.85}As$ quantum dots. It was demonstrated that the sulfide passivation results in substantial increase of photoluminescence intensity in all types of the mesa-structures studied.

1. Introduction

Compact optical emitters which are based on nanodimensional active region are promising candidates for use in optical integrated circuits. Semiconductor cavities with axial symmetry such as microdiscs and microrings allow one realizing lasing even in case cavity size is as small as few micrometers. However, because the cavity mode is close to cavity sidewalls formed by dry etching there is a problem of surface recombination. GaInAsP/InP system is characterized by low surface recombination rate [1], but lasers based on this material system cover limited optical range (1.3 – 1.6 μm) and are not temperature stable. In InGaAs/AlGaAs material system surface recombination rate is one order of magnitude higher than in GaInAsP/InP system [2] and small sized microcavities based on InGaAs quantum wells operate only at low temperatures. The problem of surface recombination can be solved by using self-organized InGaAs/AlGaAs quantum dots (QDs) as active area. Due to efficient localization of electrons and holes in QDs carrier lateral transport in QD-based structures is suppressed [3]. For InGaAs/GaAs quantum well (QWs) diffusion length at room temperature is 2.7 μm, whereas for InGaAs/GaAs QDs it is much shorter - 0.1 μm [4]. The disadvantage of QD active region is limited gain of the ground state transition. For microcavities with high external loss this may result in a switch of the lasing wavelength from the ground to excited state of QD array with temperature increase. Thus, for such microcavities InGaAs QW active region is preferable as it provides higher gain compared to InGaAs QDs. However in this case the problem of high surface recombination rate has to be solved.

As known nonradiative surface recombination can be strongly reduced by passivation. For example treatment of AlGaAs/GaAs microdisk structures with diameter from 3 to 5 μm using water solution of ammonium sulphide enabled lasing in these structures [5, 6]. However to the best of our knowledge no results about sulphide passivation of small-sized mesas formed from GaAs heterostructures with InAs/InGaAs QDs or GaAs QW active region has been reported so far. In this work we investigate the impact of surface passivation on photoluminescence (PL) intensity of mesas with diameter from 2 to 6 μm and active region based on QWs or QDs.

2. Experiment

Two structures were grown by molecular beam epitaxy on semi-insulating GaAs (100) substrate. In the first structure the active region was formed by ten 10 nm-thick GaAs QWs (MQW structure). These QWs were separated with 10 nm-thick AlAs barriers. In the second structure the active region was formed by one InAs/$In_{0.15}Ga_{0.85}As$ QD layer (QD structure). Such QDs were grown using directional migration of In adatoms [7]. In the frame of this growth approach initial islands were formed using Stransky-Krastanov growth mode by depositing 2.5 monolayers of InAs and then they were covered with 5 nm-thick $In_{0.15}Ga_{0.85}As$ layer. The density of such QDs is $5 \cdot 10^{10}$ cm^{-2}, average distance between individual QDs is 45 nm. The QWs or QDs were inserted in the middle of 200 nm thick GaAs layer. 50 nm thick $Al_{0.25}Ga_{0.75}As$ layers were deposited on the bottom and on the top of the GaAs layer to prevent carrier diffusion to the substrate and to the surface. Finally the structure was covered with a 10 nm-thick GaAs layer. Figure 1 shows schematic image of the MQW and QD structures. Note that the thin bottom $Al_{0.25}Ga_{0.75}As$ layer did not provide any optical confinement and consequently laser generation was impossible.

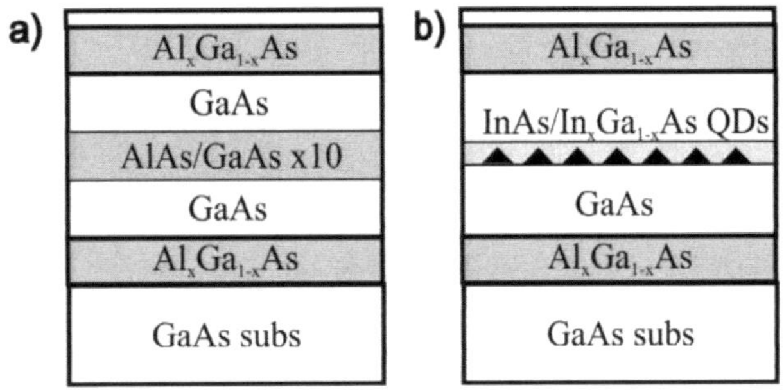

Figure 1. Schematic image of MQW (a) and QD (b) structures.

Figure 2. SEM image of the 2.8 μm in diameter microdisk mesa fabricated from a QD structure after the passivation.

MD mesas with different diameters (from 2 to 6 μm) were formed using photolithography and wet etching with $HBr:K_2Cr_2O_7:CH_3COOH$ solution. The etching was performed through the entire structures down to GaAs substrate and so the active region was exposed to air. The mesas passivation was performed with 1M water solution of sodium sulphide ($Na_2S:9H_2O$) during 4 minutes and then the mesas were washed by water and dried by air. The microphotoluminescence (micro-PL) of the mesas was investigated under CW pumping with YAG:Nd laser (λ = 527nm). The laser power was about 1 mW. Laser beam was focused in a spot of 1μm in diameter by lens Olympus LMPlan IR x100 NA0.8. The same lens was used for signal collection from the samples. Detection was conducted using monochromator FHR 1000 and multichannel cooled InGaAs-photodetector Horiba Symphony (resolution 0.05 nm).

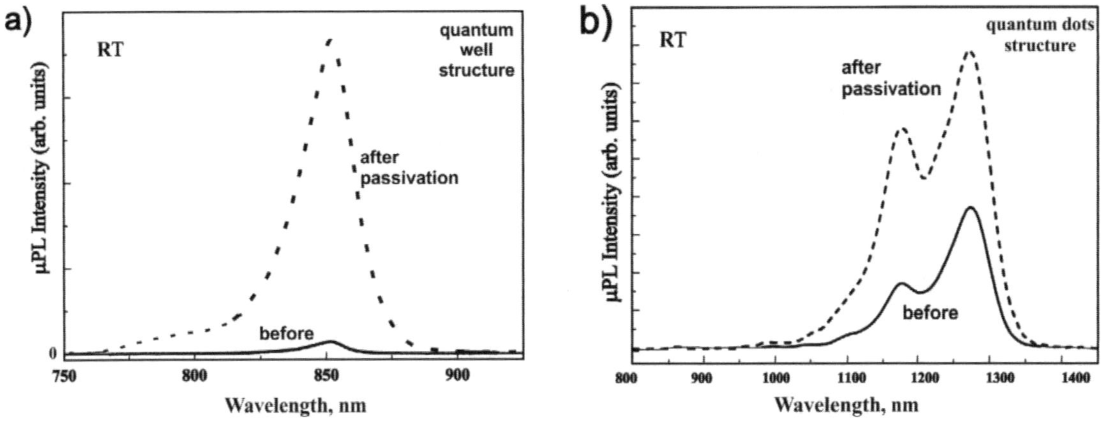

Figure 3. Micro-PL spectra of a 3.7 μm in diameter MQW microdisk mesa (a) and of a 3.9 μm in diameter QD microdisk mesa (b) at room temperature.

3. Results

Figure 2 shows scanning electron microscopy (SEM) image of the MD mesa after the passivation. We observe that the passivation does not affect mesa diameter and shape. Figure 3 shows micro-PL spectra at room temperature of mesas based on QW or QDs active region before and after the passivation. In these structures there is no vertical optical confinement so there are no narrow lines in the spectra characteristic of resonator modes. The wavelength of PL intensity maximum for the structure with QWs is 850 nm. The wavelength of PL intensity maximum for ground state of QD structure is 1280 nm. Integral micro-PL intensities increase after the passivation. For example we observe the micro-PL intensity increase by a factor of 35 for 3.7 μm in diameter MQW mesa and by a factor of 3 for 3.9 μm in diameter QD mesa (Figure 3).

Figure 4 shows the dependencies of normalized integral micro-PL intensities for both structure on the mesa diameters before and after the passivation.

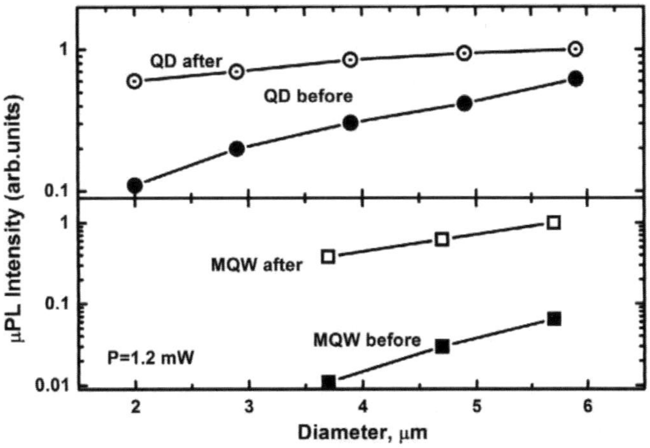

Figure 4. The normalized micro-PL intensity versus mesa's diameters for MQW (squares) and QD (circles) mesas before (solid symbols) and after (open symbols) the passivation.

The intensities were normalized on the intensity of the initial structures before mesa etching. The intensities of the both structures decrease after mesa etching. The intensity drop is stronger for the mesas with smaller diameter due to larger impact of the nonradiative recombination on the mesa sidewalls. For QD structure the intensity decreased by a factor of 1.6 for 6 μm in diameter mesa and by a factor of 9 for 2 μm in diameter mesa. For MQW structure the intensity decreased by a factor of 15.6 for 5.7 μm in diameter mesa and by a factor of 92 for 3.7 μm in diameter mesa. The weaker decrease of PL intensity for QD mesas as compared to MQW ones is due to the efficient carrier localization in QDs. Most of the carriers are located in QDs and carriers concentration in wetting and GaAs layers is small. Therefore only small amount of carriers can diffusing to mesa sidewalls. After the passivation of the structures we observed almost full recovery of emission intensity for the QD mesas. Intensity recovered to 100% for a 6 μm in diameter mesa and to 60% for a 2 μm diameter mesa. For QW structures intensity was restored to more than 60% for mesas with large diameter (4.7-5.7 μm). Integral intensity after passivation increased by a factor of 35 for 3.7 μm in diameter MQW mesas.

4. Conclusion

To conclude we show that the passivation results in increasing of microPL intensity for MQW and QD mesas of all diameters, which evidences significant decrease of surface recombination rate at passivated mesa sidewalls. The effect of passivation is more pronounced for MQW mesas due to the larger diffusion length as compares to QD structures.

Acknowledgments

The work is supported by the Skolkovo Foundation (grant agreement for Russian educational and scientific organisation no.4 dd. 25.12.2014)

References

[1] Tamura M, Kojima T, Ando T, Nunoya N, Tamura Sh and Arai S 1998 Sidewall recombination velocity in GaInAsP/InP quantum-well lasers with wire-like active region fabricated by wet-chemical etching and organo-metallic vapor-phase-epitaxial regrowth *Jpn. J. Appl. Phys.*, **37** 6569

[2] Boroditsky M, Gontijo I, Jackson M, Vrijen R, Yablonovitch E, Krauss T, Cheng C.-C, Scherer A, Bhat R and Krames M 2000 Surface recombination measurements on III-V candidate materials for nanostructure light-emitting diodes *J. Appl. Phys.* **87** 3497

[3] Moore S A, O'Faolain L, Maria Ana Cataluna, Flynn M B, Kotlyar M V and Krauss Th F 2006 Reduced Surface Sidewall Recombination and Diffusion in Quantum-Dot Lasers *IEEE Photonics Technology Letters* **18** 1861

[4] Fiore A, Rossetti M, Alloing B, Paranthoen C, Chen J X, Geelhaar L and Riechert H 2004 Carrier diffusion in low-dimensional semiconductors: A comparison of quantum wells, disordered quantum wells, and quantum dots *Phys. Rev. B* **70** 205311-1–205311-12

[5] Hobson W S, Mohideen U, Pearton S J, Slusher R E and Ren F 1993 SiNx Sulfide Passivated GaAs/AlGaAs Microdisk Lasers *Electron. Lett.* **29** 2199–2200

[6] Hobson W S, Ren F, Mohideen U, Slusher R E, Lamont Schnoes M and Pearton S J 1995 Silicon-Nitride Encapsulation of Sulfide passivated GaAs/AlGaAs Microdisk Lasers *J. Vac. Sci. Technol. A* **13** 642–45

[7] Maximov M V, Tsatsul'nikov A F, Volovik B V, Sizov D S, Shernyakov Y M, Kaiander I N, Zhukov A E, Kovsh A R, Mikhrin S S, Ustinov V M and *et al.* 2000 Tuning quantum dot properties by activated phase separation of an InGa(Al)As alloy grown on InAs stressors *Phis. Rev. B* **62** 16671-80

Ultimate modulation bandwidth of 850 nm oxide-confined vertical-cavity surface-emitting lasers

M A Bobrov[1], S A Blokhin[1], N A Maleev[1], A G Kuzmenkov[1], A A Blokhin[2,1], Yu M Zadiranov[1], S I Troshkov[1], N N Ledentsov[3], V M Ustinov[1]

[1]Ioffe Institute, 26 Polytekhnicheskaya Str, St Petersburg 194021, Russia
[2]Peter the Great St. Petersburg Polytechnic University, 26 Polytekhnicheskaya Str, St Petersburg, 195251, Russia
[3]VI Systems GmbH, Hardenbergstrasse 7, D-10623 Berlin, Germany
E-mail: bobrov.mikh@gmail.com

Abstract. Complex influence of photon lifetime (controlled by the mirror loss) and aperture size on the performance of 850 nm InGaAlAs oxide-confined vertical-cavity surface-emitting lasers (VCSELs) with fully doped AlGaAs-based distributed Bragg reflectors (DBR) was investigated. We find a tradeoff between photon lifetime and gain nonlinearity for maximizing the optical bandwidth, leading to the optimum aperture size close to 4-6 μm. In spite of the reduced photon lifetime (from 4 ps to 1 ps), the excess damping caused by the current-induced self-heating limits the ultimate modulation bandwidth for the given VCSELs design at 24-25 GHz. Further improvement in high frequency characteristics can be facilitated by decrease of the heat generation and improvement of the heat removal from the active region as well as by proper engineering of the scattering loss at the oxide aperture while keeping the low capacitance optimizing design of the oxide aperture.

1. Introduction

Vertical-cavity surface-emitting lasers (VCSELs) emitting at 850 nm are widely used as low-cost high-performance light sources, in the mode of direct current modulation, for data communication systems over short distances and for the optical interconnects in high-performance computing systems [1]. The most important factor that limits modulation response is damping of the relaxation oscillation, which directly proportional to photon lifetime in optical microcavity, gain compression and inversely proportional to differential gain. During the last few years several approaches were suggested in order to maximize the resonance frequency by increase of the differential gain of the active region and decrease of the mode volume (controlled by aperture size) as well as on minimizing the chip parasitics [2-4]. The combination of these methods allowed an increase in the modulation bandwidth of short-wavelength VCSELs up to 20 GHz [5-6]. To overcome damping limitation, the reduction of the photon lifetime was proposed [7].

In this work we have analyzed complex influence of photon lifetime together with aperture size on the high-speed performance of 850 nm InGaAlAs oxide-confined VCSELs and realize the ultimate modulation bandwidth possible for the given VCSEL design.

2. VCSEL design

A conventional VCSEL geometry with two doped distributed Bragg reflectors (DBR) was used. The VCSEL epitaxial structure consists of an n-doped GaAs contact/buffer layer, a bottom $Al_{0.15}Ga_{0.85}As/Al_{0.9}Ga_{0.1}As$ n-doped DBR, a 1.5λ-thick AlGaAs optical microcavity with a strained InGaAlAs-based active region, a top $Al_{0.15}Ga_{0.85}As/Al_{0.9}Ga_{0.1}As$ p-doped DBR, and a p-doped GaAs topmost contact layer (see figure 1). The active region consists of multiple narrow InGaAs quantum wells (QWs) with an average In-composition of ~7% sandwiched between high energy bandgap AlGaAs cladding layers to improve the differential gain of the active region and to suppress the thermal escape of nonequilibrium carriers. The QWs are placed in the antinode position of the optical field intensity to provide sufficient modal gain for lasing. The number and the thickness of the QWs as well as the barrier thicknesses were chosen to be narrow enough to reach the maximum possible optical confinement factor. To eliminate excess voltage drop and provide low resistance, compositional graded interfaces with modulated doping profiles were used in DBRs. The average doping concentration in the doped regions is about $2\cdot10^{18}cm^{-2}$ for both carrier type. To maintain reasonable internal optical losses and provide low series resistance the doping of the DBR layers is reduced near the active region, where the electric field intensity is the highest.

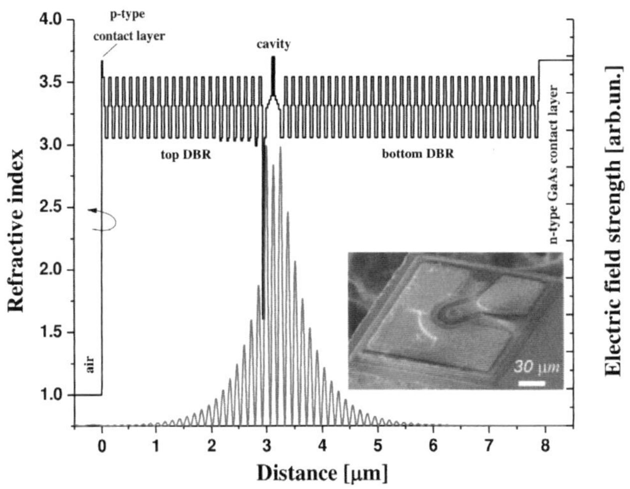

Figure 1. Refractive index and longitudinal mode profile versus the vertical coordinate for the 850 nm oxide-confined InGaAlAs VCSEL with a fully doped DBRs. Inset: the scanning electron microscopy image of the fabricated VCSEL.

VCSEL chips used in this study are a result of the joint project between VI-Systems GmbH (Federal Republic of Germany) and Connector Optics LLC (St. Petersburg, Russian Federation). The VCSEL structures were grown on an undoped GaAs substrate using metal-organic vapor-phase epitaxy. Lasers with different mesa diameters were fabricated in the high-speed design using standard fabrication techniques including: optical lithography, metal and dielectric evaporation, dry and wet etching, selective wet oxidation, etc. The schematic cross-section of the processed 850 nm oxide-confined VCSEL with a ground-source (GS) pad configuration is shown in inset in figure 1. Light power-current-voltage characteristics and small signal modulation response (S_{21}) of lasers were measured in the range current-aperture size from 1.5 μm to 10 μm.

3. Results and discussion

The photon lifetime depends on the total optical loss and the effective length of the optical microcavity. On the one hand, a decrease in the quantum efficiency and the laser speed can be expected in case of high optical loss [9]. On other hand, an increase in the microcavity length can drastically change the optical mode structure and increase the internal optical loss [10]. Since the mirror reflectivity is a quit sensitive not only to the mirror pairs, but also to the thickness of the top DBR layer, the mirror loss can be controlled by etching the top DBR.

However the reduction of top DBR reflectivity by consequent removing the mirror pairs is challenging issue in case of AlGaAs DBRs with compositional graded interfaces. Furthermore such approach requires rather deep etching which negatively affects the uniformity of carrier injection to the active region and the laser series resistance. The most promising approach is “shallow” etching of DBR surface layers. Figure 2 shows the top DBR reflectivity dependence as function of etch depth at lasing wavelength, calculated by transmission matrix method and taking into account free carrier absorption. By etching the surface of top DBR up to 70 nm, the mirror reflectivity can be adjusted within the range 96.5–99.5%. As a result, the total mirror loss can be easily changed from 0.3% up to 1.8% (at fixed bottom DBR reflectivity), reaching a local maximum at 70 nm. Note that such approach enable to investigate the static and dynamic characteristics for a fixed set of devices starting from the minimal mirror loss and ending the maximum mirror loss with step by step shallow etching of top DBR with subsequent measurements and, as a result, to eliminate the uncontrollable variations of the laser characteristics (caused by epitaxial and/or processing deviations) and to improve the validity of the experimental data.

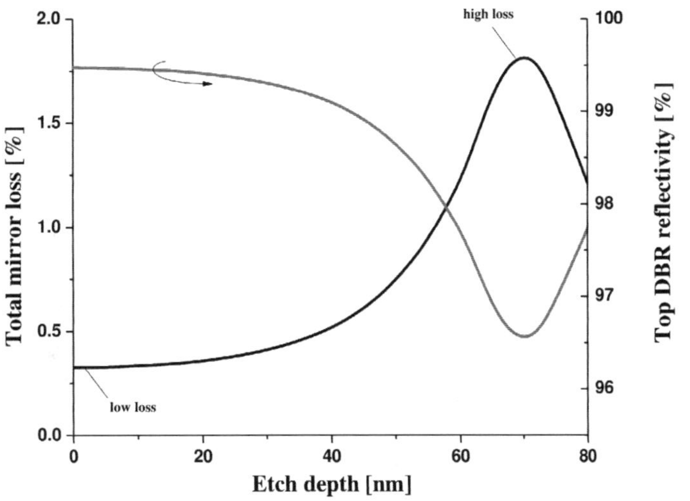

Figure 2. Calculated top DBR reflectivity and corresponding total mirror loss as a function of etch depth into the top DBR of 850 nm InGaAlAs VCSEL.

Figure 3.a shows the maximum 3dB modulation bandwidth and the rate of damping rise (so called K-factor) as function of the aperture size for VCSELs with the low mirror loss (~0.3%). As expected, the 3dB modulation bandwidth increases with reducing the aperture size, but quickly saturates at 21 GHz for aperture diameters less than 7 μm. According to the S_{11} data and the equivalent circuit model electrical, the parasitic cut-off frequency of the investigated VCSELs was found to be about 24 GHz over the entire measured current range, which means that the electrical parasitic has no significant impact on the VCSEL's high speed performance. Hence, the saturation of the 3dB modulation bandwidth can be associated to the thermal effects and/or the damping. The resulting K-factor is hardly depended on the oxide aperture and be in the range of 0.21-0.23 ns, indicating the crucial role of the photon lifetime. However the dynamic characteristic of the large aperture VCSELs are mainly limited by the self-heating effect, while the higher photon density (e.g. the higher resonance frequency) and the excess damping caused by the current-induced self-heating sets the intrinsic limit of the reachable modulation bandwidth for smaller aperture VCSELs (see figure 4.a).

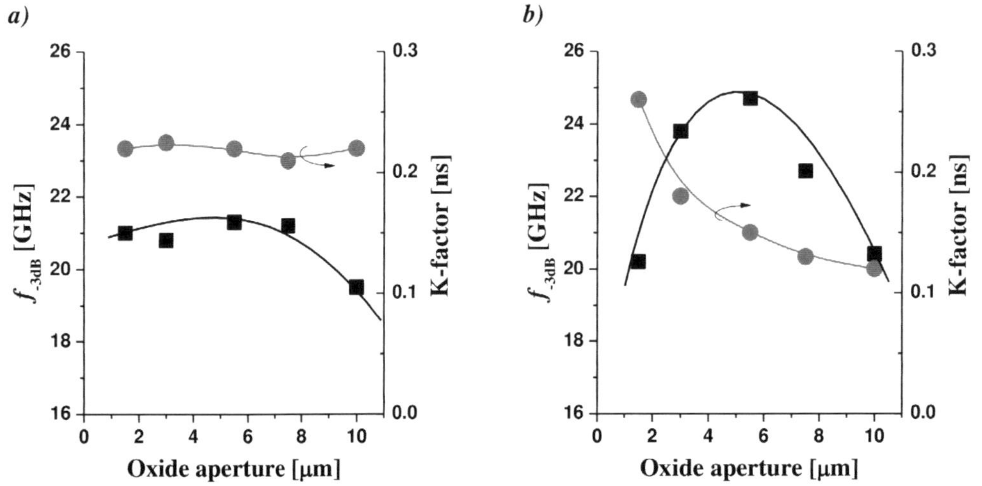

Figure 3. Measured 3dB modulation bandwidth and K-factor as function of the oxide aperture size for 850 nm InGaAlAs VCSELs with low (a) and high (b) mirror losses at 20°C.

The increase of the mirror loss up to 1.8% results in the reduction of the photon lifetime to 1-1.2 ps, however a twofold drop of the K-factor was revealed for the wide-aperture VCSELs (see figure 3.b). Moreover, the ultrasmall aperture VCSELs demonstrate a 20% increase of the resulting K-factor. It can be explained by the domination of the gain nonlinearity in the damping coefficient. The gain compression factor was estimated to be in the range $1.5–2\times10^{-17}cm^3$ for all apertures size and the mirror loss. Hence, the complex behavior of the K-factor can be attributed to the dramatic change in the differential gain. According to the dependences of the slope efficiency on the mirror loss, the internal optical loss increases significantly with the reduction of the aperture size: from 0.16% per round trip at 10 μm to 0.5% per round trip at 1.5 μm (see figure 4.b), which means that the free carrier absorption loss are dominated in the wide-aperture VCSELs, while the optical scattering loss from the oxide aperture itself occurs in the smallest devices. Moreover, as the aperture size is reduced, the current injection efficiency decreases rapidly from 73% down to 35% (due to excessive carrier leakage at high current density). The dramatic increase of the threshold current density with decreasing aperture size (from 1.1 kA/cm^2 at 10 μm to 30 kA/cm^2 at 1.5 μm) leads to the significant reduction of the differential gain (from $9.7\times10^{-16}cm^2$ at 10 μm to $4.1\times10^{-16}cm^2$ at 1.5 μm).

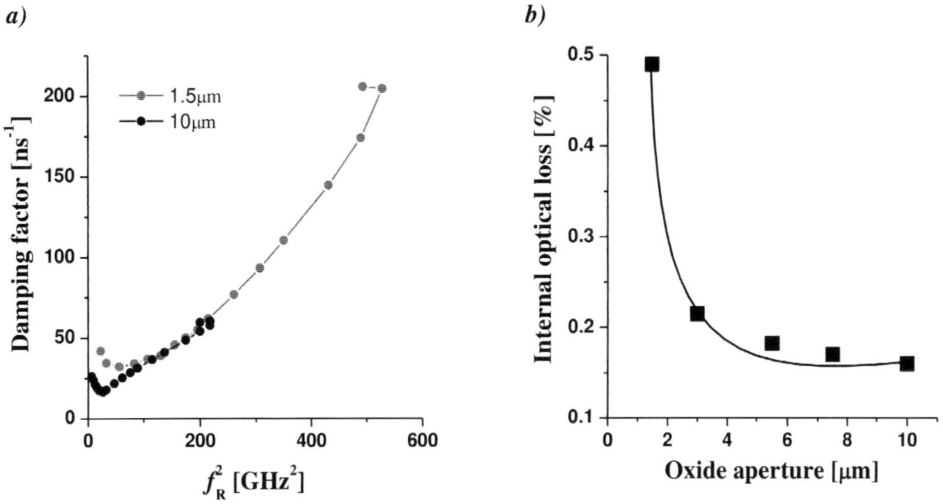

Figure 4. a) The damping coefficient versus the square of the resonance frequency for 1.5 μm and 10 μm devices with low mirror losses; b) The extracted internal optical loss of 850 nm InGaAlAs VCSELs as function of the oxide aperture size. The heatsink temperature is 20°C.

As shown in figure 3.b, the decrease of the photon lifetime leads to the increase of the maximum 3dB modulation bandwidth for the wide-aperture lasers, while the maximum 3dB modulation bandwidth of the ultrasmall aperture VCSELs drops to 20 GHz. The competing effects of the reduced photon lifetime and the reduced differential gain results in the ultimate modulation bandwidth of 24-25 GHz at oxide current aperture in the range of 4-6 μm. The further enhancement of the modulation bandwidth can be done by minimizing self-heating (low heat generation and efficient heat removal from the active region) and optimizing design of the oxide aperture (tradeoff between the optical scattering loss and the mode volume).

4. Conclusion

The combined impact of the photon lifetime and the aperture size on high-speed performance of 850 nm InGaAlAs oxide-confined VCSELs with fully doped AlGaAs DBRs was analyzed. The static physical properties of investigated lasers are determined from the static light power-current-voltage characteristics as function of the mirror loss. To clarify the limiting mechanism of high-speed performance, the dynamic parameters of each particular lasers were extracted from the small signal modulation response and microwave reflection measurements. It was found that the damping effect sets the intrinsic limit of the reachable modulation bandwidth with decreasing the aperture size. As a result, the maximum 3dB modulation bandwidth of lasers with low mirror losses quickly saturates at 21 GHz in spite of the significant reduction of the mode volume with decreasing the aperture size. However with increasing the mirror loss, the gain nonlinearity starts making a significant contribution to the K-factor. The gain compression is independent on the apertures size and the mirror loss for the given VCSEL design, while the dramatic increase of the threshold current density with decreasing aperture size results in significant reduction of the differential gain. The competing effects of the reduced photon lifetime and the reduced differential gain leads to an optimum oxide aperture range for the maximized modulation bandwidth.

5. Acknowledgments

This work was supported in part by the Ministry of Education and Science of the Russian Federation (grant № 14.607.21.0001, unique ID RFMEFI60714X0001).

References

[1] Kaminow I, Li T, and Willner A E 2013 *Optical Fiber Telecommunications Volume VIA: Components and Subsystems* (Academic Press)

[2] Healy S B, O'Reilly E P, Gustavsson J S, Westbergh P, Haglund E, Larsson A, and Joel A 2010 Active region design for high-speed 850 nm VCSELs *IEEE J. Quantum Electron.* **46** 506–512

[3] Tan F, Wu H, Feng M, and Holonyak N, Jr. 2011 Energy efficient microcavity lasers with 20 and 40 Gb/s data transmission *Appl. Phys. Lett.* **98** 191107

[4] Chang Y-C, Wang C S, Johansson L A, and Coldren L A 2006 High-efficiency, high-speed VCSELs with deep oxidation layers *Electron. Lett.* **42** 1281–1282

[5] Al-Omari A N and Lear K L 2007 Low current density, inverted polarity, high-speed, top-emitting 850nm vertical-cavity surface-emitting lasers *IET Optoelectron.* **1** 221–225

[6] Mutig A, Blokhin S A, Nadtochiy A M, Fiol G, Lott J A, Shchukin V A, Ledentsov N N, Bimberg D 2009 Frequency response of large aperture oxide-confined 850 nm vertical cavity surface emitting lasers *Applied Physical Letters* **95** 131101

[7] Westbergh P, Gustavsson J S, Haglund E, Skoeld M, Joel A, and Larsson A 2009 High-Speed, Low-Current-Density 850 nm VCSELs, *IEEE J. Sel. Top. Quantum Electron.* **15** 694–703

[8] Westbergh P, Gustavsson J S, B. Kögel, Haglund A, Larsson A, and Joel A 2010 Speed enhancement of VCSELs by photon lifetime reduction *Electron. Lett.*, **46** 938–940

[9] Blokhin S A, Bobrov M A, Maleev N A, Kuzmenkov A G, Stetsenko V V, Pavlov M M, Karachinskii L Ya, Novikov I I, Zadiranov Yu M, Egorov A Yu, and Ustinov V M 2013 Influence of Optical Losses on the Dynamic Characteristics of Linear Arrays of Near-Infrared Vertical-Cavity Surface-Emitting Lasers *Semiconductors* **47** 833–837

[10] Unold H J, Riedl M C, Mahmoud S W Z, Jäger R, and Ebeling K J 2001 Long monolithic cavity VCSELs for high singlemode output power *Electron. Lett.* **37**, 178–179

Cyclometalated complexes of platinum metals – the new luminescent sensors

M V Nikolaeva, E A Katlenok, M S Khakhalina, M V Puzyk, and K P Balashev

Department of Chemistry, Herzen State Pedagogical University, 48 Moika Embankment, St. Petersburg, 191186 Russia

E-mail: nikolaeva.mr@gmail.com

Abstract. The influence of the environment on the cyclometalated Pt(II), Pd(II), Ir(III) complexes' optical properties in the presence of various organic and inorganic compounds in solution and the gas phase is studied. The feasibility of complexes' using as optical sensors for molecular oxygen, halides ions, hydrogen and Hg(II) cations in the liquid phase, as well as for water and some organic solvents' vapor in the immobilized state in the MF-4SK membrane.

1. Introduction

Heterocyclic luminophores with high quantum yield of fluorescence and photostability [1], biological and catalytic activity [2] are attracted the attention of researches to modification of their properties. Together with introducing different substituents the modification of properties of heterocyclic phosphors occurs by metalation of their aryl substituted derivatives. It has been shown [3–7] that metalation by platinum metals leads to a significant change in the optical properties of luminophores.

In contrast to fluorescence of free luminophores, metal-complex phosphors are characterized by intense and relatively long-lived phosphorescence in solutions and outer-sphere electron transfer reversible processes. The specific structure of the cyclometalated Pt(II), Pd(II), and Ir(III) complexes with heterocyclic ligands determines the possibility of developing of these complexes-based optical sensors and chemosensors for the presence of various organic and inorganic compounds in the liquid and the gas phase, including in biosystems [7–10].

This work is aimed at studying the effect of the environment on the optical properties and the photoexcitation energy degradation efficiency of the cyclometalated Pt(II), Pd(II), and Ir(III) complexes in homogeneous solutions and which immobilized on the surface of a MF-4SK cation-exchange membrane.

The investigation subjects are the mono- and binuclear Pt(II), Pd(II), and Ir(III) complexes with metalated phenyl-substituted $(C^\wedge N)^-$ imines:

{M(C^N)} =

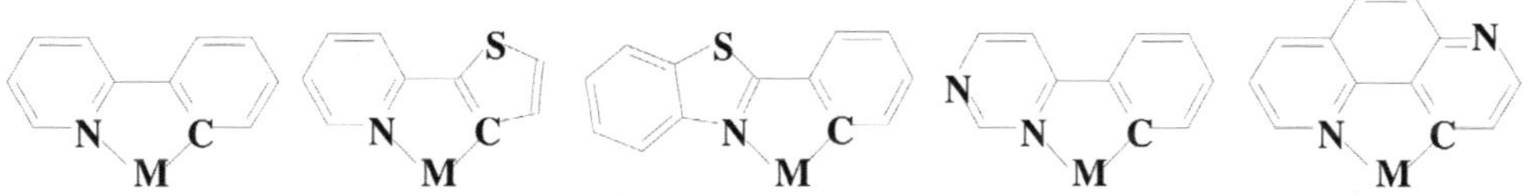

where $(C^\wedge N)^-$:	phenyl-pyridine (Ppy)	2-(2'-thienyl)-pyridine (Tpy)	2-phenyl-benzothiazole (Bt)	4-phenyl-pyrimidine (Ppm)	1,7-phenanthroline (1,7-Phen)

chelating (N–N, S–S$^-$) ligands:

{M(N–N)} = 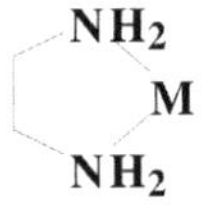 {M(S–S)} =  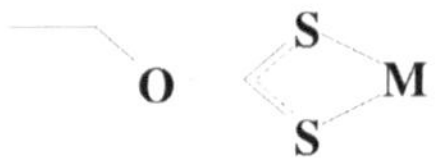

N–N: ethylenediamine (En) S–S$^-$: diethyldithiocarbamate (Dtc) ethylxanthate (Exn)

and bridging 2-mercapto-derived (μ-N–S$^-$) ligands:

{Pt(μ-N^S)Pt} = 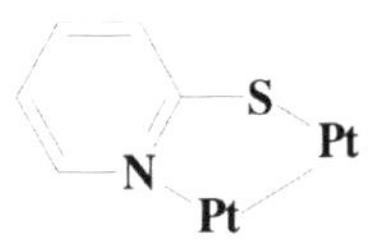 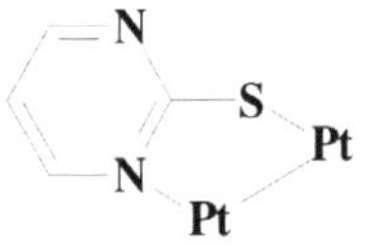 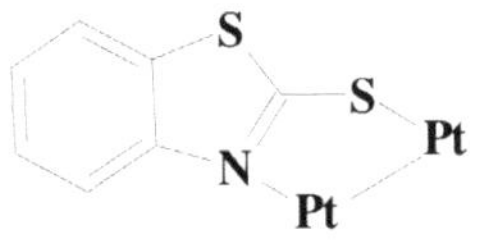 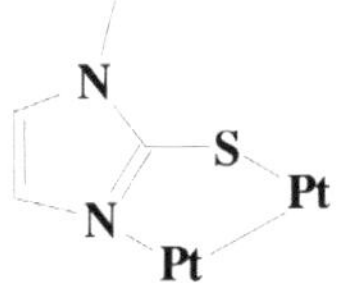

(μ-N^S$^-$):	2-mercapto-pyridine (Pyt)	2-mercapto-pyrimidine (Mpm)	2-mercapto-benzothiazole (Mbt)	2-mercapto-1-methyl-imidazole (Mim)

The cyclometalated complexes were synthesized by well-known methods and investigated by electronic absorption (SF-2000 spectrophotometer) and emission (Fluorat-02-Panorama spectrofluorimeter) spectroscopy, multinuclear one- and two-dimensional NMR spectroscopy (JEOL JNM-ECX400A spectrometer), X-ray crystallography (Agilent Technologies Xcalibur Eos single crystal X-ray diffractometer), and cyclic voltammetry (IPC-Pro MF potentiostat).

2. Sensor optical properties of cyclometalated complexes of platinum metals

2.1. Effect of halide ions on the luminescence of complexes

The luminescence quenching of complexes by halide ions (Cl$^-$, Br$^-$, I$^-$) was done in an argon-saturated ethanol solution at 298 K. In the absence of halide ions, the photoexcitation of solutions of [PtEnPpy]$^+$, [PtEnTpy]$^+$, and [PtEnBt]$^+$ complexes leads to an intense vibrationally structured luminescence in the visible spectral region (figure 1) with an exponential decay kinetics, which, within the model of localized molecular orbitals, is assigned to the spin-forbidden optical transition from the metal-to-ligand charge transfer state. The presence of halide ions ($C \leq 10^{-2}$ M) in solutions does not change the absorption spectra of complexes that points at the absence of the outer-sphere complexation of [PtEn(C^N)]Hal. At the same time, the intensity (I) and decay time (τ) of the complexes' photoluminescence with increase of halide ions concentration decrease (figure 2) [11].

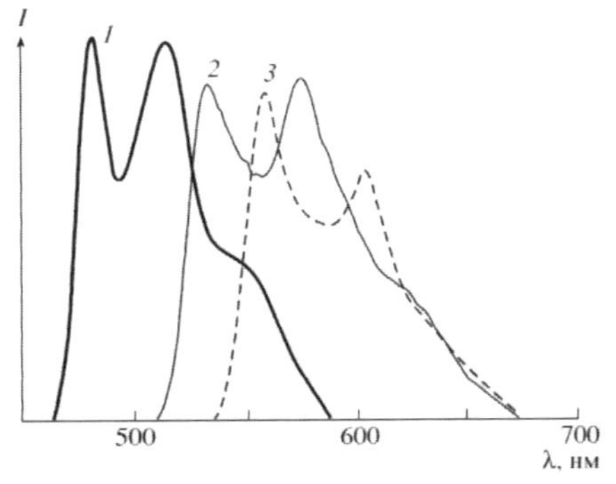

Figure 1. Luminescence spectra of (*1*) [PtEnPpy]$^+$, (*2*) [PtEnBt]$^+$, and (*3*) [PtEnTpy]$^+$ complexes in ethanol at 298 K.

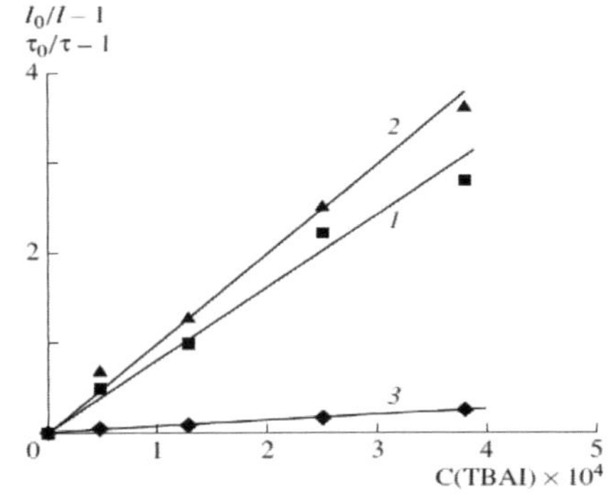

Figure 2. Dependences of the luminescence intensity and decay time for (*1*) [PtEnPpy]$^+$, (*2*) [PtEnBt]$^+$, and (*3*) [PtEnTpy]$^+$ complexes on the concentration of iodide ions in ethanol at 298 K.

The phosphorescence quenching efficiency of Pt(II) complex solutions by halide ions ($Cl^- < Br^- < I^-$) is defined by external heavy atom effect due to increasing the spin-orbit interaction constant of halide ions.

Substitution of chelated En ligand for monodentate ammonium changes fundamentally the mechanism of quenching: an effective complete reaction of replacement of both ammoniums for halogen. This is reflected in the absorption spectra and the absence of the luminescence of complex.

2.2. Effect of molecular oxygen on the luminescence of complexes

The luminescence quenching of complexes by molecular oxygen was done in an argon-deaerated water, acetonitrile, and methanol solution at 295 K. With the increase of oxygen amount in deaerated solutions of complexes, a reduction was observed in the luminescence intensity (figure 3) and lifetime. The absence of the association of oxygen molecules with complexes, as well as the dynamic character of quenching, are indicated by the absence of changes in the absorption spectra of complexes and by the coincidence of the linear dependences of $(I_0/I - 1)$ and $(\tau_0/\tau - 1)$ on the oxygen concentration (figure 4) [12].

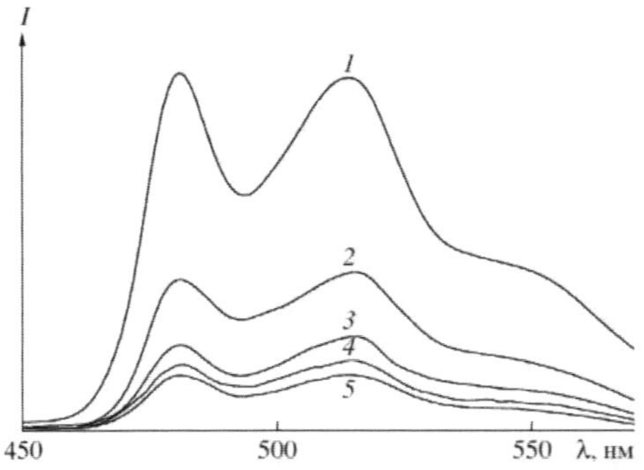

Figure 3. Luminescence intensity of $[PtEnPpy]ClO_4$ in acetonitrile at molecular oxygen concentrations of (*1*) 0, (*2*) 1.49×10^{-3}, (*3*) 2.98×10^{-3}, (*4*) 4.47×10^{-3}, and (*5*) 5.96×10^{-3} mol/l.

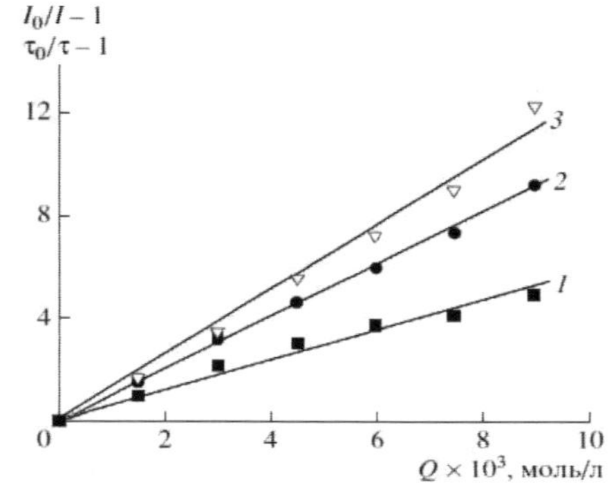

Figure 4. Quenching of the luminescence of Pt(II) complexes in acetonitrile as a function of the oxygen concentration *Q* for (*1*) $[PtEnBt]ClO_4$, (*2*) $[PtEnPpy]ClO_4$, and (*3*) $[PtEnTpy]ClO_4$.

The dynamic character of phosphorescence quenching of the complexes by molecular oxygen in solutions with the near-diffusion rate constant determines the feasibility of complexes' using as luminescent sensors for the presence of molecular oxygen in solution.

Spectral changes of the ammonia cyclometalated Pt(II) complexes in water also have their specificity: molecular oxygen causes a complete chemical reaction in which the final substance is platinum blue, i.e. polynuclear Pt(II) complexes with bridging OH^--ligand [13].

2.3. Effect of acidity on the luminescence of complexes

Protonation of nitrogen atom of cyclometalated ligands in the composition of $[Pd(NH_3)_2(Ppm)]ClO_4$ and $[Pd(NH_3)_2(1,7\text{-}Phen)]ClO_4$ complexes with the addition of perchloric acid to their ethanol solutions is shown. It leads to the phosphorescence quenching of complexes at 77 K (figure 5). The absorption spectra of complexes with the addition of acid remain almost unchanged.

The dependences of the first luminescence maximum on pH (pH = – log $[HClO_4]$) (figure 6) point at the influence of perchloric acid concentration on the luminescence spectra. The observed changes in the spectra are reversible with addition of sodium carbonate to the ethanol solution, the luminescence spectra are restored completely to their initial shapes. [14].

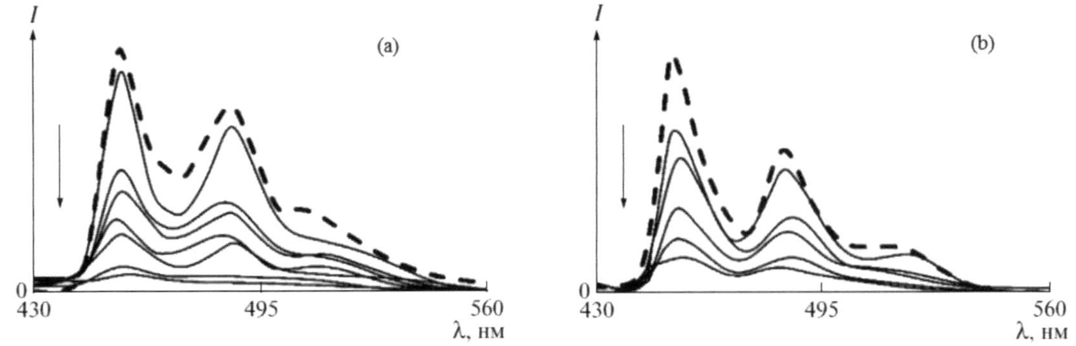

Figure 5. Luminescence spectra of (*a*) [Pd(NH_3)$_2$Ppm]ClO_4 and (*b*) [Pd(NH_3)$_2$(1,7-Phen)]ClO_4 complexes in ethanol at 77 K with increasing (in the direction of the arrow) perchloric acid concentration. The dashed lines show the initial luminescence spectra of complexes.

Figure 6. Dependence of the intensity of the first luminescence maximum of (*a*) [Pd(NH_3)$_2$Ppm]ClO_4 and (*b*) [Pd(NH_3)$_2$(1,7-Phen)]ClO_4 complexes in ethanol at 77 K on pH.

2.4. Effect of Hg(II) cation on the luminescence of complexes

Development of optical sensors for the presence of mercury, which is one of the most toxic metals, in the environment is the number one problem [15].

Addition of acetonitrile Hg(ClO_4)$_2$ solution to dichloromethane solution of binuclear [PtBt(μ-(N^S))]$_2$ complexes with a Pt–Pt chemical bond leads to a change of the optical characteristics of complexes' solutions, according to relationship between concentrations of Hg(II) and Pt(II) complex.

The optical density increase of solution in the 450–650 nm region and the phosphorescence quenching with the maximum at 670–694 nm (figure 7a), with increasing of Hg(ClO_4)$_2$ concentration to 1:1 ratio with platinum complex, result from interaction between Lewis Hg(II) acid and base – binuclear Pt(II) complex:

$$[Pt(bt)(\mu\text{-}(N\text{^}S))]_2 + Hg(ClO_4)_2 = \{Hg[Pt(bt)(\mu\text{-}(N\text{^}S))]_2\}(ClO_4)_2.$$

Further Hg(II) concentration increase leads to a decrease of solution absorption in the 450–650 nm region and an appearance (figure 7b) of the typical bands of $[Pt(bt)(CH_3CN)_2]^+$ in the luminescence spectra, with the maxima at 530 and 568 nm [16], due to adduct destruction:

$$\{Hg(ClO_4)_2[Pt(bt)(\mu\text{-}(N\text{^}S))]_2\} + Hg(ClO_4)_2 + 4\ CH_3CN = 2\ [Pt(bt)(CH_3CN)_2]ClO_4 + 2\ Hg(N\text{^}S)ClO_4.$$

It has been shown that the presence of Hg(II) cations in solutions of cycloiridated [IrBt_2Dtc] and [IrBt_2Exn] complexes leads to a substitution reaction of chelating Dtc^- и Exn^- ligands for acetonitrile solvent and is accompanied by hypsochromic shift of 20–40 nm of the phosphorescence spectra and long wavelength bands of complexes' absorption spectra.

The selective sensor optical properties of Ir(III) complex solutions as well as mono- and binuclear Pt(II) complexes with sulfur-containing donor atoms of chelating and bridging ligands are related to a specific complex formation of "soft" donor S atoms with "soft" Hg(II) ion.

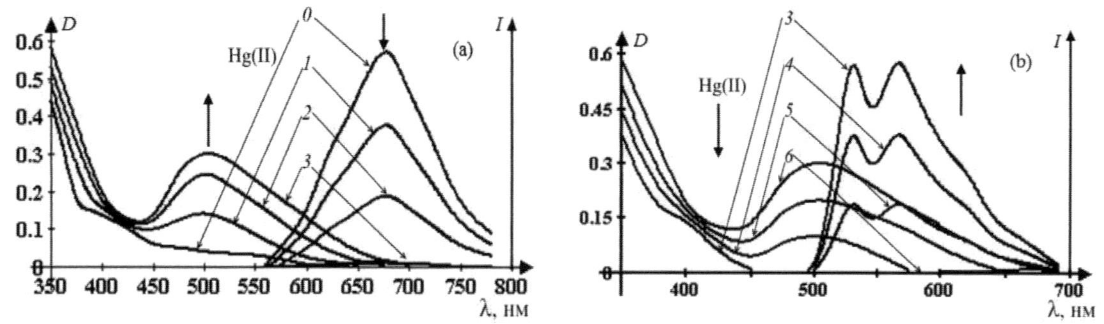

Figure 7. Optical densities and phosphorescence intensities (λ_{ex} = 400 nm) of $[PtBt(\mu\text{-}Mbt)]_2$ (C = 2 × 10^{-5} mol/l) in acetonitrile at $Hg(ClO_4)_2$ concentrations of (*0*) 0, (*1*) 5 × 10^{-6}, (*2*) 1 × 10^{-5}, (*3*) 2 × 10^{-5}, (*4*) 2.5 × 10^{-5}, (*5*) 3 × 10^{-5}, and (*6*) 3.5 × 10^{-5} mol/l.

3. Effect of vapors of water and organic solvents on the luminescence of cation-exchange MF-4SK membrane modified with Pt(II) complexes

Immobilization of cation-exchange membrane with Pt(II) complexes was done by immersing the membrane into acetonitrile solution of Pt(II) complex with concentration on the order of 10^{-5} mol/l. The efficiency of immobilization was monitored by changing the optical densities of solutions before and after immersion of membrane (figure 8), by an increase in the optical density of the modified membrane with time, and by observing bright luminescence of modified membrane in the corresponding spectral region under photoexcitation (the unmodified membrane weakly luminesces in the blue region). The sorption properties of membranes modified by platinum complexes with respect to vapors of organic solvents and water were studied by the desiccator method.

The obtained data on the sorption of vapors of water, methanol, ethanol, isopropanol, *n*-butanol, acetonitrile, and acetone (table 1) show that the modification of the cation-exchange membrane by Pt(II) complexes leads to a decrease in the molar sorption of all solvents by up to 20%. This occurs because the adsorbed cations of Pt(II) complexes decrease the volume of the membrane pores and block its sulfo groups. It was found that the sorption efficiency also depends on the nature of solvents (table 1), namely, on the size and polarity of molecules and on the presence or absence of intermolecular hydrogen bonds.

Table 1. Effect of the solvent nature on the luminescence intensity (I_0/I_{solv}) and on the molar sorption (*A*) of the initial membrane and membranes modified with Pt(II) complexes

Membrane	H_2O		CH_3OH		C_2H_5OH		*iso*-C_3H_7OH		*n*-C_4H_9OH		CH_3CN		$(CH_3)_2CO$	
	I_0/I_{solv}	*A*, mmol/g	I_0/I_{solv}	*A*, mmol/g	I_0/I_{solv}	*A*, mmol/g	I_0/I_{solv}	*A*, mmol/g	I_0/I_{solv}	*A*, mmol/g	I_0/I_{solv}	*A*, mmol/g	I_0/I_{solv}	*A*, mmol/g
Initial	–	12.2	–	14.2	–	5.5	–	3.2	–	2.5	–	2.2	–	4.6
$[PtEnPpy]^+$	2.6	10.7	8.6	13.7	4.5	4.5	2.7	3.1	2.2	2.1	3.3	2.1	4.0	4.4
$[PtEnBt]^+$	2.0	6.7	7.7	13.5	3.7	4.5	1.8	3.1	1.8	2.1	2.8	2.1	3.1	4.3
$[PtEnTpy]^+$	3.0	11.7	16.5	11.4	11.0	4.0	4.0	2.8	3.4	1.9	6.2	1.7	7.3	4.1

In the process of sorption of solvent vapors, a membrane immobilized with a Pt(II) complex sharply decreases the intensity of the vibrationally structured complex phosphorescence for the first hour spent in the atmosphere of organic solvents (figure 9). At a longer action of solvent vapors, the luminescence intensity smoothly decreases until dynamic equilibrium in the system is established. In a thermostat (up to 100°C), a membrane saturated with solvent vapors and immobilized with a Pt(II) complex restores its initial luminescence intensity, which points to reversibility of the sorption of solvent vapors and to the chemical stability of the complex in the membrane [17].

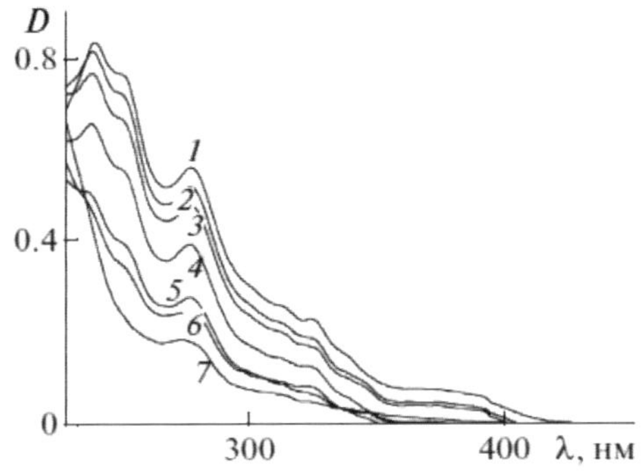

Figure 8. Optical densities of acetonitrile solutions of [PtEnPpy]ClO_4 with MF-4SK membrane immersed in them for (*1*) 0, (*2*) 20, (*3*) 40, (*4*) 60, (*5*) 120, (*6*) 180, and (*7*) 1440 min.

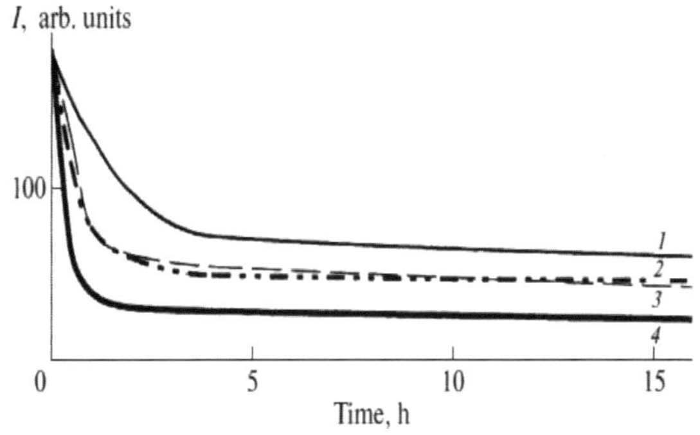

Figure 9. Dependences of luminescence intensity of membrane immobilized with $[PtEnPpy]^+$ on time of action of vapors of (*1*) acetonitrile, (*2*) acetone, (*3*) ethanol, and (*4*) methanol.

4. Conclusion

The results of this work show an application perspective of cyclometalated Pt(II), Pd(II), and Ir(III) complexes as optical sensors for the acidity and the presence of various organic and inorganic compounds, as well as for obtaining new complexes-luminophores based on reactions of coordinated ligands.

Acknowledgments

The investigation was supported by the Ministry of Education and Science of the Russian Federation, state target no. 4.131.2014K.

References

[1] Krasovitskiy B M and Bolotin B M 1984 *Organic Luminofores* (Moscow: Khimiya) [in Russian]
[2] Guzow K, Szmigiel D, Wróblewski D, Milewska M, Karolczak J, and Wiczk W 2007 *J. Photochem. Photobiol., A* **187** (1) 87
[3] Chen T-S 2008 *J. Organomet. Chem.* **693** (19) 3117
[4] Katlenok E A and Balashev K P 2014 *Opt. Spectrosc.* **116** (1) 100
[5] Katlenok E A and Balashev K P 2014 *Opt. Spectrosc.* **116** (2) 207
[6] Katlenok E A, Zolotarev A A, and Balashev K P 2015 *Rus. J. Coord. Chem.* **41** (1) 37
[7] Nikolaeva M V, Katlenok E A, Khakhalina M S, Puzyk M V, and Balashev K P 2014 *J. Phys.: Conf. Ser.* **541** 012086
[8] Zhao Q, Li F, and Huang C 2010 *Chem. Soc. Rev.* **39** (8) 3007
[9] Yang Y, Zhao Q, Feng W, and Li F 2013 *Chem. Rev.* **113** (1) 192
[10] Guerchais V and Fillaut J-L 2011 *Coord. Chem. Rev.* **255** (21–22) 2448
[11] Khakhalina M S, Rodionova O A, Puzyk M V, and Balashev K P *Opt. Spectrosc.* 2009 **106** (4) 520
[12] Khakhalina M S, Rodionova O A, and Puzyk M V *Opt. Spectrosc.* 2009 **106** (4) 529
[13] Nikolaeva M V and Puzyk M V 2013 *Opt. Spectrosc.* **114** (2) 247
[14] Nikolaeva M V and Puzyk M V 2014 *Opt. Spectrosc.* **117** (4) 589
[15] Kim H N, Ren W X, Kim J S, and Yoon J 2012 *Chem. Soc. Rev.* **41** (8) 3210
[16] Katlenok E A and Balashev K P 2013 *Opt. Spectrosc.* **114** (5) 756
[17] Khakhalina M S, Tikhomirova I Yu, and Puzyk M V 2010 *Opt. Spectrosc.* **108** (5) 703

Optical properties of multilayer films of nanocomposites based on WS_2 nanotubes decorated with gold nanoparticles

A Yu Polyakov[1], A V Nesterov[1], A E Goldt[1,2], V Zubyuk[3], T Dolgova[3], R Tenne[4] and E A Goodilin[1,2]

[1] Department of Materials Science, Lomonosov MSU, Moscow, Russia
[2] Kurnakov Institute of General and Inorganic Chemistry, RAS, Moscow, Russia
[3] Department of Physics, Lomonosov MSU, Moscow, Russia
[4] Department of Materials and Interfaces, Weizmann Institute, Rehovot, Israel

E-mail: a.yu.polyakov@gmail.com

Abstract. Multilayer films of WS_2 nanotubes decorated with gold nanoparticles are prepared for the first time using nanocomposite assemblage on the water-heptane interface and film transition onto optically transparent or semiconducting surfaces. The film morphology resembles a mosaic structure of 10-25 square micron areas with in-plane textured nanotubes. Optical properties demonstrate several features around 490, 545, and 675 nm connected either with excitonic transitions or scattering guided by high anisotropy of the nanotubes and their texturing peculiarities in the films.

1. Introduction

Nowadays, noble metal nanoparticles are widely studied due to their unique morphology-related optical properties such as surface plasmon resonance (SPR) becoming quickly of great importance in various devices. Gold nanoparticles (AuNPs) are proposed as promising optical antennas and scattering centers in metal-semiconductor composites implemented for Raman scattering and fluorescence enhancement, water-splitting hydrogen production, and light harvesting [1, 2]. Special attention is paid to AuNP-decorated carbon nanotubes (CNT) and TiO_2, $SrTiO_3/TiO_2$, sodium titanate tubular nanostructures employed in surface enhanced Raman scattering (SERS) detection of organic dyes, pollutants and explosives [3-9], SERS-based pH sensor for living cells [10], phototherapy [11], optical limiters [12-14], solar cells with improved efficiency [1, 15, 16], as well as photocatalytic systems [17-21]. A number of these applications require the development of thin films based on AuNP-nanotube composites.

After the discovery in 1992 [22], inorganic WS_2 nanotubes (INT-WS_2), found to be similar to multi-walled carbon nanotubes, have demonstrated their high effectiveness in applications either previously known for CNT, like visible light photocatalysis [23], field-effect transistors [24], or also special technologies such as heterogeneous catalysis in fuel hydrodesulfurization [25]. INT-WS_2 are synthesized readily on a large scale from WO_3 -like precursors using reduction by a H_2S + H_2/N_2 mixture in a fluidized-bed reactor [26, 27] resulting in nested S-W-S "sandwich" layers with a hexagonal graphene-like structure [28].

In the previous work [29], we developed a facile technique for decoration of INT-WS_2 with AuNPs involving a heterogeneous reaction of the disulfide with boiling $HAuCl_4$ solutions. Gold nanoparticles of tunable sizes formed near the surface defects of the nanotubes and seemed to be intimately connected to the outer sulfur atomic layer of WS_2 via Au-S bond formation. Recently, gold-decorated MoS_2 (the closest analogue of WS_2) monolayers and flake crystals were reported to show enhanced photoluminescent properties [30-32], SERS activity [32], and high-performance in solar cells [33] or photoelectrochemical cells for water splitting [34]. Inspired with these works and known applications of WS_2 films in Schottky-barrier solar cells [35] and photosensors [36], we studied optical properties of gold-decorated INT-WS_2 in the form of suspensions and films, which are useful from both the fundamental point of view and for potential applications in light harvesting, optical sensing, optical limiting, and photocatalytic technologies.

2. Experimental

WS_2 nanotubes were provided by NanoMaterials Ltd. (Israel) and had rather broad size distribution (30-100 nm in diameter and 1-20 μm in length). The $HAuCl_4$*$3H_2O$ reagent was purchased from Sigma-Aldrich (G4022-1G product, ACS reagent, ≥ 49.0 Au basis). All other reactants were of an analytical grade. All aqueous solutions and suspensions were prepared using purified water (Milli-Q RG, Millipore). Glassware and magnetic stirring bars utilized for AuNP deposition were washed by aqua regia (1:3 v/v HNO_3:HCl mixture).

INT-WS_2 were decorated with gold nanoparticles according to the previously developed and reported procedure [29]. Briefly, INT-WS_2 aqueous suspension was injected into boiling $HAuCl_4$ solutions under vigorous stirring. Boiling was continued for 3 min; then the reaction mixture was cooled down to room temperature. Typical $HAuCl_4$/MoS_2 molar ratios were of 1:56 – 1:5 resulted in 4-10 nm (in diameter) AuNPs, respectively. In order to obtain dry samples, the resulted Au-INT-WS_2 suspensions were rapidly frozen by immersion into liquid nitrogen and then dried in Labconco FreeZone freeze dry system (-50 °C, 0.03 mBar pressure).

For Au-INT-WS_2 film preparation, the composite aqueous suspension was mixed with 1/2 volume of heptane and sonicated in ultrasonic bath for 10 minutes. Within 30 min after sonication, the liquid phase segregates and Au-INT-WS_2 film forms on the liquid-liquid water-heptane interface. Such-prepared films were easily transferred onto different substrates like quartz slides and silicon wafers. Pristine INT-WS_2 films were synthesized as the reference samples using the same technique.

The films microstructure was studied by scanning electron microscopy (SEM) and optical microscopy. For SEM imaging, the films on silicon wafers were used. Micrographs were taken at 15 kV using Carl Zeiss NVision 40 microscope. For optical microscopy study, the films were deposited on quartz slides. Images were taken in a reflectance configuration using Nikon Eclipse E600 POL microscope in polarized light.

Optical properties of the Au-INT-WS_2 nanocomposites were studied by means of ultraviolet – visible – near infrared (UV-Vis-NIR) extinction spectroscopy and angle resolved reflectance spectroscopy in polarized light (for the films only). The film samples were deposited on fused quartz slides and aqueous suspensions were placed in quartz cuvettes; brief (20-30 s) sonication of suspensions before the measurement was used. UV-Vis-NIR extinction spectra were registered in a common transmission configuration using Perkin-Elmer Lambda 950 spectrophotometer (200-900 nm wavelength range, 1° step). For angle resolved reflectance spectroscopy, an original laboratory set-up was used. Reflectance spectra were registered using p-polarized light in the 400-800 nm wavelength range. The angular resolution was about 5° in 5-70° (relative to normal) angle range. Pristine INT-WS_2 films and suspensions were used as reference samples.

3. Results and Discussion

The assembly of nanoparticles on the liquid-liquid interface is widely used to produce nanocrystalline thin films of metals (especially, gold [37-41] and silver [42-46]), semiconducting chalcogenides and oxides [47, 48]. This technique is also highly effective for designing the films based on graphene sheets [49], carbon nanotubes [50-53], and their composites with plasmonic metal nanoparticles [54, 55]. Here, the assembly of pristine and gold-modified INT-WS_2 on water-heptane liquid-liquid interface was shown for the first time to result in dense composite films easily transferable onto different substrates; in particular, up to 2 cm^2 films were deposited on quartz (figure 1) and silicon.

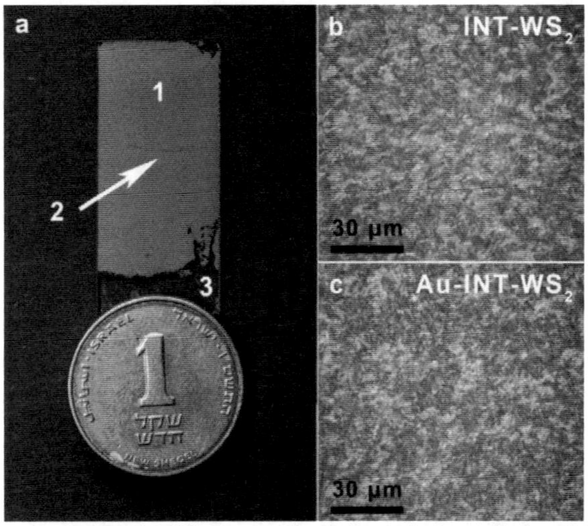

Figure 1. A general view (a) and optical images (b, c) in polarized light of WS_2 nanotubes and their composite with AuNPs. The shekel coin size in the picture (a) is 18.0 mm in diameter; 1 – textured film, 2 – a crack appeared upon film transfer from the liquid-liquid interface, 3 – quartz substrate.

According to SEM micrographs, the obtained films generally contain 1-2 layers of INT-WS_2 or gold-decorated INT-WS_2 aligned horizontally (figure 2). Only a minor fraction of larger aggregates of randomly oriented nanotubes was found. Although the complete in-plane alignment of INT-WS2 and gold-decorated INT-WS_2 along any specific direction has not yet been observed, the films were found to contain small domains (up to 25 μm^2) in which the nanotubes are textured in one direction with respect to each other. Micrographs of the composite film (figure 2d) also confirm the presence of gold nanoparticles on the INT-WS_2 surface, like reported previously [29].

Optical microscopy in a near-crossed analyzer configuration confirms the appearance of up to 5*5 μm^2 domains in which the films change the reflected light polarization identically (figure 1b, c). Since this feature was found for both INT-WS_2 and Au-INT-WS_2 films, it was ascribed to the strong light polarizing properties of INT-WS_2 observed also in [56]. It was suggested that the polarizing domains on OM images correspond to the domains with the same nanotube orientation as shown by SEM.

Figure 2. SEM micrographs of WS_2-based nanocomposites. (a) pure WS_2 nanotubes aligned onto silicon (b - a magnified view), c - Au-INT-WS_2 nanocomposite (d - its magnified view).

UV-Vis-NIR extinction spectra of INT-WS_2 and Au-INT-WS_2 films and suspensions are shown in figure 3. Pristine INT-WS_2 suspension spectrum contains three characteristic peaks at ca. 505, 575, and 675 nm. The last two features were previously attributed to B and A excitonic transitions, respectively [23]. However, as mentioned in [29], the extinction spectrum of INT-WS_2 suspension seems to contain a strong scattering impact. Interestingly, the same features were found in the extinction spectrum of INT-WS_2 film, however, their positions were red-shifted (ca. 490, 545, and 675 nm, respectively) while relative intensities are changed drastically (figure 3). Assuming the (partial) scattering origin of these features, the intensity changes can be explained by the alignment of nanotubes within the film and the red shift can be attributed to the changes of the medium refractive index (water, n=1.33, for the suspensions and air, n=1.00, for the films). However, this phenomenon needs a further study involving identification of scattered signals.

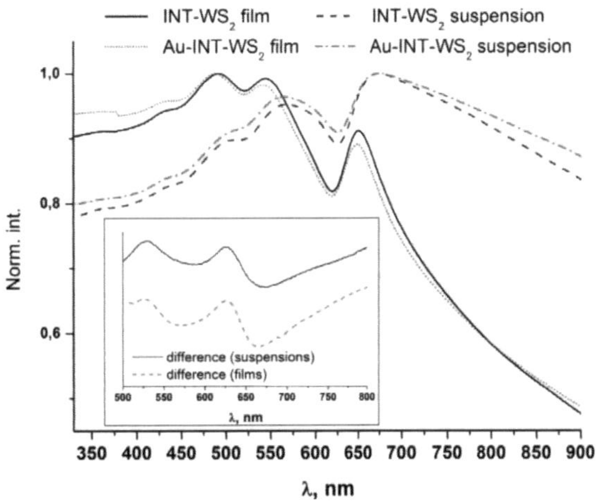

Figure 3. UV-Vis-NIR extinction spectra of INT-WS_2 and Au-INT-WS_2 films and suspensions; inset – the difference graphs between the spectra of gold-modified and pristine nanotube samples

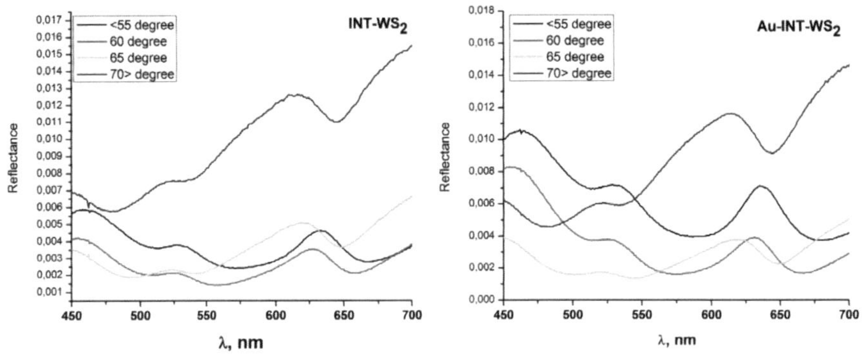

Figure 4. Reflectance spectra of INT-WS_2 and Au-INT-WS_2 films measured at different angles of incidence (relative to normal) using p-polarized light.

Similarly to the results, reported previously [29], decoration with gold nanoparticles did not result in appearance of any additional peaks in both suspension and film spectra of the nanocomposites relative to the ones of pristine nanotubes. However, subtraction of INT-WS_2 spectra from those of Au-INT-WS_2 helped to identify a weak impact of plasmonic gold nanoparticles: the difference graphs (figure 3) for suspension and film spectra contain distinct peaks at ca. 530 and 525 nm, which have to be attributed to the surface plasmon resonances of gold nanoislands (truncated spheres) grown on the disulfide surface.

Reflectance spectra of the INT-WS_2 and Au-INT-WS_2 films measured using a p-polarized beam are shown in figure 4. In the 5-55° range (relative to normal incidence), the spectra were almost unchanged. However at high angles (55-70°), the spectra change drastically. Most probably, the changes are due to the difference of excitonic and scattering contributions at low and high incidence angles due to INT-WS_2 horizontal ordering within the films; a shift of the features with angle is evidently observed. The further work on preparation of biaxially textured INT-WS_2 and Au-INT-WS_2 films is in progress.

4. Conclusions

Assembly of INT-WS_2 and their composites with AuNPs on a water-heptane interface were shown to be an effective technique for preparation of thin films based on these materials. Almost all nanotubes were found to be aligned horizontally within the 1-2 layered films. Additionally, the films contained domains, in which neighbouring nanotubes were aligned in one direction with respect to each other. Such an alignment resulted in unusual changes in extinction spectra of the films relative to the ones of the same material suspensions. For the first time, an angular dependence of INT-WS_2 film reflectance spectra in p-polarized light was shown evidencing for the possibility of new optical applications of WS_2-based materials in the future.

Acknowledgements

AYP and EAG thank The Russian Scientific Foundation for financial support (grant 14-13-00871). Authors thank Ron Tenne (Weizmann Institute) for fruitful discussions. AYP thanks Maria Mishchenko (Tsukuba University) for the help in reactant purchasing.

References

[1] Shao Z, Wang X, Ren S, Tian Z, Fan S, Sun S, Liu S and Wang Q 2013 *J. Appl. Phys.* **114** 063510
[2] Mubeen S, Lee J, Singh N, Krämer S, Stucky G D and Moskovits M 2013 *Nature Nanotech* **8** 247-51
[3] Li X, Chen G, Yang L, Jin Z and Liu J 2010 *Adv. Funct. Mater.* 2010 **20** 2815-24
[4] Roguska A, Kudelski A, Pisarek M, Opara M and Janik-Czachor M 2011 *Appl. Surf. Sci.* **257** 8182–89
[5] Yick S, Han Z J and Ostrikov K 2013 *Chem. Commun.* **49** 2861-63
[6] Jiang W F, Zhang Y F, Wang Y S, Xu L and Li X J 2011 *Appl. Surf. Sci.* **258** 1662-5
[7] Beqa L, Singh A K, Fan Z, Senapati D and Ray P C 2011 *Chem Phys Lett* **512** 237-42
[8] Zhang K, Ji J, Fang X, Yan L and Liu B 2015 *Analyst* **140** 134-9
[9] Demeritte T, Kanchanapally R, Fan Z, Singh A K, Senapati D, Dubey M, Zakarb E and Ray P C 2012 *Analyst* **137** 5041-5
[10] Chen P, Wang Z, Zong S, Chen H, Zhu D, Zhong Y and Cui Y 2014 *Anal. Bioanal. Chem.* **406** 6337-46
[11] Wang X, Wang C, Cheng L, Lee S T and Liu Z 2012 *J. Am. Chem. Soc.* **134** 7414–22
[12] Zheng C, Chen W, Ye X, Cay S, Xiao X and Wang M 2011 *Mater. Lett.* **65** 150-152
[13] Chin K C, Gohel A, Chen W Z, Elim H I, Ji W, Chong G L, Sow C H and Wee A T S *Chem. Phys. Lett.* 2005 **409** 85-8
[14] Sreeja R, John J, Aneesh P M and Jayaraj M K 2010 *Opt. Commun.* **283** 2908-13
[15] Li Y, Kodama S, Kaneko T and Hatakeyama R 2012 *Appl. Phys. Lett.* **101** 083901
[16] Chen L, He H, Zhang S, Xu C, Zhao J, Zhao S, Mi Y and Yang D 2013 *Nanoscale Res. Lett.* **8** 225
[17] Li S, Yu X, Zhang G Ma Y, Yao J, Keita B, Louis N and Zhao H 2011 *J. Mater. Chem.* **21** 2282-7
[18] Paramasivam I, Macak J M and Schmuki P 2008 *Electrochem. Commun* .**10** 71-5
[19] Pandikumar A and Ramaraj R 2013 *Mater. Chem. Phys.* **141** 629-35
[20] Kang Q, Wang T, Li P, Liu L, Chang K, Li M and Ye J 2014 *Angew. Chem. Int. Ed.* **54** 841-5
[21] Grover I S, Singh S and Pal B 2014 *RSC Adv.* **4** 51342-8

[22] Tenne R, Margulis L, Genut M and Hodes G 1992 *Nature* **360** 444-6
[23] Tsverin Y, Zak A, Livneh T, Pinkas I, Rosentsveig R and Tenne R 2012 *Nanomaterials and Energy* **2** 25-34
[24] Levi R, Bitton O, Leitus G, Tenne R and Joselevich E 2013 *Nano Lett.* ***13*** 3736–41
[25] Tsverin Y, Popovitz-Biro R, Feldman Y, Tenne R, Komarneni M R, Yu Z, Chakradhar A, Sand A and Burghaus U 2012 *Mater. Res. Bull.* **47** 1653–60
[26] Zak A, Sallacan-Ecker L, Margolin, A, Genut M and Tenne R 2009 *NANO* **4** 91−98.
[27] Zak A, Sallacan-Ecker L, Efrati R, Drangai L, Fleischer N and Tenne R 2011 *Sensors & Transducers* **12** 1−10
[28] Houben L, Enyashin A N, Feldman Y, Rosentsveig R, Stroppa D G and Bar-Sadan M 2012 *J. Phys. Chem. C* **116** 24350−7
[29] Polyakov A Yu, Yadgarov L, Popovitz-Biro R, Lebedev V A, Pinkas I, Rosentsveig R, Feldman Y, Goldt A E, Goodilin A E and Tenne R 2014 *J. Phys. Chem. C* **118** 2161−9
[30] Najmaei S, Mlayah A, Arbouet A, Girard C, Léotin J and Lou J 2014 *ACS Nano* **8** 12682–9
[31] Cao W, Pankratov V, Huttula M, Shi X, Saukko S, Huang Z and Zhang M 2015 *Mater. Chem. Phys.* **158** 89-95
[32] Singha S S, Nandi D and Singha A 2015 *RSC Adv.* **5** 24188-93
[33] Yang X, Liu W, Xiong M, Zhang Y, Liang T, Yang J, Xu M, Ye J and Chen H 2014 *J. Mater. Chem. A* **2** 14798-806
[34] Yin Z, Chen B, Bosman M, Cao X, Chen J, Zheng B and Zhang H 2014 *Small* **10** 3537-43
[35] Shanmugam M, Bansal T, Durcan C A and Yu B 2012 *Appl. Phys. Lett.* **101** 263902
[36] Perea-López N, Elías A L, Berkdemir A, Castro-Beltran A, Gutiérrez H R, Feng S, Lv R, Hayashi T, López-Urías F, Ghosh S, et al. 2013 *Adv. Funct. Mater* **23** 5511-7
[37] Kumar A, Mandal S, Mathew S P, Selvakannan P R, Mandale A B, Chaudhari R V and Sastry M 2002 *Langmuir* **18** 6478–83
[38] Lee K Y and Han S W 2005 *Bull. Korean Chem. Soc.* **26** 1306-8
[39] Smirnov E, Scanlon M D, Momotenko D, Vrubel H, Méndez M A, Brevet P-F and Girault H H 2014 *ACS Nano* **8** 9471-81
[40] Chen L-J, Ma H, Chen K, Cha H-R, Lee Y-I, Qian D-J, Hao J and Liu H-G 2011 *J. Colloid. Interface Sci.* **362** 81-8
[41] Wang D, Ma H, Chu C, Hao J and Liu H-G 2013 *J. Colloid. Interface Sci.* **402** 75-85
[42] Yamamoto S and Watarai H 2006 *Langmuir* **22** 6562-9
[43] Sathaye S D, Patil K R, Paranjape D V and Padalkar S R 2001 *Mater. Res. Bull.* **36** 1149-55
[44] Lee K Y, Kim M, Kwon S S and Han S W 2006 *Mater. Lett.* **60** 1622-4
[45] Guo S, Dong S and Wang E 2009 *Cryst. Growth Des.* **9** 372-7
[46] Lin L, Shang K, Xu X, Chu C, Ma H, Lee Y-I, Hao J and Liu H-G 2011 *J. Phys. Chem. B* **115** 11113-8
[47] Rao C N R and Kalyanikutty K P 2008 *Acc. Chem. Res.* **41** 489-99
[48] Rao C N R, Kulkarni G U, Agrawal V V, Gautam U K, Ghosh M and Tumkurkar U 2005 *J. Colloid. Interface Sci.* **289** 305-18
[49] Biswas S and Drzal L T 2009 *Nano Lett.* **9** 167-72
[50] Matsui J, Yamamoto K, Inokuma N, Orikasa H, Kyotania T and Miyashita T 2007 *J. Mater. Chem.* **17** 3806-11
[51] Matsui J, Yamamoto K and Miyashita T 2009 *Carbon* **47** 1444-50
[52] Asuri P, Karajanagi S S, Dordick J S and Kane R S 2006 *J. Am. Chem. Soc.* **128** 1046-7
[53] Matsui J, Yamamoto K, Inokuma N, Orikasa H, Kyotani T and Miyashita T 2007 *J. Mater. Chem.* **17** 3806-11
[54] Mehl H, Oliveira M M and Zarbin A J G 2015 *J. Colloid. Interface Sci.* **438** 29-38
[55] Lee K Y, Kim M, Hahn J, Suh J S, Lee I, Kim K and Han S W 2006 *Langmuir* **22** 1817-21
[56] Rafailov P M, Thomsen C, Gartsman K, Kaplan-Ashiri I and Tenne R 2005 *Phys. Rev. B* **72** 205436

Electromagnetic properties of water on GHz frequencies for medicine tasks and metamaterial applications

N D Pavlov[1] **and Y A Baloshin**[1]

[1] ITMO University, Kronverkskiy pr. 49, St.Petersburg 197101, Russian Federation

E-mail: n.pavlov@phoi.ifmo.ru

Abstract. In problems of modern radio physics and medicine it is important to know dielectric permittivity of liquids. Dispersion characteristics of water in UHF frequency band can be used to analyze the states of biological objects, and also to construct materials (metamanerials). The present work is intended to study the material properties of water in UHF frequency band based on two different techniques: Nicolson-Ross-Weir (NRW) [1] and the Active Nearfield Diagnostics [2].

1. Introduction

In problems of modern radio physics and medicine the knowledge of electromagnetic properties of liquids are very relevant. The dispersion characteristics of liquids in UHF frequency band are important to study properties of metamaterals containing water as a material of unit cells. In medical applications the non-invasive methods of biological object analysis become very relevant [3].

2. Nicolson-Ross-Weir Method

The Nicolson-Ross-Weir method [4] of measuring the material properties of substances under study is related to UHF methods. The main idea of the method is to measure the impedance on the dielectric-air boundary as shown in Fig.1, 2. This method allows not only to obtain the value of the impedance by means of S-parameter measurements, but also the phase value. These parameters allow to extract the material properties of the studied substance. One can extract the material properties (the real and imaginary parts of the dielectric permittivity) by means of the NRW method [4]. Here we consider it in details.

It is well-known that the dielectric permittivity and magnetic permeability of a substance can be represented in the following form [1]:

$$\dot{\varepsilon} = (\varepsilon' - j\varepsilon'')\varepsilon_0 \qquad \dot{\mu} = (\mu' - j\mu'')\mu_0 \tag{1}$$

where ε_0 and μ_0 are the vacuum permittivity and permeability, ε' and μ' are the real and imaginary parts of the substance permittivity and permeability, ε'' and μ'' are their imaginary

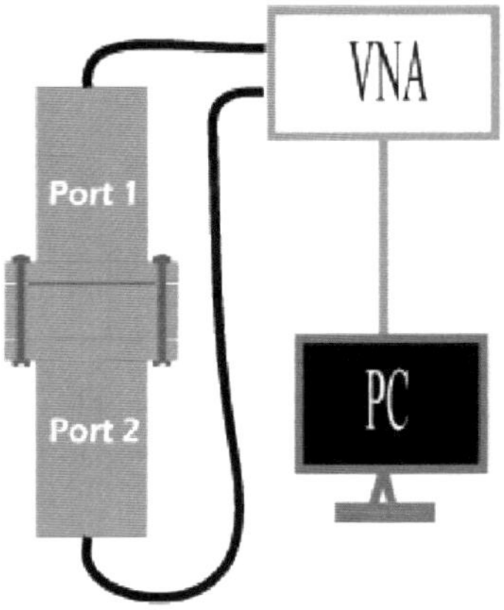

Figure 1. Schematic of the experimental setup.

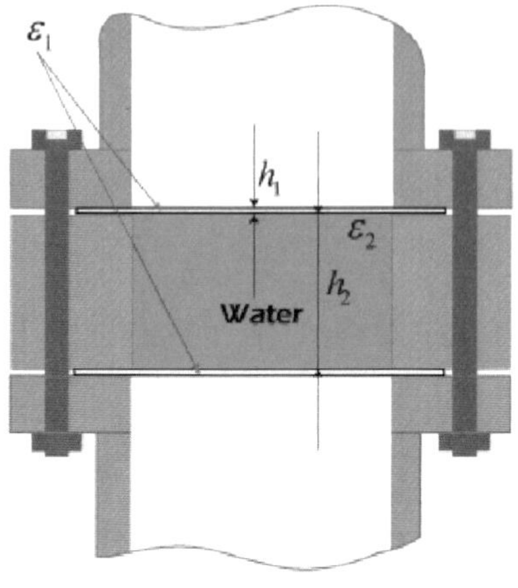

Figure 2. Waveguide section. The water is contained between of the holding films.

parts. In order to find these values one should initially calculate the wave impedance z (2) on the air-dielectric boundary [4,5].

$$z = \sqrt{\frac{(1+S_{11})^2 - S_{21}^2}{(1-S_{11})^2 - S_{21}^2}} \tag{2}$$

where S_{11} and S_{21} are S-parameters. Usually vector network analyzers can directly provide S-parameters. It is also important to take into account the phase factor $e^{\gamma d}$, which depends on the wave impedance (2):

$$e^{\gamma d} = \frac{1 - S_{11}^2 + S_{21}^2}{2S_{21}} + \frac{2S_{11}}{(z - 1/z)S_{21}} \tag{3}$$

the value γ is a propagation constant of a wave penetrating the sample, while d is a sample thickness. In measurements one obtains a set of S-parameters for the whole measured frequency band. (N values). The next step is the calculation of the complex argument ϕ_N, which can be represented in the form (4):

$$\phi_N = \phi_0 + \sum_{i=1}^{N} arg(\frac{e^{\gamma_i d}}{e^{\gamma_{i-1} d}}) \tag{4}$$

where ϕ_0 is the phase at the first frequency. Finally the refraction coefficient can be calculated by means of the following formula (5):

$$n_N = \sqrt{\frac{1}{(k_0 d)^2}(\omega_N - j\ln|e^{\gamma_N d}|)^2 - (\frac{\omega_c}{\omega_N})^2} \tag{5}$$

where ω_c is the cutoff frequency of the measurement waveguide, ω_N is the angular frequency. The next step is to calculate the magnetic permeability μ and the dielectric permittivity ϵ:

$$\dot{\mu} = z\frac{n^2 - (\omega_c/\omega_N)^2}{1 - (\omega_c/\omega_N)^2} \qquad \dot{\varepsilon} = \frac{n^2}{\dot{\mu}} \tag{6}$$

The formula (6) finally gives the parameters to be calculated ε and μ.

The important parameter in the S-parameter measurements is the phase delay of the wave propagation within the material sample. The phase delay introduces some limitations on the possible surface shape of the studied sample. Therefore is important to ensure parallel-plate shape of the sample, which does not affect the phase of the wave. The liquid state samples are harder to control the phase while measurements. Therefore we have used films with known values of the dielectric permittivity as a function of frequency. Based on those values the additional phase delay was taken into account. Results of measurements are shown in Fig.3.

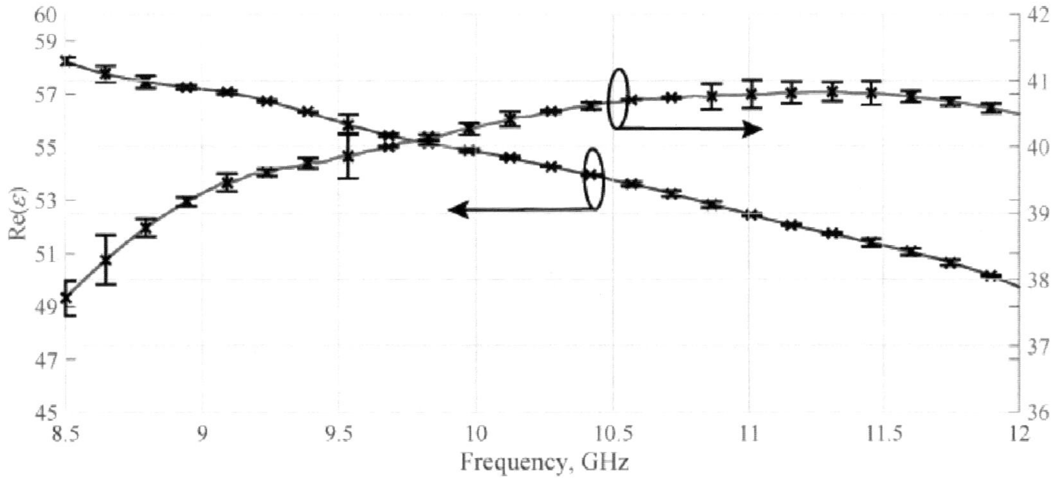

Figure 3. The extraction of the real part of the permittivity (left axis) and its imaginary part (right axis) for a distilled water at 23 °C.

3. Active nearfield diagnostics

The method of the active nearfield diagnostics allows obtaining information about the permittivity value of a liquid and its conductivity in a broad frequency band. This method is based on the measurements of an impedance of an electrically small antenna located in the vicinity of the air-liquid interface [2]. In such a case the antenna impedance is determined not only radiative part like in free space, but also by additional terms due to quasistatic components of an electric field. Therefore the essential of the method is to analyze S-parameters by means of a vector network analyzer Rohde & Schwarz, which makes possible to calculate the antenna impedance. The principal difference in radiation of such an antenna in air from its radiation on the boundary of the liquid is the presence of the imaginary part of the dielectric permittivity, which makes a part of the energy absorbed. The most of electrically small antennas have dominating capacitive part of the impedance. Therefore the matching of such antennas is extremely sensitive to the imaginary part of the substance permittivity. Consequently the small change in the imaginary part of the permittivity affects the impedance and the resonance frequency. The value of this response allows determination of the material properties, which is a physical basis of the Active Nearfield Diagnostics method. Let us note that this method can be employed not only for pure substances (e.g. a distilled water) but also mixtures, solutes and suspensions. It is important because admixtures can considerably change the losses of the main substance. It results in a change of a skin depth from several millimeters to several camtimeters on a broad frequency range. The measurements by this method include several steps. Firstly

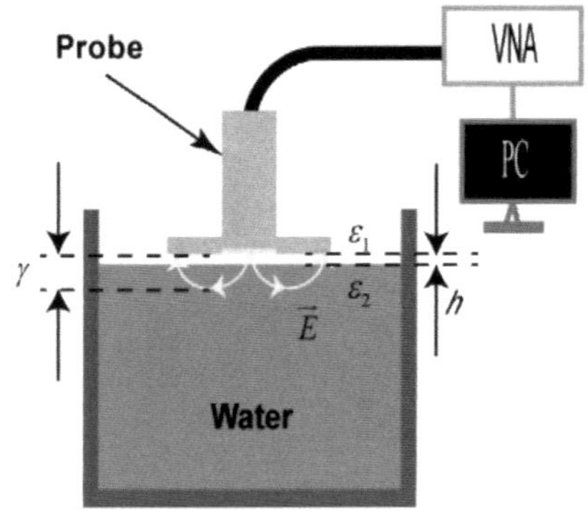

Figure 4. Schematics of the experimental setup.

one needs the information about the parameter dispersion of the calibrating liquid of a similar order of permittivity in the measured range. Thus one needs to have a prior information about the measured liquid. The second step is calibration of the whole system taking into account the effects of multiple reflections and losses in the feeding cables and connectors, the size of the probe antenna and the signal renormalization to the known material parameters of the calibrating liquid. The third step is the measurement process itself, which assumes positioning of the probe in the vicinity of the waver surface and obtaining S-parameters.

The experimental setup schematics is depicted in Fig.4. The electrically small antenna is located in air at some distance from the measured liquid's surface. The antenna sizes and the distance to the liquid are much smaller than the wavelength.Results of measurements are shown in Fig.5. In addition to water properties it is important to measure properties of salt

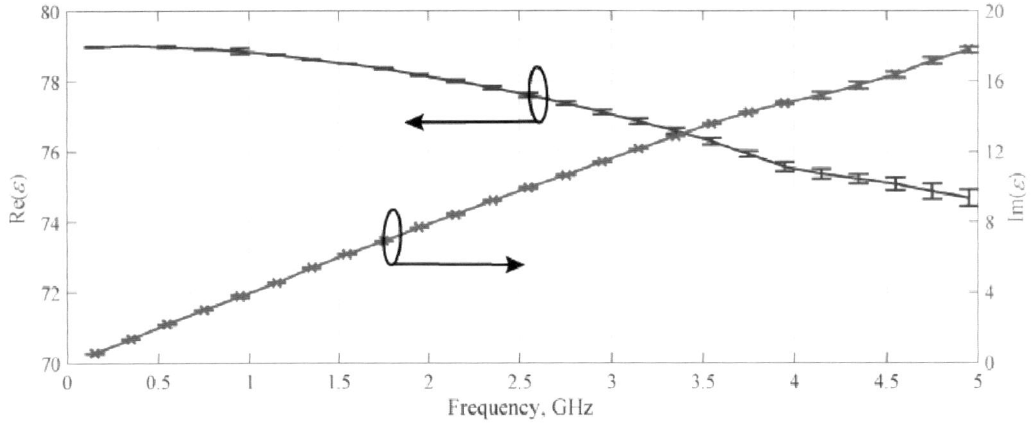

Figure 5. The extraction of the real part of the permittivity (left axis) and its imaginary part (right axis) for a distilled water at 23 °C.

solutes. Water is one of the main components of live organisms. Salt solutes can be a first order approximation of the tissues. Results of measurements are shown in Fig.6, 7.

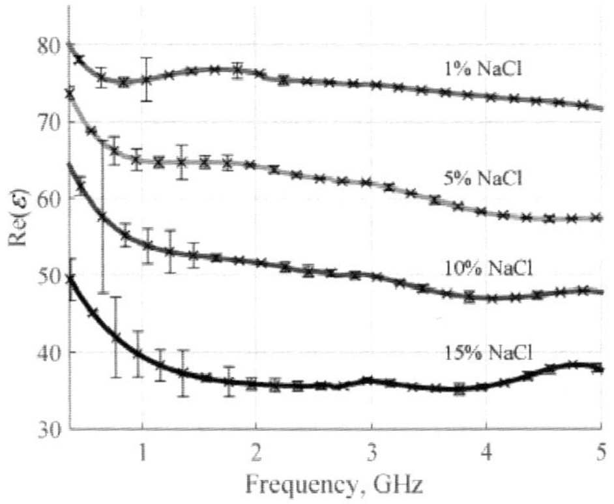

Figure 6. Extracted values of the permittivity (real part) of satl solutes with different concentration of NaCl (1, 5, 10, 15%) at 23°C.

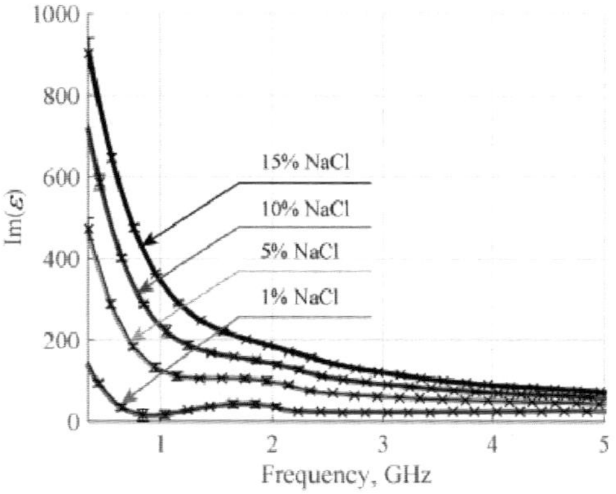

Figure 7. Extracted values of the permittivity (imaginary part part) of satl solutes with different concentration of NaCl (1, 5, 10, 15%) at 23°C.

Normally for the most of liquids the permeability in the microwave range is supposed to equal 1, and practically does not change under the influence of external factors.

4. Conclusion

In this work the values of the real and imaginary parts of the permittivity and the conductivity of water and salt solutes were measured by means of two methods: the Nicolson-Ross-Weir method and the active nearfield diagnostics method.

5. Acknowledgements

This work was supported by the Ministry of Education and Science of the Russian Federation (project 11.G34.31.0020, Zadanie No.3.561.2014/K), by Russian Foundation for Basic Research and the Dynasty Foundation (Russia).

References

[1] Luukkonen O, Maslovski S I, Tretyakov S A 2011 A Stepwise NicolsonRossWeir-Based Material Parameter Extraction Method *IEEE antennas and wireless propagation letters* vol. **10** 1295-1298
[2] Urasova N V 2006 The near-field UHF probing stratified medium *Nizhniy Novgorod* 113
[3] Samoilov V O 2004 Medical biophysics *SpecLit* 496
[4] Nicolson A M and Ross G F 1970 Measurement of the intrinsic properties of materials by time-domain techniques *IEEE Trans. Instrum. Meas.* **19** 377-382
[5] Weir W B 1974 Automatic measurement of complex dielectric constant and permeability at microwave frequencies *Proc. IEEE* **62** 33-36

Wood anomalies in the vicinity of 3D Bragg diffraction from hybrid opaline photonic crystals

V S Nikolaev[1,2], M M Voronov[1], S A Yakovlev[1], A B Pevtsov[1]
[1]Ioffe Physical-Technical Institute, Politekhnicheskaya 26, St Petersburg, 194021 Russia
[2]St. Petersburg State Electrotechnical University LETI, St. Petersburg, 197376 Russia
E-mail: vudi.vova@gmail.com

Abstract. For hybrid chalcogenide/opal photonic crystals the optical reflection spectra demonstrating the interplay of diffraction resonances such as 3D Bragg diffraction and resonant Wood anomaly have been studied. The coupling coefficient of the Wood and Bragg waves has been estimated on the basis of the coupled-waves approach and the concept of the effective refractive index of the wave-guiding surface layer of the hybrid structure. A phenomenological expression for the coefficient of reflection from a hybrid photonic crystal allowing one to investigate the interference of the Wood anomalies has been proposed.

Introduction

In recent years in the field of photonics a great deal of attention has been paid to the study of the interface radiative states to which the resonant Wood anomalies (WA) [1] can be attributed. The latter represent the reflection or transmission peaks caused by the formation of quasi-guided modes in a spatially periodic surface layer (2D photonic crystals). Since the characteristics of these spectra (e.g.,the position and half width of the resonance) depending on the parameters of the system can vary over a sufficiently wide range and the quasi-guided mode formation usually occurs in a thin surface layer, the photonic crystals can serve as basic elements of integrated photonic devices (see for example [2]).

In this work we consider the high contrast hybrid photonic crystals (HPCs) [3], which in our case are opaline photonic crystals covered with chalcogenide $Ge_2Sb_2Te_5$ (GST) films of different thickness, and study the reflection spectra in which the resonant WA and 3D Bragg diffraction can manifest themselves. We show that for certain parameters of the hybrid structure (e.g. the thickness of GST film and period of HPC) the WA maxima arise in the spectral region of 3D Bragg diffraction (the photonic band gap range) on a many-layer opal structure, which leads to the anticrossing of these spectral lines. Depending on the parameters of the system, the anticrossing may be quite strong (in case of a strong interaction between the Wood and Bragg resonances) or rather weak (in the opposite case) resembling an energy level crossing. The difference in intensity of the interaction is associated with different conditions of the propagation and scattering of light in the wave-guiding layer. In the hybrid structures being studied the above difference is determined by the thickness of the GST film because the effective parameters (the effective refractive index $\bar{n}$ and transverse wave vector component a) of the quasi-guided mode invoked to describe the angular-dependent Wood anomalies, are strongly dependent on this thickness [4]. Besides, we study the behaviour of two close modes of the

WAs and analyze the shape of the reflection coefficient spectra. In some cases (under different conditions) the WA reflection spectrum peaks corresponding to separate quasi-guided modes are symmetric Lorentzians, with their maximum positions being well described by an equation involving two effective parameters ($\bar{n}$ and a), which enables us to investigate the interplay of the Wood anomalies from the viewpoint of the interference of the light waves.

On the whole, the proposed approach using a set of the effective parameters for hybrid photonic crystals can be applied to describe some peculiarities of the light reflection spectra demonstrating the interplay of several diffraction resonances such as Wood anomalies, Bragg diffraction and Fabry-Perot oscillations.

1. The coupled-waves approach

The samples of the structures under study are formed from opal films (30 monolayers of a-SiO_2 spheres of diameter about 637 nm) grown from water suspension of a-SiO_2 particles by the vertical deposition technique on fused silica substrates [5]. The GST layers of thickness 40, 80 and 150 nm were deposited on the surface of opal films by the thermal deposition technique [6]. The schematic picture of a hybrid photonic crystal and the geometry of the experiment are shown in figure 1. Experimental reflection spectra of the opal/GST hybrid structures with GST films with the thicknesses of 40, 80 and 150 nm for different light incidence angles ($11° < \theta_0 < 66°$) are shown in figure 2. For the structures with film thicknesses of 40 and 80 nm these spectra show two kinds of features, one of which is a 3D Bragg diffraction peak moving to shorter wavelengths as the incidence angle increases, while the second is a Wood anomaly peak moving to longer wavelengths with an increase in the incidence angle. It is clearly seen that in the interaction region of the Bragg and Wood resonances the behaviour of the corresponding angular reflection spectra is different, showing a crossing of the lines (figures 2a,b) with the exception of a very narrow frequency region in one case and a strong repulsion typical for an anticrossing (figures 2e,f) in the second.

Since the experimental reflection spectra corresponding to the Wood anomaly and 3D Bragg diffraction are fitted well by Lorentzian shape functions, one can expect that their spectral-angular dependence will be described by the coupled-mode theory [7]. For this purpose we apply the two-coupled wave equation to the quasi-guided modes by analogy with the guided modes in dielectric waveguides [7]. A coupled-wave equation usually contains the wave vectors of the interacting waves, but since the wave vectors of the light propagating in the waveguide layer attributed to the Wood anomaly and Bragg diffraction correspond to the certain outgoing

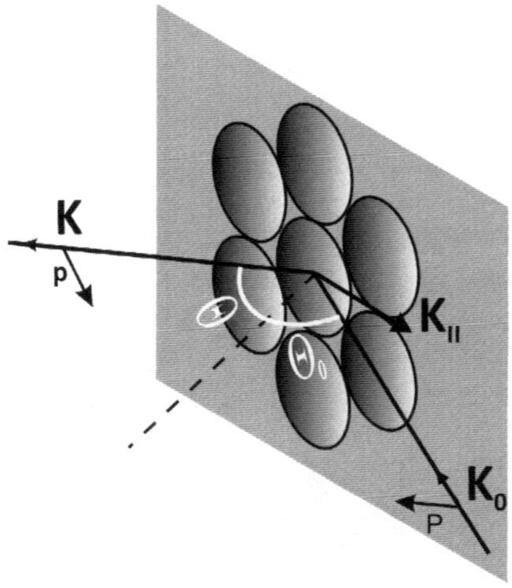

Figure 1. Schematic image of the surface of a hybrid opal/GST photonic crystal and geometry of the experiment. Here are shown the incident, outgoing and scattering wave vectors $\mathbf{k_0}$, $\mathbf{k}$ and $\mathbf{k}_{||}$, and also the incident and outgoing angles, θ_0 and θ.

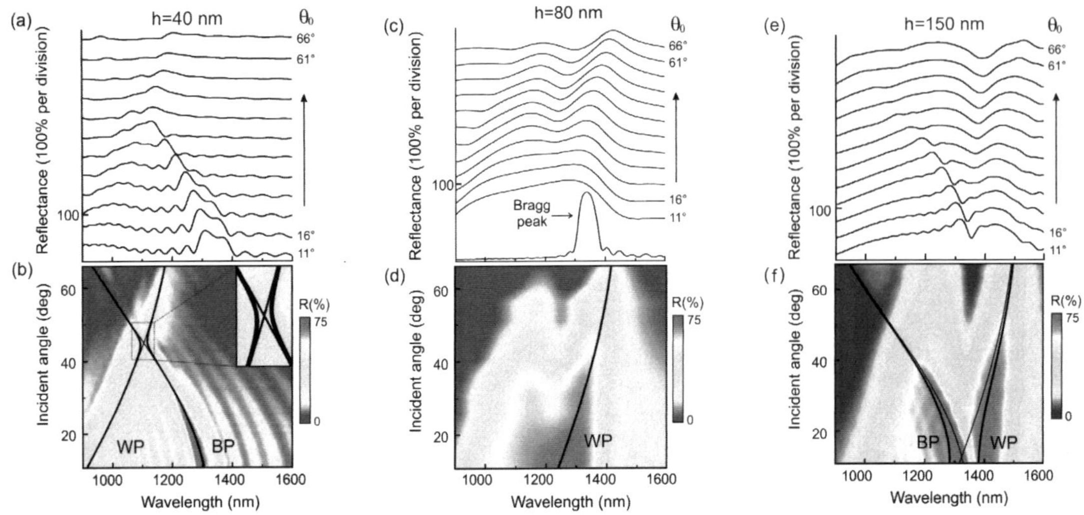

Figure 2. Experimental reflection spectra of hybrid opal/$Ge_2Sb_2Te_5$ photonic crystals for different incident angles (θ_0) and $Ge_2Sb_2Te_5$ thicknesses (h). The spectra (a,c,e panels) are shifted vertically for clarity; the color maps (b,d,f) are shown to improve visualization of the spectra. The thick full curves (b,d,f panels) passing through the spectral maxima correspond to the coupled-waves modes of 3D Bragg diffraction (BP) and the Wood anomalies (WP) and are calculated by using equation (1). The thin full curves (b,d,f panels) are calculated by expressions for $\omega_W(\theta)$ and $\omega_B(\theta)$ (without interaction between the Wood and Bragg resonances). The effective fitting parameters n and a for the quasi-guided modes calculated by equation (2) are the following: $\bar{n}$ = 1.85, a = 0.38 (h=40 nm); $\bar{n}$ = 3.11, a = 1.04 for the long-wave Wood anomaly (h=80 nm); $\bar{n}$ = 3.25, a = 1.01 (h=150 nm). The Bragg reflection spectrum (panel c) for an uncoated opal film (30 monolayers of a-SiO_2 spheres of diameter about 637 nm) is marked by arrow "Bragg peak".

angles of the light, θ, (in our case $\theta = \theta_0$) the equation for the coupled waves can be written in the form:

$$(\omega - \omega_B(\theta))(\omega - \omega_W(\theta)) = \delta\,, \tag{1}$$

where $\omega_B(\theta) = 2\pi c/\lambda_B(\theta)$ and $\omega_W(\theta) = 2\pi c/\lambda_W(\theta)$ are the frequencies corresponding to the reflection spectrum maxima for the Bragg diffraction and Wood anomaly, and δ is the effective coupling coefficient of the modes. The spectral position of the WA maximum is determined by the expression [4]:

$$\lambda_W = \frac{2\pi}{G}\frac{\sqrt{(\sin\theta\cos\varphi)^2 + (\bar{n}^2 - \sin^2\theta)(1+a)} - \sin\theta\cos\varphi}{1+a}\,. \tag{2}$$

Here φ is an angle between the projection $\mathbf{k}_{\parallel}$ of the incident wave vector $\mathbf{k}_0$, lying in the plane of the surface of a hybrid structure, and $\mathbf{G}$ is 2D vector of the reciprocal lattice formed by the top hexagonally-ordered monolayer of opal spheres of the structure. The parameter $a = (k_z/G)^2$, where $\mathbf{k}_z$ is the wave-vector $\mathbf{k}$ projection in the perpendicular direction to the surface waveguide layer.

The spectral position of the reflection coefficient maximum caused by 3D diffraction of electromagnetic waves on the planes (111) of the opal film is determined by a modified Bragg formula

$$\lambda_B(\theta) = 2d_{111}\sqrt{\bar{\varepsilon} - \sin^2\theta} + \Delta\lambda\,, \tag{3}$$

where d_{111} is the period of the opal structure in the [111] direction, $\bar{\varepsilon}$ is the average value of dielectric constant of the opal film, and $\Delta\lambda$ is the value resulting from an additional phase shift due to the thickness of the GST film and to reflection of the light in the interface area of opal-GST. Taking into account the high value ($n > 4$ in frequency wave considered) of the refractive index of GST film one can suppose that the additional phase shift, leading to the spectral shift of the Bragg diffraction peaks, will be about the same for different (but not very large, in our case $\theta < 70°$) incident angles of the light beam. This indeed takes place for hybrid structures opal/GST with the film thickness in the range of 40-150 nm. If to consider the coupling coefficient δ to be a constant (equal to its value in the crossing point of the unperturbed lines), equation (1) will be quadratic relative to the frequency ω, whose roots $\omega_1(\theta) = 2\pi c/\lambda_1(\theta)$ and $\omega_2(\theta) = 2\pi c/\lambda_2(\theta)$ describe the behaviour of the coupled Bragg and Wood modes. Generally speaking, the coupling coefficient decreases when moving off the centre of the interaction region on the frequency scale. Here we set the value of δ to a constant in the interaction region and equal to zero outside it; in fact, to simplify the final expressions we keep δ constant in the whole spectral range. This is justified by a rapid divergence of the Wood and Bragg resonances outside of the interaction region toward their ultimate values (in the absence of one of the resonances), see the thick and thin full curves for large angles θ in figures 2b,f. The value of the splitting (the minimum distance between the Bragg and Wood lines, $\lambda_B(\theta)$ and $\lambda_W(\theta)$), is equal $\Delta\omega = 2\sqrt{\delta}$ which is about 80 meV. The estimation of the coupling coefficient δ in the cases of the weak (figures 2a,b) and strong (figures 2e,f) interaction made with the help of equation (1) gives values $\sim 10^{24}c^{-2}$ and $\sim 10^{27}c^{-2}$, respectively. Such a large difference in the interaction intensity of the resonances is due to a difference in the interference conditions for 3D Bragg diffraction wave and the Wood anomaly wave in the surface waveguide layer. We note that in work [8], where 3D opal-based photonic crystals were studied under similar experimental conditions, the interference of the light waves resulted from 3D Bragg diffraction from the opal planes (111) and from disorder-induced continuous Mie spectrum was found in the transmission spectra. A similar Mie spectrum in our experimental situation (at relatively large angles θ_0) converts to the WA spectrum, which can show the interference of the light waves (and also with those resulted from 3D Bragg diffraction).

It should be noted that in the interaction region of the two resonances there are also the reflection peaks corresponding to the Fabry-Perot interference over the whole thickness of the structure (figures 2a,b and 2e,f), which are characterized by a noticeably smaller value of the reflectance and therefore we do not consider this mode (as the third wave) in the coupled-waves approach. Besides, equation (1) for the coupled modes can include the spectral-angular dependence for some other diffraction resonances, e.g. 2D Bragg diffraction (which in the frequency range studied does not appear) or Wood anomaly corresponding to another quasi-guided mode.

2. Description of the reflection spectra

Let us now explain the main features of the optical spectra of reflection from the hybrid structures with the GST film thickness $h = 50$ nm, see figures 2c,d. A peculiarity of these spectra is the absence of peaks due to 3D Bragg diffraction on the (111) planes of the opal film. Since the value of the reflection coefficient is quite small ($R < 0,1$) in the considered ranges of both wavelength λ and reflection angle θ, this situation is possible under such parameters of the structure when the total phase shift caused by the film thickness and phase change at the interfaces air - GST film and GST film - opal film leads to a condition close to that for a minimum. Therefore in a nearby wavelength range the Fabry-Perot interference can be strongly suppressed, which is indeed almost not seen (see figures 2c,d).

As is seen in figure 2c, there are two ridges of peaks, which are Wood anomalies corresponding to two different quasi-guided modes. The analysis of these spectra shows that at sufficiently large reflection angles ($\theta = \theta_0 > 30^0$) the spectra are quite well described by the Lorentzian functions whose widths increase monotonically with decreasing the angle θ. At small enough angles θ the reflection spectra are strongly deformed and take an asymmetrical form, so that it becomes impossible to differ between separate quasi-guided modes in the interaction region. The reflection coefficient peaks (for the light reflected in a small spatial angle) also monotonically increase with decreasing the angle $\theta = \theta_0$, reaching their maxima at the normal incidence ($\theta_0 = 0$). At very small angles θ the Wood anomalies join up to form a broad irregular contour. Thus, the reflectance contour of similar shape that arises at the normal incidence of light on a hybrid structure can suggest the appearance of the Wood anomalies in the reflection spectra at the inclined incidence.

We will now describe the Wood anomaly reflection spectra $R(\omega)$ for not too small angles θ when the corresponding quasi-guided modes are far enough from each other, see figures 2c,d and figure 3, shown as an example for angle $\theta_0 = 61^0$. Since the maximum of the peaks much exceeds the value of the reflection coefficient at some distance from the resonances ($\max(|r_j(\omega)|^2) \gg |r_{0j}|^2$, $j = 1, 2$), this situation resembles the one seen in the case of the light reflection from a two-dimensional periodic array of objects demonstrating a resonant response in the absence of dielectric contrast, e.g., associated with excitons in isolated or tunnel-coupled quantum dots [9],[10]. Relying on this analogy one can write the transmission coefficient for a hybrid structure as $t(\omega) = 1 + r(\omega)$, and taking into account the relation $|r|^2 + |t|^2 = 1$, one gets that $|r|^2 + Re(r) = 0$. Further, the reflection coefficient can be presented as a sum of the pole terms $r = \sum_{j=1}^{N} r_j$, $r_j = f_j/(\omega_{0j} - \omega - i\Gamma_j)$, where ω_{0j} and $\Gamma_j = \Gamma_{0j}$ are the resonant frequency and radiative damping rate of the j-th resonance. As a result, one can obtain the following expression for the reflection coefficient:

$$R = |r|^2 = \sum_{j=1}^{N} \frac{Im(f_j)\Gamma_{0j} + Re(f_j)(\omega - \omega_{0j})}{(\omega - \omega_{0j})^2 + \Gamma_{0j}^2} \, .$$

If to take into consideration the absorption in the system, the last expression can be recast into the form

$$R(\omega) = \sum_{j=1}^{N} \frac{A_j + B_j(\omega - \omega_{0j})}{(\omega - \omega_{0j})^2 + \Gamma_j^2} \, , \tag{4}$$

where A_j, B_j and Γ_j are some real constants. Alternatively, equation (4) can be derived by considering many wave interference from N resonances.

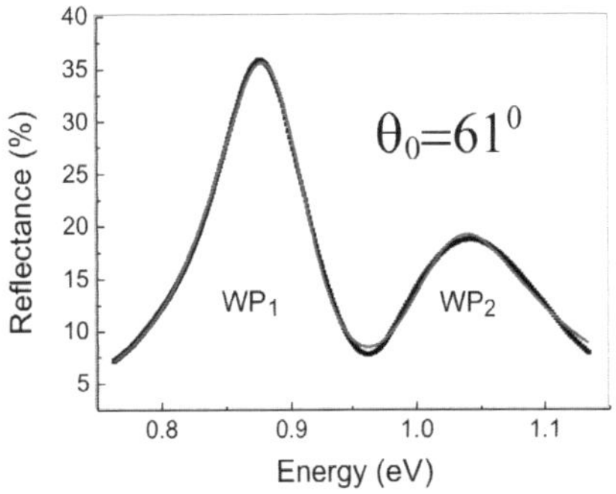

Figure 3. The optical reflection spectrum (black line) for the hybrid photonic crystal (with GST film of thickness 80 nm) for the incidence angle $\theta_0 = 61^0$, which demonstrates the Wood anomalies (WP_1 and WP_2) corresponding to two close quasi-guided modes. The extrapolation of the spectrum is made by equation (4) (red line).

The extrapolation of the reflection spectra shown in figure 3 by equation (4), where the frequency ω is expressed in terms of energy ($\omega \to E[eV]$), gives the following fitting parameters: $A_1 = 0.064$, $B_1 = -0.223$, $\Gamma_1 = 0.043$ and $A_2 = 0.072$, $B_2 = -0.024$, $\Gamma_2 = 0.066$. The resulting discrepancy between the coefficients B_1 and B_2 and the condition $B_2 = -B_1$ can be attributed to a significant difference of the spectral shape for the second Wood anomaly from the Lorentzian shape (due to a large broadening of the peak), and to the frequency dependence of the phase shift (due to a difference in the spatial locations of the quasi-guided modes in the waveguide layer).

3. Conclusion

It can be expected that the approach developed in the present work can also be used to estimate the locations of Wood anomalies in the reflection and transmission spectra of weakly disordered structures; in this case the vector's magnitude G in equation (2) is the same as in the periodic case. Despite the absence of periodicity, these structures have long-range order and therefore would provide effective quasiguided modes for propagation and scattering of light, leading to Wood anomalies in the optical spectra, albeit of less intensity but more broadened. The simplest system here is a 1D resonant grating waveguide structure with a rectangular profile composed of alternating segments with slightly different lengths to form a weakly disordered periodic structure. The study of the light reflection from such a structure can be made by a method applicable to the case of the analogous periodic structure [11] that leads to a conclusion about the possibility of appearance of resonant Wood anomalies in the reflection spectra for a weakly disordered periodic hybrid structure (and also for some other non-periodic structures having an average period, e.g., for quasicrystals).

To summarize, in the frame of the coupled-waves approach we have studied the anticrossing phenomenon of the Wood and Bragg resonances in the reflection spectra of the synthesized hybrid structures based on the opal photonic crystal. From a comparison of the experimental data with the theoretical model calculations we have determined the effective parameters of the quasi-guided mode responsible for the appearance of the resonant Wood anomalies and the interaction coefficient between the modes of the Wood anomalies and 3D Bragg diffraction. An empirical description of the Wood anomaly reflection spectra is made for the case where all other resonances in the considered frequency range are absent. It is established that in hybrid photonic crystals the interference of the light waves originating from different quasi-guided modes in the Wood anomaly reflection spectra can arise.

References

[1] Maystre D 2012 Theory of Wood's Anomalies *Plasmonics: From Basics to Advanced Topics (Springer Series in Optical Sciences vol 167)* ed S Enoch and N Bonod (New York: Springer) chapter 2 pp 39-83
[2] Eggleton B J, Luther-Davies B, Richardson K 2011 *Nature Photonics* **5**, 141
[3] Romanov S G, Regensburger A, Korovin A V, Peschel U 2011 Adv. Mater. **23**, 2515
[4] Voronov M M, Pevtsov A B, Yakovlev S A, Kurdyukov D A, Golubev V G 2014 *Phys. Rev. B* **89**, 045302
[5] Trofimova E Y, Aleksenskii A E, Grudinkin S A, Korkin I V, Kurdyukov D A, Golubev V G 2011 *Colloid J.* **73** 546
[6] Yakovlev S A, Pevtsov A B, Fomin P V, Melekh B T, Trofimova E Yu, Kurdyukov D A, Golubev V G 2012 *Tech. Phys. Lett.* **8**, 768
[7] Yariv A 1973 *IEEE J. Quantum Electron.* **9**, 919
[8] Kaplyanskii A A, Baryshev A V, Rybin M V, Sel'kin A V, Limonov M F 2012 Optical properties of low-contrast opal-based photonic crystals *Optical properties of photonic structures. Interplay of order and disorder* ed M F Limonov and R.M. De La Rue (CRC Press), section 3.5 pp 249-275
[9] Ivchenko E L, Fu Y, Willander M 2000 *Phys. Solid State* **42**, 1756.
[10] Voronov M M, Ivchenko E L 2003 *Phys. Solid State* **45**, 176
[11] Rosenblatt D, Sharon A, Friesem A 1997 *IEEE J. Quantum Electron.* **33**, 2038

Resonant plasmon-stimulated nonlinear absorption in three-level systems.

A.Zasedatelev[1], T. Dubinina[2,4], V. Krasovskii[1,3], O. Suprunova[1], L. Tomilova[2,4] and A. Chistyakov[1].

[1]National Research Nuclear University «MEPhI», Moscow, Russia
[2]Department of Chemistry, M.V. Lomonosov Moscow State University, Moscow, Russia
[3]Prokhorov General Physics Institute of the Russian Academy of Sciences, Moscow, Russia
[4]Institute of Physiologically Active Compounds, Russian Academy of Sciences, Chernogolovka, Russian Federation

E-mail: anton.zasedatelev@gmail.com

Abstract. Mechanisms of resonant nonlinear absorption in the hybrid exciton-plasmon systems based on gold nanoparticles and macroheterocyclical compounds are presented. We used novel substituted subphthalocyanines (SubPc) with 14 π-electronic system as powerful excitonic 3 – level systems, and 50 nm gold nanoparticles (AuNp) as plasmonic structures. We succeeded to define mechanisms of resonant nonlinearities and to show that local field factor (LFF) and Purcell factor (PF) lead to dramatic enhancement in SubPc nonlinearity. We suggest a theoretical model which describes exciton-plasmon interaction of porphyrin/phthalocyanine related compounds with gold nanoparticles in the resonant high power laser field.

Unique optical properties of hybrid structures based on metal nanostructures and chromophore molecules are very promising in nonlinear optics. Nowadays different hybrid plasmonic nanosystems are used as a basic part of optical modulators and switchers [1,2], nanolasers [3,4], high-sensitive sensors [5,6], harmonic generation materials [7], optical limiters etc. Tight "mixing" of plasmonic properties of metal structures with excitonic properties of chromophore is a key feature of resonant nonlinearity, it leads to dramatic changings in optical nonlinear properties in the whole exciton-plasmon system. There are two major reasons for it. The local field factor i.e. enhancement of the near field intensity would result to increase of the nonlinear responses, on the other hand enhancement of the decay rate of excited states would lead to acceleration of the rate in a radiative relaxation process. In spite of very different time scales of the processes in a uncoupled regime, when chromophore and plasmon structures don't interact, in the tightly "mixing" exciton-plasmon regime both processes can have similar time scales. Therefore relaxation and excitation rates can turn from independent values into strong correlated ones. In addition to modification of the rates of electronic transitions in plasmon-coupled excitonic systems, the impact of strong near-field on chromophores can lead to appearance of parametric nonlinearities, such as multiphoton absorption and others. Therefore it is very difficult to predict a character of nonlinear properties in general, without full description of the exciton-plasmon system's dynamics. Among promising nonlinear optical (NLO) organic materials, macroheterocyclical compounds hold a special place [8] due to their large conjugated electronic system. This is the origin of considerable nonlinear response in such compounds. High structural flexibility of the molecule allows one to construct new materials with prescribed NLO properties. According to this, researchers paid great attention to develop novel NLO devices based on different macroheterocyclical chromophores [9,10]. In particular, subphthalocyanine (SubPc) compounds are good candidates to be used in harmonic generation [11,12] and optical modulating/switching devices [13]. We believe that unique optical properties of plasmonic structures together with noticeable NLO properties of macroheterocyclical compounds are very encouraging to enhance the efficiency of existing NLO devices, as well as to reduce correspondent thresholds, and allow to discover completely new applications. The first step is a study of light absorption properties. Thus, the subject of our research is an experimental and theoretical study of resonant plasmon-stimulated nonlinear absorption in three-level chromophore systems, investigation of the involved mechanisms.

2. SYNTHESIS of nano-hybrids

An aqueous solution of gold nanoparticles (AuNp) capped with citric ion was prepared by reduction of $HAuCl_4$ with sodium citrate as described previously elsewhere [14]. The mean diameter evaluated from transmission electron microscopy (TEM) images was equal to 50±5 nm. Hexa-phenoxy substituted subphthalocyanines, bearing a halogen atom (chlorine) in axial position, were synthesized from corresponding phthalonitriles using the method described in literature [15]. Solution of ^{PhO}SubPcBCl (0.6 mL, $3\cdot10^{-5}$ M) in DMSO was mixed with aqueous solution of AuNP (0.6 mL, 50 nm gold particles), then 0.3 mL of DMSO:H_2O (1:1, V:V) was added. This mixture was stirred for 24 h for the stacking to take place. The completeness of reaction was monitored by quenching of free subphthalocyanine luminescence.

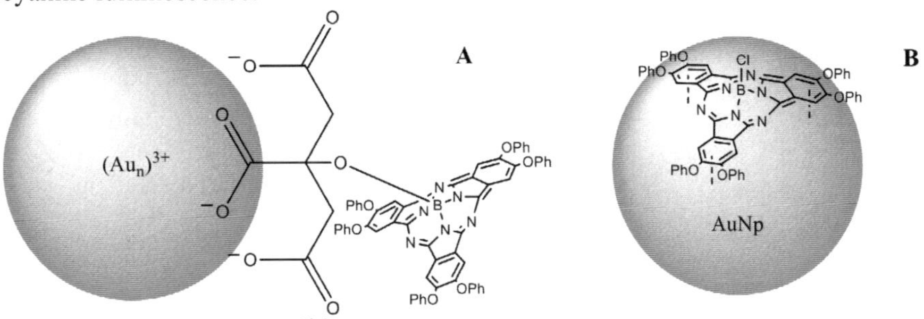

Figure 1.Scheme of possible *AuNp* - ^{PhO}*SubPcBCl* conjugate formation.

There are two approaches to formation of interaction between the target subphthalocyanines and the surface of nanoparticles. First, the formation of new B-O bond between -OH fragments of citrate and central halogen atom of subphthalocyanine (Fig.1A). It resulted from high lability of extra ligand in subphthalocyanine complexes. Second, due to the appearance of π-π interactions between subphthalocyanine macrocycle and surface of nanoparticles (Fig.1B). We suppose, that these two directions of AuNp –^{PhO}SubPcBCl conjugate formation are equally possible.

3. LINEAR properties

In order to achieve a strong exciton-plasmon interaction in hybrid structures, it's necessary to provide both excellent spectral overlap between absorption band of chromophores with plasmonic band of AuNp, and their extremely close spatial arrangement [16]. Figure 2A presents linear extinction spectra of synthesized ^{PhO}SubPcBCl, AuNp and hybrid nanostructures. From Figure 2A it follows that energy of chromophores' transitions from the ground state to the first excited state is tuned into plasmonic resonance. Linear spectrum of "hybrids" isn't a direct superposition of correspondent spectra of structures' components. Difference extinction spectrum (Fig.2B), formed by the difference between hybrid's spectrum and the sum of the ^{PhO}SubPcBCl and AuNp spectra, demonstrates a "dip". We suppose that it is Fano "dip" with the energy splitting about 50 meV. Presence of Fano "dip" in extinction spectra is a widespread proof of strong exciton-plasmon coupling in such hybrid structures. Moreover, Yoshida et al. showed that strong couplings are possible between molecules which have only been directly attached to metal surface [17], therefore absence of the "dip" in the total extinction spectrum of "hybrids" (Fig.2A) is caused by little fraction of molecules which are directly attached onto the metal surface.

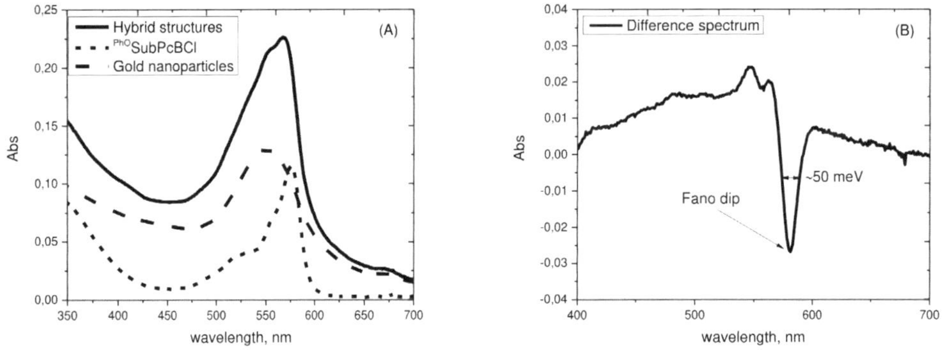

Figure 2. Extinction spectra of ^{PhO}SubPcBCl, AuNp, nano-hybrid structures (**A**). Difference spectrum (**B**).

While photoluminescence (PL) study of "free" SubPc molecules showed that PL quantum yield (QY) is equal to 0.15, PL signal of nano-hybrid structures is strongly suppressed. More detailed investigations of PL have been done using time-correlated single photon counting (TCSPC) method. PL decay rates of both structures are similar (τ_{PL}=2.6 ns). According to this, we can conclude that PL quenching is a result of coupling between radiative excitonic modes of chromophores and high-order plasmonic modes (*nonradiative*) of AuNp.

4. NONLINEAR resonant absorption

Experimental studying of resonant nonlinear absorption was carried out using open-aperture single beam Z-scan technique [18]. We utilized the second harmonic generation light of mode-locked Nd^{3+}:YAG laser operated in TEM_{00} mode with the pulse duration of 350 ps and a low repetition rate of 5 Hz (*to prevent heating processes*). The beam was tightly focused by a lens, the radius of the beam waist was equal to 25-30 μm with the pulse energy equal to 40μJ. The peak intensity in the focus was equal to $4{,}8\cdot 10^9$ W/cm^2. All measurements were done using 5 mm quartz cuvette.

According to results of our experimental and theoretical research of nonlinear absorption in "free" SubPc molecules we established that the molecules are equivalent to 3-level quantum systems, in case of short-pulse resonant excitation. Figure 3A represents fitting results of the experimental data by the 3-level model. Absorption cross-section of S_1 – S_2 transition (σ_{12}) and relaxation rate of S_2–S_1transition (τ_{21}) were determined during the fitting procedure. Nowadays, nonlinear absorption mechanisms and properties of metal nanoparticles are well-known [19], there are many papers showing that the nonlinear response of silver and gold nanoparticles in our set up conditions are well described by third order susceptibility $\chi^{(3)}$ [20,21].

An obvious feature of the nonlinear transmission of "hybrids" (Fig.3B) compared with "free" chromophores (Fig.3A) is a presence of considerable maximum near the beam focus. Since the experimental results contain both information about plasmon-stimulated absorption of 3-level systems and information about absorption in metal cores of the structures, it's impossible to determine the impact of exciton-plasmon interaction on the resonant absorption of 3-level systems and the mechanism of observed changings in Z-scan without building the physical model and separating responses of each component. In the frame of Optical Bloch formalism the theoretical model of resonant nonlinear absorption of plasmon-coupled 3-level systems was proposed. In high intensity irradiation plasmonic excitation can be considered in classical Maxwell approximation, while 3-level systems operated by Hamiltonian, can be considered in density matrix formalism.

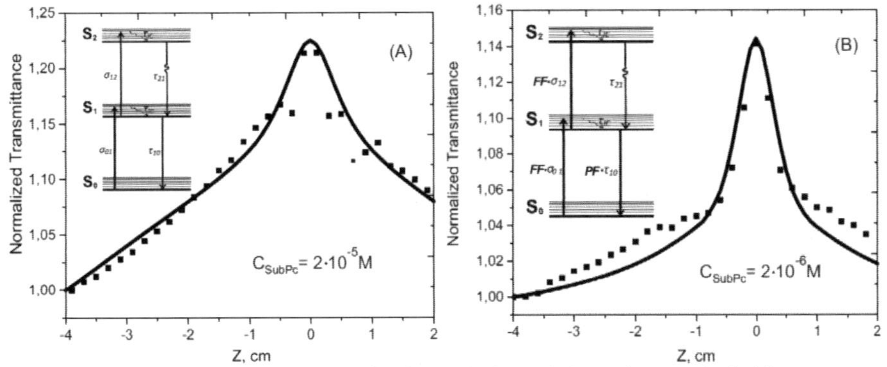

Figure 3.Open-aperture Z-scans of "free" SubPc (A) and nano-hybrid structures (B). Experimental data – dots, simulation curves – solid line.

Behavior of 3-level systems in the electromagnetic field was described using density matrix $\hat{\rho}$ with time evolution given by quantum Liouville equation [22] within phenomenological relaxation operator $\hat{\Gamma}$.

$$\frac{d\hat{\rho}}{dt} = -\frac{i}{\hbar}\left[\hat{H},\hat{\rho}\right] + \hat{\Gamma}$$

where total Hamiltonian of the system in the electromagnetic field is a sum of Hamiltonian of the system without external field (*unperturbed part*) $\hat{H}_0$ and interaction Hamiltonian (*in the dipole approximation*) $\hat{H}_I$. According to that, motion equations of the density matrix elements can be expressed as follows:

$$\dot{\rho}_{00} = \gamma_{11}\rho_{11} - \frac{i}{\hbar}(\rho_{10}\mu_{01}E_{loc}{}^* - \rho_{01}\mu_{10}E_{loc})$$

$$\dot{\rho}_{11} = -\gamma_{11}\rho_{11} + \gamma_{22}\rho_{22} + \frac{i}{\hbar}(\rho_{10}\mu_{01}E_{loc}{}^* - \rho_{01}\mu_{10}E_{loc}) - \frac{i}{\hbar}(\rho_{21}\mu_{12}E_{loc}{}^* - \rho_{12}\mu_{21}E_{loc})$$

$$\dot{\rho}_{22} = -\gamma_{22}\rho_{22} + \frac{i}{\hbar}(\rho_{21}\mu_{12}E_{loc}{}^* - \rho_{12}\mu_{21}E_{loc})$$

$$\dot{\rho}_{01} = \dot{\rho}_{10}{}^* = -(-i\omega_a + \gamma_{01})\rho_{01} - \frac{i}{\hbar}(\mu_{01}E_{loc}{}^*(\rho_{11} - \rho_{00}) - \mu_{21}E_{loc}\rho_{02})$$

$$\dot{\rho}_{12} = \dot{\rho}_{21}{}^* = -(-i\omega_a + \gamma_{12})\rho_{12} - \frac{i}{\hbar}(\mu_{12}E_{loc}{}^*(\rho_{22} - \rho_{11}) + \mu_{10}E_{loc}\rho_{02})$$

$$\dot{\rho}_{02} = \dot{\rho}_{20}{}^* = -(-i2\omega_a + \gamma_{02})\rho_{02} + \frac{i}{\hbar}(\mu_{12}E_{loc}{}^*\rho_{01} - \mu_{01}E_{loc}{}^*\rho_{12})$$

Taking into account the modification of spontaneous decay rate (*radiative*) from the 1st excited state to the ground state of 3-level systems placed in vicinity to the AuNp (*i.e. Purcell effect*), the above equations describe the dynamics of plasmon-coupled systems completely. But electromagnetic field ($\boldsymbol{E}_{loc}$) which acts on the systems, depends on the polarization of that systems. Moreover, according to Maxwell equations and boundary conditions the field is changed in space and time during the propagation through the sample. To obtain self-consistent system of differential equations we have to complete above presented motion equations by adding classical equations which bind the electric field with the induced polarization. Solving such self-consistent system of differential equations in general is a very tough issue, instead of that we can introduce some reasonable simplifications. Since the chromophore's shell thickness is about 5 nm (*it's about 10% of total "hybrid" size*) in accordance with TEM measurements, we can ignore a dependence of the near field which acts on 3-level systems on their polarization. Taking into account decay rates and transitions linewidths of 3-level systems we can also ignore rate equations of non-diagonal density matrix elements - it's another assumption. Thus, the nonlinear response of the 'hybrids' is determined only by population dynamics of 3-level systems and by the local near-field which depends on internal AuNp Kerr nonlinearity.

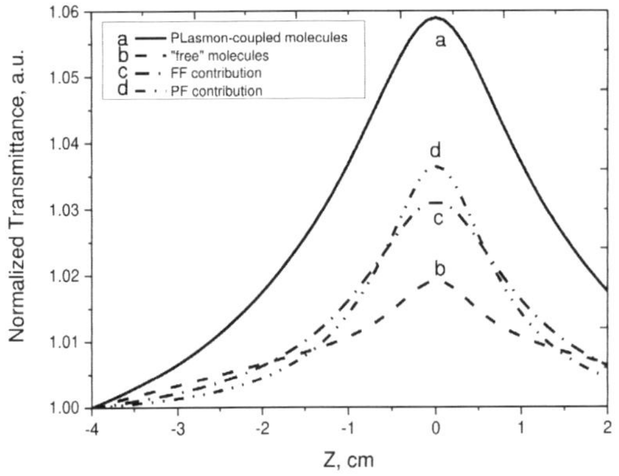

Figure 4. Simulated Z-scan curves of plasmon-coupled systems (a), "free" systems (b), coupled systems considering only field factor (c), coupled systems considering only Purcell factor (d).

Taking into account the assumptions, we numerically solved the problem using FDTD method without any adjustable parameters. The model curve is displayed by solid line in Figure 3B, it coincides well with the experimental results. In order to conduct quantative comparison of nonlinear response of plasmon-coupled 3-level systems with "free" systems we simulated curves for relevant systems with similar concentration (Figure 4a,b). We also simulated curves with just one local field factor (Figure 4c) and just one Purcell factor (Figure 4d) to determine mechanisms of changes observed. The modeling results showed that there are approximately equal contributions to the total nonlinear response - the first is caused by Purcell factor and the second by the strong near field. Difference transmittance of plasmon-coupled chromophores is three times as much than "free" molecules (Figure 4a,b), when concentration of both is equal to $2{\cdot}10^{-6}$ M. It explained by a superposition of influence of two factors (*field factor and Purcell factor*). At the same time the total nonlinear response of plasmon–coupled systems (Figure 4a) cannot be described by the direct sum of each factor's response. Analyzing dependence of the normalized populations on the intensity of laser radiation allows us to establish a reason of such dramatic differences in nonlinear transmittance. Thus, in case of "free" molecules the master mechanism which leads to the observed "saturation" effect in absorption is a decrease of ground state population especially near the beam focus. At the same time there is no significant contribution of transition from the 1st excited state to the second one to the saturation "effect". It slightly appears only in the beam focus. However, in case of plasmon-coupled systems both transitions play significant role in saturation process. Therefore both transitions $S_0 \rightarrow S_1$ and $S_1 \rightarrow S_2$, turn out to be saturated in the beam focus – it explains dramatic changings of the transmittance in plasmon-coupled systems.

5. Conclusions

This is the first time when the synthesis of hybrid nanostructures based on 3-level chromophores with high resonant interaction was implemented. We experimentally studied their linear optical properties, found out the presence of Fano "dip" in the difference absorption spectrum conforming to strong exciton-plasmon coupling between SubPc and AuNp. Relaxation rates of both free chromophores and hybrid nanostructures were measured. The experiment claims to be the first Z-scan study of resonant absorption properties of similar structures in a wide range intensity of incident radiation. Based on Optical Bloch formalism, we proposed a model that properly represents the experiment results. By applying the model we succeeded to define mechanisms of nonlinear resonant response. According to our research Purcell effect and near-field factor equally contribute to the nonlinear response. The enhancement of absorption "saturation" effect caused by Purcell effect has turned up the surprising result and, as far as we are aware, is observed for the first time.

Acknowledgements

Financial support from the Russian Foundation for Basic Research (Grant No 14-03-31579 and 15-03-05890).

References:

[1] M. Sukharev, T. Seideman, Phase and polarization control as a route to plasmonic nanodevices. Nano Lett. 6, 715–719 (2006)

[2] M. Aeschlimann, M. Bauer, D. Bayer, T. Brixner, F.J.G. Abajo de, W. Pfeiffer, M. Rohmer, C. Spindler, F. Steeb, Adaptive subwavelength control of nano-optical fields. Nature 446, 301–304 (2007)

[3] M.A. Noginov et al., Demonstration of a spaser-based nanolaser, Nature 460: 1110–1112, (2009).

[4] Noda, S. Seeking the ultimate nanolaser. Science 314, 260–261 (2006).

[5] Yurii E. Lozovik, Igor A. Nechepurenko, Alexander V. Dorofeenko, Eugeny S. Andrianov, Alexander A. Pukhov, Spaser spectroscopy with subwavelength spatial resolution, Physics Letters A378 (2014) 723–727.

[6] J.N. Anker, W.P. Hall, O. Lyandres, N.C. Shah, J. Zhao, R.P.V. Duyne, Biosensing with plasmonic nanosensors. Nat. Mater. 7, 442–453 (2008)

[7] J. Butet, G. Bachelier, I. Russier-Antoine, C. Jonin, E. Benichou, P.-F. Brevet, Interference between selected dipoles and octupoles in the optical second-harmonic generation from spherical gold nanoparticles. Phys. Rev. Lett. 105, 077401 (2010)

[8] María A. Díaz-García, Nonlinear optical properties of phthalocyanines and related compounds, J. Porphyrins Phthalocyanines 13,: 652–667, (2009)

[9] Nalwa HS and Shirk JS. In Phthalocyanines. Propertiesand Applications, Vol. 4, Leznoff CC and Lever ABP. (Eds.) VCH Publishers (LSK) Ltd.: (1996).

[10] de la Torre G, Vázquez P, Agulló-López F and TorresT. Chem. Rev., 104: 3723–3750, (2004)

[11] G. Rojo, A. Hierro, M.A. Diaz-Garcia, F. Agullo-Lopez, B. del Rey, A. Sastre, T. Torres, Appl. Phys. Lett. 70, 1802., (1997)

[12] Rojo G, Agully-Lypez F, del Rey B and Torres T.J. Appl. Phys. 84, 6507–6509. (1998)

[13] Z.Chen, C. Xia, Y. Wu, X. Zuo, Y. Song, Inorganic Chemistry Communications 9, 187–191, (2006)

[14] J. Turkevich, P. C. Stevenson, J. Hiller, "A study of the nucleation and growth processes in the synthesis of colloidal gold," Discuss. Faraday Soc., 11, 55 (1951)

[15] A. Yu. Tolbin, M. O. Breusova, V. E. Pushkarev, L. G. Tomilova, Synthesis and spectroscopic properties of new boron subphthalocyanine complexes and a heteronuclear phthalocyanine complex, Russian Chemical Bulletin, International Edition 54 (9), pp. 2083—2086, (2005)

[16] J. Bellessa, C. Bonnand, J. Mugnier and J. C. Plenet, Strong Coupling between Surface Plasmons and Excitons in an Organic Semiconductor, Phys. Rev. Lett. 93, 3, (2004)

[17] A. Yoshida, Y. Yonezawa, and N. Kometani, Tuning of the Spectroscopic Properties of Composite Nanoparticles by the Insertion of a Spacer Layer: Effect of Exciton-Plasmon Coupling, Langmuir 25(12), 6683–6689, (2009)

[18] M.Sheik-Bahae, A.A. Said, T.-H. Wei, D.J. Hagan, E.W. VanStryland, Z-Scan Measurements of Optical Nonlinearities, The Journal of Quantum Electronics 26, 760-769., (1990)

[19] CHALLENGES AND ADVANCES IN COMPUTATIONAL CHEMISTRY AND PHYSICS Vol. 1: Non-Linear Optical Properties of Matter, Edited by M. Papadopoulos, A. Sadlej, J. Leszczynski, Springer, 676 p, (2006)

[20] V.A. Karavanskii, A.V. Simakin, V.I. Krasovskii, P.V Ivanchenko, Nonlinear optical properties of colloidal silver nanoparticles produced by laser ablation in liquids, Quantum Electronics 34 (7) 644 - 648 (2004)

[21] V. I. Krasovskii and V. A. Karavanskii, Surface Plasmon Resonance of Metal Nanoparticles for Interface Characterization, Optical Memory and Neural Networks (Information Optics) 17, No. 1, pp. 8–14, (2008)

[22] P. Meystre, M. Sargent, Elements of Quantum Optics (4th edition), Springer, 507 p, (2007)

Quantum walks of photon pairs in twisted waveguide arrays

DN Vavulin[1], **A A Sukhorukov**[2]

1 ITMO University, 49 Kronverksky Ave., St.Petersburg, 197101, Russia
2 Nonlinear Physics Centre, ResearchSchool of Physics and Engineering, AustralianNationalUniversity, Canberra, ACT 2601, Australia

E-mail: dima-vavulin@mail.ru

Abstract. We consider an array of closely spaced optical waveguides, which are twisted around a central axis along the propagation direction. We derive Schrodinger-type equation of the biphoton wavefunction, taking into account the waveguide bending through the appearance of additional phase in the coupling coefficients. We present an example of the evolution of quantum photon-pair state.

1. Introduction

Quantum walks involving several particles can be used to implement quantum algorithms, which can perform faster than classical analogues. Of particular interest are quantum walks, where interference of several walkers can be used to realize various simulations, including database search [1] and quantum cryptography [2-4]. Quantum walks of photons can be implemented in arrays of coupled waveguides. In particular, coherent quantum phenomena can be potentially simulated in closed-loop arrangements of waveguides [5]. Recently, a new type of coupled waveguides with a twisted geometry were demonstrated in a photonic-crystal fiber [6]. In this work, we study the effect of twist on quantum walks of photons.

2. Initial equations, parameters and conditions

The normalized pump field profile evolution along the propagation distance z is defined through the classical coupled-mode equations [7]:

$$i\frac{dE_n}{dz} + C(\omega)(\exp[-i\Delta\varphi]E_{n+1} + \exp[i\Delta\varphi]E_{n-1}) = 0\,, \tag{1}$$

where E_n is the complex field amplitude in the nth waveguide, and E_N maps to E_0, and E_{-1} maps to E_{N-1} due to closed-loop boundary conditions [8], n is the waveguide number, N is the total number of waveguides, $\Delta\varphi = \omega d\chi/dz$ is the additional phase, ω- frequency, $d\chi/dz$ - is the periodic waveguide bending profile, $\chi(z) = \chi(z + L)$, L is the modulation period, coefficient C(ω) defines a coupling

[1] To whom any correspondence should be addressed.
[2] To whom any correspondence should be addressed.

strength between the neighboring waveguides (it characterizes diffraction strength in a straight waveguide array with $C_\chi = 0$ [9].

In our case $d\chi/dz$ = const because waveguide bending profile is the same to all values of z.

Generation of photon pairs in cubic nonlinear WGAs through SFWM in the absence of multiple photon pairs can be characterized by the evolution of a bi-photon wave function $\Psi n_s,n_i$ (z) in a Schrodinger-type equation. The equation is obtained from the Hamiltonian, and it has a form similar to that of quadratic media [6]:

$$i\frac{d\Psi_{n_s,n_i}}{dz} = -C(\exp[i\Delta\varphi]\Psi_{n_s-1,n_i} + \exp[i\Delta\varphi]\Psi_{n_s,n_i-1} + \exp[-i\Delta\varphi]\Psi_{n_s+1,n_i} + \\ + \exp[-i\Delta\varphi]\Psi_{n_s,n_i+1}) + i\gamma E^{(p)}_{n_s}(z)E^{(p)}_{n_s}(z)\delta_{n_s,n_i}\exp[i\Delta\beta^{(0)}z] \quad (2)$$

where n_s and n_i are the waveguide numbers describing the positions of the signal, and the idler photons, and $E_{ns}^{(p)}(z)$ is the pump amplitude in waveguide number ns. $\Delta\beta$ - the linear four-wave mixing phase-mismatch in a single waveguide, γ - is a nonlinear coefficient, δ - Kronecker delta [7].

3. Numerical calculation

We consider an array of closely spaced optical waveguides, which are twisted around a central axis along the propagation direction. Such structure composed of three waveguides is schematically shown in Fig. 1(a). We derive Schrodinger-type equation of the biphoton wavefunction, taking into account the waveguide bending through the appearance of additional phase in the coupling coefficients [10].

We present an example of the evolution of quantum photon-pair state in Figs. 1(b-d). We consider an input condition in the form of Einstein-Podolsky-Rosen (EPR) entangled state, where two photons are present together with equal probability in any of the waveguides, but photons cannot be in different waveguides, see correlation plot in Fig. 1(b). As the state evolves, the correlation properties can become reversed, and the photons are most likely to appear in different waveguides after a certain propagation distance, see Fig. 1(c). Interestingly, at a longer propagation distance the input state can be almost exactly restored, see Fig. 1(d). These features can be controlled by the amount of twist.

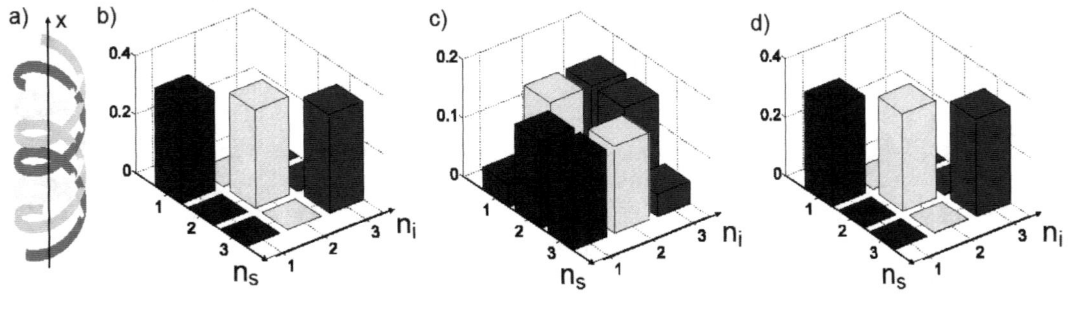

Figure 1. a). Scheme of coupled twisted waveguides. (b-d) Two-photon correlations ($|W|^2$) (W - biphoton wavefunction) between different waveguides at (b) z = 0, (c) z = L/2, (d) z = L.

We also analyse integrated photon generation and quantum walks [11], implemented through spontaneous four-wave mixing in optical fibers. Such system can further tailored to produce entangled optical-angular-momentum states, which have applications for quantum communications and imaging.

Fig.2 shows two-photon correlations ($|\Psi n_s,n_i|^2$) in the same structure as in Fig.1, but in case of photon generation by SFWM from a pump inside the array. (a) z = 0, (b) z = 0.375 L, (c) z = 0.604 L.

At z = 0 (a) there are no photons. At distance (b), the generated photons are mostly bunched: appear in the same waveguides. In (c) the photons are anti-bunched: appear at different waveguides. These features can be controlled by the amount of twist and the input pump profile.

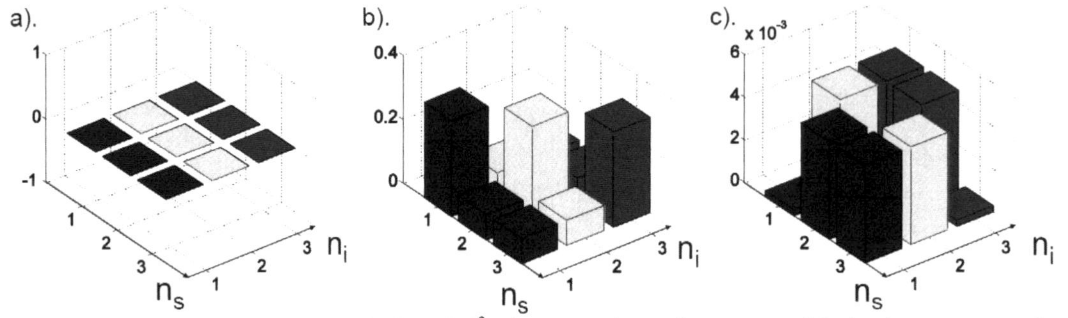

Figure 2. (a-c)Two-photoncorrelations($|\Psi|^2$) between the variouswaveguidesin the presence of photon generation throughSFWMeffect at(a) z = 0, (b) z = 0,375 L, (c) z = 0,604 L.

4. Conclusion

We predict that the correlation properties of entangled photon pairs propagating in the regime of quantum walks in closed-loop waveguide arrays can be controlled by the introduction of twist. We also analyze integrated photon generation through spontaneous four-wave mixing and predict that the input pump profile also can control the features of correlation properties of entangled photons.We also show that in both cases photons can demonstrate bunching behavior at some distances and anti-bunching behavior at other distances. These features can be controlled by the amount of twist and the input pump profile.

5. Acknowledgements

This work was financially supported by the Government of Russian Federation, Grant 074-U01.

References

[1] Hamilton C S, Kruse R, Sansoni L, Silberhorn C, and Jex I 2014 Driven Quantum Walks. *Phys. Rev. Lett.***113**, 083602-5

[2] Ekert A K, Rarity J G, Tapster P R, and Palma G M 1992Practical quantum cryptography based on two-photon interferometry. *Phys. Rev. Lett.***69**, 1293

[3] Egorov V I, Vavulin D N, Latypov I Z, Gleim A V, Rupasov A V 2013 Analysis of a sidebands based quantum cryptography system with different detector types*NANOSYSTEMS: PHYSICS, CHEMISTRY, MATHEMATICS*, **4 (2)**, pp. 190–195

[4] D N Vavulin, V I Egorov, A V Gleim, S A Chivilikhin 2014 Determining influence of four-wave mixing effect on quantum key distribution*Journal of Physics: Conference Series***541**, 012066

[5] Owens J O et al. 2011 Two-photon quantum walks in an elliptical direct-write waveguide array. *New J. Phys.***13**, 075003-13

[6] Xi X M, Wong G K L, Frosz M H, Babic F, Ahmed G, Jiang X, Euser T G, and Russell P St.J 2014 Orbital-angular-momentum-preserving helical Bloch modes in twisted photonic crystal fiber. *Optica***1**, 165-169

[7] Christodoulides D N, Lederer F, and Silberberg Y 2003 Discretizing light behaviour in linear and nonlinear waveguide lattices *Nature***424**, 817–823

[8] Markin D M, Solntsev A S, and Sukhorukov A A 2013 Generation of orbital-angular-momentum entangled biphotons in triangular quadratic waveguide arrays. *Phys. Rev. A***87**,

063814-5
[9] Garanovich I L, Longhi S, Sukhorukov A A, and Kivshar Yu S 2012 Light propagation and localization in modulated photonic lattices and waveguides. *Phys. Rep.***518**, 1-79
[10] Solntsev A S, Sukhorukov A A, Neshev D N, and Kivshar Yu S 2012 Photon-pair generation in arrays of cubic nonlinear waveguides. *OPTICS EXPRESS***20**, No. 24, 27441
[11] Jonathan C F, M and M G Thompson 2012 Quantum optics: An entangled walk of photons. *Nature***484**, 47–48

Photoluminescence of undoped InAs autoepitaxial layers

D D Firsov[1*], O S Komkov[1] and A S Petrov[2]

[1]Micro- and Nanoelectronics department, St. Petersburg Electrotechnical University "LETI", Professora Popova street 5, St. Petersburg 197376, Russia
[2]OJSC NRI "Electron", Toreza ave. 68, St. Petersburg 194223, Russia

[*]e-mail: d.d.firsov@gmail.com

Abstract. Photoluminescence (PL) properties of undoped InAs autoepitaxial layers are studied at various temperatures (9-120 K) and excitation power density (0.2-12.5 W/cm^2). The studied structures have been grown on highly doped n^{++}-InAs substrates by chloride-hydride vapour phase epitaxy method. PL spectra measurements were carried out with an FTIR spectrometer. The samples exhibit several luminescence peaks in the 2.9-3.3 μm range, which are attributed to free exciton transitions, deep donor bound excitons, and donor-acceptor pairs. A correlation is revealed between the PL intensity, and the background doping level of InAs epilayers – as supported by the free carrier concentration and mobility measurements, as well as SIMS data. A decrease of donor-bound exciton PL peak is observed for InAs sulfidized by Na_2S.

1. Introduction

Indium arsenide (InAs) autoepitaxial structures grown on highly doped substrates (n^{++}-InAs) are widely used [1] as materials for backside-illuminated infrared image detectors operating at wavelengths about λ~3 μm. Such structures can selectively cover a broad spectral range due to variable transmission of differently doped n^{++}-InAs [2], while the epitaxial n-InAs acts as an active region. To obtain high sensitivity of such photodetectors, powerful diagnostic techniques are required to control optical properties and growth quality of fabricated [3] InAs epitaxial layers. One of the available methods for contactless determination of epitaxial layers' quality is photoluminescence spectroscopy.

2. Studied samples and experimental setup

The studied structures containing undoped autoepitaxial InAs layers were grown by chloride-hydride vapour phase epitaxy method on highly doped n^{++}-InAs substrates. Epitaxial layer thickness, determined from the interference pattern in infrared (IR) reflectance spectra [4, 5], is varied from 3 μm to 10 μm. Free carrier concentration in the layers, obtained with magnetoresistance measurements, was found to be in the $6 \cdot 10^{13}$ - $6 \cdot 10^{15}$ cm^{-3} range. The parameters of the structures are listed in Table 1.

The studied samples include three series (I, II, III) differentiated by epitaxial growth parameters. Series I samples with carrier concentration less than $1 \cdot 10^{15}$ cm^{-3} were grown under the condition of In mole fraction X_{In} being greater than As mole fraction X_{As}. The shortage of arsenic leads to formation of structural defects based on As vacancies in the epitaxial InAs [6]. In combination with various dopants, these defects can result in donor or acceptor levels within the InAs band gap. A reduction of carrier mobility, observed alongside the decrease of concentration values, implies the predominant formation of acceptor-type defects in our structures.

Table 1. Sample parameters

Sample series	Sample number	Free carrier concentration at 77 K (magnetoresistance), cm^{-3}	Carrier mobility at 77 K (magnetoresistance), $cm^2 V^{-1} s^{-1}$	Comments
I	I-0	$6.2 \cdot 10^{13}$	$1.5 \cdot 10^4$	$X_{In}/X_{As} > 1$
	I-1	$2.0 \cdot 10^{14}$	$3.3 \cdot 10^4$	$X_{In}/X_{As} > 1$
	I-2	$1.3 \cdot 10^{15}$	$7.9 \cdot 10^4$	Optimal
II	II-1	$5.4 \cdot 10^{15}$	$4.3 \cdot 10^4$	After In source refilling
	II-2	$4.0 \cdot 10^{15}$	$4.7 \cdot 10^4$	In source annealed
III	III-1	$2.3 \cdot 10^{15}$	$7.3 \cdot 10^4$	Sulfidized by Na_2S
	III-2	$2.3 \cdot 10^{15}$	$7.3 \cdot 10^4$	Reference sample

The second sample series was grown to examine the influence of residual donor doping originating from the metallic In source. A concentration increase above $5 \cdot 10^{15}$ cm^{-3} was generally observed after refilling the source with a new supply of indium. A prolonged (above 30 hours) annealing of the refilled source in hydrogen atmosphere at T = 850° C results in electron mobility increase, and stabilization of carrier concentration in the (1-4)$\cdot 10^{15}$ cm^{-3} range.

Sulfidization of the sample surface by Na_2S water solution was used to remove the natural oxide layer and the surface-related impurities (sample III-1). A reference sample of non-sulfidized epitaxial InAs was taken from the same wafer.

The photoluminescence (PL) experimental setup was based upon a VERTEX 80 Fourier-transform infrared spectrometer (FTIR) operating in continuous scan mode. The spectrometer was equipped with a liquid nitrogen-cooled HgCdTe photovoltaic detector sensitive in the 1.3-16 μm range. PL measurements were carried out at various temperatures (9-120 K) using a closed-cycle He cryostat. Excitation was performed with an 809 nm laser diode of variable power (1 mW - 300 mW).

3. Results and discussion

An example of low temperature PL spectra of an InAs autoepitaxial layer (sample I-0) is shown on Fig. 1. The main PL peak at 3.0 μm (0.413 eV) is rather close to the direct bang gap of InAs (E_g = 0.415 eV at 12 K [7]). Due to the relatively small exciton binding energy of InAs ($E_{ex} \approx 1$ meV [8]), the peaks observed at 3.1 μm (0.400 eV) and 3.25 μm (0.382 eV) can not be attributed to weakly bound exciton recombination (as they are apart from the main peak by 13 meV and 31 meV, respectively). To determine their origin, we analyzed the dependence of peak intensity on laser excitation power density, as shown on Fig. 2.

The intensity of the main peak at 0.413 eV is demonstrating a linear increase proportional to laser power density, which is characteristic of free exciton recombination [9]. The long wavelength peaks exhibit a much slower change in intensity, and therefore should be connected to impurity-based states [10]. According to [11], the peak at 0.382 eV originates from donor-acceptor pairs (DAP). This is confirmed by the characteristic blue shift of the peak with increasing excitation density (Fig.1). The 0.4 eV peak, which has shown a quick decrease at higher temperatures, can be attributed to excitons bound to deep donors in InAs (presumably carbon) [12].

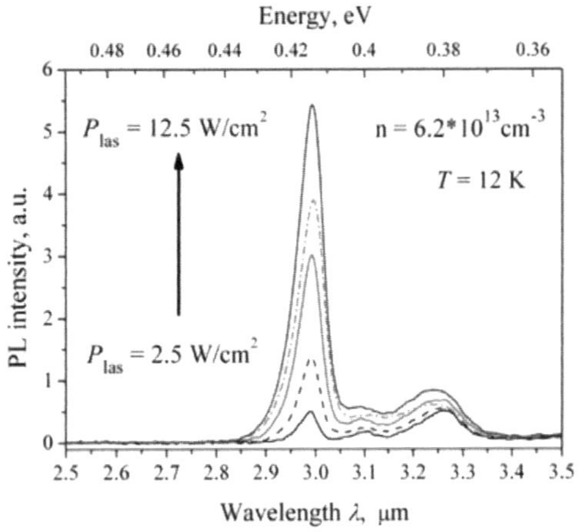

Figure 1. PL spectra of a typical InAs/n^{++}-InAs structure (Series I) at various laser excitation power densities P_{las}.

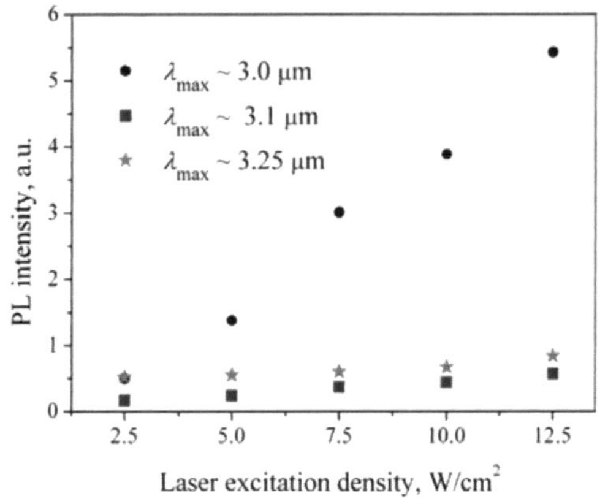

Figure 2. Dependence of PL intensity on excitation power density for 3.0 μm (dots), 3.1 μm (squares) and 3.25 μm (stars) peaks, shown on Fig. 1.

A comparison of the PL spectra for the series I samples has revealed that the donor-acceptor pair luminescence peak is more prominent for structures with lower carrier concentration in the epitaxial layers. Therefore, it can be concluded that such structures have a greater degree of impurity compensation.

The low-power PL spectra of the second series of n-InAs epitaxial layer samples are depicted on Fig. 3. Both samples demonstrate the absence of the 3.25 μm DAP peak, which implies a considerably reduced acceptor concentration. This means that the free carrier concentration in the layers should be proportional to the amount of background donor doping.

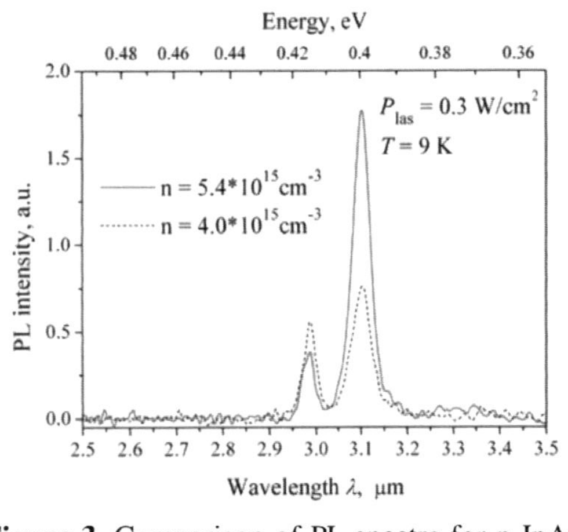

Figure 3. Comparison of PL spectra for n-InAs layers of the series II samples with different free carrier concentration.

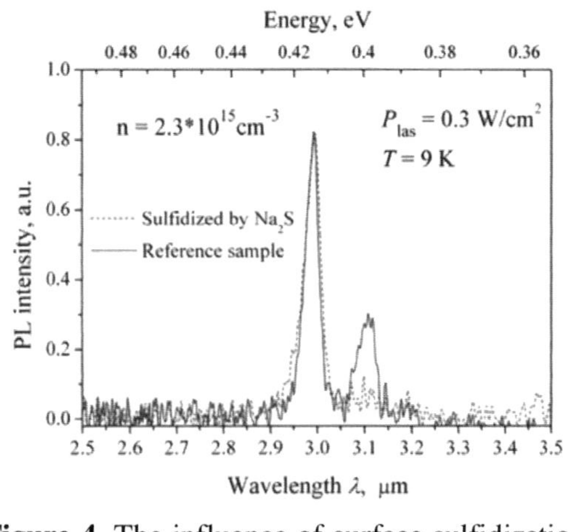

Figure 4. The influence of surface sulfidization on the PL spectra of an n-InAs/n^{++}-InAs structure (series III).

Although the intensity of the free exciton recombination peaks at 3.0 μm is roughly the same, the peak at 3.1 μm is noticeably more intensive for the sample with a higher value of free carrier concentration.

The last part of this study is dedicated to evaluating the impact of surface-based impurities on the photoluminescence of InAs epitaxial layers. Fig. 4 shows the PL spectra of an n-InAs structure treated with Na_2S water solution, and of a reference sample taken from the same wafer.

It has previously been reported [13] that the Na_2S treatment of the InAs surface should increase the main PL peak (80 K) due to removal of surface defect states promoting nonradiative recombination. However, as can be seen from Fig. 4 (9 K), the Na_2S passivation has resulted in a considerably weaker donor-bound exciton PL peak, which should suggest a decreased concentration of donors.

This can be explained if we consider the impurity profiles for n-InAs/n^{++}-InAs structures obtained by secondary ion mass spectrometry (SIMS). An order-of-magnitude increase in carbon and oxygen concentration was observed in a thin (<30 nm) layer at the sample surface. According to [14], the sulfide passivation of InAs, in addition to oxide removal, can also result in limited etching of InAs. This might lead to removal of the unintentionally doped surface layer, thus severely decreasing the donor-bound exciton luminescence – as was observed in our spectra.

In conclusion, we have observed a correlation between the PL properties of InAs autoepitaxial layers, and the residual donor and acceptor concentrations. These results can potentially be used to develop a contactless method of determining impurity concentration in InAs on the basis of photoluminescence measurements.

Acknowledgements

This work was partially supported by Russian Ministry of Education and Science (grant number 16.1307.2014/K). The authors would like to thank Saidashev I.I. (Ioffe Institute) for magnetoresistance measurements, and Ber B.Ya. (Ioffe Institute) for the SIMS data.

References

[1] Ovsyuk V N, Kuryshev G L, Sidorov Yu G et al. 2001 *Matrichnie fotopriemnie ustroistva infrakrasnogo diapazona* (Novosibirsk: Nauka)

[2] Komkov O S, Firsov D D, Kovalishina E A, Petrov A S 2014 *Prikladnaya Fizika* **4** 93

[3] Grama D M, Petrov A S, Popov S D, Stepanov R M, Chilaeva E V 2008 *Izvestiya SPbGETU* **7** 13

[4] Komkov O S, Semenov A N, Firsov D D, Meltser B Ya, Solov'ev V A, Troshkov S I, Pikhtin A N, Ivanov S V 2013 *Semiconductors* **47**(2) 292

[5] Komkov O S, Firsov D D, Kovalishina E A, Petrov A S 2014 *Izvestiya VUZov. Materialy elektronnoi tekhniki* (3)

[6] Baranov A N, Voronina T I, Gorelenok A A et al. 1992 *Soviet physics. Semiconductors* **26**(9) 905

[7] Fang Z M et al. 1990 *Journal of Applied Physics* **67**(11) 7034

[8] Tang P J P, Pullin M J, Phillips C C 1997 *Physical Review B* **55**(7) 4376

[9] Schmidt T, Lischka K, Zulehner W 1992 *Physical Review B* **45**(16) 8989

[10] Tang P J P, Phillips C C, Stradling R A 1993 *Semiconductor science and technology* **8**(12) 2135

[11] Lacroix Y, Watkins S P, Tran C A, Thewalt M L W 1994 *Appl. Phys. Lett.* **66** (9) 1101

[12] Fang Z M et al. 1991 *Applied physics letters* **59** (12) 1446

[13] L'vova T V, Sedova I V, Dunaevskiĭ M S, Karpenko A N, Ulin V P, Ivanov S V, Berkovits V L 2009 *Physics of the Solid State* **51**(6) 1114

[14] Petrovykh D Y, Yang M J, Whitman L J 2003 *Surface science* **523** (3) 231

Numerical simulation of carpet cloaking device in terahertz frequency range

V. V. Gill, A.V. Vozianova, M.K. Khodzitsky

ITMO University, Dept. of Photonics and Optical Information Technologies, 49 Kronverskiy pr. Saint-Petersburg, Russia, 197101

E-mail: viktoria.v.gill@gmail.com

Abstract. This work is devoted to the numerical calculation of the effective constitutive parameters of the carpet cloaking device and to the numerical simulation of this cloak using finite element method (FEM) for the terahertz frequency range.

1. Introduction

In recent years methods of design of composite structure called metamaterials are rapidly improving. Thereby there are unprecedented opportunities to manipulate of electromagnetic radiation. Transformation optics is new science of radiation control [1], it is very powerful instrument for emission control. This approach is based on differential geometry and tensor analysis. Transformation optics shows that space can be arbitrarily curved according to predetermined manner [2]. The rays in such space will have special trajectories. The Fermat's principle is applied to control light by changing of the medium parameters. This is accompanied by a conversion from virtual space to physical space by coordinate transformations.

One of the most interesting devices based on this principle are cloaks of invisibility. Special case of such devices is cover that crushes a cloaked object to a flat ground plane in virtual space [3, 4]. In fact there is an object and cover lying on the surface but it is perceived as the actual ground plane to the observer. This effect is achieved by coordinate transformations. The term "carpet cloak" was coined by Jensen Li and J.B. Pendry in [5] for such transformation.

The development of cloaking devices is very actual direction in spectroscopy, medicine, image analysis, etc. There are many implementations of devices based on various principles and working in the different frequency ranges [6, 7]. But the operation of the cloak in the terahertz frequency range hasn't been clearly understood. In this simulation of carpet cloaking device was proposed in terahertz frequency range. The special constitutive parameters were obtained by grid generation using difference approximation method.

2. Analytical results

For our problem the refractive index profile was calculated. The quasi-conformal mapping is used for transformation from the virtual space (with coordinates (ξ, η)) to the physical space (with coordinates (x, y)) because this mapping minimizes anisotropy [5]. The cloaking area was chosen as a symmetric domain. In the virtual space it seems as a rectangular domain divided into squares with the aspect ratio $\xi/\eta = 1$. The distortion function *f(x)* for mapping in general case [8] looks like:

$$f(\xi,\eta) = MF(\xi,\eta). \tag{1}$$

It was chosen more simply case which corresponds to the quasi-conformal mapping where distortion function has the following form:

$$f(\xi,\eta) = M. \tag{2}$$

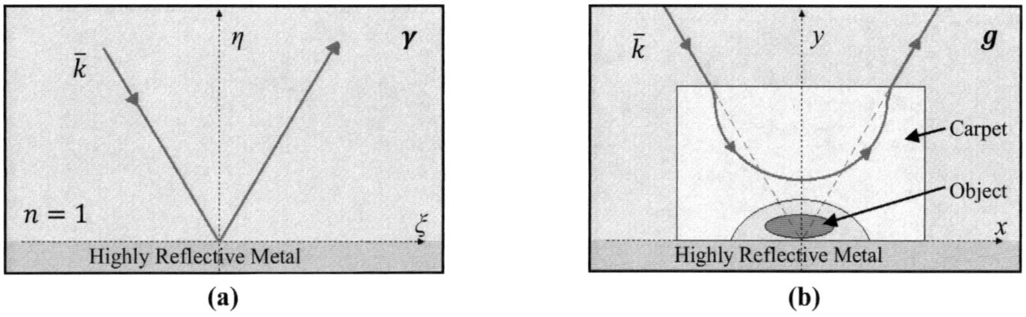

Figure 1. Propagation of wave in **(a)** virtual space; **(b)** physical space.

After the mapping the physical space will be analogously to the virtual space and it differs by only the lower boundary which is defined by the special function. This domain is divided into quadrangular cells. The refractive index profile associated with geometry has following form:

$$n^2 = \frac{1}{\sqrt{det\boldsymbol{g}}} = \frac{1}{|\xi||\eta|}, \tag{3}$$

where $\boldsymbol{g}$ is the metric tensor in the physical space. The anisotropy factor α is defined by [5]:

$$\alpha + \frac{1}{\alpha} = \frac{Tr(g)}{\sqrt{det\boldsymbol{g}}}. \tag{4}$$

According to the optimal ratio [5] between refractive index and anisotropy, the refractive index was calculated.

3. Grid generation

The two-dimensional transformation between virtual space and physical space can be obtained by grid generation. In an orthogonal grid all off-diagonal components of the metric tensor $\boldsymbol{g}$ are equal to zero:

$$g_{ij} = \bar{a}_i \cdot \bar{a}_j = \frac{\partial x}{\partial \xi}\frac{\partial x}{\partial \eta} + \frac{\partial y}{\partial \xi}\frac{\partial y}{\partial \eta} = 0, \qquad \text{with} \qquad i \neq j$$
$$\sqrt{det\boldsymbol{g}} = \frac{\partial x}{\partial \xi}\frac{\partial x}{\partial \eta} + \frac{\partial y}{\partial \xi}\frac{\partial y}{\partial \eta} = \sqrt{g_{11}g_{22}} = h_\xi h_\eta, \tag{5}$$

where g_{11}, g_{22} are diagonal elements of metric tensor $\boldsymbol{g}$; h_ξ, h_η is Lamé coefficients.

A 2-D orthogonal grid must satisfy the Beltrami equations [8]:

$$\begin{cases} f\dfrac{\partial x}{\partial \xi} = \dfrac{\partial y}{\partial \eta} \\ f\dfrac{\partial y}{\partial \xi} = -\dfrac{\partial x}{\partial \eta}. \end{cases} \tag{6}$$

Here, the distortion function f has the following form:

$$f = \frac{h_\eta}{h_\xi} = \sqrt{\frac{\left(\frac{\partial x}{\partial \eta}\right)^2 + \left(\frac{\partial y}{\partial \eta}\right)^2}{\left(\frac{\partial x}{\partial \xi}\right)^2 + \left(\frac{\partial y}{\partial \xi}\right)^2}}. \tag{7}$$

If f is the constant equal to M then the grid is quasi-conformal. In the case $M = 1$ (conformal mapping) the Beltrami equations (5) becomes the Cauchy-Riemann equations.

The discretization of typical grid cell was calculated by definition of the unknown point at the center of the control cell, as shown in Fig. 2.

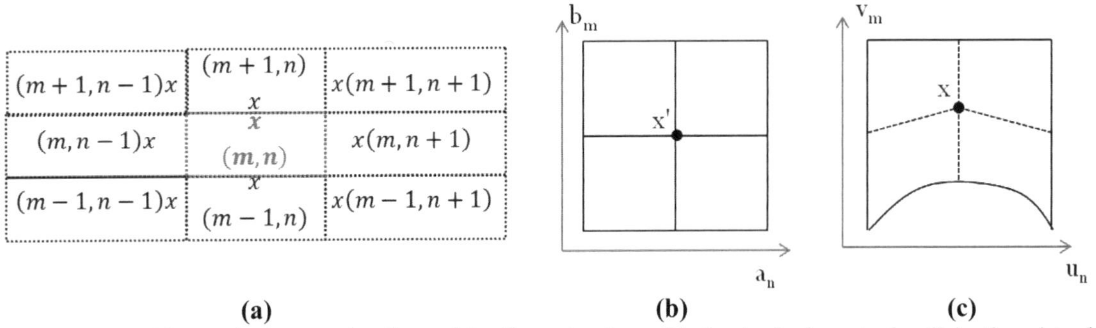

Figure 2. (a) The typical control cell used in discretization; **(b)** the typical control cell in the virtual space; **(c)** the typical control cell in the physical space.

An optimal (small enough value of anisotropy and not large enough value of refractive index) map is generated by minimizing the following functional:

$$\Phi = \iint_D \left[f\left(u_\xi^2 + v_\xi^2\right) + \frac{1}{f}\left(u_\eta^2 + v_\eta^2\right) \right] d\xi \, d\eta, \tag{8}$$

where $u(\xi,\eta)$, $v(\xi,\eta)$ describe the mapping. The difference approximation method of such variation problem [9] was used in this paper.

Two different domains in the physical space with nonlinear distortion of the low boundary and the linear one were considered. The transformed orthogonal grid in the physical space for two cases was obtained. The color maps demonstrate permittivity profile ε (Fig. 3).

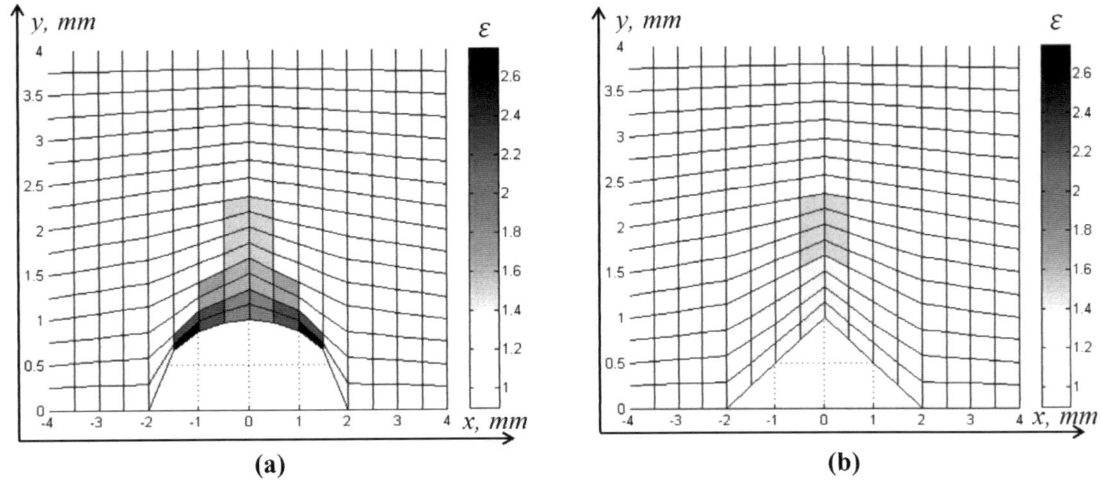

Figure 3. (a) Physical space with nonlinear distortion. **(b)** Physical space with linear distortion.

4. Numerical simulation

In this part the numerical simulation of effectiveness of the designed cloak was presented. For the simulation of wave propagation Finite Element Method (FEM) was used for solving Maxwell's equations. The cloak with sizes of 8x4 mm^2 was investigated at the frequency of 0.1 THz. The nonmagnetic cloak was considered ($\mu = 1$). The permittivity of the cloak varies from 0.99 to 2.75. The E_z-field distribution is presented in Fig. 4.

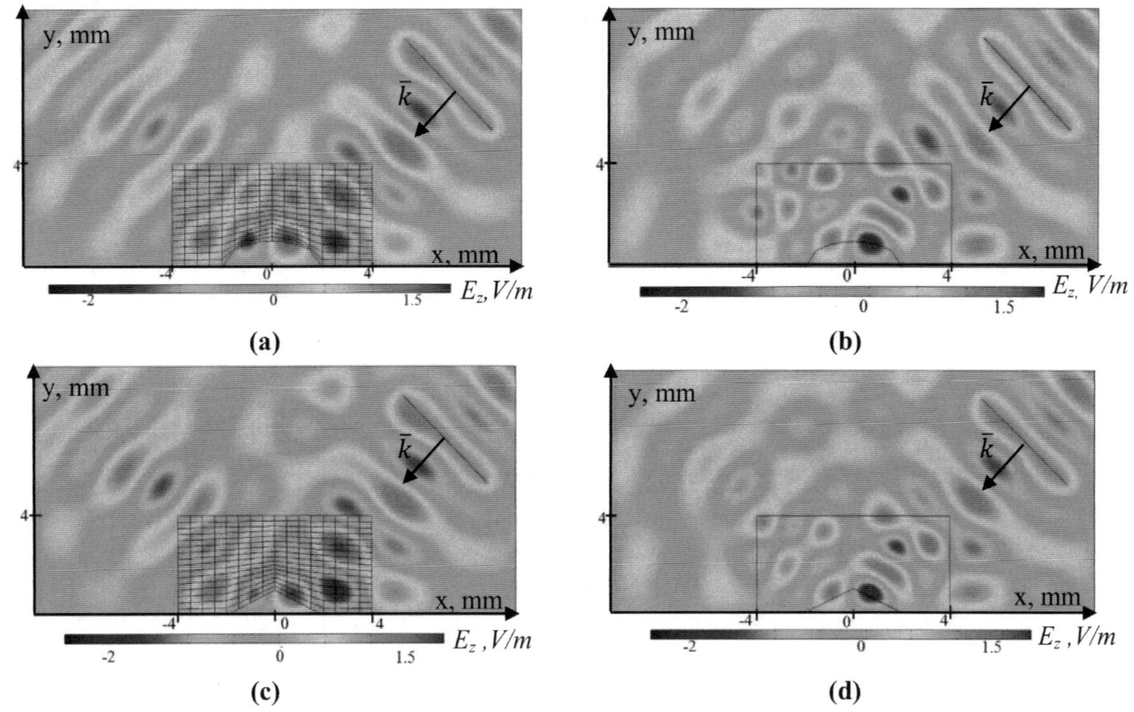

Figure 4. (a) The rectangular cloak with nonlinear boundary around the object; **(b)** the rectangular medium (ε=2.25) with the nonlinear boundary around the object; **(c)** the rectangular cloak with the linear boundary around the object; **(d)** the rectangular medium (ε=2.25) with nonlinear boundary around the object.

5. Conclusions

The simulation of carpet cloaking device in the terahertz frequency range was proposed. The effective constitutive parameters of the physical system which corresponds to wave propagation in the virtual space were calculated. The special transformation of grid was calculated by the quasi-conformal mapping using MATLAB software. Finite element method was applied for 2D electromagnetic wave simulation. The distribution of electromagnetic wave, which propagates through this cloak, was obtained for two types of lower boundaries with the different curvature (linear and nonlinear). The nonideal cloaking effect was caused by the coarse grid.

6. Acknowledgments

This work was financially supported by Government of Russian Federation, Grant 074-U01.

7. References

[1] U. Leonhardt, T. Philbin 2010 *Geometry and light: the science of invisibility* (Mineola , New-York: Dover Publication, Inc.)

[2] E. A. Gurvitz, A. V. Vozianova, M. K. Khodzitsky 2014 Simulation of beam-splitter made of

metamaterilas with angle distribution of constitutive parameters based on the transformation optics for terahertz frequency range *Journal of Physics: Conference Series* **541** 012067 (1-5)

[3] P. Zhang, M. Lobet, S. L. He 2010 Carpet cloaking on a dielectric half-space *Opt. Expr.* **18** 18158–18163

[4] H. F. Ma et al. 2009 Compact-sized and broadband carpet cloak and free-space cloak *Opt. Expr.* **12** 19947–19959

[5] J. Li, J. B. Pendry 2008 Hiding under the Carpet: A New Strategy for Cloaking *Phys. Rev. Lett.* **101** 203901-203904

[6] M. Silverinha, B. Edwards, A. Alu 2009 Experimental verification of plasmonic cloaking at microwave frequencies with metamaterials *Phys. Rev. Lett.* **103** 153901

[7] F. Zhou et al. 2011 Hiding a Realistic Object Using a Broadband Terahertz Invisibility Cloak *Scientific Reports* **1** 78

[8] J. F. Thompson, B. K. Soni, N. P. , 1999 *Weatherill. Handbook of Grid Generation* (Boca Raton: CRC Press).

[9] S. K. Godunov, G. P. Prokopov. About calculation conformal mappings (in russian) 1967 *Computational Math. and Math. Phys.* **7 (5)** 1031–1059

SPbOPEN2015 IOP Publishing
Journal of Physics: Conference Series **643** (2015) 012053 doi:10.1088/1742-6596/643/1/012053

Expressions for reflection and transmission coefficients for one-dimensional photonic quasicrystals

M M Voronov

Ioffe Physical-Technical Institute, Politekhnicheskaya 26, St Petersburg, 194021 Russia

E-mail: mikle.voronov@coherent.ioffe.ru

Abstract. We give general expressions for the light reflection and transmission coefficients for one-dimensional (1D) photonic quasicrystals in the framework of the so-called two-wave approximation. A special attention is paid to quasicrystals composed of dielectric layers, where the reflection and transmission coefficients as a function of the length of the structure take a simple form. The expression for the diffraction vectors of a 1D quasicrystal with an arbitrary number of different constituent segments is discussed as well.

Introduction

Quasicrystals (QCs) belong to a wide class of aperiodic systems possessing long-range order and allowing coherent Bragg diffraction of electron or electromagnetic waves [1,2]. In the case of light waves such structures are called resonant or active photonic quasicrystals if the constituting materials produce dipole-like excitations; such systems are, for instance, multiple-quantum-well structures (MQWs). The theory of the propagation of electromagnetic waves in MQWs in the framework of the two-wave approximation was developed in work [3] and then generalized on the case of dielectric contrast [4]. It should be noted that in distinction from 1D periodic systems for which one can obtain analytically exact solutions, for 1D non-periodic systems it generally appears to be impossible and one has to resort to some approximations, in particular to that stated here.

In this work we theoretically study the propagation of electromagnetic waves in 1D quasi-periodic media and deterministic aperiodic structures (DAS) [3,4], in particular, in quasicrystals. Firstly, we deduce a general expression for diffraction vectors of an arbitrary 1D quasicrystal. Secondly, we consider the two-wave approximation in a general form to calculate the light reflection and transmission coefficients for a one-dimensional DAS with non-zero value of the structure factor. Thirdly, we present the expressions for the reflection and transmission coefficients in the case of an all-dielectric photonic quasicrystal; the expression for the Fourier components ε_G of the dielectric function of a DAS expressed by the structure factor is also given.

1. The structure factor and diffraction vectors of 1D quasicrystals

We use the standard definition of the structure factor [2,3]:

$$f(q) = \lim_{N\to\infty} \frac{1}{N} \sum_{m=1}^{N} e^{2iqz_m} = \sum_G f_G \delta_{2q,G} \,. \tag{1}$$

Here q is the light wave vector, the coordinates z_m give the positions of the scattering elements, which in our case are interfaces between the dielectric layers, and symbol $\delta_{2q,G}$ is the Kronecker symbol for arbitrary, not necessarily integer-valued quantities $2q$ and G. It is assumed that the structure factor of the semi-infinite system can be different from 0, i.e. $f_G \neq 0$ for any G. The vectors G are diffraction vectors of a quasicrystal (or any other DAS), which can be found from a given sequence z_m determined by DAS under consideration. For example, for 1D quasicrystals constructed from two segments (A and B) the vectors G and the corresponding structure factors f_G are given by two integer indices, h and h', i.e. $G = G(h, h')$. The analogous definition of the structure factor (equation (1)) is applied to a sublattice. It is worth mentioning that vectors G of the sublattices of a quasicrystal are equal, i.e. $G_a(h, h') = G_b(h, h') = G(h, h')$. A calculation of the structure factor f_G of a quasicrystal is a separate problem and is made in [2,5].

Let us consider a one-dimensional disordered quasicrystal, with the interfaces between the segments placed along z-axis in points $z'_m = z_m + \delta z_m$, where δz_m are independent and randomly distributed variables so that $\overline{\delta z_m} = 0$, $\overline{\delta z_m^2} = \sigma^2$. The structure factor averaged over the disorder realization has a form

$$f(q) = \sum_{h,h',h''\ldots} \delta_{2q,G(h,h',h''\ldots)} f_{hh'h''\ldots} \exp(-(q\sigma)^2/2) \,.$$

The long-range correlations in the positions of the layers are still present in such a structure, and consequently the Bragg diffraction is determined by the same diffraction vectors as in the non-disordered quasicrystalline lattice. Evidently, the theory presented in the work is suitable for such correlated disordered structures as well.

We shall give a simple but not rigorous derivation of the expression for the diffraction vectors of an unlimited 1D quasicrystal composed of n types of different segments (i.e. of various constituting layers). Let us denote by a_i the length of i-type segment and by N_i the number of such segments in the structure. Thus, the full length of the quasicrystal is $L = \sum_{i=1}^{n} N_i a_i$. Using a more general definition of the structure factor $f_N(q) = N^{-1} \sum_{j=1}^{N} e^{iqx_j}$, where $N = \sum_{i=1}^{n} N_i$ is the full number of segments of the quasicrystal, and by dividing the whole quasicrystal into self-similar parts, in the limit $N \to \infty$ one can obtain that $GL = 2\pi m$, $m \in Z$. Then, by taking into account the linear independence of the solutions, one can present $m = \sum_{i=1}^{n} M_i h_i$, where $h_j \in Z$ are integers determining the diffraction vectors $G \equiv G(h_1, h_2, h_3, \ldots)$ and M_i are also integers, but not arbitrary. From the equality $G \sum_{i=1}^{n} N_i a_i = 2\pi \sum_{i=1}^{n} M_i h_i$ and independence of the sought solution for G on values N_i ($N_i \to \infty$), one can get $M_i = N_i$ ($i = 1, 2, \ldots n$). Using the notion of the average period of a quasicrystal, which is $\bar{d} = L/N$ (when $L \to \infty$, $N \to \infty$), and changing indices so that $h = h_1$, $h' = h_2 - h_1$, $h'' = h_2 - h_1$ and so on, we eventually come to the following expression for the diffraction vectors

$$G(h, h', h'', h''' \ldots) = \frac{2\pi}{\bar{d}} \left(h + \frac{h'}{t_1} + \frac{h''}{t_2} + \frac{h'''}{t_3} + \ldots \right) ,$$

where the numbers $h, h', h'', h''', \ldots \in Z$, and values of t_k are defined by the relations $t_k = 1 + \lim_{N_i\to\infty} \sum_{i=1, i\neq k}^{i=n} N_i/N_k$. In particular, the diffraction vectors of a two-segment quasicrystal are defined by expression $G(h, h') = (2\pi/\bar{d})(h + h'/t)$ [2], where $t = 1 + \lim_{N\to\infty} N_1/N_2$. The simplest and well-known example is the so-called Fibonacci quasicrystal, which can be constructed by the help of the substitution rule $A \to B$, $B \to AB$, thus, starting with a sequence $ABAABABA\ldots$. For this quasicrystal the value of t is the golden ratio, i.e. $t = (\sqrt{5} + 1)/2$.

2. Pseudo-band gaps in aperiodic systems

We consider a propagation (along z-axis) of electromagnetic waves in a 1D structure for which the modulation function $F(z)$, describing variation of the medium parameters (e.g. the dielectric permittivity) along z-direction, can be decomposed into the Fourier series $F(\omega, z) = \sum_G F_G e^{iGz}$. Besides periodic structures, which obviously satisfy this property, there are a number of deterministic aperiodic structures generating coherent Bragg diffraction, in particular, quasicrystals and incommensurate modulated crystals [1]. The frequency functions F_G depend on the system under consideration and require a separate calculation. As long as the function $F(z)$ is almost-periodic [6] (or periodic when considering a periodic structure) the electromagnetic field $E_K(z)$ with the wave vector K can be sought as a linear superposition $\sum_G E_{K-G} e^{i(K-G)z}$, and the wave equation is written as

$$-d^2 E_K(z)/dz^2 = F(\omega, z) E_K(z) \tag{2}$$

Hereafter we consider the situations where in RHS of equation (2) only the components with vectors $G = 0$ and $\pm G$ (where $G \equiv |G|$) are kept and one can take into account only two wave vectors, K and $K - G$, at the condition $|K - G/2| \ll G/2$. The last is valid if $|F_G| \ll |F_0|$, which allows neglecting the rest of diffraction vectors G. Thus, $E_K(z) \approx E_K e^{iKz} + E_{K-G} e^{i(K-G)z}$, where the wave vector magnitude K can be presented as $K = G/2 \pm Q$. Substituting the last expression for $E_K(z)$ in equation (2) and neglecting the other components except E_K and E_{K-G} we obtain that $E_{K-G} = E_K(K^2 - F_0)/F_G$ and $Q = \sqrt{(G/2)^2 + F_0 \pm \sqrt{G^2 F_0 + F_G F_{-G}}}$, where the positive sign must be ignored since $Q \ll G/2$ and the real part of F_0 is positive, consequently

$$Q = \sqrt{(G/2)^2 + F_0 - \sqrt{G^2 F_0 + F_G F_{-G}}}\,. \tag{3}$$

By neglecting the absorption in the system one can get $F_0 = F_0^*$ and $F_G = F_{-G}^*$, and if the radicand (under the big root sign) is negative we find ourselves in the region of a pseudo-band gap. The latter strongly resembles the typical band gaps in periodic structures: a high reflectivity in the pseudo-band gap region is due to the constructive interference of the light waves competing with the effect of nonperiodicity of the structure. The edges of the pseudo-band gap satisfy the condition $Q = 0$ and can easily be found from equation (3): $(G/2)^2 - F_0 \pm |F_G| = 0$. Similar to 1D quasi-periodic media, two-dimensional quasicrystals made from materials with a small dielectric contrast can also demonstrate wide pseudo-band gaps [7]; however, in this case the derivation of the explicit analytical expressions is complicated.

The above two-wave approximation can be justified by the help of the perturbation theory which allows one to evaluate the convergence of an infinite sum over all diffraction vectors except G and to determine pseudo-band gap edges placed around a Bragg frequency. Primarily, this approach was applied to the so-called 1D resonant photonic crystals for which the Bragg frequency coincides with the resonant one [3].

In the simplest cases, the Bragg condition for a DAS is defined in such a way that it would lead to the only pseudo-band gap in the energy spectrum or to two pseudo-band gaps of almost the same width. The first case is realized in 1D photonic quasicrystals with non-frequency-dependent dielectric constants of the constituent materials. Since while decreasing the dielectric contrast the value F_G converts to 0 (see section 3), the polariton wave vector becomes $Q = |G/2 - \sqrt{F_0}|$ and at a frequency $\bar{\omega}$ the Bragg condition $G/2 = \sqrt{F_0(\bar{\omega})}$ is fulfilled. The second case corresponds to a situation where the segments of a photonic quasicrystal are characterized by some frequency function symmetric about a special frequency, e.g., the exciton resonance frequency for MQWs [3]. One can show that for MQWs the calculation of the corresponding Fourier-components leads to the following expressions:

$$F_0 = \left(\frac{\omega}{c}\right)^2 \bar{\varepsilon} - \frac{2ir}{1+r}\frac{\omega\sqrt{\varepsilon_a}}{c\,\bar{d}}\,, \quad F_G = \left[\left(\frac{\omega}{c}\right)^2 (\varepsilon_a - \varepsilon_b) - \frac{2ir}{1+r}\frac{\omega\sqrt{\varepsilon_a}}{c\,a}\right] \frac{f_{-G}}{\pi G} \sin\frac{Ga}{2}\,. \tag{4}$$

Here we have used the following notations: ε_a and ε_b are the background dielectric constants of the quantum well and the barrier, respectively, a is the well width, $\bar{d}$ is the mean period defined as the average distance between quantum wells, and r is the amplitude reflection coefficient of light for a single quantum well (reduced to the plane in the centre of the well), calculated for the case $\varepsilon_a = \varepsilon_b$. In addition, equation (4) contains the structure factor f_G which for quasi-crystalline sequences is equal to $2\pi f_{hh'}/\bar{d}$, and $f_{-G} = f_G^*$. The optical spectra and pseudo-band gaps for light propagation in MQWs were studied in [3,4,5]. In this case the resonant Bragg condition is $\omega_0 n/c = G/2$ (where $n \approx \sqrt{\varepsilon_b}$) and in the energy spectrum there are two pseudo-band gaps which in the reflection spectra manifest themselves as two "shelves" of almost the same width being located at about equal distances from the frequency ω_0.

Now we will find the general expressions for the amplitude reflection and transmission coefficients (at normal light incidence) for a structure of the length L with an arbitrary quasi-crystalline sequence of the constituent layers of different types. For this purpose we define the field $E(z)$ as

$$E_0 e^{iqz} + E_r e^{-iqz} \quad (z < 0)\,; \qquad E_t e^{iq(z-L)} \quad (z > L)$$

$$E_{K_1} e^{iK_1 z} + E_{K_1 - G} e^{i(K_1 - G)z} + E_{K_2} e^{iK_2 z} + E_{K_2 - G} e^{i(K_2 - G)z} \quad (0 < z < L)$$

where $K_1 = G/2 - Q$, $K_2 = G/2 + Q$, and $q = (\omega/c)\sqrt{\varepsilon_{out}}$ is the wave vector magnitude of the plane light wave incident from a uniform medium surrounding the quasicrystal. By using the boundary conditions for the field $E(z)$ and its derivative $dE(z)/dz$ at the interfaces $z = 0$ and $z = L$ and introducing the notations $\xi_{1,2} = (K_{1,2}^2 - F_0)/F_G$,

$$A(K_{1,2}) = q - K_{1,2} + (q + K_{2,1})\xi_{1,2}$$

$$B(K_{1,2}) = q + K_{1,2} + (q - K_{2,1})\xi_{1,2}$$

$$C(K_{1,2}) = K_{1,2} - q - (q + K_{2,1})\xi_{1,2} e^{-iGL}$$

the reflection coefficient $r_L = E_r/E_0$ can be reduced to the form

$$r_L = \frac{A(K_2)C(K_1) - A(K_1)C(K_2)e^{2iQL}}{B(K_2)C(K_1) - B(K_1)C(K_2)e^{2iQL}} \tag{5}$$

and the transmission coefficient $t_L = E_t/E_0$ to

$$t_L = 2qe^{iK_2 L} \frac{C(K_1)(1 + \xi_2 e^{-iGL}) - C(K_2)(1 + \xi_1 e^{-iGL})}{B(K_2)C(K_1) - B(K_1)C(K_2)e^{2iQL}}\,. \tag{6}$$

The analysis of expressions (5) and (6) under condition $|F_G| \to 0$ shows that if $q^2 = F_0$, then $r_L = 0$ and $t_L = e^{iqL}$, therefore the dielectric constant of the surrounding medium must be equal to $\varepsilon_{out} = F_0(c/\omega)^2$. This equality can be strictly satisfied for any frequency only for purely dielectric photonic quasicrystals because in this case $F_0 = (\omega/c)^2\bar{\varepsilon}$ (see section 3), nevertheless it can be approximately used in other cases also. In fact, the dielectric constant ε_{out} of the surrounding medium should be set to $\bar{\varepsilon} = L^{-} \int_0^L \varepsilon(\tilde{\omega}, z)dz$, where $\tilde{\omega}$ must satisfy the condition $|F_G(\tilde{\omega})| \ll 1$. We pay attention that the above expressions for r_L and t_L can be exploited only in the frequency ranges for which the criterion for the validity of the two-wave approximation is satisfied, and for each individual system these expressions should be checked numerically. In the case where the value of the dielectric constant of the medium surrounding a quasicrystal is arbitrary the reflection and transmission coefficients, instead of expressions (5) and (6), can be found by using the transfer-matrix technique, see [4].

In the pseudo-band gap region, in the limit as $L \to \infty$, $|r_\infty| = 1$ and from equation (5) one can obtain the dispersion relation between the frequency ω and imaginary part Q'' of the polariton wave vector Q, which is just equation (3). Let us introduce the following notations: $\eta = \sqrt{G^2F_0 + |F_G|^2}$, $X = 2GQ''F''_G + 2|F_G|^2 + G^2F'_G - 2\eta F'_G$, where F'_G and F''_G denote the real and imaginary parts of F_G, respectively, and $Q'' = \sqrt{\eta - (G/2)^2 - F_0}$. Then the reflection coefficient of light from a semi-infinite structure can be written as $r_\infty = e^{i\phi} = Re(r_\infty) + iIm(r_\infty)$ (where ϕ is the reflection coefficient phase), and its real and imaginary parts after some transformations may be reduced to the following forms:

$$Re(r_\infty) = \frac{(q^2 + F_0)X + 2|F_G|^2(F'_G - \eta)}{(q^2 - F_0)X + 2|F_G|^2(\eta - F'_G)}, \quad Im(r_\infty) = \frac{2q[2Q''(F'_G\eta - |F_G|^2) + GF''_G(2F_0 - \eta)]}{(q^2 - F_0)X + 2|F_G|^2(\eta - F'_G)} \tag{7}$$

These expressions can be used to calculate the phase change ϕ on reflection from the interface between the structure and the surrounding medium and also to state the conditions under which the phase of the light wave changes by a given value ϕ. The latter has a direct relation to different types of 1D topological insulators, e.g., 1D dielectric topological insulator [8]. Since for such systems $F_0 = q^2$, the frequency ω at which $\phi = \pi$, as is seen from the first equation (7), can be found from the condition $X = 0$, which can be simplified to the form $2GQ'' \sin\psi + 2|\varepsilon_G|(\omega/c)^2 + (G^2 - 2\eta)\cos\psi = 0$, where ψ is the phase of $\varepsilon_G = |\varepsilon_G|e^{i\psi}$.

3. The dielectric photonic quasicrystals

If a 1D photonic quasicrystal consists of layers characterized with non-frequency dependent dielectric susceptibilities (and magnetic permeability $\mu = 1$), one can find the Fourier-components $F_0 = (\omega/c)^2\varepsilon_0$ and $F_G = (\omega/c)^2\varepsilon_G$, where ε_0 and ε_G are Fourier-components of the dielectric profile function $\varepsilon(z) = \sum_G \varepsilon_G e^{iGz}$. The function $\varepsilon(z)$ can be represented in terms of the unit step functions, from where one can obtain the expression for ε_G which for a two-segment quasicrystal is

$$\varepsilon_G = \varepsilon_a f^{(a)}_{-h,-h'} \frac{(1 - e^{-iGa})}{iG\bar{d}_a} + \varepsilon_b f^{(b)}_{-h,-h'} \frac{(1 - e^{-iGb})}{iG\bar{d}_b}, \tag{8}$$

where $\bar{d}_a = \lim_{N\to\infty}(N_a a + N_b b)/N_a$ and $\bar{d}_b = \lim_{N\to\infty}(N_a a + N_b b)/N_b$ are the mean periods of sublattices of the initial quasi-crystalline lattice, and $f^{(a)}_{hh'}$ and $f^{(b)}_{hh'}$ are the structure factors related to the corresponding sublattices. The generalization of equation (8) on an arbitrary number of segments is trivial.

It is useful to mention the second simplest case of a dielectric photonic QC, when a deterministic aperiodic structure consists of dielectric layers A with the width a incorporated into a dielectric medium with the dielectric constant ε_b in such a way that their centres, lying along z-axis, form a sequence given by the expression $z_m = z_0 + m\bar{d} + \Delta\{m/t + \varphi\}$. Here z_0 is an arbitrary constant, Δ, t, φ are the parameters of the structure, $\bar{d}$ its average period, and $\{x\}$ denotes the fractional part of a number. In this case the Fourier components F_0 and F_G can be found from equation (4) by setting $r \to 0$.

The edges of a pseudo-band gap calculated on the basis of the two-wave approximation are found from equation (3) under condition $Q = 0$: $\omega_\pm = (Gc/2)/\sqrt{\varepsilon_0 \pm |\varepsilon_G|}$. The value of the wave vector of the light propagating in the surrounding medium, q, must be chosen equal to $\omega\sqrt{\bar{\varepsilon}}/c$ because it leads to zero reflection coefficient when $F_G \to 0$; $\varepsilon_0 = \bar{\varepsilon}$ is the average dielectric constant and L is the length of a quasicrystal. After some transformations and by dividing the numerators and denominators of the expressions for r_L (equation (5)) and t_L (equation (6)) by $Z(\omega) = F_G e^{iGL/2} + (Q^2 - (q + G/2)^2)e^{-iGL/2}$ the amplitude reflection coefficient of light can be reduced to the following form

$$r_L = \frac{(F_G + Q^2 - (q + G/2)^2)(Q^2 - (q - G/2)^2)}{(F_G + Q^2 - (q - G/2)^2)((G/2)^2 - q^2 - Q^2 - 2iqQ\cot QL)} \tag{9}$$

and the amplitude transmission coefficient to

$$t_L = \frac{-2iqQ[F_G e^{iGL/2} + (Q^2 - (q - G/2)^2)e^{-iGL/2}]}{(F_G + Q^2 - (q - G/2)^2)((G/2)^2 - q^2 - Q^2 - 2iqQ\cot QL)\sin QL}. \tag{10}$$

A further simplification of the above expressions for r_L and t_L leads to the following answers:

$$r_L = \frac{F_G(qG - \sqrt{q^2G^2 + |F_G|^2}) + |F_G|^2}{(F_G + qG - \sqrt{q^2G^2 + |F_G|^2})(\sqrt{q^2G^2 + |F_G|^2} - 2q^2 - 2iqQ\cot QL)}, \tag{11}$$

$$t_L = \frac{-2iqQ[F_G e^{iGL/2} + (qG - \sqrt{q^2G^2 + |F_G|^2})e^{-iGL/2}]}{(F_G + qG - \sqrt{q^2G^2 + |F_G|^2})(\sqrt{q^2G^2 + |F_G|^2} - 2q^2 - 2iqQ\cot QL)\sin QL}. \tag{12}$$

The equations (11) and (12) can be used when $|\varepsilon_G| \ll \bar{\varepsilon}$, in particular, for low-contrast photonic quasicrystals. Since these expressions contain explicitly the length L of a structure, it allows one to determine the length at which the formation of pseudo-band gaps occurs in different 1D deterministic aperiodic structures.

A typical reflection spectrum $|r_L(\omega)|^2$ for a quasicrystal calculated by equation (11) in the vicinity of frequency $\bar{\omega} = Gc/2\sqrt{\bar{\varepsilon}}$ is almost symmetrical about $\bar{\omega}$ and resembles in much the reflection spectrum for an analogous periodic structure. Since the equation $Z(\omega) = 0$ determines some of the eigenfrequencies of the system and the function $Z(\omega)$ is reduced in expressions for r_L and t_L (equation (5) and (6)), one can conclude that the set of eigenfrequencies and the reflection and transmission spectra for 1D purely dielectric photonic quasicrystals are much poorer than those for photonic quasicrystals composed of constituent layers with a frequency-dependent dielectric function.

We note that in the pseudo-band gap region in the limit as $L \to +\infty$ the product $Q\cot QL$ gives Q'', where $Q'' = \sqrt{\eta - (G/2)^2 - q^2}$, and $|\sin QL| \to +\infty$ leading to zero transmission coefficient ($t_\infty = 0$). Another possible approximation is $qG - \sqrt{q^2G^2 + |F_G|^2} \approx -|F_G|^2/(2qG)$, which is valid for very low-contrast structures. With these simplifications, equations (11) and (12) for the optical coefficients r_L and t_L can be considerably simplified, especially in the cases of the pseudo-band gap edges (at frequencies $\omega_\pm$) and the gap "centre" (at frequency $\bar{\omega}$).

In conclusion, we have given the general expression for the diffraction vectors of 1D quasicrystals and generalized the theory of light propagation in quasicrystals (and in other deterministic aperiodic structures) developed in works [3,4]. An analysis of the obtained expressions allows one to optimize the parameters of different 1D photonic quasicrystals to get pseudo-band gaps in the frequency range of interest. Also the developed theory can be generalized to the case of inclined light incidence or the propagation of acoustic waves in phononic quasicrystals, and as a particular case it can give answers for the corresponding periodic systems.

Acknowledgments
The author is grateful to M.M. Glazov and E.L. Ivchenko for useful discussions.

References

[1] Fujiwara T and Ishii Y 2008 *Quasicrystals* (Amsterdam: Elsevier)
[2] Valsakumar M C and Kumar V, 1986 *Pramana* **26** 215
[3] Poddubny A N, Pilozzi L, Voronov M M and Ivchenko E L 2009 *Phys. Rev. B* **80** 115314
[4] Poddubny A N and Ivchenko E L 2010 *Physica E* **42** 1871
[5] Werchner M, Schafer M, Kira M, Koch S W, Sweet J, Olitzky J D, Hendrickson J, Richards B C, Khitrova G, Gibbs H M, Poddubny A N, Ivchenko E L, Voronov M and Wegener M 2009 *Opt. Express* **17** 6813
[6] Corduneanu C 1989 *Almost-periodic functions* (New York: Chelsea)
[7] Rechtsman M C, Jeong H-C, Chaikin P M, Torquato S and Steinhardt P J 2008 *Phys. Rev. Lett.* **101** 073902
[8] Poshakinskiy A V, Poddubny A N, Hafezi M 2015 Preprint at http://lanl.arxiv.org/abs/1501.07425

Simplified approach for modeling ZnO luminescence dependence on excitation level with and without surface plasmon resonance

A P Tarasov[1,2], Ch M Briskina[1], M V Ryzhkov[1], S I Rumyantsev[1], V M Markushev[1]

[1]Kotel'nikov Institute of Radio Engineering and Electronics of Russian Academy of Sciences, Moscow 125009, Russian Federation
[2]Moscow Institute of Physics and Technology (State University), Dolgoprudny, Moscow Region 141700, Russian Federation

E-mail: tarasov.ap@phystech.edu

Abstract. The research of excitation level dependence of ZnO UV and visible luminescence was carried out. To describe the behavior of radiation intensity there was proposed the model based on the set of rate equations. The simultaneous equations include the set of parameters which characterizes processes participating in luminescence. The values of parameters were estimated using experimental dependence of luminescence intensity on excitation level. Proposed model can give the way for optimization of samples structure to maximize the effect of surface plasmon resonance influence on luminescence intensity.

1. Introduction

Research of the SPR influence on the recombination luminescence intensity of *ZnO* films covered by *Ag* island films was carried out. Different excitation sources were used. In the case of He-Cd laser luminescence enhancement (compared to the same *ZnO* sample without *Ag* cover) was observed. However, while using the Nd:YAG laser for the excitation of the same samples the luminescence quenching was observed, and enhancement could be achieved only with significant decreasing of excitation level. It was shown that the dependence of luminescence intensity on pumping level in presence of *Ag* cover is significantly nonlinear, while the dependence without *Ag* cover is, as a rule, linear (see for example [1]). This nonlinearity leads to the following effect: while the excitation increases, the luminescence enhancement decreases and finally turns into quenching.

It is suggested simplified approach for description of *ZnO* luminescence and the SPR influence on it. This approach bases on the set of rate equations (SRE), which describes processes taking part in the luminescence. As is known, luminescence of *ZnO* consists of near-UV part caused by exciton recombination and visible part caused by defects which form energy levels in the band-gap. Proposed approach gives an opportunity for modeling the dependence of radiation intensity on excitation level and to estimate the values of parameters in SRE.

The analyze of luminescence of various *ZnO* films with *Ag* cover using SRE allows to estimate variation of different characteristics, for example, losses, closeness of luminescence maximum wavelength to SPR and others. In order to use suggested SRE it is necessary to consider that all processes participants have corpuscular nature.

Using SRE we describe time dependence of concentrations for: electrons in conduction band (n_1), holes in valence band (n_3), excitons (n_4), photons caused by exciton recombination and forming plasmons (n_2), photons which simplistically describe field produced by plasmons in nanoscale region between *Ag* cover and *ZnO* (n_5) and phonons caused by heating (n_6).

It should be noted that a part of this investigation has already been reported by authors in [2]. But in that article phonons caused by heating of the *ZnO/Ag* system were not taken into account. In course of investigation it was discovered that maximum of spectra shifts to long wavelength region with the increase of pumping energy. Most likely, that the shift is caused by band-gap narrowing with temperature growth. The value of the shift equals approximately 4-5 nm (~40 meV). Taking into account that band-gap decreases by ~ 0.35 meV with 1 K heating [3] in the case of T > 200 K it is possible to estimate that *ZnO/Ag* system temperature rises up to ~380 K. This results in necessity of addition of the equation for phonons in SRE.

2. UV radiation

The set of rate equations for recombination (UV) radiation is following:

$$
\begin{cases}
\frac{dn_1}{dt} = Pe^{\frac{-(t-10)^2}{26,4}} - (M+B)n_1n_3 \\
\frac{dn_2}{dt} = Hn_4 + A(1-L_An_6)n_5n_4 - Cn_2 - Sn_2 \\
\frac{dn_3}{dt} = Pe^{\frac{-(t-10)^2}{26,4}} - (M+B)n_1n_3 \\
\frac{dn_4}{dt} = Mn_1n_3 - Rn_4 - A(1-L_An_6)n_5n_4 \\
\frac{dn_5}{dt} = N_0(1-L_Nn_6)Sn_2 - N_0(1-L_Nn_6)Sn_2W(1+L_Wn_6)n_2 - Tn_5 \\
\frac{dn_6}{dt} = \alpha Pe^{\frac{-(t-10)^2}{26,4}} + Vn_2 + Sn_2W(1+L_Wn_6)n_2 - En_6
\end{cases}
$$

Here t refers to time in nanoseconds, n_i refers to concentration in 10^{18}cm^{-3}, P refers to excitation level, multiplier by P is time dependence of excitation pulse that corresponds to experiment. Rough estimation shows that to the energy density of excitation pulse 1 mJ/cm^2 corresponds to $P = 3.3$.

Let us fix on the system parameters in more detail. In equations for n_1 and n_3 parameter B refers to interband recombination probability, parameter M refers to exciton formation probability. Those two equations coincide and difference of results are caused only by different initial conditions.

In the equation for excitons (n_4) parameter R is probability of exciton spontaneous recombination, parameter A refers to probability of exciton recombination under the influence of resonance field formed by plasmons, parameter L_A refers to the influence of system heating on parameter A.

In the equation for photons produced by exciton recombination (n_2) parameter H is radiation exciton recombination probability ($H < R$), parameters C and S are probabilities of photons escape from the system and interaction of photons with *Ag* cover with plasmons formation respectively.

In the equation for photons produced by plasmon field (n_5) parameter *N0* is a coefficient that refers to plasmons field increasing against the field of photons that form plasmons, parameter L_N refers to the influence of system heating on parameter *N0*, parameter T is the characteristics of plasmons radiation decay, parameter W is the characteristics of Joule's losses in *Ag* cover that are determined by *Ag* electrical resistivity, parameter L_W refers to temperature dependence of *Ag* electrical resistivity.

In the equation for phonons (n_6) parameter α refers to relation of the part of photoexcitation converted into heat to the part of photoexitation that was used to *ZnO* excitation, parameter E is the probabilty of phonon attenuation, parameter V is the probability of direct conversion of photon into phonon.

Consider the term of equation that determines Joule's losses in more details. Photons caused by exciton recombination get inside *Ag* cover. The rate of their input equals $\gamma = Sn_2^0$ (per unit time). They produce electron vibrations (plasmons). Those vibrations actually appear to be an electrical current. The second power of electrical current is determined by the concentration of input photons – $I^2 \sim n_2$. Losses with power of $I^2R \sim n_2$ arise in *Ag* cover. Dividing this value by the energy of photon ($h\nu$) it is possible to estimate photon lifetime before its conversion into heat: $1/\tau = I^2R/h\nu \sim n_2$. To estimate the amount of photons which will be converted into heat after they form plasmons let us assume $d(Sn_2)/dt = -(Sn_2)/\tau$. From this relation it follows: $Sn_2 = Sn_2^0 exp(-t/\tau) \approx Sn_2^0(1 - t/\tau)$ if $t/\tau << 1$ which appears to be correct. So, Joule's losses equal $Sn_2^0Wn_2^0$ and significant nonlinearity appears in the equation. Hence, the field in the narrow region at the interface between *ZnO* and *Ag*, simplistically described by photons n_5, will be formed by remained plasmons. The rate of input of such plasmons is proportional to $Sn_2 \approx Sn_2^0(1-Wn_2^0)$. As is known, electric field strength in the interface is much higher than field strength inside the metal [4] in the equation for n_5 term which describes plasmon input rate should be multiplied by $N0 > 1$. As expression $-Sn_2Wn_2$ describes heat losses it appears in the equation for phonons (n_6) with positive sign.

The set of rate equations was solved using MathCAD by Runge-Kutta numerical method. The result of calculation is the sum of photon concentrations$I_2 = \Sigma_t n_2^t$ at all (50000) points of time, in which the calculation was performed. This value is proportional to the integrated luminescence intensity. Calculation time was defined by the duration of photo excitation laser pulse.

For the uncovered part of the film, parameters which correspond to exciton luminescence lifetime ~100 ps and efficiency ~5% were taken. Those values are rather close to experimental ones. The values equal: $R = C = 10$, $H = 0{,}5$. We also set B and M to 20.

Consider estimations of parameters that are taking into account the influence of heating (phonons) on the process in consideration. As was mentioned in the Introduction the heating of the system *ZnO/Ag* in the experiment was ~100 degrees. By virtue of the fact that the temperature dependence of the resistance is $R = R_0[1+\beta(T-T_0)]$, where for *Ag* $\beta = 0.0037$ the change of resistance must be around 40%. That is why the value of L_W equals 0,3. Parameters L_A and L_N should take into account widening of the *Ag* extinction spectrum and, as a result, resonance degradation. Because the width of this spectrum is big even at low temperature, heating does not play significant role in this process. Solutions of the set of rate equations using values of L_A and L_N equal 0,3 and 1 correspondingly gave the same result for I_2 so for those parameters the value 0,3 was taken. For parameters α, E and V more or less arbitrary values were taken ($\alpha = 10$, $E = 1000$, $V = 1$). Preliminary calculations have shown that the variation of those parameters values (staying in constant order of magnitude) does not affect the result significantly.

For the comparison of experimental dependencies of luminescence intensity on excitation level *I(P)* with the calculated ones those dependences should be plotted in the same scales. To get the same scale on *X* axis we should take into account that, as it was mentioned above, energy density of excitation pulse of 1 mJ/cm^2 corresponds to $P = 3.3$. To make the same scale along *Y* axis we should add one more simplification: we should assume that calculated linear dependence *I(P)* obtained using parameters for uncovered part of the film coincides with experimental dependence.

Suggested approach was applied to 2 types of *ZnO* films with different manufacturing methods. The first film was prepared by thermal evaporation; the second film was prepared by magnetron sputtering. On the first film there were 3 regions: uncovered *ZnO* and *ZnO* covered with 10 nm and 20 nm *Ag* film (effective thickness was determined using the suggestion of uniform *Ag* distribution on semi-sphere). On the second film there were two regions: uncovered *ZnO* and *ZnO* covered with ~10 nm *Ag* film.

The comparison between experimental and calculated dependences *I(P)* for the first *ZnO* film is shown on Fig.1. Calculated curves for 10 nm and 20 nm *Ag* approximate experimental data rather well. Those calculated curves differs one from the other only by the value of parameter *W* (determines losses in *Ag* cover) and *S* (probability of interaction of photons with *Ag* cover with plasmon formation). All other parameters coincide for both curves. It turns out that with the increase of *Ag* film thickness losses and probability of interaction of photons with *Ag* cover with plasmon formation become higher.

a b

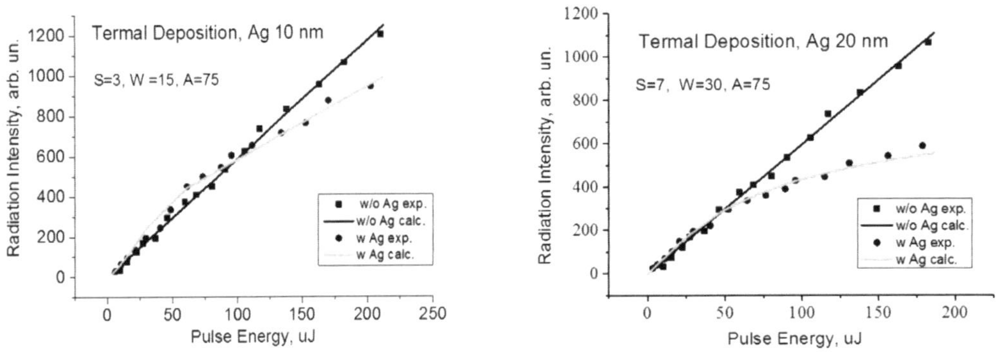

Fig. 1. Comparison of experimental and calculated dependences of luminescence intensity on excitation level for the first *ZnO/Ag* sample (thermal deposition): (a) *Ag* thickness is 10 nm; (b) *Ag* thickness is 20 nm

On Fig.2 the dependence *I(P)* for the second film is shown. It can be seen that in this case significantly greater enhancement of luminescence intensity is possible. In comparison with the first film here noticeable greater value of parameter *A* that characterize probability of exciton recombination under the influence of resonance filed of plasmons and lower value of losses (*W*) was obtained.

So, such approach to analysis of dependence of luminescence intensity on excitation level gives a possibility to optimize samples structure for the highest possible luminescence intensity enhancement due to SPR.

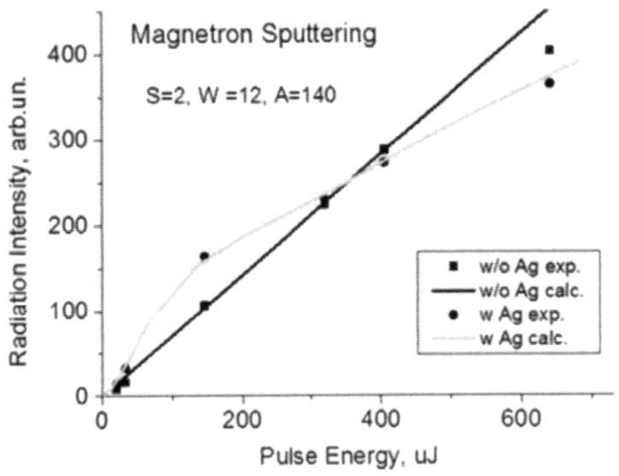

Fig. 2. Comparison of experimental and calculated dependences of luminescence intensity on excitation level for the second *ZnO/Ag* sample (magnetron sputtering)

3. Visible ZnO radiation

The suggested approach can be used for zinc oxide visible radiation consideration along with UV radiation. For this purpose the set of rate equation is supplemented with the rate equations for the population of energy levels in band-gap, caused by defects, for photons of visible radiation, caused by transitions between levels mentioned above and/or valence band, and (in the presence of SPR) for photons of visible radiation which we use for description of field produced by plasmons in nanoscale region between *Ag* cover and *ZnO*. As a result, in the case of single defect level in *ZnO* band-gap, SRE consists of 9 equations. In the case of existence of donor-acceptor pairs (DAP) in *ZnO*, in other words in case of two energy levels in *ZnO* band gap SRE consists of 10 equations.

In [1] it was experimentally demonstrated that for radiation of *ZnO* the dependence of visible radiation intensity on excitation level with high accuracy could be approximated by following relation: $I(P) \sim P^{1/3}$.

Mathematical modelling of the dependence of visible radiation intensity on excitation level using suggested SRE revealed that the dependence like $I(P) \sim P^{1/3}$ can be obtained only for the case of DAP existence i.e. two energy levels existence in band-gap (Fig. 3b). If in band gap only one defect level exists the dependence $I(P)$ looks like $I(P) \sim P^k$, where $1/3 < k < 1$ (Fig. 3a).

a

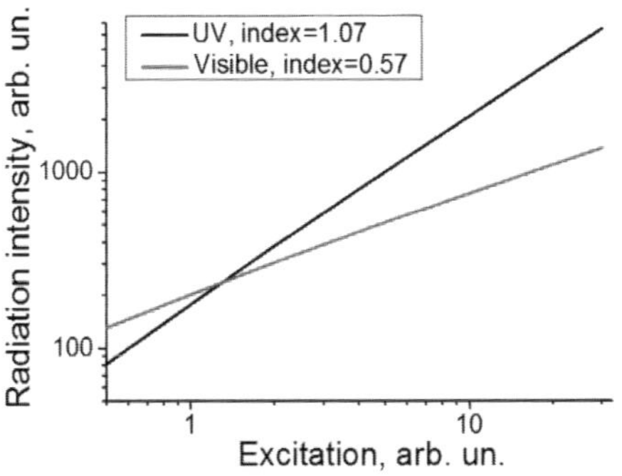

b

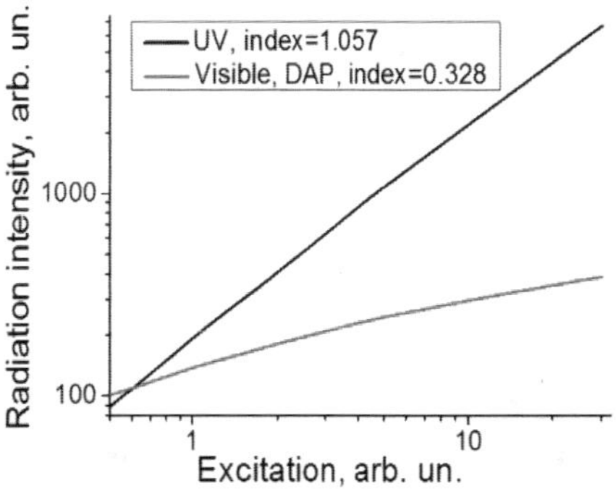

Fig. 3. Calculated dependences of luminescence intensity on excitation level for: (a) one donor energy level in band-gap; (b) two energy levels in band-gap (DAP)

So, utilization of suggested simplified approach to mathematical modeling in combination with experimental dependences $I(P)$ can give additional information about the origin of *ZnO* visible radiation.

4. Conclusion

For the analysis of the influence of SPR on recombination radiation intensity of *ZnO* films and origin of visible luminescence of such films simplified approach was suggested. This approach is based on the system of rate equations (SRE) which gives the possibility to calculate the dependence of radiation intensity on excitation level. SRE includes the set of parameters that describe processes determining luminescence.

Comparison of experimental and calculated dependences of UV radiation intensity on excitation level allows estimating parameters that characterize *ZnO* films with or without *Ag* cover. Those parameters could give a possibility to optimize sample characteristics, for example, for higher intensity and enhancement. In particular, it was shown that while thickness of *Ag* cover increases (from 10 to 20 nm), losses also increase.

It was found out that it is possible to make a conclusion about visible radiation origin on the base of the analysis of the dependence of radiation intensity on excitation level. Namely, if visible radiation is caused by existence of DAP in *ZnO* the dependence will look like $I(P) \sim P^{1/3}$; in other cases index of power will be bigger than 1/3 (but less than 1). In conclusion, it is worth noting that the proposed approach can be applied not only to *ZnO* but to other materials also.

References

[1] Studenkin S A and Cociverab M 2002 *J. Appl. Phys.* **91** 5060
[2] Ryzhkov M V, Markushev V M, Briskina Ch M, Rumyantsev S I, Tarasov A P and Lyaskovskii V L 2014 *J. Nanoelectron. Optoelectron.* **9** 1
[3] Wang L and Giles N C 2003*J. Appl. Phys.* **94** 973
[4] Maier S A 2007 Plasmonics: Fundamental and Applications. Springer Busness and Science Media LLC

384x288 readout integrated circuit for MWIR and LWIR HgCdTe based FPA

A.V. Zverev[1], Yu.S. Makarov[2], E.A. Mikhantiev[3], I.V. Sabinina[4], G.Yu. Sidorov[5], S.A. Dvoretskiy[6]
[1] Research scientist, Rzhanov Institute of Semiconductor Physics SB RAS, Russia
[2] PhD Student, Rzhanov Institute of Semiconductor Physics SB RAS, Russia
[3] Engineer, Rzhanov Institute of Semiconductor Physics SB RAS, Russia
[4] Senior research scientist, Rzhanov Institute of Semiconductor Physics SB RAS, Russia
[5] Research scientist, Rzhanov Institute of Semiconductor Physics SB RAS, Russia
[6] Laboratory head, Rzhanov Institute of Semiconductor Physics SB RAS, Russia

E-mail: makarov.zgr@gmail.com

Abstract. The review of architecture and characteristics of 384x288 silicon readout integrated circuit with 25 μm pixel pitch for MWIR and LWIR HgCdTe based FPA is presented.

1. Introduction

The unique physical properties of the mercury-cadmium-telluride (MCT, HgCdTe) solid solution allow to realize infrared (IR) detectors with high sensitivity in the mid-wavelength infrared (MWIR) band (3 – 5 μm) and long-wavelength infrared (LWIR) band (8 – 12 μm). The most attractive method for fabrication multi element photovoltaic focal plane array (FPA) is fabricating the MCT diode in pixel of readout integrated circuits (ROIC). Some attempt to fabricate so-called monolithic IR detectors on the basis of MCT growth onto ROIC was carried out [1-5]. But it is clear that the development of monolithic detectors on the basis of HgCdTe can not used as a basic technology for development and fabricating multi element up to megapixel format FPA. Today all developers are used hybridization photo diode sensitive array and ROIC by «flip-chip» through indium bumps. Figure 1 shows the scheme of FPA fabrication which obtained by «flip-chip» hybridization. Usually MWIR and LWIR HgCdTe photodiode detectors are operated at liquid nitrogen temperature [6]. It is necessary to provide high detector sensitivity for registration low radiation fluxes in MWIR and LWIR spectral ranges in presence high background ones.

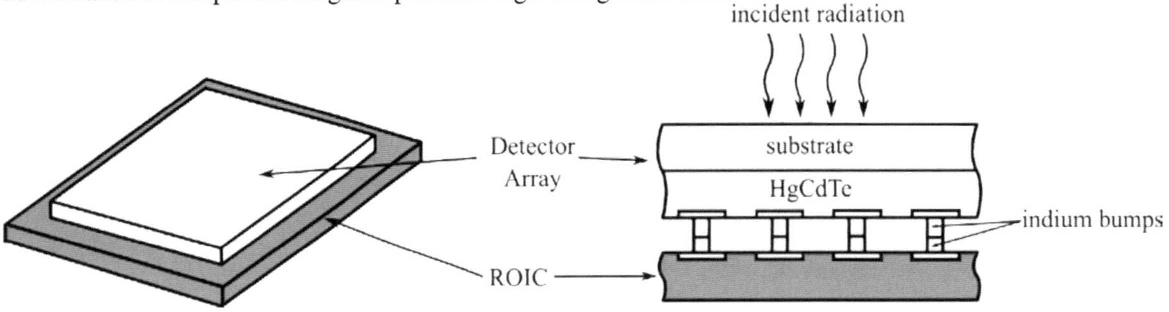

Figure 1. The scheme of hybrid HgCdTe FPA

The main purpose of a ROIC is a registration of photocurrents that are generated by FPA photodiodes and then a serial output of the accumulated information for imaging. The modern IR ROICs are fabricated by CMOS technology, which gradually replaced the CCD technology [7]. Operation at liquid nitrogen temperature requires as low as possible power dissipation of ROIC because of limitation of micro cryogenic cooler in refrigerating capacity. The modern ROIC should support the following features: integration then read (ITR) mode, integration while read (IWR) mode and window modes. Also, important requirements is frame rate, which is low in most serial available ROIC. The work objective was design modern high-speed low power ROIC with minimal number of control and reference signal pads.

The infrared flux of useful information is depending on the wavelength range and from the surrounding background. Usually it is necessary to detect small useful signal information from thermal object in MWIR and LWIR bands at high background flux intensity. Direct injection pixel cell architecture is suitable for the MWIR and LWIR bands. The electron capacity is the main characteristic of such cell. The increase of electron capacity leads to increase of image integration time and dynamic range [6]. The large dynamic range gives better temperature resolution. A temperature resolution is minimum temperature difference between an object that can be detected by focal plane array (FPA). Another important parameter that determines the FPA speed is frame rate. Frame rate is limited by information output time and thus depends on a number of pixels, i.e. matrix format, a number of signal outputs and a maximum frequency of the output information from the ROIC. High frame rate allows to provide external signal integration, that improves the temperature resolution of FPA. In order to improve these characteristics in modern FPA ROIC the following techniques are used:

- Subtract of DC component of the photocurrent. This method allows to increase the dynamic range, especially where the proportion of the useful signal is small relative to the total flow of the IR radiation, detected by photodiodes array. But it does not give better temperature resolution.
- The possibility of discrete switching of the charge capacity that allows to use the ROIC in conditions with different intensity of infrared radiation flux.
- Using the window mode. This method allows to significantly increase the temperature resolution by reading information from reduced area of photosensitive elements array (PSEA). Reduction of the readout area reduces output information time, thereby increasing the frame rate.
- Multiplexing of nearest pixels in order to increase charge capacity while reducing the spatial resolution. Spatial resolution determines the number of elements that build the image. This method reduces the number of readout pixels, reducing the output time information.
- Increasing the number of analog outputs. The greatest contribution to power consumption of ROIC make analog output amplifiers. Therefore increasing their numbers imposes additional requirements to the cooling cryogenic coolers. Also in this case, the number of wires crossing the border of the cryostat increases, which increases the heat gain.
- Increasing the frequency of the output information from ROIC. Increasing the frequency output of the analog signal is limited by difficulty of transmission without distortion.
- Using the built-in analog-to-digital conversion and digital outputs. It allows to increase the frequency of the output information, but requires the use of more expensive ROIC production process.
- Digital integration directly into the pixel unit cell. It can significantly increase the dynamic range and, as a consequence, the temperature resolution. This type ROIC requires a very small ROIC production process and consequently more expensive [8].

2. Key features

The 384x288 ROIC with 25 μm pixel pitch for MWIR and LWIR HgCdTe photodiodes FPA was developed and tested. The samples were made by XC06 production process of XFab foundry. Figure 2 shows developed ROIC die photo. The column bus partition architecture is used in ROIC. Charge from the unit cell is transferred to the column bus and input column followers. This method allows to increase the area under the charge capacity, but requires further amplification of the attenuated signal [9]. The signal from the column followers multiplier for preamplifiers with adjustable gain and then transmitted to the output buffer. The decoders which are controlled by digital block perform selection of rows and columns.

Figure 3 shows a schematic of the ROIC direct injection pixel cell. The ROIC operates in the snapshot mode. ROIC can operate in ITR mode and also in IWR mode. The PMOS transistors VTC1 and VTC2 are for accumulating charge during integration in both ITR and IWR modes. It is possible to disable the capacity of transistor VTC1 for reducing image integration time. It preserves the dynamic range where the average flux intensity is low enough, for example, when the ROIC is connected to MWIR FPA. PMOS transistor VTC3 used to read the accumulated charge in both ITR and IWR modes, and is also used for accumulating charge during integration in the ITR mode. In IWR mode charge accumulated at previous frame is transferred to capacity of transistor VTC3 and closed transistor VT6 prevents discharge of VTC3 capacity during the accumulation of photocurrent onto VTC1 and VTC2. NMOS transistor VT1 is used for Antiblooming and also can be used as an imitator of the current to test an array of pixel cells. It allows to evaluate performance of the ROIC before hybridization with a photodiode array. Pin PHDK of the cell is connected via indium bump to the cathode of the FPA photodiode with common anode. NMOS transistor VT3 is direct injection transistor which used to set the offset voltage of photodiode.

Inside the ROIC 12-bit digital-to-analog converters (DAC) are installed. DACs can be used to generate a voltage for gates of direct injections transistors, and also a voltages for gates and drains of Antiblooming transistors (VT1, Fig. 3). The use of DACs allows to reduce the number of wires crossing the border of the cryostat. Also the signals generated by the DACs directly in ROIC are more noise-immune. It is possible to disable the DACs and use the external contact pads for supplying this signals.

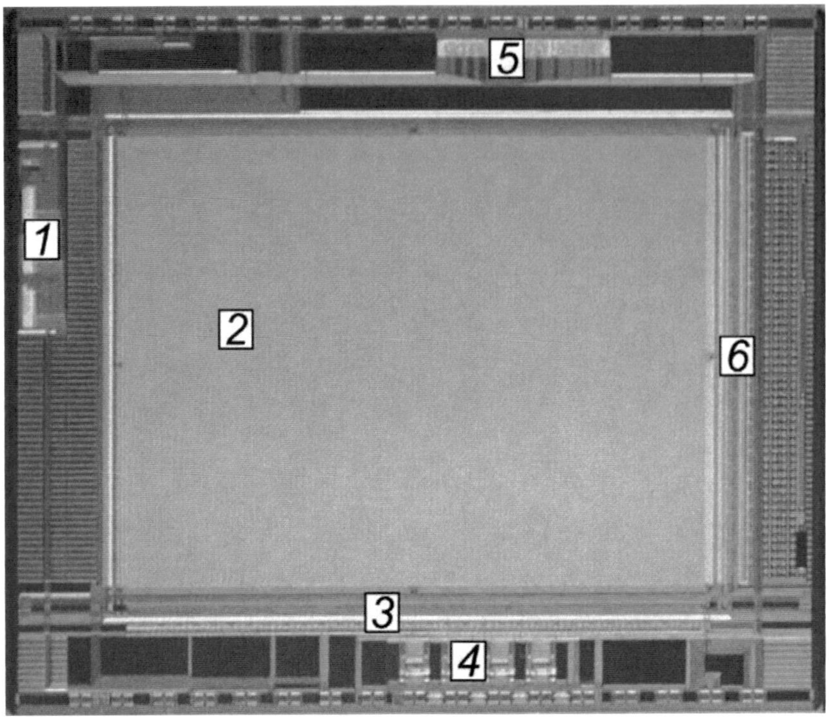

Figure 2. The ROIC die photo. 1 – DACs block, 2 – array of pixel cells, 3 – column followers and decoder, 4 – preamplifiers and output buffers, 5 – digital block, 6 – row decoder.

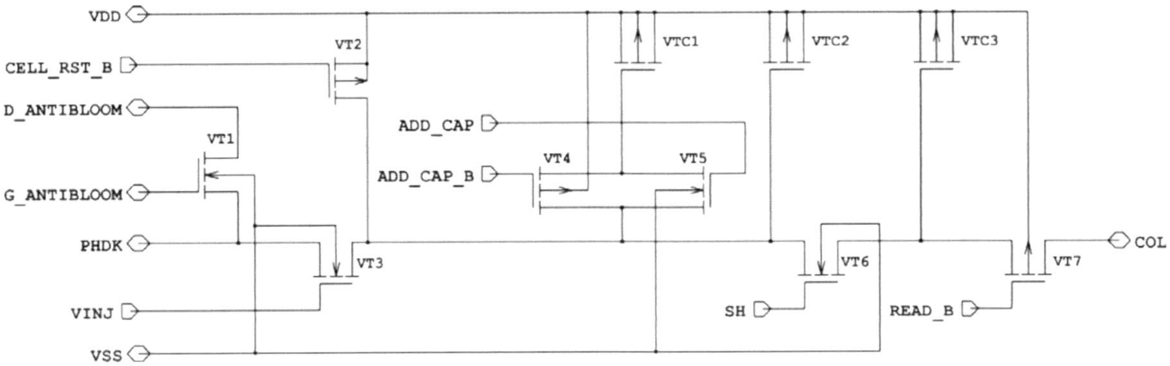

Figure 3. The schematic of the ROIC pixel cell. C_{VTC1}=0,6 pF, C_{VTC2}=0,2 pF, C_{VTC3}=0,2 pF.

The designed ROIC allows to readout accumulated video signal from the restricted area of PSEA, the so-called window mode. The minimum size of the window is equal to 4x4, which allows to detect moving target with an increased frame rate. In ROIC the serial interface is used. The operating modes of ROIC, the values build-in DACs, format and location of window, the number of signal outputs, preamplifier's gain, current consumption of the output buffers are determined by the configuration that loaded in internal registers via this serial interface. It allows to minimize the number of control pads of ROIC. The loading of registers is performed using the same input that is used to control the process of image integration. New values of registers can be loaded at the beginning of each frame. Thus, each frame can change the window size from which information is read. Thus the window size can be changed for each frame. Figure 4 shows ROIC timing diagrams of the in the ITR and IWR modes.

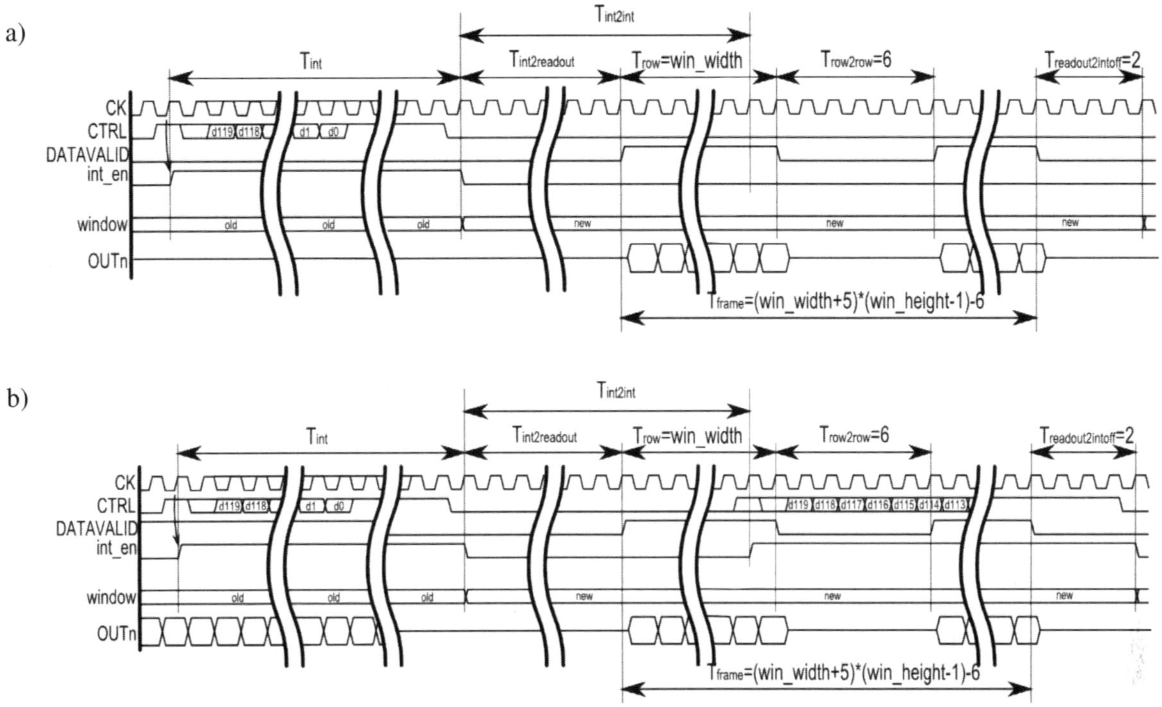

Figure 4. ROIC timing diagrams: a) ITR mode; b) IWR mode.

Figure 5 shows the thermal images obtained with LWIR FPA based on the developed ROIC. The noise equivalent temperature difference lower than 30 mK was obtained at integration time 600 μs and FOV 30 degrees.

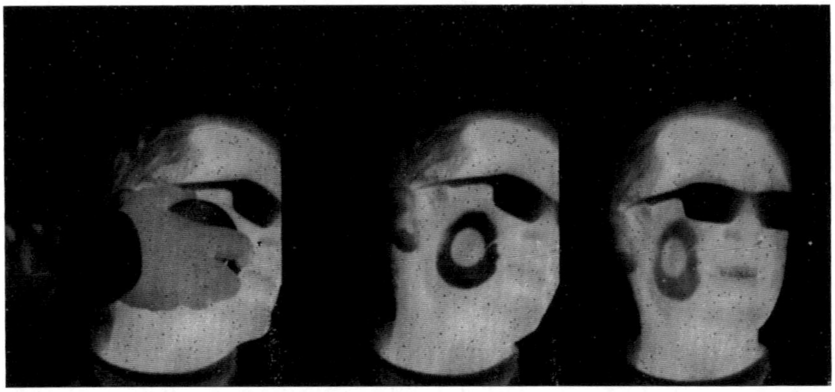

Figure 5. The thermal images HgCdTe LWIR FPA.

3. Conclusion

Table 1 shows some characteristics of the developed ROIC.

Table 1. The charateristics of designed 384x288 ROIC.

Parametrer	Value
Design rules	0,6 μm
Pixel pitch	25 μm
Detector type	N on P
Operating voltage	5 V
Pixel charge capacity	> 21 Me⁻
Output voltage swing	3,4 V (ITR) 2,6 V (IWR)

Number of outputs	1, 2, 4
Max. pixel rate per output	20 MHz
Max. full window frame rate	700 frames/sec
Max. power dissipation	< 100 mW
Operating temperature	from 70 to 300 K

Summing up, we list the basic functional features of the ROIC:

- Provides built-in digital-to-analog converters (DAC), which can be used to set the offset voltage of photodiodes and to control Antiblooming and build-in testing system of the pixel cells.
- Provides a serial interface for controlling the functionality of the ROIC.
- Provides reception and accumulation of electric charge from each FPA photodiode simultaneously for a predetermined time by direct injection technique.
- Enables reading signaling information simultaneously on four video outputs, also using only one or two video outputs.
- Operates in the ITR or IWR modes.
- Allow selection of rectangular submatrix of any size and location which will be readout video information.
- Provides the ability to adjust the bias circuit of the analog part of the ROIC.
- Allow selection the value of the charge capacity of the unit cell between two fixed values using the serial control interface.

4. References

[1] K. Zanio, R. Mattson 1992 *HgCdTe on Si fpr monolithic focal plane arrays* (Proc. SPIE, v. 1683) **179-190**

[2] J.W. Cairns, L. Buckle, G.J. Pryce, J.E. Hails, J. Giess, M.A. Crouch, D.J. Hall, A. Hydes, A.Graham, A.J. Wrigth, C.J. Hollier, D.J. Lees, N.T. Gordon, T. Ashley 2006 *Integrated infrared detectors and readout circuits* (Proc. SPIE, v. 6206, p. 620614-1–620614-9)

[3] G.B. Dalton, P.N. Dennis, D.J. Lees, D.J. Hall, J.W. Cairns, N.T. Gordon, J.E. Hails, J. Giess 2008 *Development of non-hybridised HgCdTe detectors for the next generation of astronomical instrumentation* (Proc. SPIE, v. 7021, p. 702101-702105)

[4] S. Velicu, T.S. Lee, R. Ashokan C.H. Grein, P. Boieriu, Y.P. Chen, J. Dinan, D. Lianos 2003 *Monolithically integrated HgCdTe focal plane arrays* (Proc. SPIE, v. 5209) **14-32**

[5] Maxim V. Yakushev, Sergei A. Dvoretsky, Alexander I. Kozlov, Irina V. Sabinina, Yuri G. Sidorov, Alexander V. Sorochkin, Boris I. Fomin, and Alexander L. Aseev 2010 *HgCdTe monolithic infrared detector* (Phys. Status Solidi C 7, No. 6) **1681-1683**

[6] Rogalski A 2011 *Infrared Detectors – 2nd ed.* **662, 744** 876

[7] Michel Zécri, Patrick Maillart, Eric Sanson, Gilbert Decaens, Xavier Lefoul, Laurent Baud 2008 *Advanced ROICs design for cooled IR detectors* (Proc. of SPIE Vol. 6940, 69402X)

[8] Kenneth I. Schultz, Michael W. Kelly, Justin J. Baker, Megan H. Blackwell, Matthew G. Brown, Curtis B. Colonero, Christopher L. David, Brian M. Tyrrell, and James R. Wey 2014 *Digital-Pixel Focal Plane Array Technology* (Lincoln Laboratory Journal v. 20, No. 2) **36-51**

[9] Mottin E., Pantigny P., Boch R. 1996 *An improved architecture of IRFPA readout circuits* (Proc. SPIE v. 3061) **118-124**

Dual-wavelength method and optoelectronic sensor for online monitoring of the efficiency of dialysis treatment

A.M. Vasilevsky[1], G.A. Konoplev[1], O.S. Stepanova[1], A.Yu. Zemchenkov[2], R.P. Gerasimchuk[2], A. Frorip[3]

[1] St. Petersburg Electrotechnical University, St. Petersburg, 197376 Russia
[2] Mariinsky Hospital, St. Petersburg, 194104 Russia
[3] LDIAMON AS, Tartu, 51014 Estonia

oksana_lopatenko@mail.ru

Abstract. The absorption spectra of effluent dialysate in the ultraviolet region were investigated. A novel dual-wavelength spectrophotometric method for uric acid determination in effluent dialysate and an optoelectronic sensor based on UV LED were developed. Clinical trials of the proposed sensor were carried out in the dialysis unit of St. Petersburg Mariinsky Hospital. The relative error of measurement for the concentration of uric acid does not exceed 10%.

1. Introduction

Hemodialysis (HD) is an extracorporeal method of blood purification by means of a dialysis machine for patients suffering from renal failure. During this life-supporting medical procedure low molecular weight waste products of metabolism and excessive water are removed from patient's blood circulating through a dialyzer to dialysis fluid (dialysate) via semi-permeable membranes. HD treatment is carried out 3 times a week; every session lasts 3-5 hours. The adequacy (efficiency) of HD treatment is usually estimated by taking blood samples before and after an HD session for laboratory analysis. The concentrations of uremic markers (urea, creatinine, uric acid etc.) in these sampled are measured and so-called dialysis indexes are calculated using special formulae. To make HD more safe and efficient it is necessary to control the concentrations of uremic markers in patients` blood or effluent dialysate online by means of special HD monitoring systems.

Three types of such devices have been developed and partly introduced to clinical practice: electrochemical sensors for urea monitoring, conductometric systems (ionic dialysance sensors) and optical spectral systems, which measure the spectral transmittance of effluent dialysate in the ultraviolet (UV) region at one or more wavelengths [1-4]. Spectral systems have some advantages over other devices including the possibility to measure the concentrations of several substances simultaneously without having to use any disposables or chemical reagents. In recent years, deep UV light-emitting diodes (LEDs) which are based on aluminum-gallium nitride AlGaN have been developed. These LEDs can emit optical radiation in the spectral range of 240 ... 350 nm, which allows creating compact and low-cost optoelectronic spectral sensors for HD monitoring instead of using large and expensive UV spectrophotometers.

2. Experiment

Previous experimental studies of the UV absorption spectra of spent dialysate have proved the validity of optical spectrophotometry based methods for online HD monitoring [2-3]. However, additional research work is required to make possible the quantitative control of uremic toxins using UV spectrophotometry.

The maximum of spectral absorption of urea lies at the wavelengths shorter than 200 nm (the spectral region that is not accessible by conventional equipment); the absorption maxima of uric acid (UA) are located at the wavelengths 238 and 293 nm, creatinine - 217 and 234 nm. Therefore UA and creatinine were chosen as marker substances for HD monitoring by optical spectral methods.

Comprehensive analysis of the profiles of the absorption spectra of the dialysate samples from the large group of patients (more than 300) revealed that using of the single wavelength of 293 nm for quantitative analysis of UA in dialysate (so-called single wavelength method) leads to significant errors, since at this wavelength not only UA make significant contribution to dialysate absorption but also other substances. Therefore it is necessary to introduce additional working wavelengths and to evaluate the contribution of other components.

Depending on the character of the UV absorption curve in the region of 260 ... 290 nm the spectra of spent dialysate can be classified into three types (Fig.1): type «A» – absorption spectra that have the maximum at 290 nm (~ 15% of total patients); type «B» - spectra that have "plateau" without distinct maxima (~ 70% of the total number of patients); type «C» - spectra with a slope in the mentioned range (~ 15% of total patients). It should be noted that the profile of dialysate absorption curve for each patient remained virtually unchanged throughout the whole period of the study (more than two years).

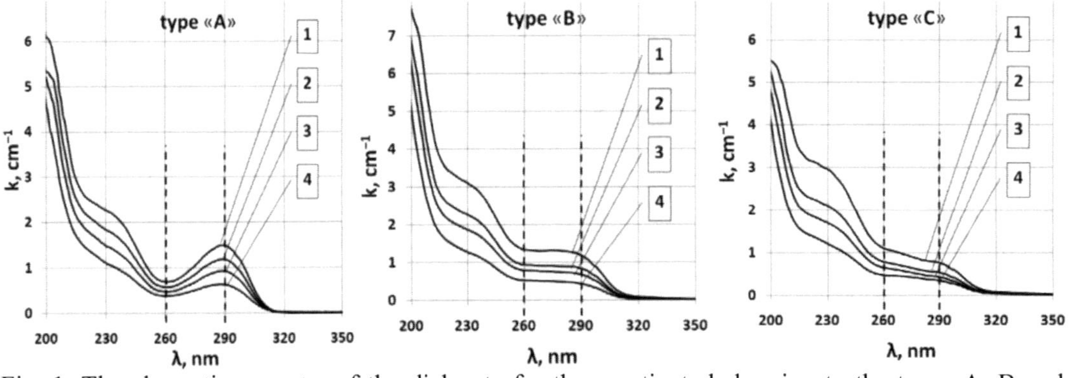

Fig. 1. The absorption spectra of the dialysate for three patients belonging to the types A, B and C, respectively: 1 - at the beginning of the procedure, 2 - after an hour from the start of the procedure, 3 - two and a half hours from the beginning of the procedure, 4 - at the end of the procedure.

Further researches showed that in the wavelength range of 260 ... 290 nm the spectral absorption of dialysate is dominated by UA and the set of identified and unidentified components, which include tryptophan, hippuric acid, pseudouridine, adenosine and other substances associated apparently with individual metabolic processes [5]. The individual concentration of every such component is low, but their combination, conventionally called pseudocomponent (pseudo-NK), determines the characteristic shapes of absorption curves. Spectral analysis in the wavelength range 260 ... 320 nm of dialysis fluid samples taken from patients assigned to different types, and simultaneous biochemical tests made it possible to find the profile of the absorption spectrum for the pseudo-NK which can be used for UA quantitative determination during HD online monitoring.

3. Results

The results of experimental studies and mathematical modeling of spent dialysate absorption spectra in the range of 260…350 nm showed:

- The spectral absorbance of dialysate in 285 ... 295 nm is mainly due to UA;

- At shorter wavelengths absorption is due to both UA and pseudo-NK. The region of 260 ... 270 nm is optimal for estimating of the absorption of pseudo-NK, where the absorption of UA is minimal.
- A novel dual-wavelength spectrophotometric method for UA determination in effluent dialysate and an optoelectronic sensor based on UV LED were developed [6].

The mathematical basis of the method is the law of Bouguer-Lambert for the two-component medium:

$$\begin{cases} k_{\lambda_1} = \varepsilon_{\lambda_1}^{UrAc} \cdot C^{UrAc} + \varepsilon_{\lambda_1}^{NK} \cdot C^{NK} \\ k_{\lambda_2} = \varepsilon_{\lambda_2}^{UrAc} \cdot C^{UrAc} + \varepsilon_{\lambda_2}^{NK} \cdot C^{NK} \end{cases} \quad (1)$$

where k_{λ_1} and k_{λ_2} – absorption coefficients at the respective wavelengths λ_1 and λ_2, $\varepsilon_{\lambda_1}^{UrAc}$ and $\varepsilon_{\lambda_2}^{UrAc}$ – molar spectral absorption coefficients of UA at the respective wavelengths λ_1 and λ_2, $\varepsilon_{\lambda_1}^{NK}$ and $\varepsilon_{\lambda_2}^{NK}$ – molar spectral absorption coefficients of pseudo-NK at the respective wavelengths λ_1 and λ_2.

In solving a system of equations (1) for the C^{UrAc} we find the concentration of UA by the values of the absorption coefficients at wavelengths λ_1 and λ_2:

$$C^{UrAc} = \frac{\varepsilon_{\lambda_2}^{NK} k_{\lambda_1} - \varepsilon_{\lambda_1}^{NK} k_{\lambda_2}}{\varepsilon_{\lambda_1}^{UrAc} \varepsilon_{\lambda_2}^{NK} - \varepsilon_{\lambda_1}^{NK} \varepsilon_{\lambda_2}^{UrAc}} \quad (2)$$

The proposed method is implemented as a compact optical sensor using UV LEDs at the wavelengths 262 and 287 nm, which was designed and manufactured by LDIAMON AS (Tartu, Estonia) in collaboration with Saint-Petersburg Electrotechnical University (Saint-Petersburg, Russia).

The sensor allows measuring the time profiles of UA concentration with extremely high temporal resolution (down to 5 sec).

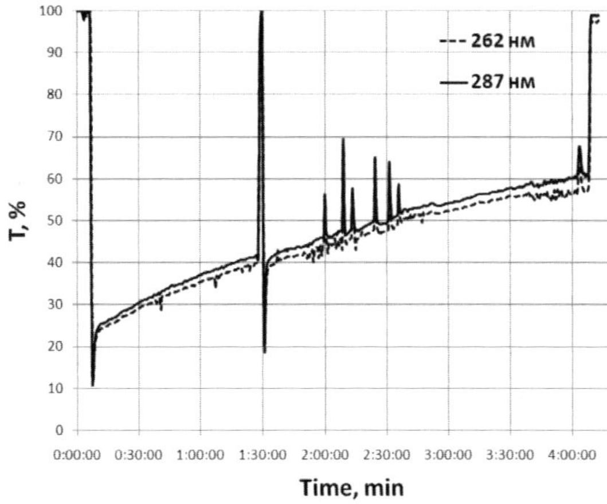

Fig. 2. The spectral transmittance of the dialysate for one of the patients in two narrow spectral ranges (262 nm and 287 nm)

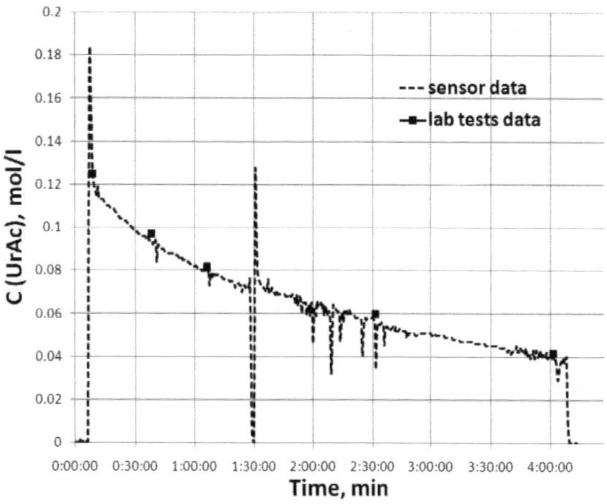

Fig. 3. The time profiles of UA concentration in effluent dialysate and results of the lab tests for UA

Clinical trials of the proposed sensor carried out in the dialysis unit of St. Petersburg Mariinsky Hospital. The time profiles of UA concentration were measured for the group of 10 patients with the help of the sensor. The shapes of the absorption spectra of the patients' dialysate belong to different types of classification proposed earlier. For validation the samples of effluent dialysate were taken at 0, 30, 60, 150 and 240 min of HD and lab tests for UA were carried out.

The time profiles of effluent dialysate spectral transmittance in two narrow spectral ranges (262 nm and 287 nm) for one of the patients are presented in Fig. 2. Fig. 3 shows the time profiles of UA concentration in effluent dialysate and results of the lab tests for UA. It should be noted that, the sensor allows early detecting of adverse events during HD (see Fig.3 at the middle of the HD session).

4. Conclusion

Clinical trials of the developed sensor were carried out in the dialysis unit of St. Petersburg Mariinsky Hospital. The results of spectral measurements are presented as time profiles of dialysate spectral transmission at two wavelengths and UA concentration in the course of a HD session. The relative error of measurement for the concentration does not exceed 10% (for more than 100 samples).

The correlation coefficient between the UA concentrations is 0.987, standard error of estimate is 4.4 umol/l. The sensor allows early detecting of adverse events during HD.

References

[1] Daugirdas J T, Tattersall J E, Kidney Int. 78, 833(2010)
[2] Fridolin I, Lauri K, Jerotskaja J, Luman M, Estonian Journal of Engineering. 14, 177 (2008)
[3] Vasilevsky A, Konoplev G, Journal of Biomedical Optics, 10, 044026 (2005).
[4] Umimoto K, Tatsumi Y, Jokei K, Nephrol. Dial. Transplant. 22, 127 (2007)
[5] Vanholder R C, De Smet R, Kidney Int. 63, 1932 (2003)
[6] Vasilevsky A, Konoplev G, Stepanova O, Rubinsky A, Gerasimchuk R, Frorip A, Nephrol. Dial. Transplant. 29, 461 (2014)

Photoluminescence of Gd_2O_3:Er - based materials for conversion of solar energy

Yu A Kuznetsova, A F Zatsepin
Ural Federal University named after the first President of Russia B.N.Yeltsin, Yekaterinburg, Russia

E-mail: kuznetsova.bess@mail.ru

Abstract. The spectral and luminescence properties of Er^{3+} activated Gd_2O_3 compacted powders have been investigated. There are several excitation channels of Er^{3+} luminescence: interband transitions, the energy transfer from Gd^{3+} ions and intracenter excitation. The quantum efficiency of the Er^{3+} ions luminescence under different excitation was calculated.

1. Introduction

Problems of modern energy, in particular, the growth of energy consumption and limited fossil resources, contribute to the development of alternative energy sources. Nowadays solar energy has broad prospects, therefore the majority of studies are focused on improving the efficiency of photovoltaic cells. It is known that the efficiency limit of most common photovoltaic cells based on single crystalline Si (c-Si) is estimated to be 32% (the Schockley–Queisser limit) due to mismatch between the solar spectrum and the Si absorption spectrum [1]. There are energy losses in Si solar cell such as thermalization (when photons with energy greater then Si band gap are absorbed and excess energy is lost as heat) and transmission (when photons with energy less then Si band gap are not absorbed). These losses can be reduced by using luminescent solar concentrators containing optically active centers [2]. The basic idea is better exploitation of the solar spectrum through the down-conversion (one high energy photon is converted into two or more lower-energy photons) and up-conversion (two or more low-energy photons combine to create one higher-energy photon) mechanisms. Luminescent solar concentrators represent the additional layers above (for down-conversion) or below (for up-conversion) solar cells (figure 1).

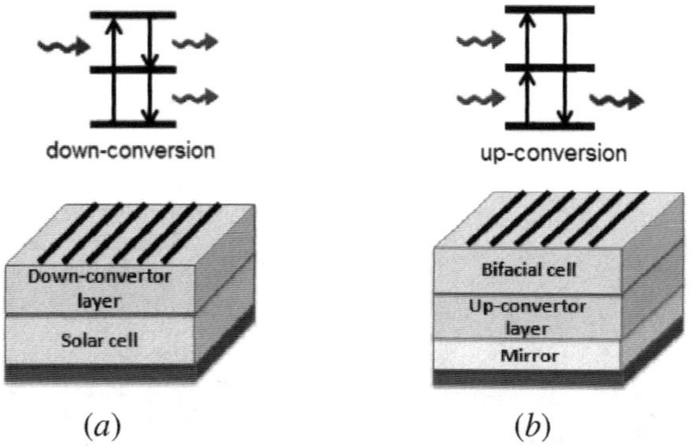

(*a*) (*b*)

Figure 1. Basic principle and application of down-converter (*a*) and up-converter (*b*).

Rare-earth ions are of interest as optical centers in luminescent energy converter materials due to specialty of their energy levels structure in the visible, ultraviolet and infrared regions of the spectrum. Rare earth oxides are used as the host materials because of their optical and chemical properties (broad band gap, low phonon energy) and the ability of being easily doped with rare earth ions in high concentrations. It is known that solar energy conversion efficiency is determined by the quantum efficiency of optically active ions photoluminescence [3]. In present work the quantum efficiency of the Er^{3+} ions luminescence in host lattice Gd_2O_3 under different excitation was determined.

2. Samples and Methods

Compacted powder samples of Gd_2O_3: Er^{3+} were prepared by sol-gel method at the Institute of Physics Polish Academy of Sciences, Warsaw, Poland. Crystal phase of Gd2O3 was identified as monoclinic with space group C2 / m (lattice parameters a = 14.09 A; b = 3,57 A; c = 8,76 A) by powder XRD pattern. The photoluminescence and excitation spectra and luminescence decay kinetics were obtained at room temperature using a fluorescence spectrometer PerkinElmer LS 55. The temperature luminescence intensity dependence was investigated using a spectrometer McPherson.

3. Results and discussion

3.1 Emission and excitation spectra

Upon excitation at 220 nm (interband transition in host lattice Gd_2O_3) the emission spectrum of Gd_2O_3: Er^{3+} demonstrates the typical Er^{3+} ions emission peaks at 522 and 544 nm originated from $^2H_{11/2}$ and $^4S_{3/2}$ to $^4I_{15/2}$ transitions respectively (figure 2) [4]. In addition, a lot of narrow lines with lower intensity are observed in the ultraviolet and visible regions of the emission spectrum. So the line photoluminescence spectrum is characteristic of TR^{3+} 4f-4f transitions that indicates uncontrolled TR^{3+} traces presence in Gd_2O_3: Er^{3+} powder. In particular, emission bands at 487 nm (Pr^{3+}), 579 and 588 nm (Tb^{3+}), 612 and 621 nm (Eu^{3+}) were identified according to [5].

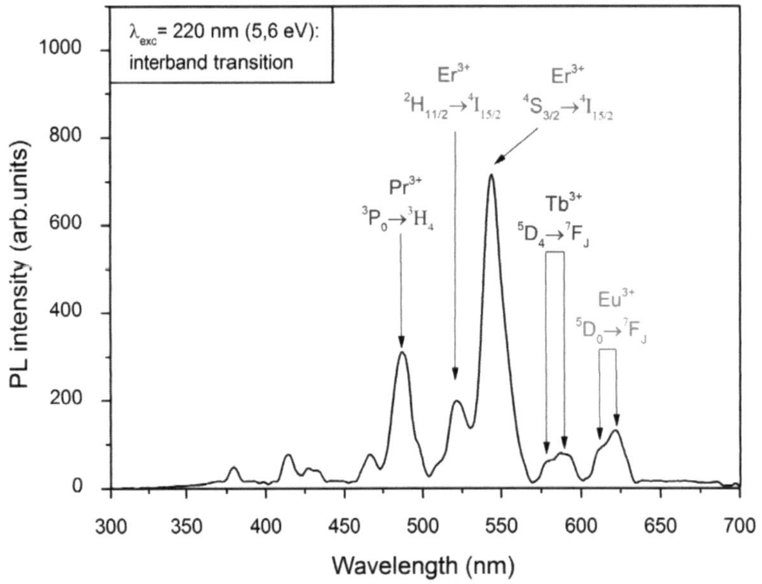

Figure 2. Emission spectrum of Gd_2O_3: Er^{3+} upon excitation at 220 nm. The arrows indicate lines corresponding to Er^{3+} transitions and Pr^{3+}, Tb^{3+}, Eu^{3+} uncontrolled traces transitions. Unmarked lines interpretation is not available presently.

Excitation spectra monitored at 544 nm emission of Er^{3+} at different wavelength ranges are shown in figure 3. The lines at 409, 380 и 367 nm are assigned to the $^4I_{15/2}\rightarrow{}^2H_{9/2}$, $^4G_{11/2}$, $^2G_{9/2}$ intra-4f-4f transitions of Er^{3+}, respectively [6]. Excitation spectrum in the short-wave region demonstrates lines at 220, 247, 277 и 312 nm corresponded to the interband transition and inta-4f-4f transitions of Gd^{3+} from the ground state $^8S_{7/2}$ to the excited states ^6D_J, ^6I_J and ^6P_J [4]. Excitation of Gd^{3+} ions through the mentioned above transitions in the region of Gd_2O_3 band gap indicate that such ions cannot be located at regular lattice positions and may be associated with intrinsic defects of host lattice Gd_2O_3.

Intracenter excitation of Er^{3+} luminescence is dominant but the presence of Gd^{3+} transition lines in the excitation spectrum points out Gd^{3+} to Er^{3+} energy-transfer. So there are several excitation channels of Er^{3+} luminescence: interband transitions, the energy transfer from imperfect Gd^{3+} ions and intracenter excitation.

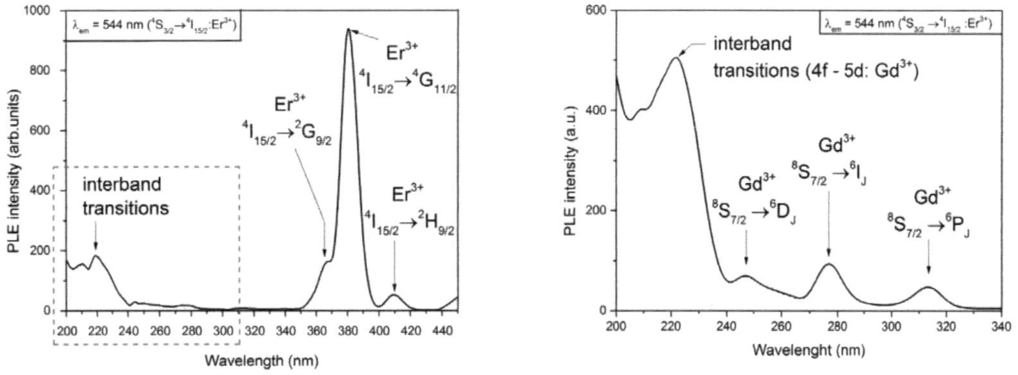

(*a*) (*b*)

Figure 3. Excitation spectra at emission 544 nm ($^4S_{3/2}\rightarrow{}^4I_{15/2}$: Er^{3+}) in the wavelength range from 200 to 450 nm (*a*) and from 200 to 340 nm (*b*).

3.2 Emission decay curves and energy transfer

The emission decay curves of Er^{3+} $^4S_{3/2}\rightarrow{}^4I_{15/2}$ transition under different excitation are presented in figure 4. All the decay curves are described by double-exponential equation (1) [3]:

$$I = A_1\exp(-t/\tau_1) + A_2\exp(-t/\tau_2), \quad (1)$$

where *I* is the luminescence intensity; A_1 and A_2 are constants; *t* is the time; τ_1 and τ_2 are decay time for exponential components.

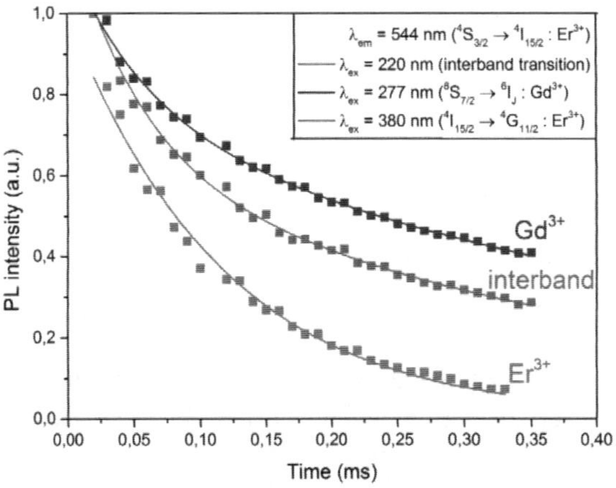

Figure 4. Luminescence decay of Er^{3+} $^4S_{3/2}\rightarrow{}^4I_{15/2}$ transition under intracenter excitation, interband excitation and excitation by imperfect Gd^{3+} ions. All curves are obtained at room temperature and approximated by double-exponential.

Because of non-exponential behavior of Er^{3+} luminescence decay, the average lifetime (τ) of Er^{3+} ion could be calculated using following equation (2):

$$\tau = \frac{A_1\tau_1^2 + A_2\tau_2^2}{A_1\tau_1 + A_2\tau_2} \quad (2)$$

The parameters of decay curves fitting and average decay times of Er^{3+} luminescence are listed in table 1.

Table 1. The parameters obtained from fitting data and calculated average lifetimes of Er^{3+} under different excitation.

Excitation channel	A_1	A_2	τ_1 (ms)	τ_2 (ms)	τ (ms)
Intracenter excitation	0.85559	0.73017	0.13111	0.01943	0.11857
Interband transitions	0.65838	0.57281	0.40587	0.05204	0.37036
Excitation by Gd^{3+} ions	0.38634	0.76739	0.05541	0.53693	0.51308

It is found that the decay time of Er^{3+} luminescence increases under interband excitation and excitation by Gd^{3+} ions. Possible mechanism of excitation energy transfer from Gd^{3+} to Er^{3+} are represented using energy level diagram (figure 5). The population of the Er^{3+} radiative level $^4S_{3/2}$ is carried out in several stages under indirect excitation. This fact confirm the increase decay time of Er^{3+} luminescence under this excitation type.

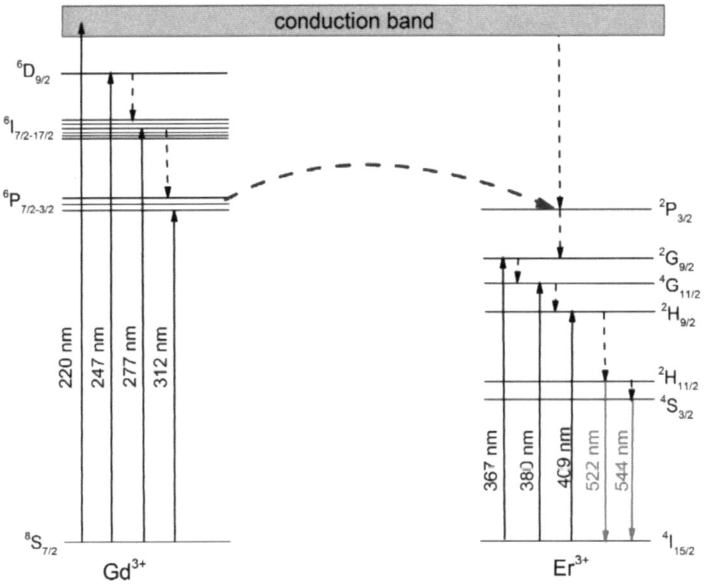

Figure 5. Schematic diagram of Gd^{3+} and Er^{3+} energy levels demonstrating the mechanism of energy transfer from the Gd_2O_3 host lattice and imperfect Gd^{3+} ions to the luminescent Er^{3+} centers.

3.3 Quantum efficiency calculation

Traditionally, the quantum efficiency is defined as ratio of the luminescence decay time to the its radiation component [7]. Using the interrelation between the decay time and its radiation and non-radiative components the analytical expressions for calculating the quantum efficiency of Er^{3+} ions luminescence under intracenter excitation (3) and excitation by Gd^{3+} ions (4) are obtained.

$$\eta_1 = (1 + \frac{C}{e^{E/kT}})^{-1}, \quad (3)$$

where η_1 – quantum efficiency under intracenter excitation;
C – quenching constant;
E – activation energy.

$$\eta_2 = \gamma_{Gd-Er} \cdot \eta_1, \quad (4)$$

where η_2 – quantum efficiency under excitation by Gd^{3+} ions;
$\gamma_{Gd\text{-}Er}$ – efficiency of the energy transfer from Gd^{3+} to Er^{3+}.

The parameters of temperature quenching of the Er^{3+} luminescence are determined as a result of quenching curve fitting according to Mott's law (E = 0.0685 eV, C = 14.907). The Gd^{3+} to Er^{3+} energy transfer efficiency is estimated to be 80% as a typical value in the case of 1% acceptor rare-earth ions concentration [8].On the basis of the obtained data the quantum efficiency of the Er^{3+} ions luminescence under intracenter excitation (η_1=0,5) and excitation by Gd^{3+} ions (η_2=0,4) are calculated.

4. Conclusion

The results indicate that the efficiencies of direct and indirect excitation of Er^{3+} ions luminescence are characterized by the same order. It is found that the most decay time of the Er^{3+} ions luminescence is observed under excitation by Gd^{3+} ions, however, the quantum efficiency in this case is less. It means that the major losses under indirect excitation take place at intermediate stages. So there is a reserve for minimizing these losses and the quantum efficiency of the Er^{3+} ions luminescence can be controlled by changes the defectiveness of host lattice Gd_2O_3.

Acknowledgments

This research was carried out in the framework of state assignment of Russian Federation Ministry of Science and Education (№ 3.2016.2014/K) with partial support of the RFBR (project № 13-08-00568).

References

[1] Sandra F. H. Correia, Ver´onica de Zea Bermudez, Sidney J. L. Ribeiro et all 2014 *J. Mater. Chem. A* **2** 5580–5596
[2] Strumpel C, McCann M, Beaucarne G et all 2007 *Solar Energy Materials & Solar Cells* **91** 238-249
[3] Jiayue Sun, Guangchao Sun, Yining Sun and Liu Han 2014 *Optical Materials* **36** 1097-1100
[4] Gai S., Yang P., Wang D., Li C. 2011 *CrystEngComm* **13** 5480-5487
[5] Gordon W.O., Carter J.A., Tissue B.M. 2004 *Journal of Luminescence* **108** 339-342
[6] Baryshnikov V.I., Krivorotova V.V. 2008 *Physics of the Solid State* **50** 1664–1666
[7] C. Gorller-Walrand, K. Binnemans, K.A. Gschneidner et all 1998 *Handbook on the Physics and Chemistry of Rare Earths* **25** Chapter 167
[8] Xinmin Zhang, Fangui Meng, Wenlan Li et all 2013 *Journal of Alloys and Compounds* **578** 72-76

Subminiature eddy current transducers for studying semiconductor material

S F Dmitriev[1], V N Malikov[1], A M Sagalakov[1], L I Shevtsova[2], A O Katasonov[1]
[1] Altai State University, Lenina avenue, 61,656049,Barnaul, Altai region
[2] Novosibirsk State Technical University, 20 K.Marx Prospekt, Novosibirsk Oblast

E-mail: osys11@gmail.com

Abstract. Based on an eddy-current transducer (ECT), a probe has been designed to study the semiconductor material. The structural diagram of the probe is given and the basic technical data are stated (the number of windings is 10–130 turns, and the value of the initial permeability of the core μ max = 500). Due to the high confinement of the magnetic field, it is possible to acquire data from small areas of semiconductor plates. Response versus ECT positioning diagram is also provided.

1. Introduction

The demand for improvement and development of semiconductor material production technology brings up a concern of efficient quality control. Production of materials for microelectronics requires implementation of high-precision measuring methods and instruments in the industry, adoption of modern nondestructive inspection techniques[1].

Today, diagnostics of semiconductor plates is mainly performed using the 4-point probe technique. This technique has a number of disadvantages. The nondestructive inspection (NDI) using the eddy-current method (ECM) is based on registration of variations in the eddy-current electromagnetic field, which are induced by the exciting coil of eddy-current transducer (ECT) in the electrically conducting unit under inspection. The advantage of electromagnetic inspection is the possibility to perform measurement without a contact between the transducer and the object of measurement. Interaction between them usually takes place at narrow intervals (0.1-2mm). That's why this technique can yield good results, when the unit under inspection is moving.

One more advantage of ECM is the relative simplicity of the transducer's design. Coils are placed into a protective case and then encapsulated in a compound. As a result, they are resistant to mechanical and weather stresses, and they are able to work harsh environments with wide ranges of temperature and pressure.

2. The main part

In general the scheme of the sensor is presented in Fig.1. In this scheme there are actuating (AW) and measuring (MW) windings, placed in the VCT.

A digital signal comes from the virtual generator to the input of the digital analog transformer (DAT) of the sound card and then it is transformed into the analog one. After passing though the power-amplifier (PA) the analog signal is sent to the actuating winding (AW) of the transformer. While passing through the actuating winding of the VCT, the sinewave signal makes the electromagnetic field, that creates EMF in the measuring winding (MW) of the VCT. The tension is received by the microphone input of the sound card and, after coming though the preamplifier (PA), it goes to the input of the analog digital transformer (ADT) sound card. The analog signal is transformed into the digital one and is sent to the treatment and operation unit of the software. The treatment and control unit fixes the level of the digital signal in conventional units that correspond to the voltage values in the measuring winding[2].

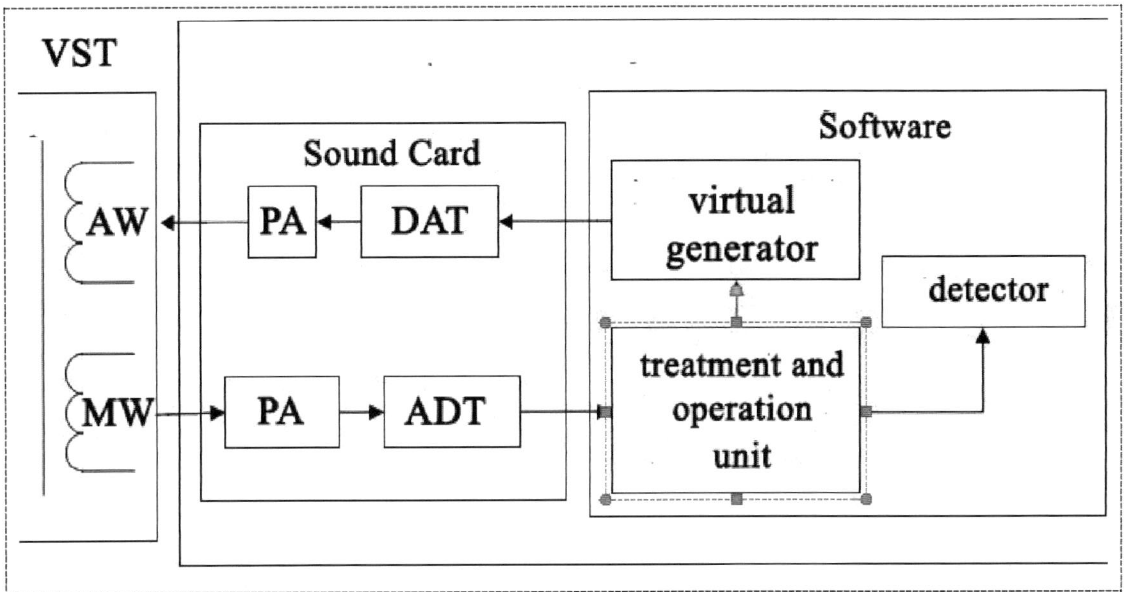

Figure 1. The scheme of sensor operation

This level is accepted as a zero one that corresponds to the voltage level on the measuring winding without the control object. Without the control object the indicator shows zero, it corresponds to the zero value of conductivity.

Along with the considerable price reduction of the device, the virtualization of the generator and signal receiver makes it possible to mark out the frequency range, at which the process of scanning of the control object gets very effective. The effectiveness is reached due to the quick varying of the current frequency, given by the virtual generator to the VCT. Thus, researchers can define the frequency, at which changing the amplitude of the signal on the measuring winding (when it is passing through the imperfect area) is at a maximum. Besides, frequency varying enables us to find the depth of the defect hiding, because the depth of the electromagnetic field distribution of the VCT directly depends on the frequency of the field sent to the actuating winding [3].

The principle of operation of the VCT is based on the alternating magnetic field, localized in the controlled object with the help of the oxide pyramid shaped core. The shape of the core is conditioned with the necessity to localize the magnetic flux from the actuating (generator) coil. The core of this model was made of a midrange metadiscipline ferrite NM3, chosen according to the value of the maximum initial magnetic conductivity 500. The actuating winding of the subminiature transformer consists of 10 winds and its diameter is 0.12-0.13 mm. The measuring winding includes 130 winds and its diameter is 0.05-0.08 mm. The compensation winding is included into the scheme to minimize the influence of the actuating winding on the received signal. The compensation winding is connected with the measuring winding so that the researcher could subtract the tension of the actuating winding. It consists of 20 winds. A copper wire, that is 0,005 mm thick, is used to reel the winds. The windings are wound round the pyramid shaped core.

The electromagnetic field of the generating winding gives an impulse to eddy currents in the electroconductive controlled object. Eddy current density in the object depends on the geometric and electromagnetic values of the object and on the back-to-back location of the measuring VCT and the object. The magnetic field of the eddy currents is opposed to the primary magnetic field of the generating winding, as a consequence the resulting field depends on the electromagnetic characteristics of the controlled object and on the distance between the VCT and the object, as the distribution of the eddy current density depends on these factors. There is EMF in the measuring winding of the VCT; it serves as a signal, hands over the information about the object into the measuring unit.

In this case EMF of the measuring winding decreases due to the opposite magnetic field produced by the eddy currents.

3. The results of the experiment and their discussion.

An experimental installation for electromagnetic diagnostics of semiconductor plates through eddy-current method was created at the premises of the Altai State University. Information gathering system in the form of eddy-current probe is the installation's main module. In order to test the device, a series of measurements was performed for electrical conductivity of reference semiconductors with a-priori known conductivity values. Response versus ECT positioning diagram is shown in Fig. 2.

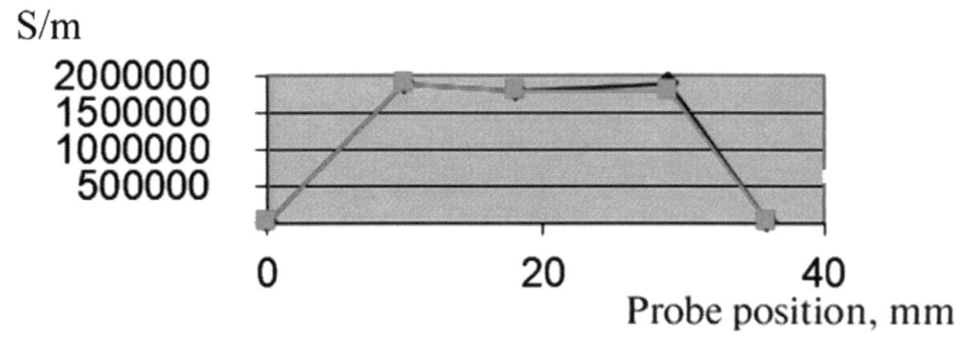

Figure 2. Response versus ECT positioning diagram. InAs

By means of reference semiconductor plates with known values of specific conductivity, the response was plotted against specific conductivity. Gallium-arsenide plate conductivity versus the probe's position above the plate is plotted in Fig. 3.

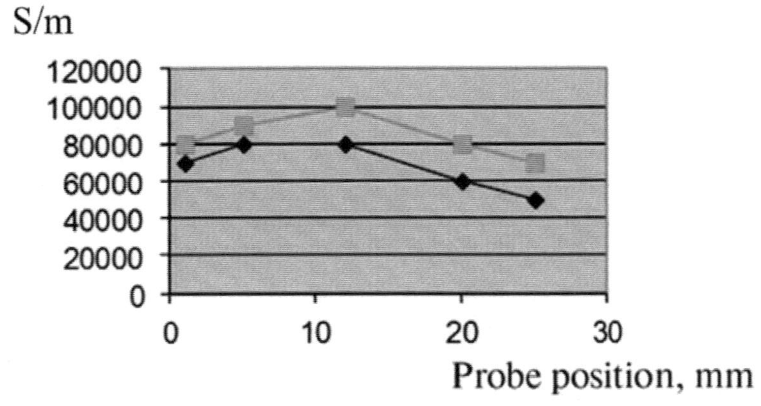

Figure 3. Response versus ECT positioning diagram. GaAs

On the presented diagram, we can note sufficient variations in the conductivity depending on the probe's position above the object. This fact provides the evidence of the semiconductor's structural discontinuity and (consequently) of the low quality.

4. Conclusions.

Semiconductor conductivity measurement results gathered during the research show the potential of the eddy-current method for quality control of semiconductor materials.

Due to the high confinement of the magnetic field, it is possible to acquire data from small areas of semiconductor plates, which allows for the conclusion about semiconductor impurity distribution in small areas. Therefore, this method offers exciting possibilities for evaluation of quality of semiconductor plates.

References

[1] S. F. Dmitriev, A. V. Ishkov, V. N. Malikov, A. M. Sagalakov Subminiature eddy current transducers for studying metal-dielectric junctions. *Instruments and Experimental Techniques*. November 2014, **Volume 57, Issue 6**, pp 751-754

[2] Polyakov V V, Dmitriev S F, Ishkov AV, Kolubaev EA and Malikov VN. Non-destructivetesting of aluminium alloys by using miniature eddy-current flaw transducers. Advanced Materials Research. 2014. **Vol. 880.** Pp. 105-108.

[3] S F Dmitriev, A V Ishkov, V N Malikov, A M Sagalakov and A O Katasonov. Non-destructive testing of the metal-insulator-metal using miniature eddy current transducers. IOP Conference Series: Materials Science and Engineering. **V. 71**

Features of collimation and focusing of few-cycle optical beams

Danila Puzyrev and Sergey Kozlov
Department of Photonics and Optical Information Technology, ITMO University, Saint-Petersburg 197101, Russia
E-mail: puzurdan@gmail.com

Abstract. The paper reports theoretical features of collimation and focusing of few-cycle optical beams in transparent isotropic dielectric medium. It was showed, that a spatial collimation of such short waves results in a peculiar shaping of spatio-temporal structure with the central part moving faster than its outlying areas and collimated beam loses with distance half cycle of temporal electromagnetic field oscillation. In focus of focusing mirror, number of oscillations in temporal field structure of beams depends on only temporal profile of radiation source, when focusing collimated wave packet.

1. Introduction

Development of the methods of generation terahertz radiation made it possible to obtain few-cycle (less than 10 field oscillations) electromagnetic waves [1]. Such radiation is widely applicable in spectroscopy, in detection systems of drugs and explosive materials and for medical diagnostics [2]. The optimization of optical systems of emission, collimation and focusing initially few-cycle waves is important problem in this area [3].

Laws of paraxial difraction and collimation of electric field spatio-temporal structure, which is formed in far field region of diffraction of initially Gaussian one-sycle wave are well studied in the present time [3-4]. In this paper, we study collimation and focusing of initially Gaussian few-cycle waves propagation in homogeneous isotropic medium.

2. Paraxial radiation dynamics of spatio-temporal spectrum

Paraxial radiation dynamics of spatio-temporal spectrum

$$g(\omega, k_x, k_y, z) = \int_{-\infty}^{+\infty}\int_{-\infty}^{+\infty}\int_{-\infty}^{+\infty} E(t, x, y, z) \exp\left(-i\left(\omega t + k_x x + k_y y\right)\right) d\omega dk_x dk_y \qquad (1)$$

in an isotropic homogenous optical medium can be described by the expression [4]:

$$g(\omega, k_x, k_y, z) = g_{x,y}(\omega, k_x, k_y) \exp\left(-ik(\omega) z \left(1 - \frac{k_x^2 + k_y^2}{2k(\omega)^2}\right)\right), \qquad (2)$$

where x, y, z are Cartesian axes coordinates (z axis coincides with wave propagation direction), C_x and C_y are assumed to be known spectral components of radiation at $z = 0$,

ω is the temporal frequency of radiation, k_x and k_y are the spatial frequencies, $k = \dfrac{\omega n(\omega)}{c}$ is the wave number, $E(t, x, y, z)$ is the spatio-temporal distribution of electric field strength in optical beam, $n(\omega)$ is the refractive index of the medium, c is the speed of light in a vacuum.

Diffraction of the initial Gaussian beams at the distance z_0 can be presented in the form, which is equivalent to expression in [4]

$$G(\omega, x, y, z_0) = z_R(\omega) \frac{z_R(\omega) + iz_0}{z_R^2(\omega) + z_0^2} \exp\left(-\frac{x^2+y^2}{\rho^2} z_R(\omega) \frac{z_R(\omega) + iz_0}{z_R^2(\omega) + z_0^2} - ik(\omega) z_0\right) G_0(\omega) \quad (3)$$

Where $z_R(\omega) = \frac{\rho^2 n(\omega)\omega}{2c}$ is the Rayleigh range, ρ is the transverse beam width at $z = 0$.

We set collimating mirror with reflection function $F_{col}(\omega, x, y) = \exp\left(ik(\omega)\left(x^2+y^2\right)/2f_{col}\right)$ at distance $z = 0$ and focal length $f_{col} = z_0$. Hence, spatial distribution of temporal spectrum immediately after reflection can be described by

$$G_{col}(\omega, x, y, z_0) = G(\omega, x, y, z_0) \times F_{col}(\omega, x, y), \quad (4)$$

where $G(\omega, x, y, z_0)$ is defined by expression (3), in contrast to paper [3], where collimation mirror has been placed in far field diffraction region,

$$\begin{aligned} G_{col}(\omega, x, y, z_0) &= iz_R(\omega) \frac{iz_R(\omega) - z_0}{z_R(\omega)^2 + z_0^2} \times \\ &\times \exp\left(-\frac{\left(x^2+y^2\right)}{\rho^2} \frac{z_R(\omega)^2}{z_0} \left(\frac{iz_R(\omega) - z_0}{z_R(\omega)^2 + z_0^2}\right) - ik(\omega) z_0\right) G_0(\omega). \end{aligned} \quad (5)$$

Spatio-temporal spectrum is determined by Fourier transform

$$g(\omega, k_x, k_y, z) = \int_{-\infty}^{+\infty}\int_{-\infty}^{+\infty} G(\omega, x, y, z) \exp\left(-i\left(k_x x + k_y y\right)\right) dk_x dk_y. \quad (6)$$

Inserting of expression (5) in formula (6) gives us spatio-temporal spectrum

$$g_{col}(\omega, k_x, k_y, z_0) = i\pi\rho^2 \frac{z_0}{z_R(\omega)} \exp\left(-\frac{k_x^2 + k_y^2}{4}\rho^2 z_0 \frac{z_0 + iz_R(\omega)}{z_R(\omega)^2} - ik(\omega) z_0\right) G_0(\omega). \quad (7)$$

For boundary condition (7) expression (2) takes form

$$g_{col}(\omega, k_x, k_y, z) = i\pi\rho^2 \frac{z_0}{z_R(\omega)} \exp\left(\begin{array}{l} -\frac{k_x^2+k_y^2}{4} \frac{\rho^2}{z_R(\omega)} \left(\frac{z_0^2 - iz_R(\omega)(z-z_0)}{z_R(\omega)}\right) \\ -ik(\omega)(z_0 + z) \end{array}\right) G_0(\omega). \quad (8)$$

Spatial distribution of temporal spectrum is described by Fourier transform formula $G(\omega, x, y, z) = \frac{1}{(2\pi)^2} \int_{-\infty}^{+\infty}\int_{-\infty}^{+\infty} g_{col}(\omega, k_x, k_y, z) \exp\left(i\left(k_x x + k_y y\right)\right) dk_x dk_y$, therefore the dynamics of electro-magnetic field after collimation mirror, are described by expression

$$G_{col}(\omega, x, y, z) = \frac{iz_0\left(z_{col}(\omega) + i(z - z_0)\right)}{\left(z_{col}(\omega)\right)^2 + (z - z_0)^2} \exp\left(\begin{array}{l} -\frac{x^2+y^2}{\rho^2} \frac{z_R(\omega)\left(z_{col}(\omega) + i(z-z_0)\right)}{\left(z_{col}(\omega)\right)^2 + (z-z_0)^2} \\ -ik(\omega)(z_0 + z) \end{array}\right) G_0(\omega), \quad (9)$$

where $z_{col}(\omega) = \frac{z_0^2}{z_R(\omega)}$.

For far field region $z - z_0 \gg z_{col}(\omega)$ relation (9) is simplified to form

$$G_{col}^{Far}(\omega, x, y, z) = -\frac{z_0}{(z - z_0)} \exp\begin{pmatrix} -\frac{x^2+y^2}{\rho^2}\frac{z_R(\omega)(z_{col}(\omega)+i(z-z_0))}{(z-z_0)^2} \\ -ik(\omega)(z_0 + z) \end{pmatrix} G_0(\omega). \quad (10)$$

Let is consider spatio-temporal structure of collimated beam. Such structure may occur in, for example, initially single-cycle waves on condition that refractive index is present in the form $n(\omega) = N = const$. At distances given in this article atmospheric air for THz radiation may be an example of such media [5].

Spectrum of single-cycle wave on the source can be described by the expression

$$G_0(\omega) = -\frac{\sqrt{\pi}}{2} E_0 \tau_p^2 i\omega \exp\left(-\left(\frac{\tau_p \omega}{2}\right)^2\right), \quad (11)$$

where τ_p is the duration of radiation at $z = 0$.
Using Fourier transform formula

$$E(t, x, y, z) = \frac{1}{2\pi} \int_{-\infty}^{+\infty} G(\omega, x, y, z) \exp(i\omega t)\, d\omega \quad (12)$$

we can take expression that describes spatio-temporal structure of collimated beam at the focus of collimation mirror ($z = z_0$)

$$E_{col}(t', x, y, z) = E_0 \frac{\rho^2 N}{2cz_0} \frac{\tau_p^2}{\tau_\partial^3} \left(1 - 2\left(\frac{t'}{\tau_\partial}\right)^2\right) \exp\left(-\left(\frac{t'}{\tau_\partial}\right)^2\right), \quad (13)$$

where is $t' = t - 2\frac{z_0 N}{c}$, which is called "delay time" [4] and duration $\tau_\partial = \tau_p \sqrt{1 + \frac{x^2+y^2}{\rho^2}\left(\frac{\rho^2 N}{\tau_p z c}\right)^2}$, which is directly proportional to the distance from beam axis.

Radiation in far field region can be described by the Fourier transform (12) of function (10) with the radiation spectrum of the emitter (11):

$$E_{col}^{far}(t'', x, y, z) = -E_0 \frac{z_0}{(z - z_0)} \frac{t''}{\tau_p} exp\left(-\frac{x^2 + y^2}{\rho^2} \frac{z_0^2}{(z - z_0)^2}\right) \exp\left(-\left(\frac{t''}{\tau_p}\right)^2\right), \quad (14)$$

where delay time is $t'' = t - \frac{N}{c}\left(z_0 + z + \frac{x^2+y^2}{2(z-z_0)}\right)$, while duration has become the constant in cross section of wave packet.

Figure 1 illustrates expressions for collimation of initial one-cycle wave packet. Red and blue areas correspond to maximum positive and minimum negative field strength in two-dimensional representation. Here we take $\rho/\lambda_{\max} = 10$, where ρ is the initial transverse beam width, $\lambda_{\max}$ is the wavelength corresponding to the maximum of input wave packet spectrum. For example, for terahertz single-cycle waves this value is usually $\lambda_{\max} = 300 \mu m$.

Figure 1 and analysis of equation (9) show that in the focus of collimation mirror the central part of spatio-temporal structure of collimated beam moves faster than its outlying areas and that radiation have extra half cycle oscillation of temporal electromagnetic field (figure 1 (b)), which fades with distance in far field diffraction region (figure 1 (c)).

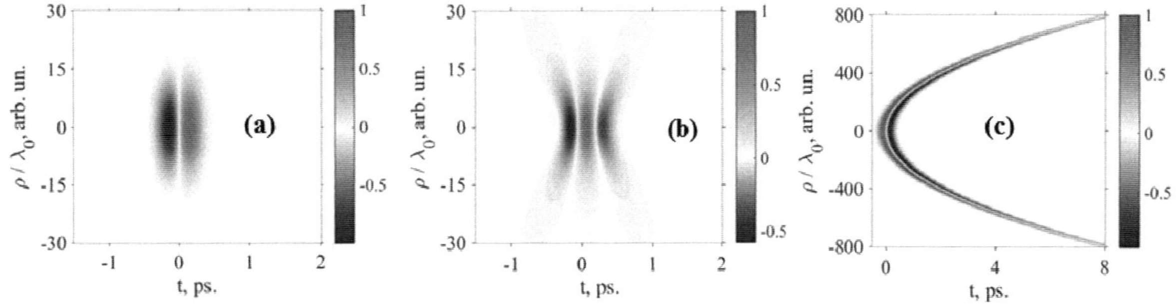

Figure 1. Spatio-temporal structure of electric field at the source of radiation (a); after collimation at the distance z_0 (b) build by expression (13); after collimation at the distance $80z_0$ (c) build by expression (14); where z_0 is the focusing distance of collimation mirror.

3. Focusing of collimated radiation

Focusing collimated radiation (9) by using the mirror with reflection function $F_{foc}(\omega, x, y) = \exp\left(ik(\omega)\left(x^2+y^2\right)/2f_{foc}\right)$ can be described by expression

$$G_{foc}(\omega, x, y, z_f, f_{foc}) = \frac{iz_0\left(z_{col}(\omega)+i\left(z_f-z_0\right)\right)}{\left(z_{col}(\omega)\right)^2+\left(z_f-z_0\right)^2}\times$$
$$\times\exp\left(\begin{array}{l} -\frac{x^2+y^2}{\rho^2}z_R(\omega)\frac{z_{col}(\omega)+i\left(\left(z_f-z_0\right)\left(1-\frac{z_f-z_0}{f_{foc}}\right)-\frac{z_{col}(\omega)^2}{f_{foc}}\right)}{\left(z_{col}(\omega)\right)^2+\left(z_f-z_0\right)^2} \\ -ik(\omega)\left(z_0+z_f\right)\end{array}\right)G_0(\omega). \tag{15}$$

where z_f is the distance from collimation mirror to focusing, f_{foc} is focal length.

By analogy with the collimated radiation we have derive the expression that describes spatial distribution dynamics of the temporal spectrum

$$G_{foc}(\omega, x, y, z, z_f, f_{foc}) = \frac{iz_0}{z_{col}(\omega)\left(1-\frac{z}{f_{foc}}\right)+i\left(\left(z_f-z_0\right)\left(\frac{z}{f_{foc}}-1\right)-z\right)}\times$$
$$\times\exp\left(\begin{array}{l} -\frac{x^2+y^2}{\rho^2}\frac{z_R(\omega)}{f_{foc}}A\left(z, z_f, , z_0, f_{foc}\right) \\ -ik(\omega)\left(z_0+z_f+z\right)\end{array}\right)G_0(\omega). \tag{16}$$

Where

$$A\left(z, z_f, z_0, f_{foc}\right) = \frac{z_{col}(\omega)\left(\left(1-\frac{z}{f_{foc}}\right)\left(\left(z_f-z_0\right)-\left(z_f-f_{foc}-z_0\right)\right)+z\right)}{\left(z_{col}(\omega)\left(1-\frac{z}{f_{foc}}\right)\right)^2+\left(\frac{z}{f_{foc}}\left(z_f-z_0-f_{foc}\right)-\left(z_f-z_0\right)\right)^2}+$$
$$+\frac{i\left(\left(z_f-f_{foc}-z_0\right)\left(\left(z_f-z_0\right)\left(\frac{z}{f_{foc}}-1\right)-z\right)-z_{col}(\omega)^2\left(1-\frac{z}{f_{foc}}\right)\right)}{\left(z_{col}(\omega)\left(1-\frac{z}{f_{foc}}\right)\right)^2+\left(\frac{z}{f_{foc}}\left(z_f-z_0-f_{foc}\right)-\left(z_f-z_0\right)\right)^2} \tag{17}$$

is the complex factor, which determine duration and wavefront curvature of focusing radiation.

In focus of focusing mirror expression (16) can be simplified and we can find analytical solution of Fourier integral (12), which can be represented in the form

$$E_{foc}^{z=f}(\omega, x, y, z_f, f_{foc}) = -E_0\frac{z_0}{f_{foc}}\frac{t'}{\tau_p}\exp\left(-\left(\frac{t'}{\tau_p}\right)^2-\frac{x^2+y^2}{\rho^2}\frac{z_0^2}{f_{foc}^2}\right) \tag{18}$$

where delay time is $t' = t + \frac{x^2+y^2}{2c}\frac{N\left(z_f-f-z_0\right)}{f_{foc}^2} - \frac{N}{c}\left(z_0+z_f+z\right)$.

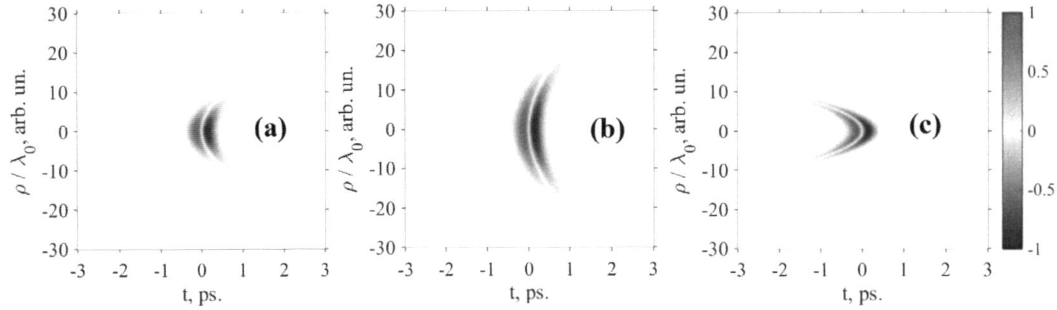

Figure 2. Spatio-temporal structure of electric field at focus of focusing mirror:
(a) distance between mirrors $z_f = f_{col}$ and their focal length correlated as $f_{foc} = 0.5 f_{col}$,
(b) distance between mirrors $z_f = 0.5 f_{col}$ and their focal length correlated as $f_{foc} = f_{col}$,
(c) distance between mirrors $z_f = 5 f_{col}$ and their focal length correlated as $f_{foc} = 0.5 f_{col}$.

Analysis of equation (16) shows that number of electric field oscillations is determined only by temporal profile of radiation source and that electric field has phase shift relative to initial wave at the emitter at focus of focusing mirror.

4. Conclusion

We obtained analytical expressions, which describe paraxial features of spatio-temporal electric field structure that is forming due to collimation and focusing of initially few-cycle waves. We have demonstrated that duration of wave packet is directly proportional to the distance from beam axis and that the spatio-temporal structure of electric field has extra half cycle of oscillation at the focus of collimation mirror. In far field region, collimated beam loses that field oscillation and wavelength becomes constant in cross section of wave packet. In focus of focusing mirror electric field has phase shift relative to initial wave at the emitter. Curvature of wavefront depends on distance, which the radiation passed, and focal length of collimation and focusing mirror. These features can be used for optimization of optical systems using THz radiation.

Acknowledgments

This work was financially supported by Government of Russian Federation, Grant 074-U01.

References

[1] Bespalov V G, Gorodetski A A, Denisyuk I Y, Kozlov S A, Krylov V N, Lukomski G V, Petrov N V and Putilin S E 2008 J. Opt. Technol. 75 636-642
[2] Sakhnov S N, Leksutkina E V, Smolyanskaya O A, Usov A V, Parakhuda S E, Grachev Y V and Kozlov S A 2011 Optics and Spectroscopy 111 257-261
[3] Puzyrev D., Drozdov A.A. Collimation and focusing of initially single-cycle paraxial optical beams. Journal of Physics: Conference Series. 2014. Vol. 541. No. 1. pp. 012065.
[4] Ezerskaya A A, Ivanov D V, Kozlov S A and Kivshar Y S 2012 J. Infrared Millim. Terahertz Waves 33 926{942
[5] Kulya M S, Grachev Y V, Bespalov V G and Kujanpaa V P 2013 Nanosystems: Physics, Chemistry,Mathematics 4 206{213

Theoretical investigation of the correlation between perturbations of quantum optical circuit parameters and its performance

Anton V Kozubov and Sergey A Chivilikhin

ITMO University, Kronverkskiy pr. 49, St.Petersburg 197101, Russian Federation

E-mail: kozubov.anton@gmail.com

Abstract. In this paper described the operability of some quantum optical circuits which depends on their constructions, such as reflectivity of beam splitters in the linear optical controlled-NOT gate in the coincidence basis.

1. Introduction

In the present there is an active development of such direction of quantum information as quantum teleportation [1]. Experimental and theoretical research of quantum teleportation, and its practical application, is one of the main directions in the field of quantum information. As a result, there were quite a number of experimentally realized schemes, ranging from those using photons [2] and the atoms [3], various hybrid systems and ensembles of particles [4]. The CNOT gate is one of the most important elements in quantum information theory and particularly in quantum computing. There are different realizations of this gate, both linear and non-linear. But all of them have one major problem – it operates with certain probability. However, furthermore its work is also influenced by inaccuracies of the constructions which may have a significant impact on the performance of the gate.

2. Methods

In this work we investigate how these quantum optical circuits work with some perturbed parameters such as reflectivity (transmittivity) of beam splitters. In the linear optical CNOT gate in the coincidence basis [5] circuit there are five beam splitters – two of them have the reflectivity of one-half and others have reflectivities of one-third.

In spite of this we want to see how it operates with the arbitrary values of reflectivity of the beam splitters. By using the Heisenberg equation for beam splitters

$$
\begin{aligned}
a_{out} &= \sqrt{\eta}a_{in} + \sqrt{1-\eta}b_{in}, \\
b_{out} &= \sqrt{1-\eta}a_{in} - \sqrt{\eta}b_{in},
\end{aligned}
\tag{1}
$$

where η and $(1-\eta)$ is the reflectivity and transmittivity of the beam splitters, we can obtain the relation of control and target input modes to their corresponding output mode operators. It is necessary for calculating the output state of the system in the Schrödinger picture. We can calculate it through substituting input mode operators for the output operators, which we already derived. These operators are

$$c_{H_0}=\sqrt{\eta_1}c_H+\sqrt{1-\eta_1}v_c,$$

$$c_{V_0}=-\sqrt{\eta_2}c_V+\sqrt{\eta_3(1-\eta_2)}t_H+\sqrt{(1-\eta_3)(1-\eta_2)}t_V,$$

$$t_{H_0}=\sqrt{\eta_4(1-\eta_2)}c_V+\sqrt{\eta_3(1-\eta_2)\eta_4+(1-\eta_3)(1-\eta_5)(1-\eta_4)}t_H+\sqrt{(1-\eta_4)(1-\eta_5)}v_T, \tag{2}$$

$$t_{V_0}=\sqrt{(1-\eta_4)(1-\eta_2)}c_V+\sqrt{\eta_3(1-\eta_5)\eta_4+(1-\eta_3)(1-\eta_2)(1-\eta_4)}t_V-\sqrt{\eta_4(1-\eta_5)}v_T.$$

Thereby we obtain the function of the output state for the circuit with the arbitrary beam splitter ratios, substituting these parameters (2) in next equation [5]:

$$|\varphi_{out}\rangle=(\alpha c_{H_0}^{\dagger}t_{H_0}^{\dagger}+\beta c_{H_0}^{\dagger}t_{V_0}^{\dagger}+\gamma c_{V_0}^{\dagger}t_{H_0}^{\dagger}+\delta c_{V_0}^{\dagger}t_{V_0}^{\dagger})|0000\rangle|00\rangle. \tag{3}$$

After substituting we obtain several new summands, which reduced when the optimal parameters used. These summands have an important influence on the circuit's operability, due to the fact that with them this circuit operates as identity operator but not like the CNOT gate. Thus we can analyze how this gate operates and what the error probabilities are.

Also in this work we have calculated and plotted density matrices for certain values of the deviations of the reflectivity in the beam splitters from the optimal values by using next equation:

$$\rho=|\varphi_{out}\rangle\langle\varphi_{out}|. \tag{4}$$

Here presented density matrices of output state for different qubits with optimal values of reflectivity in the beam splitters and certain deviations from them.

α=1, β=γ=δ=0

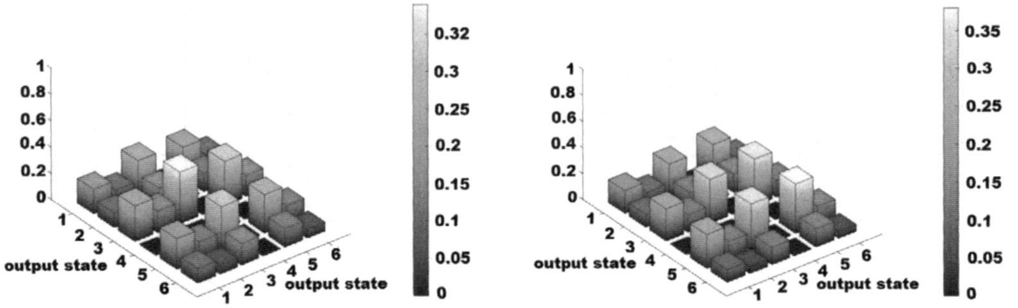

a b

Figure1. a. Output state density matrix for optimal values for beam splitters $\eta_1=\eta_2=\eta_5=0.33$ and $\eta_3=\eta_4=0.5$. b. Output state density matrix for arbitrary deviations of values for beam splitters $\eta_1=\eta_2=\eta_5=0.37$ and $\eta_3=\eta_4=0.58$.

β=1, α=γ=δ=0

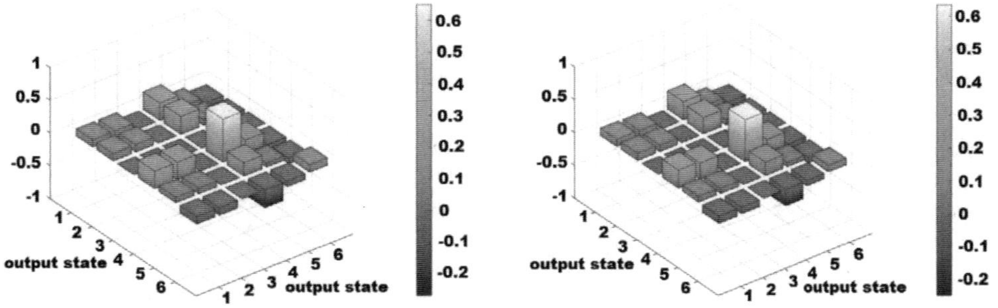

a b

Figure 2. a. Output state density matrix for optimal values for beam splitters $\eta_1 = \eta_2 = \eta_5 = 0.33$ and $\eta_3 = \eta_4 = 0.5$. b. Output state density matrix for arbitrary deviations of values for beam splitters $\eta_1 = \eta_2 = \eta_5 = 0.37$ and $\eta_3 = \eta_4 = 0.58$.

γ=1, α=β=δ=0

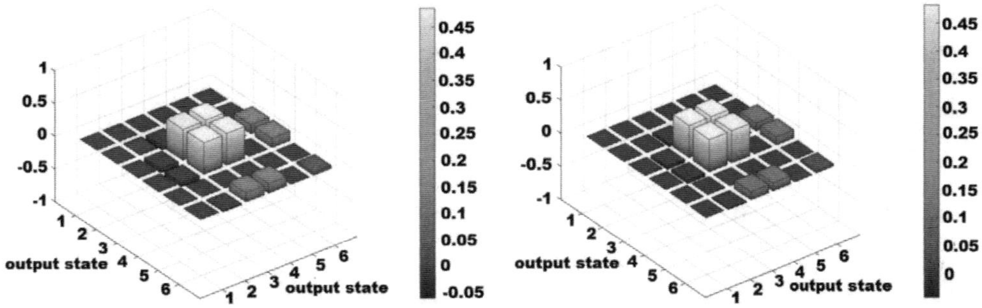

a b

Figure 3. a. Output state density matrix for optimal values for beam splitters $\eta_1 = \eta_2 = \eta_5 = 0.33$ and $\eta_3 = \eta_4 = 0.5$. b. Output state density matrix for arbitrary deviations of values for beam splitters $\eta_1 = \eta_2 = \eta_5 = 0.37$ and $\eta_3 = \eta_4 = 0.58$.

δ=1, α=β=γ=0

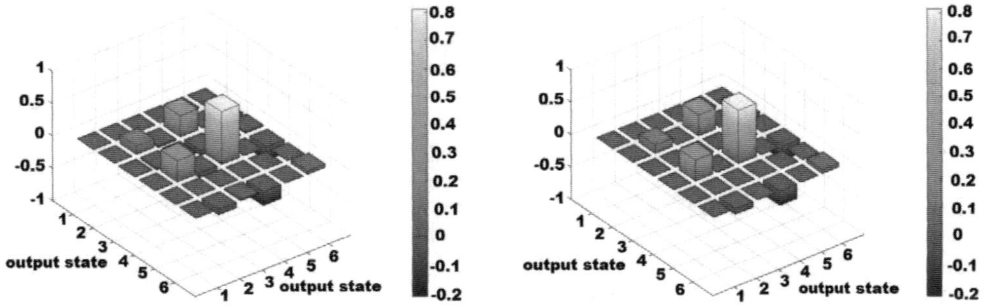

a b

Figure 4. a. Output state density matrix for optimal values for beam splitters $\eta_1 = \eta_2 = \eta_5 = 0.33$ and $\eta_3 = \eta_4 = 0.5$. b. Output state density matrix for arbitrary deviations of values for beam splitters $\eta_1 = \eta_2 = \eta_5 = 0.37$ and $\eta_3 = \eta_4 = 0.58$.

In order to have a better representation of this difference we plotted some graphics to show it.

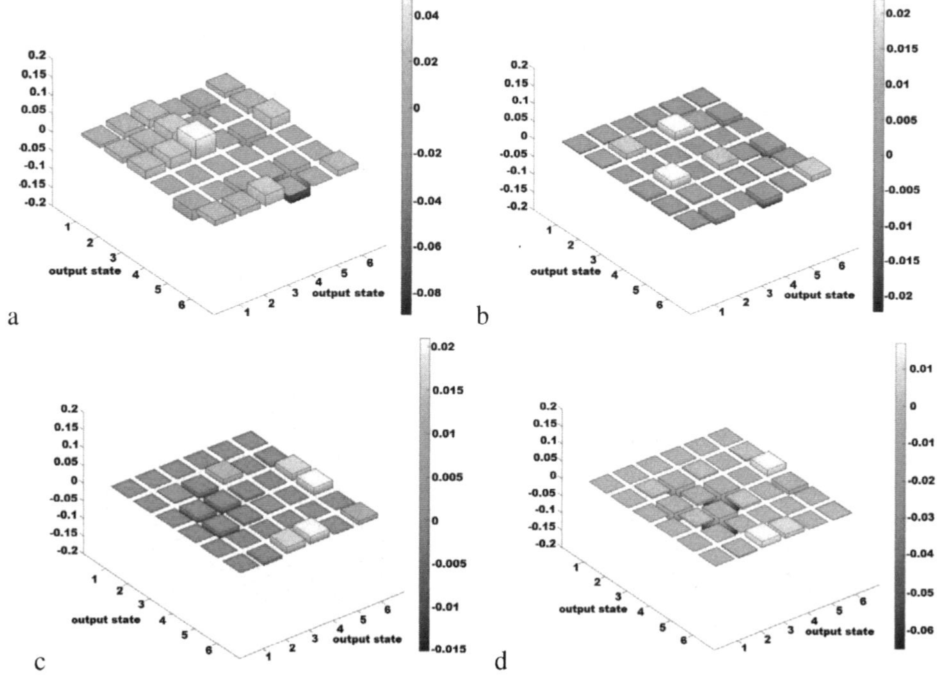

Figure 5. Difference between density matrices of output state with $\eta_1 = \eta_2 = \eta_5 = 0.33$ and $\eta_3 = \eta_4 = 0.5$ and density matrices of output state with $\eta_1 = \eta_2 = \eta_5 = 0.37$ and $\eta_3 = \eta_4 = 0.58$. a. For $\boldsymbol{\alpha}$**=1**, $\boldsymbol{\beta}$=$\boldsymbol{\gamma}$=$\boldsymbol{\delta}$=**0**. b. For $\boldsymbol{\beta}$**=1**, $\boldsymbol{\alpha}$=$\boldsymbol{\gamma}$=$\boldsymbol{\delta}$=**0**. c. For $\boldsymbol{\gamma}$=**1**, $\boldsymbol{\alpha}$=$\boldsymbol{\beta}$=$\boldsymbol{\delta}$=**0**. d. For $\boldsymbol{\delta}$=**1**, $\boldsymbol{\alpha}$=$\boldsymbol{\beta}$=$\boldsymbol{\gamma}$=**0**.

There are different realizations of this gate with various type and amount of elements, in particular beam splitters, but despite this we can apply similar technique and methods to obtain the function of output states for these circuits.

3. Results and discussion

In the presented work, we have derived the analytical expression for the output state function of the system for realized circuit of control-NOT gate, such as linear optical CNOT gate in the coincidence basis. Furthermore we have calculated output state function for this circuit with arbitrary deviations in beam splitter ratios. Also we have obtained and plotted density matrices for certain values of the deviations of the reflectivity in the beam splitters from the optimal values. Thus it was investigated, that with $\eta_1 = \eta_2 = \eta_5 = 0.33 \pm 0.03$ and $\eta_3 = \eta_4 = 0.5 \pm 0.05$ the circuit operates good and the error probability is near 1 per cent, but starting from $\eta_1 = \eta_2 = \eta_5 = 0.33 \pm 0.04$ and $\eta_3 = \eta_4 = 0.5 \pm 0.08$ the probability to find photon on some outputs can be very different from optimal values.

Acknowledgments
This work was financially supported by the Government of Russian Federation, Grant 074-U01.

References

[1] Bennett C Brassard G 1993 *Phys. Rev. Lett.* **70** 1895
[2] D Bouwmeester J-W Pan K Mattle M Eibl H Weinfurter and A. Zeilinger 1997 *Nat.* **390** 575
[3] M D Barrett J Chiaverini T Schaetz J Britton W M Itano J D Jost E Knill C Langer D Leibfried R Ozeri & D J Wineland 2004 *Nat.* **429** 737
[4] Alexey N Pyrkov and Tim Byrnes 2014 *Phys. Rev.* A **90** 062336
[5] T C Ralph N K Langford T B Bell and A G White 2002 *Phys. Rev.* A **65** 062324

Mode switching in InN microresonators

D R Kazanov*, V H Kaibyshev, V U Davydov, A N Smirnov, V N Jmerik, N V Kyznetsova, S V Ivanov, T V Shubina

Ioffe Physical-Technical Institute of the RAS, St.-Petersburg, 194021 Russia

*E-mail: kazanovdr@gmail.com

Abstract. InN optical resonators, which can support whispering-gallery modes of low orders up to room temperature, were made by MBE on patterned substrates. The observed effect of mode switching with a temperature rise was ascribed to the change of optical parameters induced by the shift of an absorption edge and modification of its shape. The results of modelling taking into account variation of refraction index reproduce typical distribution of the intensity of electromagnetic field within the resonators.

1. Introduction to optical resonators with different optical modes

Resonant stationary waves, that are so called whispering gallery modes (WGM) following Rayleigh [1], appear in closed optical systems using effect of total internal reflection. The necessity of amplification of very low signals of quantum oscillators by means of optical resonators of different types stimulates increasing interest in them. The most explored systems are planar Bragg structures with dielectric or semiconductor mirrors. However, difficulties of their creation are significant. At the same time, modern technologies allow to make optical resonators with WGMs in the shape of microdisks, microcolumns, *etc.* [2]. We can point two basic methods of creation of such types of these resonators: 1) epitaxial growth, that allows to make an active region inside the resonator at the same growth cycle [3], and 2) lithography with dry or wet etching [4]. During the development of methods and technology of forming WGM microresonators, theoretical research of mode variety in such structures was done [5-8].

We have recently shown the way of creation of the optical microresonators that were formed during InN growth by molecular beam epitaxy (MBE) on patterned substrates [9]. Such microresonators could be promising in order to create nanolasers and sources of single photons, since they can amplify very weak emission intensity of two-level system by means of the Purcell effect. Generally, people use high order WGMs because of their higher quality factor in resonators. This reason sets some restrictions on the size of crystals, which should be quite large. In this paper, we investigate the application of the low-order WGMs, excited in the microcrystals of InN with a cup-like shape. In particular, we analyze the effect of temperature switching of dominant type of the modes. We demonstrate that this effect is related to a shifted absorption edge and respective temperature-induced changing refractive index in microcrystals.

2. Creation and optical properties of samples

Structures were fabricated by plasma-assisted MBE on cone-shaped patterned c-Al_2O_3 substrates, where the half-spherical cones have a diameter 3 and height 1.5 microns. Microcrystals have very specific shape of inverted truncated hexagonal pyramids, which results from different lateral and vertical growth velocities. They are formed on about 15% of all cones (Fig. 1). As shown by micro-Raman studies, structural quality of microcrystals was higher than in the area around them. Their plane tops have smooth surface [8]. In addition, the results of energy-dispersive X-ray microanalysis (EDX) showed decreasing In/N ratio in the crystals that excludes the Purcell effect associated with plasmonic resonance of metallic In particles.

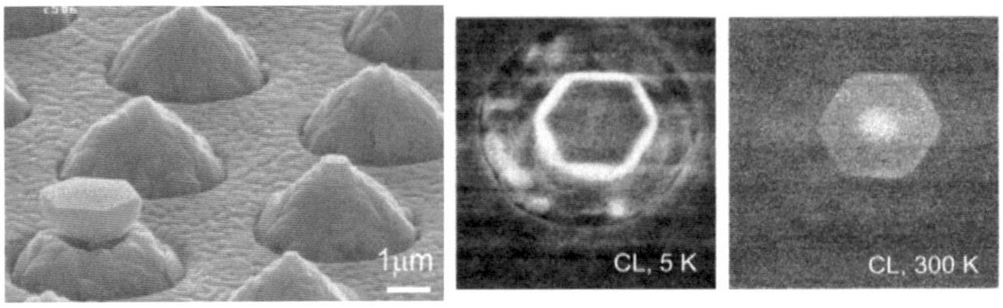

Figure 1. (from left to right). Image of InN structure with microcrystals taken by scanning electron microscopy (SEM), cathodoluminescence (CL) images taken at 5K and 300K.

In order to clarify the mode content and its temperature change we used micro-photoluminescence (μ-PL) spectroscopy. The wavelength of a laser used for excitation was 532 nm. The laser beam was focused in a spot with a diameter about 1 □m; thus a spectrum μ-PL can be measured from a single crystal. Spectrum of transmission and reflection were taken using a tungsten lamp; they characterize integral properties of area about 1-2 mm^2, which includes both the crystals and an InN planar layer. Figure 2 shows the results of these studies.

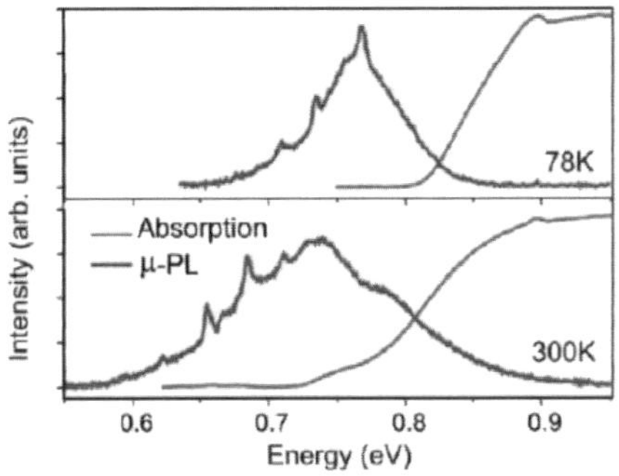

Figure 2. Spectra of μ-PL (red lines), measured in the same crystal with the diameter of ~2.2 μm, and absorption spectra of InN lstructure (blue lines) at different temperatures.

Illustrations of distribution of electromagnetic energy in microresonators, obtained by micro-cathodoluminescence (μ-CL), are shown in Fig. 1. These pictures show that the azimuthal type of mode is dominant at low temperature with highest intensity on the edges of the crystal. However, the type of modes changes while temperature rises (at ~80-100K) and mostly become radial at 300 K, which corresponds to highest electromagnetic energy intensity in the center of the crystal. It is worth saying that changing optical properties of semiconductor is the reason of different mode content. In the simplest cylindrical approximation, it is well known that the frequency of a resonator mode of *m*-order appears to be $\omega \propto mc/Dn$, where *D* - diameter, *n* - refractive index, *c* - speed of light. Therefore, while *D* is constant, only *n* can change the type of modes.

3. Modelling of optical modes in microcrystals InN

In order to derive the temperature dependence of the refractive index, which is important for computer modelling, the measurements of the integral transmission and reflection spectra were done. Then, we calculated complex dielectric function by applying Kramers-Kronig relations to these spectra:

$$\varepsilon_1(\omega) - 1 = \frac{2}{\pi}\int_0^{\infty} \frac{x\varepsilon_2(x)}{x^2 - \omega^2}dx\,,\ where \begin{cases} \varepsilon_1 = n^2 - k^2 \\ \varepsilon_2 = 2nk \end{cases}$$

Figure 3 shows the result of the determination of the temperature dependent refractive index for three energies that fall within the emission band of InN.

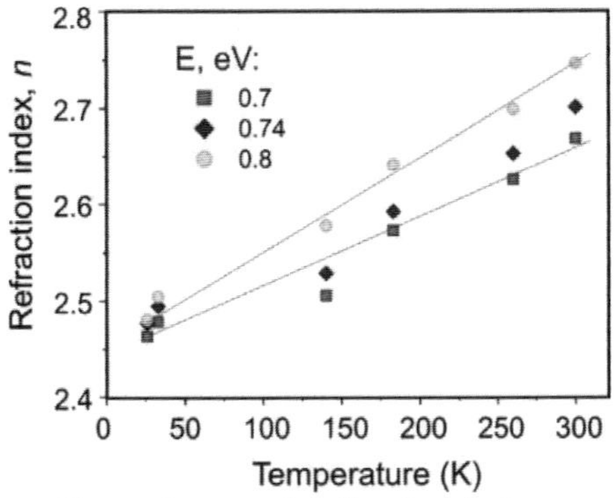

Figure 3. Dependence of refractive index on temperature for three different energies in InN emission band

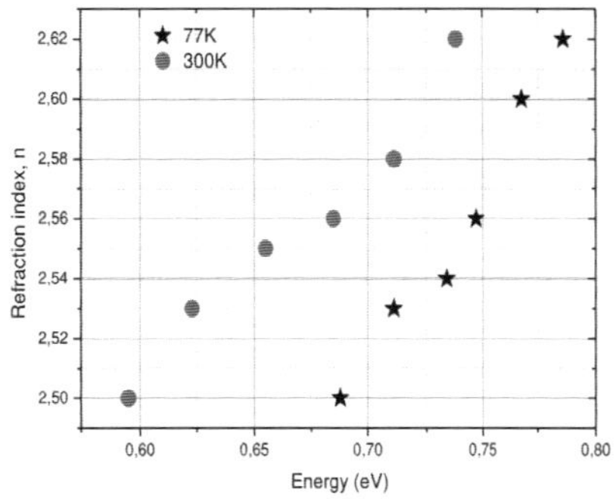

Figure 4. Dispersion of refractive index at low (77 K) and room (300 K) temperatures derived from modelling the resonator modes in InN microcrystals.

We have done modelling of optical modes in InN microcrystals using the spectral dependence of the refractive index. Calculation was made in Comsol Multyphysics software that allows to find the frequencies and distribution of intensity of electromagnetic waves inside the resonator by solving the Maxwell's equations. In the first approximation, we used refractive index dispersion that was found earlier. Inverted truncated hexagonal pyramids were replaced by inverted truncated cones, which size (height, and top/bottom radiuses) was obtained by scanning electron microscopy. We chose boundary conditions in order to make rapid decay of wave outside the crystal. After this iteration, more accurate analysis of frequencies was made by varying refractive index, so that experimental and calculated data were in quite agreement with each other. We considered the modes lying between 0.65-0.8 eV, which is inside the InN emitting band. Table 1 presents the optical mode energies, both calculated and detected from μ-PL experiment, along with the reconstruction of the field intensity distributions. Figure 4 shows the values of the refractive index characterizing the crystal material. They increase in the vicinity of the absorption edge, like those derived from analysis of the optical spectra. Besides, we investigated the quality factor of the WGM cavities, which is estimated to be around 200.

77 K		300 K	
Energy (eV) experim./calc	Field distribution	Energy (eV) experim./calc	Field distribution
0.688/ 0.690	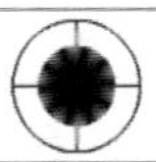	0.623/0.625	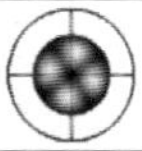
0.711/0.709	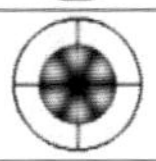	0.655/0.673	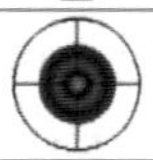
0.734/0.721	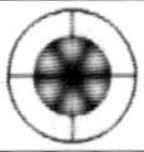	0.685/0.685	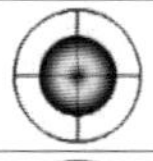
0.747/0.756	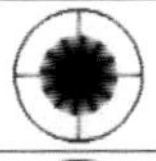	0.711/0.709	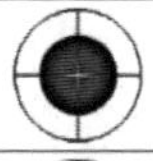
0.768/0.762	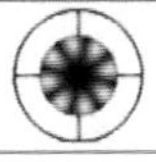	0.738/0.739	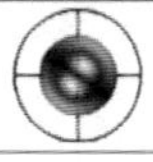

Table.1. Experimental and theoretical frequencies of optical modes in microresonators with 2.2 μm in diameter at different temperatures shown together with electromagnetic field intensity distributions.

This theoretical modelling shows that the azimuthal type of the WGMs prevails against the radial at low temperatures. This situation inversely changes at room temperature. We have observed exactly the same in the experiments. The simulations show that the main reason of that associated with the shift of absorption edge towards low-energies when temperature rising. It induces changing the refractive index resulting in the variation of the mode content and the modification of the spatial energy distribution. Besides, the increasing absorption suppresses the modes of higher orders. The modeling reveals also a large gap between the existent modes at the short-wavelength side of the emission band.

Furthermore, we need to draw attention to refractive index in figures 3 and 4. They are a little bit different. The reason of that is the fact that in Fig. 3 the data describe structure as a whole, including microcrystals and a layer between them. As we know, InN is very imperfect material, so the integral spectra have the information not only about absorption edge but the "tail" of states, which are typical for strongly compensated semiconductors [10, 11]. The data in Fig. 4 displays only the optical properties of the InN microcrystals. As we said before, their quality is much higher than those of the area around them. Thus, the second method of establishing refractive index by modelling optical modes in resonators could be more precise to describe their material.

4. Summary

Our data testify the possibility of amplification of weak signals in the near IR by means of low-order optical modes in InN microresonators. It is shown that changing the dielectric function of material with temperature affects the type of electromagnetic energy distribution and the frequencies of optical modes. This effect should be considered while designing nanophotonics devices, where selective amplification by WGMs determines working frequencies. Besides, we demonstrate that modelling of optical modes is one of the suitable ways to derive the material dispersion of the refractive index in microcrystals.

Acknowledgement

This work has been supported by Russian Science Foundation, grant № 14-22-00107

References

[1] Rayleigh, Lord, The problem of whispering gallery, Phil. Mag., v. 20, 1001-1004, 1910
[2] K. J. Vahala, Nature, **424**, 839 (2003).
[3] M.Burger *et al.*, Appl. Phys. Lett. **103** , 021107 (2013),
[4] Y.Zhang *et al.*, Appl. Phys. Lett. **104**, 221106 (2014),
[5] T. Nobis *et al.*, Phys. Rev. Lett. **93,** 103903 (2004)
[6] T. Nobis and M. Grundmann, Phys. Rev. A **72,** 063806 (2005)
[7] M.A. Kaliteevski *et al.*, Phys. Rev. B **75,** 233309 (2007)
[8] J V Alekseenko *et al.*, Журнал технической физики, **79**, 72 (2009)
[9] T V Shubina et al., XVIII Symposium "Nanophysics & Nanoelectronics", 686 (2014)
[10] A.L. Efros and B.L. Shklovskii, Electronic of Doped Semiconductors (Springer, Heidelberg, 1989)
[11] T V Shubina *et al.*, Phys. Stat. Sol. c 4, 2474 (2007).

Extracting S-parameters of bilateral electro-optic network for lightwave component analyzer calibration

D Frolov, A Levchenko, K Korotkov

Kuban State University, 149 Stavropolskaya st., Krasnodar, Russia

E-mail: frolov.drf@gmail.com

Abstract. A new method for extracting E/O and O/E S-parameters of a bilateral electro-optic network (BEON) is theoretically proposed. It is based on measuring reflection coefficients from three optical loads: an absorber and two mirrors. This technique includes two series of reflections measurements: first when loads are connected to optical port of BEON directly and second when loads are connected in series with optical waveguide of fixed length. Using two BEONs and this calibration technique allows to make calibrated lightwave measurements with a standard microwave network analyzer without using additional electro-optical equipment such as lightwave component analyzer or optical heterodyne techniques.

1. Introduction

A bilateral electro-optic network (BEON) and its use for lightwave network analysis measurements was first reported by Pollard [1]. Unlike majority E/O and O/E components that are not reciprocal in nature BEON is a network in which forward (electrical to optical) and reverse (optical to electrical) transmission is allowed. Using BEONs in lightwave network analyzers instead of single laser diodes and photodetectors allow to use an optical analogue of various microwave two-port calibrations such as TRL. This allows to measure full S-matrix of O/O components. However it sacrifices the ability to perform E/O and O/E measurements.

A two-tier calibration of the electro-optic analyzer was also reported by Pollard [2]. However in this calibration BEON is treated as a blackbox whose S-parameters are measured with a HP8703A or other lightwave component analyzer, that uses the optical heterodyne technique [3] to measure the frequency response $S_{21}^{O/E}$ of its photodetector. This technique uses two Nd:YAG lasers, one passing through the modulator, the other functioning as a local oscillator to convert one of the sidebands of the modulated carrier to IF frequency. This IF frequency can be swept by sweeping the wavelength of laser beam, thus the wavelength can be controlled by varying the temperature of YAG crystal.

However these measurements are conducted with complex optoelectronic system that includes lasers, spectrum analyzer, temperature control system, different lenses and polarization control devices. This can be a serious disadvantage in case if one needs to characterize an unknown photodetector or a BEON that is operated in O/E mode because it is impossible to make such measurements with a standard lightwave component analyzer like HP8703A.

2. Description of measurement procedure

Due to the fact that BEON has low bandwidth because it contains circulator [1], for calibration we need some optical devices that will be equivalent to electrical loads: open, short, matched load, and the optical equivalent of airline. It will be shown that in the proposed method it is needed to know exactly only parameters of short equivalent optical load (mirror) that is not hard to calculate for specific optical connector and fabrication of such mirror for a fixed optical wavelength has no difficulties. An equivalent of matched load – an optical absorber can also be easily done.

The optical equivalent of airline for this calibration should be long enough to fit necessary number of points in frequency domain. A piece of optical fiber can be used and due to the small value of loss in it (0.3 dB/km, at 1310 nm wavelength) comparing to RF lines (1.6 dB/m) this optical fiber can be ten times longer than ordinary coaxial RF line. Exactly this condition allows to do this calibration at narrow bandwidth and it is possible to adjust this bandwidth for specific BEON by changing the length of optical fiber (waveguide).

Unfortunately, it is rather difficult to fabricate an equivalent open optical load that will provide 180° RF phase difference between forward and reflected optical waves that are modulated with RF signal (in the optical fiber). But accordingly to low bandwidth of calibration it is possible to fabricate this load as a quarter RF wave optical waveguide with a mirror on the end. The quarter RF wave optical waveguide should be calculated for the average RF frequency of the calibration bandwidth for specific BEON.

The process of extracting scattering parameters of BEON can be done with a standard VNA. First it is needed to calibrate vector network analyzer (VNA) with one of known techniques [4,5,6]. After calibration, BEON electrical port should be connected to one of VNA's ports and six S11 measurements with optical loads should be done as it is shown in figures 1 and 2.

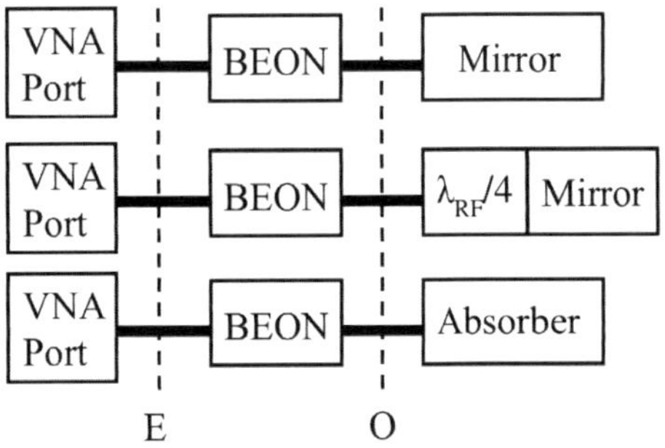

Figure 1. VNA measures complex reflection coefficients of BEON's electrical port with three type of loads: first mirror (equivalent to short load at RF), second mirror placed $\lambda_{RF}/4$ far (open load) and absorber (matched load at RF).

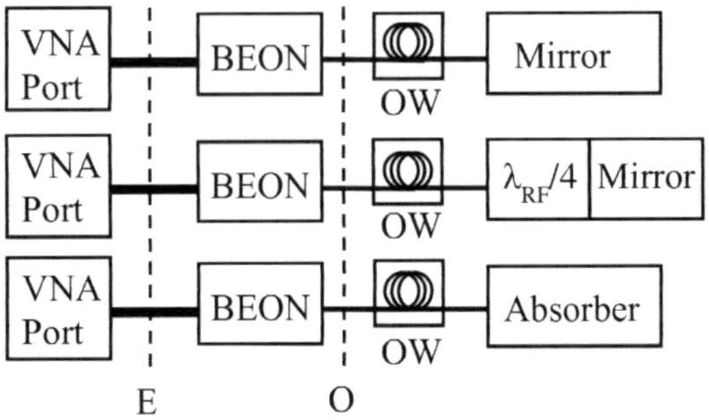

Figure 2. VNA measures complex reflection coefficients of BEON's electrical port with the same loads, but connected in series with optical waveguide (OW) of calibrated length (equivalent to air line at RF).

As a result we have six complex reflection coefficients that can be presented as a system of six equations – three equations (1-3) for reflections without optical waveguide and three with it (4-6):

$$S_S = S_{11}^{E/E} + \frac{S_{21}^{E/O} S_{12}^{O/E} \Gamma_S}{1 - S_{22}^{O/O} \Gamma_S} \tag{1}$$

$$S_O = S_{11}^{E/E} + \frac{S_{21}^{E/O} S_{12}^{O/E} \Gamma_O}{1 - S_{22}^{O/O} \Gamma_O} \tag{2}$$

$$S_L = S_{11}^{E/E} + \frac{S_{21}^{E/O} S_{12}^{O/E} \Gamma_L}{1 - S_{22}^{O/O} \Gamma_L} \tag{3}$$

$$S_S^{/} = S_{11}^{E/E} + \frac{S_{21}^{E/O} S_{12}^{O/E} \Gamma_S^{/}}{1 - S_{22}^{O/O} \Gamma_S^{/}} \tag{4}$$

$$S_O^{/} = S_{11}^{E/E} + \frac{S_{21}^{E/O} S_{12}^{O/E} \Gamma_O^{/}}{1 - S_{22}^{O/O} \Gamma_O^{/}} \tag{5}$$

$$S_L^{/} = S_{11}^{E/E} + \frac{S_{21}^{E/O} S_{12}^{O/E} \Gamma_L^{/}}{1 - S_{22}^{O/O} \Gamma_L^{/}} \tag{6}$$

Using analogue with (1-6) for $\Gamma_S^{/}$, $\Gamma_O^{/}$, $\Gamma_L^{/}$ we also have three equations:

$$\Gamma_S^{/} = S_{11}^{OW} + \frac{S_{21}^{OW} S_{12}^{OW} \Gamma_S}{1 - S_{22}^{OW} \Gamma_S} \tag{7}$$

$$\Gamma_O^{/} = S_{11}^{OW} + \frac{S_{21}^{OW} S_{12}^{OW} \Gamma_O}{1 - S_{22}^{OW} \Gamma_O} \tag{8}$$

$$\Gamma_L^{/} = S_{11}^{OW} + \frac{S_{21}^{OW} S_{12}^{OW} \Gamma_L}{1 - S_{22}^{OW} \Gamma_L} \tag{9}$$

where S_{11}^{OW}, S_{12}^{OW}, S_{21}^{OW} and S_{22}^{OW} are S-parameters of optical waveguide.

Values S_S, S_O, S_L, $S_S^{/}$, $S_O^{/}$, $S_L^{/}$ are known from the measurements, all other values: $S_{11}^{E/E}$, $S_{21}^{E/O} S_{12}^{O/E}$, $S_{22}^{O/O}$, Γ_O, Γ_L, S_{11}^{OW}, $S_{21}^{OW} S_{12}^{OW}$, S_{22}^{OW} are unknown. So we have system with six equations and eight unknown values. By solving analytically system of equations in fixed frequency points where S_{11}^{OW} and S_{22}^{OW} parameters compensate each other and their influence will be very small, because in these fixed frequency points the equivalent RF length of optical waveguide will provide 180° difference between RF waves reflected from the ends of the optical waveguide, we shall find twelve solutions. Using the condition that reflection from open equivalent load has limits $0.5 \le |\Gamma_O| \le 1$ and reflection from matched equivalent load (absorber) has limits $0 \le |\Gamma_L| \le 1$ we can leave only two solutions. In the first, the phase of open load will increase with frequency increasing and in the second – the phase will decrease. Accordingly to the physical meaning, we can leave only one solution. Wherein because of the high length of the optical waveguide in narrow bandwidth it is possible to get many points with known derivatives that makes possible to use Hermite polynomials to get values of Γ_O, Γ_L, and by substituting them to (1-3) we can extract $S_{11}^{E/E}$, $S_{21}^{E/O} S_{12}^{O/E}$, $S_{22}^{O/O}$ parameters of BEON under test.

3. Conclusion

Theoretically is proposed a simple principle for extracting full S-matrix of a BEON both in E/O and O/E directions that is based on using only one port of standard microwave VNA three optical loads and a light waveguide of fixed length. The benefits of this calibration process are:

1) It is not needed fabricate several precision optical load standards and measure their parameters. Only one load (mirror) is needed to be precisely characterized.
2) It is possible to calculate inaccuracy of calibration and it is not needed to calibrate optical ports of BEON with some other lightwave component analyzer.
3) Using the proposed calibration technique it is possible to make lightwave components measurements by using only an ordinary VNA and a pair of BEONs.

Journal of Physics: Conference Series **643** (2015) 012062 doi:10.1088/1742-6596/643/1/012062

References

[1] Iezekiel S, Elamaran B and Pollard R 2000 *Eng. Sci. and Educ. J.* **12** 247
[2] Pollard R 2002 *IEE Proc.-Sci. Meas. Technol.* **149** **6** 345
[3] Tan T, Jungerman R, Elliott S 1989 *IEE Trans. Microw. Theory Tech.* **37** **8** 1217
[4] Kruppa W and Sodomsky K 1971 *IEEE Trans. Microw. Theory Tech* **19**, 122–123
[5] Rehnmark S 1974 *IEEE Trans. Microw. Theory Tech.* **22** 457–458
[6] Heuermann H and Schiek B 1994 *IEEE Trans. Instrum. Meas.* **43** 18–23

IR spectra of ICPCVD SiNx thin films for MEMS structures

G Rudakov[1] and I Reshetnikov[2]

[1] SMC Technological Centre, Zelenograd, Moscow, Russian Federation
[2] National Research University of Electronic Technology (MIET), Zelenograd, Moscow, Russian Federation

e-mail: grigory.rudakov@gmail.com

Abstract. Optical properties of non-stoichiometric silicon nitride (SiN_x) films for thermo sensitive membranes of microelectromechanical systems (MEMS) and microoptomechanical systems (MOMS) has been studied applying infrared (IR) spectroscopy. For the structures SiN_x/Si and (thin metal layer)/SiN_x/Si transmission and reflection spectra in the region of wave numbers of 500-7000 cm^{-1} has been investigated. For the investigated structures analysis of optical properties observed in the IR spectra both in the form of selective absorption bands and interference modulation of a baseline was conducted.

1. Introduction

Silicon nitride is widely used in microelectronics as a material for hard mask, dielectric and passivation layers [1] formation. Silicon nitride is characterized by highly thermal stability, good dielectric and mechanical properties and chemical resistance in various aggressive liquid or gaseous environments even at high temperatures. The main advantage of silicon nitride for application in solar cells and various thermal-imaging structures lays in its quite intensive absorption band of IR spectra of about 12 um wavelength that corresponds to the IR window of the atmosphere.

Stoichiometric silicon nitride (Si3N4) films produced by high temperature CVD method are characterized by great internal stress which causes to their destruction, and can't be applied in suspended membranes or consoled in 3D multi-element devices of MEMS or MOMS systems. In order to solve these problems, low temperature PECVD methods are applied by decomposing silane, ammonia or nitrogen at temperature lower than 300°C to form layers of a non-stoichiometric silicon nitride (SiN_x). PECVD methods allow to generate SiN_x films with controlled inner stresses and good mechanical properties [2,3]. For successive development of the MEMS/MOMS devices based on the SiN_x layers, detailed studies of optical and structural properties of films produced from that material are required.

2. Experiment

2.1. Samples fabrication

SiN_x thin films were deposited on (100) n-Si double-polished substrates and performed on an inductive coupled plasma chemical vapor deposition (ICPCVD) system (PLASMALAB 133, Oxford Instruments, Oxford, UK). The ICPCVD process ran at 250°C, under 10 mTorr pressure, 1200W RF power (13.56 MHz) and gas flow ratio (SiH4:N2) - 1:3. Prior to deposition the wafers was RCA cleaned to remove organics and particles. Metal reflecting layers were magnetron sputtered. Detailed samples characteristics are listed in table 1. All films were deposited on frontal surface.

Table 1. Samples information surface parameters

No.	SiN_x thickness, nm	Reflecting layer	Index	Maximum change of height, nm	Roughness, nm	Average lateral grain size, nm	Average height of grain, nm
1	497	n/a	2.12	4.8	0.6	73	2.1
2	402	Al (200 nm)	2.12	77	10	98	51
3	280	n/a	2.13	n/a	n/a	n/a	n/a
4	288	Ni (5 nm)	2.11	n/a	n/a	n/a	n/a
5	236	n/a	2.10	3.2	0.4	56	1.5
6	254	NiCr (50 nm)	2.12	2.4	0.3	61	1.3
7	128	n/a	2.10	1.9	0.2	48	0.8
8	128	Al (200 nm)	2.11	52	7	86	34
9	0	n/a	n/a	n/a	n/a	n/a	n/a

2.2. Thin film analysis

The film surfaces were examined using atomic force microscopy Smena™ VV (NT-MDT, Russia) with silicon cantilever NSG10S (NIIFP, Russia). The AFM results were obtained using tapping mode and scan speed of 4 nm/sec, which provide high resolution. In all cases, the roughness and grain size was measured for scanned area of 2x2 um^2.

Transmission and reflection experiments were conducted using a Perkin-Elmer Spectrum 100 IR Fourier Spectrometer. The IR spectra were taken at wave number range of 500-7800 cm^{-1} with 4 cm^{-1} spectral resolution. For each sample, 16 series of the measured spectra were performed for good signal-to-noise ratio determination. The transmission spectra (T) were obtained for normal incidence of non-polarized IR radiation. The reflection spectra (R) were measured using PIKE Technologies VeeMAX™ reflector and KRS-5 polarizer attached to the IR Fourier Spectrometer. The reflection of the p-polarized IR radiation at 30° were registered both on the frontal surface of the sample (R_f) and on its back surface (R_b).

The film thickness and reflective index were measured using spectral ellipsometer Senduro (Sentech, Germany). Table 1 shows the film parameters at wavelength of 632.8 nm.

3. Results and discussion

3.1. Surface morphology

The AFM examination of the SiN_x film surface revealed that the surface roughness increases by increasing thickness. Detailed surface parameters are shown in table 1. The data given in table 1 show that 50 nm of NiCr coating results in a slight change of surface roughness. However, for 200 nm Al the surface roughness is higher than that of the initial surface, especially clearly seen for thin SiN_x film.

3.2. Spectral analysis

Figure 1 and figure 2 show transmission IR spectra for samples with different SiN_x thickness. As it is seen from the figures, in the wave number 700-1200 cm^{-1} appear both selective absorption bands of the SiN_x and modulation the baseline. This is due to the interference of the IR in the film and in fact is Fabry-Perout modes. Peaks selective absorption bands correspond to minimum value of the transmission spectra. Identification of absorption bands is based on the [5, 6].

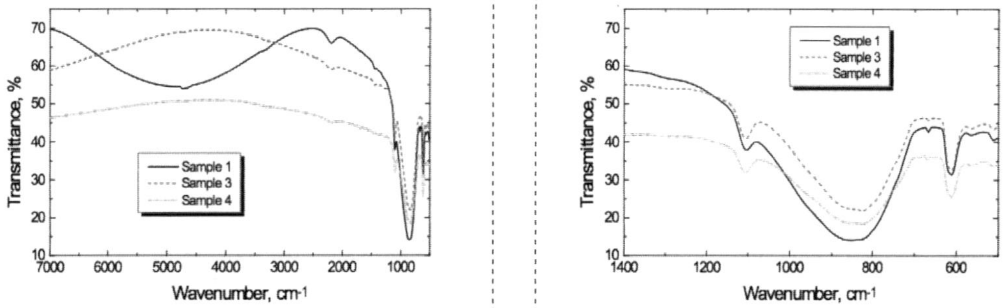

(a) (b)

Figure 1. IR transmission spectra for samples with different SiN_x thickness for wave numbers 500-700 cm^{-1} (a) and fragment of these spectra in the range wave numbers 500-1400 cm^{-1} (b).

The most intensive, quite wide and complex absorption band within the range of wave numbers of 700-1020 cm^{-1} with a minimum in the range of 820-860 cm^{-1} corresponds to absorption bands of stretching vibration of Si-N groups. It is worth noting that this absorption band detected may cause an interest in silicon nitride as a heat-sensitive material which is suitable for application in MEMS and MOM devices due to its window of spectral transparency of the atmosphere [7].

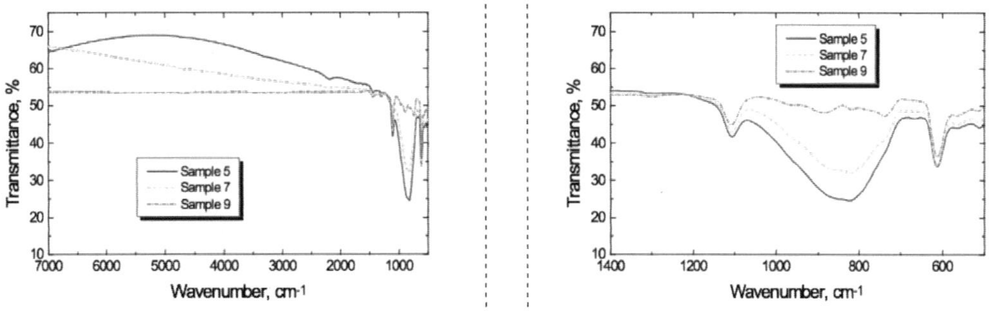

(a) (b)

Figure 2. IR transmission spectra for samples with different SiN_x thickness for wave numbers 500-700 cm^{-1} (a) and fragment of these spectra in the range wave numbers 500-1400 cm^{-1} (b).

Interference extremes in the modulated spectral baseline for various samples in figure 1 and figure 2 are observed at different wave numbers associated with different SiNx layer thickness. In the transmission spectra of the most thickest SiNx film the neighboring broad interference maxima in the spectra of the modulated baseline appear at wave numbers of ~2580 cm-1 and of ~7520 cm-1. An approximate estimate of intensity of selective absorption bands of stretching vibrations of Si-N at 860 cm-1 with a minimum transmittance ~14% and transmittance at baseline (background level) ~49% showed the value of 0.544 for layer thickness of 497 nm. Baseline level ~49% of transmittance in the IR spectrum at the wave number of ~860 cm^{-1} is determined basically by nonselective reflection and interference modulation baseline.

Figure 3 shows reflection spectra for samples with metal reflector. The spectrum of bare silicon is also given to compare. The identification method of IR reflectance spectra allows to qualitative estimate the dependence of selective absorption on the SiN_x thickness. For samples with IR semitransparent NiCr modulation baseline is clearly seen. Obviously that is due to the interference of the radiation in the SiN_x film. However, selective absorption bands SiNx practically are not visible because the angle 30° does not provide sufficient sensitivity method for recording reflection spectra of thin films bordered with the metal surface [8, 9].

On the other hand, samples with Al and different SiNx thickness are characterized by huge differences. For 128 nm SiN_x film modulation baseline is actually absent and absorption bands stretching vibration Si-N in the range 820-860 cm^{-1} is extremely weak. But in case with 402nm SiN_x film district interference modulation of the baseline and rather intense selective absorption band Si-N in the 820-860 cm^{-1} are observed. This obvious difference in the selective absorption of the reflection spectra of SiN_x films with different thickness is due to the nonlinear dependence of the intensity of the absorption band bordering on the metal surface of a thin dielectric film on the film thickness in the optical system at low angles of incidence of IR radiation at the metal - dielectric interface [8, 10].

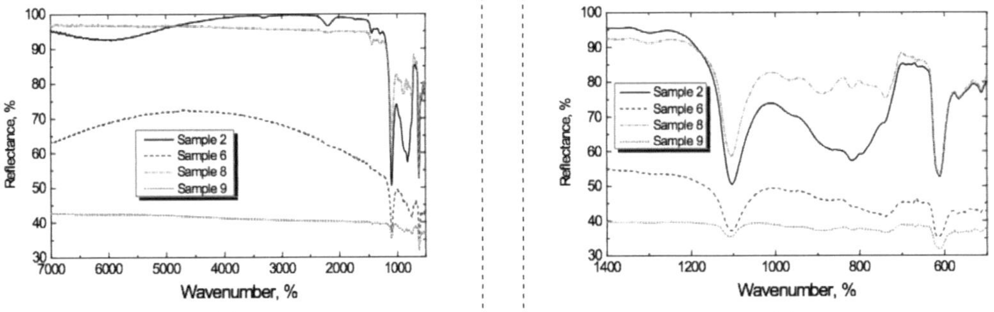

(a)

(b)

Figure 3. IR reflection spectra for samples with metal reflector for wave numbers 500-700 cm^{-1} (a) and fragment of these spectra in the range wave numbers 500-1400 cm^{-1} (b).

4. Conclusion

For the non-stoichiometric silicon nitride films of thickness range 120-500 nm in the SiN_x/Si and (thin metal layer)/SiN_x/Si structures designated for using in MEMS/MOMS, IR transmission and reflection spectra were investigated for IR region. One possible way to increase the amount of light energy absorbed by the SiN_x film in the window of spectral transparency of the atmosphere is the selection of SiN_x film thickness so the interference peak occurs in the region wave numbers 700-1100 cm^{-1}. Taking into consideration the analysis of SiN_x selective absorption bands and interference modulation of the base line in IR spectra, the optimal thickness of SiN_x layer was confirmed to be about of 1.0-1.5 um.

Acknowledgements

The presented work was carried out in National Research University «MIET» with a financial support of the Ministry of Education and Science of the Russian Federation, unique number RFMEFI57814X0009. We are grateful to A.A. Sugarev for his contribution to the analysis of results.

References

1. Dange M D, Lee J Y, Sooriakumar K 1991 New applications of low temperature PECVD silicon nitride films for microelectronic device fabrication *J. Microelectronics* **22** 19-26
2. Scheeper P R, Voorthuyzen J A and Bergveld P 1991 PECVD silicon nitride diaphragms for condenser microphones *Sensors and Actuators* B **4** 79-84A reference
3. Winchester K J and Dell J M 2001 Tunable Fabry-Perot cavities fabricated from PECVD silicon nitride employing zinc sulphide as the sacrificial layer *J. Micromech. Microeng.* **11** 589
4. Yakovlev V F, Vinogradov E F, Novikova N N, Mattei G, Godinho V, Fernandez-Ramos C, Fernandez A and Delplancke-Ogletree M P 2008 *Phys. Stat. Sol.* C **7** 2460-2463
5. Butarret E, Bensonda M, Habrad M, Bruyere J, Paulin S, Gujrathi S 1988 Configurational statistics in a a-SixNyHz alloys: A quantitive bonding analysis *Phis. Rev.* B **38/12** 8171-8184
6. Batan A, Franquet A, Vereechen J, Reniers F 2008 Characterisation of the silicon nitride thin films deposited by plasma magnetron *Surface and Interface Analysis* **40** 754-757
7. Yu X, Yi Y, Ma S, Lui M, Liu X, Dong L and Zhao Y 2008 Design and fabrication of high sensitivity focal array for uncooled IR imaging *J. Micromech. Microeng.* **18** 57001
8. Greenler R G 1975 Design of a reflection-absorption experiment for studying the IR spectrum of molecules adsorbed on a metal surface *J. Vac. Sic. Techno.* **12** 1410-1417
9. Greenler R G 1969 Reflection method for obtaining the infrared spectrum of a thin layer on a metal surface *J. Chem. Phys.* **50** 1963-9168
10. Rabolt J P, Jurich J, Swalen J D 1985 Infrared reflection-absorption studies of thin films at grazing incidence *Appl. Spectr.* **39** 269-272

Elimination of light coupling negative effect on the accuracy of external quantum yield determination in a MJ SC

S A Kozhukhovskaia, E D Filimonov, M A Mintairov and M Z Shvarts

Ioffe Institute, 26 Politekhnicheskaya, St Petersburg 194021, Russian Federation

E-mail: svetlana07.87@mail.ru

Abstract. This paper presents a new technique allows to eliminate negative effect of light coupling and to determine true value of external quantum yield of a multijunction solar cell.

1. Introduction

The light coupling (LC) observed inside monolithic semiconductor heterostructures and discussed actively during recent years, which results in appearance of induced current in a p-n junction (subcell) owing to absorption of luminescence from the adjacent wideband p-n junction, can lead to such positive effects as the rise of the short circuit current, of the open circuit voltage and, as a result, of the efficiency of a GaInP/GaAs/Ge multijunction solar cell (MJ SC). The processes mentioned above complicate methodology of the experimental study of MJ SCs and can substantially affect the accuracy in determining spectral and I-V characteristics [1-6].

It is known that, in investigating MJ SC spectral characteristics, an abnormal photoresponce out of typical spectral sensitivity range of the subcell under study is fixed. So a Ge subcell can demonstrate a high sensitivity in the short wavelength spectrum region, radiation from which is absorbed completely in photoactive GaInP and GaAs layers located above. In this case, the absolute external quantum yield (EQY) values of the Ge subcell being registered appear to be low. In the majority of cases, the registered abnormal photoresponce (out of main photoresponce range) of Ge subcell appears to be as a result of a complex negative effect of fundamental (light coupling) processes and of properties of p-n junctions (low shunting resistance) inherent in the semiconductor structure on the photosensitivity spectral dependences being determined by standard methods [2-5, 7]. This highlighted the need to improve corresponding experimental procedures.

In standard procedures for obtaining the spectral dependence of the EQY, the current signal from a subcell under study is registered at frequency-modulated monochromatic light in a required spectral range. Influence of other p-n junctions being not investigated at the moment is eliminated by constant broadband illumination creating conditions of current limitation from the side of a tested subcell, and shift of the tested subcell to the short circuit current condition is ensured by additional bias voltage [7].

Depicting photocurrents generated in the subcells in the form of bar chart (Fig.1), it is easy to see that, at intense illumination of the wideband (WB) subcell, in the narrowband (NB) p-n junction, an induced current J_0^N can arise caused by luminescent radiation, value of which will be proportional to internal current J_0^{pn} flowing through the WB p-n junction (Fig. 1a). In adding photocurrent J_λ^N caused by chopped monochromatic light into the NB subcell (Fig. 1b), an instantaneous drop of internal current through the WB p-n junction down to J_1^{pn} occurs. As a result, the reaction on the monochromatic radiation, i.e. registered "alternative" photoresponce ΔJ, will be less than the value J_λ^N, and, hence, the magnitude of the EQY for the NB subcell will be underestimated:

$$QE = \frac{\Delta J}{J_{ref}} \cdot QE_{ref}, \tag{1}$$

where ΔJ – registered increment of the photocurrent caused by monochromatic light and used in standard measurement procedures for determining the EQY of the photoresponce (expression 2); J_{ref} – photocurrent of a reference SC at its illumination by chopped monochromatic light; QE_{ref} – external quantum yield of the reference SC at chosen wavelength.

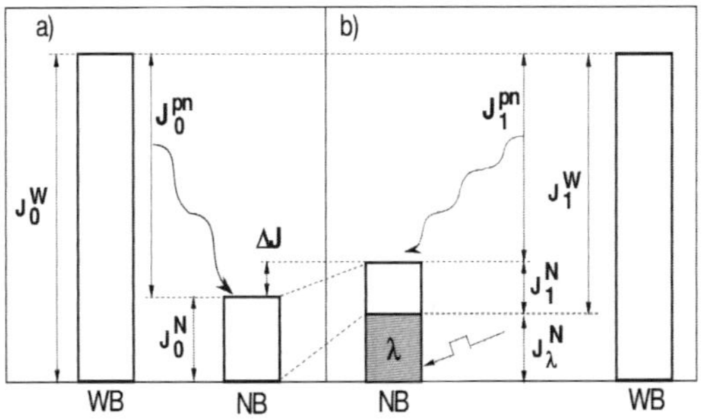

Figure 1. Bar chart of photocurrents of WB and NB subcells in conditions of standard procedure for recording spectral dependence of the EQY of photoresponce.

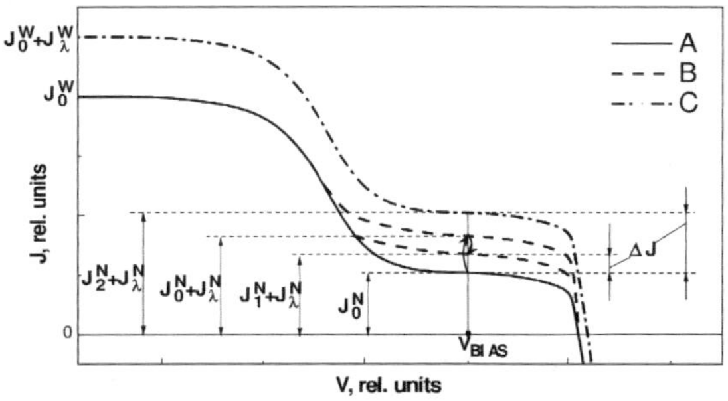

Figure 2. Graphical representation of photocurrent values for I-V curves, obtained under various illumination conditions: initial (A), in adding chopped monochromatic light for NB (B) and WB (C), correspondingly.

2. Experimental details

In the work, a new approach to determining true EQY values of NB subcells in a MJ SC, based on study of spectral characteristics only with application of sources of frequency-modulated radiation ensuring also a bias light for the uninvestigated WB subcells, is presented.

The use of sources of only alternative (frequency-modulated) radiation leads to that registered signals will be noticeably different from those obtained in a standard measurement procedure. At absence of monochromatic illumination, an induced alternative photocurrent J_0^N will be registered, but in adding it - $J_1^N+J_\lambda^N$ (see Fig. 1b). At a chosen level of illumination of WB subcell, the following relation for photocurrent values at the bias voltage V_{BIAS} (Fig.2) is valid:

$$\Delta J + J_0^N = J_\lambda^N + J_1^N, \tag{2}$$

where the desired value J_λ^N is associated with currents of the WB subcell as $J_0^W - J_1^W = J_\lambda^N$ (see Fig.1b) and can be determined graphically by the subtraction method.

For this it is necessary, first of all, to determine experimentally the dependence of the photocurrent in the NB subcell (J_{LC}^N), induced by luminescence, on photocurrent generated in the adjacent irradiating subcell (J^W) (Fig, 3). In setting initial conditions in experiment (J_0^W, J_0^N), one can, in deviating consistently to the left (zone 1 in Fig.3) of the value J_0^W by some value J_λ^N, search for a pair of values (J_1^N, J_λ^N), at which equation (2) will be fulfilled.

The proposed method for determining values of J_λ^N appears to be rather sensitive to both initial conditions (J_0^W, J_0^N) and value J_λ^N itself depending on the level of illumination with monochromatic radiation.

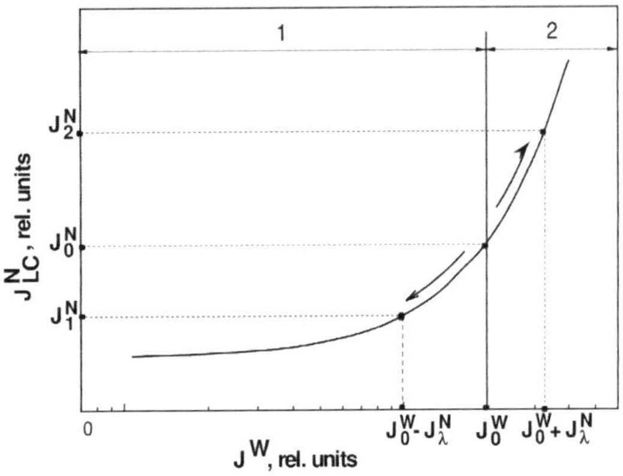

Figure 3. The dependence of the photocurrent induced by luminescence in the NB subcell (J_{LC}^N) on the photocurrent generated in the neighboring WB subcell (J^W): 1 and 2 designate the zones for J_λ^N determination, when λ is within the NB (1) or WB (2) subcell photosensitivity range.

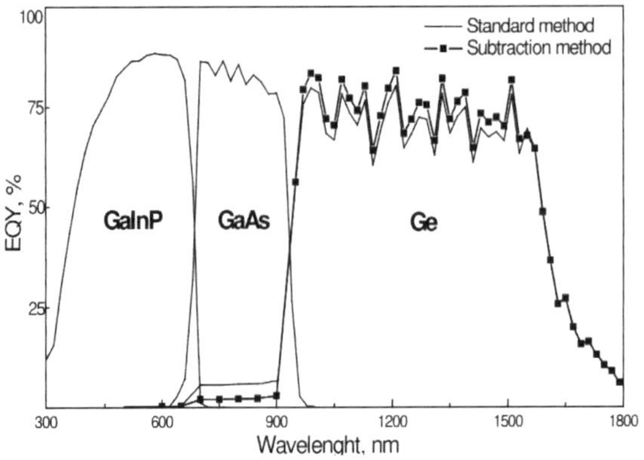

Figure 4. Spectral dependence of the external quantum yield of photoresponce obtained in two ways: standard method and subtraction method).

Since the magnitudes of WB subcell photocurrent J_0^W and of induced current J_0^N are bound through the luminescent coupling effectiveness γ_0, as well as J_1^N and J_1^W through γ_1 [1]:

$$\gamma_0 = \frac{J_0^N}{J_0^{pn}} = \frac{J_0^N}{J_0^W - J_0^N}, \tag{3}$$

$$\gamma_1 = \frac{J_1^N}{J_1^{pn}} = \frac{J_1^N}{J_1^W - J_1^N} = \frac{J_1^N}{(J_0^W - J_\lambda^N) - J_1^N}, \tag{4}$$

then, at low effectiveness of luminescent coupling, the $J_{LC}^N(J^W)$ dependence will be gently sloping ($\gamma_0 \approx J_0^N / J_0^W$; $\gamma_1 \approx J_1^N / J_1^W \approx J_1^N/(J_0^W - J_\lambda^N)$), which will make the difference between values of J_0^N и J_1^N extremely minor (compared with the accuracy of registration of the pointed out values) even at high levels of illumination with monochromatic light and large J_λ^N. It is apparent that at considerable luminescent coupling the J_{LC}^N (J^W) dependence will be steeper, and more "comfortable" initial conditions (J_0^W, J_0^N) can be preset in experiment. Requirements to ensuring high illumination from the monochromatic source are also reduced. Nevertheless, in allowing for probable rather strong change of monochromatic illumination at scanning the optical range (inherent in the systems based on monochromators and xenon/halogen light sources), which will affect the value of J_λ^N, it should be preferred to such initial conditions (point (J_0^W, J_0^N) in Fig.3), in which the tangent to the J_{LC}^N (J^W) dependence will have a slope of about 45 angular degrees.

It should be emphasized that the dependence $J_{LC}^N(J^W)$ for each pair of subcells and for any MJ SC is individual and will be determine by the luminescent coupling effectiveness. In case of high luminescent coupling effectiveness and in using standard measuring methods, an error in determining values of the EQY of photoresponce of a NB (absorbing luminescent radiation) subcell can be substantial (up to 20 rel.%) [8]. For this reason, the described above multistage procedure to obtain correct (without "artifacts") spectral characteristics of MJ SCs is proposed to be used:

- construction of the $J_{LC}^N(J^W)$ dependence and determination of the initial conditions (J_0^W, J_0^N);
- scanning the spectral sensitivity range with recording values of $J_1^N+J_\lambda^N$ for subcell under test and corresponding to them values of reference photodetector photocurrent (J_{ref}) at illumination with chopped monochromatic radiation;

- calculation of J_λ^N values for each wavelength of monochromatic radiation and using them in constructing the spectral dependence of external quantum yield of photoresponce (EQY (λ)) in correspondence with the formula (1), where ΔJ is substitute for J_λ^N.

3. Conclusion

The proposed procedure has been realized on an installation for recording spectral dependencies of MJ SC [9], in which an optical chopper was placed in front of the tested specimen thereby modulating fluxes of monochromatic radiation and bias light.

Even the first experiments have shown that the proposed multistage method gives the increase to the photocurrent of a germanium subcell greater than in 1% within its spectral sensitivity range and practically completely excludes the effect of anomalous photoresponce in the short wavelength part of spectrum (Fig.4).

Further development and application of the proposed measuring procedure will allow determining more correctly output and metrological characteristics of MJ SCs at research practice and life tests associated with action of different damaging factors on the SC semiconductor structure.

Acknowledgments

Support for this work comes from the Russian Foundation on Basic Research (Grant 14-08-01222-a).

References

[1] Shvarts M Z, Mintairov M A, Emelyanov V M, Evstropov V V, Lantratov V M and Timoshina N Kh 2013 Method for direct measurements of luminescent coupling efficiency in concentrator MJ SCs *AIP Conf. Proc* **1556** 147-151

[2] Siefer G, Baur C, Bett AW 2010 External quantum efficiency measurements of germanium bottoms subcells: measurement artifacts and correction procedures *Proceedings of the 35th IEEE PVSC* (Honolulu) 704–707

[3] Lim S H, Li J J, Steenbergen E H and Zhang Y H 2011 Luminescence coupling effects on multi-junction solar cell external quantum efficiency measurements *Progress in PV: Research and Applications* vol 21 p 344

[4] Steiner M A, Geisz J F, Moriarty T E, France R M, McMahon W E, Olson J M, Kurtz S R and Friedman D J 2013 Measuring IV Curves and Subcell Photocurrents in the Presence of Luminescent Coupling IEEE J. Photovoltaics vol 3 pp 879–887

[5] Li J J, Lim S H, Allen C R, Ding D and Zhang Y H 2011 Combined effects of shunt and luminescence coupling on external quantum efficiency measurements of multi-junction solar cells *IEEE J. Photovoltaics* vol 1 pp 225–230

[6] Kozhukhovskaia S A, Filimonov E D and Shvarts M Z 2015Experimental model of multi-junction solar cell: IV curve tailoring and photoresponce artifact simulation *Journal of Physics: Conference Series* **586** 012014

[7] Meusel M, Baur C, Le´tay G, Bett A W, Warta W and Fernandez E 2003 Spectral Response Measurements of Monolithic GaInP/Ga(In)As/Ge Triple-Junction Solar Cells: Measurement Artifacts and their Explanation *Progress in PV: Research and Applications* **11(8)** pp 499-514

[8] Li J J and Zhang Y H 2013 Elimination of measurement artifacts in external quantum efficiency of multijunction solar cells using a pulse light bias *IEEE J. Photovoltaics* vol 3 pp 364-369

[9] Shvarts M Z, Chalov A E, Ionova E A, Larionov V R, Malevskiy D A, Rumyantsev V D, Titkov S S 2005 Indoor characterization of the multijunction III-V solar cells and concentrator modules *Proceedings of the 20th European Photovoltaic Solar Energy Conference* (Barcelona, June) pp 278-281

Magnetic system design for investigation of magneto-optical properties of objects in strong magnetic field in terahertz frequency range

S E Azbite, A K Denisultanov, M K Khodzitsky

Department of Photonics and Optical Information Technologies, ITMO University, 49 Kronverksky av., St. Petersburg, 197101 Russian Federation

E-mail: azbite@mail.ru, alaudi.denisultanov@gmail.com, khodzitskiy@yandex.ru

Abstract. A strong localized magnetic field hemispherical system at room temperature was proposed for application in terahertz magnetooptics, magneto-resonance spectroscopy and magneto-tunable devices. The most effective and suitable structural and material parameters of the system were estimated.

1. Introduction

As known, many advanced scientific experiments need a strong magnetic fields usage [1-4]. For example, properties graphene, which has a huge potential in practice applications [5-9], strongly depend on influence of external temperature and magnetic field [10-12]. For generation of strong magnetic fields electromagnets and superconductor materials are often used, but these systems are not always suitable and energy-consuming for their usage in experimental setups. Therefore, this paper is dedicated to static magnetic system design with the strong magnetic field for the experimental investigation of magneto-dependent graphene properties such as conductivity in terahertz frequency range.

2. Magnetic system parameters and configurations

In the paper [13] the several magnetic system design types are investigated and according to the results hemispherical magnetic system has the strongest magnetic field. This hemispherical system is constructed by equal magnetic segments and a cone underneath the structure. This system is shown in Figure 1.

Magnetic field strength of the system can be found by the following expression:

$$H(\alpha, \phi_1, \phi_2) = [2\pi\sigma_{av} \sin 2\alpha + 2n \cdot 2M_s(\sigma_1 \sin\alpha + \sigma_2(1 - \sin\alpha))] \cdot \ln(R/r)\,, \qquad (1)$$

where R is a magnetic system radius, $r = (x^2 + y^2)^{1/2}$ is the distance of a point from the origin to the observer point, M_s is a saturation magnetization, α is a angle at the cone vertex,ϕ_1 and ϕ_2 is angles between vector $\mathbf{M_s}$ and cone axis, n is a segments number, the average density of

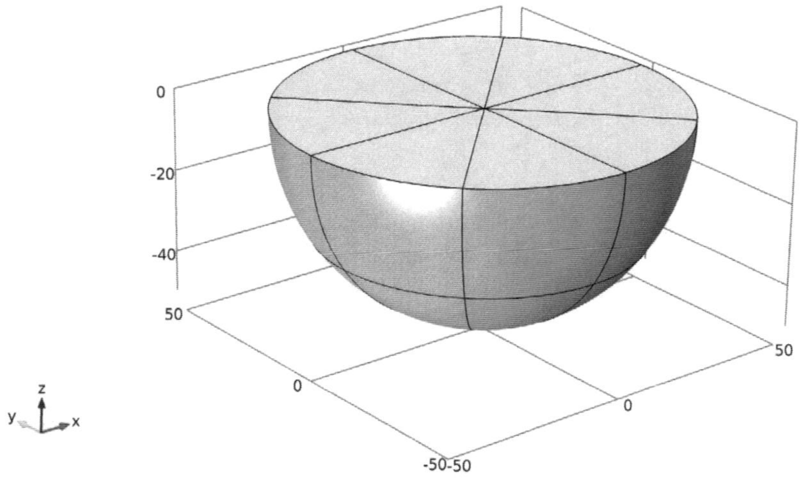

Figure 1. Hemispherical static magnetic system. Coordinate scale in *mm* units.

charges is

$$\sigma_{av} = M_s[\arccos(\sin(\alpha+\phi_1) - \frac{(\pi/n)^2}{6}\cos\alpha\sin\phi_1) - \arccos(\sin(\alpha+\phi_2) - \frac{(\pi/n)^2}{6}\cos\alpha\sin\phi_2)] \quad (2)$$

the charge densities at the boundaries between the magnet sectors $\sigma_1 = \sin(\pi/n)\sin\phi_1$ and $\sigma_2 = \sin(\pi/n)\sin\phi_2$. For the segments number n=8 the parameters $\alpha = 54°$,$\phi_1 = -107°$ and $\phi_2 = 14°$ allow to obtain the most effective system design [13]. For the hemisphere radius R=50 mm, the segments number n=8 and the saturation magnetization $M_s = 1280$ Gs the magnetic field H dispersion has the following form (Figure 2):

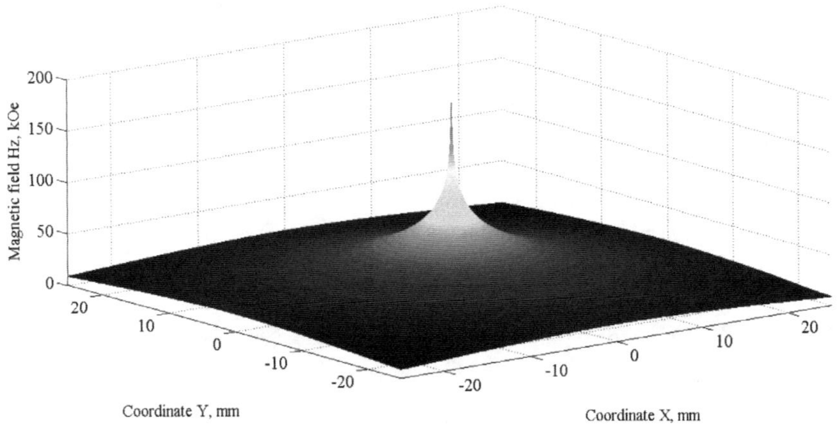

Figure 2. The magnetic field mapping for the investigated system.

2.1. Material
For the simulation the material with the highest magnetization value was chosen. It was $Nd_2Fe_{14}B$ with magnetization $M_s = 1280$ Gs [14]. The magnetic systems consisted of this material can be made by industrial firms on request. In another words, nowadays it is no problem to get this system for experimental proposes.

2.2. Segments number
The segments number n of the magnetic system is able to impact on the magnetic field strength. The dependance of the magnetic field H on the coordinate X was simulated at the coordinate $Y = 0$. The comparison of the impact of different segments number is demonstrated in Figure 3:

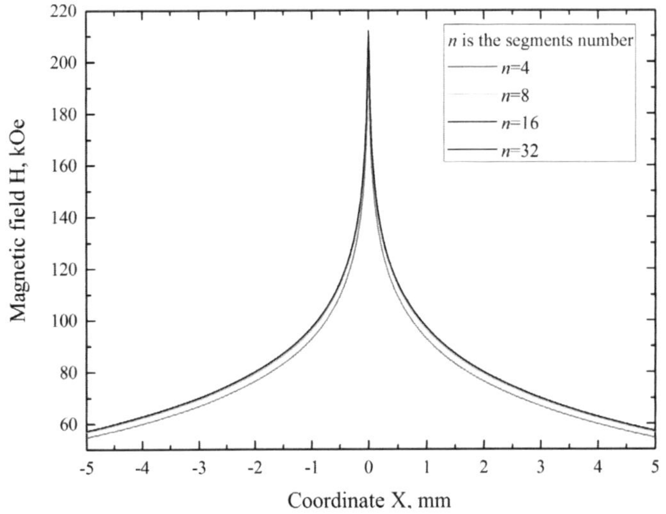

Figure 3. Dependence of the magnetic field strength on the segments number.

As seen in Figure 3, by increasing of the segments number the magnetic system saturation comes after $n = 8$. It is worth noting, the assembling of 16 segments is harder than 8, so $n = 8$ is the most optimal segments number.

2.3. Magnetic field for various radius
The next step is the definition of the system radius influence on the magnetic field distribution. In simulation segments number n=8 and magnetization $M_s = 1280$ Gs were used. In Figure 4 and Figure 5 the magnetic field distribution for various radius are shown.

By the magnetic field distribution behavior in Figure 5 it can be concluded, that by increasing of the system radius the magnetic field distribution smooths. Too much smoother field is more suitable for a research, because in every point more precise magnetic field value is obtained. But we are limited by experimental setup proportions, therefore the radius of magnetic system should be agree with the system compactness.

3. Conclusions
In this paper analytical simulation of hemispherical magnetic system with the strong magnetic field was investigated. The influence of the magnetic system configuration parameters such as the system radius R, the magnetization M_s and the segments number n on the magnetic

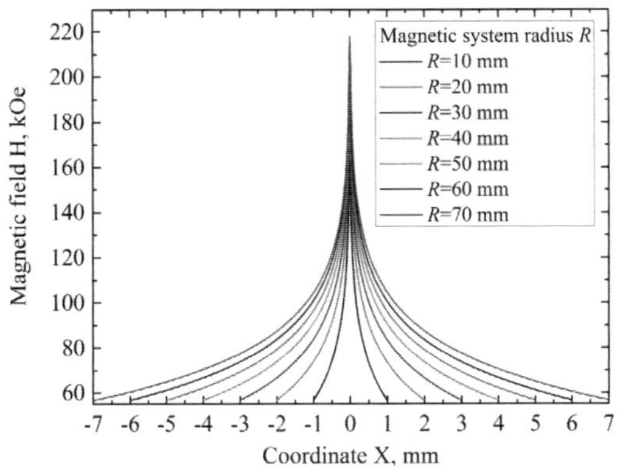

Figure 4. Magnetic field strength for various R.

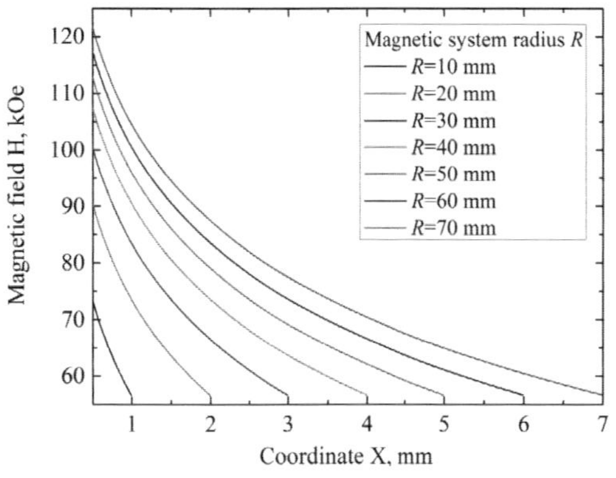

Figure 5. Magnetic field strength for various R in zoom mode.

field distribution was demonstrated. This type magnetic system with most suitable parameters (R = 50 mm, M_s = 1280 Gs and n=8) will be used in terahertz spectrometer setup for experiments due to simplicity in using (a sample is shifted along the magnetic system surface) and availability of magnetic field map, that allows to know magnetic field value in every point.

Acknowledgments
This work was financially supported by Government of Russian Federation, Grant 074-U01.

4. References

[1] Oka T, Tanaka K, Kimura T, Mimura D, Fukui S, Ogawa J, Sato T, Ooizumi M, Yokoyama K and Yamaguchi M 2010 *Physica* C **470** 17991803
[2] Hu C, Sakka Y, Tanaka H, Nishimura T and Grasso S 2011 *J. Am. Ceram. Soc.* **94** 410415
[3] Ginossar E, Glazman L I, Ojanen T, Felix von Oppen, Shanks W E, Bleszynski-Jayich A C and Harris J G E 2010 *Phys. Rev.* B **81** 155448
[4] Denisultanov A, Azbite S and Khodzitsky M 2014 *J. Phys.: Conf. Ser.* **541** 012058
[5] Postma H W Ch 2010 *Nano Lett.* **10** 2
[6] Tokman M, Yao X and Belyanin A 2013 *Phys. Rev. Lett.* **110** 077404
[7] Gurvitz E, Vozianova A and Khodzitsky M 2014 Tunable frequency converter for terahertz frequency range based on space-time transformation *Proc. SPIE* vol 9160 (San Diego: USA) p 91602C
[8] Wang Y, Shao Y, Matson D W, Li J and Li Y 2010 *ACS Nano* **4** 4 pp 17901798
[9] Warner J H 2013 *Graphene. Fundamentals and emergent applications* (Amsterdam)
[10] Goerbig M O 2011 *Rev. Mod. Phys.* **83** 1193
[11] Denisultanov A, Azbite S and Khodzitsky M 2014 Influence of magnetic field on the surface waves properties in the photonic crystal/graphene structure for terahertz frequency range *Proc. SPIE* vol 9160 (San Diego: USA) p 91602T
[12] Raza H 2012 *Graphene Nanoelectronics. Metrology, Synthesis, Properties and Applications* (New York)
[13] Samofalov V N, Belozorov D P, Ravlik A G 2006 *Phys. of Metals and Metallography* **102** 5 pp. 494-505
[14] Skomski R and Sellmyer D J 2006 *Handbook of Advanced Magnetic Materials* vol 1 (New York)

Dielectric permittivity of artificial periodic structure and local field effects

M N Anokhin, A A Tishchenko and M N Strikhanov

National Research Nuclear University MEPhI, Moscow, Russia

E-mail: MNAnokhin@mephi.ru, tishchenko@mephi.ru

Abstract. In this work we study dielectric properties of an artificial periodic structure on the base of the local field theory. We generalize Clausius-Mossotti relation for periodic structures and demonstrate that in this case a spatial dispersion of dielectric permittivity takes place.

1. Introduction

Propagation of the electromagnetic waves in periodic structures based on different kinds of materials is of great interest for numerous applications [1-4]. Periodicity plays a key role from point of view of propagation of electromagnetic waves in such structures [5, 6], which is of importance for applications in resonators, signal dividers, printed antennas include higher antenna gains [7] playing a vital part in microwave electronics. It is also very important that a new class of materials with unusual properties, so called metamaterials, is realized on the base of the periodic structures [8].

In this paper we obtain a dielectric permittivity of a periodic structure *ab initio*, with help of direct solution of Maxwells equations, and without numerical calculations. We also take into account the so called local field effects, which is very important for analysis of resonant conditions in periodic structures.

2. Dielectric properties

Let us consider the infinite periodic structure composed of anisotropic particles with polarizability $\alpha_{ij}(\omega)$. The particles can be both usual atoms or molequles and nanoparticles, quantum dots or another objects. Let the external field $\mathbf{E}^0$ act upon this material.

Using the dipole approximation it is not hard to obtain the Fourier transform of the solution of Maxwell's equations in a medium (see, e.g., [9]):

$$E_i^{mic}(\mathbf{r},\omega)=E_i^0(\mathbf{r},\omega)+\frac{1}{2\pi^2}\int d^3lS_{ij}(\mathbf{l},\omega)\alpha_{jk}(\omega)\sum_b E_k^{mic}(\mathbf{R}_b,\omega)\exp\{-i\mathbf{l}(\mathbf{R}_b-\mathbf{r})\}, \quad (1)$$

where

$$\alpha_{jk}(\omega)=\alpha_\perp(\omega)(\delta_{jk}-e_je_k)+\alpha_\parallel(\omega)e_je_k, \quad (2)$$

$$S_{ij}(\mathbf{l},\omega)=\frac{k^2\delta_{ij}-l_il_j}{l^2-k^2-i0}, \quad (3)$$

$$k=\omega/c. \quad (4)$$

One can solve this system only approximately because $N\gg 1$. Below we shall follow the method developed in our previous research [9-12].

For $\mathbf{r}=\mathbf{R}_a$ one can easily obtain from Eq. (1) a system of N equations for acting on the a-th particle field:

$$E_i^{mic}\left(\mathbf{R}_a,\omega\right)=E_i^0\left(\mathbf{R}_a,\omega\right)+\frac{1}{2\pi^2}\int d^3lS_{ij}\left(\mathbf{l},\omega\right)\alpha_{jk}\left(\omega\right)\sum_b E_k^{loc}\left(\mathbf{R}_b,\omega\right)\exp\left\{-i\mathbf{l}\left(\mathbf{R}_b-\mathbf{R}_a\right)\right\}, \quad (5)$$

where the local field acting upon the particle in $\mathbf{R}_a$ can be obtained by averaging over the location of all other particles located in points $\mathbf{R}_b \neq \mathbf{R}_a$:

$$\begin{aligned}&E_i^{loc}\left(\mathbf{R}_a,\omega\right)=E_i^0\left(\mathbf{R}_a,\omega\right)+\\&+\frac{1}{2\pi^2}\int d^3lS_{ij}\left(\mathbf{l},\omega\right)\alpha_{jk}\left(\omega\right)\int d^3R_{ba}Nw\left(\mathbf{R}_{ba}\right)E_k^{loc}\left(\mathbf{R}_a+\mathbf{R}_{ba},\omega\right)\exp\left\{-i\mathbf{l}\mathbf{R}_{ba}\right\},\end{aligned} \quad (6)$$

where $w\left(\mathbf{R}_{ba}\right)$ is the probability density of finding the b-th particle at the distance $\mathbf{R}_{ba}=\mathbf{R}_b-\mathbf{R}_a$ $\left(k=\overline{1,N-1}\right)$:

$$w\left(\mathbf{R}_{ba}\right)=\frac{1}{N}\sum_{m=1}^{N}\delta\left(\mathbf{R}_{ba}-\mathbf{R}_m\right). \quad (7)$$

Averaging Eq. (1) over the coordinates of all the particles we get the conventional macroscopic field:

$$E_i\left(\mathbf{R},\omega\right)=E_i^0\left(\mathbf{R},\omega\right)+\frac{1}{2\pi^2}\int d^3lS_{ij}\left(\mathbf{l},\omega\right)\alpha_{jk}\left(\omega\right)\left\langle\sum_b E_k^{loc}\left(\mathbf{R}_b,\omega\right)\exp\left\{-i\mathbf{l}\left(\mathbf{R}_b-\mathbf{R}\right)\right\}\right\rangle, \quad (8)$$

where

$$\begin{aligned}&\left\langle\sum_b E_k^{loc}\left(\mathbf{R}_b,\omega\right)\exp\left\{-i\mathbf{l}\left(\mathbf{R}_b-\mathbf{R}\right)\right\}\right\rangle=\\&=\frac{N}{V}\int d^3\mathbf{R}'E_k^{loc}\left(\mathbf{R}',\omega\right)\exp\left\{-i\mathbf{l}\left(\mathbf{R}'-\mathbf{R}\right)\right\}=n(2\pi)^3E_k^{loc}\left(\mathbf{l},\omega\right)\exp\left\{i\mathbf{l}\mathbf{R}\right\}\end{aligned}. \quad (9)$$

Thus, we obtain the macroscopic field in form:

$$E_i\left(\mathbf{R},\omega\right)=E_i^0\left(\mathbf{R},\omega\right)+4\pi nS_{ij}\left(\mathbf{R},\omega\right)\alpha_{jk}\left(\omega\right)E_k^{loc}\left(\mathbf{R},\omega\right). \quad (10)$$

Using Eq. (6) and (10) one can express the macroscopic field $\mathbf{E}$ trough the local field $\mathbf{E}^{loc}$:

$$\begin{aligned}&E_i\left(\mathbf{R},\omega\right)=E_i^{loc}\left(\mathbf{R},\omega\right)+4\pi nS_{ij}\left(\mathbf{R},\omega\right)\alpha_{jk}\left(\omega\right)E_k^{loc}\left(\mathbf{R},\omega\right)-\\&-\frac{1}{2\pi^2}\sum_{m=1}^{N}S_{ij}\left(-\mathbf{R}_m,\omega\right)\alpha_{jk}\left(\omega\right)E_k^{loc}\left(\mathbf{R}+\mathbf{R}_m,\omega\right)\end{aligned}. \quad (11)$$

Eq. (11) in the Fourier coordinates $\left(\mathbf{q},\omega\right)$ takes the form:

$$\begin{aligned}&E_i\left(\mathbf{q},\omega\right)=E_i^{loc}\left(\mathbf{q},\omega\right)+4\pi n\,S_{ij}\left(\mathbf{q},\omega\right)\alpha_{jk}\left(\omega\right)E_k^{loc}\left(\mathbf{q},\omega\right)-\\&-\frac{1}{2\pi^2}\sum_{m=1}^{N}S_{ij}\left(-\mathbf{R}_m,\omega\right)\alpha_{jk}\left(\omega\right)\exp\left\{i\mathbf{q}\mathbf{R}_m\right\}E_k^{loc}\left(\mathbf{q},\omega\right)\end{aligned}. \quad (12)$$

This allows to obtain the equation connecting the Fourier transforms of the local and macroscopic fields:

$$t_{ik}\left(\mathbf{q},\omega\right)E_k^{loc}\left(\mathbf{q},\omega\right)=E_i\left(\mathbf{q},\omega\right), \quad (13)$$

where

$$t_{ik}\left(\mathbf{q},\omega\right)=\delta_{ik}+4\pi n\,S_{ij}\left(\mathbf{q},\omega\right)\alpha_{jk}\left(\omega\right)-\frac{1}{2\pi^2}\sum_{m=1}^{N}\exp\left\{i\mathbf{q}\mathbf{R}_m\right\}S_{ij}\left(-\mathbf{R}_m,\omega\right)\alpha_{jk}\left(\omega\right), \quad (14)$$

and

$$S_{ij}\left(-\mathbf{R}_m,\omega\right)=\int d^3lS_{ij}\left(\mathbf{l},\omega\right)\exp\left\{-i\mathbf{l}\mathbf{R}_m\right\} \quad (15)$$

If one neglects the anisotropy of individual particles, i.e.:

$$\alpha\left(\omega\right)\equiv\alpha_{\perp}\left(\omega\right)\equiv\alpha_{\parallel}\left(\omega\right) \quad (16)$$

the Eq. (15) can be represented in the form:

$$\int d^3 l S_{ij}(\mathbf{l},\omega)\exp\{-i\mathbf{l}\mathbf{R}_m\} = a(R_m)\delta_{ij} + b(R_m)\frac{R_i^m R_j^m}{R^2}, \tag{17}$$

where

$$\begin{aligned} a(R_m) &= -2\frac{\pi^2 k}{R_m^2}\sin(kR_m) - 2\frac{\pi^2}{R_m^3}\cos(kR_m), \\ b(R_m) &= 4\frac{\pi^2 k^2}{R_m}\cos(kR_m) + 6\frac{\pi^2 k}{R_m^2}\sin(kR_m) + 6\frac{\pi^2}{R_m^3}\cos(kR_m) \end{aligned}. \tag{18}$$

Thus, tensor $t_{ik}(\mathbf{q},\omega)$ can be found the form:

$$\begin{aligned} t_{ik}(\mathbf{q},\omega) &= \delta_{ik} + 4\pi n\alpha(\omega)\frac{k^2\delta_{ik} - q_i q_k}{q^2 - k^2 - i0} - \\ &- \frac{1}{2\pi^2}\sum_{m=1}^{N}\left[a(R_m)\alpha(\omega)\delta_{ik} + b(R_m)\alpha(\omega)\frac{R_{mi}R_{mk}}{R_m^2}\right]\exp\{i\mathbf{q}\mathbf{R}_m\} \end{aligned}. \tag{19}$$

Finally, we obtain the dielectric permittivity expressed through the reciprocal tensor $t_{jk}^{-1}(\mathbf{q},\omega)$:

$$\varepsilon_{ik}(\mathbf{q},\omega) = \delta_{ik} + 4\pi n\alpha_{ij}(\omega)t_{jk}^{-1}(\mathbf{q},\omega) \tag{20}$$

The dependence of dielectric permittivity on the vector $\mathbf{q}$ means the spatial dispersion, i.e. non-local relationship between electromagnetic field and corresponding response of the structure. The spatial dispersion proves to be immanent property of periodic structure considered here.

3. Discussion

Eq. (20) describes all dielectric properties of the periodic structure consisting of particles of arbitrary nature in the dipole approximation. The most important feature here is that dielectric permittivity demonstrates the spatial dispersion. Also, the poles of function $t_{jk}^{-1}(\mathbf{q},\omega)$ express the local field effects, which is very promising for the further analysis.

4. Acknowledgment

This work was supported by the Ministry of Education and Science of the Russian Federation, Agreement No. 14.575.21.0002, ID RFMEFI57514X0002.

References

[1] Datta S, Chan C T, Ho K M and Soukoulis C M Effective dielectric constant of periodic composite structures 1993 Phys. Rev. B **48** 14936.

[2] Silin R A Electromagnetic waves in artificial periodic structures 2006 Phys. Usp. **49** 542.

[3] Zhang Z, Satpathy S Electromagnetic wave propagation in periodic structures: Bloch wave solution of Maxwell's equations 1990 Phys. Rev. Lett. **65** 2650.

[4] Stefanout N, Karathanosf V and Modinosf A Scattering of electromagnetic waves by periodic structures 1992 J. Phys: Condens. Matter **4** 7389.

[5] Economou E N and Sigalast M M Classical wave propagation in periodic structures: Cermet versus network topology 1993 Phys. Rev. B **48** 13434.

[6] Pisarenco M, Setija I D On the complexity of aperiodic Fourier modal methods for finite periodic structures 2014 Journal of Computational Physics **261** 130.

[7] Yang H.-Y.D. Wang J Surface waves of printed antennas on planar artificial periodic dielectric structures 2001 IEEE Transactions on Antennas and Propagation **49** 444.

[8] Koschny Th, Markoš P, Economou E N, Smith D R, Vier D C and Soukoulis C M Impact of inherent periodic structure on effective medium description of left-handed and related metamaterials 2005 Phys. Rev. B **71** 245105.

[9] Ryazanov M I and Tishchenko A A Clausius-Mossotti-type relation for planar monolayers 2006 JETP **103** 539 [ZhETF **130** 621].

[10] Ryazanov M I, Strikhanov M N and Tishchenko A A Local field effect in diffraction radiation from a periodical system of dielectric spheres 2008 Nucl. Instr. and Meth. B **266** 3811.

[11] Anokhin M N, Tishchenko A A, Ryazanov M I and Strikhanov M N Permittivity of anisotropic dielectric near surface with local field effects 2014 Journal of Physics: Conference Series **541** 012023.

[12] Porvatkina O V, Tishchenko A A and Strikhanov M N Local field effects in anisotropic metamaterials 2014 Journal of Physics: Conference Series **541**, 012024.

Enhancement of spontaneous emission in metal-dielectric multilayer structures accounting losses

A. R. Gubaydullin[1], M. A. Kaliteevski[1,2]

[1]St. Petersburg Academic University, Khlopina 8/3, 194021, St. Petersburg, Russia

[2]ITMO University, Kronverkskiy pr. 49, 197101, St. Petersburg, Russia

E-mail: gubaydullin.azat@gmail.com

Abstract. We study the emission rate enhancement of the dipole emitter centred in the stratified metal-dielectric metamaterial, characterized by the hyperbolic isofrequency surface. We find out a limited enhancement of the Purcell factor in the layered metamaterial. We demonstrate that the radiative decay rate is strongly depends on a ratio of the thickness of layers and is affected by the level of losses in metal.

1. Introduction

Metamaterials constructed from periodic metal-dielectric multilayer system are generally known as a simple realization of the so-called hyperbolic media [1]. Such hyperbolic metamaterial (HMM) systems have attracted great scientific interest over the past decade due to their special optical properties emanating primarily from hyperbolic dispersion of isofrequency surface. When one of the principal components of the effective permittivity tensor ε_{eff} of the diagonal form is opposite in sign to the other two principal components [2]:

$$\varepsilon_{eff} = \begin{pmatrix} \varepsilon_{\perp} & 0 & 0 \\ 0 & \varepsilon_{\perp} & 0 \\ 0 & 0 & \varepsilon_{||} \end{pmatrix}, \tag{1}$$

where components perpendicular and parallel to the anisotropy axis are given by:

$$\varepsilon_{\perp} = \frac{d_m \varepsilon_m + d_d \varepsilon_d}{d_d + d_m}, \quad \varepsilon_{||} = \frac{(d_m + d_d)\varepsilon_d \varepsilon_m}{d_m \varepsilon_d + d_d \varepsilon_m}. \tag{2}$$

The hyperbolic dispersion relation for eigenmodes of such structures is given by:

$$\frac{\mathrm{k}_x^2 + \mathrm{k}_y^2}{\varepsilon_{||}} + \frac{\mathrm{k}_z^2}{\varepsilon_{\perp}} = \left(\frac{\omega}{c}\right)^2 \tag{3}$$

and depends strongly on frequency, and can be obtained only for a certain ratio of the layer thicknesses.

Hyperbolic metamaterials have attracted great scientific attention as an application for engineering the spontaneous emission. The concept of the radiative decay enhancement in a cavity first proposed by Purcell in 1946 [3] within the framework of nuclear magnetic resonance and is known as the Purcell effect. This effect is described by the Purcell factor, which in the broad sense is defined by the ratio

$$\mathrm{F_p} = \tau_{\mathrm{bulk}}/\tau_{\mathrm{cav}} \tag{4}$$

between lifetimes of spontaneous emission of a point light source in the infinite homogeneous medium to that inserted into a resonant cavity. According to the Fermi golden rule, the radiative decay rate $1/\tau$ in a homogeneous lossless medium is defined by:

$$\frac{1}{\tau}=\frac{2\pi}{h}\sum_{\boldsymbol{k},\sigma}|\langle f\ |H_{int}(\boldsymbol{k},\sigma)|i\rangle|^2\ \delta(h\ \omega_{\boldsymbol{k},\sigma}\ - h\omega) \quad (5)$$

where $\langle f\ |H_{int}(\boldsymbol{k},\sigma)|i\rangle$ is the matrix of the perturbation between the final state $|f\rangle$ and the initial state $|i\rangle$, $h\omega$ is the transition energy with the emission of the photon $(\boldsymbol{k},\sigma)$, with the wave vector $\boldsymbol{k}$ and polarization σ.

The density of photonic states in HMM diverges [4], affording an enhancement of the spontaneous emission [5]. The Purcell effect provides a means to tune the emission properties of the light source, potentially useful for a variety of applications. Recently, it was predicted theoretically, that for a point dipole-emitter embedded in a metamaterial structure constructed with thinner dielectric layers the Purcell factor rapidly increases [6]. Therefore, it is interesting to study how the enhancement of the Purcell factor depends on the thickness of layers. In addition, structures with metallic layers are also characterized by losses, which should be accurately taken into account. Therefore, we need to clarify how losses affect the electromagnetic properties of the multilayered HMM and the Purcell factor.

2. Results and discussion

We study a silver-dielectric stratified structure shown schematically in figure 1, the dielectric constant of silver is described by the Drude model: $\varepsilon_m=\varepsilon_\infty-\frac{\omega_p^2}{\omega(\omega+i\gamma)}$, which was fitted to a particular frequency range of the Johnson and Christy bulk dielectric data [7] with the following parameters: $\varepsilon_\infty \approx 4.96$, the plasma frequency $\omega_p \approx 8.98$ eV, and γ=0.018 eV is the damping rate. The dielectric constant ε_d=1, the structure period $D=d_m+d_d$=60 nm, and the ratio of the thickness of silver and dielectric layers is varied $\eta=\frac{d_m}{d_d}$.

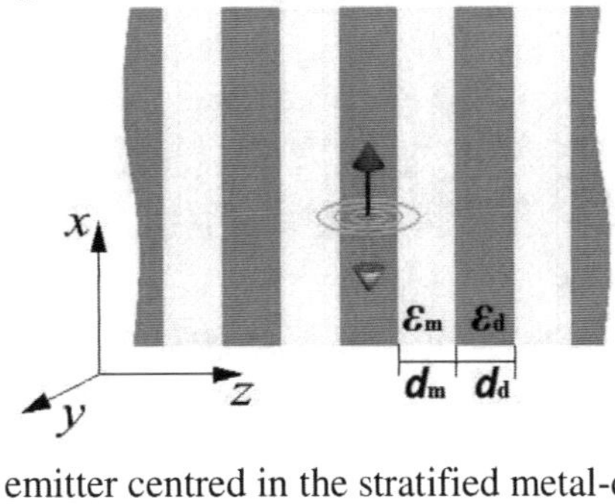

Figure 1. The dipole emitter centred in the stratified metal-dielectric metamaterial.

We study the Purcell factor for a point dipole emitter embedded in the central dielectric layer of the metal-dielectric stratified media. We assume that the dipole-source is enclosed between two identical mirrors formed by semi-infinite metal-dielectric periodical structures. Therefore the system can be treated as a cavity, and to calculate the Purcell factor we apply an approach based on the quantum theory of the spontaneous emission (SpE) from an active microcavity [8]. Reflectivity and transmittance through layered media were calculated by applying the transfer matrix approach [9].

Figure 2 illustrates the dependence of the Purcell factor over the in-plane component of the wavevector $k_{||}$. We have calculated separately the contribution of two orientations of the dipole: perpendicular (figure 2 (a,b,c)) and parallel (figure 2 (d,e,f)) to the interface of layers.

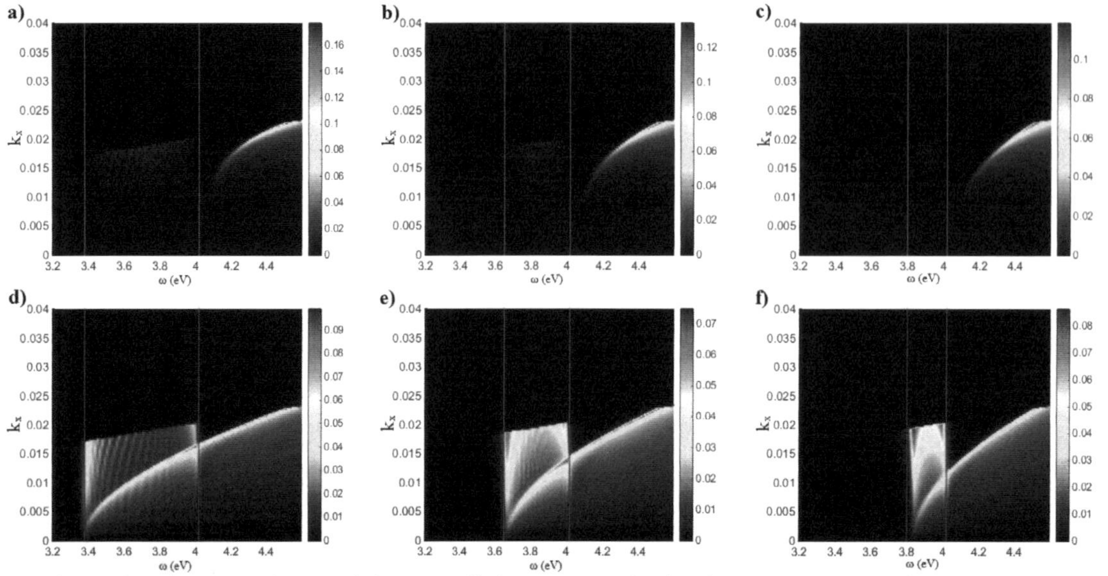

Figure 2. The dependence of the Purcell factor over the in-plane component of the wave vector. With the dipole orientation respectively perpendicular (a,b,c) and parallel (d,e,f) to the interface of semi-infinite periodic mirrors. The period of the structure is fixed D=60 nm, while the ratio of the thickness of layers $\eta = \frac{d_m}{d_d}$ is varied: $\eta = 0.5$ (a, d), $\eta = 1$ (b, e), and $\eta = 2$ (c, f).

When the dipole is perpendicular to the interface of layers, only the case corresponding to TM polarization significantly contribute to the Purcell factor. The surface waves exist at interfaces between materials with opposite signs of the real part of their dielectric permittivity, $Re[\varepsilon_m] = -\varepsilon_d$, defining the resonant limit of a single surface plasmon polariton (SPP). We observe a sharp maximum of the Purcell factor at the frequency of the SPP. Additionally, there are two main branches (indicated by red vertical lines in figure 2) contributing to the Purcell factor: first branch related to the odd modes ω_+, which have frequencies higher than the respective frequency for a single interface SPP; second branch corresponds to the even modes ω_-, which have lower frequencies [10]. Decreasing metal layers thickness we can see an interesting property of odd modes: the confinement of the coupled SPP to the metal layers decreases, as a result the mode evolves into a plane wave propagating in the homogeneous dielectric layer. In contrast, for the even modes the confinement to the metal increases while decreasing thickness of metal layers, as a consequence the propagation length is reduced.

In case when the dipole is parallel to the interface of layers, both the TE and TM polarizations contribute to the Purcell factor. We present the evolution of the Purcell factor as a function of $k_{||}$ in figure 3 (a,d), calculated independently for TE and TM polarization respectively. For TE polarization, we observe an enhancement of spontaneous emission at the edge of the stop-band, when the Bloch wave vector becomes imaginary. The dependence of the Bloch wave vector K_b is defined by:

$$\cos(K_b D) = \cos(k_{1z}d_1)\cos(k_{2z}d_2) - \frac{1}{2}\left(\frac{Z_1^{s,p}}{Z_2^{s,p}} + \frac{Z_2^{s,p}}{Z_1^{s,p}}\right)\sin(k_{1z}d_1)\sin(k_{2z}d_2), \tag{6}$$

where $k_{(1,2)z} = \left(\varepsilon_{(1,2)}\left(\frac{\omega}{c}\right)^2 - k_{||}^2\right)^{\frac{1}{2}}$, $Z_i^s = k_{iz}$ for TE and $Z_i^p = \frac{\varepsilon_i}{k_{iz}}$ for TM polarization. The dependence of the K_b over the in-plane component of the wave vector $k_{||}$ is illustrated in figure 3, independently real and imaginary parts of K_b for TE and TM polarizations respectively. For TM polarization we observe the enhancement of the spontaneous emission at the frequencies of the SPP, and both branches related to odd and even coupled SPP modes (indicated by black vertical lines). We note that the contribution of the SPP exhibit only for TM polarization.

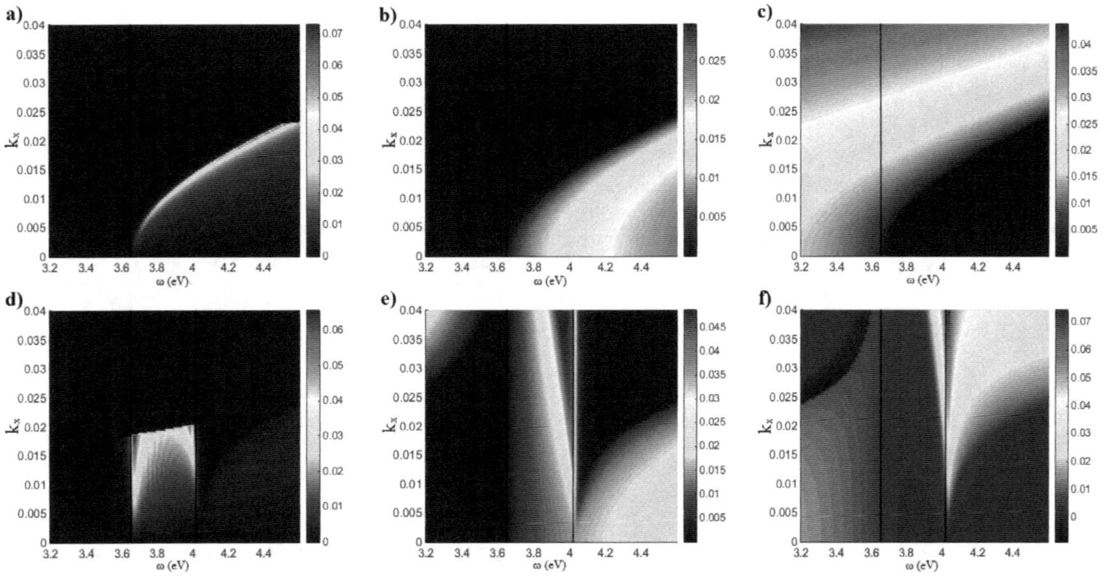

Figure 3. The dependence of the Purcell factor over the in-plane component of the wave vector when the dipole is parallel to the layers. We present TE-(a) and TM-(d) polarizations independently. The dependence of the real (b,e) and imaginary (c,f) parts of the Bloch wave vector over the in-plane component of the wave vector, for TE-(b,c) and TM-(e,f) polarization correspondingly. The thickness of metal and dielectric layers: $d_m = d_d = 30$ nm.

3. Conclusions

In summary, we have calculated the spontaneous emission rate of the dipole emitter embedded in the stratified metal-dielectric structure. Our modeling has demonstrated the increase of the molecular decay rates at the frequencies related to the SPP, and both odd and even coupled SPP modes. We have shown results introducing realistic level of losses into the problem. We have observed that the Purcell factor significantly depend on the ratio of the layers thicknesses.

Acknowledgements.
The work was supported by FP7 ITN NOTEDEV network.

References

[1] Smolyaninov I I and Narimanov E E 2010 *Phys. Rev. Lett.* **105** 067402
[2] Smith D and Schurig D 2003 *Phys. Rev. Lett.* **90** 077405
[3] Purcell E M 1946 *Phys. Rev.* **69** 681 Abstract B10
[4] Jacob Z, Kim J, Naik G V, Boltasseva A, Narimanov E E, Shalaev V M 2010 *Applied Phys. B: Lasers and Optics* **100** 215
[5] Poddubny A N, Belov P A, and Kivshar Y S 2011 *Phys. Rev.* A **84** 023807
[6] Iorsh I, Poddubny A, Orlov A, Belov P, and Kivshar Y S 2012 *Physics Letters A* **376** 185187
[7] Johnson P B, Christy R W 1972 *Phys. Rev.* B **6** 4370
[8] Martini F, Marrocco M, Mataloni P, Crescentini L, Loudon R 1991 *Phys. Rev.* A **43** 2480
[9] Yariv A and Yeh P 2002 *Optical Waves in Crystals: Propagation and Control of Laser Radiation* Wiley Classics Library
[10] Maier S A 2007 Plasmonics: Fundamentals and Applications Springer (New York: Springer)

Spectral sensitivity characteristics simulation for silicon p-i-n photodiode

S.U. Urchuk[1], S.A. Legotin[1], U.V. Osipov[1], D.S. Elnikov[1], S.I. Didenko[1], V.P. Astahov[1], O.I. Rabinovich[1], V.P. Yaromskiy[2] and K.A. Kuzmina[1]

[1]National university of science and technology “MISiS” 119049, Leninskiy pr., bl.4, Moscow, Russia
[2]JSC "Scientific-manufacturing plant for testing equipment", 141070, Pioneer st., bl. 2, Korolev, Moscow region, Russia

E-mail: rawork2008@mail.ru

Abstract. In this paper the simulation results of the spectral sensitivity characteristics of silicon p-i-n-photodiodes are presented. The analysis of the characteristics of the semiconductor material (the doping level, lifetime, surface recombination velocity), the construction and operation modes on the characteristics of photosensitive structures in order to optimize them was carried out.

1. Introduction

In equipment types which use the yielding the radiation detector, in most cases, defines the basic parameters of optical systems. Modern development of optical receivers is characterized by the further parameters improvement and radiation characteristics detectors: sensitivity, speed, the spectral sensitivity range, reliability and etc. [1]

Silicon photodiodes have a long life usage, mechanical strength and compact size. P-I-N photodiodes have high bandwidth at low bias voltage, which makes them ideal detectors for use in high speed photometry of optical communication lines [2].

At the same time, the effective multilayer photodetector structures creation requires consideration of factors that affect the photosignal formation. Simultaneous optical radiation absorption consideration, the transfer of photogenerated charge carriers and their recombination, including recombination at the layer boundaries and surface requires computer simulation that allows a detailed analysis of the structure design influence and electrical characteristics of individual regions with the lowest material cost.

2. Results and discussion

In order to optimize physical and topological structure of multilayer silicon structures simulation program for spectral characteristics was developed. Simulating photosensitivity silicon of photosensitive structures was performed by solving the basic system of equations: Poisson equation, the continuity and the electrons and holes transfer [3-6]. The simulation program was developed using the language Borland Delphi 7. The program allows recreating the simulated layers structure,

[1] To whom any correspondence should be addressed.

specifying the layers quantity, the material, the conductivity type, doping level and the each layer thickness.

For determination of the spectral dependence of the photosensitivity it was given the spectral range and in every area of monochromatic spectral range the photocurrent was determined. Moreover, in order to analyze the physical processes occurring in the structure, and then optimize the design of the structure separately the photocurrents generated separately p- and n- regions and the space charge region were determined. As a result, the spectral sensitivity distribution of the units in A/W was achieved. As an initial structure for simulation it was adopted the structure shown in figure 1.

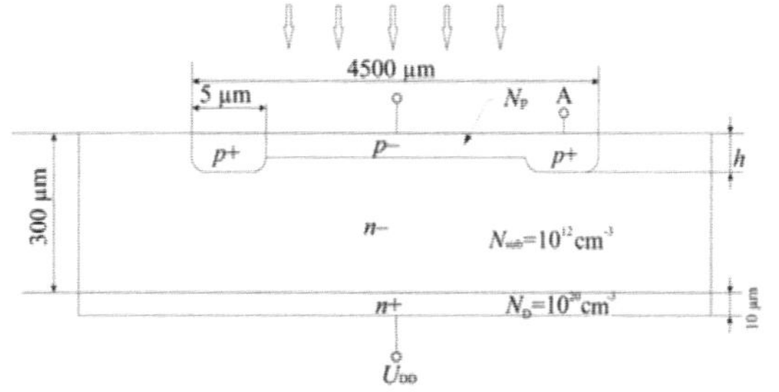

Figure 1. The initial silicon p-i-n-structure chosen for simulation

Figure 2 shows an example of calculated p-i-n-structure spectral characteristics.

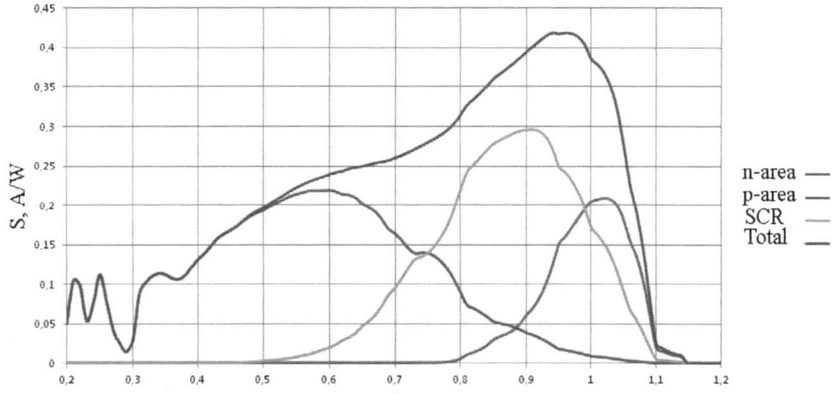

Figure 2. Estimated multilayer p-i-n structure spectral characteristics

The program also allows to estimate the lifetime impact, recombination rate, thickness and doping level in various areas, consider the effect of changing the doping level without changing the lifetime and mobility. It may seem strange, but the total spectral sensitivity is not too dependent on the doping level of the lightly doped n-region (if not to take into account the change in mobility and lifetime) (fig. 3). Although the space charge region varies considerably.

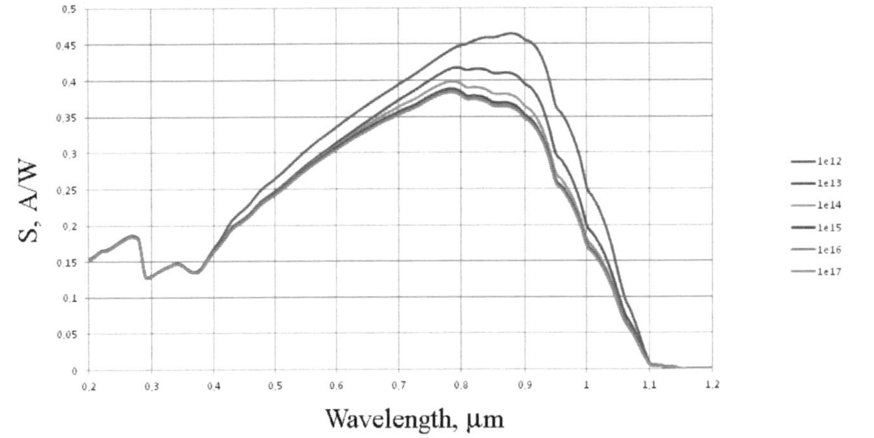

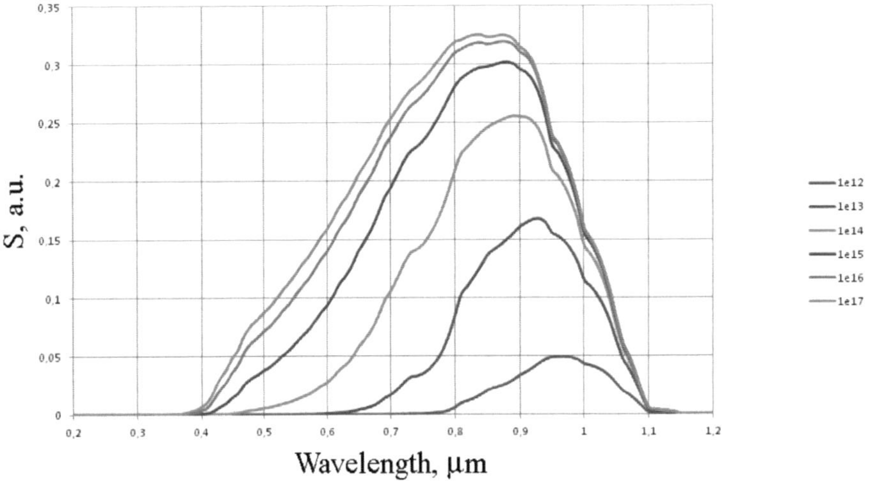

Figure 3. Effect of the doping level of the lightly doped n-region on the spectral sensitivity characteristics (top) and contribute to the spectral sensitivity characteristics of the space charge region (bottom) ($N_d = 10^{12} \div 10^{17} cm^{-3}$).

At the same time, an increase in the p-region doping level does not lead to a significant change in the photosensitivity spectral characteristics, but there is a slight increase in sensitivity of the long waves part. This is due to the fact that the space charge region is expanded slightly by increasing the contact potential difference, which leads to an increase in the contribution of the space charge region. The n-region contribution is slightly reduced due to a decrease in the area where photons are absorbed, by expanding the space charge region (SCR). The top p-layer contribution in the spectral response curve is not changed (taking into account reducing the lifetime and the electron mobility with increasing doping level the p-region contribution should be decreased).

To check the results reliability which was obtained by the developed simulation program, additional calculations were carried out in a specialized software ISE TCAD. This software package gives opportunity fully enough to reproduce all the process steps of creating the final structure and analyze the device operation in electrical mode. One advantage of TCAD is the possibility to create two-dimensional structures, which currently in the developed program is not implemented. Also considers only the generation-recombination processes directly within the active structure and allows to save time required to adjust processes.

As the initial structure has been selected the structure shown in figure 1 with the thermal oxide (thickness - 100 nm). Figure 4 shows the calculated spectral characteristics obtained in the developed program and in TCAD program for maximum concentration in a low doped p - layer 10^{16} cm^{-3}. The results were normalized by the final structure area. There is good results correlation. Spectral peak at 0.5 microns wavelength is due to maximum transparency caused oxide (thickness- 100 nm).

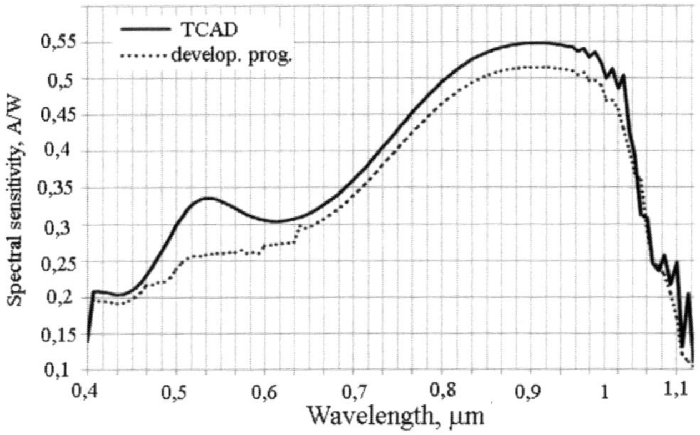

Figure 4. p-i-n structure spectral characteristics

To improve the structure sensitivity in the short-wavelength spectrum region it is necessary to reduce the concentration in the low doped p-region to approach the SCR surface. Another way to increase the SCR width in the subsurface region is to create a mesh structure in which the high doped p-region alternating with the original substrate parts. Figure 5 shows the spectral characteristics of the mesh structure with a 5 micron pitch between p-type regions and used for comparison the structure (impurity concentration in the p - region - 10^{16} cm^{-3}). It is important that even with a sufficiently small distance between the p-type region it is possible to achieve a significant sensitivity increase in the blue and green spectrum regions.

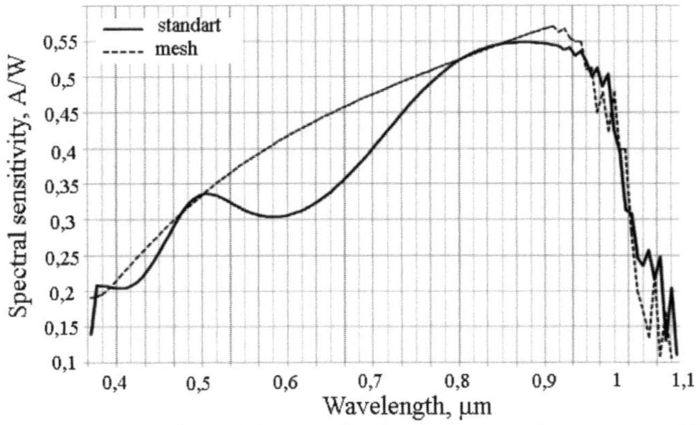

Figure 5. The spectral characteristics of the standard structure and structure with mesh p- region

For the initial substrate parameters the voltage of full depletion U_{dep} is about 80 V, which allows the p-i-n-diode using in mode when the applied voltage is 1.5 - 2 times higher than the U_{dep}. The maximum operating voltage will be determined by photodiode breakdown characteristics. This mode makes it possible not only to increase the overall output level, but also to raise the response.

3. Conclusions

It was carried out the various factors influence: the electrons and holes lifetime and mobility, surface recombination velocity on the spectral sensitivity characteristics of p-i-n-structures. In order to optimize the photodetector structure parameters it was calculated the parameters when changing the top layer depth of p-doping levels and changing doping areas of the structure. To expand the spectral sensitivity characteristics of silicon photodetector structures toward shorter wavelengths it is required the maximum achieving space charge region to the surface. At the same time, action is required to ensure the improvement of the surface properties to reduce the surface recombination velocity and the simultaneous use of anti-reflection coatings.

Acknowledgements
This study was supported by the Federal Targeted Program "Research and development on priority directions of scientific-technological complex of Russia for 2014—2020, state contract № 14.575.21.0018 (unique identifier for applied scientific research (project) RFMEFI57514X0018).

References

[1] Baranochnikov M.L. 1985 *IR photodetectors*. (M. Review)
[2] Volkov D.L., Karmanov D.E., Murashev V.N., Legotin S.A., Mukhamedshin R.A, and Chubenko A.P. 2009 *Instruments and Experimental Techniques*, **52(5),** 655
[3] MOS SBIS. 1988 *Simulation of elements and technological processes*. (M.: Radio & communication)
[4] Murashev, V.N. , Legotin, S.A. , El'nikov, D.S. , Didenko, S.I. and Rabinovich, O.I. 2015 *Journal of Nano- and Electronic Physics*, **7(1)**, 01009
[5] Murashev V.N., Yurchuk S.Y., Legotin S.A., Yaromskiy V.P., Osipov Y.V., Astahov V.P., El'nikov D.S., Didenko S.I., Rabinovich O.I., and Kuz'mina K.A., 2015, Journal of Nano- and Electronic Physics, **7(2),** 1-5
[6] Legotin S.A., Murashev V.N., Didenko S.I., Rabinovich O.I., Elnikov D.S., Krasnov A.A., Bazalevsky M.A., Koltsov G.I. and Kuzmina K.A. 2014,, Journal of Nano- and Electronic Physics, **6(3),** 03020

SPbOPEN2015 IOP Publishing
Journal of Physics: Conference Series **643** (2015) 012069 doi:10.1088/1742-6596/643/1/012069

Analysis of magnetic systems with high-anisotropy localized magnetic field for terahertz-radiation generation

S E Azbite A Kh Denisultanov and M K Khodzitsky

Department of Photonics and Optical Information Technologies, ITMO University, 49 Kronverksky av., St. Petersburg, Russia

E-mail: azbite@mail.ru

Abstract. For enhancement of terahertz radiation generation in terahertz devices four permanent magnetic systems by strong localized magnetic field were considered. Their different configurations were used for simulation and comparative analysis and choice compact system with maximum value of localized magnetic field at the room temperature.

1. Introduction

At the present time scientific researches in terahertz frequency range attract more interest. Terahertz radiation may be applied for biomedicine, non-destructive examination, security system, spectroscopy, etc. There are various terahertz (THz) radiation generation systems which use magnetic field for efficiently generation. Rungsawang *et al.* [1] theoretically described the THz electromagnetic radiation originating from spins in CdMnTe diluted magnetic semiconductor quantum wells. Kumar *et al.* [2] examined terahertz generation at the beat frequency of two lasers in the presence of transverse magnetic field. Also an optimization of magnetic field is an important aspect [3]. Weiss *et al.* [4] applied an external magnetic field for increasing of the generated THz radiation power in semiconductor materials. Currently, superconductor and permanent magnetic systems are used. Strong magnetic field superconductor systems need cooling and power source whereas permanent magnetic systems operate at room temperatures and produce small magnetic fields [5]. This paper represents the research of compact permanent magnetic systems with strong magnetic field at room temperature.

2. Description of magnetic systems under study

For this research four magnetic systems were considered. These systems have various segmental configuration: system of two magnets, Halbach cylinder, four-segment cylinder and hemisphere with cone. Each of these systems was calculated for NdFeB that is a material with a giant magnetic anisotropy, its magnetization, M_s, is 1280 Gs. Due to use this material a strong localized magnetic field can be obtained at room temperature. The magnetic field strength distribution depends on magnet geometrical parameters. Theoretical investigation was performed in the papers [6], [7]. The first system consists of two parallelepipeds without a gap, which have with the opposite magnetization direction. 3D model is shown in Fig.1. The vertical

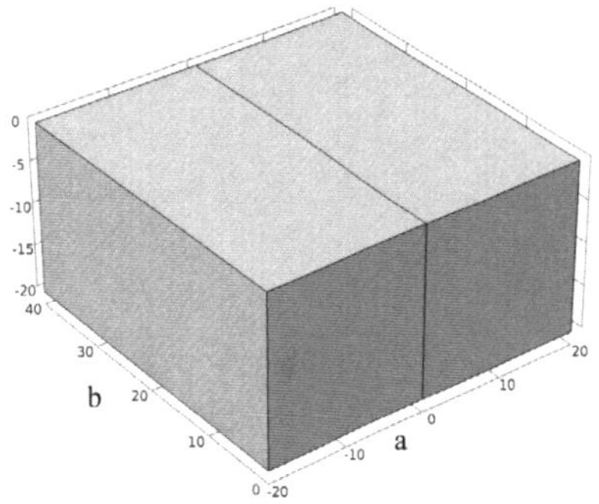

Figure 1. Two magnets with the opposite magnetization direction; width a is 20 mm, length b is 40 mm.

component of stray field, H_z, is:

$$H_z(x,y) = 2M_s(\arctan\frac{a+x}{z} - \arctan\frac{a-x}{z} - 2\arctan\frac{x}{z}), \quad (1)$$

It is seen that $H_z(x,z) < 2\pi M_s$ and it yields a small contribution in a total field $H = (H_x^2 + H_y^2)^{1/2}$. H_x component of stray magnetic field at $b \gg a$ is determined by the expression:

$$H_x(x,y) = 2M_s[ln(a^2+z^2+2ax+x^2) - 2ln(x^2+y^2) + ln(a^2+z^2-2ax+x^2)], \quad (2)$$

In the case $x < 0.01a$ and $z = 0$ we can write $H_x(x) \simeq 4M_s\ln(a/x)$ The second system consists of four triangular prisms it has a cavity near point O. Practically, it is the simplest Halbach cylinder. 3D model is shown in Fig.2 On the gap edges the singular points exist. Their

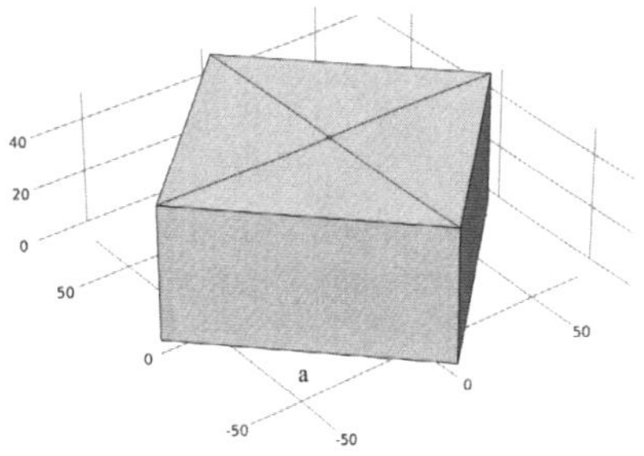

Figure 2. Halbach cylinder, lateral side a is 100 mm, gap width 2δ is 2 mm.

coordinates are $z = 0$ and $x = \pm\delta$. In these points magnetic field attains the largest value. The expression for H_x has the form

$$H_x(x,z) = M_s[\ln((a+\delta+x)^2+z^2)+\ln((a+\delta-x)^2+z^2)-\ln((\delta+x)^2+z^2)-\ln((\delta-x)^2+z^2)] \quad (3)$$

After an algebraic transformation we can write for field strength in narrow gap

$$H_x(n) = 8M_s\ln\frac{R}{r}sin\frac{\pi}{2n} \quad (4)$$

The third system is a cylinder that consists of four equal segments. It is shown in Fig.3 For system, which consists of several charged surfaces, the magnetic field attains the largest value. In paper [7] it was shown that for obtaining maximum value of magnetic field it is necessary magnetization vector is centrally directed along bisector. Charge density at both surface is

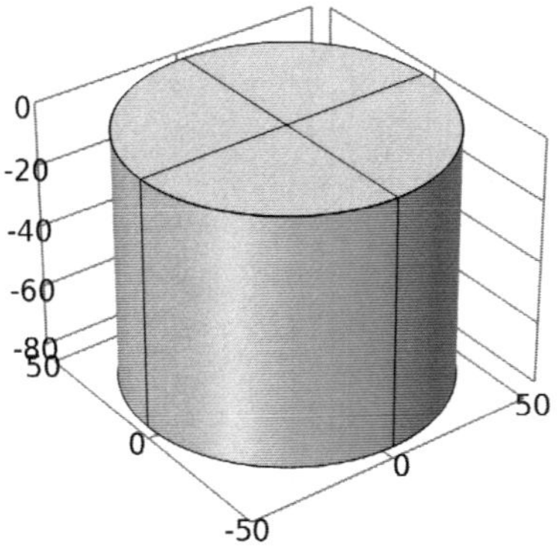

Figure 3. Cylinder made as four equal segments, radius R is $50mm$, magnetization vector is centrally directed along bisector.

$\sigma_{xz} = \sigma_{xy} = \sqrt{2}M_s$. Total field is sum of linear fields. It possesses maximum value near point O and may be described by equal

$$H_z \simeq 2nM_s \sin\frac{\pi}{n} \ln\frac{h}{r} \tag{5}$$

$r = (x^2 + y^2 + z^2)^{1/2}$ is the distance of a point from the origin to observer point, h is the cylinder height. For four sectors we obtain

$$H_z \simeq 4\sqrt{2}M_s \ln\frac{h}{r} \tag{6}$$

The fourth system is cone in hemisphere, which is divided by radial planes passing through the cone axis (Fig. 4) In each of the sectors the vector Ms lies in the plane passing through the

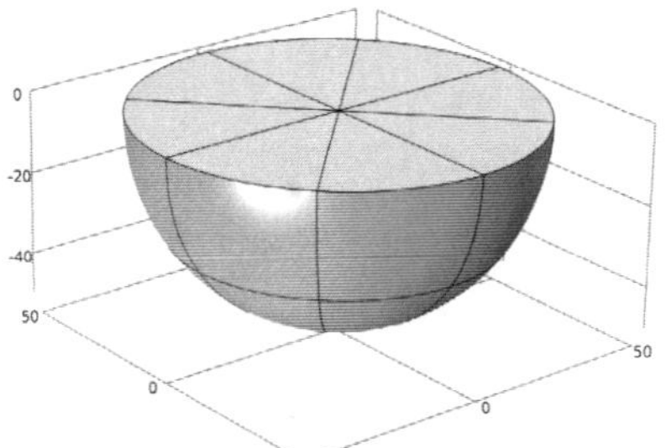

Figure 4. Hemisphere is divided by radial planes it has a hole with cone; radius R is 50 mm, vertex cone angle α is 54°. Magnetization vector lies in the bisector plane for each magnet and makes an angle with axis z equal $\varphi_1 = -107°$ for cone and $\varphi_2 = 14°$ for segments of hemisphere.

bisector and it is directed along to the normal to cone surface [7]. H_x component of magnetic field is

$$H_x \simeq 0.77M_s[\sin\varphi_1 \sin\alpha + \sin\varphi_2(1 - \sin\alpha)] \ln\frac{R}{r}, \tag{7}$$

where $\alpha = 54°$, $\varphi_1 = -107°$, $\varphi_2 = 14°$. That system is source of strong magnetic field of point type. In all points inside the magnet strong magnetic field is at right angle to magnetization vector, therefore possibility of magnet demagnetization by generic field is lower. One can use this system for obtaining of strong stray field.

3. Estimation and simulation

Optimum parameters was matched for these systems, like size, quantity of segments and direction of magnetization vector for composite system. The dimensional parameters were calculated for an operating terahertz pulsed spectrometer consistent with design compactness and strong localized magnetic field capability system for obtainment of known beforehand magnitude and magnetic field strength distribution. The dependence of magnetic field strength on magnet characteristics (like magnetization, magnet dimension, gap between plates) was analytically calculated by Matlab. Each of four systems was simulated by COMSOL MultiPhysics software using AC/DC Module. Two magnets with the opposite magnetization direction have a strong localized magnetic field in the region from -2 till 2 mm, it is above 2×10^4 Oe (Fig. 5) This system has magnetic field along line of contact of both magnets. Halbach cylinder has a strong

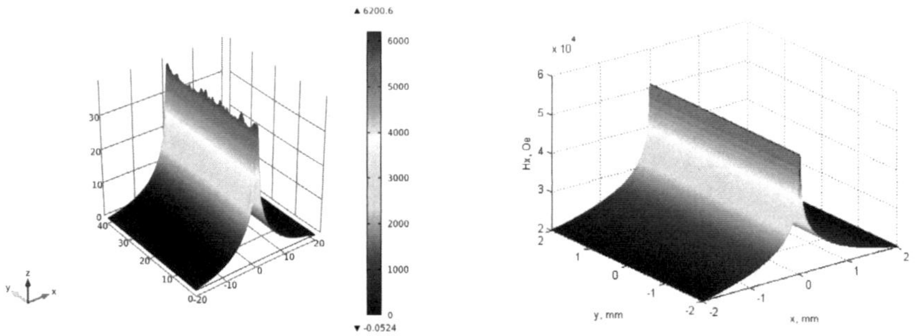

Figure 5. Distribution $H_x(x, y)$ at $z = 0$ in the two magnets system: a) in all xy-surface (simulated in COMSOL); b) in the region of $-2 < x < 2$ mm (simulated in Matlab).

localized magnetic field in singular points on gap edges. We can obtain magnetic field value above 2.3×10^4 Oe in over the range 2 till 2 mm (Fig. 6). Four-segment cylinder has a strong

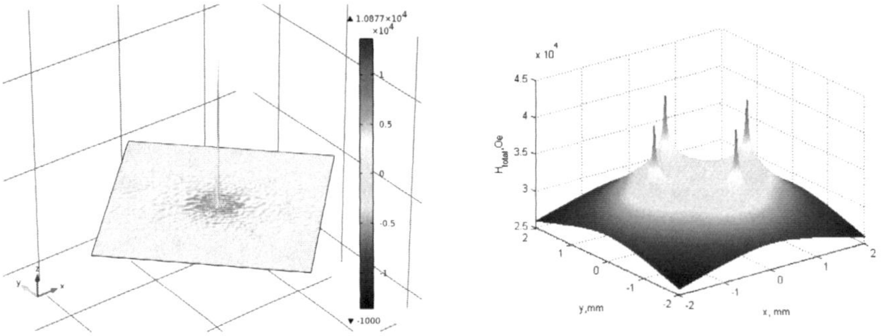

Figure 6. Distribution $H_{total}(x, y)$ at $z = 0$ in the Halbach cylinder system: a) in all xy-surface (simulated in COMSOL); b) in the region of $-2 < x < 2$ mm (simulated in Matlab).

localized magnetic field near point O, in the region from -2 till 2 mm, it is above 2.5×10^4 Oe. (Fig. 7) The cylinder is source of strong magnetic field of point type. Hemisphere, divided by radial planes, with cone has a strong localized magnetic field near point O, it is above 3×10^4 Oe within the interval from -4 till 4 mm. (Fig. 8). Also the hemisphere is source of point type magnetic field. The system consisting of two same uniaxial magnets with opposite magnetic field strength direction allows obtaining twofold magnetic field strength compared to one magnet case.

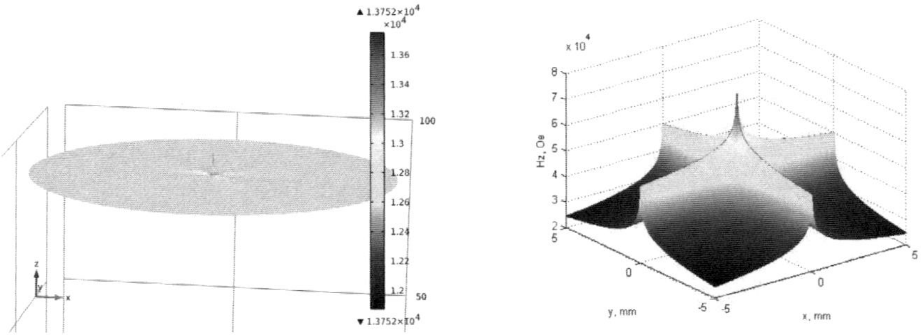

Figure 7. Distribution $H_z(x, y)$ at $z = 0$ in the four-segment cylinder system: a) in all xy-surface (simulated in COMSOL); b) in the region of $-2 < x < 2$ mm (simulated in Matlab).

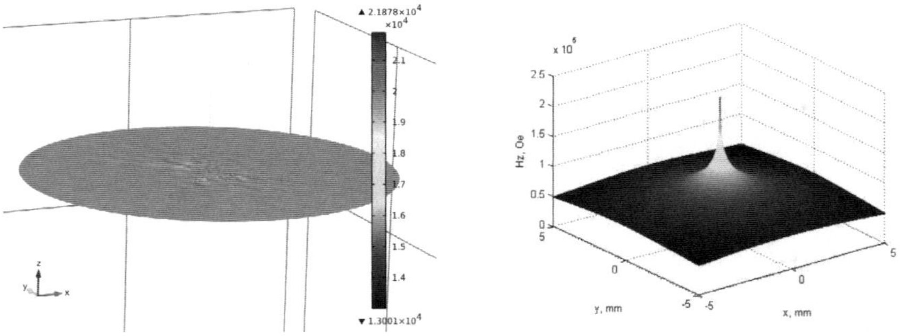

Figure 8. Distribution $H_z(x, y)$ at $z = 0$ in the hemisphere, divided by radial planes, with cone system: a) in all xy-surface (simulated in COMSOL); b) in the region of $-5 < x < 5$ mm (simulated in Matlab).

Comparison of the two magnets system and the Halbach cylinder was made (Fig. 9). It is seen that magnetic field of the Halbach cylinder is lower than field of two magnets system, it is above 2.3×10^4 Oe in region from -2 till 2 mm and practically uniform field from -0.5 till 0.5 mm. Two magnets system magnetic field is above 2.5×10^4 Oe on the interval from -1 till 1 mm. Thus the two magnet systems can be used for a measurement near axis $x = 0$, and the Halbach cylinder system can be useful for enhancement of THz generation using small square samples. On the results of comparative analysis of the four-segment cylinder and the hemisphere, divided

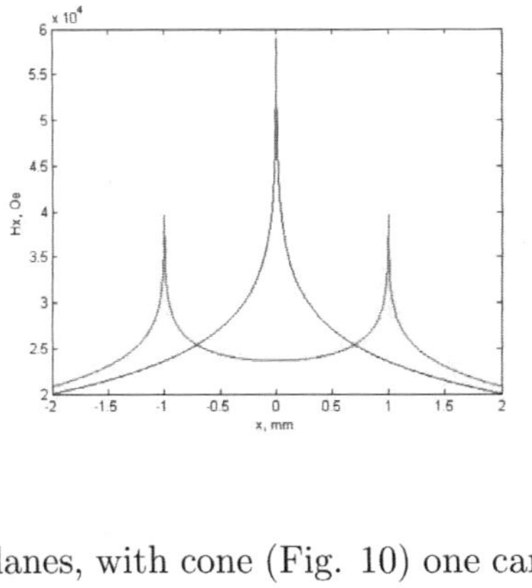

Figure 9. Comparison of the two magnets system (blue line) and the Halbach cylinder (red line) at $z = 0, y = 0$.

by radial planes, with cone (Fig. 10) one can see that magnetic field of the hemisphere is higher and reaches value H above 3×10^4 Oe in the interval from -3 till 3 mm, whereas the magnetic field of the cylinder is above 2.5×10^4 Oe in this region. The hemisphere yields the highest magnetic field strength while it has elaborate design. On the results of analytical estimation, simulation and comparison the hemisphere with cone was selected as the most effective compact system.

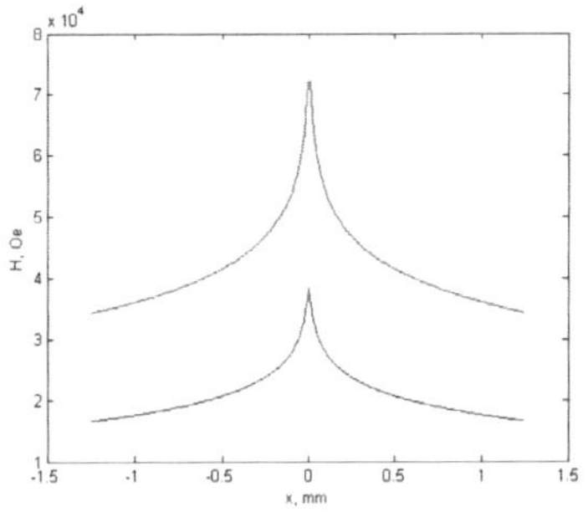

Figure 10. Comparison of the four-segment cylinder (blue line) and the hemisphere with cone (red line) at $z = 0, y = 0$.

4. Conclusions

In conclusion, the magnetic systems with high localized magnetic field were investigated. Due to these systems made from NdFeB that is material with giant magnetic anisotropy, the high magnetic field is possible. Each of magnetic systems was estimated and simulated. The two magnets system and the simplest Halbach cylinder yield low magnetic field strength and may be used for enhancement of THz generation using small oblong or square samples respectively. By the four-segment cylinder and hemisphere, divided by radial planes, with cone we can obtain higher magnetic field strength. The cylinder is easy to develop, but its magnetic field strength localizes in a narrow area. The hemisphere has an elaborate design and produces strong magnetic field in relative wide area. As the result of this research the hemisphere with cone was selected for obtaining strong magnetic field above 3×10^4 Oe in the region from -3 till 3 mm.

Acknowledgments

This work was financially supported by Government of Russian Federation, Grant 074-U01.

5. References

[1] Rungsawang R *et al.* 2013 *Phys. Rev. Lett.* **110** 177203
[2] Kumar *et al.* 2012 *J. of Phys. and Chemis. of Solid.* **73** pp 269-274
[3] Malik A *et al* 2012 *Phys. Rev. Lett.* **85** 016401
[4] Weiss C, Wallenstein R and Beigang R 2000 Appl. Phys. Lett. **77** pp 4160-4162
[5] Bespalov V *et al* 2008 *J. of Opt. Tech.* **75** pp 636-642
[6] Samofalov V, Belozorov D and Ravlik A 2006 *The Phys. of Met. and Metallogr.* **102** pp 494-505
[7] Samofalov V, Belozorov D and Ravlik A 2013 *Physics -Uspekhi* **56** pp 269-288

Journal of Physics: Conference Series **643** (2015) 012070 doi:10.1088/1742-6596/643/1/012070

Photoionization of Rydberg States by Ultrashort Wavelet Pulses

S Yu Svita, V A Astapenko
Moscow Institute of Physics and Technology, 141700, Institutskii per.9, Dolgoprudnyi, Russia

E-mail: sergey.svita@gmail.com

Abstract. Photoionization of Rydberg atomic state by the ultra-short wavelet pulses (USWP) is calculated in the frame of perturbation theory with the use of Kramers ionization cross-section. Two simplest types of USWP are considered, namely, cosine and sine pulses. Analytical expressions containing scaling functions for photoionization probability are derived for above-mentioned USWP.

1. Introduction

Atoms in highly excited (Rydberg) states are important nanoobjects both from theoretical point of view and for different applications in nanophysics particularly in nanophotonics. Rapid development of ultra-short pulse generation technology makes it urgent to investigate the peculiarities of the USWP-Rydberg atom interaction.

2. General formulas for radiation absorption probability

The present work is devoted to theoretical description of atomic Rydberg states photoionization by USWP within the frame work or perturbation theory. We consider two types of USWP cosine- and sine ones. Fourier transforms of these pulses are given by the following expressions:

$$E_{\cos}(\omega) = 2\sqrt{\frac{2}{3}}\sqrt[4]{\pi}\, E_0\, \omega^2 \tau^3 \exp\left(-\omega^2 \tau^2/2\right), \tag{1}$$

$$E_{\sin}(\omega) = 2i\sqrt[4]{\pi}\, E_0\, \omega\tau^2 \exp\left(-\omega^2 \tau^2/2\right), \tag{2}$$

here E_0 is the amplitude of electric field strength in the pulses and τ is pulse duration parameter.

For the calculation of Rydberg state photoionization probability during all time of USWP action we use the formula obtained in paper [1]

$$W_{ph} = \frac{c}{4\pi^2} \int\limits_{I_{th}/\hbar}^{\infty} \sigma_{ph}(\omega') \frac{\left|E(\omega')\right|^2}{\hbar\omega'} d\omega' \tag{3}$$

here $\sigma_{ph}(\omega)$ is photoionization cross section for which we use Kramers formula:

$$\sigma_n(\omega) = \frac{8\pi}{3\sqrt{3}} \frac{Z_{eff}^4}{n^5 c\, \omega^3}. \tag{4}$$

Here n is principal quantum number of Rydberg state and Z_{eff} is effective charge of atomic core. Using equalities (1), (3) – (4) we obtain the following expression for photoionization probability of Rydberg states by cosine pulse ($I_{nZeff} = Z_{eff}^2/2n^2$ is ionization potential of Rydberg state):

$$W_n^{(\cos)}(\tau) = \frac{2^8}{9\sqrt{3}}\left[\frac{E_0}{E_{sc}}\right]^2 f_c\left(\tilde{\tau} = \tau I_{nZeff}\right), \quad (5)$$

$$E_{sc}\left(n, Z_{eff}\right) = \frac{Z_{eff}^3}{n^{5/2}}, \quad (6)$$

$$f_c(\tilde{\tau}) = \tilde{\tau}^5\, erfc(\tilde{\tau}). \quad (7)$$

Here $f_c(\tilde{\tau})$ is scaling function of dimensionless time $\tilde{\tau} = \tau I_{nZeff}$, $erfc(x)$ is complimentary error function ($erfc(x) = 1 - erf(x)$).

For sine pulse we have

$$W_n^{(\sin)}(\tau) = \frac{2^8}{3\sqrt{3}}\left[\frac{E_0}{E_{sc}}\right]^2 f_s\left(\tilde{\tau} = \tau I_{nZeff}\right), \quad (8)$$

here

$$f_s(\tilde{\tau}) = \tilde{\tau}^5\left[\frac{\exp(-\tilde{\tau}^2)}{\sqrt{\pi}\,\tilde{\tau}} - erfc(\tilde{\tau})\right]. \quad (9)$$

It follows from the definitions of scaling functions (7), (9) that they have maxima at

$$\tilde{\tau}_{\max}^{(c)} \cong 1.457, \; \tilde{\tau}_{\max}^{(s)} \cong 1.175 \quad (10)$$

and corresponding maximum values:

$$f_{c,\max} \cong 0.258, \; f_{s,\max} \cong 0.054. \quad (11)$$

3. Results and conclusions

Plots of function (7) and (9) are shown in figure 1.

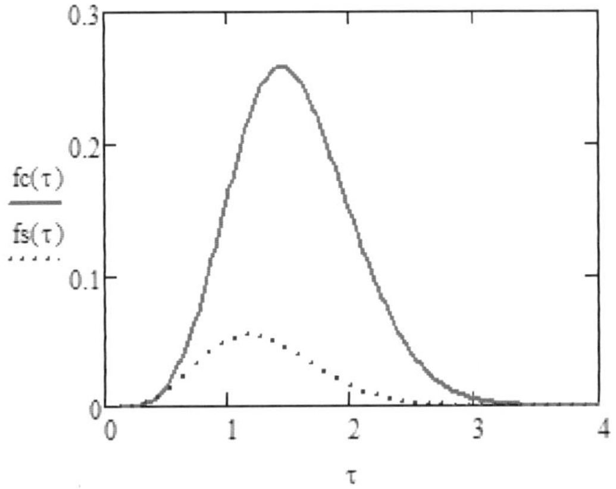

Figure 1. Solid curve – cosine USWP, dotted curve – sine USWP, abscissa is dimensionless time.

Figure 2 shows the photoionization probability of Rydberg state with given principal number (n = 10, Z_{eff} = 1) by cosine- and sine wavelet pulses as function of pulse duration for electric field amplitude $E_0 = 10^{-6}$ at.u. Note that such small value of electric field enable one to neglect the mixing of Rydberg states due to Stark effect.

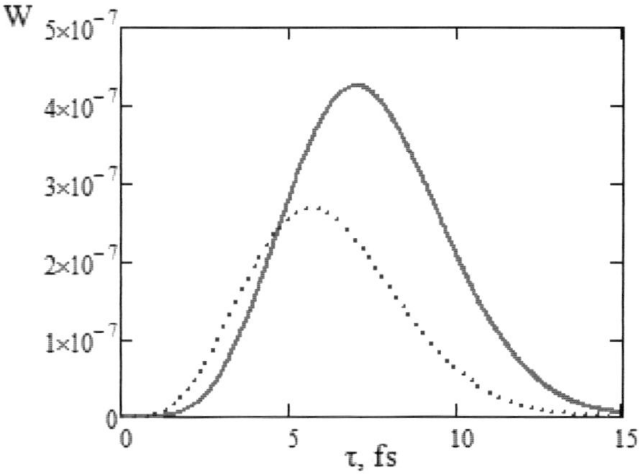

Figure 2. Photoionization of Rydberg state (n = 10, Z_{eff} = 1) by cosine (solid curve) and sine (dotted curve) wavelet pulses.

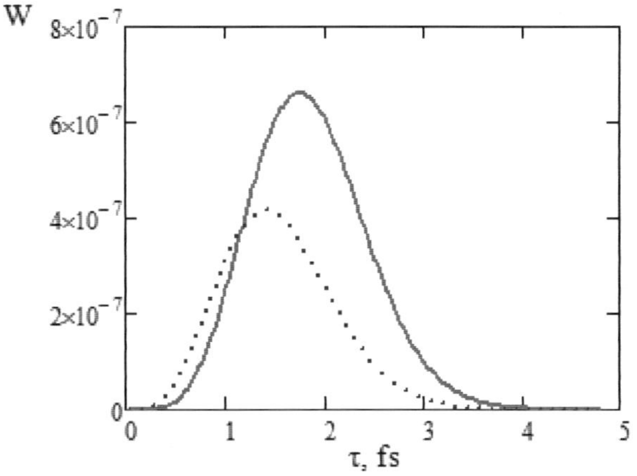

Figure 3. Just the same as in figure 2 for Z_{eff} = 2 and $E_0 = 10^{-5}$ at.u.

Let us consider the photoionization of Rydberg state with given principal n and orbital l quantum numbers. Corresponding cross section is given by the following expression

$$\sigma_{nl}(\omega) = \frac{32\pi}{3\sqrt{3}} \frac{(l+1/2)Z_{eff}^2}{n^3 c\,\omega^2} \exp\left(-\frac{2\omega(l+1/2)^3}{3Z_{eff}^2}\right). \tag{12}$$

This formula can be obtained in the frame of quasi-classical approach as it shown in [3], [4].

Substituting formulas (1), (12) in expression (3) after simple algebra we have:

$$W_{nl}^{(\cos)}(\tau,\omega) = \frac{2^9(l+0.5)}{9\sqrt{3\pi}} \left(\frac{E_0}{E_{sc}}\right)^2 f_{cl}(\tilde{\tau},\zeta), \tag{13}$$

here

$$f_{cl}(\tilde{\tau},\zeta)=\tilde{\tau}^4\left\{\exp(-\tilde{\tau}^2-2\zeta)-\sqrt{\pi}(\zeta/\tilde{\tau})\exp(\zeta^2/\tilde{\tau}^2)erfc[\tilde{\tau}+\zeta/\tilde{\tau}]\right\} \quad (14)$$

is appropriate scaling function for cosine-pulse and parameter ζ is given by equality:

$$\zeta=\frac{(l+0.5)^3}{6n^2}. \quad (15)$$

Analogously for photoionization probability of Rydberg state *nl* by sine-pulse (2) one can find

$$W_{nl}^{(\sin)}(\tau,\omega)=\frac{2^9(l+0.5)}{3\sqrt{3\pi}}\left(\frac{E_0}{E_{sc}}\right)^2 f_{sl}(\tilde{\tau},\zeta), \quad (16)$$

here

$$f_{sl}(\tilde{\tau},\zeta)=\tilde{\tau}^4\int_1^\infty \exp\left(-2\zeta w-w^2\tilde{\tau}^2\right)\frac{dw}{w} \quad (17)$$

is scaling function for Rydberg state (with given principal and orbital quantum numbers) ionization by sine wavelet pulse.

Photoionization probabilities of Rydberg states with given principal quantum number (n = 10) and different orbital numbers l by cosine wavelet pulse with amplitude E_0 = 10^{-6} at.u. as a function of pulse duration are presented in figure 4

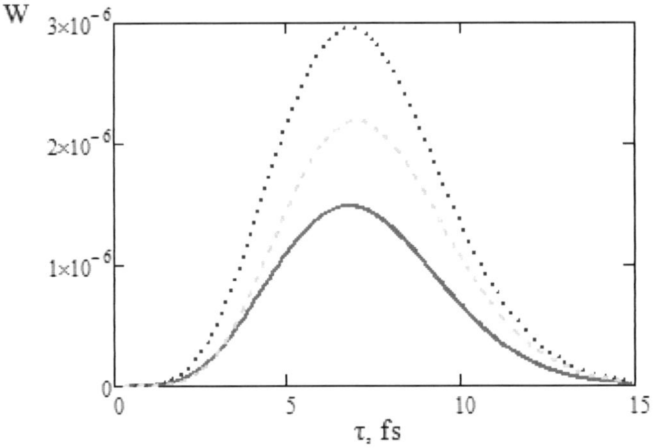

Figure 4. Photoionization probabilities of Rydberg states (n = 10) by cosine wavelet pulse as a function of pulse duration for different l numbers: solid line – l = 1, dotted line – l = 3, dashed line – l = 6

One can see from this figure that the photoionization probability dependence upon orbital quantum number has non-monotonic character.

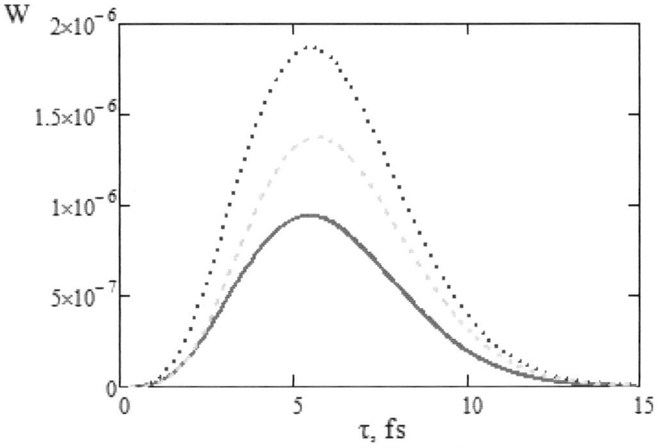

Figure 5. Just the same as in figure 4 for sine wavelet pulse.

One can see from figures 2-5 that photoionization probability dependencies upon wavelet pulses duration are the curves with maxima and go to zero for long pulses. The last conclusion is a specific feature of photoionization by wavelet pulses without carrier frequency. So in the considered situation we have no transition to the monochromatic limit as in the case of atom photoionization by Gaussian pulses [5] when the probability of the process is linear function of pulse duration for sufficiently long pulse.

Thus in the framework of perturbation theory and quasi-classical approach we obtain general expressions for total photoionization probability of Rydberg states by cosine and sine wavelet pulses for two cases, namely, for (I) Rydberg state with given principal quantum number n and for (II) Rydberg state with given principal n and orbital l quantum numbers. Obtained expressions describe the photoionization probability as a function of wavelet pulse duration via scaling functions in universal manner and can be used for simple description of considered process.

The work has been executed with the financial support of the Ministry of Education and Science of Russian Federation within the framework of the basic part of Government order (No2014/120-1940).

References

[1] Astapenko V A 2010 *Phys. Lett. A* **374** 1585
[2] Sobelman I I 1972 *Introduction to the theory of atomic spectra* (Oxford, New York: Pergamon Press) p 609
[3] Bureyeva L A, Lisitsa V S 2000 *A Perturbed Atom* (New York: Gordon and Beach)
[4] Goreslavskii S P, Delone N B, Krainov V P 1982 *Sov. Phys. JETP* **55**, 1032
[5] Astapenko V A , Svita S Yu 2014 *JETP* **119**, 811

Tunable narrowband filters with cross-shaped resonators for THz frequency band

E.A. Sedykh, V.Y. Soboleva and M.K. Khodzitsky
Department of Photonics and Optoinformatics, ITMO University, 49 Kronverksky Ave. 197101, Saint-Petersburg, Russia
E-mail: egor.a.sedykh@gmail.com

Abstract. This work presents simulation of transmission spectra of tunable narrowband and broadband filters for terahertz frequency range comprised of cross-shaped metallic resonators both free standing and deposited on semiconductor substrate.

1. Introduction

Terahertz radiation is an intensely researched field which strives to find applications in our everyday life: from medicine and security to nondestructive evaluation [1]. Special interest is paid to terahertz spectroscopy and communication at terahertz frequencies. It is also important to develop elementary base for terahertz frequency communication devices. Such elements as low-pass, high-pass, bandstop [2, 3] and bandpass filters are required to control the frequency of generated THz radiation, providing the opportunity to implement frequency modulation in THz frequency range. Another important field is terahertz spectroscopy, which can be used to study and identify organic molecules which have energy transition with energies that of photon of terahertz frequency associated with rotation and vibration degrees of freedom. The goal of this work was to estimate different designs of X-shaped bandpass filters which can be used for THz frequency domain. A lot of papers devoted to the topic have been continuously published since the first papers of Ulrich[4, 5] considering square shaped resonators and complementary slot structures for far infrared frequency domain and the following papers considering arrays with different slot geometries [6]-[11].

2. Simulated structures

Present work shows the results of numerical simulation of transmission spectra of the following structures: the first being arrays of X-shaped slots of different shapes (with straight and mitred edges) cut out in the layer of metal, the second being X-shaped slot placed on tunable substrate (simulated for values of $\epsilon = 10.24, 10.89, 11.56$), the third being X-shaped slot with varying slope of slot sides and the forth being the X-shaped slot with width varying alongside the length of a bar. Transmittance spectra of all structures were simulated for sets of parameters listed in Tables 1-3. All the slot arrays were simulated using unit cell model with periodic boundary conditions backed by 30 μm layers of air followed by PML layers on both sides of the plane . Ports were set at boundaries between a layer of air and PML.

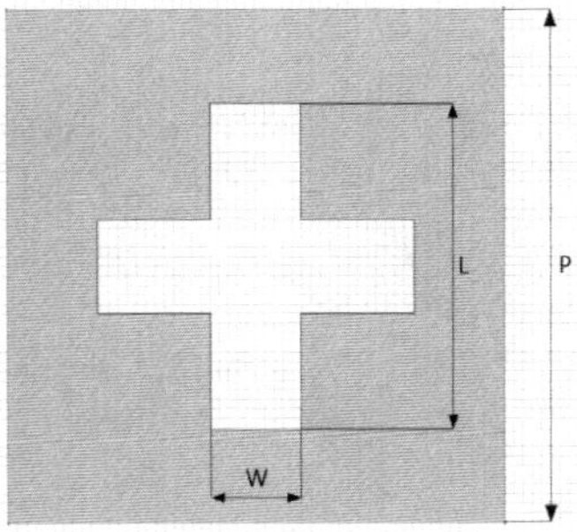

Figure 1. Unit cell of the grid composed of X-shaped slots.

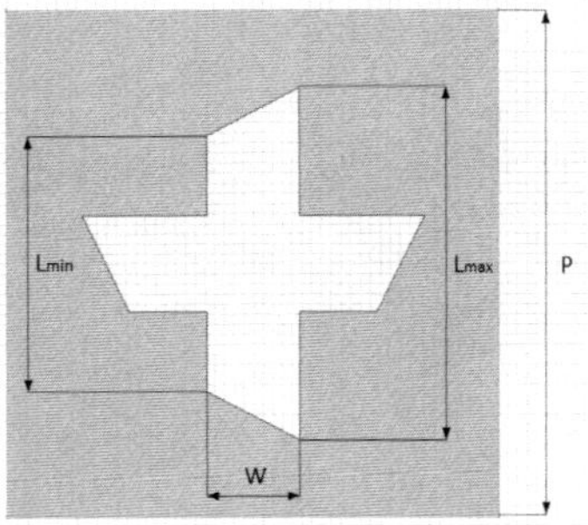

Figure 2. Unit cell of the grid composed of X-shaped slots with length varying across the width of the slot.

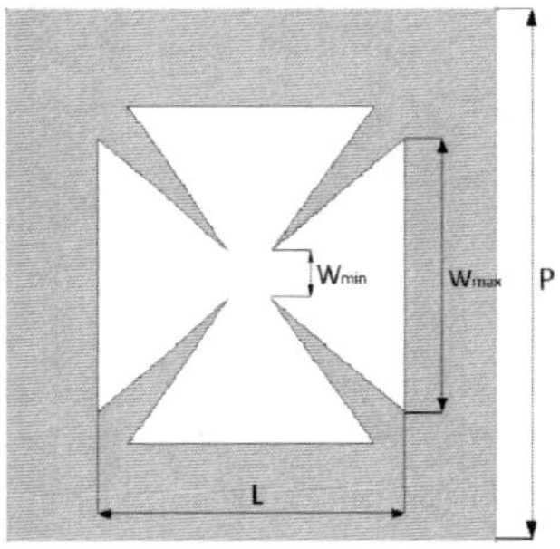

Figure 3. Unit cell of the grid composed of X-shaped slots width varying across the length of the slot.

Table 1. Simulation parameters for grid of X-shaped slots with straight edges with respective resonant frequencies

Resonant frequency, THz	W, um	L, um	P, um
0.27	163	574	901
0.49	89	313	492
0.76	57	203	318

Table 2. Simulation parameters for grid of X-shaped slots with edges mitred at 45 deg with respective resonant frequencies

Resonant frequency, THz	W, um	L min, um	L max, um	P, um
0.27	163	492	656	901
0.49	89	268	358	492
0.76	57	174	231	318

Table 3. Simulation parameters for grid of X-shaped slots with bar width varying along its length

Resonant frequency, THz	W min, um	W max, um	L, um	P, um
0.27	82	232	574	901
0.49	44	127	313	492
0.76	29	82	203	318

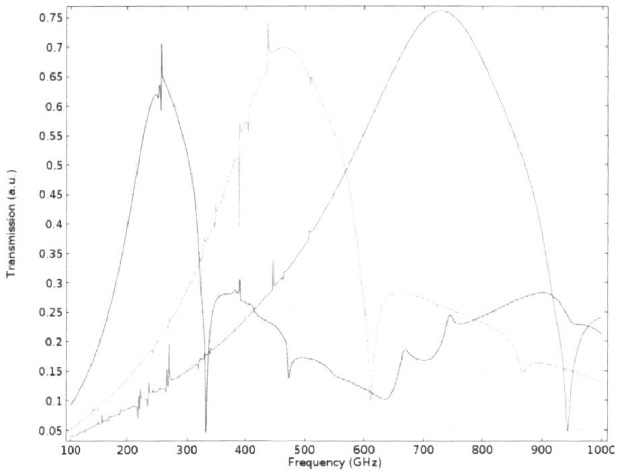

Figure 4. Simulated transmission spectra of arrays of X-shaped slots for parameter sets (peak transmission frequencies equal to 0.27, 0.49, 0.76 THz for blue, green and red curves respectively) listed in Table 1.

2.1. X-shaped slots with straight and mitred sides

The structure composed of X-shaped slots (Fig.1) exhibits narrow transmission band with central frequency determined by its dimensions. The central wavelength of the transmission peak for X-shaped slot depends on its surface area to bar width relation [6, 7] which gives:

$$\lambda_r = 2L - W,$$

where L is the length of cross bar and W is its width, see Fig.1.As mentioned in [6] this estimation can be made more accurate using results of numerical simulation of several grid structures which gives the following empiric equation:

$$\lambda = 1.8L - 1.35W + 0.2P,$$

where P is the period of the structure, see Fig.1.

It was demonstrated by numerical simulation of transmission spectra that metasurface composed of the X-shaped slots with sides mitred at 45 ° (length of the slot varies over its width, see Fig.2) and having the same surface area for the each parameter set listed in Fig.1 exhibits transmission peaks with higher quality (Fig.5) (e.g. 126 GHz (straight sides) versus 107 GHz (mitred sides) using FWHM criteria for the 1st set of parameters) which increases with for higher freuqencies. The choice of particular combination of L, W and P parameters satisfying this criteria is determined by the desired quality Q of the grid [9].

2.2. X-shaped slots with variable slope of its sides

The central frequencies of the resonant transmission peaks can be varied by altering the width of the cross bar alongside its length or vice versa, assuming that the surface of the slot remains unchanged. Simulated transmission spectra for the case where width of the slot varies along the length are shown in Fig.6, and Fig.7 shows the case where the width of the slot varies across the length of a bar. Transmission spectra of the structure show in Fig.6 was simulated for the same sets of parameters and surface area values as X-shaped slots with straight sides in order to make a meaningful comparison. It is shown that transmission peak can be shifted by varying arm width of a slot alongside the length of the bar of the cross. The spectra of the structure shown in Fig.7 appears to have low frequency transmission peak with high Q-factor and high frequency transmission peak with low Q-factor.

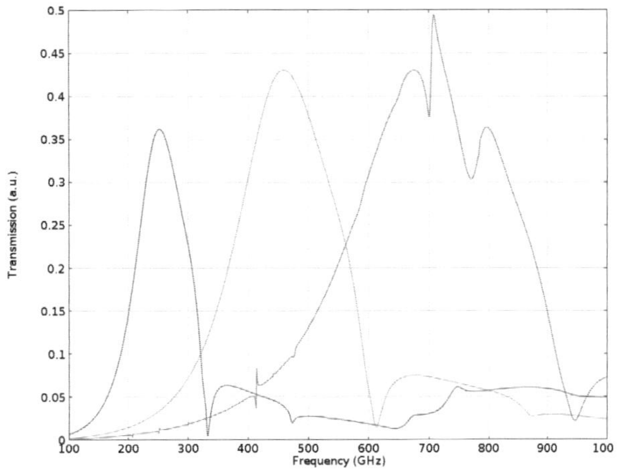

Figure 5. Simulated transmission spectra of arrays of X-shaped slots with sides mitred at 45 for the parameter sets listed in Table 2 (peak transmission frequencies equal to 0.27, 0.49, 0.76 THz for blue, green and red curves respectively) .

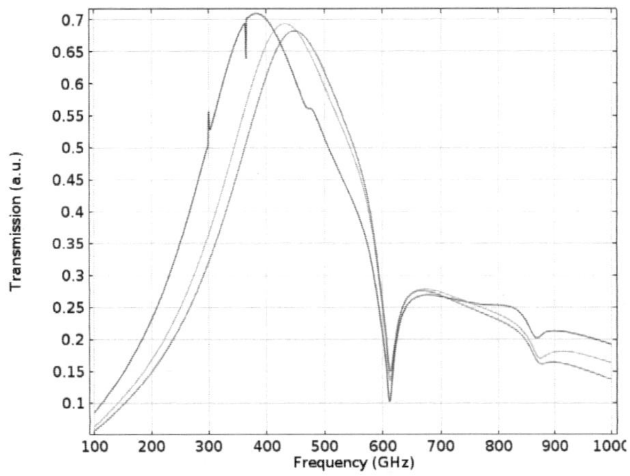

Figure 6. Simulated transmission spectra of arrays of X-shaped slots with sides mitred at 30° (blue), 45° (green), 60° (red) for the 2nd parameter set listed in Table 1.

2.3. Influence of substrate on the transmission properties of the filter
By changing substrate permittivity (e.g. by applying electric field to the silica film) the transmission peak of the structure is shifted alongside the spectrum (increasing the ϵ redshifts the transmission peaks, see Fig.8). The presence of the substrate also gives a set of resonant peaks in the transmission spectra attributed to Fabri-Perot resonances, which are not desirable for passband filters.

3. Conclusion
In the present work the transmission spectra of X-shaped slots with straight bars, with bar width varying over length, with bar length varying over width and with substrate of variable were numerically simulated. It is shown that X-shaped slots with mitred sides have narrower passbands than straight ones but significantly less transmission. Variation of slope angle of

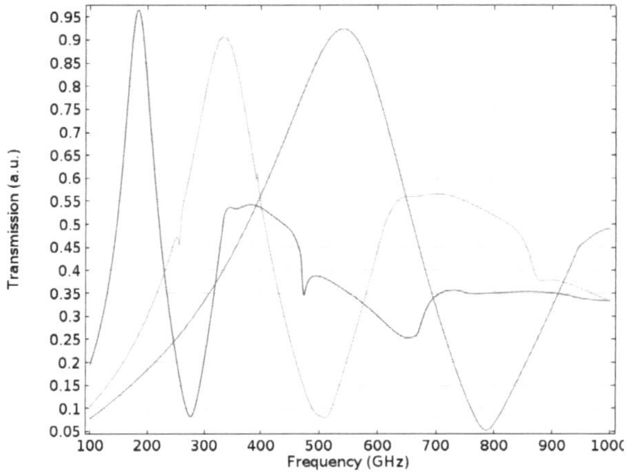

Figure 7. Simulated transmission spectra of the grid composed of X-shaped slots with width varying alongside the bar length of the cross. Blue, green and red curves correspond to the same parameter sets listed in Table 3 (peak transmission frequencies of 0.27, 0.49, 0.76 THz respectively).

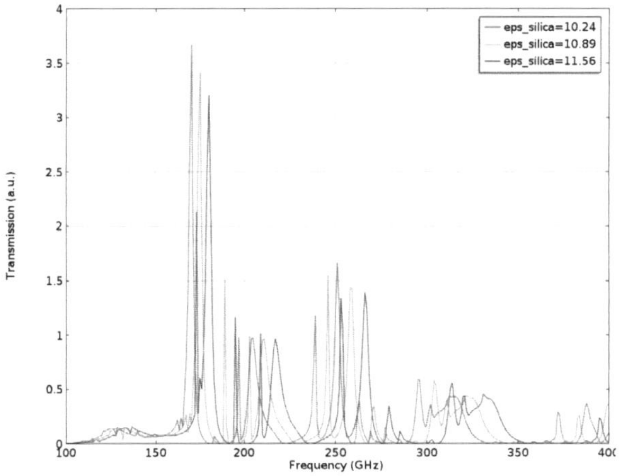

Figure 8. Simulated transmission spectra of the grid composed of X-shaped slots placed on $500\mu m$ thick Si substrate which permittivity ϵ takes the values of $10.24, 10.89, 11.56$. The parameters of the slot are that for resonant frequency 0.49 THz listed in Table1

the sides of the cross while preserving its surface area redshifts the resonant transmission peak. Spectra of the X-shaped slots with width varying along the length of a bar show two passbands in their spectra: one low frequency transmission peak of high quality and wide, low quality, transmission peak at higher frequencies. Finally the position of the transmission peak of the slot grid can be altered by varying the substrate permittivity: increasing its value redshifts the center frequency of the passband. These features can be used to develop narrowband filters with high Q-factor.

4. Acknowledgements

This work was financially supported by Government of Russian Federation, Grant 074-U01

References

[1] Lee Yun-Shik 2009 Principles of Terahertz Science and Technology
[2] Terekhov Y et al. 2013 The influence of period between U-shaped resonators on metasurface response at terahertz frequency range (SPIE NanoScience+ Engineering, 88062Q-88062Q-7)
[3] Girich A et al. 2011 Experimental Analysis of Metamaterials Spectra to Design Tunable THz-GHz Passive Devices (Terahertz and Mid Infrared Radiation) pp. 159-164
[4] Ulrich R et al. 1967 Far-infrared properties of metallic mesh and comlementary structure (Infr. Phys. Vol. 7) pp. 37-55
[5] Ulrich R et al. 1967 Effective low-pass filter for infrared frequencies (Infr. Phys. Vol. 7) pp. 65-74
[6] Moller K et al. 1999 Cross-shaped bandpass filters for the near- and mid-infrared wavelength regions (Appl. Opt. Vol. 35 No.31) pp. 6210-6215
[7] Moller K et al. 1999 Thin and thick cross shaped metal grids (Infr. Phys.Tech. Vol. 40) pp. 475-485
[8] Chase R 1983 Resonant array bandpass filters for the far infrared (Appl. Opt. Vol. 22 No. 11) pp. 1775-1779
[9] Tarasov M et al. 2009 Fabrication and Characteristics of Mesh Band-Pass Filters (Instrum. Exp. Tech. Vol. 52, No. 1) pp. 74-78
[10] Voisiat B et al., 2011 Band-pass filters for THz spectral range fabricated by laser ablation (Appl. Phys. A Vol. 104) pp. 953-958
[11] Melo A et al. 2012 Cross-Shaped Terahertz Metal Mesh Filters : Historical Review and Results, (Adv. Opt. Tech. Vol. 2012, Article ID 530512)

Journal of Physics: Conference Series **643** (2015) 012072 doi:10.1088/1742-6596/643/1/012072

Reaction-diffusion optoelectronics based on dispersed semiconductors

O V Gradov[1] **and M A Gradova**[2]
[1] Institute for Energy Problems of Chemical Physics, Russian Academy of Sciences, Leninsky pr., 38/2, Moscow, Russia
[2] Semenov Institute of Chemical Physics, Russian Academy of Sciences, Kosygin Str., 4, Moscow, Russia

E-mail: o.v.gradov@gmail.com

Abstract. Since many dispersed semiconductors are capable of light energy conversion and possess photocatalytic and luminescent properties, and any discreet light-sensitive medium can be applied for the positional-sensitive light flux registration (similar to pixels and voxels in semiconductor-based image recording), the use of chemically active dispersed semiconductors allows to perform a direct signal / image registration based on light-sensitive reaction-diffusion redox systems without conventional CCD / CMOS devices. The image capturing in this case will correspond to the formation of the metastable dissipative structures in the active medium, with their morphological properties determined by the flux gradient and provided by the corresponding dispersed semiconductor medium sensitivity.

1. Prerequisites

Many dispersed semiconductors are known to be effective light energy converters, photocatalysts and photoluminescent nanoparticles [1-4]. At the same time a dispersed, i.e. a discreet photosensitive medium can be used in a position-sensitive light flux registration, and hence can be applied for the image recording similar to the CCD with the pixel function performed by the discrete semiconductor elements. The above mentioned chemical activity of the dispersed semiconductors allows to perform the charge transport and the image fixation based on the chemical principles. There are several light-sensitive autocatalytic reactions, such as Kuhnert reaction [5] which is a special case of the self-oscillatory Belousov-Zhabotinsky reaction where the redox processes occur in accordance with the light flux gradient. There also have been developed a number of advanced systems for image registration and neural network processing based on light-sensitive BZ-type reactions and other reaction-diffusion systems, including the biomolecular ones [6-10]. However, the lack of the semiconductor properties of the active medium prevents from the direct image capturing, directional data recording and transduction in such systems without using conventional optoelectronic devices (CCD or CMOS positional-sensitive sensors).

2. Our proposals

We propose to use self-oscillatory systems with the redox-catalysis / photocatalysis on the surface of the dispersed semiconductor nanoparticles involved as the redox-agents in the autocatalytic and reaction-diffusion processes. The image recording process in this case will correspond to the formation of the metastable dissipative structures in the dispersed medium, with their morphological properties determined by the flux gradient and provided by the corresponding dispersed semiconductor medium light-sensitivity, as well as by the reversibility of the photoredox processes similar to those occurring in the photographic process where the silver halide micro / nanoparticles are also both dispersed semiconductors and photocatalysts [11]. The charge nature of the processes in the active medium makes it possible to perform a direct registration according to the principles of the "direct optoelectronics" without using CCD or CMOS active pixel sensors as the secondary image converters in the framework of the above approach [12,13].

Moreover, it is possible to perform registration of non-optical parameters using the appropriate converters of the arbitrary signals into the optical ones, fixed at the optical path in front of the reaction-diffusion medium. In particular, reaction-diffusion dispersed semiconductor registering "matrixes" allow to perform visualization of a number of important physical parameters of the sample, such as:

a) magnetic field (using magnetic film indicators [14-18] or “flux-detectors”; either with or without the transition through the Curie point under the local laser irradiation)
b) electrochemical parameters (using planar potentiometric / pH / pX indicators). A similar method with the local diffusion cell walls allows to perform the mapping of the local chemical composition (similar to the reagent indicator paper methods [19-22] but with the transparent carrier), where the light beam with the spectral parameters corresponding to those of the substance under investigation passes through the analyte, located on the optical path in front of the registering reaction-diffusion medium;
c) laser beam transmission outside the visible spectral range (using the doped solid matrices) [23-26];
d) distribution of the emitting regions in autoradiography (using the scintillation detectors with different quenching factors, from the simplest CsI(Tl) or CaI_2 [27,28] and other inorganic thin scintilllators [29], including rare-earth activated sol-gel films [30], to organic plastic scintillators [31]. The above technology is particularly similar to the Scintillator Pixel Sensors [32], with the solid state optoelectronic components substituted by the reaction-diffusion medium which is sensitive in the optical spectral region of the scintillator emission under the penetrating radiation or particles. This is of special interest in comparison with the beam action on the oscillation reactions in the active medium without a scintillator, which have been comprehensively studied earlier [33-35]. This approach can be also applied to the electron beam-excitation assisted optical microscopy [36] and thin layer fluorography on the mesoscopic scale [37].
e) polarization characteristics and the angular fluorescence polarization (using position-sensitive polaroids instead of a technique described in [38]; position-sensitive polarization systems can be also designed for both X-rays and gamma rays, but this is a more complex problem [39,40])
f) the local temperature of the sample at different points (using thermochromic liquid crystals [41-43] or other specially prepared for a certain temperature and spectral emission range thermochromic mixtures and coatings for temperature fields measurements [44,45]).

Thus, reaction-diffusion optoelectronics based on dispersed semiconductors is perspective not only for the optical flux visualization, but also for the colocalization analysis of different physical parameters of the sample, drug or analyte, which can be visualized using various pre-converters of the physical parameter into the optical signal registered by the photosensitive reaction-diffusion medium. Of special interest are the works by Vanag on the self-oscillating reactions in heterogeneous media [46-48] containing a kind of pixels, since their implementation on the dispersed semiconductor medium allows one to perform a position-sensitive analysis without the external controlling factors of the beam structuring, but directly in the self-oscillating system due to the inherent properties of the medium which can be in this case interpreted as an adaptive dynamic system of sensitive elements (sensels).

3. References

[1] Vasil'tsova O V and Parmon V N 1999 *Kinet. Catal.* **40** 62-70.

[2] Savinov E N, Gruzdkov Yu A and Parmon V N 1989 *Int. J. Hydrogen Energy* **14** 1-9.

[3] Gruzdkov Yu A, Savinov E N, Korolkov V V and Parmon VN 1988 *React. Kinet. Catal. Lett.* **36** 395-400.

[4] Bavykin D V, Savinov E N and Parmon V N 1999 *Langmuir* **15** 4722-27.

[5] Kuhnert L 1986 *Nature* **319** 393-4.

[6] Rambidi N G, Kuular T O O and Machaeva E E 1998 *Adv. Mater. Opt. Electron.* **8** 163-71.

[7] Rambidi N G and Maximychev A V 1996 *Optical Memory Neural Networks* **5** 129-43.

[8] Gritsenko O V, Sidelnikov D I, Simonova A P, Chernavskii D S and Rambidi N G 1991 *J. Mol. Electron.* **7** 155-66.

[9] Rambidi N G and Maximychev A V 1998 *Proc. of SPIE* **3402** 434-40.

[10] Rambidi N G 2005 *Int. J. Unconvent. Comput.* **1** 101-21.

[11] Jiang J, Li H and Zhang L 2012 *Chemistry* **18** 6360-9.

[12] Gradov O V and Gradova M A 2012 *3rd Int. Symp. "Molecular Photonics"*, Saint Petersburg, Russia, p 156.

[13] Gradov O 2015 *Reaction-diffusion optical soft matter lab-on-a-chip*. Preprint (in press).

[14] Grechishkin R, Chigirinsky S, Gusev M, Cugat O and Dempsey N M 2008 *In "Magnetic Nanostructures in Modern Technology (Spintronics, Magnetic MEMS and Recording)"* Eds. B Azzerboni, G Asti, L Pareti and M Ghidini, Dordrecht: Springer pp 195-224.

[15] Polyanskii A A, Cai X Y, Feldmann D M and Larbalestier D C 1999 *NATO Science Series* **72** 353-70.

[16] Polyanskii A A 2014 *Rus. Phys. J.* **56** 1007-12.

[17] Mikheenko P, Yurchenko V V, Cardwell D A, Shi Y H and Johansen T H 2013 J. Supercond. N. Magn. **26** 1499-502.

[18] Nikitenko V I, Gornakov V S, Dedukh L M, Kabanov Y P, Khapikov A F, Bennett L H, Chen P J, McMichael R D, Donahue M J, Swartzendruber L J, Shapiro A J, Brown H J and Egelhoff W F 1996 *IEEE Transactions on Magnetics* **32** 4639-41.

[19] Amelin V G, Kolodkin I S and Irinina Y A 2000 *J. Anal. Chem.* **55** 374-7.

[20] Ostrovskaya V M, Davidovskii N V, Prokopenko O A and Man'shev D A 2004 *J. Anal. Chem.* **59** 882-4.

[21] Ostrovskaya V M, Reshetnyak E A, Nikitina N A, Panteleimonov A V and Kholin Y V 2004 *J. Anal. Chem.* **59** 995-1001.

[22] Gaiduk O V, Pantaler R P, Grebenyuk N N and Ostrovskaya V M *J. Anal. Chem.* 2009 **64** 201-5.

[23] Yanes A C, Del-Castillo J, Méndez-Ramos J and Rodríguez V D 2011 *J. Nanopart. Res.* **13** 7295-301.

[24] Oleinik A S and Fedorov A V 2011 *Meas. Tech.* **53** 1331-6.

[25] Singh V, Rai V K, Al-Shamery K, Haase M and Kim S H 2013 *Appl. Phys. A* **113** 747-53.

[26] Lyapin A A, Ryabochkina P A, Ushakov S N and Fedorov P P 2014 *Quantum Electron.* **44** 602–5.

[27] Novosad S S 1998 *Tech. Phys.* **43** 956-8.

[28] Avdeyev S P, Karnaukhov V A, Kuznetsov V D, Petrov L A, Rodionov V K, Karcz W, Janitcki M and Oeschler H 2001 *Instrum. Exp. Tech.* **44** 634-87.

[29] Berdnikov V V, Gurov Y B, Dolgoshein B A, Dmitrenko V V, Zadneprovskii B I, Kantserov V A, Sosnovtsev V V, Tikhomirov V O and Shmeleva A P 2013 *Instrum. Exp. Tech.* **56** 146-50.

[30] Garcia-Murillo A, LeLuyer-Urlache C, Dujardin C, Pédrini C and Mugnier J 2003 *J. Sol-Gel Sci. Tech.* **26** 957-60.

[31] Tolstukhin A I, Somov A S, Somov S V, Bolozdynya A I 2014 *Instrum. Exp. Tech.* **57** 658-61.

[32] Allier C P, Hollander R W, Van Eijk C W E, De Boer M J, Gardeniers J G E, Sarro P M and Czirr J B 2001 Proc. Sensor Technology Conference, Enschede, Netherlands pp. 191-6.

[33] Artamonov D N, Lebedev V M, Priselkova A B, Spassky A V and Trukhanov K A 2008 *Bull. Russ. Acad. Sci: Phys.* **72** 988-91.
[34] Usanov D A and Rytik A P 2013 *Russ. J. Phys. Chem.* **87** 872-5.
[35] Priselkova A B, Artamonov D N, Ermakov A N, Lebedev V M, Spassky A V, Trukhanov K A and Kruglov O S 2009 *Bull. Russ. Acad. Sci.: Phys.* **73** 494-7.
[36] Inami W, Fukuta M, Masuda Y, Nawa Y, Ono A, Lin S, Kawata Y and Terakawa S 2015 *Opt. Rev.* **22** 354-8.
[37] Luthi U and Waser PG 1995 *Adv. Tracer Methodol.* **3** 149-55.
[38] Gilles H, Girard S and Hamel J 2002 *Opt. Lett.* **27** 1421-3.
[39] Xu D, He Z and Zhang F 2005 *IEEE Trans. Nuc. Sci.* **52** 1160-4.
[40] Stohlker T, Spillmann U, Banas D, Beyer H F, Dousse J C, Chatterjee S, Hess S, Kozhuharov C, Kavcic M, Krings T, Protic D, Reuschl R, Szlachetko J, Tashenov S and Trotsenko S 2007 *J. Phys. Conf. Ser.* ***58*** 411–4.
[41] Sabatino D R, Praisner T J and Smith C R 2000 *Exp. Fluids* **28** 497-505.
[42] Smith C R, Sabatino D R and Praisner T J 2001 *Exp. Fluids* **30** 190-201.
[43] Ménard V, Reulet P, Nörtershäuser D, Le Masson S and Millan P 2006 *J. Visualization* **9** 131.
[44] Ma Y, Zhu B and Wu K 2000 *J. Coat. Technol.* **72**, 67-71.
[45] MacLaren D C and White M A 2005 *J. Mat. Sci.* **40** 669-76.
[46] Epstein I R, Vanag V K, Balazs A C, Kuksenok O, Dayal P and Bhattacharya A 2012 *Acc. Chem. Res.* ***45*** 2160-8.
[47] Vanag V K and Epstein I R 2009 *Phys. Chem. Chem. Phys.* **11** 897-912.
[48] Vanag V K and Epstein I R 2007 *Chaos* **17** 037110.

Temperature dependence of photoluminescence of GaAs/AlGaAs quantum rings

Yu.D. Sibirmovsky, I.S. Vasil'evskii, A.N. Vinichenko, I.S. Eremin, O.S. Kolentsova, N.I. Kargin and M.N. Strikhanov

National Research Nuclear University "MEPhI", 115409, Moscow, Russia

E-mail: sibirmovsky@gmail.com

Abstract. Samples with arrays of GaAs/AlGaAs quantum rings (QRs) of different shapes were grown by molecular beam epitaxy in droplet epitaxy mode. Photoluminescence (PL) spectra of the samples were measured at 20 – 90 K and 300 K, intense peaks attributed to the QR layers were observed. The peaks were identified by comparison of as-grown and selectively etched samples and by the calculation of the ground state energy for charge carriers in GaAs QRs. FWHM narrowing with the temperature increase was observed below 70 K.

1. Introduction

Quantum rings (QRs) are novel nanostructures similar to quantum dots but with unique electronic and optical properties, especially in magnetic fields [1]. Photoluminescence spectra of GaAs/AlGaAs quantum rings were studied in several works [2-5], though the temperature dependence of the spectra has not been investigated in detail. In addition, there are some difficulties with identification of additional PL peaks at higher energy (some authors attribute them to AlGaAs barrier layer, while others – to the GaAs wetting layer). Also, QR layers are usually overgrown and annealed before the PL experiments, so the spectrum of as-grown surface layers of QRs has not been studied so far.

2. Experiment and results

Two samples were investigated in the present work, containing one layer of single QRs (A1) and two layers of double QRs (A2), formed by droplet epitaxy [6]. Both samples were grown on GaAs(001) substrate with GaAs buffer layer and $Al_{0.28}Ga_{0.72}As$ barrier layer. For A2 sample one QR layer was overgrown by $Al_{0.28}Ga_{0.72}As$, and another (identical) QR layer was grown on the surface. For more information on the growth procedure see [7].

The mean diameter and height of the rings were measured using atomic force microscopy and scanning electron microscopy. The outer/inner diameter of the QRs were $53\pm12/22\pm6$ nm for A1, $56\pm9/28\pm3$ nm for A2 inner rings and $140\pm24/100\pm23$ nm for A2 outer rings. The mean height of QRs was 4.5 ± 0.5 and 4.3 ± 0.3 nm for A1 and A2, respectively.

For PL measurements 488 nm Ar^+ laser was used with intensity of 100 mW/cm^2. Spectra were measured at 20-90 K and at room temperature. PL peaks of QR layers were identified by comparing their energy with GaAs band gap (1.52 eV at 20 K) and $Al_{0.28}Ga_{0.72}As$ band gap (1.87 eV at 20 K). The A2 low temperature PL spectrum has an additional peak at lower energy

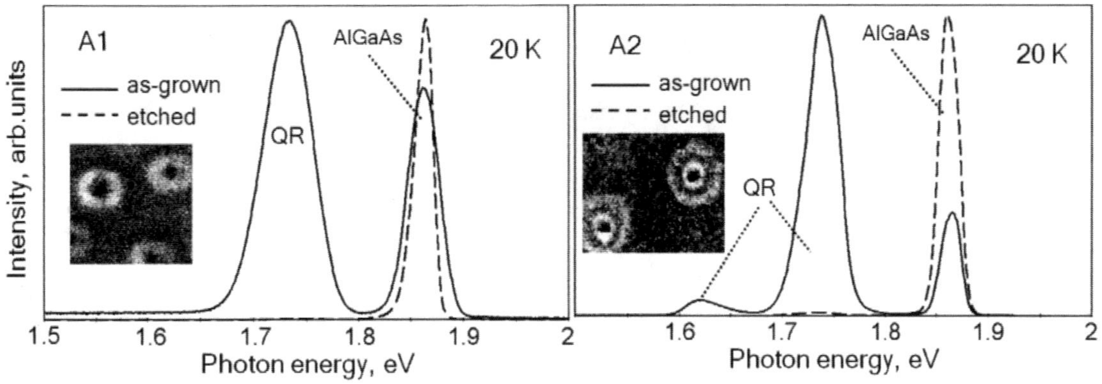

Figure 1. Normalized PL spectra of the samples A1 (left) and A2 (right) at T=20 K. As-grown (solid line) and etched (dashed line) samples are compared.

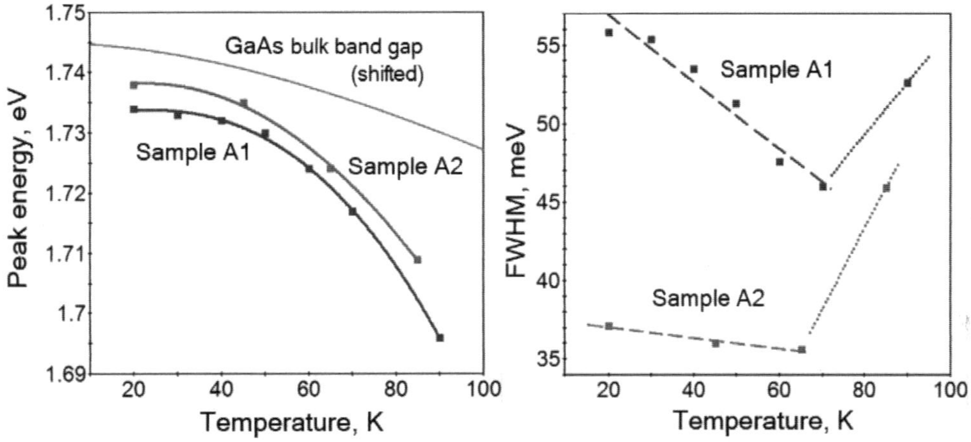

Figure 2. Temperature dependence of the energy (left) and FWHM (right) of QR peaks for samples A1 and A2. GaAs bulk band gap (Varshni relation) is shown shifted to higher energy for comparison.

which is attributed to the buried QR layer. The reference spectra of the etched samples without QR layers were also measured (Fig. 1).

Room temperature (RT) spectra were measured in accumulation mode. There was no PL from the $Al_{0.28}Ga_{0.72}As$ layer at RT, but there was a peak at 1.42 eV (which is GaAs bandgap at 300 K) from the substrate/buffer layer for both samples in addition to the peaks from GaAs quantum rings. For reference, we also measured the PL spectra of the substrates, where only bulk GaAs peak at 1.42 eV was found.

3. Discussion

The temperature dependences of QR peak energy and full width at half-maximum (FWHM) are presented in Fig. 2. The FWHM narrowing with the temperature increase was observed below 70 K. This phenomenon has been noticed before in disordered systems such as quantum dots [8,9] and rough quantum wells [10] and is usually explained by thermal assisted tunnelling and relaxation of the carriers into global minima of the potential (larger QRs in this case) intead of local minima. The peak energy also decreases more rapidly with the temperature increase than GaAs bulk band gap (Fig. 2), which is in good agreement with the proposed model. At temperatures higher than 70 K electron-phonon scattering starts to play major role, leading to the peak broadening.

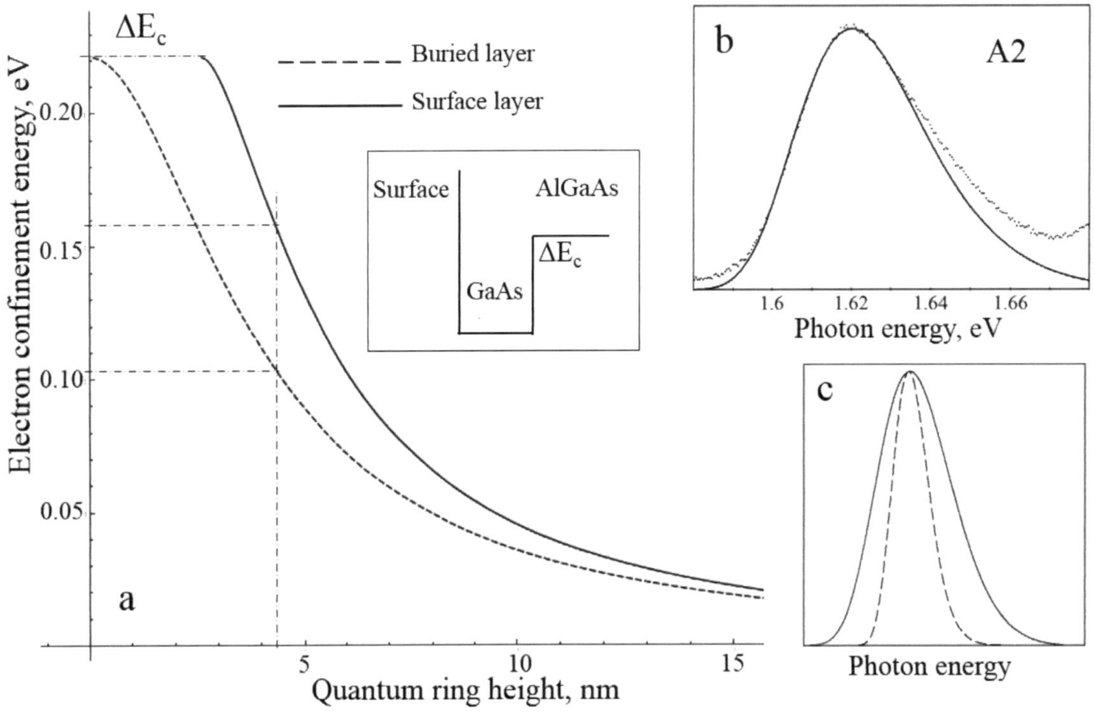

Figure 3. (**a**) Electron confinement energy dependence on the QR height for surface and buried layers, the considerable difference is shown for 4.3 nm (sample A2) (**b**) Peak shape approximation for the buried QR layer of A2 at T=20 K, (**c**) Peak shapes for surface and buried QR layers with the same height distribution.

Finally, the ground state energy of QRs was calculated. We use one-particle two-band model here for electrons and heavy holes. Because the height of QRs is much smaller than their diameter, simple 2D quantum well model is sufficient for estimating the quantum confinement energy. This model shows that the ground state energy is much higher for surface QR layer than for buried QR layer (Fig. 3a) which is confirmed by the spectrum of sample A2. We can also model the low temperature shape of the peaks, determined by the QRs size distribution (Fig 3b). We consider gaussian distribution for QR height and model the peak shape using energy dependence on the 2D quantum well width. The calculated FWHM is smaller for the buried layer than for the surface layer with identical height distribution (Fig. 3c).

For more accurate calculation of the confinement energy for 3D quantum rings numerical method described in [11] was implemented. We expand the envelope function into a complete orthonormal set of eigenfunctions of a rectangular potential well with infinite barrier. Then we transform the Schrodinger equation into matrix form which is solved numerically. The size of the well for any number of the basis functions can be chosen so as to minimize the ground state energy. The method was tested by comparing with the analytical results for exactly solvable problems, such as quantum well and two-dimensional quantum ring. For final calculation we used the basis set of 11 functions in each dimension. The results for the peaks from the surface QR layers are shown in the Table 1. They are in good agreement with the experiment.

For the buried QR layer of the sample A2 the calculated energy is higher than the experimental value which can be explained by interdiffusion with the $Al_{0.28}Ga_{0.72}As$ barrier layer during overgrowth. In that case the height distribution would also be different compared to the surface layer, which can explain the large FWHM of the peak.

Table 1. Experimental and calculated values of e1-hh1 transition energy for the samples.

Sample	Experimental	Calculated
A1	$\Delta E_{exp} = 0.217 \pm 0.028$ eV	$\Delta E_{calc} = 0.20 \pm 0.03$ eV
A2	$\Delta E_{exp} = 0.220 \pm 0.018$ eV	$\Delta E_{calc} = 0.21 \pm 0.01$ eV – inner ring
		$\Delta E_{calc} = 0.20 \pm 0.01$ eV – outer ring
A2	$\Delta E_{exp} = 0.102 \pm 0.041$ eV	$\Delta E_{calc} \approx 0.12$ eV

Acknowledgements
The authors would like to thank Dr Denis M. Zhigunov for performing the photoluminescence measurements (Multiple-Access Center, Faculty of Physics, Lomonosov Moscow State University).

This work was supported by the Ministry of Education and Science of the Russian Federation, Complex project no. 2013-218-04-050 and was performed using the equipment of the Multiple-Access Center "Heterostructure Microwave Electronics and Physics of Wide Band-Gap Semiconductors" of the National Research Nuclear University "MEPhI".

References
[1] F. Ding, N. Akopian, B. Li et al. 2010 *Phys. Rev.* B **82** 075309
[2] T. Mano and N. Koguchi 2005 *J. Cryst. Growth* **278** 108
[3] T. T. Kuroda, T. Mano, T. Ochiai et al. 2005 *Phys. Rev.* B **72** 205301
[4] Z.Y. AbuWaar, Y.I. Mazur, J.H. Lee. 2007 *J. Appl. Phys.* **101** 024311
[5] J. S. Kim 2008 *J. Korean Phys. Soc.* **53** 2138
[6] N. Koguchi, S. Takahashi, T. Chikyow. 1991 *J. Cryst. Growth* **111** 688
[7] Yu. D. Sibirmovskii, I. S. Vasil'evskii, A. N. Vinichenko et al. 2014 *Bull. Lebedev Phys. Inst.* **41** 243
[8] D. I. Lubyshev, P. P. Gonzlez-Borrero, E. Marega Jr. et al. 1996 *Applied Physics Letters* **68** 205
[9] Y. T. Dai, J. C. Fan, Y. F. Chen et al. 1997 *Journal of Applied Physics* **82** 4489
[10] M. Grassi Alessi, F. Fragano, A. Patane, and M. Capizzi. 2000 *Phys. Rev.* B **61** 16
[11] M. Califano and P. Harrison. 2000 *Phys. Rev.* B **61** 10959
[12] Yu. D. Sibirmovskii, I. S. Vasil'evskii, A. N. Vinichenko et al. 2015 *Semicond.* **49** 638

Local field effects and metamaterials based on colloidal quantum dots

O V Porvatkina, A A Tishchenko and M N Strikhanov

National Research Nuclear University MEPhI, Moscow, Russia

E-mail: OVPorvatkina@mephi.ru, Tishchenko@mephi.ru

Abstract. Metamaterials are composite structures that exhibit interesting and unusual properties, e.g. negative refractive index. In this article we consider metamaterials based on colloidal quantum dots (CQDs). We investigate these structures taking into account the local field effects and theoretically analyze expressions for permittivity and permeability of metamaterials based on CdSe CQDs. We obtain inequality describing the conditions when material with definite concentration of CQDs is metamaterial. Also we investigate how the values of dielectric polarizability and magnetic polarizability of CQDs depend on the dots radius and properties the material the quantum dots are made of.

1. Introduction

Metamaterials demonstrate new properties, some with no analogue in conventional materials, for example, artificial magnetism and near-field focusing. Unusual electromagnetic response of these materials could lead to a range of applications including organic light emitting diodes (OLEDs), ultrafast LEDs and nanoscale lasers [1]-[3]. In our work we consider metamaterials based on spherical CQDs. CQDs are organic nanocrystals with semiconductor core (for example, CdSe) that have organic ligands bound to surface. CQDs offer promise in light sensing, energy conversion etc. CQDs have tunability and atomic-like optical emission, therefore they are very interesting for studying interaction between light and metamaterial based on them.

2. Dielectric and magnetic properties of metamaterials based on colloidal quantum dots

We investigate metamaterial consisting of N anisotropic CQDs. Properties of the quantum dots are described by dielectric polarizability tensor $\alpha_{ij}^{e}(\omega)$ and magnetic polarizability tensor $\alpha_{ij}^{m}(\omega)$:

$$\alpha_{ij}^{e}(\omega)=\alpha_{\perp}^{e}(\omega)\left(\delta_{ij}-e_i e_j\right)+\alpha_{l}^{e}(\omega)e_i e_j, \quad (1)$$

$$\alpha_{ij}^{m}(\omega)=\alpha_{\perp}^{m}(\omega)\left(\delta_{ij}-e_i e_j\right)+\alpha_{l}^{m}(\omega)e_i e_j. \quad (2)$$

In our previous work [4] we obtained the dependence of permittivity tensor $\varepsilon_{ik}(\omega)$ on dielectric polarizability tensor $\alpha_{ij}^{e}(\omega)$ and permeability tensor $\mu_{ij}(\omega)$ on the magnetic polarizability tensor $\alpha_{ij}^{m}(\omega)$:

$$\varepsilon_{ik}(\omega)=\varepsilon^{\perp}(\omega)\left(\delta_{ik}-e_i e_k\right)+\varepsilon^{l}(\omega)e_i e_k, \quad (3)$$

$$\mu_{ij}(\omega)=\mu^{\perp}(\omega)\left(\delta_{ij}-e_i e_j\right)+\mu^{l}(\omega)e_i e_j, \quad (4)$$

where

$$\varepsilon^{\perp}(\omega)=\frac{1+(8\pi/3)n\alpha_{\perp}^{e}(\omega)}{1-(4\pi/3)n\alpha_{\perp}^{e}(\omega)}, \quad \varepsilon^{l}(\omega)=\frac{1+(8\pi/3)n\alpha_{l}^{e}(\omega)}{1-(4\pi/3)n\alpha_{l}^{e}(\omega)}, \quad (5)$$

$$\mu^{\perp}(\omega)=\frac{1+(8\pi/3)n\alpha_{\perp}^{m}(\omega)}{1-(4\pi/3)n\alpha_{\perp}^{m}(\omega)},\quad \mu^{l}(\omega)=\frac{1+(8\pi/3)n\alpha_{l}^{m}(\omega)}{1-(4\pi/3)n\alpha_{l}^{m}(\omega)}. \tag{6}$$

In case of isotropic CQDs our results are identical to those obtained by another method in the article [5]. These equations make it possible to create metamaterials with desirable characteristics taking into account the properties of constituent particles.

We do our calculations with using Maxwell's equations. We also take into consideration the so called local field effects. The local field is an effective field calculated by averaging over the positions of all CQDs except one [6]. Adjacent quantum dots have more strong influence on the local field acting upon the given quantum dot, so the value of the local field is produced in a finite volume near the selected quantum dot. Because of it we use a long-wave limit for our computation:

$$c/\omega >> L >> d, \tag{7}$$

where d is the lattice constant, L is the linear dimension of the crystal.

It is well known that permittivity and permeability are complex values (for metamaterials real parts of permittivity and permeability are negative $\varepsilon'(\omega)<0,\ \mu'(\omega)<0$):

$$\varepsilon(\omega)=\varepsilon'(\omega)+i\varepsilon''(\omega), \tag{8}$$

$$\mu(\omega)=\mu'(\omega)+i\mu''(\omega). \tag{9}$$

Taking into account that dielectric and magnetic polarizability are also complex values, using the results of work [4] we obtain equations for real parts of permittivity and permeability:

$$\varepsilon'(\omega)=\frac{1-32/9\pi^2n^2\left(\left(\alpha_e'(\omega)\right)^2+\left(\alpha_e''(\omega)\right)^2\right)+4/3\pi n\alpha_e'(\omega)}{1+16/9\pi^2n^2\left(\left(\alpha_e'(\omega)\right)^2+\left(\alpha_e''(\omega)\right)^2\right)-8/3\pi n\alpha_e'(\omega)}, \tag{10}$$

$$\mu'(\omega)=\frac{1-32/9\pi^2n^2\left(\left(\alpha_m'(\omega)\right)^2+\left(\alpha_m''(\omega)\right)^2\right)+4/3\pi n\alpha_m'(\omega)}{1+16/9\pi^2n^2\left(\left(\alpha_m'(\omega)\right)^2+\left(\alpha_m''(\omega)\right)^2\right)-8/3\pi n\alpha_m'(\omega)}. \tag{11}$$

Combining conditions $\varepsilon'<0,\ \mu'<0$ with Eqs. (10), (11) one can obtain limits for the production of the concentration of CQDs and dielectric/magnetic polarizability of CQDs:

$$0.1<n\alpha_{e,m}(\omega)<0.16. \tag{12}$$

The resulting Eq. (12) allows defining the concentration of CQDs which is necessary to create metamaterials (and vice versa). Possibility of controlling concentration of CQDs could be used in various fields such as OLEDs, optical filter and switching device [7].

We consider metamaterial consisting of CQDs made of CdSe ($\varepsilon=10,\lambda=733{,}6$nm). Using equations for real and complex parts of dielectric and magnetic polarizability from works [8],[10] we obtain dependence of the dielectric and magnetic polarizability of the QDs on the radius of quantum dots. Note that we can control the color of the light emitted by the dots by changing their size. Therefore, OLEDs with metamaterials based on QDs can produce any color of light.

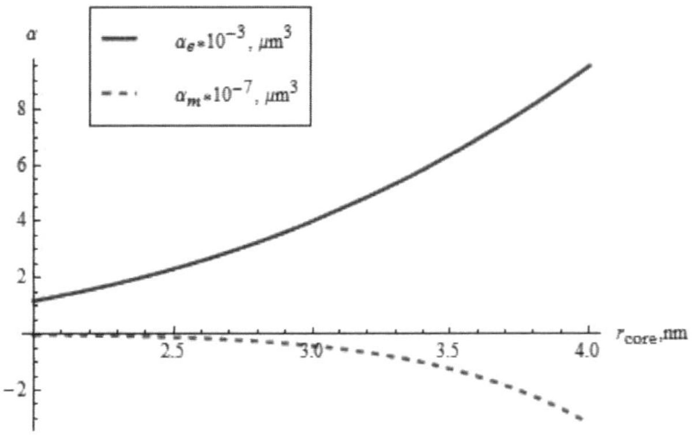

Figure 1. Dependence of the dielectric and magnetic polarizability of the QDs on the radius of QDs for $\varepsilon = 10$, $\lambda = 733{,}6$ nm.

Dependence of the dielectric and magnetic polarizabilities of the quantum dots on the radius of quantum dots (Figure 1) allows us to predict metamaterial state for material based on CQDs made of CdSe with a specified radius. For this reason OLEDs with metamaterials based on CQDs can produce any color of light, including white [8]. Also such metamaterials may consist of nanorods. Equations for dielectric and magnetic polarizabilities for such materials were obtained in [9].

3. Discussion

Permittivity and permeability of metamaterials based on colloidal QDs were investigated taking into account the local field effects. These effects caused by interaction between single QDs are proved to play a key role in the problem of originating of metamaterial condition. Eqs. (10), (11) allow creating metamaterials with simultaneously negative permittivity and permeability. Eq. (12) describes the conditions when the amorphous structure consisting of QDs can manifest the properties of metamaterial. Metamaterials are in the center of modern electrodynamics now. Such materials can also operate in the terahertz range [11],[12]. This fact makes them more attractive for investigations. Also nanostructured thin film metamaterials can be useful in light-emitting devices (OLEDs, solar battery etc). Nanostructured thin film metamaterials derived from nanostructured layers which consist of three or more CQDs (semiconductor nanocrystal) populations can perform charge transport as well as charge recombination in light emitting devices, or exciton creation and charge separation in light absorbing devices such as solar cells and photodetectors. Liquid crystal colloidal nanoparticle dispersions provide a powerful platform for self-assembly-based tunable optical metamaterial fabrication and may allow their potential applications to come within reach, for example, for a broad range of breathtaking applications that range from nanoscale diffraction-unlimited optical imaging with "perfect" lenses to invisibility cloaking [13].
This work was supported by Ministry of Education and Science of Russian Federation, Agreement No. 14.575.21.0002, ID RFMEFI57514X0002.

References

[1]Pendry J B Negative refraction makes a perfect lens 2000 *Phys. Rev. Lett.* **85** 39

[2]Veselago V G The electrodynamics of substances with simultaneously negative values of ε and μ 1968 *Physics-Uspekhi* **10** 517 [UFN **92** 517]

[3]Agranovich V M and Gartstein Yu N Spatial dispersion and negative refraction of light 2006 *Physics-Uspekhi* **49** 1029 [UFN **176** 1051]

[4]Porvatkina O V, Tishchenko A A and Strikhanov M N Local field effects in anisotropic metamaterials 2014 *Journal of Physics: Conference Series* **541** 012024

[5]Kästel J and Fleischhauer M Local-field effects in magnetodielectric medium: Negative refraction and absorption reduction 2007 *Physical Review A* **76** 062509

[6]Ryazanov M I and Tishchenko A A Clausius-Mossotti-type relation for planar monolayers 2006 *JETP* **103** 539 [*ZhETF* **130** 621]
[7]Boatman E M, Lisensky G C and Nordell K J A Safer, Easier, Faster Synthesis for CdSe Quantum Dot Nanocrystals 2005 *Journal of Chemical Education* **82** 1697-99
[8]Niu Y, Kazlas P T Patent WO 2013103440 (A1) 2013
[9]Shalaev V M, Wenshan C, Chettiar U K, H-K Yuan, Sarychev A K, Drachev V P and Kildishev A V Negative index of refraction in optical metamaterials 2005 *Optics Letters* **30** 33
[10]Morokhov I D, Petinov V I, Trusov L I, Petrunin V F Structure and properties of fine metallic particles 1981 *Sov. Phys. Usp.* **24** 295 [*UFN* **133** 653]
[11]Yen T J, Padilla W J, Fang N, Vier D C, Smith D R, Pendry J B, Basov D N, and Zhang X Terahertz Magnetic Response from Artificial Materials 2004 *Science* **303** 56
[12]Linden S, Enkrich C, Wegener M, Zhou J, Koschny T and Soukoulis C Magnetic response of metamaterials at 100 Terahertz 2004 *Science* **306** 1351
[13]Gardner D F, Evans J S and Smalyukh I I Towards reconfigurable optical metamaterials: colloidal nanoparticle self-assembly and self-alignment in liquid crystals 2011 *Mol. Cryst. Liq. Cryst.* **545** 1227

Hot exciton relaxation in multiple layers CdSe/ZnSe self-assembled quantum dots separated by thick ZnSe barriers

M Eremenko, G Budkin and A Reznitsky
Ioffe Institute, Politekhnicheskaya 26, St. Petersburg, 194021, Russia

E-mail: erjomenko@gmail.com

Abstract. We have studied PL and PLE spectra of two samples (A and B) of MBE grown CdSe/ZnSe asymmetric double quantum wells with different amount of deposited CdSe layers separated by 14 nm ZnSe barrier. It has been found that PLE spectra of the states forming short wavelength side of the PL spectra of both deep and shallow QWs of the sample A as well as that of deep QW of the sample B demonstrate oscillating structure in the spectral ranges corresponding to exciton states of self-assembled quantum dots only. Meanwhile PLE spectra of the short wavelength states of shallow QW the sample B revealed pronounced oscillating structure with energy period of ZnSe LO phonon under excitation with photons in a wide energy range both in the regions of quantum-dot states and in that of free states in the ZnSe barrier. In these spectra creating of excitons with kinetic energies more than 0.3 eV was observed which considerably exceed the exciton binding energy as well as LO phonon energy (both appr. 0.03 eV). It has been concluded that oscillating structure of the PLE spectra arises due to cascade relaxation of hot excitons. We discuss the model which explains these experimental findings.

1. Introduction

The numerous structural and spectroscopic studies carried out so far have shown that, under certain conditions, epitaxy of several CdSe monolayers into a ZnSe matrix leads to formation of a quantum well (QW) from an inhomogeneous ZnCdSe solid solution with nanosize objects, many properties of which are of 0D nature. Depending on the growth conditions, and primarily on temperature, such self-organized quantum dots (QD) may be of 2D or 3D nature. The most complex structure is observed for 2D objects obtained at T = 280°C. At this temperature the coherent CdSe insertions in ZnSe with CdSe nominal thickness in the range of 1-3 ML are known to generate planar CdSe-enriched QDs incorporated into the body of ZnCdSe QW [1, 2].

In this communication, we discuss the optical spectra of two ZnSe/CdSe/ZnSe double heterostructures grown by MBE at growth T = 280°C which includes two CdSe quantum wells with different nominal thicknesses separated by a ZnSe barrier with a thickness of 14 nm. The nominal thickness of the first CdSe deposited layer was 1.4 and 1.8 ML (samples B and A, respectively). The thickness of the second layer was t_{CdSe}=2.7 ML in both samples. Other details of the growth conditions as well as design of the samples under study and experimental details can be found in [3].

As it was shown previously, in multilayer ZnSe/CdSe/ZnSe samples with thicknesses of ZnSe barriers exceeding 4 nm there is (i) no correlation of the CdSe QDs in the adjacent layers [4] and (ii) at low temperatures there is no evidence of exciton tunneling between QD states in adjacent QW layers [5].

2. Experimental results and discussion.

Previous study has shown that the main features of the PLE spectrum of heterostructures with planar QDs are the following: (i) the oscillating character of PLE spectra for states on the short-wavelength side of the PL band reflecting exciton cascade relaxation with emission of LO phonons, which practically disappears from the PLE spectrum above a certain photon energy E_{ME}, which is related to the region of transition from states localized within QDs to those distributed over all the QW plane; and (ii) increasing of the PL signal more than an order of magnitude under excitation by photons with energy exceeding the energy gap of the ZnSe barrier material [3]. As one can see from figure 1 a-c both above-mentioned features of PLE spectra can be seen in PLE spectra of both deep and shallow QWs of the sample A as well as that of deep QW of the sample B.

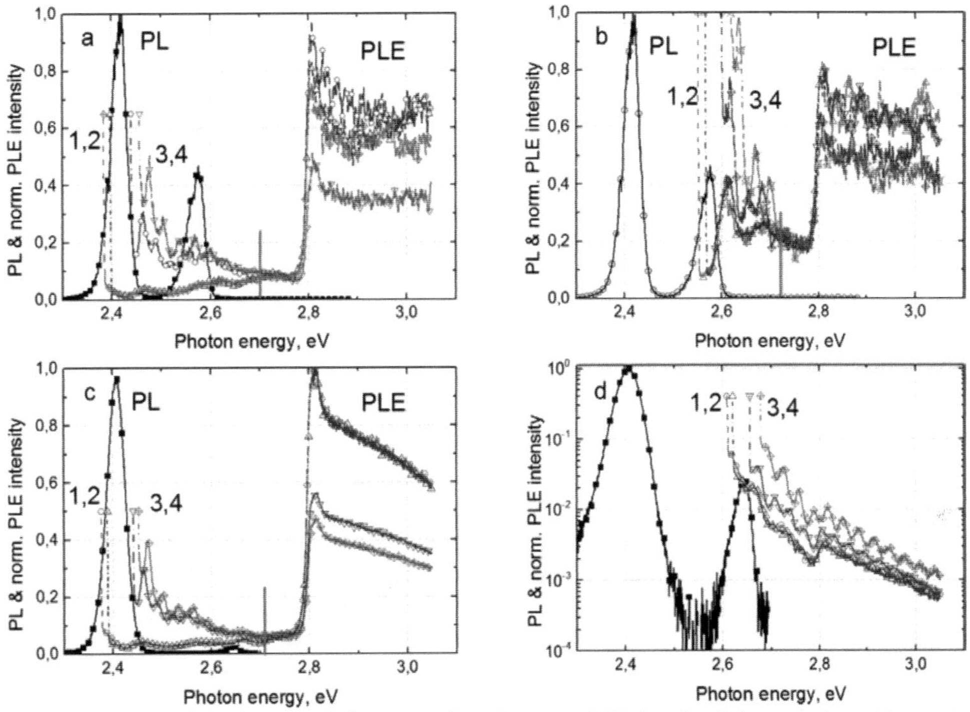

Figure 1. PLE spectra of the long- and short-wavelength parts of PL bands of the samples with nominal CdSe contents t_{CdSe} = 2.7 and 1.8 ML (sample A, panels a and b) and 2.7 and 1.4 ML (sample B, panels c and d), respectively. Each panel shows two PLE spectra for each of two detector positions on the long- and short-wavelength parts of the PL band. All the excitation spectra in panels a, b and c are normalized to the amplitude of a spectrum at energy of 2.75 eV. The PL and PLE spectra of the sample B present in panel d in a semilog scale. All PLE spectra of short-wavelength parts of a shallow PL band of sample B are shifted along the vertical axis for clarity. The vertical lines in panels a, b, and c show the position of the exciton percolation threshold E_{ME}

Unlike abovementioned character of PLE spectra, shown in panels a, b and c, distinct oscillating structure of PLE spectra was found for the states forming short-wavelength side of the PL spectrum of the sample B (see figure 1, panel d) not only within QW states but also at the spectral range above ZnSe fundamental absorption edge where we have detected up to 10 maxima in PLE spectra separated by the energy of ZnSe LO-phonon. It worth pointed out that energy relaxation with emission of LO phonons is a characteristic feature of hot exciton relaxation process [6].

Hot exciton nature of the structure observed in PLE spectra is the most clearly manifested when the excitation spectra are detected in polarized light. Figure 2 shows two PLE spectra obtained under excitation by linearly polarized light and recording in the parallel (solid curve Y1) and crossed (dash curve Y2) orientations of the polarizer and analyzer. The same figure presents a spectrum of the degree of polarization $Y3=(Y1-Y2)/(Y1+Y2)$ (curve Y3 with symbols).

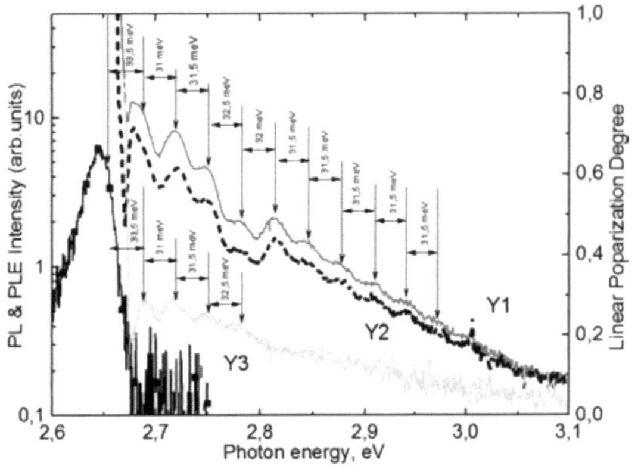

Figure 2. Black solid line with symbols shows the PL spectrum of sample B. Solid line Y1 and dashed line Y2 represents linearly polarized PLE spectra of the same sample in polarizations with parallel and crossed orientations of polarizer and analyser, respectively. Curve Y3 indicates the spectrum of the linear polarization degree (see text).

It can be concluded that both above-mentioned features of PLE spectra reflect the role of two main channels of the population of the QD emitting states, namely, (i) direct capture of excitons from the barrier states to the QD states and (ii) initial capture of electronic excitations in QW states with the following diffusion along QW states and consequent capture in QD states. The simple estimation of the relative role of these two ways of populations of QD emitting states is given by total square of QDs in QW plane. For 10 nm of mean lateral QD size and QD density states of 10^{11} cm^{-2} it gives 10 percent for direct capture from the barrier states and 90 percent for the capture to QW states followed by diffusion and capture in QD states. So, in normal conditions direct capture of excitons from the barrier states gives a small correction to the population of QD states.

Nevertheless if one supposes that due to some reasons excitation channel of the QD emitting states through QW states is suppressed, then the only possible way for population of QD states would be direct capture of excitons from the barrier states. In this case cascade relaxation of hot excitons with the emission of LO phonons is dominant mechanism of the population of the emitting QD states (the direct capture of free e-h pairs from the barrier to QD states can be neglected due to built-in electric fields available in the samples under study [5]).

Thus, the clue point in the discussion of the differences in the shapes of PLE spectra of the samples A and B is the understanding of the negligible role of QD states population via the states of shallow QW in the sample B.

One of the possible ideas was suggested in paper [3] according to which all amount of CdSe in shallow QW of the sample B (1.4 ML) accumulates into QDs. In this case QDs will be surrounded by ZnSe barrier material in all three dimensions and thereby QDs are spatially isolated from each other. To check this idea we have studied the PLE spectra of specially grown sample ZnSe/CdSe/ZnSe with single CdSe QW with the same nominal thickness t_{CdSe} = 1.4 ML and have found that the shape of its PLE spectra looks very similar to that shown in Figure 1 a-c. This result indicates that any possible explanation should take into account multilayer character of the samples under study. In the following we discuss the model which is based on this idea and explains available experimental findings.

3. Model description

3.1. Quantum well states in growth direction.

High resolution transmission electron microscopy of the samples grown in similar conditions shows [1, 2] that the Cd distribution in the growth direction with a good accuracy can be approximate by the function

$$C(z) = \frac{4C}{\left[\exp(-z/l) + \exp(z/l)\right]^2}, \quad (1)$$

where the point of reference for z is taken in the middle point of distribution. A parameter l defines the width of the Cd distribution, and C is the Cd concentration in the center monolayer of the QW at some point of the lateral QW plane. The function $C(z)$ in Eq.(1) is considered as a continuous function and presents the averaged two dimensional Cd concentration.

The knowledge of the Cd distribution allows us to construct the potential profile of the QW in growth direction of heterostructure. We consider the limit of strong confinement of exciton in z direction [7] and a large lateral size of island in comparison with exciton radius a_B in the plane of QW. Using $C(z)$ we can present the averaged potential profile $U_e(z)$ for electrons and $U_h(z)$ for holes of the QW as

$$U_e(z) = -C(z)\Delta_e, \quad U_h(z) = -C(z)\Delta_h, \quad (2)$$

where z is the particle coordinate in the growth direction, Δ_e and Δ_h are the conduction and valence bands offsets, respectively. The quantization levels for exciton in the QW with potential profile given by $U_e(z)$ and $U_h(z)$ and corresponding wave functions can be found as eigenvalues and eigenfunctions of the Schrödinger equation.

3.2. Numerical characteristic of energy transfer rates between asymmetrical QW pairs.

Potential profile in growth direction of samples A and B with asymmetrical quantum wells based on optical spectroscopy data are shown in Figure 3. Wavy and thick arrows indicate energy transfer from the ground QD and QW states to corresponding states in deep QW (we suppose that rates of all exciton energy relaxation processes in barrier and quantum well states are much exceed that of recombination and energy transfer processes).

Some important conclusions can be made from the data presented in Figure 3:

(i) in sample A energy transfer from QD ground states of shallow well is possible only to QD states of deep well but this process looks improbable due to random spatial occupation of QDs in adjacent QW layers;

(ii) similar process in sample B includes as final states continuous spectra of QWs and probability of this process is determined by overlap integral between corresponding wave functions only.

Features listed in (i) and (ii) explain differences in relative intensities of emission bands from shallow and deep QWs in PL spectra of samples A and B

In order to explain different shapes of PLE spectra of the samples A and B shown in Figure 1 *b* and *d*, we should be mostly interesting in comparison of the energy transfer rates from ground state of shallow QW and corresponding states of deep QW in these samples shown by thick arrows in Figure 3. We estimate probability of transition from the ground state of shallow QW to a state of deep QW which is proportional to the overlap integral between corresponding wave function:

$$I_{1,2} = \int \Psi_{QW_1} \Psi_{QW_2} dr. \quad (3)$$

Transition rates between two layers of QWs in samples A and B can be compared by parameter

$$\beta = \left|I_1^2\right| / \left|I_2^2\right|.$$

We have shown that in the samples under study this parameter is $\beta << 1$ for $\Delta_e/\Delta_h = 3$ [8], which explains the peculiarities of the energy transfer in the samples A and B.

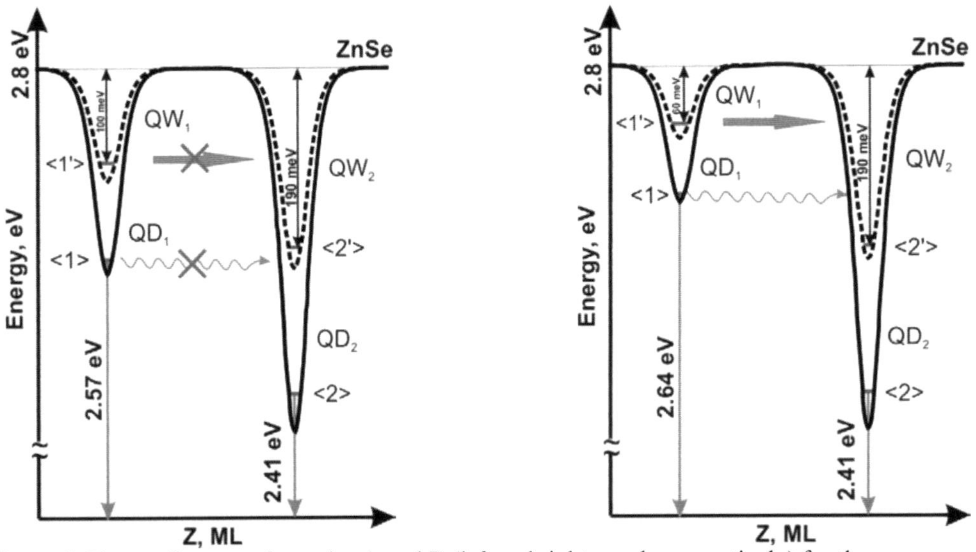

Figure 3. Energy diagram of samples A and B (left and right panels, respectively) for the average potential in the growth direction (*Z*). Dashed and solid lines indicate QW potential profile in *Z* direction between and within quantum dots, respectively. <1'> and <2'> - percolation thresholds of shallow and deep QWs, <1> and <2> - emitting states of QDs of shallow and deep QWs, respectively, (ground states of QWs and QDs in valence bands set as an energy points of reference).

3.3. Estimations of exciton capturing rates

The PLE spectra of shallow QW of the sample B at the spectral range above ZnSe fundamental absorption edge shows pronounced oscillation and exponential decay. To explain these phenomena we consider cascade relaxation of hot excitons with consequent direct capture from the barrier states to the QD states. During single LO phonon emission there are two competing processes: exciton transition as whole particle from 1s state to 1s state with lower kinetic energy and non-radiative exciton decay on free electron and hole, which are then separated by the built-in electric field. Therefore, the higher photon excitation energy the higher probability of exciton decay in the process of cascade relaxation.

The details of calculations are as follows. Wavefunction of the exciton initial state is given by

$$\Psi_{1s,\mathbf{k}}(\mathbf{r_e},\mathbf{r_h}) = exp(i\mathbf{k}\ \mathbf{R})\ exp(-r/a_B)/(\pi\ Va_B^{\ 3})^{1/2}, \quad (4)$$

where $\mathbf{r_e}$ and $\mathbf{r_h}$ are electron and hole coordinates respectively, $\mathbf{r} = \mathbf{r_e}\text{-}\mathbf{r_h}$, $\mathbf{R}$ is coordinate of the center of mass, V is normalizing volume, $\mathbf{k}$ is exciton wavevector. The final state is described either by the wavefunction of 1s with smaller momentum or by wavefunction $\Psi_{\mathbf{k_e},\mathbf{k_h}}$ of free electron and hole with wavevectors $\mathbf{k_e}$ and $\mathbf{k_h}$, respectively. The rates of the transitions are given by Fermi's golden rule

$$\begin{aligned} W_c &= (2\pi/\hbar)\ \Sigma\left|<\Psi_{\mathbf{k_e},\mathbf{k_h}}\right|H_{Fr}\left|\Psi_{1s,\mathbf{k}}>\right|^2 \delta\left(E_f + \hbar\Omega_{LO} - E_i\right), \\ W_{1s} &= (2\pi/\hbar)\ \Sigma\left|<\Psi_{1s,\mathbf{k}}\right|H_{Fr}\left|\Psi_{1s,\mathbf{k}}>\right|^2 \delta\left(E_f + \hbar\Omega_{LO} - E_i\right), \end{aligned} \quad (5)$$

where W_c is the rate of exciton decay, W_{1s} is the rate LO phonon emission by exciton as whole particle, summation is taken over all final states, δ is Dirac delta function, E_f and E_i are energies of final and initial states, respectively, and the exciton-phonon interaction is described by Fröhlich mechanism

$$H_{Fr} = i\, e \left[2\pi\, \hbar\Omega_{LO} / V \, \left(\varepsilon_{\infty}^{-1} - \varepsilon_{0}^{-1} \right) \right]^{1/2} \hat{c}_{\mathbf{q}}^{\dagger} \left[exp\left(i\mathbf{q}\, \mathbf{r}_{\mathbf{e}} \right) \; - \; exp\left(i\mathbf{q}\, \mathbf{r}_{\mathbf{h}} \right) \right] \; + \; c.c. \quad (6)$$

where e is electron charge, Ω_{LO} is LO phonon frequency, **q** is phonon wavevector, $\hat{c}_{\mathbf{q}}^{\dagger}$ is phonon creation operator, ε_{∞} and ε_0 are the high- and low-frequency dielectric constants. Straightforward calculations yield the probability of bulk exciton decay in the process of single LO phonon emission $W_c / \left(W_c + W_{1s} \right) \approx 0.4$, which is in a good agreement with experimental findings.

4. Summary and conclusions.

We have studied energy transfer between electronic states in MBE grown double CdSe/ZnSe asymmetrical quantum wells with self-assembled quantum dots in dependence of nominal thicknesses of deposited CdSe in adjacent layers. It was found that in the samples of proper design the situation can be realized when the rate of energy transfer from ground state of shallow QW to all states of deep QW is much larger than energy transfer rate between QW and QD states within the shallow QW itself. We have shown that in this case the only channel of population of QD exciton states participated in the emission spectrum of shallow QW is direct capture of free excitons from ZnSe barrier through the mechanism of cascade relaxation of hot excitons. We put forward the model which explains the shape of PLE spectra of the samples with such features.

Acknowledgements

The authors thank S. V. Ivanov, I. V. Sedova and S. V. Sorokin for the samples studied in this communication and S. A. Tarasenko for useful discussion. The work is partly supported by the program P220 of the Goverment of Russia (project #14.Z50.31.0021, leading scientist M. Bayer) and the "Dynasty" Foundation.

References

[1] Peranio N, Rosenauer A, Gerthsen D, Sorokin S, Sedova I, Ivanov S 2000 *Phys. Rev.* B **61** 16015
[2] Schikora D *et al.* 2000 *Appl. Phys. Lett.* **76** 418
[3] Reznitsky A, Eremenko M, Sedova I, Sorokin S, Ivanov S 2015 accepted for publication in *Phys. Status Solidi B*
[4] Litvinov D, Gerthsen D, Rosenauer A, Preis H, Kurtz E and Klingshirn C 2001 *Phys. Status Solidi B* 224 **147**
[5] Reznitsky A, Klochikhin A, Permogorov S, Korenev V, Sedova I, Sorokin S, Sitnikova A and Ivanov S 2009 *Phys. Status Solidi* C **9** 2695
[6] Permogorov S 1975 *Phys. Status Solidi B* **68** 9
[7] Heller O, Bastard G 1996 *Phys. Rev.* B **54** 5629
[8] Maksimov M, Krestnikov I, Ivanov S, Ledentsov N and Sorokin S 1997 *Semiconductors* **31** 800

Refraction index modulation induced with transverse electric field in double tunnel-coupled GaAs/AlGaAs quantum wells

A A Shumilov[1], M Ya Vinnichenko[1], R M Balagula[1], L E Vorobjev[1], D A Firsov[1], M M Kulagina[2], A P Vasil'iev[2], C A Duque[4], A Tiutiunnyk[3,4], V Akimov[3,4], R L Restrepo[4,5], V N Tulupenko[3,4], A L Ter-Martirosyan[6,7]

[1] Peter the Great St. Petersburg Polytechnic University, St. Petersburg 195251, Russia
[2] Ioffe Physical-Technical Institute, St. Petersburg 194021, Russia
[3] Donbass State Engineering Academy, Kramatorsk 84313, Ukraine
[4] Grupo de Materia Condensada, Universidad de Antioquia, Medellín, Colombia
[5] Escuela de Ingeniería de Antioquia, Medellín, Colombia
[6] JSC ATC-Semiconductor Devices, St. Petersburg, Russia
[7] ITMO University, St. Petersburg, Russia

E-mail: mvin@spbstu.ru

Abstract. Modulation of refraction index under transverse electric field was studied in structures with multiple tunnel-coupled GaAs/AlGaAs quantum wells in the spectral range corresponding to intersubband light absorption. The change of refraction index in electric field was calculated using Kramers-Kronig relation and experimentally determined spectra of intersubband light absorption in equilibrium conditions and under transverse electric field.

1. Introduction

Quantum cascade lasers are efficient radiation sources. They cover wide spectral range, and can be used in data transmission systems in infrared spectral range. In order for their performance to be efficient, an effective and easy to implement modulation system should be introduced [1]. In case of GaAs/AlGaAs-based quantum cascade lasers there is a way to create frequency modulation system by introducing the layers with tunnel-coupled quantum wells (QW) into the laser waveguide. Application of transverse electric field to the tunnel-coupled QW can induce intersubband absorption modulation, which, in turn, will produce refraction index modulation. Lattice temperature variation can have the same effect in the lesser extent. Here we present the results of the investigation of transverse electric field influence on intersubband light absorption in GaAs/AlGaAs tunnel-coupled QWs and calculation of refraction index modulation in these conditions.

2. Sample and experiment

The investigated structure contains 100 pairs of tunnel-coupled GaAs/$Al_{0.25}Ga_{0.75}As$ QWs. It was grown using molecular beam epitaxy on semi-insulating GaAs substrate. The narrow and wide quantum wells have widths of 4.5 and 6.6 nm, respectively; the barrier between them is 3 nm thick. Structure was selectively doped in the wider QWs up to the surface electron concentration of $4\cdot10^{11}$ cm^{-2} (potential profile of the conduction band is shown in figure 1). The energies of the levels of size quantization were calculated using the self-consistent solution of the Schrödinger and Poisson equations. The space charge emergence due to partial ionization of the donor impurity and the electron redistribution between the wide and narrow QWs with temperature change was taken into account. The investigated structure contains three electron levels of size quantization. The energy of the *e*1-*e*3 transition is about 125 meV (corresponding to wavelength approximately of 10 μm), energy of *e*2-*e*3 transition is about 95 meV at liquid nitrogen temperature. Third electron level is close to the continuous spectrum, which leads to a significant modification of its wave function under transverse electric field and causes the absorption modulation increase. In order to apply transverse electric field mesa structure with 7x10 mm dimensions was etched on the surface of structure at the stage of post-growth processing. $Al_{0.9}Ga_{0.1}As$ stop layer was grown right above the doped bottom layer to stop the selective etching of the structure. For detailed information about postgrowth processes (mesa etching, contacts coating, sample processing in multipass geometry) see [2]. Multipass geometry was formed in such manner that light passed through the QW layers 4 times.

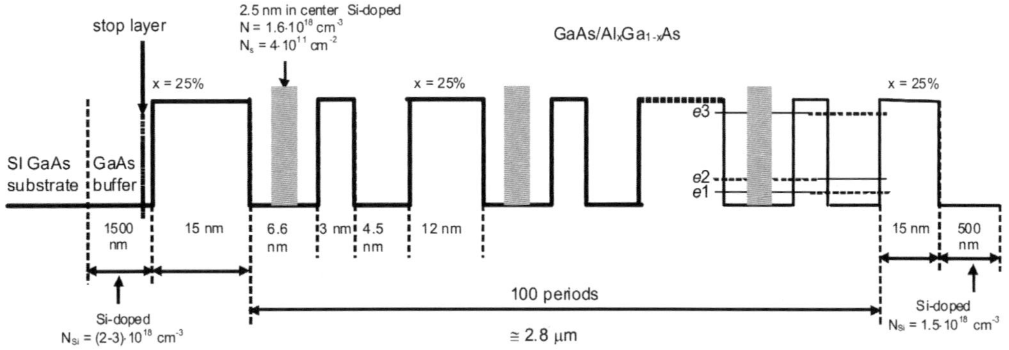

Figure 1. Potential profile of the conduction band of structure.

Structure characterization was performed using photoluminescence (PL) spectra obtained with Horiba Jobin Yvon FHR640 monochromator with CCD-camera Symphony II as a detector. Interband pumping was performed with 532 nm YAG laser. Vacuum Fourier transform spectrometer Bruker Vertex 80v with different photodetectors (MCT, pyroelectric, silicon bolometer) was used to obtain equilibrium intersubband absorption spectra. Closed cycle cryostat Janis PT407RM with operational temperature range of 4–320 K and a temperature setting accuracy of 0.1 degrees was used in the experiment. Modulation spectra under transverse electric field were obtained in step-scan spectrometer mode with boxcar integrator and current preamplifier added to the experimental setup. We used two ADC inputs in the measurements: one to record interferogram of modulation spectra, and the other one to register interferogram and phase of equilibrium spectra. Due to conjugation properties of Fourier transform, positive and negative modulation signals are indistinguishable in the interferogram. In order to determine true signal sign the phase correction of modulation spectra was performed using the phase spectra of interferogram of equilibrium spectra which is positive in the spectral region in question.

3. Results and discussions

Photoluminescence spectra were measured for original structure. Analysis shows that there is a significant spectral shift of the PL maximum from the position of the GaAs band gap. The value of spectral shift corresponds to the Fermi level of the doped GaAs top layer of the structure. Thus we can conclude that the observed PL originates in the bulk GaAs surface layer. Mechanical removal of this heavily doped top layer allowed us to obtain true PL spectrum. The positions of the peaks in PL spectrum of the structure are in good agreement with the theoretical calculation of electron transition energies at room temperature: *e*1 - *hh*1 = 1.486 eV and *e*2 - *hh*2 = 1.525 eV. Peak *e*2 - *hh*2 disappears with the temperature decrease. It is associated with variation of the carrier distribution function and decrease in electron concentration on the level *e*2.

In earlier work [2] we presented the results of intersubband absorption investigation without electric field and studied the influence of temperature variation (from 4 to 300 K) on the absorption spectra. Intersubband absorption related only to quantum wells was estimated by dividing the spectrum of *p*-polarized light transmission by the transmission spectrum for *s*-polarized light (see curve 1 in figure 2 at liquid nitrogen temperature). These spectra contained one distinct peak at the photon energy corresponding to the *e*1-*e*3 electron transitions. This peak widened and slightly shifted with the temperature variation. We associate the observed variation of absorption with the redistribution of electrons between first two electron levels (*e*1 and *e*2) and with a change in the matrix element of *e*1-*e*3 transition. Slight broadening of the *e*1-*e*3 absorption peak with the temperature increase can be attributed to the thermal effects.

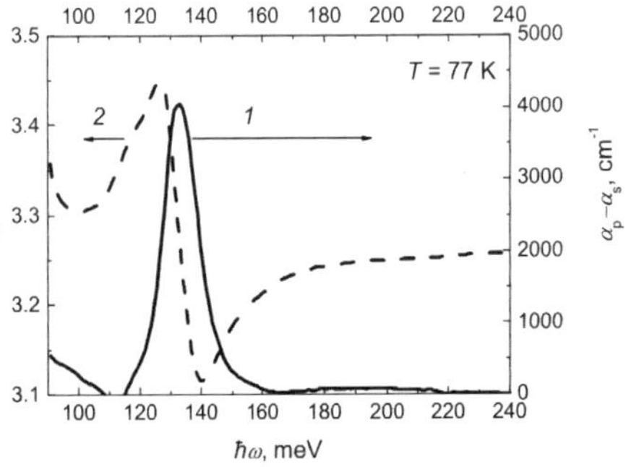

Figure 2. Spectra of intersubband absorption (1) and refraction index *n* (2) at $T = 77$ K.

It stands to mention that interference influence should also be considered in intersubband absorption studies. It can be observed in case of standing wave occurrence in the waveguide geometry as a result of the interference between the incident and totally reflected light at the semiconductor–air interface [3]. The structure described in the present paper has highly doped surface GaAs layer and metal coating which together should result in standing wave with antinode for *p*-polarized light and node for *s*-polarized light at the structure surface. However, the thickness of the active layer containing 100 pairs of tunnel-coupled QWs is about 3 μm which results in the averaging of interference effects for several periods of standing wave located there [4].

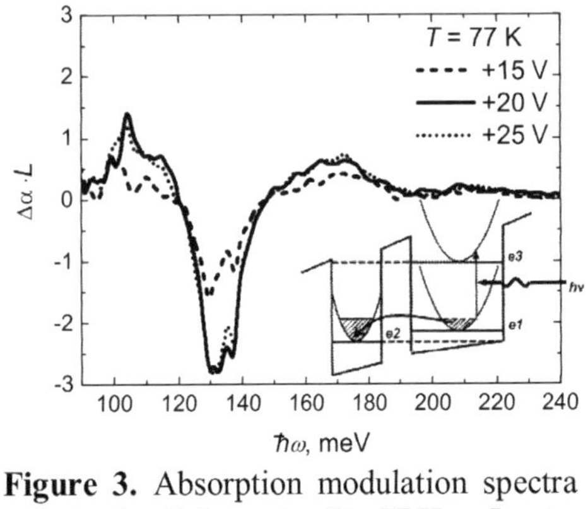

Figure 3. Absorption modulation spectra for *p*-polarized light at $T = 77$ K. Inset shows schematically electron redistribution under transverse electric field. *L* - optical length.

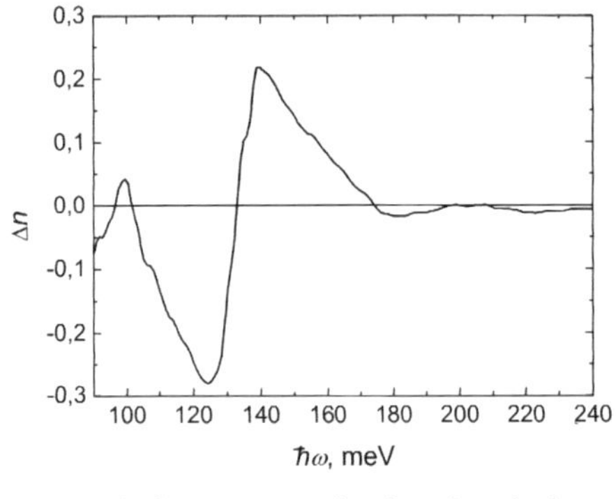

Figure 4. Spectrum of refraction index variation Δn under transverse electric field corresponding to applied voltage $U = 20$ V at $T = 77$ K.

The figure 3 shows measured intersubband absorption modulation spectra $\Delta\alpha(\omega)$ for *p*-polarized light at $T = 77$ K (in case of *s*-polarized light there are no changes in QW-related absorption in multipass geometry) under transverse electric field. It is also important to mention that electric field influences light absorption as well as the reflection at the interface between active region and substrate. It will be found that electric field changes the value of refraction index n on approximately 0.3. Providing that refraction index of bulk GaAs is 3.27 it could be calculated that electric field changes reflection coefficient R on 0.2% thus its effect can be neglected.

The dip in modulation spectra (see figure 3) corresponds to the decrease in absorption. Its energy position is in good agreement with $e1$-$e3$ transition energy. Observed absorption decrease is associated with electron redistribution between first two subbands $e1$ and $e2$ (see inset in figure 3). One can see that spectra for voltage $U = 20$ V and $U = 25$ V are almost identical. It can indicate that $U = 20$ V corresponds to complete depletion of $e1$ subband. This also corresponds to our numerical estimations. The electric field of the other polarity does not lead to a significant increase in the population of first subband, therefore intersubband absorption change in this field was not observed.

Spectra of refraction index change $\Delta n(\omega)$ in electric field were calculated from the obtained intersubband absorption modulation spectra using Kramers-Kronig relation:

$$\Delta n(\omega) = \frac{c}{\pi}\,\text{v.p.}\int \frac{\Delta\alpha(\omega')}{\omega'^2 - \omega^2}\,d\omega' \qquad (1)$$

where ω is light frequency, c is light speed, $\Delta\alpha(\omega)$ stands for the changes of intersubband absorption (see figure 3). Results of calculation (see figure 4) show that maximal change of refraction index under transverse electric field is approximately 10% of the whole refraction index of GaAs ($n = 3.27$). Spectrum of refraction index without electric field was also calculated (see curve 2 in figure 2). It is associated with intersubband absorption in equilibrium conditions. It is clear that modulation of infrared radiation in electric field is efficient and effective in structures with tunnel-coupled QWs.

4. Conclusion

The obtained results confirm the feasibility of tunnel-coupled QW-based infrared modulators. Experimental results show that low voltage value of transverse electric field is enough to significantly change the energy spectrum of structure and provide the electron distribution between QW. Thus we have confirmed significant changes in refraction index of mid-IR radiation. Besides, redistribution of the charge carriers between lowest electronic states in tunnel-coupled QW structures under transverse electric field can also affect the refractive index in near infrared spectral range, namely the part of refractive index related to virtual interband transitions induced by the radiation with photon energy slightly less then the effective energy gap. The corresponding spectral range is of interest for application in commercially developed medical laser systems, for example 980 nm lasers for stomatology.

Acknowledgments

This work has been supported by the Russian Foundation for Basic Research (Projects № 14-02-31489, 14-02-00336), the Ministry of Education and Science of the Russian Federation, and partially by the project on the development of cooperation between Russian universities and industrial enterprises (contract № 02.G25.31.0064).

References

[1] Luryi S, Gouzman M 2006 *International journal of high speed electronics and systems* vol 16(2), pp 559-566.

[2] Rybalko D A, Vinnichenko M Ya, Vorobjev L E, Firsov D A, Balagula R M, Panevin V Yu, Kulagina M M, Vasil'iev A P 2014 *Journal of Physics: Confernce Series* vol 541, p 012081.

[3] Vodopyanov K L , Chazapis V , Phillips C C , Sung B , Harris J S 1997 *Jr. Semicond. Sci. Technol.* vol 12, p 708.

[4] Schneider H, Schönbein C, Walther M, Koidl P, Weimann G 1999 *Appl. Phys. Lett.* vol 74(1), 16.

Dynamics of mid-infrared light absorption related to photo-excited charge carriers in Ge/Si quantum dots

O I Kirilenko[1], R M Balagula[1], A N Sofronov[1], V Yu Panevin[1], L E Vorobjev[1], D A Firsov[1] and A A Tonkikh[2,3]

[1] Peter the Great St. Petersburg Polytechnic University, Polytechnicheskaya 29, St. Petersburg 195251, Russia

[2] Max Planck Institute of Microstructure Physics, Weinberg 2, Halle (Saale) 06120, Germany

[3] Institute for Physics of Microstructures RAS, GSP-105, Nizhny Novgorod 603950, Russia

Abstract. Mid-infrared optical absorption spectra of Ge/Si quantum dot structures in polarized light were obtained under conditions of interband optical excitation. Mid-infrared absorption changes in comparison to equilibrium conditions for certain light polarization in spectral range of 0.25 – 0.6 eV. The sign of the effect is found to be different in various spectral ranges. Transients of photoinduced absorption contain fast and slow components. The fast decay component of the absorption is related to the direct recombination of localized holes and electrons in quantum dots while a slow component is determined by significantly less probable processes of indirect in real space recombination.

1. Introduction

Nanostructures with quantum dots are being studied by a number of scientific groups around the globe. Intensive studies of optical and photoelectric properties of Ge/Si quantum dots are motivated by a possibility to develop mid- and near-infrared optoelectronic devices based on these structures. Lately a noticeable progress has been made in technology aspects of Ge/Si quantum dots growth [1]. There is a number of papers dedicated to studies of near-IR photoluminescence dynamics [2,3], and the results contain a lot of original information explaining recombination processes in quantum dots. The present work is focused on the optical properties of Ge/Si quantum dots in the mid-infrared spectral range under conditions of non-equilibrium population of the dot states with photo-excited charge carriers and also presents recent results on photoinduced absorption dynamics in GeSi quantum dot structures

2. Experimental setup and samples

Structure with MBE-grown Ge/Si quantum dots on silicon (100) substrate was studied. Structure contains 10 periods of quantum dot layers formed by deposition of 7 Ge monolayers with 0.14 ML/s rate at 600^0C. Quantum dot layers were separated from each other with 15 nm Si interlayer. Structure was isolated from substrate with epitaxial 100 nm silicon buffer layer and was covered with 100 nm top layer. A surfactant (Sb) was used during the growth process in order to increase the density of the QD array as described in [4]. No doping operations were performed during structure growth.

QD structure was characterized using transmission electron microscopy and atomic force microscopy. The first method allowed us to determine the shape of quantum dots as a truncated pyramid with average height of 2.7 nm, and average base size of 14 nm. Results of atomic force microscopy of specially grown samples with no surface silicon layer showed the QD density at the level of $2 \cdot 10^{11}$ cm^{-2}. The material of the quantum dot is Ge_xSi_{1-x} solid solution with approximate germanium content 60-65%.

Optical measurements in mid-IR were performed with the samples prepared in multi-pass geometry with opposite sample sides ground at 45^0 angle so the light beam could pass the QDs layers multiple times due to total internal reflection in the sample. This geometry allows one to study optical transmission of the sample for different light polarizations with respect to growth direction.

Charge carriers in the undoped sample were created only using interband illumination of the top surface of the sample with solid-state 532 nm laser.

Experimental setup for steady-state photoinduced absorption spectra measurements is the variation of steady state pump-probe technique. Absorption spectra in mid-infrared range were measured with automated prismatic monochromator. Globar was used as a source of probe radiation. Intensity of the probe radiation after the sample was measured with InSb photodetector. Sample was mounted in the liquid nitrogen-cooled cryostat with ZnSe optical windows. ZnSe metal grid polarizer was placed in front of the entrance window. Both probe and pump radiation were mechanically chopped at different frequencies, and photodetector signal was measured simultaneously with two SR830 Lock-in amplifiers. One of them measured a signal proportional to the equilibrium optical transmittance T of the sample, while the second one measured a signal proportional to the variation of the optical transmittance ΔT, related only to non-equilibrium photo-excited charge carriers.

Dynamics of photoinduced absorption was registered with digital oscilloscope after a short 10 ns pump radiation pulse at the spectral maximum of p-polarized light absorption.

3. Experimental results

Typical measured steady-state spectra of photo-induced absorption related to photo-excited charge carriers captured to the QDs are shown in Figure 1 (a) for p- and s-polarization of probe radiation at lattice temperature $T_L = 79$ K. Figure 1(b) presents spectra for probe light polarized along the growth direction (z) at two different temperatures. These spectra were calculated from measured data taking into account that polarization vector for *p*-polarization contains equal *xy*- and *z*-components (while *s*-polarization is in fact 100% *xy*).

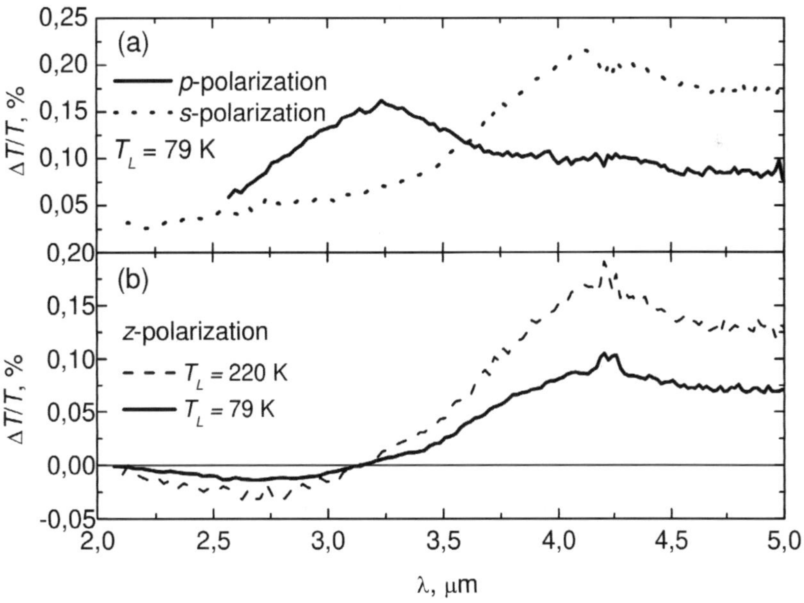

Fig.1. Typical spectra of photo-induced absorption for p- and s- infrared light polarization at $T_L = 79$ K (a); calculated spectra for probe light polarized along the growth direction at 79 K and 220 K (b). Pump intensity $I = 20$ mW

Since the optical response of the sample related to the quantization of the hole energy spectrum in the growth direction (which is the main confinement factor due to small "height" of the QDs with respect to its lateral dimensions) is sensitive only to z-polarized light, one can treat the spectral features presented in Figure 1(b) as related to optical transitions of the non-equilibrium holes between the energy states of the QD. The noticeable feature of the $\Delta T/T$ spectrum for z-polarized probe is the change of the sign of the ΔT.

Positive value of ΔT corresponds to the increase of absorption at the certain spectral range close to 4 μm. Absorption increase is related to the contribution of optical transitions of non-equilibrium holes from the quantum dot ground states to the continuum of states, and the energy of these transitions is about 300 meV. Negative value of ΔT corresponds to the decrease of absorption at the presence of non-equilibrium charge carriers at the spectral range close to 2.75 μm. This fact can be associated with the suppression of interband-like processes such as generation of hole inside the dot and bound electron outside of the dot under conditions of full occupation of appropriate discrete states with charge carriers photo-generated by pump radiation. Figure 2 shows the optical transition scheme. Transitions marked with solid arrow 1 are responsible for absorption increase at long-wavelength part of $\Delta T/T$ spectrum. Suppression of transitions with energy of about 450 meV marked with dashed arrow 2 is responsible for absorption change drop below zero.

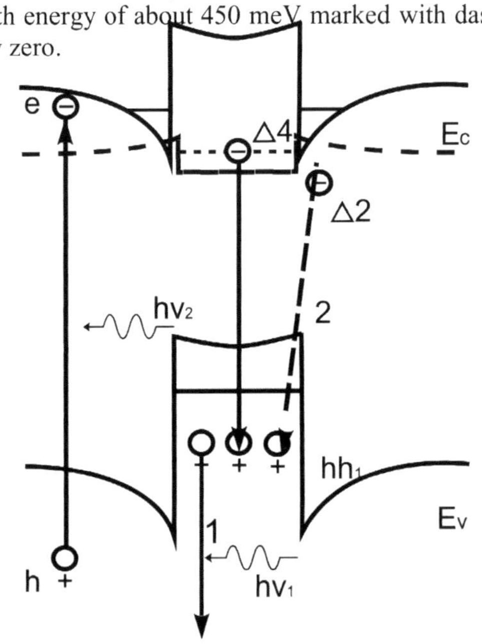

Fig.2. Band diagram of the Ge/Si QD and optical transition scheme.

The dynamics of the processes involving transitions from ground states was investigated at the probe radiation wavelength corresponding to the maximum of the positive part of $\Delta T/T$ spectrum. The absorption relaxation curves were obtained in the wide lattice temperature range T_L = 90…290 K. Typical decay curve at the intermediate temperature of 190 K is presented in Figure 3.

Observed $\Delta T/T$ decay curves demonstrate two-stage behavior with a pronounced fast and slow components with a characteristic decay times that differ by almost one order of magnitude at 90 K. Heating of the sample leads to the decrease of the slow component decay time so at room temperature it can not be isolated from the fast one.

Such two-stage kinetics was observed earlier in interband photoluminescence studies [2,3]. Since the measured $\Delta T/T$ decay curves in essence reflect the dynamics of the population of the QD ground state with non-equilibrium charge carriers, the microscopic mechanism of the photoinduced absorption relaxation should be similar to the one of the interband photoluminescence. Non-uniform mechanical strain distribution in the QD structure can reduce the degeneracy of the electron Δ-valleys in silicon around the QD [7] and form a complicated potential profile at the quantum dot interface similar to the one presented in Figure 2. Two Δ2 valleys stretched along z axis form a potential well for electrons at the silicon side of the QD Ge/Si interface while four Δ4 valleys stretched in the xy-plane can lead to weak electron localization inside the quantum dot. Thus, two different ways of recombination are possible in such structures. Direct recombination of localized in Δ4 states electrons and holes in quantum dots can be associated with a fast decay component. Indirect in real space recombination of electrons in Δ2 states with holes inside the dot has significantly less probability due to weak overlap of the involved states and can be responsible for a slow decay component.

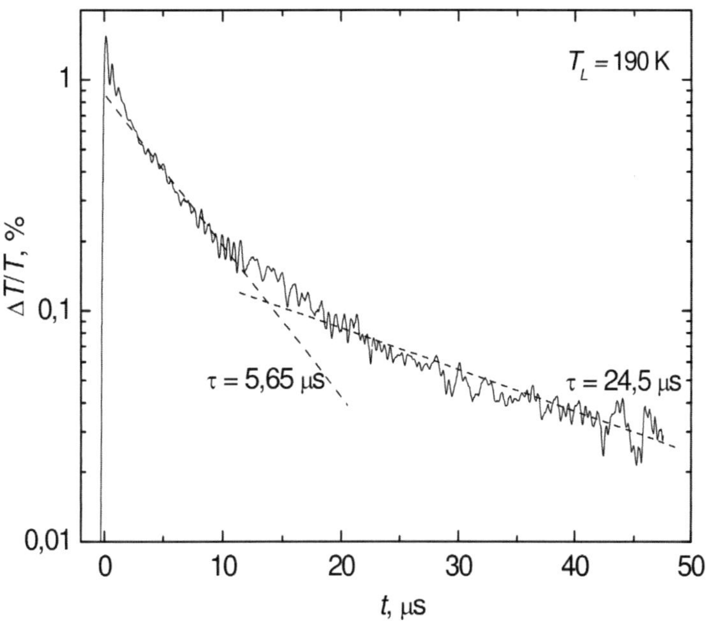

Fig. 3. Typical photoinduced absorption relaxation curve at $T_L = 190$ K. Characteristic time values derived from the experimental data are shown for each decay component.

4. Conclusion

In conclusion, in this work the photo-induced mid-infrared absorption spectra were measured in Ge/Si quantum dot structures. The analogue of the Burstein-Moss effect is observed. Dynamics of photoinduced light absorption was studied under conditions of pulse interband optical excitation. Two distinct temporal components were isolated. The underlying mechanisms were proposed.

Acknowledgments

This work has been supported by the Russian Foundation for Basic Research (Project № 13-02-12203) and the Ministry of Education and Science of the Russian Federation.

References

[1] Tonkikh A A, Zakharov N D, Novikov A V, Kudryavtsev K E, Talalaev V G, Fuhrmann B, Leipner H S and Werner P 2012 *Thin Solid Films* **520** 3322

[2] Julsgaard B, Balling P, Lundsgaard Hansen J, Svane A and Nylandsted Larsen A 2011 *APL* **98** 093101

[3] Kamenev B V, Tsybeskov L, Baribeau J-M, Lockwood D J 2005 *Phys. Rev.* B **72** 193306-1-4

[4] Tonkikh A A, Zakharov N D, Talalaev V G, Werner P 2010 *Phys. Status Solidi* RRL **4**, 224.

[5] Panevin V Yu, Sofronov A N, Vorobjev L E, Firsov D A, Shalygin V A, Vinnichenko M Ya, Balagula R M, Tonkikh A A, Werner P, Fuhrman B, Schmidt G. 2013 *Semiconductors*, **47**, 1574.

[6] Talochkin A B, Chistokhin I B 2011 *Semiconductors*, **45**, 907

[7] Kurdi M El, Sauvage S, Fishman G and Boucaud P 2006 *Phys. Rev.* B **73** 195327..

Mid-infrared photoluminescence from structures with InAs/GaSb type II quantum wells

A V Selivanov[1], I S Makhov[1], V Yu Panevin[1], A N Sofronov[1], D A Firsov[1], L E Vorobjev[1] and L. Shterengas[2]

[1] Peter the Great St. Petersburg Polytechnic University, Politechnicheskaya str., 29, St. Petersburg, 195251, Russia

[2] State University of New York at Stony Brook, Stony Brook, New York 11794, USA

Email: a_selivanov@mail.ru

Abstract. The results of experimental investigations of mid-infrared interband photoluminescence spectra from quantum well nanostructures based on $InAs/GaSb/Al_{0.35}Ga_{0.65}As_{0.03}Sb_{0.97}$ type II quantum wells in the wide temperature range from 10 K to the room temperature are presented. The photoluminescence line spectral positions are compared with the effective quantum well bandgap values obtained from the results of calculations of electron and hole energy states in quantum well. Analysis of the temperature dependence of the interband emission intensity allows to assume the presence of the traps near the middle of the effective energy gap. It is found that interband radiative recombination rate and nonradiative Shockley-Read-Hall rate are comparable.

1. Introduction

Type II heterojunctions have recently attracted considerable attention. The distinctive feature of these structures is the overlap of the one semiconductor valence band with conduction band of other one. The overlap of bands results in strong hybridization of states of one's conduction band and valence band of other one. Type II heterojunctions can be formed by InAs/GaSb alternating layers. Quantum well (QW) for electrons would be formed in InAs layer and quantum well for holes would be formed in GaSb layer. Quantum wells and superlattices based on InAs/GaSb type II heterojunctions allow to create the photodetectors operating in mid-infrared spectral range [1-3], which are compete with HgCdTe-based devices. In the type II QWs, the spectral position of the interband absorption edge (or the luminescence line) is determined by the effective bandgap Eg^{eff} - energy gap between the ground electron and hole states, which are separated in the real space. Using the calculations of the electron and hole energy levels, one is able to design the necessary band profiles for devices operating in mid-infrared spectral range. Unfortunately, often the total error in Eg^{eff} corresponding to mid-infrared range could be the same order of magnitude as Eg^{eff} itself because of large dispersion of band structure parameters of the hetero-pair components in the reference data. In these conditions, the experimental investigation of the effective bandgap is very important task. In the present work, the effective bandgap value for InAs/GaSb type II QW is obtained from the study of the mid-infrared interband photoluminescence (PL) spectra at the different lattice temperatures. Additionally, temperature dependence of the PL intensity allows one to obtain information about ratio between radiative and nonradiative lifetimes which define the performance of the many optical devices.

2. Samples and experimental setup

Sample for optical studies with type II QWs was MBE grown on GaSb substrate after a 500 nm GaSb buffer layer. Sample contains 20 periods GaSb/InAs/GaSb layers forming QWs for electrons and holes, separated with 20 nm $Al_{0.35}Ga_{0.65}As_{0.03}Sb_{0.97}$ lattice-matched barriers. Each period contains 4 nm InAs layer, forming QW for electrons. 25 nm GaSb layers are located on both sides of InAs layer forming QWs for holes. Finally, the structure covered by 25 nm undoped GaSb layer.

The potential profiles of the conduction band bottom and valence band top which determine the energy spectrum of electrons and holes in QWs along growth axis are shown in figure 1. The energy spectrum for our structure was calculated using the Schrödinger equation for the envelope function and transfer matrix method taking into account the mechanical stress in InAs layer. The values of material parameters were taken from paper [4], interpolation formula for parameters of alloys was taken from paper [5]. The calculated values of the band structure parameters for different crystal lattice temperatures T used in our energy spectrum calculations are shown in table 1. As it follows from table 1 the effective energy gap Eg^{eff} for our structure decreases with the temperature increase. According to energy spectrum calculations only the one state for electrons (*e1*) and one state for heavy holes (*hh1*) are belong to our type II QW. These states are shown with the dashed lines on the figure 1.

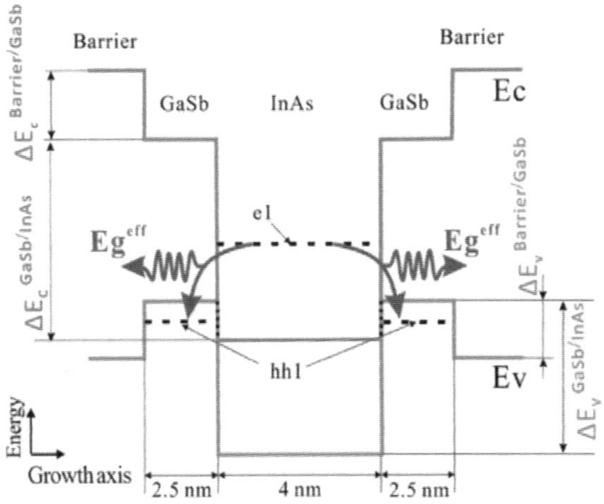

Figure 1. Band diagram of QW and optical transitions scheme. Electron and heavy hole levels marked $e1$ and $hh1$, respectively.

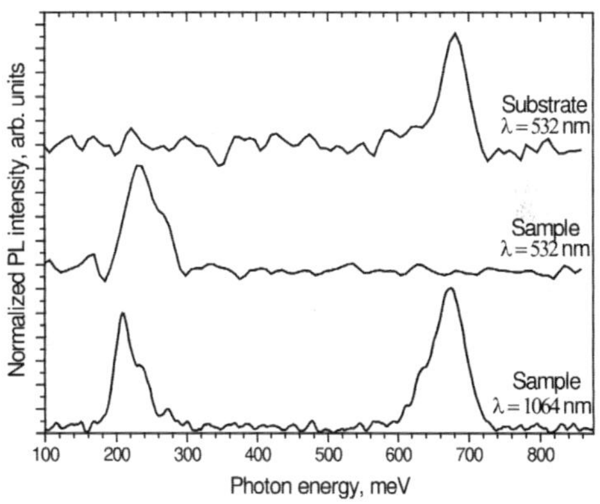

Figure 2. Mid-infrared PL spectra of the sample and it's substrate in the photon energy range of 100-875 meV for the different optical excitation wavelengths: $\lambda = 1.064$ µm and $\lambda = 532$ nm. Crystal lattice temperature $T = 80$ K. Curves are shifted along the vertical axis for the clear view.

For experimental studies the sample was mounted into the low vibration closed cycle cryostat based on the pulse tube thermodynamic cycle. Sample temperature can be varied from 10 K to 320 K. Interband optical excitation of the sample was attained through fused silica window by means the laser with $\lambda = 1.064$ µm, pulse duration 0.25 µs, repetition rate 8 kHz, maximum peak power about 30 W or by means the laser with $\lambda = 532$ nm, pulse duration 10 ns, repetition rate 10 Hz, peak power about 140 kW. The PL spectrum of the substrate was studied on a part of sample surface with the track from MBE holder leg. That part of the surface is left not covered by QW layers.

Table 1. Band offset parameter values for our structure at different crystal lattice temperatures.

	GaSb/InAs		AlGaAsSb		ΔEg_{eff}
T, K	ΔE_C (meV)	ΔE_V (meV)	ΔE_C (meV)	ΔE_V (meV)	(meV)
10	967	561	246	204	226
77	967	561	246	204	217
300	945	561	246	204	160

The mid-infrared photoluminescence spectra were studied using vacuum Fourier transform infrared spectrometer Bruker Vertex 80v operating in step scan mode. The photoluminescence intensity was measured by liquid nitrogen cooled HgCdTe-photodiode with cut-off wavelength about 12.5 μm equipped with the preamplifier. Spectrometer, cryostat and HgCdTe-photodiode were optically coupled using ZnSe windows. In order to block the pumping light, the high resistance Ge optical filter (transmittance of about 40 % in the mid-infrared spectral range) was set in the front of the HgCdTe-photodiode. The photoresponse of HgCdTe-detector was measured by gated integrator SR250.

3. Experimental results and discussion

The photoluminescence spectra of the sample and its substrate studied in the spectral range 100-875 meV for the crystal lattice temperature $T = 80$ K (under optical excitation with wavelengths $\lambda = 1.064$ μm and $\lambda = 532$ nm) are shown in figure 2 (see the plot to recognize the curves). It should be noted that real emission spectra can slightly differ from spectra presented in our paper because of spectral dependency of photodetector sensitivity, windows, KBr-beamsplitter and Ge-filter transmission using in our experimental setup. In the sample PL spectrum under optical excitation with wavelength $\lambda = 1.064$ μm (please, see the bottom spectrum presented in figure 2) two peaks of radiation can be observed. The first peak shows a luminescence band lying in the spectral range from 180 - 260 meV near the effective bandgap calculated value ($Eg^{eff} \approx 217$ meV for the 80 K). We associate above mentioned luminescence band with the *e1-hh1* indirect in the real space radiative recombination of photoexcited electrons and holes in type II InAs/GaSb QWs. These optical transitions schematically shown on figure 1 with the red (online) arc arrows. The second peak with the photon energy about 700 meV could be connected with photoluminescence from bulk GaSb substrate. The high-energy part of this photoluminescence peak is determined by absorption in Ge-filter. Really, this peak does not observed on the PL spectrum for the sample under optical excitation with wavelength $\lambda = 532$ nm (please, see the middle curve on the figure 2). To understand that strong dependence of PL spectra from optical excitation wavelengths let us consider the sample design. The thickness of active layers containing QWs is 0.62 μm and about 70% of them consist of $Al_{0.35}Ga_{0.65}As_{0.03}Sb_{0.97}$ barrier layers. The energy gap for this barrier layer at $T = 77$ K is 1.25 eV, so for the radiation with wavelength $\lambda = 1.064$ μm (~ 1.16 eV) barrier layers are almost transparent (the absorption coefficient is very small). From the other side, for the radiation with $\lambda = 532$ nm (~ 2.33 eV) the absorption coefficient in barrier is much greater. Thus radiation with $\lambda = 532$ nm doesn't penetrate to the substrate and is absorbed in the nanostructure layers. When we excite with $\lambda = 532$ nm directly the substrate, the same luminescence band near 700 meV observing in the PL spectrum (please, see the top curve in figure 2).

The photoluminescence spectra of the sample under optical excitation with wavelength $\lambda = 1.064$ μm at different crystal lattice temperatures are presented in figure 3. The comparison of photoluminescence spectra for the different crystal lattice temperatures demonstrates the weak shift (within 40 meV) of the luminescence band to the lower photon energies region with the temperature increase from 10 K to 300 K. The shift of the luminescence band with temperature could be connected with the change of the band offsets on the heterointerfaces of the QW with the temperature change. The spectral positions of PL peaks for different crystal lattice temperatures are in a good agreement with effective energy gap calculations for our type II QW structure presented in table 1. Corresponding to the different crystal lattice temperatures values of Eg^{eff} are marked on figure 3 by arrows. Difference between the calculated effective energy gap values and the PL peak spectral positions can be explained by large dispersion of band structure parameters of the hetero-pair components for different temperatures in the reference data.

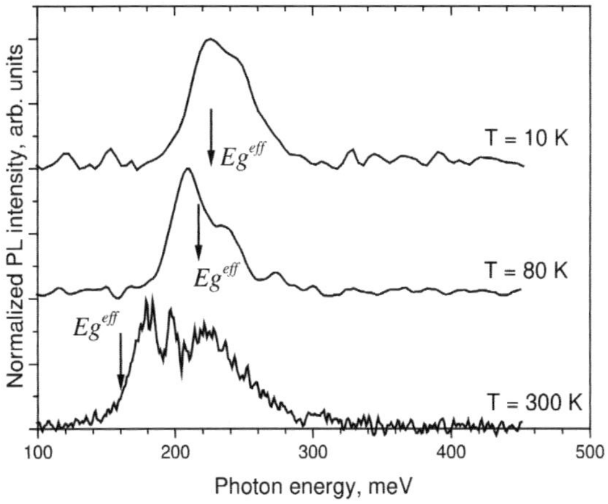

Figure 3. Mid-infrared PL spectra for different crystal lattice temperatures in the photon energy range of 100-450 meV. Optical excitation with the wavelength λ = 1.064 μm. The curves are measured under different excitation intensities.

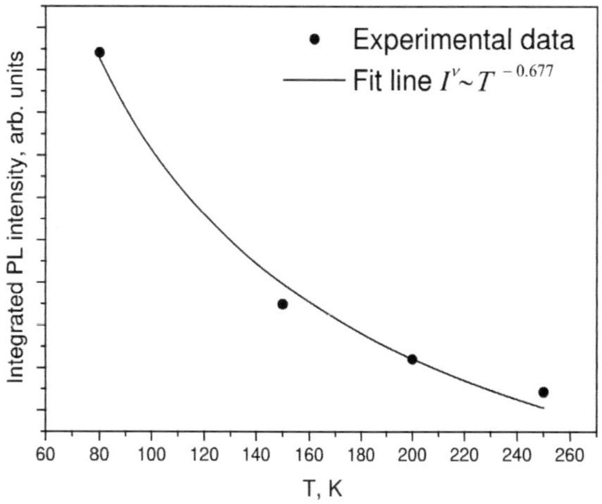

Figure 4. Temperature dependence of the integrated luminescence intensity for the interband transition $e1$-$hh1$ in QW and its approximation (to recognize the curves see comments on the plot).

The temperature dependence of the integrated emission intensity I^{ν} related to e1-hh1 optical transitions in QW under excitation with the wavelength λ = 1.064 μm is presented in figure 4 (points – the experimental data, line – approximation $I^{\nu} \sim T^{-m}$). The decrease of the interband luminescence intensity could be connected with the internal quantum efficiency decrease with the temperature increase. From fit line we can observe that spontaneous emission intensity I^{ν} varies as $I^{\nu} \sim T^{-m}$ (where $m \approx 0.7$). If τ_{ν} – the charge carriers lifetime relative to nonradiative recombination, then $I^{\nu} = \frac{\Delta p}{\tau_{\nu}} \cong \frac{I_{pump} \cdot \tau}{\tau_{\nu}}$, where $\tau = \frac{\tau_{SRH} \cdot \tau_{\nu}}{\tau_{SRH} + \tau_{\nu}}$. Here I_{pump} – optical excitation intensity, Δp – nonequilibrium charge carrier concentration, τ_{SRH} – nonradiative lifetime related to carrier capture on trap levels (Shockley-Reed-Hall lifetime (SRH) [6, 7, 8]), τ - charge carriers lifetime. If $\tau_{\nu} \ll \tau_{SRH}$ then I^{ν} doesn't depend on temperature, and if $\tau_{\nu} \gg \tau_{SRH}$ then $I^{\nu} = \frac{I_{pump} \cdot \tau_{SRH}}{\tau_{\nu}}$. According to ref. [8] $\tau_{\nu} \sim T$ and $\tau_{SRH} \sim T^{-\frac{1}{2}}$ (when capture cross section σ does not depend on temperature ($\tau_{SRH}^{-1} = \sigma \cdot M \cdot \sqrt{k_b \cdot T} \sim T^{-\frac{1}{2}}$, where M – trap concentration, k_b – Boltzmann constant)). τ_{SRH} varies like $T^{-\frac{1}{2}}$ if trap energy level for holes lies higher on energy than Fermi level for holes [6, 7, 8]. In this case $I^{\nu} \sim T^{-\frac{3}{2}}$. If τ_{ν} and τ_{SRH} are comparable, then the dependence of the spontaneous emission intensity I^{ν} on temperature T could be weaker and will better fit the experimental data, according to which $I^{\nu} \sim T^{-0.7}$. Evaluation of the Fermi energy for the acceptor impurity concentration about $0.5 \cdot 10^{16}$–$1 \cdot 10^{16}$ cm^{-3}, ionization energy $E_A \approx 0.03$ eV, hole effective mass about $m_h \approx 0.4\ m_0$ (where m_0 – free electron mass in vacuum) gives the trap energy level value about 0.1 eV. Thus, from the analysis of the temperature dependence of the spontaneous emission intensity it should be noted that lifetimes τ_{SRH} and τ_v are comparable, and traps are lying near middle of the effective energy gap Eg^{eff}.

Summary

In the present work, the interband photoluminescence spectra from quantum well nanostructures based on $InAs/GaSb/Al_{0.35}Ga_{0.65}As_{0.03}Sb_{0.97}$ type II QWs in mid-infrared spectral range and its temperature dependence were investigated. The quantum well nanostructure effective bandgap calculated value is in satisfactory agreement with the spectral position of interband photoluminescence line. The wavelength of emission line for the room temperature is about 180 meV (~ 6.8 μm). By means of the temperature dependence of the QW emission intensity analysis, the trap levels deepness and ratio between radiative and nonradiative (SRH) lifetimes were estimated. This work is partially supported by RFBR (Grant 13-02-12203) and the Ministry of Education and Science of Russian Federation.

References

[1] Haugan H J, Brown G J, Szmulowicz F, Grazulis L, Mitchel W C, Elhamri S, Mitchell W D2005 *Journal of. Crystal Growth* **278** 198;
[2] Rehm R, Walther M, Schmitz J, Fleißner J, Fuchs F, Ziegler J, Cabanski W 2006 *Opto-Electronics Review* **14** 19;
[3] Plis E A 2014 *Advances in Electronics 2014* 246769;
[4] Vurgaftman I, Meyer J R, Ram-Mohan L R 2001 *Journal of Applied Physics* **89** (11). 5815;
[5] Donati G P, Kaspi R, Malloy K J 2003 *Interpolating Journal of Applied Physics* **94**(9) 5814;
[6] Shockley W., Reed W. T., Jr. 1952 *Physical Review* **87** 835;
[7] Hall R. N. 1952 *Physical Review* **87** 387;
[8] Vorobjev L. E., Danilov S. N., Zegrya G. G., Firsov D. A., Shalygin V. A., Yassievich I. N., Beregulin E. V. 2001 *Photoelectric phenomena in semiconductors and dimensionally quantized structures* ed. Il'in V. I. and Shik A. Ya. (Saint-Petersburg, Nauka) P. 236 (in russian).

Ballistic thermopower of suspended semiconductor Hall bars with two dimensional electron gas

E Yu Zhdanov[1,2], A G Pogosov[1,2], M V Budantsev[1], D A Pokhabov[1,2], A K Bakarov[1] and A I Toropov[1]

[1] Institute of Semiconductor Physics, SB RAS, Novosibirsk, Russia
[2] Novosibirsk State University, Novosibirsk, Russia

E-mail: zhdanov@isp.nsc.ru

Abstract. We study ballistic electron transport in suspended semiconductor nanostructures containing high mobility two dimensional electron gas structured with periodical square lattice of artificial scatters — antidots. Thermopower demonstrates magnetic field commensurability oscillations resulting from geometrical resonances similar to those earlier observed in magnetoresistance, thus indicating the retaining of the ballistic regime in thermopower measurement on suspended structures and the validity of the Mott rule in this case. In spite of peculiarities of the heat transport in suspended structures leading to the observed anomalies in non-linear effects, the amplitude of thermopower oscillations remains unchanged after the suspension. This can be explained by short Thouless time compared to the time of electron-phonon interaction.

1. Introduction

Recently there has been growing interest in nanoelectromechanical systems. Conducting nanostructures with two-dimensional electron gas (2DEG) detached from the substrate are one of the examples of such systems. The study of these structures showed them to demonstrate a series of interesting peculiarities [1,2]. It was found that additional mechanical degrees of freedom drastically influence electron-transport properties of suspended membranes [3,4]. Moreover, it was shown that the heat transport in suspended structures essentially differs from that in non-suspended ones [5-7]. However electron transport in ballistic regime in suspended structures still remains poorly investigated [8].

In the present work we investigate ballistic electron transport, measuring conductance and thermopower in suspended 2DEG structured with periodical square lattice of artificial scatters — antidots.

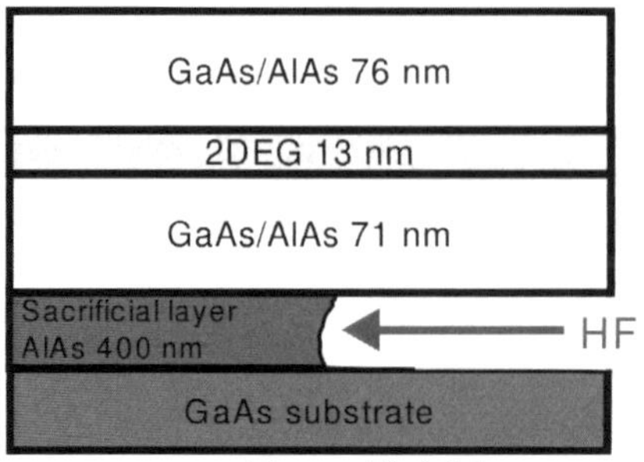

Figure 1 The GaAs/AlGaAs heterostructure with sacrificial layer.

2. Experimental details

Experimental samples were fabricated from AlAs/GaAs heterostructures containing high-mobility 2DEG grown by molecular beam epitaxy. The 2DEG density and electron mobility measured at liquid helium temperature were $n_s = 5\times10^{11}$ cm^{-2} and $\mu = 2\times10^6$ cm^2/Vs correspondingly. The described heterostructures were grown over a 400 nm-thick $Al_{0.7}Ga_{0.3}As$ sacrificial layer (see Figure 1). After photolithography the obtained samples represented themselves Hall bars with typical length of $L = 15$ µm, and typical width of 3 µm. A periodical square lattice of antidots was created on these Hall bars by means of electron-beam lithography followed by anisotropic plasmachemical etching. The antidots diameter varied from 100 to 200 nm while the lattice period varied from 500 to 800 nm for different samples (see Figure 2). The sacrificial layer was removed from-under the Hall bars by means of selective wet etching in HF water solution (see Figure 1). Magnetotransport measurements were performed at low temperatures (0.5÷4.2 K) by standard low-frequency lock-in technique in linear regime. Non-linear effects were studied by passing low-amplitude ac current together with dc current through the sample and measuring their resistance on the ac signal. We performed identical measurements on the same samples before and after the suspension.

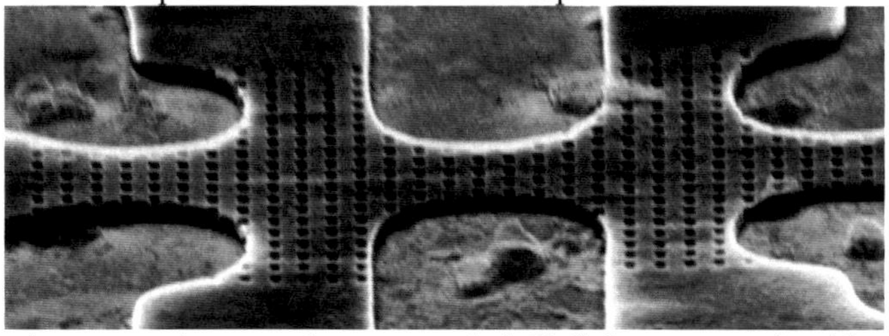

Figure 2 Electron-microscope image of a suspended Hall bar covered with periodical lattice of anditots captured at an angle of 60°.

3. Experimental results

Solid lines in Figures 3(a) and 3(b) show magnetoresistance of non-suspended and suspended samples correspondingly (the antidots diameter is a = 200nm, the lattice period is d = 600 nm). We found that the magnetoresistance of the samples with antidots lattice demonstrates commensurability oscillations in both cases. Earlier it was shown that, in the case of non-suspended samples, these oscillations originate from geometrical resonances [9]. We demonstrate that detaching of the samples from the substrate has almost no influence on the magnitude of the commensurability oscillations which means that electron mean free path remains practically unchanged ($l > 2$ µm) after the suspension. The suppression of commensurability oscillations by dc current demonstrates different behavior for suspended and non-suspended samples (see Figure 3 (a) and 3 (b)) which can be explained by poor heat coupling of suspended Hall bars to the substrate. More comprehensive understanding of non-linear effects requires further theoretical and experimental studies.

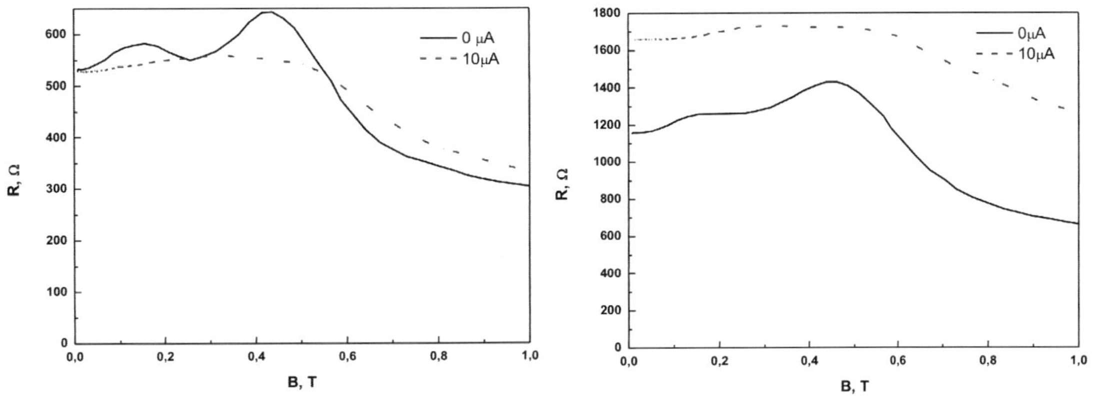

Figure 3. The influence of applied dc current (values are shown in the Figure) on the magnetoresistance of the Hall bar with antidots lattice (the antidots diameter is $a = 200$nm, the lattice period is $d = 600$ nm). (a) Non-suspended sample; (b) suspended sample.

Figure 4 shows well pronounced "S-shaped" peculiarities of ballistic thermopower measured on non-suspended and suspended samples. They are observed at magnetic fields corresponding to commensurability oscillations of magnetoresistance. The characteristic form of the peculiarities is explained by the fact that, according to the Mott rule, thermopower is proportional to the logarithmic derivative of conductivity on the Fermi energy [10]. This makes thermopower measurements more sensitive compared to magnetoresistance and explain earlier beginning Shubnikov – de Haas oscillations. One could expect some additional features in thermopower caused by modified phonon spectrum in thin suspended sample. However, the commensurability oscillations in thermopower are quite similar for suspended and non-suspended Hall bars. In our case, we measure the diffusion thermopower without the contribution of phonon drag because the electron-electron interaction time is much less than electron-phonon interaction time ($\tau_{e-e} \ll \tau_{e-ph}$). Hence, the thermopower signal is determined by electron temperature gradient. The observed similarity of the thermopower curves shows that electron temperature distributions both in suspended and in non-suspended samples are similar. This is possible if electron-phonon relaxation time τ_{e-ph} is much greater than the Thouless time τ_{Th}, i.e. the time an electron spends in the sample. The described behavior is the same for all the periods of the investigated lattices, except the number of commensurability peaks varies. For example, in Figure 5 and Figure 6 one can see four peaks of commensurability oscillations of magnetoresistance and corresponding four "S-shaped" features in thermopower of the suspended sample with 800 nm period of the antidot lattice.

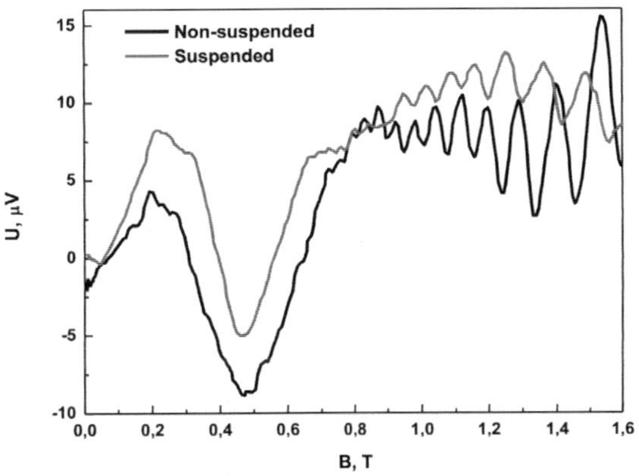

Figure 4 Ballistic thermopower of non-suspended and suspended Hall bars with antidot lattice(diameter of antidots $a = 200$nm, period of the lattice $d = 600$ nm).

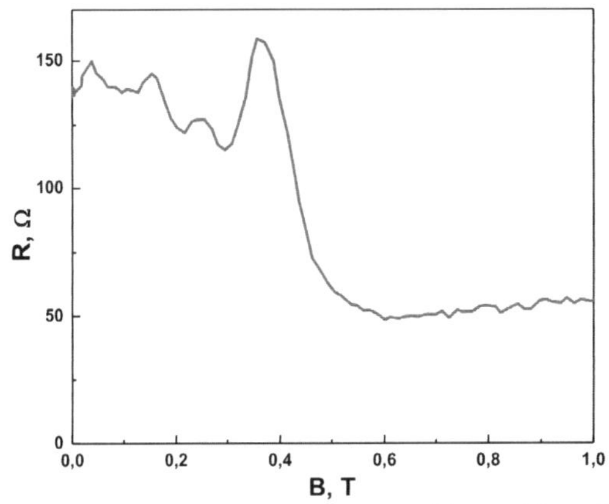

Figure 5 Magnetoresistance of the suspended Hall bar with antidots lattice (diameter of antidots a = 200nm, period of the lattice d = 800 nm).

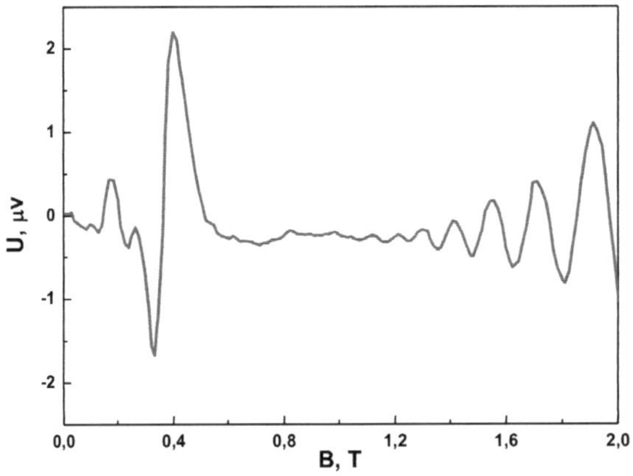

Figure 6 Ballistic thermopower of the suspended Hall bar with antidots lattice (diameter of antidots a = 200nm, period of the lattice d = 800 nm).

4. Conclusions

Magnetic field dependence of thermopower in suspended 2DEG with square lattice of antidots demonstrate commensurability oscillations originating from geometrical resonances similar to those observed in magnetoresistance. This indicates that the suspension of the 2DEG does not suppress ballistic regime in thermopower measurements. While the non-linear effects in suspended and non-suspended structures demonstrate difference, caused by the peculiarities of the heat transfer in suspended samples, the thermopower does not demonstrate any additional features related to the suspension. This can be explained by short Thouless time in our samples compared to electron-phonon relaxation time ($\tau_{Th} \ll \tau_{e\text{-}ph}$).

Acknowledgements

The work is supported by RFBR (grants 13-02-01153, 14-02-31585) and Russian Academy of Sciences.

References

[1] Cleland A N and Roukes M L 1998 *Nature* **392** 160.
[2] Nishiguchi N 2003 *Phys. Rev. B* **68** 121305.
[3] Shevyrin A A, Pogosov A G, Budantsev M V, Bakarov A K, Toropov A I et al. 2012 *Appl. Phys. Lett.* **101** 241916.
[4] Shevyrin A A, Pogosov A G, Budantsev M V, Bakarov A K, Toropov A I, Ishutkin S V 2013

Appl. Phys. Lett. **103** 131905.

[5] Rössler C, Herz M, Bichler M, Ludwig S 2010 *Solid State Communic.* **150** 861.

[6] Pogosov A G, Budantsev M V, Zhdanov E Yu, Pokhabov D A, Bakarov A K and Toropov A I 2012 *Appl. Phys. Lett.* **100** 181902.

[7] Zhdanov E Yu, Pogosov A G, Budantsev M V and Pokhabov D A 2013 *AIP Conf. Proc.* **1566** 211.

[8] Pogosov A G, Budantsev M V, Lavrov R A, Plotnikov A E, Bakarov A K, Toropov A I, Portal J C 2006 *JETP Letters* **83** 122.

[9] Baskin E M, Gusev G M, Kvon Z D, Pogosov A G, Entin M V 1992 *JETP Letters* **55** 678 (1992).

[10] Pogosov A G, Budantsev M V, Lavrov R A, Plotnikov A E, Bakarov A K, Toropov A I 2005 *JETP Letters* **81** 462.

Journal of Physics: Conference Series **643** (2015) 012080 doi:10.1088/1742-6596/643/1/012080

The luminescence properties of colloidal quantum dots

I I Mikhailov[1], S A Tarasov[1], I I Lamkin[1], M Y Andreev[1], A V Solomonov[1]
[1]Saint Petersburg Electrotechnical University «LETI» (ETU)
E-mail: iimihalov@gmail.com

Abstract. Luminescent properties of systems containing lead sulfide (PbS) nanoparticles deposited on substrates of porous silicon (p-Si), oxidized porous silicon (OPS) and have been studied. It was shown that structure of the matrix and its composition are of crucial importance affecting luminescence spectra of colloidal quantum dots by setting their radiation wavelength.

1. Introduction

For the last decade colloidal quantum dots (CQDs) have become an object of particularly intensive studying. CQDs are semiconductor particles of nanometer scale synthesized in liquid media and characterized by unique properties which substantially differ from bulk material properties. Important advantages of CQDs are simplicity of size tunability through technological parameters variation (time, temperature etc.) and that synthesis procedures do not require expensive equipment and materials. Placing the quantum dots in porous matrixes lead to additional advantages due to stabilization of their size and increasing density of emitting centers. Such structures are promising materials for a high-performance light-emitting diode (LED) and laser structures, as well as photosensitizers in solar cells or luminophores in micro-devices for optical tomography of biological tissues.

2. Materials and methods

In present report we investigate luminescence of lead sulfide (PbS) CQDs synthesized via pyrolysis of metalorganic compounds with organic systems containing a source of sulfur [1-6]. As a result of synthesis CQDs with various sizes were obtained. After measurements of photoluminescence spectra of CQDs in an solution QDs CQDs were deposited on various silicon substrates [7-9]. Monocrystalline silicon (mono c-Si), purified porous silicon (por-Si) and porous silicon oxidized in air for 14 days were employed as substrates [10-12]. Porous silicon was obtained by anodic electrochemical etching in a one-compartment cell in electrolyte based on hydrofluoric acid supplemented with isopropyl alcohol. Size of the pores ranged from 10 to 40 nm.

In order to investigate properties of CQDs, methods of non-invasive optical spectroscopy were applied. The photoluminescence spectra was investigated with a laboratory setup based on a double monochromator spectrometer SDL-1 [13,14]. A Nd:YVO4 diode-pumped solid-state laser with doubled frequency ($\lambda = 532$ nm) and a semiconductor laser diode with the emission wavelength of $\lambda = 445$ nm were used as sources of exciting radiation.

3. Results

In a first phase, we studied luminescent properties of PbS CQD samples in the solution. The PbS CQD photoluminescence spectra are shown in Figure 1. For further investigation we chose two types of samples with a maximum of their spectral response at 950 nm (PbS-1) and 1000 nm (PbS-2) Preliminary theory-based estimate of CQD size gave values of 4-5 nm. Besides, we observed a significant spectral broadening (the half-width exceeded 100 nm) indicating a considerable spread of particle sizes. Altering of particle sizes was carried out via variation of process parameters of CQD synthesis such as temperature and time.

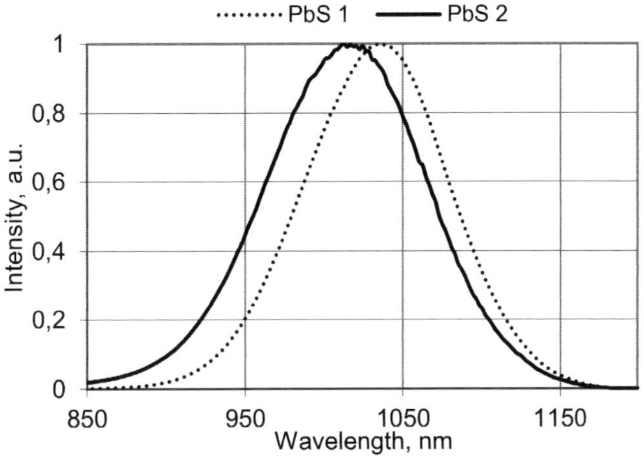

Figure 1. PL spectra of PbS CQDs in the solution

We also studied luminescent properties of porous silicon substrates. In contrast to bulk silicon crystals such materials shows a significant emissivity because of photon emission from nanorods and nanowalls [6], and can partially be caused by surface states. PL spectra of studied samples are shown in Figure 2. The maximum of spectral response was at the wavelength of 660 nm. The spectrum half-width was about 150 nm due to the spread of particle sizes and other parameters of nanosystems in the substrate.

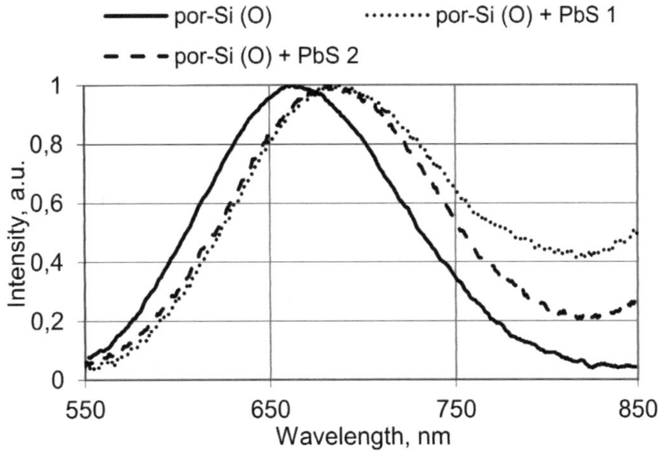

Figure 2. The luminescence spectra of the oxidized substrate with deposited quantum dots and without.

Porous substrates were found to significantly influence PL spectra of PbS CQDs, while the use of mono-Si blank substrates did not result in major changes of luminescent properties of the nanoparticles. When particle settling onto the porous substrates was carried out immediately after anodic treatment, modification of PL spectrum was relatively small (Figure 3). However significant shift of the maximum to shorter wavelengths was reached with porous silicon substrates oxidized in air during 14 days.

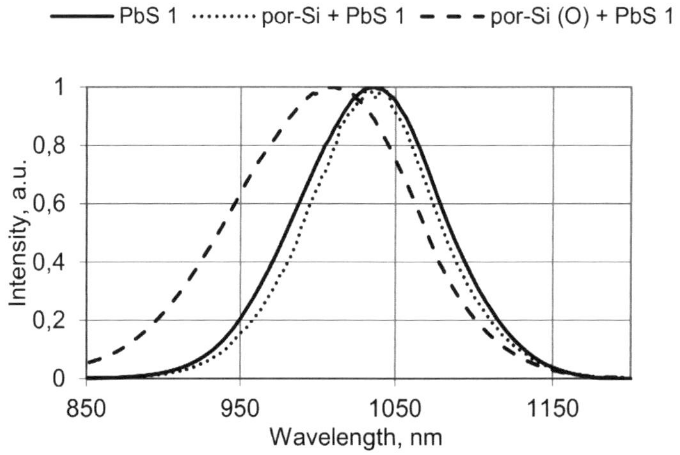

Figure 3. PL spectra of PbS CQD deposited on the porous substrate or in solution

The initial colloidal solution contains QDs of various sizes. In case of unoxidized substrate, pore diameter could be oversized to considerably affect the size of particles entering into the pores. When using the substrate based on oxidized porous Si, the pore size reduces due to oxide formation. As a result smaller nanoparticles enter into the pores, that is selection takes place. When the nanoparticles are being applied and distributed over the substrate surface, decomposition of larger particles takes place which followed by selection of newly formed smaller particles. These phenomena result in decreasing the sizes of the emitting particles, and, consequently, in increasing the energy of the emitted photons.

The presence of particle selection is confirmed by the fact that initial size of the particles practically had no effect on the emission spectrum of the particles precipitated into the oxidized matrixes (Figure 4).

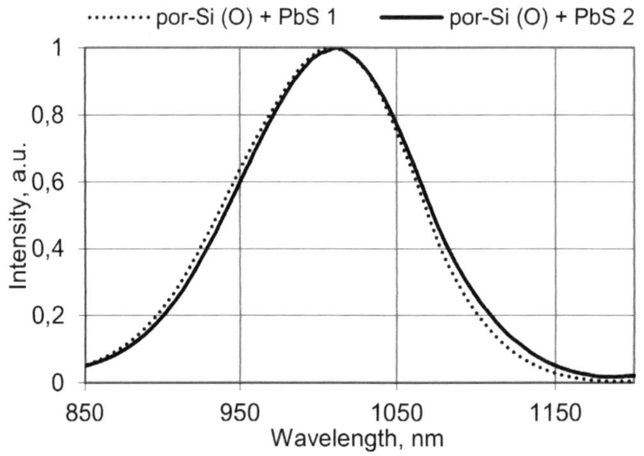

Figure 4. PL spectra of PbS CQD deposited on the substrate.

We also found that deposition PbS CQDs onto surface of porous Si results in alteration of substrate PL spectra. PL spectra of porous Si samples without nanoparticles and structures of oxidized por-Si are shown in Figure 2. When depositing of QD they are distributed in pores and in immediate vicinity from the nanosystems inside the porous matrixes. It changes wave functions of the Si nanostructures and their energy level positions resulting in splitting or broadening of the levels. And with this, the effective band gap energy reduces, and the PL peak shifts into longer wavelengths which is similar to interaction of coupled quantum wells. When the barrier is very thin, this system transforms into a quantum well where the width is twice as much as the initial well, and the levels of the new well are lower. We estimated the width of such a QW and level positions and demonstrated a satisfactory fit between the simulation and experimental results.

4. Discussion and conclusion

Thus, it has been found out that interactions between quantum dimensional elements of the porous Si-PbS QDs nanosystem result in a shift of por-Si PL spectra and a change in the photoluminescent properties of PbS QDs. We established that PbS CQDs produced by chemical deposition onto por-Si matrixes have a luminescence in the infrared spectrum region but with the spectrum maximum shifted to shorter wavelengths. It was shown that matrix structure and its composition are of crucial importance affecting luminescence spectra of CQDs by setting their radiation wavelength

5. Acknowledgements

The work was supported by the Ministry of Education of the Russian under project of the state task in the field of scientific activity, project number 16.1307.2014K.

6. References

[1] L Bakueva, S Musikhin, M Hines, T Chang, M Tzolov, G Scholes, E. Sargent, .Applied physics letters **82**, 2895(2003).

[2] O Alexandrova, A Maximov, E Maraeva, L Matyushkin, V Moshnikov, S Musikhin, S Tarasov, Нано- и микросистемная техника **2**, 19 (2013).

[3] S Tarasov, O Alexandrova, A Maximov, E Maraeva, L Matyushkin, E Men'kovich, V Moshnikov, S Musikhin, Известия высших учебных заведений. Электроника **3**, 28 (2013).

[4] I Mikhailov, S Tarasov, A Presnyakova and D Romanovsky, Izvestiya SPbGETU «LETI» **5**, 22 (2013).

[5] I Mikhailov, S Tarasov, A Solomonov, L Matyushkin, D Mazing, Functional materials 21, 142 (2014).

[6] S Tarasov, O Aleksandrova, A Maksimov, E Maraeva, L Matyushkin, E Men'kovich, V Moshnikov, S Musikhin, Semiconductors **48**, 1729 (2014).

[7] A Lenshin, V Kashkarov, Yu Spivak, V Moshnikov, Materials Chemistry and Physics **135**, 293 (2012).

[8] A Lenshin, V Kashkarov, Yu Spivak, V Moshnikov. ФХС **38**, 383 (2012).

[9] V Moshnikov, I Gracheva, V Kuznezov, A Maximov, S Karpova, A Ponomareva, Journal of Non-Crystalline Solids **356**, 2020 (2010).

[10] V Levickiy, A Len'chin, A Maximov, E Maraeva, V Moshnikov, Сорбционные и хроматографические процессы **12**, 725 (2012).

[11] S Tarasov, I Gracheva, K Gareev, O Gordyushenkov, I Lamkin, E Men'kovich, V Moshnikov and A Presnyakova, Известия высших учебных заведений. Электроника **94**, 21 (2012).

[12] S Tarasov, I Gracheva, K Gareev, O Gordyushenkov, I Lamkin, E Men'kovich, V Moshnikov and A. Presnyakova, Semiconductors **46**, 1584 (2012).

[13] E Menkovich, S Tarasov, I Lamkin, Functional Materials **19**, 233 (2012).

[14] D Romanovskiy, S Tarasov, G Galiev, S Pushkarev, Journal of Physics: Conference Series **541**, 012080 (2014).

Estimation of the effective electron-capture cross section in the emission processes from arrays of vertically coupled InAs quantum dots in the n-GaAs matrix

A Baklanov[1], A Gutkin[2], P Brunkov[1,2]

[1] St. Petersburg State Polytechnical University, Institute of Physics, Nanotechnology and Telecommunications, St. Petersburg, 195251 Russia
[2] Ioffe Physical-Technical Institute, Russian Academy of Sciences, St. Petersburg, 194021 Russia

E-mail: baklanov@mail.ioffe.ru

Abstract. In the paper we investigate the processes of electron emission from arrays of vertically coupled InAs quantum dots in the space charge region of the Schottky barrier in the matrix of n-type of GaAs. We developed a model of thermally activated tunneling of charge carriers, which allows to calculate the parameters that characterize the processes of electron emission. The temperature dependence of the activation energy of the carriers, the ground state energy carrier in the quantum dot at different reverse bias voltage of the Schottky barrier, and the dependence of the electron-capture cross section on the binding energy of the quantum dot ground state were obtained. The comparison of the values performed for samples with different number of layers of quantum dots: 3, 6 or 10 layers.

1. Introduction

The study of the emission processes of charge carriers in heterostructures that contain arrays of self-assembled quantum dots (QDs) is an important part in the design and optimization of devices and active elements which are based on these structures [1,2]. Therefore it is important to investigate the mechanisms of this emission and different the factors affecting the value of its rates. Such the task is interesting due to the practical and fundamental importance. The emission electron from multilayer assemblies vertically coupled quantum dots was investigated in [3,4]. In [3] the ground state energy carriers in QD was estimated by an analysis of the capacitance-voltage characteristics of structures, as in [4] a mechanism of electron emission process is thermally activated tunneling involving virtual state was proposed and the effective thermal activation energies at different temperatures were evaluated without including the effect of tunnelling in the analyse. In addition to calculating the characteristic energies is important to estimate the effective carrier capture cross-section [5], assuming that electron emission from arrays of vertically coupled QDs to conduction band of the semiconductor matrix is in the form of thermally activated tunneling process with the involvement of virtual states. For such the process the model was implemented that describes the experimental data of the temperature dependence of the emission rate which obtained by admittance spectroscopy. In [6] the dependencies of the effective carrier capture cross sections on the ground-state energy of the electrons in the quantum dots were determined with this model for samples containing a single layer of InAs QDs in n-GaAs matrix.

However, it is interesting to estimate the capture cross sections for samples containing different numbers of layers of QDs and comparison of their values with the number of layers.

2. Samples and experimental data

The investigated structures were Schottky barriers based on epitaxial layers of *Si*-doped GaAs ($n \approx 2 \cdot 10^{16}$ cm^{-3}), which contained 3, 6 or 10 layers of InAs QDs [3] (Fig. 1d). Figure 1 (a-c) shows the bright field TEM-images of cross-sectional structures demonstrating the number of QDs layers in the structure (3, 6 or 10), as well as the vertical ordering of quantum dots over each other. Arrays of QDs were embedded into the matrix so that they could be included into the space charge region of the Schottky barrier with changing the reverse-bias voltage (V_{rev}). The emission process of electron occurred only from the last layer of QDs, because the electrons from remaining layers rapidly dumped to the last layer due to tunneling. Layers of QDs probed when changing the bias voltage. The emission rate of charge carriers from the QDs was determined by the method of admittance spectroscopy at different reverse-bias voltages [3,4,7,8]. The QDs with different binding energies can be probed by use of different V_{rev}. The frequency (f) of the probe AC voltage was varied within the limits of 10^4 – 10^6 Hz. Whereas the differential conductance reaches a maximum peak when the emission rate is $e_n \approx \pi f$ [7,8]. Thus the temperature dependence of the rate of emission at different reverse bias voltage to the Schottky barrier were obtained (Fig. 2).

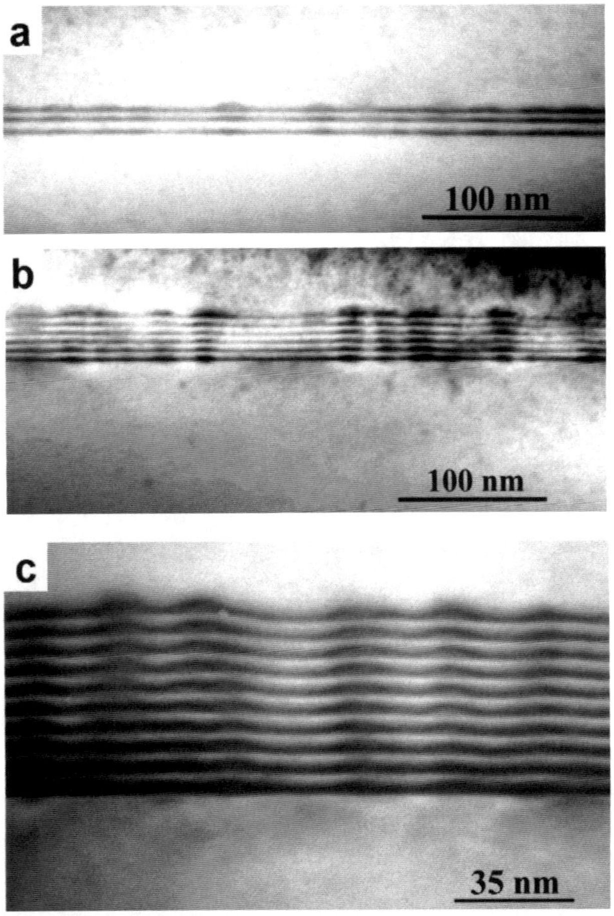

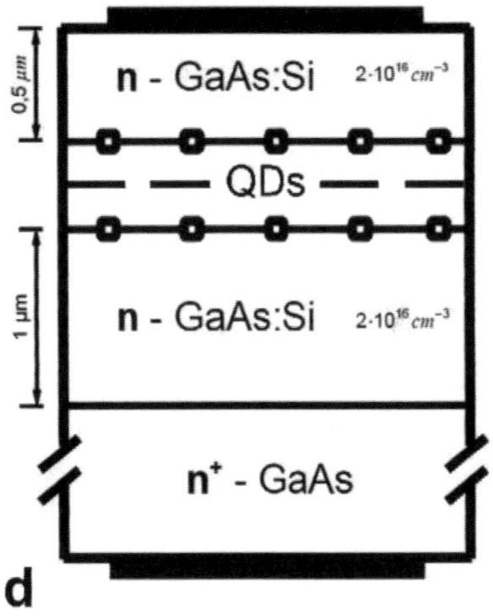

Figure 1. TEM-images of structures with arrays of vertically coupled QDs InAs/GaAs containing: a – 3, b – 6, c – 10 layers; d – schematic view of the samples.

In the case of thermally activated tunneling emission of an electron, process consists of thermal jump to a virtual state (E^*) and tunneling through the barrier [6]:

$$e_n = A(T)\cdot\exp\left(-\frac{E^*}{k_B T}\right)\cdot\exp\left(-\frac{2}{\hbar}\int_0^{x_0(E^*)}\left(2m_n^*\left[E(x)-E^*\right]\right)^{1/2}dx\right) \tag{1}$$

$$A(T)=\langle V_n\rangle\,\sigma_n^{\infty}N_c;\quad \langle V_n\rangle=\left(\frac{8k_BT}{\pi m_n^*}\right)^{1/2};\quad N_c=2\cdot\left(\frac{m_n^* k_B T}{2\pi\hbar^2}\right)^{3/2} \tag{2}$$

where $E(x)$ is energy barrier shape similar to [6], $x_0(E^*)$ – the tunneling length, σ_n – the capture-cross section, N_C – the effective density of states in the conduction band, m_n^* – the effective mass of electron in GaAs matrix, $<V_n>$ – the thermal velocity of free electrons; q – the charge, ε – the permittivity of GaAs, N_d – the donor concentration, The value of e_n has a maximum at E^* is equal to the activation energy E_a, which characterizes the process of thermally activated tunneling. Determining the activation energy E_a from this condition, we can obtain the equation that relates the barrier height E_0 (E_0 *is* $E(x)$ at $x=0$), which is the binding energy of probed electron state in QD, and activation energy E_a [6]:

$$E_a(T)=4\exp\left(\frac{2\beta}{T}\right)\left(1+\exp\left(\frac{2\beta}{T}\right)\right)^{-2}(E_0-E_F)+E_F,\quad \beta=\frac{q\hbar}{k_B}\left(\frac{\pi N_d}{\varepsilon m_n^*}\right)^{1/2} \tag{3}$$

The experimental dependence were approximated by the expression (1-3), (approximation parameters: the capture cross section of electrons and the binding energy) with using the proposed model in [6], which describes the process of thermally activated tunneling of charge carriers (Fig. 2).

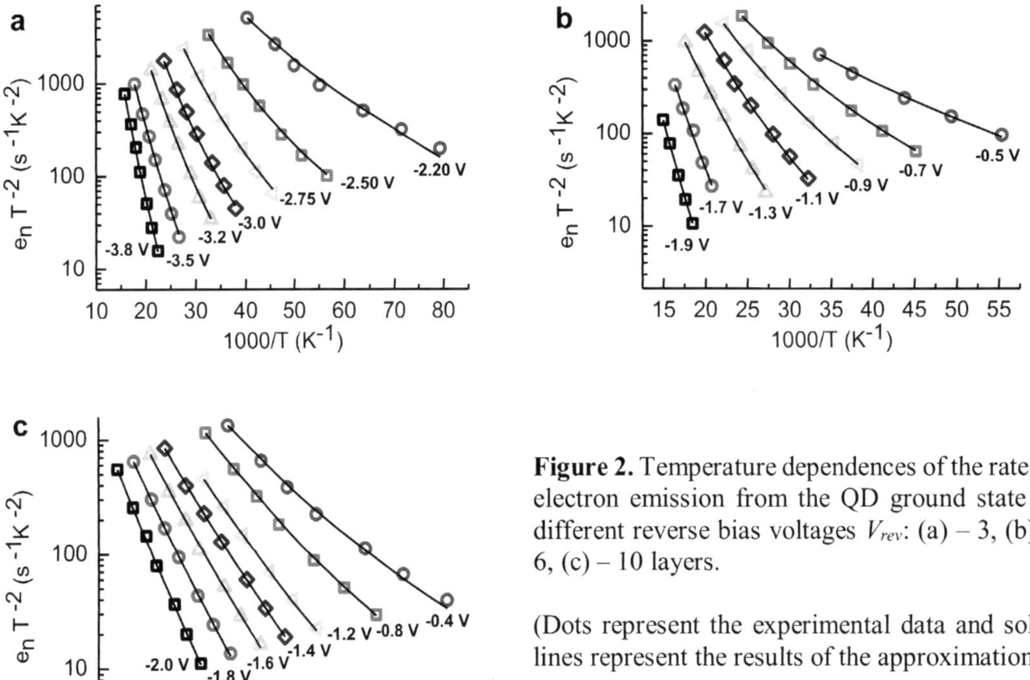

Figure 2. Temperature dependences of the rate of electron emission from the QD ground state at different reverse bias voltages V_{rev}: (a) – 3, (b) – 6, (c) – 10 layers.

(Dots represent the experimental data and solid lines represent the results of the approximation)

3. Conclusions

As a result of analysis of the experimental temperature dependence of the emission rate at different reverse-bias voltages were calculated the binding energy of the carriers in QDs (Fig. 3) and activation energy using the expression (3) (Fig. 4).

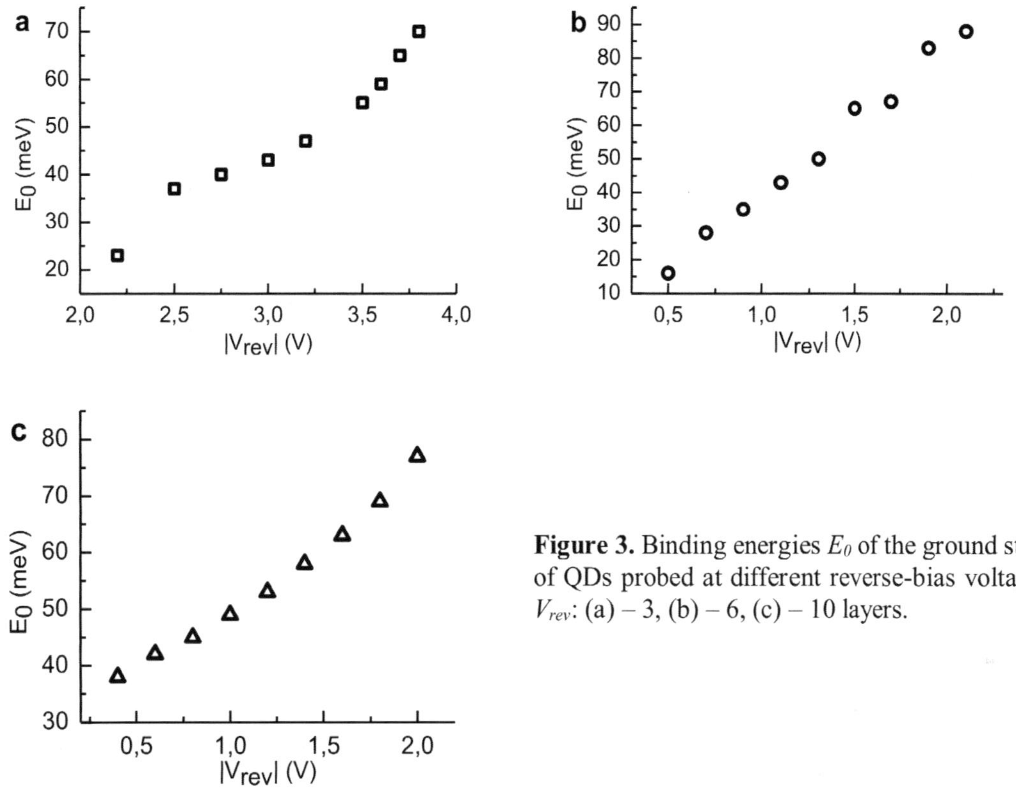

Figure 3. Binding energies E_0 of the ground state of QDs probed at different reverse-bias voltages V_{rev}: (a) – 3, (b) – 6, (c) – 10 layers.

The binding energy of electrons in the QD E_0 probed at different reverse-bias voltages increases with value of the voltage on the Schottky barrier, as expected in this case. The maximum value of E_0 is 70 meV (Fig. 3a), 88 meV (Fig. 3b) and 77 meV (Fig. 3c), respectively for samples containing 3, 6 and 10 layers of QDs. Such an increase in the binding energy for samples with 3 and 6 layers of QDs has been in [3,4]. However, in this paper, for the sample with 10 layers of QDs, this value is less than the 6-layer (the reverse situation), which may be associated with an increasing influence in the emission process of indirect mechanisms (additional mechanical stress, piezoelectric effects, the geometric parameters of the dots). Wherein the values of the activation energy increases with temperature, as well as the difference between the energies ($E_0 - E_a$), see Figure 4.

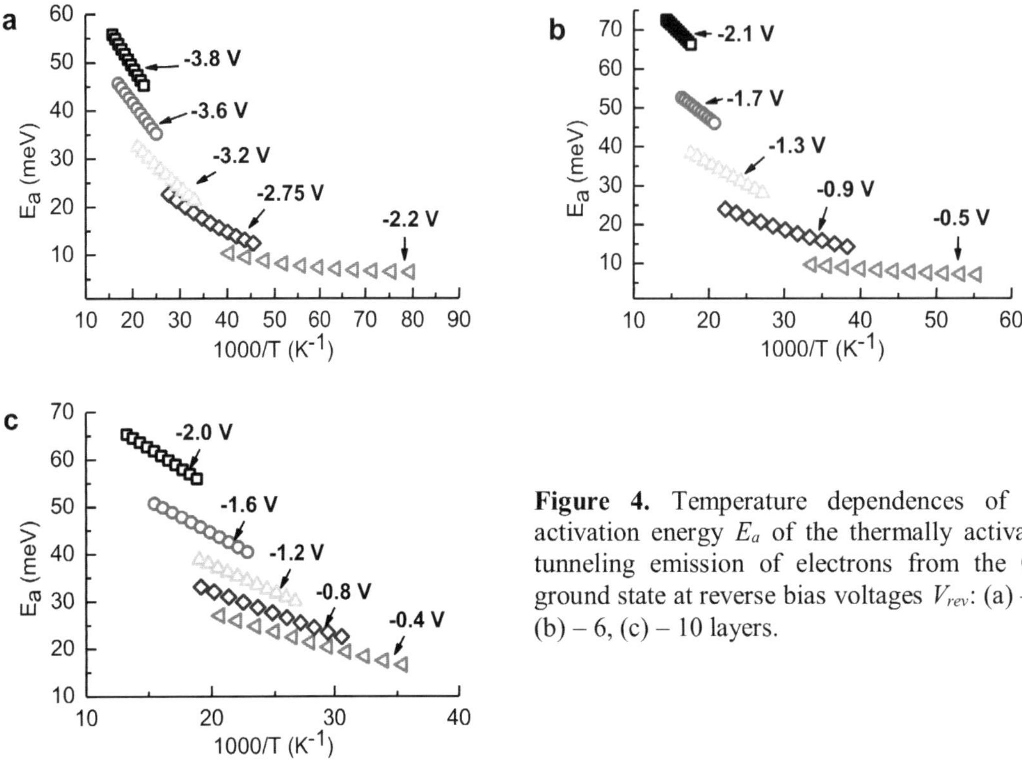

Figure 4. Temperature dependences of the activation energy E_a of the thermally activated tunneling emission of electrons from the QD ground state at reverse bias voltages V_{rev}: (a) – 3, (b) – 6, (c) – 10 layers.

Finally, were also evaluated the effective electron-capture cross section (Fig. 5): the capture cross section increases with the binding energy of the electron in the QD, i.e. with an increase in the size of QD. Thus it is possible to obtain estimates for the effective electron-capture cross section by analyzing experimental data. The effective electron capture cross section varies from $1 \cdot 10^{-14}$ cm^2 to $60 \cdot 10^{-14}$ cm^2, while the real mean lateral size of the quantum dots is about 20 nm, see Figure 1 (a-c); i.e. the effective capture cross section is smaller than quantum dots. And the cross section is constant for every particular binding energy (Fig. 5).

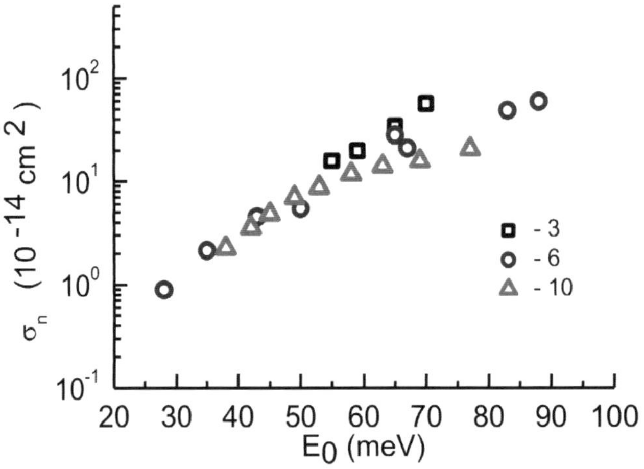

Figure 5. Dependence of the effective electron-capture cross section on the binding energy E_0 of the QD ground state for all samples.

The obtained dependence of the capture cross section on the electron binding energy has an exponential form. In some works, this curves is called as the rule of Meyer-Neldel, which allows to extract the characteristic energy of the task, that could be related to number and energy of phonons involved in capture process, see e.g. [9,10].

Acknowledgements
This studies were performed using the equipment of the regional Joint Research Centre "Material Science and Characterization in Advanced Technology", Ioffe Physical–Technical Institute, Russian Academy of Sciences.

References
[1] Zhiming M Wang 2012 *Quantum Dot Devices* (Berlin: Springer)
[2] Bimberg D 2008 *Semiconductor Nanostructures* (Berlin: Springer)
[3] Gutkin A et al 2007 *Semiconductors* **41** 1335
[4] Gutkin A et al 2008 *Semiconductors* **42** 1104
[5] Abakumov V N, Perel V I, Yassievich I N 1991 *Nonradiative Recombination in Semiconductors* (North-Holland: Elsevier)
[6] Baklanov A et al 2014 *Semiconductors* **48** 1155
[7] Losee D 1975 *J. Appl. Phys.* **46** 2204
[8] Brunkov P et al 1999 *J. Electron. Mat.* **28** 486
[9] Yakimov A et al 2004 *JETP Lett.* **80** 321
[10] Bloshkin A et al 2014 *Semiconductors* **48** 1036

Ion-beam synthesis of GaN in silicon

V A Sergeev, D S Korolev, A N Mikhaylov, A I Belov, V K Vasiliev, A E Smirnov, D E Nikolitchev, S I Surodin, D V Guseinov, A V Nezhdanov, A S Markelov, V N Trushin, A V Pirogov, D A Pavlov and D I Tetelbaum

Lobachevsky State University of Nizhni Novgorod, 23/3 Gagarin prospect, Nizhni Novgorod, 603950, Russia

E-mail: v.a.sergeyev94@gmail.com

Abstract. The structure and composition of a subsurface silicon layer subjected to a dual implantation of Ga and N ions with subsequent annealing have been investigated using X-ray photoelectron spectroscopy, electron spin resonance, X-ray diffraction, Raman microscopy, transmission electron microscopy. The results indicate a possibility of ion-beam synthesis of GaN composite nanostructures in silicon-based materials.

1. Introduction

New development stage of electronic technologies is characterized by the transfer from conventional microelectronics to optoelectronics and integrated optics. This requires synthesis of light emitters and photodetectors on a single chip for different ranges of wavelengths, including the ultra-violet range. A key material of conventional microelectronics – silicon – cannot be used as an efficient light source because of its indirect band gap. This fact has stimulated the search for other light-emitting materials compatible with silicon technology. Such a compatibility would allow the usage of tremendous long-term achievements of silicon technology and could give high economic effect. For example, the cost of a silicon wafer is seven times lower than that of a GaAs wafer of the same area.

It is known that LEDs, lasers and photodetectors on the basis of III-nitrides, in particular gallium nitride, have excellent performance as well as high resistance to radiation, chemically aggressive environment and temperature. However, there is a problem of high cost of GaN and difficulties of its integration into a silicon or other low-cost substrates. Fabrication of high-quality epitaxial III-nitride layers on silicon is difficult due to the high lattice mismatch and differences in thermal expansion coefficients of the materials [1]. It results in a high dislocation density (more than $1 \cdot 10^6$ cm^{-2}) and the degradation of device performance.

One of the perspective solutions to this problem is the ion-beam synthesis of III-V (especially GaN) nanocrystals by dual (sequential) ion implantation into silicon or into dielectric films (SiO_2, Si_3N_4, Al_2O_3, HfO_2, etc.) deposited on silicon substrate. An advantage of ion implantation over the other synthesis methods (like epitaxy) is in a high controllability of the implanted layer composition and concentration of implanted atoms (not restricted by the thermodynamic solubility limit), an ability of local phase synthesis with focused implantation, and perfect compatibility of this method with the existing CMOS-technology. Using ion implantation with subsequent annealing permit to synthesize not only continuous layers, but semiconductor nanocrystals in different matrices, including amorphous matrices.

The possibility of ion-beam synthesis of GaN nanocrystals is based on the fact that nanocrystals of some semiconductor compounds of a number of III-group elements have been already synthesized in silicon, SiO_2 and Al_2O_3 and, in some cases, the expected light emission has been demonstrated [2,3,4,5,6]. However, the literature on GaN ion synthesis in silicon or in Si-compatible matrices is almost absent. Only one research group [7,8] has reported on the synthesis of light-emitting GaN crystallites in SiO_2 and Al_2O_3 matrices after co-implantation of gallium and nitrogen ions and subsequent annealing in ammonia atmosphere.

In addition to the applications in optoelectronics and integrated optics, GaN nanocrystals in different matrices could be used as ultra-violet detectors, solar panels with high efficiency (owing to the conversion of shortwave part of solar radiation into electric energy), and possibly as semiconductor ferromagnetic material for spintronic devices, non-classical photon emitters in quantum communication devices, in biosensors, etc. GaN-based nanostructures are also important for biomedical applications because GaN is almost non-toxic material.

Thus, the problem of GaN nanocrystals ion synthesis in different matrices, especially silicon-compatible materials, is very relevant. In this paper, the possibility of GaN synthesis in Si matrix by dual implantation of gallium and nitrogen ions with subsequent annealing is studied.

2. Experimental

The investigated GaN-nanostructures were obtained by dual implantation in silicon of Ga^+ ions with energy of 80 keV and N_2^+ ions with energy of 40 keV. Implantation of molecular N_2^+ ions instead an N_1^+ ions was used to shorten the duration of implantation (in this case, every N_2^+ ion splits into two ions with two time lower energy at the first collision with target atoms). Ion implantation was carried out in two ways: first, Ga^+ ions were implanted with the dose of $5\cdot10^{16}$ cm^{-2} or $1\cdot10^{17}$ cm^{-2}, and then, N_2^+ ions with the corresponding doses of $2.5\cdot10^{16}$ cm^{-2} or $5\cdot10^{16}$ cm^{-2} were implanted and vice versa. Energy and doses of ions were chosen to obtain the correspondence of Ga and N depth distributions according to the SRIM calculation [9].

The ion-irradiated samples were annealed at 900°C (30 min) in dry nitrogen atmosphere. The structure, chemical composition of implanted layers and state of the implanted atoms were analyzed by several techniques: electron spin resonance (ESR), X-ray photoelectron spectroscopy (XPS) with depth profiling by Ar^+ ion etching, X-ray diffraction (XRD) in Bragg-Brentano geometry and confocal Raman spectroscopy/microscopy.

Some of the samples before and after annealing were investigated by the cross-section transmission electron microscopy (XTEM) using the JEOL JEM-2100F microscope operated at accelerating voltage of 200 keV. The electron beam was focused to the diameter of 0.7 nm in the bright-field scanning transmission electron microscopy (STEM) mode, which was used also for the analysis of local chemical composition by the INCA Energy TEM 250 X-MAX X-ray energy-dispersive spectrometer (EDS). The cross-sections were prepared by standard techniques on the Gatan equipment.

3. Results and discussion

The ESR data (figure 1) show that ion implantation into the Si surface layer transforms it into the amorphous state (typical dangling bonds with g = 2.0055 are detected with the concentration in the range of $5\cdot10^{17}$-$6\cdot10^{18}$ cm^{-3} depending on the order and dose of ion implantation). After annealing at 900 °C, the defect concentration is reduced, but remains still noticeable. It is related to the non-ideal crystalline structure – disordered areas such as grain boundaries and mosaic blocks in the implanted layer.

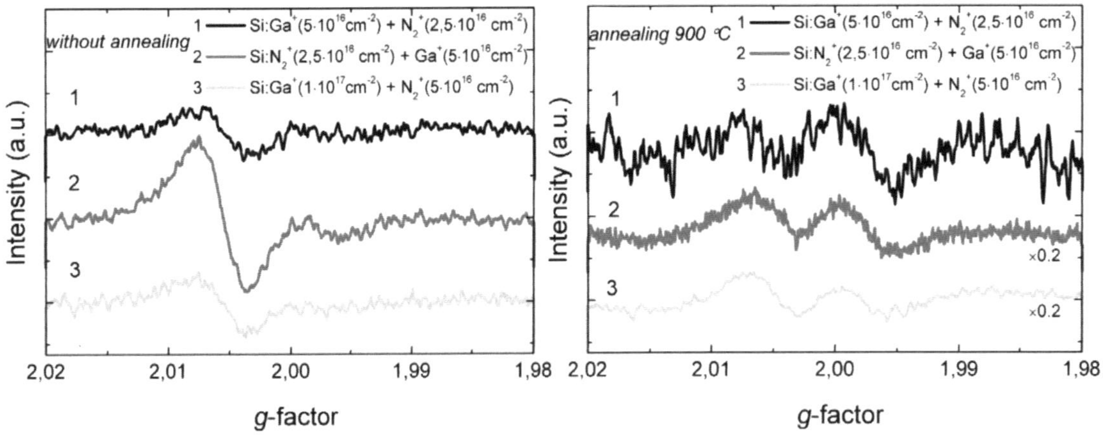

Figure 1. ESR spectra of silicon samples subjected to irradiation with gallium and nitrogen ions: before (a) and after (b) annealing at 900°C.

XRD data (figure 2) show that there is not any crystalline phase in the silicon subsurface layer after ion implantation. After annealing, the recrystallized Si is formed at the depth where amorphization has occurred. The formed crystallites do not have preferential orientation, this leads to the appearance of Si (111) XRD line for the lower dose. For the higher dose, Si line has lower intensity. The XRD lines from GaN are not revealed probably due to a small volume fraction of GaN.

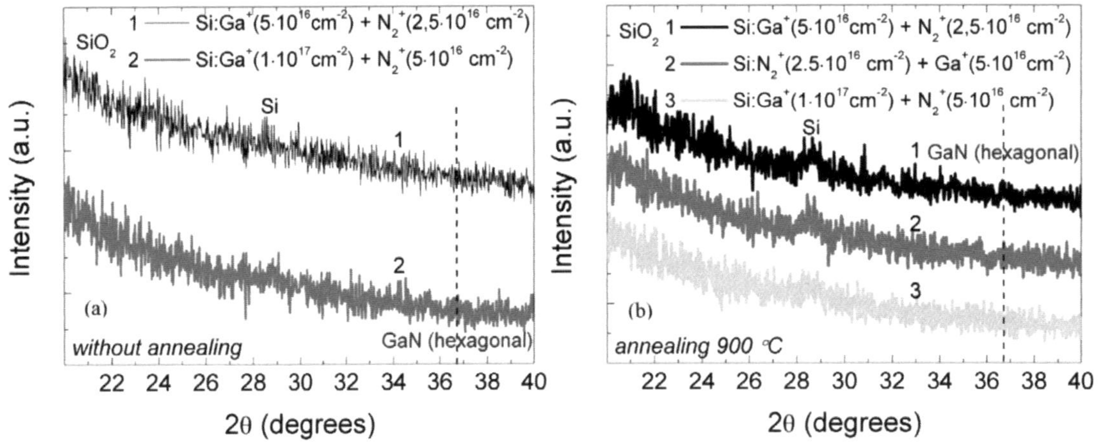

Figure 2. Diffraction patterns of Si samples irradiated by N and Ga ions with different sequence and doses, before annealing (a) and after annealing (b) at 900°C.

On the Raman spectrum (figure 3) for Si implanted by Ga^+ ($5\cdot10^{16}$ cm^{-2}) and N_2^+ ($2.5\cdot10^{16}$ cm^{-2}), then annealed at 900°C, we can clearly observe several Si-related peaks. The main peak (521 cm^{-1}) is asymmetric due to the existence of Si nanocrystallites (nc-Si) in the implanted layer. Inclusions with the sizes about hundred nanometers are identified on image of a part of the surface obtained by scanning with primary laser beam. Raman scattering in the synthesized GaN phase is revealed as a weak peak, the small intensity of which can be also related to low volume fraction of GaN phase.

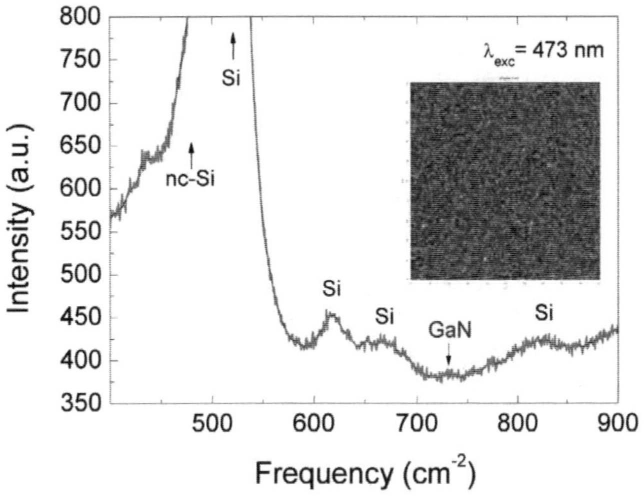

Figure 3. Raman spectrum of Si implanted by Ga^+ ($5·10^{16}$ cm^{-2}) and N_2^+ ($2.5·10^{16}$ cm^{-2}) after annealing at 900°C. Surface morphology is illustrated on the inset (frame size 50×50 μm).

The XTEM data show (figure 4), that there is a clear edge between the irradiated layer and underlying substrate before annealing. The depth of the edge agrees well with the ion projected range. Electron diffraction has not revealed any crystallites in the irradiated layer, but the existence of small nanoclusters could be observed as dark heterogeneity on the XTEM image.

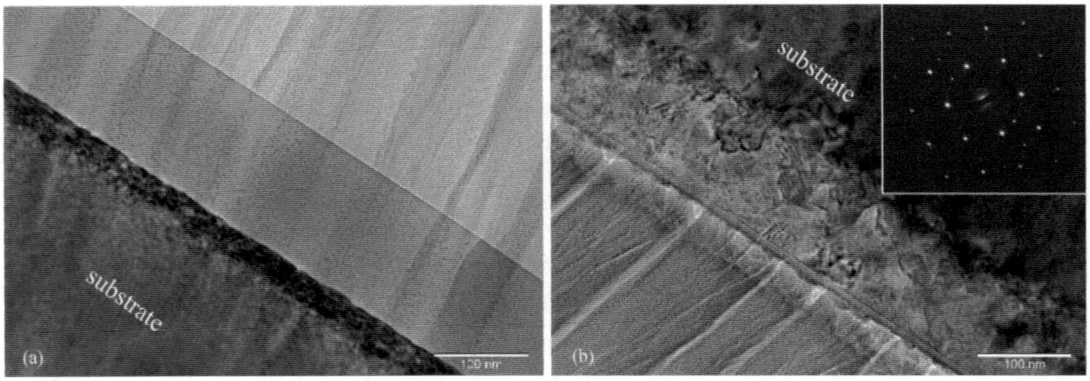

Figure 4. XTEM images of Si sample irradiated with Ga^+ ($5·10^{16}$ cm^{-2}) and N_2^+ ($2.5·10^{16}$ cm^{-2}) before (a) and after (b) annealing at 900°C. Electron diffraction picture for the implanted layer after annealing is shown on the inset.

There is a lot of disoriented nanocrystals in the implanted layer after annealing, which are identified with silicon. A dark layer is located closer to the edge of the implanted layer; presumably, this is a recrystallized Si layer coherent to matrix. The corresponding diffraction picture is shown on the inset. Interpretation of this diffraction picture doesn't reveal any crystalline phases except silicon. However, the EDS analysis shows that nanometer-sized areas with the most pronounced dark contrast contain a high concentration of Ga (up to 25 at.%) and correspond to the Ga-rich nanoclusters in silicon matrix.

The XPS data have given an integral information about the chemical and phase composition of Si samples irradiated with Ga^+ ($5·10^{16}$ cm^{-2}) and N_2^+ ($2.5·10^{16}$ cm^{-2}) ions in a different order, without and after annealing at 900°C. The depth distributions of species and their chemical bonding have been analyzed based on the position, shape and intensity of the corresponding photoelectron lines.

The depth distributions of gallium and nitrogen in the implanted Si layer before and after annealing are demonstrated in figure 5. The SRIM-calculated ion profiles are also shown. The distribution of gallium bonded to nitrogen is shown in figure 6. From the analysis of these data and data for distribution of nitrogen bonded to gallium or silicon (not shown), we can come to the following conclusions.

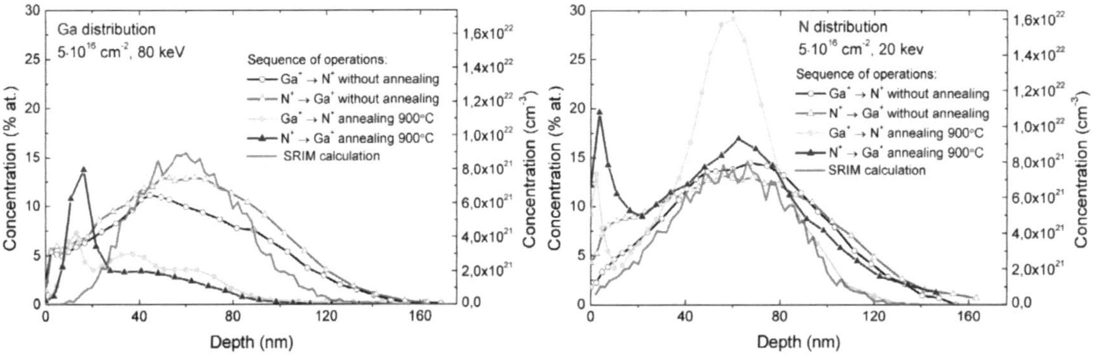

Figure 5. The depth distribution profiles of Ga and N for different sequences of implantation, before and after annealing at 900°C.

The excess of nitrogen concentration over the gallium concentration indicates that a part of implanted gallium ions is lost during implantation; this conclusion should be considered as preliminary. The distribution of implanted atoms (gallium and nitride) before annealing is wider than estimated. This could be explained by the radiation enhanced diffusion during the implantation process, which isn't a trivial fact and can be explained by disordering of the structure caused by ion implantation of high doses of heavy ions (Ga^+).

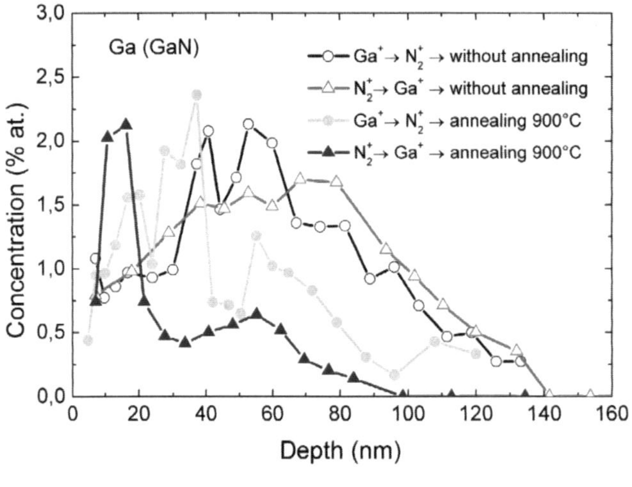

Figure 6. The depth distribution profiles of Ga bonded to N.

As a result of annealing Ga atoms migrate to the surface. Some of them leave the sample, others form elementary gallium and some part of Ga atoms (up to 2 at.%) reacts with nitrogen. Migration of impurity atoms to the surface was detected before, when Ga^+ and N^+ were implanted in SiO_2 film [8].

This was a result of diffusion enhancement because of radiation defects. When Ga^+ is implanted before N_2^+, the migration of gallium atoms to the surface after annealing is more significant than for the reverse sequence. Those nitrogen atoms which have not reacted with gallium form bonds with silicon (SiN_x, Si_3N_4).

4. Conclusions

Ion synthesis of gallium nitride is difficult because of the diffusion and evaporation of Ga atoms forming phases of elementary Ga and silicon nitride. At the used synthesis conditions, we are able to synthesize Ga-N bonds with the concentration about several atomic percents. Further optimization would allow improving the conditions of gallium nitride synthesis. For that aim, we need to change annealing regime and irradiation dose. Also we should use oxygen or nitrogen coating layer to prevent the evaporation of gallium atoms.

The work is supported by the Ministry of Education and Science of the Russian Federation (RFMEFI58414X0008).

References

[1] Kumar M, Roul B, Bhat T N, Rajpalke M K, Misra P, Kukreja L M, Sinha N, Kalghatgi A T and Krupanidhi S B 2010 *Materials Research Bulletin* **45** 1581

[2] White C W, Budai J D, Zhu J G, Withrow S P and Aziz M J 1996 *Appl. Phys. Lett.* **68** 2389

[3] White C W, Budai J D, Zhu J G, Withrow S P, Zuhr R A, Hembree D M, Henderson D O, Ueda A, Tung Y S, Mu R and Magruder R H 1996 *J. Appl. Phys.* **79** 1876

[4] Kanemitsu Y, Tanaka H, Fukunishi Y, Kushida T, Min K S and Atwater H A 2000 *Phys. Rev. B* **62** 5100

[5] Komarov F, Vlasukova L, Milchanin O, Mudryi A, Dunetz B, Wesch W and Wendler E 2012 *Phys. Stat. Sol. A* **209** 148

[6] Prucnal S, Facsko S, Baumgart C, Schmidt H, Liedke M O, Rebohle L, Shalimov A, Reuther H, Kanjilal A, Mucklich A, Helm M, Zuk J and Skorupa W 2011 *Nano Letters* **11** 2814

[7] Borsella E, Garcia M A, Mattei G, Maurizio C, Mazzoldi P, Cattaruzza E, Gonella F, Battaglin G, Quaranta A and D'Acapito F 2001 *J. Appl. Phys.* **90** 4467

[8] Borsella E, de Julian Fernandez C, Garcıa M A, Mattei G, Maurizio C, Mazzoldi P, Padovani S, Sada C, Battaglin G, Cattaruzza E, Gonella F, Quaranta A, D'Acapito F, Tagliente M A and Tapfer L 2002 *Nucl. Instr. Meth. B* **191** 447

[9] Ziegler J F, Ziegler M D and Biersack J P 2010 *Nucl. Instr. Meth. Phys. Res. B.* **268** 1818

Spin-dependent energy gap oscillations in the ultra-short carbon nanotube (5, 5)

A V Tuchin, S V Popov, G I Glushkov, L A Bityutskaya

Voronezh State University, 394006, Universitetskaya pl.1, Voronezh, Russia

E-mail: a.tuchin@bk.ru

Abstract. Results of the numerical simulation of an electronic structure of an ultrashort single-walled carbon nanotube (5, 5) at singlet and triplet states were presented. An antiphase energy gap oscillation on the length of the ultrashort nanotube (5, 5) at singlet and triplet states was revealed. It was found that the ground state of the nanotube is singlet, herewith the energy of the singlet-triplet transition corresponds to the energy value of visible and IR-radiation.

1. Introduction

In recent years a significant advance was achieved in the field of the ultra-short single-walled carbon nanotubes (us-SWCNT) synthesis with controlled both chirality and length [1–3]. Sanchez-Valencia et al. [1] synthesized single-chirality and essentially defect-free us-SWCNTs (6, 6) from a flat molecular precursors on a Pt (111). During the surface-catalyzed cyclodehydrogenation process the precursor transforms to the singly capped us-SWCNT (6, 6) seed with the length 4.5 Å. A manage of a concentration of a carbon feedstock gas at the time of an epitaxial elongation allows to control the length of us-SWCNTs up to few Å. The solubility of ultra short nanotubes in organic solvents, acids and water at the level of 2 weight % was revealed [4–6] that is an important achievement for the creation of functional nanomaterials. The experimental success opens up a way for usage of us-SWCNTs for applications such as light detectors, photovoltaics, field-effect transistors and sensors with highly optimized characteristics [1, 3, 7–9]. Lots of theoretical researches of carbon nanotubes demonstrate that the decrease their length less than 10 nm causes of great changes in the electronic structure and fundamental parameters such as the energy gap (E_{LH}) between the lowest unoccupied (LUMO) and the highest occupied (HOMO) molecular orbitals, ionization potential (IP), electron affinity (EA), work function [10–17]. It is known that infinite armchair carbon nanotubes must have zero energy gaps because the difference of chirality indexes equals to zero [18–20]. Rochefort et. al [12], Buonocore et al. [13], Yumura et al. [17], Cioslowski et al. [11] showed that us-SWCNTs (6, 6) and (5, 5) with the length L<2.5 nm have nonzero energy gaps with oscillation dependences on the length of nanotubes. Energy gap oscillations on the length for a range of open-ended armchair us-SWCNTs with the chiral index n=3–15 were revealed in works [14, 15]. Numerous investigations of the electronic structure of armchair us-SWCNTs devoted to the singlet state of nanotubes [10–15] however other spin states hitherto not been studied. The investigation of the spin-dependent properties of us-SWCNTs is important for spintronics, optoelectronics and molecular electronics. Electric and magnetic fields can be used for the spin manipulation through the spin-orbit interaction in carbon nanotubes [21, 22]. It is expected that properties and fundamental parameters of us-SWCNTs strongly depend on their spin states. The aim of this work is the theoretical study of the electronic structure of us-SWCNTs (5, 5) at singlet and triplet states.

2. Computational details

The numerical simulations of the electronic structure of the capped armchair us-SWCNT (5, 5) were carried out using hybrid DFT (density functional theory) method of Becke [23] and Lee, Yang, Parr [24] (B3LYP) and 6-31G basis set [25] with the help of the Gaussian09 program package [26] in the Supercomputing center of Voronezh State University. This level of theory widely used for the investigation of the electronic structure of carbon nanomaterials at its ground and excited states and allows to achieve a good agreement of the calculated energy gap, ionization potential, electron affinity with experimental data [11, 27–29].

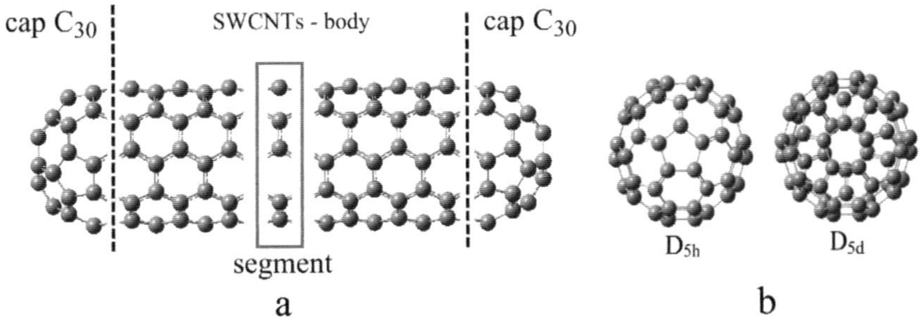

Figure 1. The structure of the ultrashort single-walled carbon nanotube (5, 5) (a) and the mutual orientation of C_{30} caps of nanotubes D_{5h} and D_{5d} symmetries

A stoichiometric formula of the capped SWCNT (5, 5) is C_{60+10j}, where j denotes the number of segments which forms the body of the nanotube while caps essentially are halves of the fullerene C_{60} dissected perpendicularly to the C_5 axis (figure 1) [10, 11]. The sequential increment of j causes to the rotation of caps of the us-SWCNT (5, 5) on the $\pi/5$ angle relatively to each other, thus the number of segments j determines symmetry of nanotubes: D_{5h}(i=2m+1) or D_{5d}(i=2m) m=0, 1, 2 ... The increment of j leads to the nanotube length increase on 1.23 Å. The diameter of the us-SWCNT (5, 5) equals to the 7 Å.

3. Results and discussion

3.1. The energy gap oscillation of the capped us-SWCNT (5, 5) in its singlet state

The dependence of the energy gap E_{LH} on the length of the us-SWCNT (5, 5) shows clear oscillations (solid line in the figure 1). The value of E_{LH} significantly exceeds the energy gap of the infinite-length SWCNT (5, 5) Eg=30meV [15, 30, 31]. The quantum confinement of the electron along the tube axis leads to the non-monotonic change of the energy gap from 2.8 to 0.7 eV in the range of the number of segments j=0–13. Hence, the single nanotube (5, 5) in the range of ultra small lengths forms a whole family of semiconductor nanomaterials. Three types of us-SWCNT (5, 5) with the maximal, intermediate and minimal energy gaps can be distinguished. These include nanotubes with the number of segments j=3m+1, 3m, 3m+2 (m=0, 1, 2…).

The existence of three types of armchair nanotubes is due to the electron density redistribution of LUMO and HOMO [12, 14, 15] and rights for capped, open-ended and single capped us-SWCNTs [32]. Thereby, the electronic structure of us-SWCNTs (5, 5) almost doesn't depend on the boundary condition and primarily determines by the number of segments. That also confirms by the proximity of oscillation amplitudes of us-SWCNT (5, 5) and graphene nanoribbon with the same structure [33].

3.2. The electronic structure of the capped us-SWCNT (5, 5) in its triplet state

At the triplet state of us-SWCNTs (5, 5) the number of electrons with the spin "up" more on two than with the spin "down". Thereby spins are not compensated and energy gaps for two spin channels $E_{LH}^{\uparrow}$ and $E_{LH}^{\downarrow}$ are different. The dependence of these energy gaps on the length of us-SWCNTs (5, 5) are nonmonotonic (figure 2). In the range of the number of segments j=0–13 energy gaps $E_{LH}^{\uparrow}$ and $E_{LH}^{\downarrow}$ belong to the interval from 0.5 to 2.0 eV. At the triplet state nanotubes of three types with the number of segments j=3m+1, 3m, 3m+2 have minimal, intermadiate and maximal energy gaps. Thereby, dependences of energy gaps on the number of segments of us-SWCNTs(5, 5) at singlet and triplet states are antipase (figure 2).

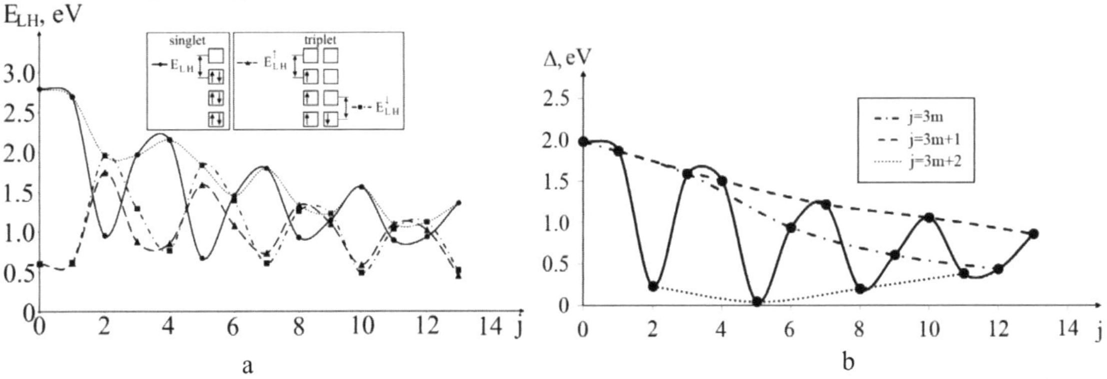

Figure 2. The Enegry gap E_{LH} of singlet (solid line) and spin-dependent $E_{LH}^{\uparrow}$ (dash line) and $E_{LH}^{\downarrow}$ (dot-dash line) of triplet states of us-SWCNT(5, 5) versus the number of segments j (a). The maximum gap E_{LH} is shown by the dot line. The energy of the singlet-triplet transition of us-SWCNT (5, 5) (b)

The energy gap is a measure of the stability of molecular systems. At this rate, at the singlet state nanotubes with the number of segments j=(3m+1) are more stable, than nanotubes of the (3m+2) type and vice versa at the triplet state. The figure 2 shows that energy gaps at the triplet state are more than at the singlet state of us-SWCNTs (5, 5) with the number of segments j=3m+2. Hence, the ground state of this type of nanotubes is triplet and singlet for 3m and 3+1 types. This results agrees with Fowler [34, 35] who shows that among isomers of carbon clusters C_n (n=60+6k, k>1) and clusters with the five-fold symmetry axis C_n (n=70+30k or n=84+36k, k>0) at least one isomer with the closed electron shells exists. In the case of capped us-SWCNT (5, 5) the first and second type of carbon clusters are nanotubes with the number of segments j=3m and 3m+1. The third type of clusters at k=5j+1 (j=0, 1, 2…) are nanotubes of the 3m type. Which implies, that nanotubes with j=3m+2 have open electron shells at the ground state.

With the aim of the investigation of singlet-triplet transition total energies E_{tot} of us-SWCNTs (5, 5) at two spin states S=0 and S=1 were compared. Remarkably, the difference of the total energies $\Delta=E_{tot}(S=0)-E_{tot}(S=1)$ is positive in the all over range of the number of segments (figure 2b). Hence, the ground state of us-SWCNTs (5, 5) is singlet, which contradicts to Fowler et al. [34, 35]. The energies of the singlet-triplet transitions of three types of nanotubes correspond to near (for 3m and 3m+1) and far (for 3m+2) infrared (IR).

4. Conclusions

The size confinement in the capped armchair us-SWCNT (5, 5) causes to the non-zero size-dependent energy gap between lowest unoccupied and the highest occupied molecular orbitals. There are three types of us-SWCNTs (5, 5) with the number of segments j=3m, 3m+1, 3m+2 that have the intermediate, maximal and minimal energy gap at the singlet state and the intermediate, minimal and maximal energy gap at the triplet state of nanotubes. It was revealed that the ground state of us-SWCNTs (5, 5) is the state with closed electron shells. The energy of the singlet-triplet transition corresponds to the energy value of visible and IR-radiation that makes us-SWCNTs promise materials for optoelectronic and spintronic devices.

Acknowledgments

This work was supported by RFBR, research projects N 14-02-31315 mol_a and by Targeted Federal Program "Research and development in priority directions of scientific-technological complex of Russia in 2014-2020 years", government contract No 14.574.21.0112.

References

[1] Sanchez-Valencia J R, Dienel T, Gröning O, Shorubalko I, Mueller A, Jansen M, Amsharov K, Ruffieux P, Fasel R 2014 *Nat. Lett.* **512** 61
[2] Kato T, Hatakeyama R 2010 *ACS Nano* **4** 7395
[3] Liu L, Yang C, Zhao K, Li J, Wu H-C 2013 *Nature Comm.* **4** 3989
[4] Price B K Lomeda J R, Tour J M 2009 *Chem. Mater.* **21** 3917
[5] Chen Z Kobashi K, Rauwald U, Booker R, Fan H, Hwang W-F, Tour J M 2006 *J. Am. Chem. Soc* **128** 10568
[6] Chattopdhyay J. Cortez F J, Chakraborty S, Slater N K H, Billups W E 2006 *Chem. Mater.* **18** 5864
[7] Jariwala D, Sangwan V K, Lauhon L J, Marks T J, Hersam M C 2013 *Chem. Soc. Rev.* **42** 2824
[8] Pascale-Hamri A, Perisanu S, Derouet A, Journet C, Vincent P, Ayari A, Purcell S T 2014 *Phys. Rev. Lett.* **112** 126805
[9] Island J O, Tayari V, Yiğen S, McRae A C, Champagne A R 2011 *Apll. Phys. Lett.* **99** 243106
[10] Tuchin A V, Nestrugina A V, Bityutskaya L A, Bormontov E N 2014 *J. of Phys.: Conf. Ser.* **541** 012008
[11] Cioslowski J, Rao N and Moncrieff D 2002 *J. Am. Chem. Soc.* **124** 8485
[12] Rocherfort A, Salahub D R and Avouris 1999 *J. Phys. Chem.* B **103** 641
[13] Buonocore F, Trani F, Ninno D, Matteo A Di, Cantele G and Iadonisi G 2008 *Nanotech.* **19** 025711
[14] Wang B-C, Wang H-W, Lin I-C, Lin Y-S, Chou Y-M and Chiu H-L 2002 *J. Chin. Chem. Soc.* **50** 939
[15] Tuchin A V, Ganin A A, Zhukalin D A, Bitytskaya L A and Bormontov E N 2014 *Recent Adv. In Biomedical & Chem. Eng. and Mat. Sc.* **1** 40
[16] Lu D, Li Y, Rotkin S V, Ravaioli U, Schulten K 2004 *Nano Lett.* **4** 2383
[17] Yumura T, Hirahara K, Bandow S, Yoshizava, Iijima S 2004 *Chem. Phys. Lett.* **386** 38
[18] Saito R, Fujita M, Dresselhaus G and Dresselhaus M 1992 *Appl. Phys. Lett.* **60** 2204
[19] Minitmire J 1995 *Carbon* **33** 893
[20] Dresselhaus M S, Dresselhaus G and Saito R 1995 *Carbon* **33** 883
[21] Klinovaja J, Schmidt M J, Braunecker B and Loss D 2011 *Phys. Rev.* B **84** 085452
[22] Ingerslev Jørgensen H I, Grove-Rasmussen K, Wang K-Y, Blackburn A M, Flensberg K, Lindelof P E, Williams D A 2008 *Nat. Phys.* **4** 536
[23] Becke A D 1993 *J. Chem. Phys.* **98** 5648
[24] Lee C, Yang W, Parr R G 1988 *Phys. Rev. B* **37** 785
[25] Ditchfield R, Hehre W J, Pople J A 1971 *J. Chem. Phys.* **54** 724
[26] http://www.gaussian.com
[27] Wu J, Hagelberg F 2010 *Phys. Rev. B* **81** 155407
[28] Parker S F, Bennington S M, Taylor J W, Herman H, Silverwood I, Albers P, Refson K 2011 *Phys. Chem. Chem. Phys.* **13** 11192
[29] Tuchin A V, Bityutskaya L A, Bormontov E N 2015 *Eur. Phys. J. D* **69** 1
[30] Bormontov E N, Ganin A A and Bitytskaya L A 2012 *Semiconductors* **13** 1573
[31] Bormontov E N, Ganin A A and Bitytskaya L A 2012 *Proc. of SPIE* **8700** 870011
[32] Tuchin A V, Bityutskaya L A, Bormontov E N 2014 *Proc. of Voronezh State University. Series: Physics. Mathematics* **3** 59
[33] Son Y W, Cohen M L, Louie S G 2006 *Phys. Rev. Lett.* **97** 216803
[34] Fowler P, Batten R, Manolopoulos D 1991 *J. Chem. Soc. Faraday Trans.* **87** 3103
[35] Fowler P W 1991 *J. Chem. Soc. Faraday Trans* **87** 1945

Purification non-aqueous solution of quantum dots CdSe-CdS-ZnS from excess organic substance-stabilizer by use PE-HD membrane

K Kosolapova[1], A Al-Alwani[2], I Gorbachev[2, 3] and E Glukhovskoy[2]

[1]Faculty of Nano- and Biomedical Technologies, Saratov State University, Russia

[2]Department of Educational and Research Institute of Nanostructures and Biosystems, Saratov State University , Russia

[3]Department of Colloids and Interfaces, Max Planck Institute of Colloids and Interfaces, Germany

E-mail: feelg8d@yandex.ru

Abstract. Recently, a new simple method for the purification of CdSe-CdS-ZnS quantum dots by using membrane filtration, the filtration process, successfully separated the oleic acid from quantum dots through membranes purification after synthesis; purification of quantum dots is a very significant part of post synthetical treatment that determines the properties of the material. We explore the possibilities of the Langmuir–Blodgett technique to make such layers, using quantum dots as a model system. The Langmuir monolayer of quantum dots were then investigated the surface pressure-area isotherm. From isotherm, we found the surface pressure monolayer changed with time.

1. Introduction

The interest of nanoparticles – semiconductor quantum dots, – it was continued to increase in last decades. Quantum dots have a scientific interest due to unique physical properties. Purification of quantum dots – it's very important process after synthesis. Purified quantum dots (QDs) are the foundation for the creation of light-emissive devices.

Nanoparticles rectify it's very difficult due to small size (2-5 nm). As well it's difficult due to a small selection of suitable solvents, irreversible adsorption with other materials, aggregation. Aggregation nanoparticles lead to a change in the size of the system elements. Change in the size of the system to make difficult selection of membrane with pore to a certain size.

Is now used many methods of purification [1]. In this work presented the results of studied that show the possibility of purification by membrane filtration solutions QDs in chloroform and toluene. Present results of experimental use one of low-cost materials for the membrane. This filter with nanopores - its high density polyethylene. It was used for separation of organic molecules of the stabilizer (oleic acid) from initial solution QDs.

2. Experiment

Purification of QDs, stabilized oleic acid, was carried out with using a specially assembled unit (see. Fig. 3) for 168 hours. Isotherm compression of the external solution were removing for observation and diffusion control during this time.

A series of pilot experiments to select a suitable membrane. There were three options materials – cellulose dialysis membrane, plastic films of low and high density (LD-PE and HD-PE). As a result of the experiments was chosen high density polyethylene.

Cellulose dialysis membrane and low density polyethylene film were excluded for various reasons. Firstly, they have pores too large: there through QDs permeate. Secondly, the dialysis membrane is most convenient for the filtration of aqueous solutions. In organic solvents, dialysis membrane retains fragility, resulting in the formation of microcracks at the bend and the loss of functionality.

The solution of colloidal quantum dots stabilized oleic acid in chloroform to about 10^{-6} M concentration of 125 μl was placed in a dialysis bag which was immersed in the box with chloroform. The internal volume of the solution in the package was 125 μl (Fig. 1a), the volume of the external solution in a bank – 50 ml (Fig. 1b).

As a vessel for carry through a membrane filter was taken jar reagent plastic screw cap 100 ml.

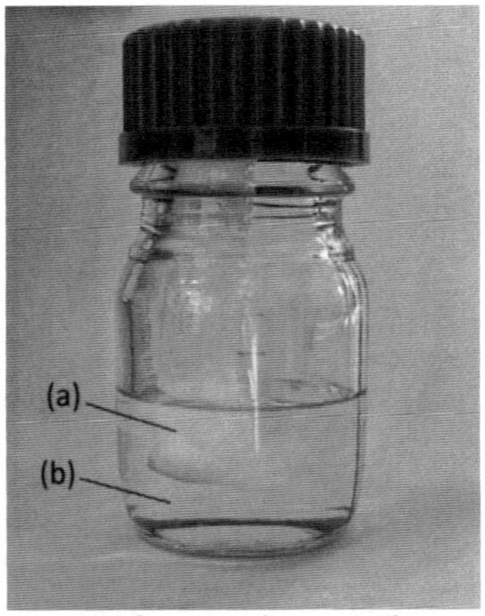

Figure 1. Chemical reaction apparatus for carrying out the process of separation: QD membrane solution bag (a), chloroform (b).

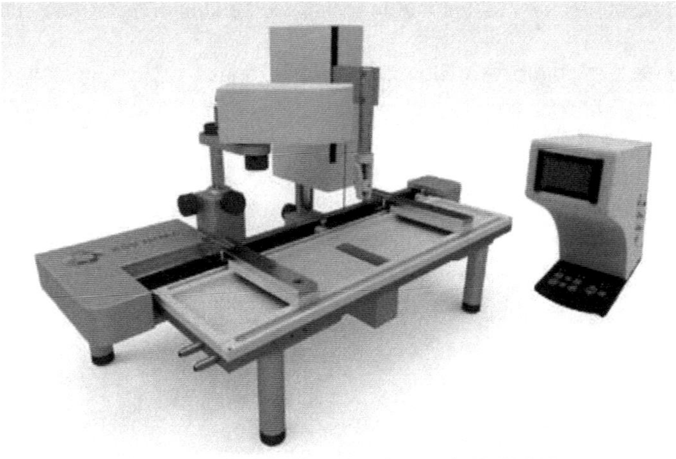

Figure 2. KSV Nima LB Trough KN2002.

Control and monitoring the diffusion of the surfactant through the membrane was done by measuring the surface pressure (isothermal compression) of the external solution. To measure the surface pressure setting used KSV Nima LB Trough KN2002 (Fig. 2) [2].

3. Results

Compression isotherms were recorded every 24 hours. For cleanliness before each measurement surface degreasing bath and purified with chloroform. As subphase used deionized water with a resistivity of 18 MOhm × cm. On the surface of the water each time make the same amount of material from the outer volume (of the solution in a glass container) – 800 μl.

After evaporation of the solvent (8 minutes) and the monolayer was compressed isothermally movable barriers at a constant rate – 8 cm^2/min.

The data obtained during the compression isotherm measurement shown in (Fig. 3). It is noticeable that the increases of maximum surface pressure with increasing filtration time, (Fig. 4). Obviously the Reducing of length of gas phase of external solution is due to increase the number of surfactant molecules in the external volume of the solution. This growth is due to the process of diffusion of unbound molecules of oleic acid through polyethylene membrane.

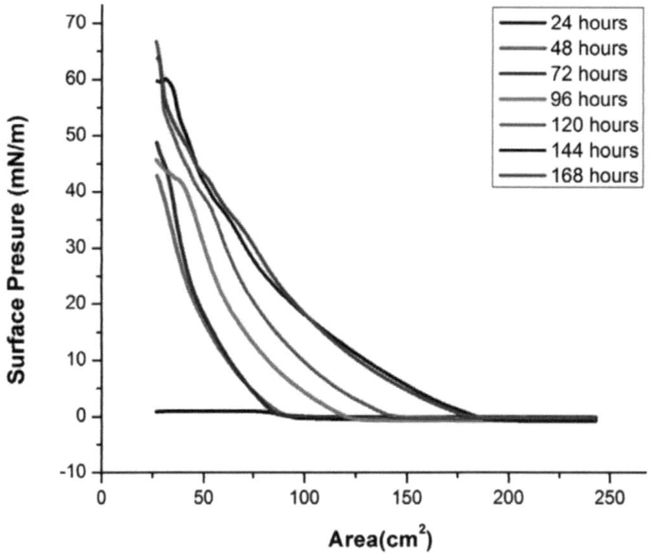

Figure 3. Compression isotherms surfactants of external solution.

When comparing images obtained by the method of Brewster angle microscopy, it can be seen that with increasing filtration time increases the amount of surfactant in the external volume. Monolayers were formed from solutions past 48 hours (Fig. 5) and 168-hour filtration (Fig. 6).

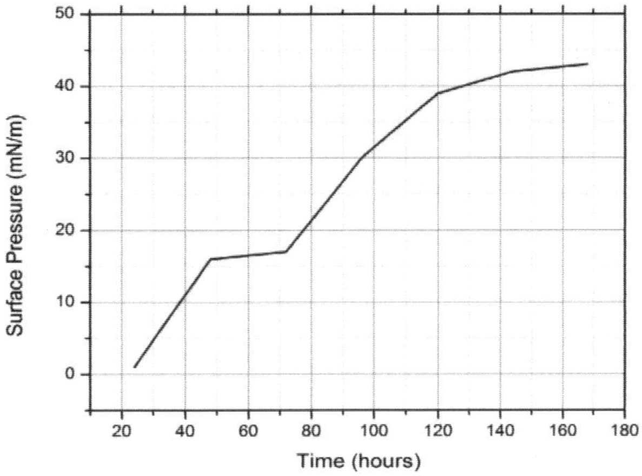

Figure 4. The growth of surface pressure over time.

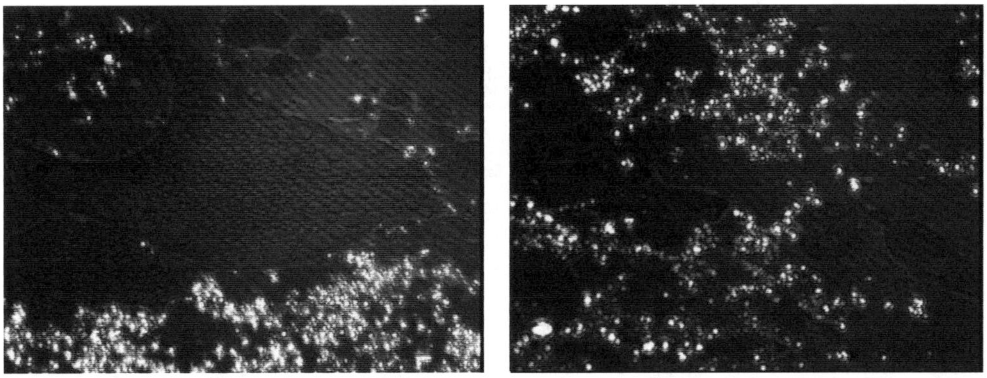

Figure 5. Images BAM outer monolayer of the solution after filtration of the solution to RT for 48 hours.

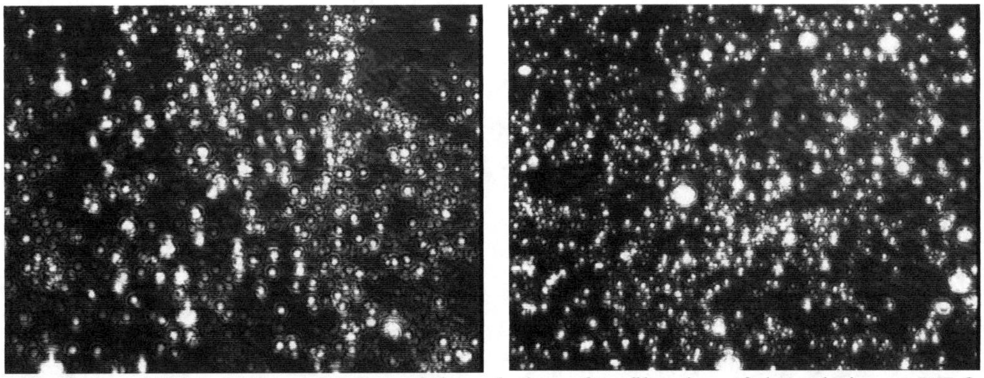

Figure 6. Images BAM outer monolayer of the solution after filtration of the solution at RT for 168 hours (right).

Thus, a method of purifying a solution of quantum dots, CdSe-CdS-ZnS by membrane filtration of unbound molecules from the excess oleic acid [2,3].

4. Conclusions

To investigate the changes in the number of surfactant in the external volume of the solution used Langmuir-Blodgett technique. In the process of compression isotherms were taken the outside solution during the membrane filtration of the solution of quantum dots. In the solution the unbound surfactant molecules leaving from a stock solution of QDs. Isotherms obtained at significantly increasing the maximum surface pressure and reduce the length of the gas phase in monolayers with increasing time filtering solution QDs scan, which shows an increase in surfactant concentration on the surface of the water surface.

Gratitude

The work is supported by grant № 14-12-00275 of Russian Science Foundation and Saratov State University

References

[1] Kowalczyk B., Lagzi I., Grzybowski B.A., Nanoseparations: Strategies for size and/or shape-selective purification of nanoparticles // Current Opinion in Colloid & Interface Science. 2011. Vol. 16. P. 135–148.

[2] Kim T., Organic-solvent resistant ultrafiltration and nanofiltration membrane modules for separation and purification of nanoparticles // Diss. PhD in Chemistry. Oregon State University. 2011. 107 p.

[3] Munir A., Dead end membrane filtration. / Instructor: Dr. Syed A. Hashsham // Michigan State University. Laboratory Feasibility Studies in Environmental Engineering. ENE 806. - Spring, 2006, 36 p.

Increasing of AlGaAs/GaAs quantum well robustness to resonant excitation by lowering Al concentration in barriers

I A Solovev, V G Davydov, Yu V Kapitonov, P Yu Shapochkin, Yu P Efimov, V A Lovcjus, S A Eliseev, V V Petrov and V V Ovsyankin

St. Petersburg state university, Ulyanovskaya 1, Petrodvorets, St. Petersburg, 198504, Russia

E-mail: Solivan2007@yandex.ru

Abstract. The robustness of AlGaAs/GaAs quantum well to strong resonant excitation is studied. It was found that lowering Al concentration in barriers to 3% does not influence the measured radiative linewidth of exciton resonance in the sample at low intensities. At the same time parasitic broadening of the resonance by an additional resonant illumination is strongly suppressed as compared to the quantum well with 30% of Al in barriers.

1. Introduction

III–V semiconductor heterostructures with quantum wells (QW) are very attractive for research because of theoretical possibility to use them as an all-optical logic element[1] due to the strong resonant interaction of excitons in QW with light. We have demonstrated that the molecular beam epitaxy technique (MBE) allows one to produce high-quality samples with the inhomogeneous broadening comparable to the radiative linewidth of excitonic resonance for both $In_{0.02}Ga_{0.98}As/GaAs$[2] and $GaAs/Al_{0.3}Ga_{0.7}As$[3] QWs.

There are limitations of direct use of $GaAs/Al_{0.3}Ga_{0.7}As$ structures in optical information processing due to the reversible bleaching which manifests itself as an additional broadening of exciton's spectral line under strong resonant excitation. This effect accumulates with characteristic times in microseconds scale and longer, and eventually vanishes after termination of the resonant illumination.

In this work we have shown that the bleaching effect can be reduced through lowering aluminum concentration in barriers from 30% to 3% while maintaining the same spectral quality of the sample.

2. Experimental

The sample T670 is grown by MBE technique. It contains a 14 nm GaAs QW between $Al_{0.03}Ga_{0.97}As$ barriers.

We use the reflection spectroscopy to investigate coherent optical properties of excitons with heavy holes (HH-exciton). The *p*-polarized probe light is incident on the sample surface at the Brewster's angle[2, 3] in order to eliminate nonresonant reflection off the surface (Fig. 1).

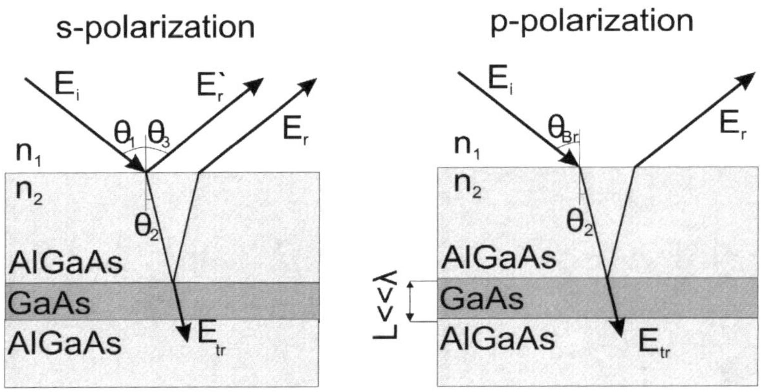

Figure 1. The scheme of propagation polarized light through a sample.

In this case the reflection spectrum could be described by the Lorentzian:

$$K_R(\omega) = \frac{\Gamma_R^2}{(\omega - \omega_0)^2 + (\Gamma_R + \Gamma_{NR})^2}, \tag{1}$$

where ω_0 is the resonant frequency, Γ_R is the radiative linewidth, Γ_{NR} denotes all nonradiative broadening of various kinds combined together and K_R is the experimentally determined absolute reflection coefficient.

Sample temperature was 8K. Probe intensity was weak enough to guarantee the linear response. Spectra were recorded using femtosecond Ti:sapphire laser as a spectrally broad source of probe light and 0.5 m spectrograph with multichannel CCD-detector.

In order to study the bleaching effect we use an additional CW monochromatic semiconductor laser which was carefully tuned to the HH-exciton resonance frequency. Fig. 2a illustrates the dependence of the resonant reflection (RR) spectra $K_R(\omega)$ on the pump intensity I_{pump}. It should be noted that there is no significant shift of spectral line, so thermal broadening is negligible (fig. 3). By quantitative fitting the spectra with Eq. (1) (modified by addition of a weak nonresonant background) we obtain both quantities Γ_R and Γ_{NR}.

3. Discussion

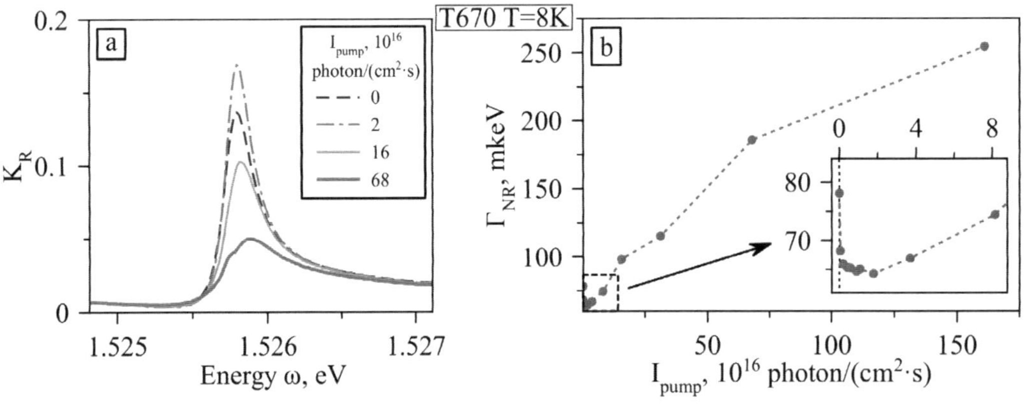

Figure 2. Reflection spectra of the T670 sample at different intensities of the resonant pump (a) and dependence of the nonradiative linewidth on that intensity (b).

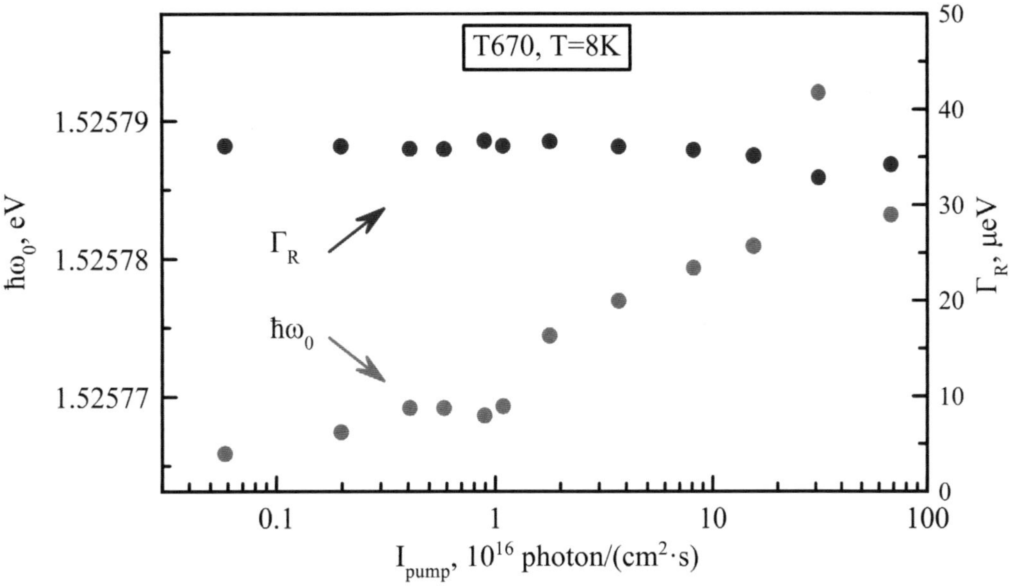

Figure 3. Dependence of the spectral position of excitonic resonanse $\hbar\omega_0$ and radiative width Γ_R on the pump intensity $I_{\rm pump}$.

It has been experimentally established that radiative width remains constant in wide range of pump intensity and equals $36 \pm 3\,\mu$eV (Fig. 3). This value is similar to the Γ_R obtained for the sample E296 which has been investigated in a previous work[2] and contains a 13 nm GaAs QW between $Al_{0.3}Ga_{0.7}As$ barriers. Observed ratio $\Gamma_{NR}/\Gamma_R = 0.46$ in linear limit $I_{\rm pump} \to 0$ proves the high quality of the sample T670 (for comparison: in the sample E296 the ratio Γ_{NR}/Γ_R is approximately the same).

However considerable nonradiative broadening appears only for rather strong pump intensities approaching 10^{18} photons/cm^{2}s which are two orders of magnitude higher than that in E296 sample(Fig. 2b).

We attribute such behavior to the weaker effects of free charge carriers in this system, which are believed to cause strong scattering of the excitons and thus deteriorating their optical resonance.

An indirect proof of this fact is the absence of noticeable spectral features at energies slightly lower than the HH-exciton resonance that correspond to the creation of bounded charged excitons (trions) by excitation of an exciton in the vicinity of the free charge carrier. This trion lines were readily observed in E296 sample. Nevertheless we couldn't totally rule out charge effects from consideration due to the observed small decrease of Γ_{NR} at low pump intensities (see inset Fig. 2b) due to the charge sign reversal, which is typical for photovoltaic phenomena in GaAs quantum wells[4].

4. Conclusion

We have investigated resonant reflection spectra from QW with additional resonant pumping. Quantitative analysis demonstrate that lowering Al concentration in barriers to 3% lets us to decrease the bleaching effect, while spectroscopic quality of the sample being as high as that of classical sample with 30% aluminum.

Acknowledgments
The reported study was partially supported by RFBR, research project No 15-52-12016 nnio_a. This work was carried out on the equipment of the SPbU Resource Center "Nanophotonics" (http://photon.spbu.ru)

References

[1] Gerlovin I Ya, Ovsyankin V V, Stroganov B V and Zapasskii V S 2000, *Nanotechnology* **11**, 383–6.
[2] Poltavtsev S V, Efimov Yu P, Dolgikh Yu K, Eliseev S A, Petrov V V and Ovsyankin V V 2014 *Solid State Communications* **199**, 47–51.
[3] Poltavtsev S V and Stroganov B V 2010, *Physics of the Solid State* **52**, 1899–905.
[4] Glasberg Sh, Finkelstein G, Shtrikman H and Bar-Joseph I 1998 *Phys. Rev.* B **59**, R10425–8.

Measurements of doping density in InAs by capacitance-voltage techniques with electrolyte barriers

D Frolov, G Yakovlev and V Zubkov

Department of micro- and nanoelectronics, St. Petersburg Electrotechnical University "LETI", Prof. Popov str. 5, St. Petersburg 197376, Russia

E-mail: frolovds@gmail.com

Abstract. The doping densities in n-InAs structures were studied by means of capacitance-voltage technique using electrolyte to form Schottky-like contact. It was shown that in heavily doped InAs ($> 10^{18}$ cm^{-3}) the depletion approximation can be used to obtain the true doping concentration. Concentration in low doped InAs can be estimated by simulation (using modified Thomas-Fermi approximation). Measured doping densities were compared with concentration obtained by Hall measurements. The difference between CV and Hall results in undoped samples was explained.

1. Introduction

Capacitance-voltage (CV) methods are widely used for characterization of semiconductor materials and structures [1–3]. Conventional CV measurements are performed by forming metal Schottky contact on semiconductor surface. However, in some semiconductor materials like InAs and InN the formation of reliable Schottky contact is difficult due to carrier accumulation at the surface, in this case an electrolyte can be used to form rectifying contact [4, 5]. Another advantage of electrolyte barriers, used in electrochemical capacitance-voltage (ECV) profiling technique, is the ability to perform controlled dissolution of the semiconductor [6, 7].

The interval of applied biases, in which depletion capacitance dominates, is very important for determination of doping densities by CV method with electrolyte barrier. In wide bandgap materials ($E_g > 2$ eV) this interval usually spans over more than 1 V and can be much higher for many semiconductors like Si, SiC, GaAs and GaSb at lower doping levels (approaching up to more than 6 V) [8].

The InAs, as a narrow band gap material, is very different in this regard. Gopal et al. [9] have reported on the ECV profiling of n-InAs. They noted that measured carrier concentration in highly doped InAs corresponds well to Hall measurements. For undoped InAs epi-layers it was noted that the measured CV concentration was much higher in comparison to Hall concentration, but no explanation was given.

The purpose of the present work is to explain the mismatch between CV and Hall results in undoped InAs samples and to determine the optimal parameters of CV measurements for doping density extraction in n-InAs.

Journal of Physics: Conference Series **643** (2015) 012086 doi:10.1088/1742-6596/643/1/012086

2. Experimental and Simulation technique

Undoped InAs epi layers were grown by HVPE on n^+-InAs substrates. Doping densities were studied in undoped layers and in substrates. Capacitance-voltage measurements were performed using ECVPro profiler (Nanometrics). LCR meter Agilent E4980A was additionally connected directly to electrochemical cell to perform capacitance measurements in a wider frequency range. The aqueous solution of 0.2 M H_2SO_4 was used to form electrolyte barrier with nominal contact area of 0.1 cm^2.

CV characteristics were simulated by modelling potential and carrier distributions using Poisson equation with modified Thomas-Fermi approximation (MTFA) [10,11]. MTFA was used as an alternative approach to the self-consistent solution of Poisson and Schrödinger equations. To calculate capacitance the potential distribution must be found first. The potential as function of depth z must satisfy 1D-Poisson equation:

$$\frac{d^2\varphi(z)}{dz^2} = -\frac{q}{\varepsilon\varepsilon_0}\left[N_D^+ - N_A^- - n(z) + p(z)\right], \tag{1}$$

where ε is the dielectric constant of InAs, $n(z)$ and $p(z)$ are density of free electrons and holes. The donor (N_D) and acceptor (N_A) concentration were assumed constant throughout the space charge span range.

The distribution of electron density within MTFA was calculated as:

$$n(z) = \int_0^\infty \rho_c(z,E) f_{FD}(E) f_{MTFA}(z,E)dE, \tag{2}$$

where $f_{FD}(E)$ is the Fermi-Dirac function and $\rho_c(z,E)$ is density of state (DOS) function, which selected taking into account the conduction band nonparabolicity [12, 13]:

$$\rho\left(z,E\right) = \frac{1}{2\pi^2}\left(\frac{2m_e^*}{\hbar^2}\right)^{3/2}\sqrt{E}\cdot\sqrt{1+\alpha E}\cdot(1+2\alpha E). \tag{3}$$

Here $\alpha = (1 - m_e^*/m_0))^2/E_g$ is the nonparabolicity coefficient, and m_e^* is the effective mass of electrons.

The $f_{MTFA}(z,E)$ function in equation (2) is the MTFA correction function used to account boundary condition for the wave function at the semiconductor surface:

$$f_{MTFA}(z,E) = 1 - sinc\left(\frac{2z}{L}\left(\frac{E}{k_BT}\right)^{1/2}(1+\alpha E)^{1/2}\right), \tag{4}$$

where $L = \hbar/\left(2m_e^* k_B T\right)^{1/2}$ is the thermal length.

We numerically solved Poisson equation (1) using Dirichlet boundary conditions with $\varphi(\infty) = 0$. The capacitance per unit area was determined from charge increment over two applied voltages which differ by small step ΔV:

$$C = \frac{\Delta Q}{\Delta V} = \frac{\varepsilon\varepsilon_0 \Delta F_s}{\Delta V},$$

where $F_s = d\varphi(z)/dz|_{z=0}$ is the surface electric field.

3. Results and discussion

Experimental and simulated capacitance-voltage curves are shown in figure 1. All voltages in electrochemical cell were applied with respect to the reference Pt electrode. The main feature of the measured characteristics in comparison to ones for the many other materials (e.g. GaAs in figure 2c) is a strong deviation of C^{-2} vs. V plot from linear, which behave very similar

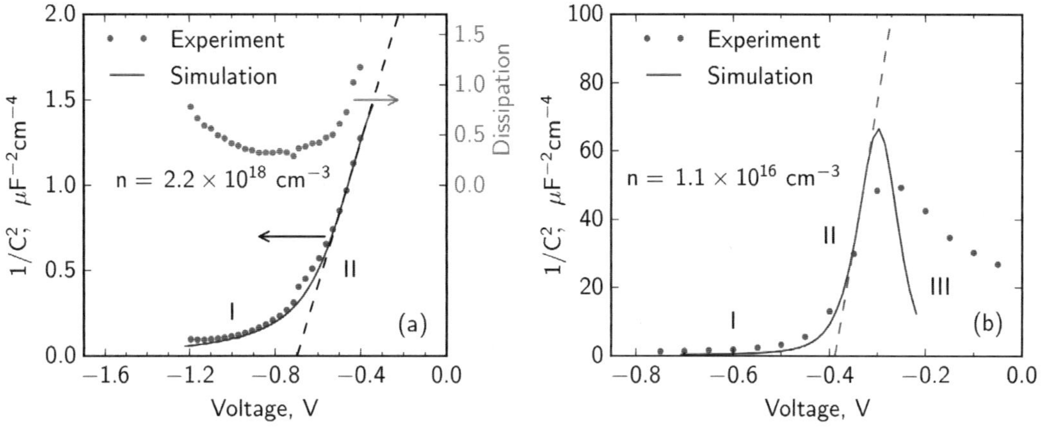

Figure 1: Mott-Schottky plot of experimental data and simulation for: (a) n-InAs substrate with $N_D = 2 \times 10^{18}$ cm^{-3} and (b) unintentionally doped InAs epi-layer ($n = 1 \times 10^{15}$ cm^{-3}). The electron densities calculated from linear fitting at the region II (dashed line) are also shown. The Roman numerals depicts: I – accumulation, II – depletion, III – inversion.

to CV characteristic of metal-oxide-semiconductor (MOS) structures. At lower biases, in the region of low positive slope of Mott-Schottky plot (region I in figure 1), the surface potential bends downwards which causes the formation of accumulation layer and increase of free carrier density. At higher positive voltages a depletion takes place. Here (region II) the slope of the Mott-Schottky plot is constant and proportional to the doping density. Further increase of bias leads to inversion which is characterized by a negative slope in C^{-2} vs. V curve (region III in 1b). The width in Volts of the transition region from depletion to inversion affects the accuracy of doping density measurements.

Another feature of InAs is its band position inside aqueous solution of electrolyte depicted in figure 2a. In this figure the band edge position of InAs is determined experimentally from Mott-Schottky plot as the difference between open circuit potential and flat-band potential

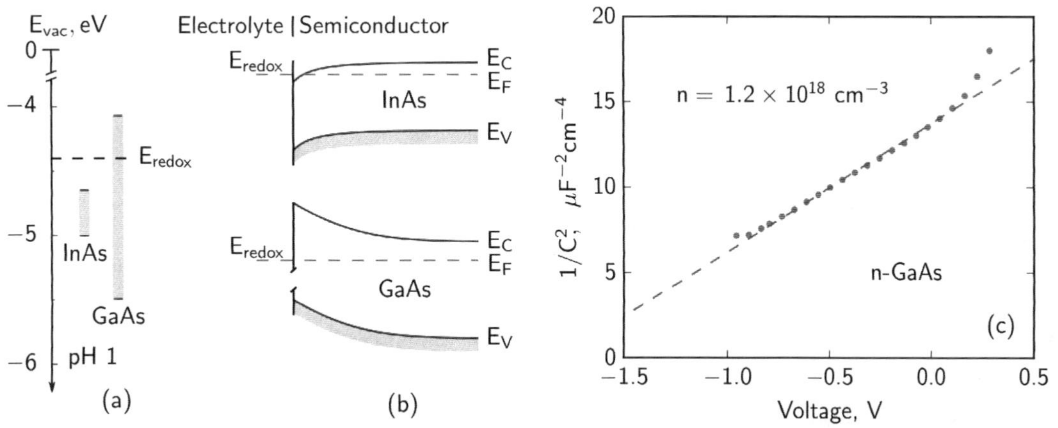

Figure 2: (a) Position of energy bands at the surface of InAs and GaAs in aqueous solution; (b) band diagrams of n-InAs and n-GaAs in equilibrium; (c) Mott-Schottky plot for n-GaAs.

relative to redox potential of electrolyte, the GaAs band edge position is given elsewhere [8]. It can be noted that in InAs the bottom of conduction band is located below the electrochemical potential (E_{redox}) of electrolyte which causes formation of accumulation layer in equilibrium (for $N_D < 10^{18}$ cm^{-3}), whereas in GaAs depletion arises (figure 2b) at any concentration.

Doping densities used in the simulation were obtained from magnetoresistance (for epi structure) and Hall effect measurements (for substrate) performed at 77 K. For n$^+$-InAs substrate the measured CV has a good agreement with simulation results (figure 1a). The dopant concentration of 2×10^{18} cm^{-3} was used for heavily doped n-InAs in simulation. Extraction of doping densities is usually carried out in voltage range, where dissipation factor is minimal. In our case it will give overestimated value of electron density (see figure 1a). To make accurate extraction of doping densities the voltage region, corresponded to depletion, should be used. The electron concentration calculated from linear approximation of C^{-2} vs. V curve in the depletion region gives 2.2×10^{18} cm^{-3}.

The difference between the dopant concentration used in the simulation and the calculated electron density is due to the small bias range of the depletion region. This difference is much higher for epi-layer with doping level of 1×10^{15} cm^{-3} (figure 1b) because the Fermi level shifts toward the center of forbidden energy gap, and the inversion starts earlier.

4. Conclusions

The measurements and simulations have shown that in heavily doped n-InAs ($n > 10^{18}$ cm^{-3}) the range of biases, in which the depletion occurs, is wide enough to use the depletion approximation for accurate doping density estimation. At the lower doping levels standard analysis of Mott-Schottky plot does not give the correct value of dopant concentration. In this case simulation of capacitance-voltage characteristics should be used to estimate doping density. The main reason of the mismatch between the true doping concentration and the concentration observed in CV measurement is the small band gap of InAs, leading to premature start of inversion. In general, to get accurate doping concentration in n-InAs one should choose the CV parameters, which results in formation of depletion region in wider voltage span range.

Acknowledgments

The investigation were carried out at Resource Center for Solid State Physics of SPbETU. In addition, authors would like to thank Dr. A.S. Petrov for fruitful discussion.

References

[1] Barnes P 2002 Capacitance–Voltage (C-V) Characterization Of Semiconductors *Characterization of Materials* (John Wiley & Sons, Inc.) pp 579–589
[2] Kucherova O V, Zubkov V I, Tsvelev E O, Yakovlev I N and Solomonov A V 2011 *Inorg. Mater.* **47** 1574–1578
[3] Hurni C A, Kroemer H, Mishra U K and Speck J S 2012 *J. Appl. Phys.* **112** 083704
[4] Yim J W L, Jones R E, Yu K M, Ager J W, Walukiewicz W, Schaff W J and Wu J 2007 *Phys. Rev. B* **76** 41303
[5] Vasheghani Farahani S K, Veal T D, Mudd J J, Scanlon D O, Watson G W, Bierwagen O, White M E, Speck J S and McConville C F 2014 *Phys. Rev. B* **90** 155413
[6] Blood P 1986 *Semicond. Sci. Technol.* **1** 7–27
[7] Zubkov V, Kucherova O, Frolov D and Zubkova A 2013 *Phys. Status Solidi C* **10** 342–345
[8] Memming R 2001 *Semiconductor electrochemistry* (John Wiley & Sons)
[9] Gopal V, Chen E H, Kvam E P and Woodall J M 2000 *J. Electron. Mater.* **29** 1333–1339
[10] Veal T D, Piper L F J, Schaff W J and McConville C F 2006 *J. Cryst. Growth* **288** 268–272
[11] Paasch G and Übensee H 1982 *Phys. Status Solidi B* **113** 165–178
[12] Ariel-Altschul V, Finkman E and Bahir G 1992 *IEEE Trans. Electron Devices* **39** 1312–1316
[13] Bebb H B 1971 *J. Appl. Phys.* **42** 3189–3194

Dependence of microwave-excitation signal parameters on frequency stability of caesium atomic clock

Petrov A.A., Davydov V.V., Vologdin V.A., Zalyotov D.V.
Peter the Great Saint-Petersburg Polytechnical University, Saint-Petersburg, Russia

E-mail: Alexandrpetrov.spb@yandex.ru

Abstract. New scheme of the microwave – excitation signal for the caesium atomic clock is based on method of direct digital synthesis. The theoretical calculations and experimental research showed decrease step frequency tuning by several orders and improvement the spectral characteristics of the output signal of frequency synthesizer. A range of generated output frequencies is expanded, and the possibility of detuning the frequency of the neighboring resonance of spectral line that makes it possible to adjust the C-field in quantum frequency standard is implemented. Experimental research of the metrological characteristics of the quantum frequency standard on the atoms of caesium - 133 with new design scheme of the microwave – excitation signal showed improvement in daily frequency stability on $1.2*10^{-14}$.

1. Introduction

Most of modern technologies based on the use of precision measuring instruments. The most accurate measurements are frequency measurements. Therefore, modernization of existing and development of new frequency generation devices are some of the most relevant physical and technical problems of modern science [1-3]. High-precision atomic clocks have a high importance in science and technology and a vast area of their application. So these devices are requirements for high accuracy and reliability of measurements.

A quantum frequency standards on the atoms of caesium ^{133}Cs have a special place among the atomic clocks. These standards are used as a clock generators in the communications equipment and in a data transmission devices, applied in the satellite navigation systems GLONASS and GPS as a clock generators and in the various metrological services. Also these standards perform a role of the reference signals with high precision and stability in radio equipment [3 - 5]. Due to the high accuracy and reliability quantum frequency standards on atoms of caesium ^{133}Cs are applied in the positioning systems of underwater devices.

With the development of scientific - technical progress operating conditions of caesium atomic clocks are constantly changing. Therefore new requirements for measurement accuracy, reliability and weight - dimensional characteristics are produced to frequency standard. This leads constantly upgrade existing and develop new models of caesium atomic clocks. For this aims it is reasonable to carry out research, develop new methods and find a new design solutions based on the latest appearing electronic components and discovered physical phenomena.

Development and commissioning of new atomic clock are very long and costly process, which in most cases is not enough funds and time. Therefore, in most cases, for specific tasks related to the operating conditions of frequency standards modernization is occur. The process of modernization of frequency standards includes various directions: change the weight and dimensions, reduced energy consumption, improved metrological characteristics. And for frequency standards characterized by the fact that modernization may not be for the whole construction and may be for individual units or blocks.

In present work one of the directions of modernization of the cesium atomic clock is considered. A new implementation of a digital frequency synthesizer for atomic clocks is presented. Improved microwave – excitation signal parameters leads to improvement of characteristics of quantum frequency standard in particular the frequency stability.

2. Frequency synthesizer of the cesium atomic clock

The work of a caesium atomic clock is based on the principle of adjustment of the frequency of a highly stable crystal oscillator to quantum transition frequency of atoms of cesium-133 [3]. Figure 1 shows a block diagram of a caesium atomic clock

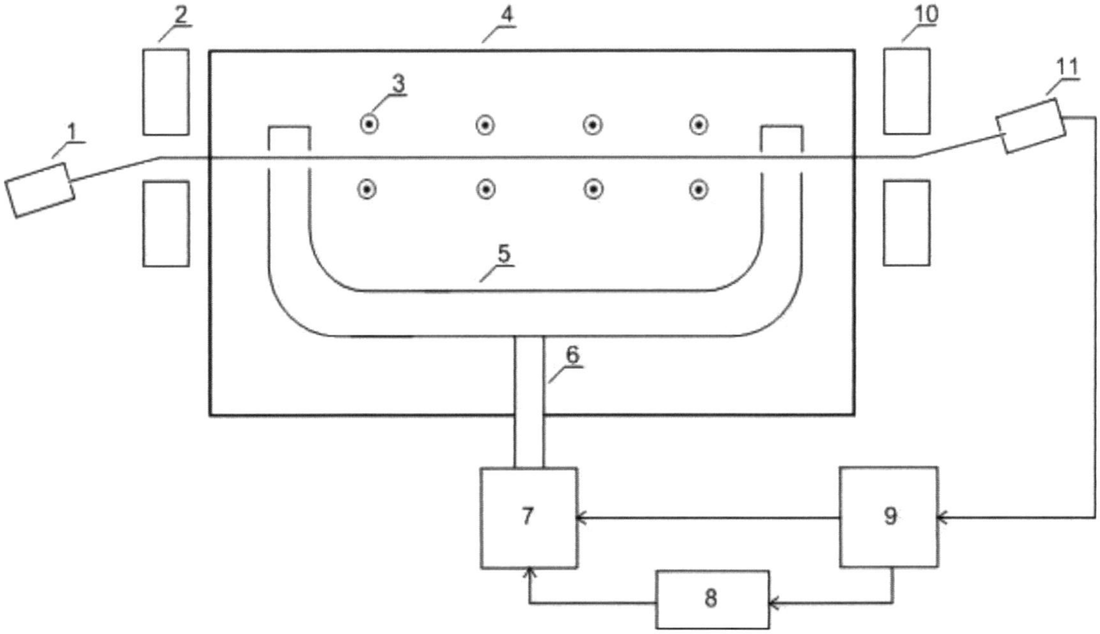

Figure 1. Block diagram of a caesium atomic clock. 1 - a source of caesium atoms, 2 - magnet polarizer 3 - magnetic field 4 - magnetic shield 5 - Ramsey resonator 6 - the waveguide 7 - frequency converter 8 - crystal oscillator 9 - automatic frequency control system 10 - magnet analyzer, 11 – detector

The output signal frequency of 5MHz of the crystal oscillator 8 is supplied to the frequency converter 7. Frequency converter consists of the frequency synthesizer, the mixer signals and the multiplier signals. In the frequency synthesizer input signal frequency of 5MHz is converted to the signal frequency of 12,631772 MHz and supplied to the input of mixer signals. In the multiplier signals input signal frequency of 5 MHz is multiplied to the frequency of 270 MHz and then to frequency of 9180 MHz. This signal frequency of 9180 MHz is also supplied to the input of mixer signals. As a result, the output signal of the frequency converter is the signal of ultrahigh frequency of 9192,631772 MHz. This signal is supplied in the waveguide 6 and then on the input of atomic beam tube.

In caesium atomic clock with the help of magnet polarizer 2 the atoms are prepared such that they are either in the F=4, m_F=0 or in F=3, m_F=0 state. Afterwards the atoms interact with an electromagnetic field that induces transitions into the former unoccupied state. The atoms in this state are detected and allow one to determine the frequency of the interrogating field where the transition probability has a maximum. The observed transition frequency is corrected for all known frequency offsets that would shift the transition frequency from the unperturbed transition and is used to produce a standard frequency or pulse per second every 9192631772 cycles [3].

Scanning the frequency ν of the atomic resonance leads to a detector current like the one shown on the figure 2. The signal shows the Ramsey resonance structure on a broader, so-called, Rabi pedestal.

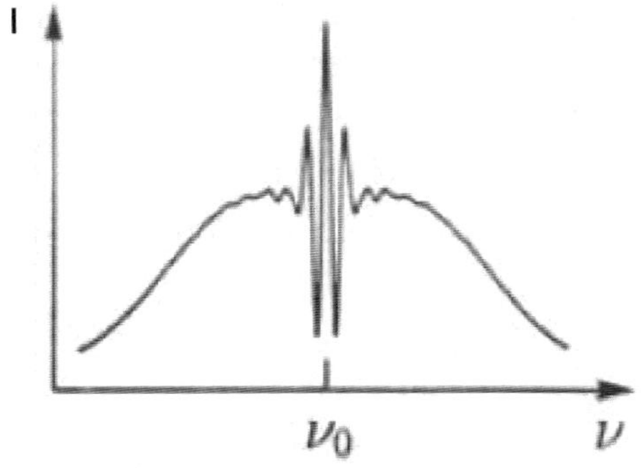

Figure 2. Ramsey resonance structure on the Rabi pedestal

The central feature with the maximum at the transition frequency ν_0 is used to stabilize the frequency of the crystal oscillator to the atomic transition frequency. To this end, the frequency from the synthesizer is modulated across the central peak. The signal from the detector is phase-sensitively detected in the automatic frequency control system, integrated and this signal is used for stabilising the frequency of the crystal oscillator.

Improvement the accuracy of the signal frequency of 12,631770 MHz results in better accuracy of the resonant frequency of the atomic transition [3-5].

The main characteristic of the frequency synthesizer is its ability to impact the characteristic of frequency stability of the output signal of quantum frequency standard. Frequency instability introduced by the synthesizer is determined by the lateral discrete spectrum components of the signal that occurs in dividing, multiplying, mixing frequency signals, the accuracy of the generated frequency, and the impact on the signal of natural and technical noise.

Experimental study showed up that the present method of generating the output signal of the frequency synthesizer needs to increase the accuracy. The large resolution of step frequency is necessary. New scheme of the frequency synthesizer is designed using direct digital synthesis (DDS - Direct Digital Synthesis) [6]. This method allows to generate the output signal of the synthesizer with accuracy more than 10^{-6} Hz.

The implementation of our proposed method enables us to control the frequency of the output signal frequency synthesizer in real time. This ensures a high rate of frequency tuning, which makes it possible to more efficiently adjust the crystal oscillator frequency in contrast the previously used scheme.

The application of direct digital synthesis gave the possibility of obtaining the generated frequencies in a wide range (0-3MHz), in contrast to previous schemes, where this feature was absent [7, 8].

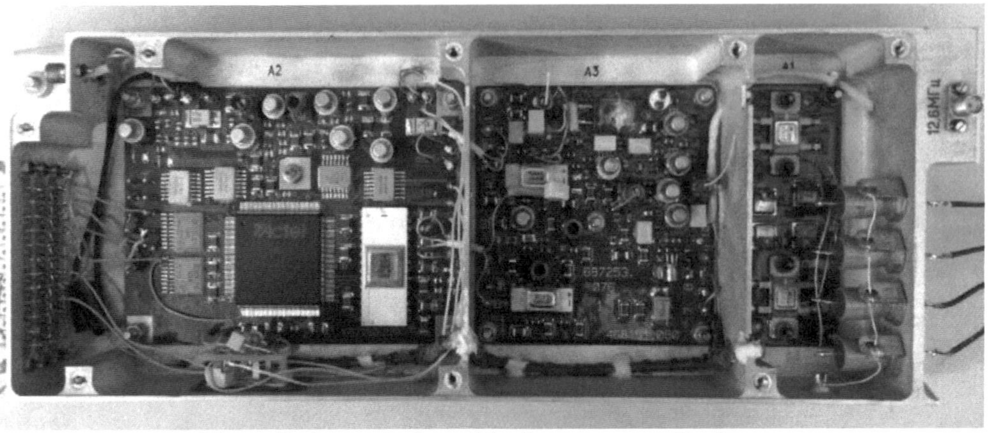

Figure 3. A new design of the frequency synthesizer

In figure 3 a new design of the frequency synthesizer is presented.

3. Experimental results and discussion

Experimental study of the output signal parameters of a new scheme frequency synthesizer, confirmed the simulation results and showed great advantages in comparison with previously used scheme.
In figure 4, as an example, oscillograms measured in the band of 1 kHz of the output signal of a new scheme (a) and a previously used (b) of the frequency synthesizer are presented.

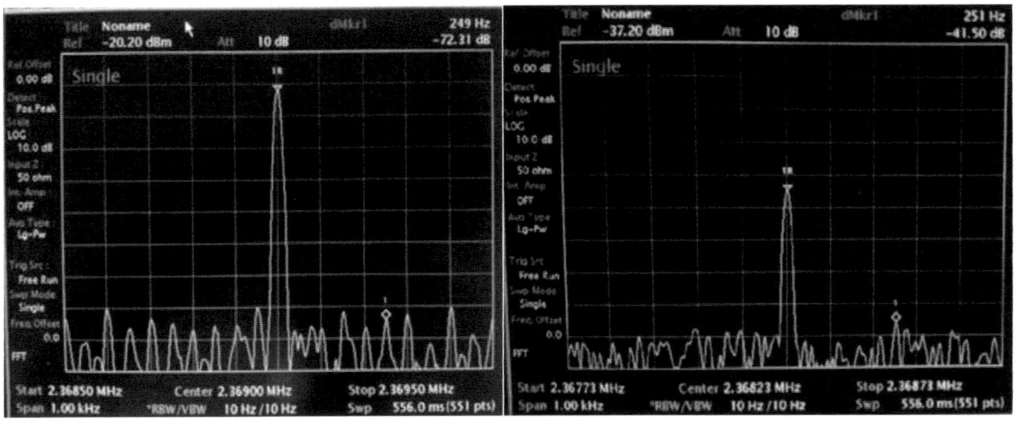

a) b)

Figure 4 Suppression of the amplitude of the lateral components in the band of 1 kHz.

The experimental results show that the suppression of lateral discrete components in the spectrum of microwave-excitation signal in the band of 1 kHz is improved by 75%.
The shift of the peak of the line resonance atomic transition Δf is arising due to there are lateral discrete components in the microwave-excitation signal spectrum. The shift is defined by the following expression [3]:

$$\frac{\Delta f}{f} = \frac{A}{I}\frac{(\delta f)^2}{f(f-f_s)}, \quad (1)$$

where A is the amplitude of the lateral components, I is the amplitude of the carrier, δf is the width of the spectral line, $(f - f_s)$ is the detuning of the lateral components relative to peaks of the spectral line.

In the developed scheme of the frequency synthesizer the suppression of lateral components is not less than -72 dB, spectral line width is equal 500 Hz and the band of the lateral components is equal 1 kHz. Then relative frequency shift of the atomic transition is equal $\frac{\Delta f}{f}$ =2.2.10^{-14}. Previously obtained result is equal $\frac{\Delta f}{f}$ =1.62.10^{-13} Hz.

By reducing the frequency shift of the atomic transition Δf the more fine-tuning on the center of the resonance line is occur. This leads to a more accurate determination of the value of the nominal output frequency of frequency standard and, consequently, improves the long-term frequency stability of a cesium atomic clock.

4. Conclusion

Experimental study of frequency synthesizer showed improvement parameters of the microwave-excitation signal, such as the step of frequency tuning, time of the frequency tuning, range of generated frequencies, phase stability and spectral characteristics. Experimental research of the metrological characteristics of the quantum frequency standard on the atoms of caesium - 133 with new scheme of the microwave – excitation signal showed improvement in daily frequency stability on 1.2*10^{-14}.

Reference

[1] Davydov V V, Dudkin V I and Karseev A U 2013 *Optical Memory & Neural Networks (Information Optics).* **112–117** 237
[2] Davydov V V and Semenov V V 1999 *Radiotekhnika I Elektronika.* **1528–1531** 1965
[3] Riehle F 2004 *Frequency standards. Basics and applications.* 496
[4] Oduan K and Gino B 2002 *Chronometry and basics of GPS.* 400
[5] Dudkin V I and Pachomov L N 2012 *Quantum electronics. Devices and their applications.* 422
[6] Ridiko L I 2005 *Components and Technology.* **27-39** 83
[7] Petrov A A.and Davydov V V 2014 *J. Proc of the International youth scientific forum "LOMONOSOV-2014".* **187–188** 232
[8] Petrov A A.and Davydov V V 2014 *Proc of the 52nd international scientific student conference "ISSC - 2014"* **15** 101

Study of colloidal particles Fe_mO_n-SiO_2 synthesized by two different techniques

K G Gareev[1], S A Ionin[1], D V Korolev[2], V V Luchinin[1], V A Moshnikov[1,3], M F Panov[1] and N V Permyakov[1]

[1]Saint Petersburg Electrotechnical University "LETI",
5th Prof. Popova Str., Saint-Petersburg, 197376, Russia
[2]Institute of Experimental Medicine, Federal North-West Medical Research Centre,
15th Parkhomenko Str., Saint-Petersburg, 194156, Russia
[3]Saint Petersburg Polytechnic University,
29th Politechnicheskaya Str., Saint-Petersburg, 195251, Russia

kggareev@yandex.ru, agorange@mail.ru

Abstract. Colloidal solutions Fe_mO_n-SiO_2 were obtained by two different techniques based on sol-gel process: silica cores partially coated by iron oxide particles and iron oxide cores coated by silica. Dried layers were investigated by atomic force microscopy and size distribution was plotted. Chemical contain of dried specimen was analysed by FTIR-spectroscopy. Magnetic properties of colloidal particles obtained by both the techniques were measured by VSM. Resistivity's change under the influence of magnetic field was measured using LCR-meter.

1. Introduction

Composite magnetic colloids could be applied for solving biomedicine problems such as MRI diagnostics [1] or blood vessel controlled occlusion [2], including oncology [3]. Synthesis of superparamagnetic nanoparticles may be provided by sol-gel technique [1]. Using this technological route authors obtained aggregate stable specimens starting to settle under the influence of constant magnetic field [4]. Aggregation and sedimentation processes were studied by MRI relaxation times' and AC signal resistivity measurements [1, 4]. Authors of [5] investigated process of iron oxide phase transformation from magnetite/maghemite to hematite initiated by laser radiation at wavelength of 532 nm. Also oxidation of magnetite/maghemite particles may be caused by the absence of the outer coating so using of one-stage technique based on the sol-gel process should give better result. However the biological inertness of two-stage derived colloids is probably much higher because of silica heat treatment at 300 °C for 30 minutes in comparison to the presence of 0.04 vol. % untreated tetraetoxisilane in the one-stage derived colloids.

Intensive heating of iron oxide nanoparticles by the influence of laser radiation is used in biomedical application. Photothermal effects of laser heating iron oxide nanoparticles in cartilaginous tissues are studied in work [6]. Presence of magnetite nanoparticles allowed to achieve about 41 degrees Celsius in 1.5 seconds. Wavelength of the heating laser was determined by the absorbance spectra of an aqueous dispersion of nanoparticles. Thus using the absorbance spectra of the synthesized colloidal particles it is possible to choose the optimal laser wavelength. As it was shown in the work [7] UV, visible and near-infrared spectra of colloidal particles Fe_mO_n-SiO_2 obtained by two-stage technique doesn't have any intensive absorbance lines. In accordance to the work [8] authors tried to find some appropriate wavelength in IR range. Despite of higher absorbance by the biological tissue than, for example, at the terahertz range, IR lasers now are more widespread and can generate much higher optical power.

2. Experimental methods

In present work authors investigated colloidal solutions Fe_mO_n-SiO_2 derived by two different ways. The first one is based on magnetite precipitation in the presence of colloidal silica as described in [1] and the second one assumes magnetite precipitation followed by coating iron oxide particles by tetraetoxisilane at the alkaline aqueous medium. Estimated concentration of synthesized solutions equaled 3 mM/L. Analysis of particles' size distribution was made by NTEGRA TERMA AFM (NT-MDT). The average diameter was calculated using Gwyddion software basing on AFM images. Magnetic properties were investigated measuring DC-hysteresis curves using vibrating sample magnetometers Lake Shore 7410 and MPMS SQUID (Quantum Design). Resistivity's change under the influence of constant magnetic field 500 A/m at 1 kHz AC signal was measured by LCR-meter E7-20 (MNIPI). To analyze chemical contain of dried specimen FTIR-spectroscopy using Nikolet 6700 spectrometer was applied.

3. Experimental results

FTIR spectra of dried specimen obtained by one-stage process (figure 1) indicated the presence of magnetite. Peaks about 1100, 1400, 1600 and 3400 cm^{-1} are in conformity to the work [8] which describes FTIR-spectroscopy analysis of magnetite particles covered by oleic acid. The most intensive absorbance line at 1100 cm^{-1} (9.1 μm) may be used as the laser wavelength for heating the particles. Thus quantum cascade lasers and interband cascade lasers with the radiation range of 3.42…9.60 μm (2924…1042 cm^{-1}) may be used.

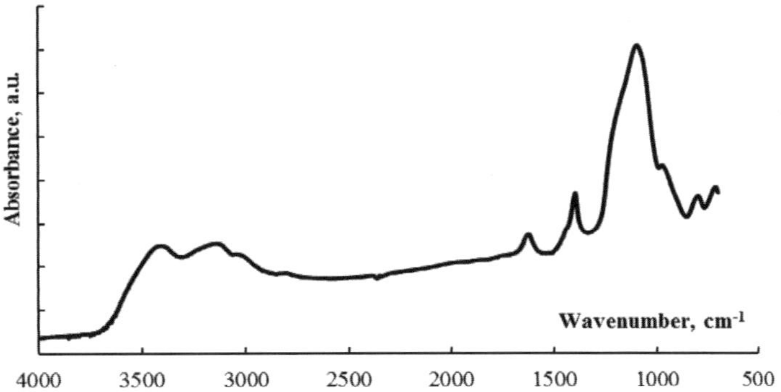

Figure 1. FTIR-spectra of dried specimen obtained by one-stage process

AFM images of the colloidal particles obtained by one and two-stage processes are shown at Figure 2.

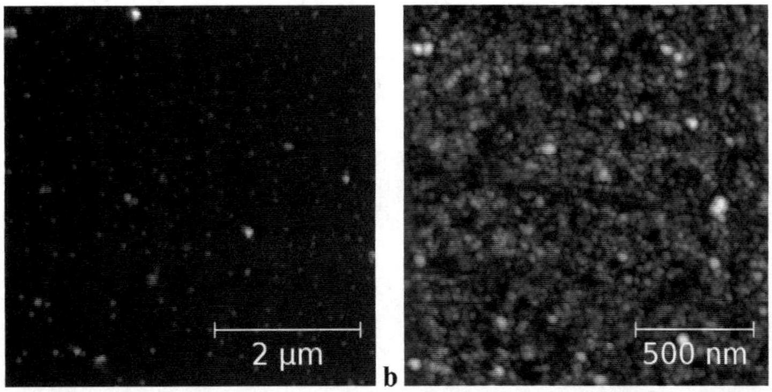

a b

Figure 2. AFM image of dried specimens obtained by one (a) and two-stage (b) processes

To obtain the image of standing alone particles initial specimens of 0.3 mM/L were diluted to 0.015 mM/L solutions and dried at room temperature. Size distribution graphs plotted using AFM images are shown at figure 3. The average particles diameter is about 15 nm for both the techniques although the one-stage process allows to obtain colloidal particles with narrower size distribution and to avoid any additional filtration.

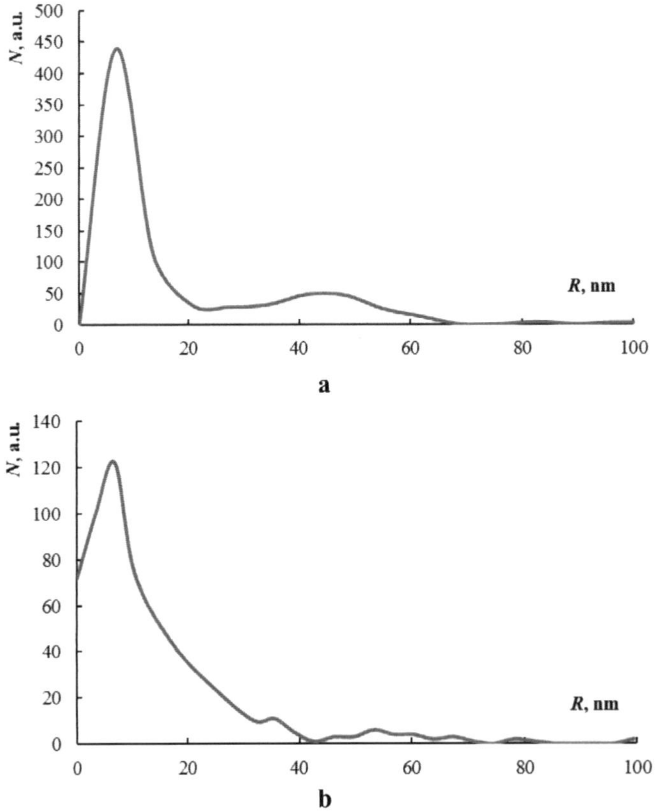

Figure 3. Size distribution for particles based on AFM image of dried specimen obtained by one (a) and two-stage (b) processes

DC-magnetization curves are shown at figure 4. Saturation magnetization for the one-stage process derived specimen is about five times higher probably because of inhibition of oxidation process in the solution. As it was stated in the work [5] colloids consisted of silica cores partially coated by iron oxide particles could intensively oxidize even without any heat treatment and therefore significantly decrease in magnetization.

Phase transformation of magnetite/maghemite to hematite could be one of the reasons leading to an increase of protons spin-spin relaxation time T_2 that is described in the work [1] but the main disadvantage of colloids obtained by two-stage process is intensive aggregation of particles under the influence of constant magnetic field. Aggregates provoke rapid sedimentation process and loss of colloidal stability. Nanoparticles Fe_mO_n-SiO_2 obtained by the one-stage process could be more chemically and aggregate stable due to the silica coating. As it can be stated from figure 4 relative resistivity change is slightly higher for the specimen obtained by the two-stage process. The following restoration of resistivity after the release of magnetic field is different for the specimens and its rate is much higher for the one-stage derived colloid.

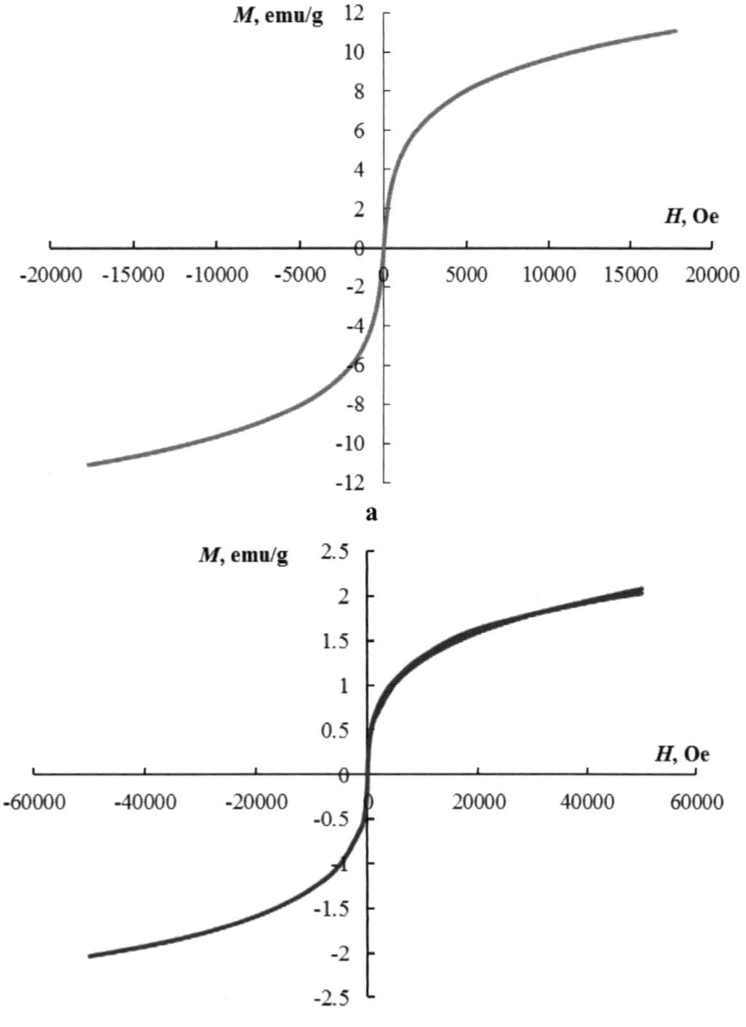

b

Figure 4. Magnetization curves of dried specimens obtained by one (a) and two-stage (b) processes

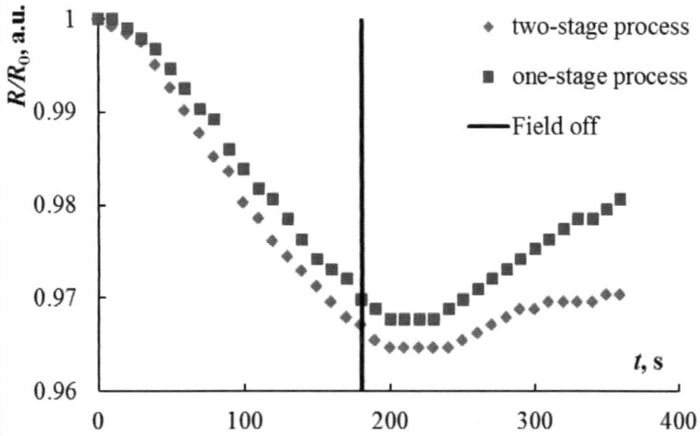

Figure 5. Time dependence of resistance under the influence of magnetic field for solutions obtained by one and two-stage processes

4. Conclusions

Colloidal particles Fe_mO_n-SiO_2 obtained by one and two-stage techniques have similar average diameter but the other properties differ significantly. One-stage process allows to achieve narrower size distribution of particles and higher saturation magnetization. According to the results of FTIR-spectroscopy silica coated iron oxide particles consist in the main of magnetite. The silica shell leads to higher aggregate stability as it was stated using the resistivity change measurements. However further experiments are needed to determine the toxicity of specimens obtained by both the techniques.

Authors thank A.I. Romanychev (director of Resource Centre "Innovative technologies of composite nanomaterials", Saint-Petersburg State University) and S.V. Lebedev (director of Centre for diagnosis of functional materials in medicine, pharmacology and nanoelectronics, Resource Centre of Saint-Petersburg State University) for providing VSM measurements of dried specimens Fe_mO_n-SiO_2 derived by one-stage and two-stage techniques accordingly.

The study was partially supported by RFBR, research project No. 14-03-31534.

References

[1] Bogachev Yu V, Chernenco Ju S, Gareev K G, Kononova I E, Matyushkin L B, Moshnikov V A, Nalimova S S 2014 Appl. Magn. Reson. **45** 329-337

[2] Smolkova I S, Kazantseva N E, Makoveckaya K N, Smolka P, Saha P and Granov A M 2015 Mat. Sci. and Eng. C **48** 632–641

[3] Kawai N, Kobayashi D, Yasui T, Umemoto Y, Mizuno K, Okada A, Tozawa K, Kobayashi T and Kohri K 2014 Vascular Cell **6** 15

[4] Gareev K G, Kononova I E, Levitckii V S, Moshnikov V A and Nalimova S S 2014 J. of Phys.: Conf. Ser. **572** 012027

[5] Al'myashev V I, Gareev K G, Ionin S A, Levitskii V S, Moshnikov V A, Terukov E I 2014 Physics of the Solid State **56**, *11* 2155–2159

[6] Gulyaev P Yu, Kotvanova P Yu, Pavlova P Yu, Sobol' E N, Omel'chenko A I 2012 Nanotechnologies in Russia **7**, *3-4* 127-131

[7] Bogachev Yu V, Gareev K G, Matyushkin L B, Moshnikov V A, Naumova A N 2013 Physics of the Solid State **55**, *12* 2431–2435

[8] Liu C, Shan Y, Zhu Y, Chen K 2009 Thin Solid Films **518** 324–327

Cu/Diamond composite heat-conducting shims

E N Galashov[1], A A Yusuf[1] and E M Mandrik[1]

[1] Department of Applied Physics, Scientific Research Department Novosibirsk State University, 2 Pirogov's Street, Novosibirsk 630090, Russia

E-mail: chessplanet86@yandex.ru

Abstract. Composite material with high thermal conductivity was obtained by the method of thermal sintering of a diamond (50 - 75 %) with a size of 20 to 250 μm in a matrix of copper.Coefficient of thermal conductivity of copper diamond composite materials was measured and is 450 – 650 $W \cdot m^{-1} \cdot K^{-1}$. The coefficient of thermal expansion CTE was measured and is 5.5 – 7.5 $\cdot 10^{-6}$/°C. The obtained copper diamond composite materials are promising objects for use in THz and microwave devices.

1. Overview

For the manufacture of heat sinks working under extreme heat loads in various areas of technology (electronics, nuclear energy, particle accelerators), materials with high thermal conductivity are required [1]. This problem can be solved through the creation of composite materials based on diamond because diamond has a thermal conductivity 1000-2600$W \cdot m^{-1} \cdot K^{-1}$ [2].The manufacturing of products entirely of diamond is technically difficult, and they have a very high cost [3]. So now diamond-containing composite materials with dispersed diamond filler, in particular metal matrix Ag, Cu, Al, are being developed actively [4].Diamond-metal composites with particles of a diamond of the 20-250 μm size with copper matrix have a high conductivity 600 – 950 $W \cdot m^{-1} \cdot K^{-1}$.

Along with the conductivity in the semiconductor device CTE is an important parameter. Semiconductor devices have a short lifespan if the heat sink substrate has a low thermal conductivity. Heat-conducting copper-diamond composites have CTE5.5 – 10 10^{-6}/°C and they have a long lifespan.

2. Fabrication

We have developed the method of obtaining diamond-copper composites with high thermal conductivity. The diamonds are covered with tungsten coating by the method of vapor deposition of tungsten carbonyl. The thickness of the coating is 0.5 – 1 μm. The coating composition may vary within W-W_2C-WC.Carbides provide adhesion of the coating to the diamond. Metallic tungsten and tungsten carbide are well moistened copper. Uncoated diamond is very poorly moistened by copper. In addition, tungsten and carbide are practically not dissolved in the melt of copper. Therefore, the tungsten coating on the diamond provides a copper-diamond composite by thermal sintering diamond in the matrix of copper.

Diamonds and copper were mixed and merged into a pill. The tablet was placed in the vacuum chamber of the furnace of the electrical resistance. The sintering temperature was 970 °C. The sintering time was 1 hour.

2.1. Equipment for coating of diamonds

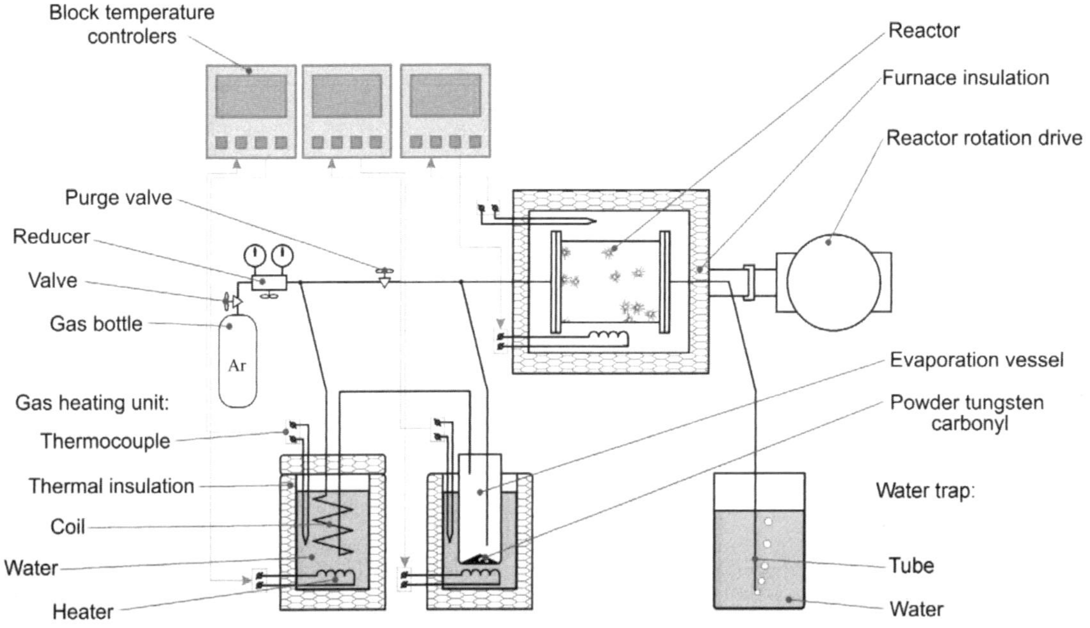

Figure 1. Diagram of the apparatus for applying coatings of tungsten.

Argon passes through the tube to the coil. Coil is heat by heater and water. Heated argon enters in evaporation vessel, where gas and vapor of tungsten carbonyl are mixed. Then argon with vapor of tungsten carbonyl moves in reactor, where diamonds are sited. Surface of diamonds are cleared from dirt due to pre-heating. Reactor with diamonds is rotated by reactor rotation drive and diamonds are coated composition of W-W_2C-WC. Then unnecessary gas residues are retired in water trap.

2.2. Equipment for synthesis of Cu/Diamond composite shims

Copper and diamonds coated with tungsten are mixed and pressed in steel press mold by mechanical press.

Figure 2. Steel press mold.

Figure 3. Pressed Cu/Diamond shims.

3. Results

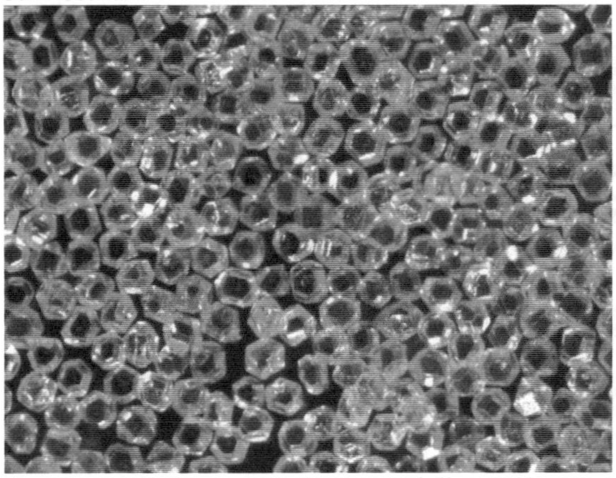

Figure 4. The original uncoated diamonds.

Figure 5. Covered diamonds.

By electron microscopy analysis was determined that covered diamonds have developed morphology. The coating has two types. The first is continuous. On top of a continuous coating islet coating is located.

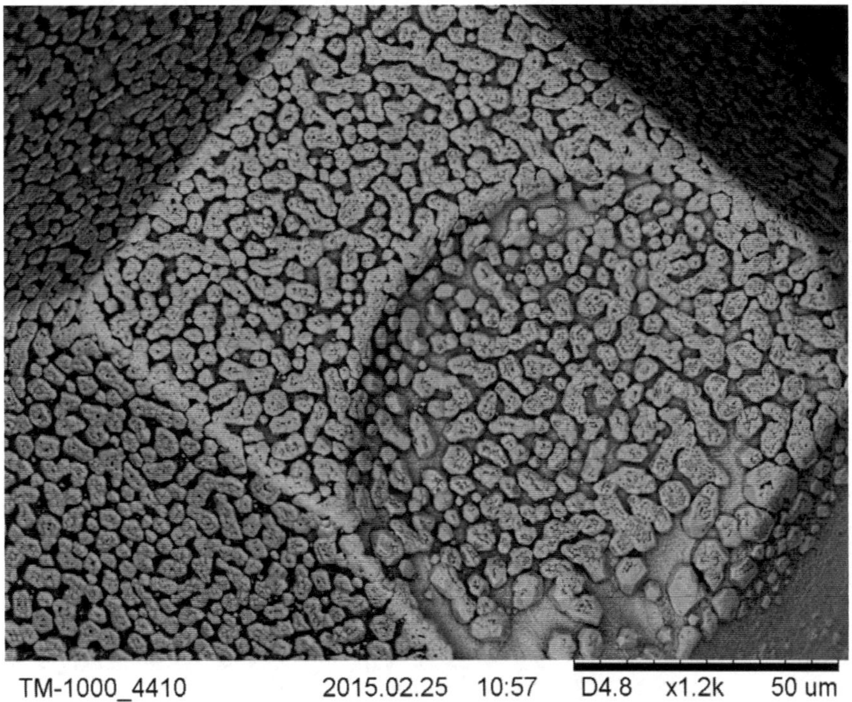

Figure 6. Electron microscopy of coated diamond with size 230 μm.

By X-ray analysis was determined that covered diamonds have diamond phase, tungsten carbide phase and tungsten phase. According to electron microscopy (Fig. 6) and X-ray diffraction (Fig. 7) is suggested that continuous coating has tungsten carbide phase and islet coating has tungsten carbide phase.

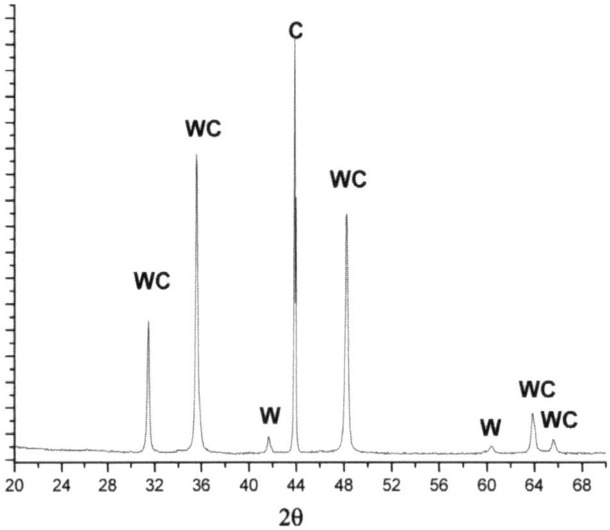

Figure 7. Diffraction pattern of coated diamond with size 230 μm.

Copper diamond tablet was obtained. Pores are practically absent, the surface is smooth. The thickness of the tablets is 1 mm. Coefficient of thermal conductivity of copper diamond composite materials was measured and is 450 – 650 W·m^{-1}·K^{-1}.The coefficient of thermal expansion CTE was measured and is 5.5 – 7.5 10^{-6}/°C.

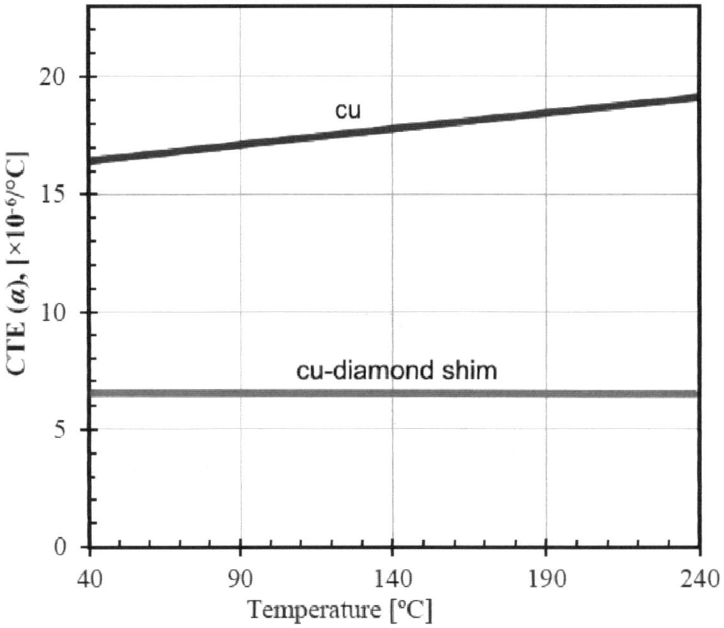

Figure 8. CTE of Cu and Cu-Diamond shim with diamond size 230 μm

References

[1] S.V. Kidalov, F.M. Shakhov 2009 Materials **2**, p2467
[2] J.S. Goela, S.L. Shinde 2006 High Thermal Conductivity Materials, Springer, N.Y., p 271
[3] Synthetic CVD diamond products from Element Six http://www.e6cvd.com
[4] ExtreMat —heat sink materials http://www.extremat.org/heat-sink-materials?Edition=en

The high-frequency conductivity of a thin semiconductor wire in the longitudinal magnetic field

O V Savenco[1], I A Kuznetsova[2], A A Yushkanov[3]

[1,2]Demidov Yaroslavl State Yniversity, ul. Sovetskaya 14, Yaroslavl, 150000, Russia
[3]Moskow State Regional University, ul. Radio 10a, Moskow, 105005, Russia

E-mail: [1]cryak92@mail.ru, [2]kuz@uniyar.ac.ru

Abstract. The problem of high-frequency conductivity of a thin semiconductor cylindrical wire at arbitrary temperature in the longitudinal magnetic field is solved by the kynetic method. The charge carrier reflection from inner wire surface is assumed to be diffuse. The ratio between the mean free path of charge carriers and the wire radius is supposed to be arbitrary. The results obtained for limiting cases of degenerate and nondegenerate semiconductor were subjected to comparative analysis.

1. Introduction

The electrical properties of conductive materials with the linear size comparable to the mean free path of charge carriers differ substantially from the corresponding properties of bulk materials [1]. Particularly a nontrivial dependence of conductivity on ratio R/λ (R is the wire radius, λ is the mean free path of charge carriers) appears.

In many conventional semiconductors at room temperature the mean free path of charge carriers $\lambda = 10 \div 1000$ nm, and de Broglie wavelength is $\lambda_B \sim 10$ nm. In high-conductivity metals we have $\lambda = 10 \div 100$ nm, and de Broglie wavelength is of the order of the interatomic distance: $\lambda_B \approx 0.3$ нм [2, 3]. Modern technology allows one to prepare the materials with the size of the order of several dozen nanometers. Therefore the situation when we can disregard the quantum confinement effects and take into account the classical size effects is actually occurs. It should be noted that in recent years because of rapid development of micro- and nanoelectronics the problems of thin wire conductivity become much more actual.

In work [1] the problem about the influence of surface scattering on the conductivity and the magnetic properties of thin films and wires is considered for the case of constant electrical and magnetic fields. In some works [4 – 7] the problems of the conductivity of thin wire with a rectangular [4] and circular [5 – 7] cross-section are solved. A metal wire was considered in works [4 - 7] and a semiconductor wire was considered in work [7]. In [8 – 10] the problem of the conductivity of a thin metal wire in longitudinal magnetic field was studied.

In this paper the expression of high-frequency conductivity of a thin semiconductor wire with a circular cross section in a longitudinal magnetic field at arbitrary temperature is obtained by the kynetic method. Charge carriers are assumed to be reflected from the wire boundary diffusely.

2. Conductivity calculation

We consider a cylindrical semiconductor wire with radius *R*, which is much smaller than length *L*. This wire located in a magnetic field with induction *B*. An alternating electrical voltage with frequence ω is applied to its ends. The directions of electrical and magnetic fields coinsides with the wire symmetry axis. A skin effect isn't considered (we assume that $R < \delta$, where δ is the skin layer depth).

A homogeneous time-periodic electric field

$$\mathbf{E} = \mathbf{E}_0 \exp(-i\omega t) \tag{1}$$

affects on the charge carriers and induces the deviation of the distribution function of those f_1 from the equilibrum Fermi-Dirac's distribution function f_0:

$$f(\mathbf{r},\mathbf{v},t) = f_0(\varepsilon) + f_1(\mathbf{r},\mathbf{v},t) = f_0(\varepsilon) + f_1(\mathbf{r},\mathbf{v})\exp(-i\omega t), \tag{2}$$

where $\varepsilon = m\upsilon^2/2$ is the kinetic energy of an electron (a hole) for the case of spherically symmetric energy band, **v** and *m* are the velocity and the effective mass of an electron (a hole), **r** is the radius vector of an electron (a hole).

The distribution function f_1 is found from the solution of the Boltzmann kinetic equation in the relaxation time approximation and in the linear approximation in the external field [11]:

$$-i\omega f_1 + \mathbf{v}\frac{\partial f_1}{\partial \mathbf{r}} + e\mathbf{vE}\frac{\partial f_0}{\partial \varepsilon} + \frac{e}{m}[\mathbf{v},\mathbf{B}]\frac{\partial f_1}{\partial \mathbf{v}} = -\frac{f_1}{\tau}, \tag{3}$$

where τ is the relaxation time.

We assume the diffuse reflection of charge carriers from the wire inner surface as a boundary condition [11]:

$$f_1(\mathbf{r}_\perp, \mathbf{v}_\perp, \mathbf{v}_z) = 0 \quad \text{at} \quad \begin{cases} |r| = R, \\ \mathbf{r}_\perp \mathbf{v}_\perp < 0, \end{cases} \tag{4}$$

where $\mathbf{r}_\perp$ and $\mathbf{v}_\perp$ are respectively the components of the electron (hole) radius vector $\mathbf{r}$ and velosity $\mathbf{v}$ in the plane perpendicular to the wire symmetry axis; $\mathbf{v}_z$ is the electron (hole) velocity component along the wire symmetry axis.

The kinetic equation (3) is solved by the method of characteristics [12]:

$$f_1(t') = -\frac{e(\mathbf{vE})}{\nu}\frac{\partial f_0}{\partial \varepsilon}[1 - \exp(-\nu t')], \tag{5}$$

where $\nu = \tau^{-1} - i\omega$ is the complex scatterring frequence, $t' = \varphi\delta/\upsilon_\perp$ is the electron (hole) movement time from the boundary, in which reflection happens, to the current point. The trajectory of charge carrier movement in the magnetic field represents a spiral line. This trajectory may not cross the wire boundary (in this case $t' \to \infty$). The projection of the trajectory in the plane perpendicular to wire symmetry axis looks as the arc of a circle. The arc radius is $\delta = m\upsilon_\perp/eB$, φ is the central angle respectiving to this arc.

Find the wire integral conductivity by the mean similar to the work [8] for the case of a metal wire:

$$G = G_0 P(x_0, y_0, \mu_0), \qquad G_0 = \frac{3ne^2R^3}{m\upsilon_1 L}, \tag{6}$$

$$P(x_0, y_0, \mu_0) = \frac{1}{3I_0 z_0} \int_0^1 \int_0^\infty \int_0^\infty \int_0^{2\pi} \xi \sqrt{U_z} \frac{\exp\left(U_z + U_\perp - U_\mu\right)}{\left[\exp\left(U_z + U_\perp - U_\mu\right) + 1\right]^2} \left[1 - \exp\left(-\frac{\varphi z_0}{\mu_0}\right)\right] d\alpha dU_\perp dU_z d\xi,$$

$$\varphi = 2\,\mathrm{arcctg}\left[\left(2\,\xi \frac{\sqrt{U_\perp}}{\tilde{\upsilon}_1}\cos\alpha - \sqrt{D}\right) \Big/ \mu_0\left(\xi^2 - 1\right)\right],$$

$$D = 4\frac{U_\perp}{\tilde{\upsilon}_1^2}(1 - \xi^2 \sin^2\alpha) - 4\xi\mu_0 \frac{\sqrt{U_\perp}}{\tilde{\upsilon}_1}\sin\alpha(1 - \xi^2) - \mu_0^2(1 - \xi^2)^2 .$$

Nondimensional parameters x_0, y_0, z_0, ξ, I_0, $U_\perp$, U_z, U_μ, also parameters υ_1, $\tilde{\upsilon}_1$ are definited in the work [7], $\mu_0 = eRB / m\upsilon_1$ is the nondimensional magnetic field induction, α is the angle in the velocity space.

3. Limited cases

3.1. The case of degenerate semiconductor

Let us consider the case of degenerate semiconductor ($U_\mu >> 1$) that is equal to the case of a metal wire. The expression of the dimensionless integral conductivity will have the form:

$$P(x_0, y_0, \mu_0) = \frac{1}{z_0} \int_0^1 \int_0^1 \int_0^{2\pi} \xi\rho\sqrt{1 - \rho^2} \left[1 - \exp\left(-\frac{\varphi z_0}{\mu_0}\right)\right] d\alpha d\rho d\xi; \tag{7}$$

$$\varphi = 2\,\mathrm{arcctg}\left[\frac{2\,\xi\,\rho\cos\alpha - \sqrt{4\rho^2(1 - \xi^2\sin^2\alpha) - 4\xi\mu_0\rho\sin\alpha(1 - \xi^2) - \mu_0^2(1 - \xi^2)^2}}{\mu_0\left(\xi^2 - 1\right)}\right]; \tag{8}$$

where the designation $\rho = \upsilon_\perp / \upsilon_F$ is entered, υ_F is Fermi velocity.

3.2. The case of nondegenerate semiconductor

Let us consider the case of degenerate semiconductor ($U_\mu \to -\infty$). The expression of the dimensionless integral conductivity will have the form:

$$P(x_0, y_0, \mu_0) = \frac{5}{3z_0} \int_0^1 \int_0^\infty \int_0^{2\pi} \xi\rho \exp\left(-\frac{5}{2}\rho^2\right) \left[1 - \exp\left(-\frac{\varphi z_0}{\mu_0}\right)\right] d\alpha d\rho d\xi. \tag{9}$$

Here the angle φ is determined by the expression (8), if ρ is implied by the following expression $\rho = \upsilon_\perp / \upsilon_T$. $\upsilon_T = \sqrt{5k_0 T / m}$ is the average thermal velocity.

4. Analysis of obtained results

Fig. 1 (a, b) shows the dependences of the dimensionless integral conductivity module $|P|$ on the dimensionless induction of the external magnetic field μ_0 for the case of stationary electric field ($y_0 = 0$) for the limited cases of a metal wire (7) and of a semiconductor wire made of nondegenerate semiconductor (9). The values of the dimensionless reciprocal mean free path of charge carriers are 0,1 (Fig. 1(a)) and 1 (Fig. 1(b)). In these figures the contributions to the total integral conductivity from the charge carriers that do not collide with the wire boundary (the volume part of the total dimensionless integral conductivity) (curves 2, 5), and from the charge carriers colliding with the wire boundary (the surface part of the total dimensionless integral conductivity) (curves 3, 6) are considered. The behaviour of this curves is explained by the fact that for the small values of dimensionless induction of the external magnetic field $\mu_0 << 1$ the charge carriers colliding with the wire boundary contribute generally to the total integral conductivity. With increasing of magnetic field the amount of charge carriers moving along a trajectory not crossing with the wire boundary increases, and the contribution from those to the total integral conductivity grows. Therefore the curves 1 and 2 (for the case of a metal wire) and the curves 4 and 5 (for the case of a semiconductor wire) converge asymptotically each other, and the curves 3 and 6 smoothly converge to zero. The converging velocity of the curves 1, 4 and 2, 5 and the velocity of the dropping to zero of the curves 3 and 6 for the case of the semiconductor wire are greater than those for the case of the metal wire. For sufficiently great values of the dimensionless reciprocal mean free path of charge carriers (Fig. 1(b)) the dimensionless integral conductivity discontinues to depend on the dimensionless induction of the magnetic field because of the small contribution from charge carrier surface scattering to the conductivity.

In Fig. 2 (a, b) the dependences of the dimensionless integral conductivity module |P| (a) and argument arg(P) (b) on the dimensionless induction of the external magnetic field μ_0 for the case of the metal (7) and the semiconductor (9) wire for the dimensionless electric field frequence $y_0 = 1$ and for different values of the dimensionless reciprocal mean free path of charge carriers are built. In Fig. 2(a) we see that the dimensionless integral conductivity module has a minimum more pronounced for the case of the metal wire. For the wire with the determined radius with charge carrier mean free path increasing the wire conductivity grows. With dimensionless induction of the external magnetic field increasing the dimensionless integral conductivity argument (Fig. 2(b)) monotonically grows and converges to $\pi/2$. Thus the conductivity argument value for the case of the semiconductor wire is greater than that for the case of the metal wire. This difference increases with charge carrier dimensionless reciprocal mean free path growing.

In Fig. 3 (a, b) the dependences of the dimensionless integral conductivity module |P| (a) and argument arg(P) (b) for the case of the metal (7) and the semiconductor (9) wire on the dimensionless electric field frequence y_0 for the dimensionless external magnetic field induction value $\mu_0 = 1$. The conductivity module value for the case of the semiconductor wire is greater than for the case of the metal wire. At that with charge carrier mean free path increasing this difference increases. For great values of y_0 the dimensionless integral conductivity module decreases and all curves merge. This is due to the fact that the free charge carriers system not keeping up with electric field intensity oscillating behaves as bound charge collection not contributed to the conduction current. With electric field frequence increasing the conductivity argument grows and converges to $\pi/2$ for large frequences, that is the wire conductivity becomes the purely imaginary value.

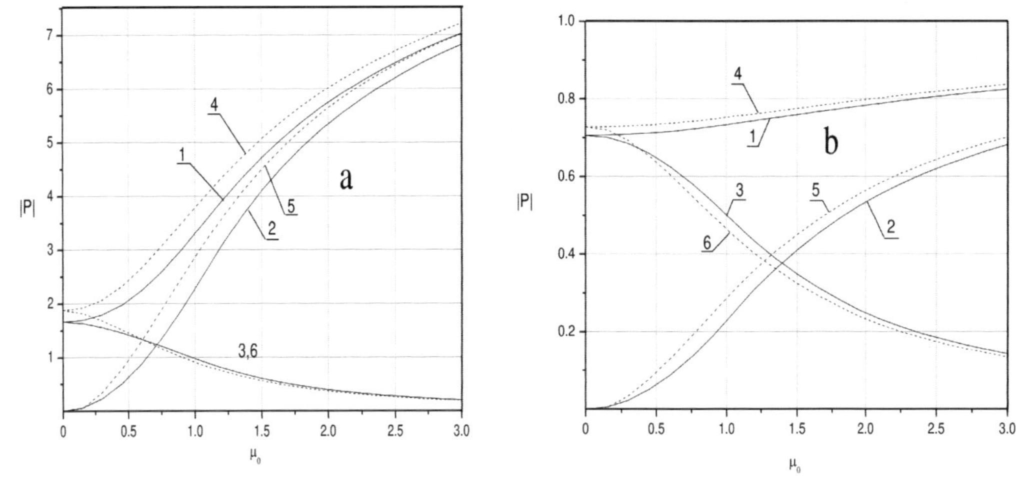

Figure 1. The dependence of the dimensionless integral conductivity module |*P*| on the external magnetic field induction μ_0 at $x_0 = 0.1$ (a), $x_0 = 1$ (b) and $y_0 = 0$ for the cases of a metal wire (solid curves 1 – 3) and a semiconductor wire (dashed curves 4 – 6): 1, 4 are the total dimensionless integral conductivity module; 2, 5 are volume integral conductivity module; 3, 6 are surface integral conductivity module.

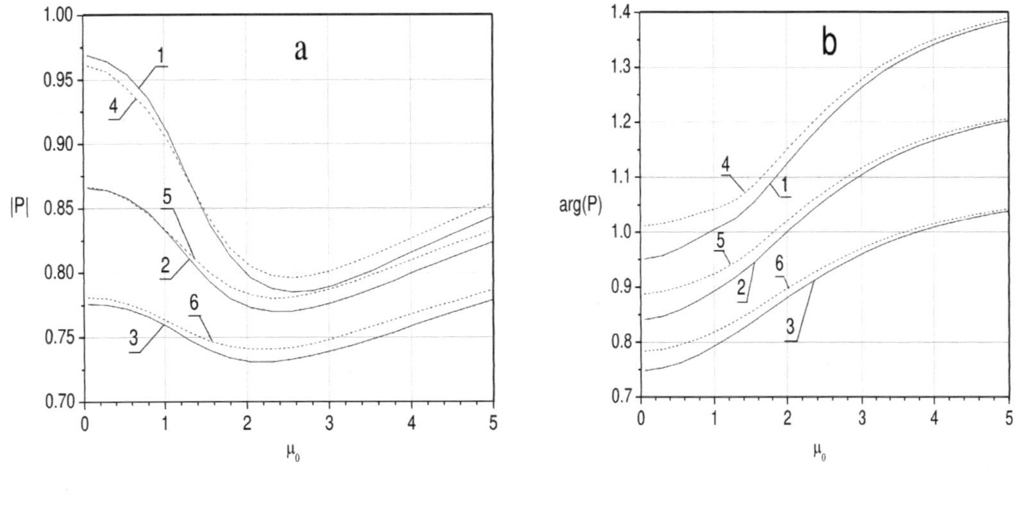

Figure 2. The dependence of *a*) module |*P*| and *b*) argument arg(*P*) of the wire dimensionless integral conductivity on the external magnetic field induction μ_0 at $y_0 = 1$ for the cases of a metal wire (solid curves 1 – 3) and a semiconductor wire (dashed curves 4 – 6): 1, 4 – $x_0 = 0.1$; 2, 5 – $x_0 = 0.3$; 3, 6 – $x_0 = 0.5$.

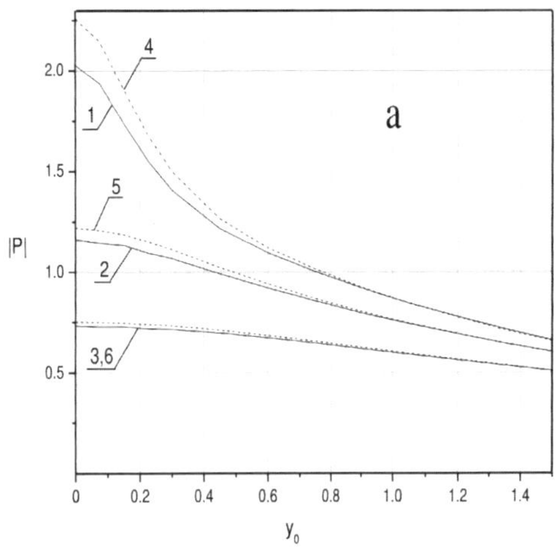

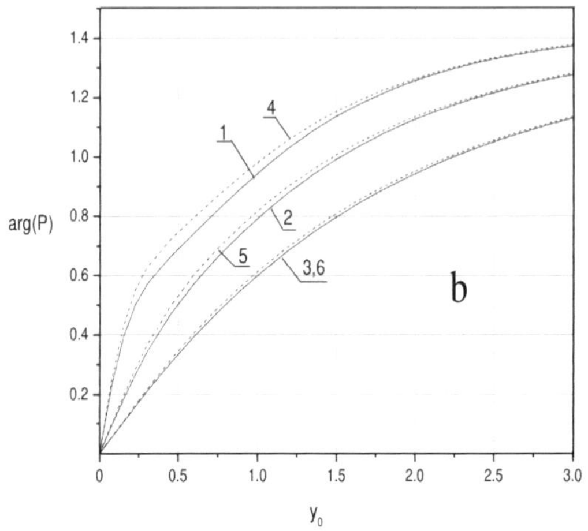

Figure 3. The dependence of *a*) module $|P|$ and *b*) argument arg(*P*) of the wire dimensionless integral conductivity on the dimensionless electric field frequence y_0 at $\mu_0 = 1$ for the cases of a metal wire (solid curves 1 – 3) and a semiconductor wire (dashed curves 4 – 6): 1, 4 – $x_0 = 0.2$; 2, 5 – $x_0 = 0.5$; 3, 6 – $x_0 = 1$.

5. Conclusion

In this work the results obtained demonstrate that magnetic field influences substantially on a thin wire conductivity (for example the conductivity minima were discovered for determined dimensionless external magnetic field induction values). Also we showed the differences in the behaviour of the metal and semiconductor wire conductivity. Thus for specified parameter values (the electric field frequence, the charge carrier mean free path) the conductivity module and argument for the case of the semiconductor wire are greater than those for the case of the metal wire.

References

[1] Sondheimer E H 2001 *Advances in Physics* **50** 499
[2] Anselm A I 1978 *Introduction to Semiconductor Theory* (Moscow: Nauka)
[3] Lifshitz I M, Azbel M Ya, Kaganov M I 1971 *Electron Theory of Metals* (Moscow: Nauka)
[4] Zavitaev E V, Yushkanov A A 2006 *JETP* **129** 938
[5] Zavitaev E V, Yushkanov A A 2008 *Microelectronics* **37** 429
[6] Kuznetsova I A, Chapkin A V, Yushkanov A A 2011 *Microelectronics* **40** 45
[7] Kuznetsova I A, Yushkanov A A, Khadchukaev R R 2009 *Phys. and Tech. of Semiconductors* **43** 645
[8] Zavitaev E V, Yushkanov A A 2006 *JETP* **130** 887
[9] MacDonald D K C 1950 *Proc. Phys. Soc. A* **63** 290
[10] Chambers R G 1952 *Proc. Phys. Soc. A* **65** 458
[11] Ziman J M 1962 *Electrons and Phonons* (Moscow: IL) chapt. II
[12] Courant R 1964 *Partial Differential Equations* (Moscow: Mir) chapt. II

MEMS switch with the active contact breaking mechanism

I V Uvarov, V V Naumov, O M Koroleva and I I Amirov
Yaroslavl Branch of the Institute of Physics and Technology, Institution of Russian Academy of Sciences, 150007, Universitetskaya Street 21, Yaroslavl, Russia

E-mail: ilnik88@mail.ru

Abstract. Electrostatically actuated MEMS switch with resistive contact is presented. Movable electrode of the switch is a beam suspended by torsion springs. Low stiffness of springs allows to achieve the low values of the actuation voltage. Main feature of the switch is a mechanism of the active contact breaking, allowing to solve the problem of adhesive sticking of the beam to the signal electrode. The active opening of the switch is realized by the presence of two driving electrodes. The theoretical analysis, finite element simulation and experimental investigation of the switch characteristics are performed.

1. Introduction
Electrostatically actuated MEMS switches are widely used in various radio frequency and microwave systems [1]. They have advantages of low power consumption, small insertion losses in the closed state, high isolation in the opened state, low harmonic distortion, high radiation resistance, wide operating temperature range and the possibility of integration with CMOS circuits [2].

The one of main disadvantages of MEMS switches is the high actuation voltage, which is a few tens of volts [3]. The most effective way to make the actuation voltage lower is to reduce the stiffness of the movable electrode. However, in the vast majority of conventional switches the breaking of a contact between the movable electrode and the signal electrode is carried out by the force of elasticity (passive opening). Reducing the stiffness of the movable electrode increases the probability of its stiction to the signal electrode by forces of adhesion, and hence reduces the reliability of the switch. The design of the switch must contain mechanisms, which allow, if necessary, to overcome the adhesion forces and open the electrodes (active opening).

2. MEMS switch design and principle of operation
Design of the switch is schematically shown on figure 1. Movable electrode is a metallic beam, attached in the middle of its length to the torsion suspension. Driving and signal electrodes (also metallic) are situated under each arm of the beam symmetrically with respect to the clamping point. Each driving electrode occupies the space from the clamping point of the beam almost to its free end, thereby providing a large area of an electric field. Beam has the contact bumps on its bottom side. Bumps localize the area of contact of the beam with the signal electrode and prevent the contact of the beam with the gate electrode when triggered.

Initially the beam is in a horizontal position (figure 1). When the voltage is applied to one of driving electrodes, beam bends towards the electrode under the electrostatic force and comes in contact with the signal electrode (figure 2, state 1). Elastic force is generated in the torsion hinges, which tries to return the beam to its initial position. The stiffness of hinges is selected to be small in order to provide low values of actuation voltage. Elastic force is insufficient to overcome the adhesion forces and to detach the beam from the signal electrode. Therefore, when the driving voltage is removed, switch retains its state. To break the contact, driving electrode located under the raised arm of the beam is used. Voltage is applied to it, the beam is tilted to the opposite direction and comes in contact with another signal electrode (figure 2, state 2). Thus, two states of the switch, shown on figure 2, are realized during operation. The horizontal state of the beam is not reached. This design provides a mechanism of active contact breaking, which is realized by the presence of the driving electrode under the each arm of the beam. It is important that the beam should be sufficiently stiff and should not bend when trying to detach it from the signal electrode.

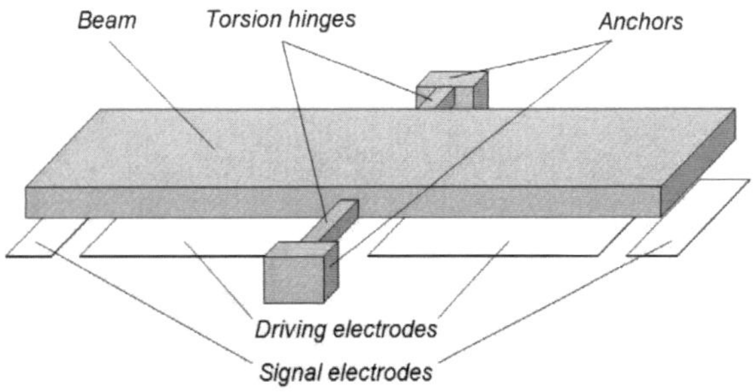

Figure 1. MEMS switch design.

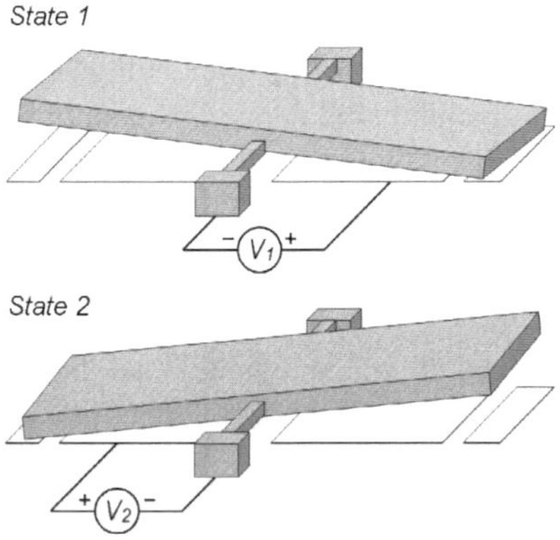

Figure 2. Possible states of the switch.

3. Calculation of the pull-in voltage

Switch design with the designation of dimensions is shown on figure 3. The beam has a length L_{beam} and a thickness t_{beam}. Torsion hinges have a length L_{hinge}, width and thickness w_{hinge} and t_{hinge}. The gap between the beam and the electrodes is g. The beam and the electrodes have the same width w. The driving electrode is located at a distance a_1 from the center of the switch, and has a length of a_2-a_1.

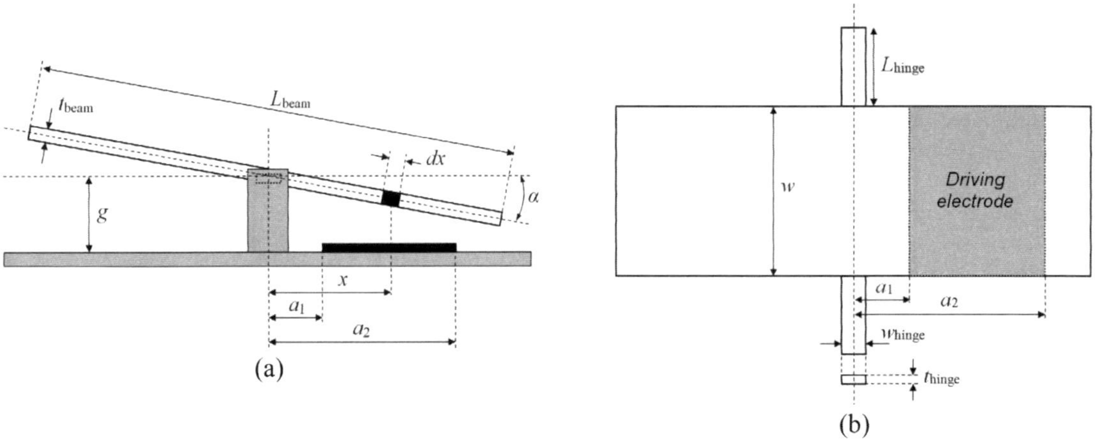

Figure 3. Design of the switch with the designation of dimensions: (a) side view, (b) top view.

Let the beam of the switch in a horizontal position. DC voltage V is applied between the beam and one of the driving electrodes. Electrostatic force bends the beam at an angle α towards the electrode (figure 3a). Assuming small angle α, an element of the beam length dx and a part of driving electrode, located under it, can be considered as two parallel plates. Neglecting fringing fields, we write the electrostatic force acting on a small element of the beam [4]:

$$dF = \frac{\varepsilon_0 w V^2 dx}{2(g - x\alpha)^2}, \tag{1}$$

where ε_0 is the vacuum permittivity. Full torque generated by the electrostatic force is given by the expression:

$$M_e = \int_{a_1}^{a_2} x dF = \int_{a_1}^{a_2} \frac{\varepsilon_0 w V^2 x}{2(g - x\alpha)^2} dx. \tag{2}$$

Doing the integration in the equation (2), we obtain [5]:

$$M_e = \frac{\varepsilon_0 w V^2}{2\alpha^2}\left[\frac{g}{g - a_2\alpha} - \frac{g}{g - a_1\alpha} + \ln\left(\frac{g - a_2\alpha}{g - a_1\alpha}\right)\right]. \tag{3}$$

The torque generated by the elastic force is given by the expression $M_{mech} = K\alpha$, where K is the stiffness of torsion hinges [6]:

$$K = \frac{G w_{hinge} t_{hinge}^3}{8 L_{hinge}}\left[\frac{16}{3} - 3.36\frac{t_{hinge}}{w_{hinge}}\left(1 - \frac{t_{hinge}^4}{12 w_{hinge}^4}\right)\right], \tag{4}$$

where G is the shear modulus of the material of hinges.

For a given driving voltage, the position of beam is determined by the condition of balance between the electrostatic and mechanical torques: $M_e = M_{mech}$. In expanded form:

$$\frac{\varepsilon_0 w V^2}{2\alpha^2}\left[\frac{g}{g - a_2\alpha} - \frac{g}{g - a_1\alpha} + \ln\left(\frac{g - a_2\alpha}{g - a_1\alpha}\right)\right] = K\alpha. \tag{5}$$

Equation (5) represents a dependence of the tilt angle α on a driving voltage V. With the gradual increase of V, the increasing electrostatic force is no longer balanced by the elastic force, and the beam sharply tilts and touches the signal electrode (electrostatic pull-in). The voltage at which this occurs is called the pull-in voltage, V_{PI}. The expression for V_{PI} is derived from the equation (5) using numerical methods [7]:

$$V_{PI} = \sqrt{\frac{0.8275 K g^3}{\varepsilon_0 w a_2^3}}\left[1 - \left(\frac{a_1}{a_2}\right)^2\right]^{-0.5}\left[1 + 0.6735\left(\frac{a_1}{a_2}\right)^{1.931}\right]^{-0.3244}. \tag{6}$$

Pull-in voltage values, calculated using equation (6), and the nominal values of switch dimensions are shown in the table 1. Four types of switch design were considered, having different width of the beam and the electrodes w. The material of torsion hinges was chromium ($G = 115$ GPa), the material of beam was aluminum. According to the equation (4), stiffness of hinges is equal to $3.9\cdot10^{-9}$ N·m. Pull-in voltage of the switch should be between 4.1 V in the case of $w = 32$ µm and 8.2 V for $w = 8$ µm.

4. FEM modeling of the switch

The proposed design of the switch should be modified to facilitate its production and improve the performance. Firstly, the beam should have etching holes (perforation). They allow to speed up the removal of the sacrificial layer from under the beam, and partially take off the residual stresses in the beam material. Secondly, only twist of the torsion hinges should occur during operation of the switch, the beam should not bend. To this end, the beam must have sufficient stiffness. Effective method of increasing the stiffness of the beam is the formation of ribs on its surface. Ribs can increase the stiffness without increasing the thickness of the beam material, and hence avoid the increasing of the beam mass and reduce the bending of the beam under influence of residual stress [8].

Table 1. Dimensions of the switch and results of analytical calculation and FEM modeling of V_{PI}.

Switch type	L_{beam} (μm)	t_{beam} (μm)	L_{hinge} (μm)	w_{hinge} (μm)	t_{hinge} (μm)	a_1 (μm)	a_2 (μm)	G (μm)	w (μm)	V_{PI} (V) analytic	V_{PI} (V) FEM
1	102	1	1.5	3	0.3	4	44	0.5	8	8.2	12.8
2									16	5.8	9.5
3									24	4.7	7.9
4									32	4.1	7.0

The influence of perforation and ribs on the properties of the switch can be considered using finite element method (FEM) simulation. The simulation was performed using validated FEM software. Ribs having 98 μm length, 4 μm width and 1 μm height were located on the upper surface of the beam. Etching holes having area of 2×2 μm^2 were located between the ribs.

FEM model of each type of the switch included a beam, the one driving and one signal electrode. Dimensions of the switch, used in simulations, are indicated in table 1. The ends of the torsion hinges were fixed. The beam was grounded. Electric potential was applied to the driving electrode. Potential was gradually increased until the pull-in occurrence. The dependence of the distance between the end of the beam and the signal electrode on the applied potential, calculated for the four switch types, is shown on figure 4. Pull-in voltage was determined from this dependence.

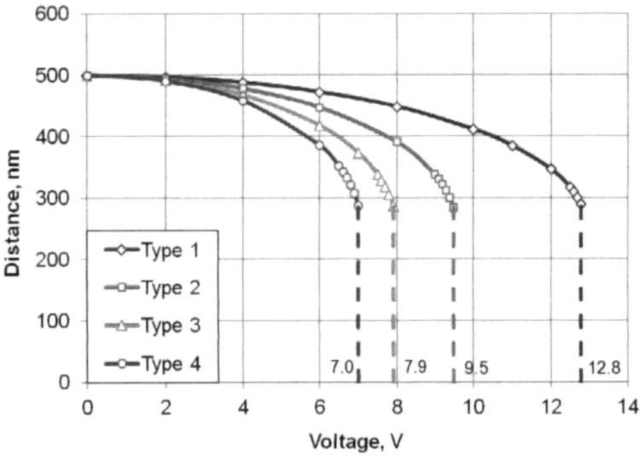

Figure 4. Dependence of the distance between the end of the beam and the signal electrode on the driving voltage.

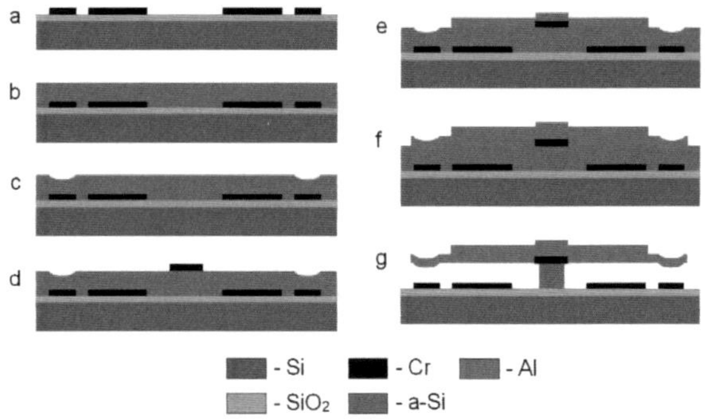

Figure 5. Procedure of MEMS switch fabrication.

Values of V_{PI}, obtained by FEM simulation, are presented in table 1. Simulated values exceeded ones calculated analytically at 1.5-1.7 times. The reason of such an excess could be etching holes, which reduced the overlap area of the beam and the driving electrode and therefore increased the actuation voltage. According to the simulation results, the lowest value of the pull-in voltage should be 7 V.

5. Experimental results

Switches were fabricated by surface micromachining. The main steps of the manufacturing process are shown on figure 5. At the first step the dielectric layer was formed on the silicon wafer. The wafer was thermally oxidized in wet oxygen to a SiO_2 layer thickness of 1 μm. Next, driving and signal electrodes of the switch were performed on SiO_2 layer (figure 5a). For this purpose, chromium layer with 50 nm thickness was deposited on the wafer by magnetron sputtering, followed by contact photolithography and wet etching of chromium.

At the next step the wafer was coated by a sacrificial layer of amorphous silicon (a-Si, figure 5b). In order to obtain the contact bumps of the beam, the dimples with nearly 200 nm depth were formed on the surface of the sacrificial layer (figure 5c). To do this, the round windows of 2 μm in diameter were formed in the photoresist layer. Through these windows the a-Si layer was etched in SF_6 plasma.

Further, torsion hinges and a beam of the switch were formed on the sacrificial layer. Cr layer with 0.3 μm thickness was deposited, anchors and hinges were made from it using contact lithography and wet etching (figure 5d). Next, ribs of the beam were formed (figure 5e). For this purpose, Al layer with 1 μm thickness was deposited, followed by lithography and wet etching. Another Al layer with 1 μm thickness was deposited over the ribs, and the beam was formed from it (figure 5f). The final stage of fabrication was the removal of the a-Si sacrificial layer from under the beam by etching in SF_6 plasma (figure 5g). Fabricated switch of type 3 is shown on figure 6.

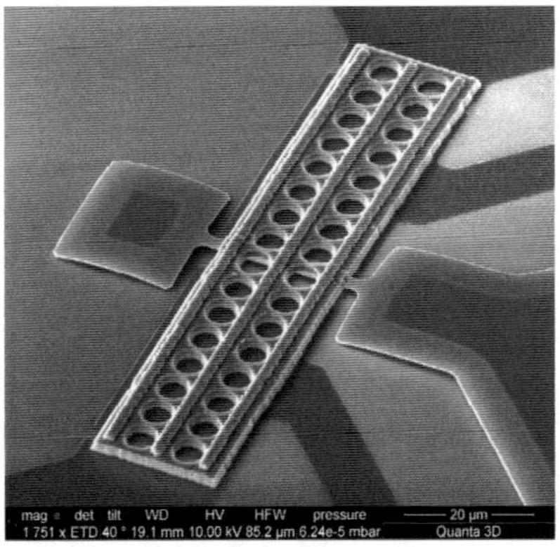

Figure 6. SEM image of the fabricated switch (type 3).

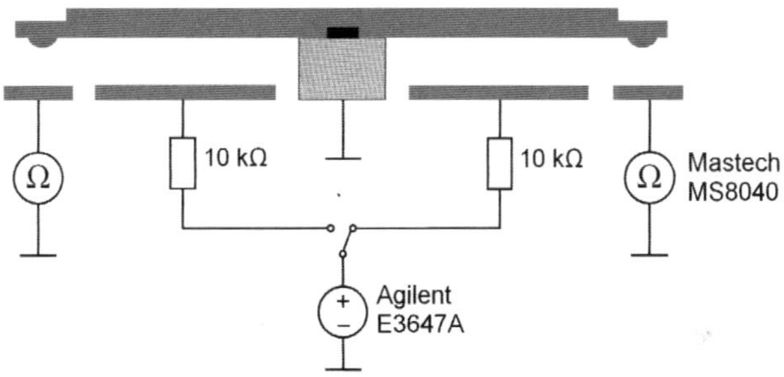

Figure 7. Scheme of connection of the measurement equipment to the switch.

The switch had five contact pads with dimensions of 1.4×1.4 mm^2, which were connected to the beam and electrodes by means of metallic lines with 200 μm width. Probes were placed to the contact pads. Measurement equipment was connected to the probes, wiring diagram is shown on figure 7. All the measurements were performed in air under normal conditions [9].

Experimentally obtained values of the pull-in voltage are shown in table 2. The switches with the widest beam had lowest pull-in voltage. The minimum value of V_{PI} was 6.0 V. On one hand, the experimental values should exceed simulated ones, because the etching of aluminum when forming the beam caused a significant change of its lateral dimensions (about 1.5 μm per side). It can be clearly seen at the perforation holes with a diameter of 5 μm (see figure 6), while the nominal dimensions of the holes is 2×2 μm^2. Reduction of the lateral dimensions of the beam and enlargement of perforation led to decrease in the overlap area of the beam and the driving electrode, and hence, reduced the electrostatic force. On the other hand, the reduction of lateral dimensions was also observed at the step of forming the chromium hinges (about 300 nm per side). This effect led to a narrowing of hinges and caused the decrease of its stiffness, which should lower V_{PI}. Additionally, suspensions bent under the weight of the beam, resulting in a decrease of the distance between the beam and electrodes. It also had to reduce the pull-in voltage. Apparently, the factors listed above compensated one another so that there was good agreement between the experimental data and the results of FEM simulation (table 1).

Table 2. Experimentally obtained values of the pull-in and recovery voltage, and the resistance of the switch in a closed state.

Switch type	V_{PI} (V)	V_{REC} (V)	R (Ω)
1	13.0 ± 0.7	27.7 ± 5.1	344 ± 41
2	10.6 ± 1.1	32.0 ± 8.3	318 ± 33
3	9.0 ± 1.5	26.7 ± 3.9	336 ± 95
4	6.5 ± 0.5	24.6 ± 5.2	317 ± 20

Experimentally obtained values of recovery voltage are shown in table 2. The lowest value of V_{REC} was 19.4 V. Recovery voltage exceeded the pull-in voltage at 3-4 times. There are two main reasons for the excess. Firstly, the air gap between the raised arm of the beam and the driving electrode exceeded the initial value *g*. Secondly, to open the contact, the adhesion forces between the beam and the signal electrode had to be overcome. Since the magnitude of these forces is difficult to calculate, the calculation of V_{REC} is also difficult. At the same time, it is possible to reverse the path: knowing the recovery voltage (from the experiment) and the switch geometry, it is possible to determine the value of adhesion forces. However, this is a separate research task. It is worth noting that during the switch operation V_{REC} is the actuation voltage, since the horizontal position of the beam will not be achieved. At the initial stage of the work, recovery voltage was relatively high (20-40 V). In the future we plan to reach lower values by reducing the gap between the beam and the electrodes and, possibly, by modifying the beam design.

The resistance of the switch in a closed state was 250-400 Ω (table 2). The decrease of resistance with the increase of the beam width was not observed. Although the wider the beam, the less its own resistance, and it contained the more contact bumps. The absence of dependence of the resistance on the width of the beam can be explained by the fact that much of the resistance was caused by metal lines, connecting the contact pads with the beam and electrodes. The contact resistance of the switch was difficult to determine from the total resistance. Presented switches had Al-to-Cr contact. It is planned to use the MEMS contact materials such as gold or platinum, in order to achieve low contact resistance and reliable contact of the beam with the signal electrodes.

6. Conclusion

Electrostatically actuated MEMS switch with resistive contact was designed and fabricated. The main feature of the switch was the mechanism of active opening, solving the problem of adhesive stiction of the beam with the signal electrodes. Active opening was realized by the presence of two driving electrodes. The fabrication procedure of the switch was developed using standard processes of microelectronics. The experimental values of the pull-in voltage were in agreement with the results of finite element modeling. Recovery voltage of the switch was 20-40 V. In the future we plan to get lower values, since the switch design allows to achieve it.

The reported study was supported by RFBR research project No. 14-07-31156 mol_a, by Facilities Sharing Centre "Diagnostics of Micro- and Nanostructures" and by the Ministry of education and science of Russian Federation.

References

[1] Rebeiz G M 2003 *RF MEMS: Theory, Design, and Technology* (Hoboken, New Jersey: John Wiley & Sons, Inc.)
[2] Dai C-L and Chen J-H 2006 *Microsyst. Technol.* **12** 1143
[3] Sharma A K, Gautam A K, Farinelli P, Dutta A and Singh S G 2015 *J. Micromech. Microeng.* **25** 035014
[4] Younis M I 2011 *MEMS linear and nonlinear statics and dynamics* (New York: Springer Science + Business Media, LLC)
[5] Degani O, Socher E, Lipson A, Leitner T, Setter D J, Kaldor S and Nemirovsky Y 1998 *J. Microelectromech. Syst.* **7** 373
[6] Timoshenko S P and Goodier I N 1970 *Theory of Elasticity* (New York: McGraw-Hill)
[7] Degani O and Nemirovsky Y 2002 *J. Microelectromech. Syst.* **11** 20
[8] Gupta A, Barron L, Brainin M and Lee J-B 2014 *J. Micromech. Microeng.* **24** 065023
[9] Uvarov I V, Naumov V V and Selyukov R V 2014 *Proc. SPIE* **9440** 94400W

Fabrication technology of CNT-Nickel Oxide based planar pseudocapacitor for MEMS and NEMS

E.A. Lebedev[2], E.P. Kitsyuk[1], I.M. Gavrilin[2], D.G. Gromov[2], N.E. Gruzdev[2], S.A. Gavrilov[2], A.A. Dronov[2], A.A. Pavlov[3]
[1]Scientific Manufacturing Complex "Technological Centre", 124498, Russia, Moscow, Zelenograd, Bld. 5, Pas. 4806
[2]National Research University of Electronic Technology, 124498, Russia, Moscow, Zelenograd, Bld. 5, Pas. 4806
[3]Institute of Nanotechnologies of Microelectronics Russian Academy of Sciences, 119991, Russia, Moscow, Leninsky Prospekt, 32A

E-mail: dr.beefheart@gmail.com

Abstract. Fabrication technology of planar pseudocapacitor (PsC) based on carbon nanotube (CNT) forest, synthesized using plasma enhanced chemical vapor deposition (PECVD) method, covered with thin nickel oxide layer deposited by successive ionic layer adsorption and reaction (SILAR) method, is demonstrated. Dependences of deposited oxide layers thickness on device specific capacities is studied. It is shown that pseudocapacity of nickel oxide thin layer increases specific capacity of the CNT's based device up to 2.5 times.

1. Introduction

A rapid development of miniature electronics devices such as micro- and nanoelectromechanical systems, robots and implantable medical devices has significantly increased the demand for effective micro-sized energy storage devices. Electrochemical capacitors and batteries are the most popular solutions for these purposes in the "macro world" today, which are fundamentally different from each other in accumulation and energy storage mechanisms. With one hand, in the electrical double layer capacitors (EDLC) the charge accumulation occurs due to the formation of the electric double layer at the electrode-electrolyte interface. On the other hand, electric energy accumulation in batteries occurs as a result of ion intercalation into bulk electrode material. The mechanism of charge accumulation and storage determines the physical properties of the device – electrochemical capacitors are powerful and capable for rapid recharging, and batteries are high capacity sources [1].

However, in recent years a growing interest have been confined to pseudocapacitors (PsC) or supercapacitors. These devices accumulate and store energy using a reversible redox reactions (Faradaic process) in thin surface layer of the electrode, have a higher capacity compared to EDLC and demonstrate a much higher power density compared to batteries, combining the advantages of these two types of energy storage devices.

Transition metal oxides (Co_3O_4, RuO_2 and NiO) have significant specific pseudocapacitance (3560, 2200 and 2584 F/g respectively [2]) compared to the values for carbon (about 150 F/g [3]). To create the PsC an electrodes with a high specific surface area are covered with thin metal oxide layers using various methods. From this point of view methods of layer-by-layer deposition, such as atomic layer deposition from gas-phase (ALD) [4], are of the greatest interest. Such approaches provide the most effective usage of the entire electrode area due to the conformal deposition, and allow to control the deposited layers thickness with high precision. Another promising method is SILAR [5] (successive ion layer adsorption and reaction) - a stepwise chemical deposition technique based on immersion of the substrate into separately placed cationic and anionic precursors and rinsing after every immersion cycle with deionized water to avoid homogeneous precipitation. The possibility of growing good quality thin films at room temperatures and normal pressure is the main advantage of SILAR as compared with the gas-phase techniques. Moreover, SILAR deposition equipment is simple and inexpensive, since vacuum systems are not required.

In this work we report on the planar pseudocapacitor based on CNT forest covered by nickel oxide using SILAR method - promising electrochemical energy storage device, which are in one hand combines significant energy and power density and long life cycle, and on the other hand the integral fabrication technology allows to implement on-chip energy sources directly on substrates for "micro world" applications.

2. Experimental data

Planar PsC based on the structure which consists of two metal combs nested into each other on the substrate surface. This structure allows to place two electrodes in the same planar surface and do not use separator. All experimental samples were prepared using standard integral technology and equipment.

The fabrication process schematically presented on Figure 1 and starts with wet cleaning and thermal oxidation of a bare silicon wafer to isolate the substrate from the electrodes (1). After that, Ti, TiN and Ni layers were evaporated onto the substrate using magnetron sputtering system with a thickness of 300 nm, 25 nm and 5 nm, respectively (2). Ti acts as the current collector material, TiN as the diffusion-barrier layer and Ni as the common catalyst for CNT forest growth. Then deposited layers were covered by photoresist (3) and lithography was made to pattern two comb-like electrodes and pads (4). Wet metal etching (5) and photoresist lift-off (6) followed to remove the metals on unwanted areas and photoresist mask.

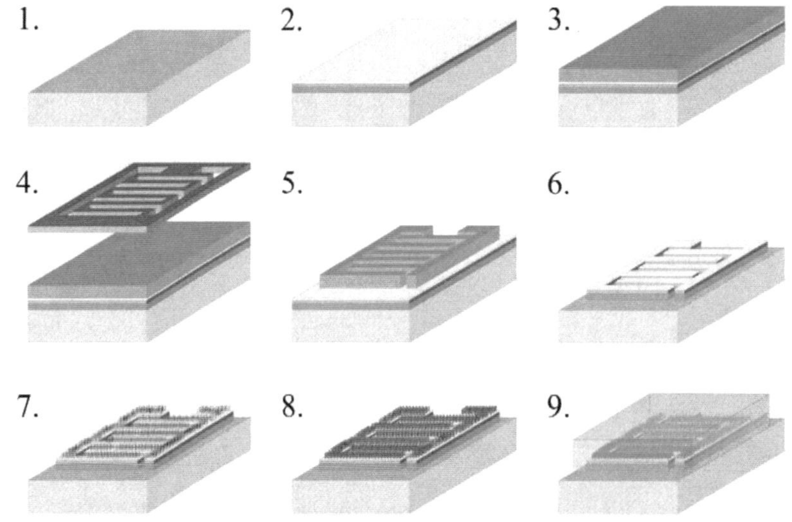

Figure 1. Basic steps of planar PsC fabrication.

Vertically-aligned multi-wall carbon nanotubes were directly synthesized on the substrate surface by PECVD method (7) on Oxford PlasmaLabSystem 100 (Nanofab 800 Agile) according to the author's described procedure [6]. For CNT arrays growing it was needed to form a catalyst nanoclusters, that was performed by thermal oxidizing of samples at 280°C in oxygen environment and subsequent annealing at 450°C in H_2 atmosphere. Synthesis of multi-wall carbon nanotubes was carried out in gas mixture of C_2H_2, H_2 and NH_3 at 2 Torr operating pressure. Additional plasma assistance allow to reduce process temperature down to 500°C. Temperature of substrate holder was controlled by heat sensor. Plasma energy generator creates the radio frequency (RF) – 13,56 MHz radiation with power of 20 W and low frequency (LF) – 500 Hz radiation with power of 30 W. The growth rate of CNT arrays was 600 nm per minute.

To increase the PsC specific capacity CNT arrays were covered with a nickel oxide thin layer (8). Prior to the NiO deposition the CNTs surface was preliminary functionalized in Ar-O_2 plasma at low incident power of 50 W for 120 seconds on NIITM MVU TM Plasma RIT-T. The deposition of NiO onto the aligned CNT forest was done using the SILAR method. The cationic precursor for this study was a mixture of 0.01 M $Ni(NO_3)_2$*$6H_2O$ and aqueous ammonia. Initially, precipitate of $Ni(OH)_2$ was formed from 0.01 M $Ni(NO_3)_2$*$6H_2O$, which was dissolved upon addition of ammonia to adjust the pH to 12, thereby forming hexaamminonickel (II) complex. The anionic precursor source was a hot water bath. When the substrate was immersed in the precursor solution (0.01 M $Ni(NO_3)_2$*$6H_2O$) at room temperature for 30 s, nickel ions were adsorbed onto the substrate surface. The substrate was transferred to the hot water bath at a temperature of 90°C for another 30 s to convert it into NiO, after which it was rinsed in deionized water for 30 s to remove the loosely bound or excess nickel hydroxyl ions. The SILAR steps were repeated for different number of cycles. After deposition, the samples were heat treated at 250 °C for 2 h in air. In the last step 1 M aqueous solution of KOH was used as the electrolyte (9).

Study of the deposited NiO layer thickness influence on the PsC capacity was performed using bulk samples consisting of two stainless steel electrodes on which surface the CNT arrays were synthesized and covered with NiO layer using SILAR method; porous polypropylene separator impregnated with electrolyte was placed between the electrodes.

Testing and measurement of electro-physical charge-discharge characteristics and cyclic voltammetry curves using potentiostat Elins P-30J were made. Charge and discharge current varied from 0.5 to 2 mA, and the voltage changed from 0 to 2 V. The scan rate during cyclic voltammetry ranged from 10 to 50 mV/sec.

CNT arrays structure and morphology were investigated and controlled using scanning electron microscope Agilent 8500 FE-SEM.

3. Results and discussion

During the research it was found that the synthesized CNT arrays after a short-term air exposure lose their hydrophilic properties and cease to be water wettable. This become a serious problem because SILAR method uses aqueous solutions and hydrophobic means the inability of the layer deposition. However, several methods of carbon nanotubes surface activation are referred to in earlier publications [7]. We chose plasma method, which uses a gas mixture of argon and oxygen – each component provides respectively the physical and chemical activation mechanism. Experiments with different values of incident power and the process duration were conducted and low power (50W) and short process duration (maximum 120 seconds) were selected. The increase of the incident power causes significant etching of nanotubes – arrays become sparse and lower in height. Before plasma treatment the water formed droplets on the array surface and no CNT wetting was observed. But after the treatment the water is promptly went into the arrays and wetted carbon nanotube surface.

The CNT arrays height varied in the range from 3 to 5 μm. Such heights interval was determined by the depth of nickel oxide depth deposition. At higher arrays altitudes the delivery of cationic and anionic solutions to the reaction zone becomes difficult, because of limited depth of plasma functionalization, and, therefore, large CNT surface area will not be involved in the charge accumulation and storage process.

Figure 2 shows SEM images of the morphological features of clean CNT (2a) arrays and NiO/CNT (2b) composite after 15 SILAR cycles (NCc-15).

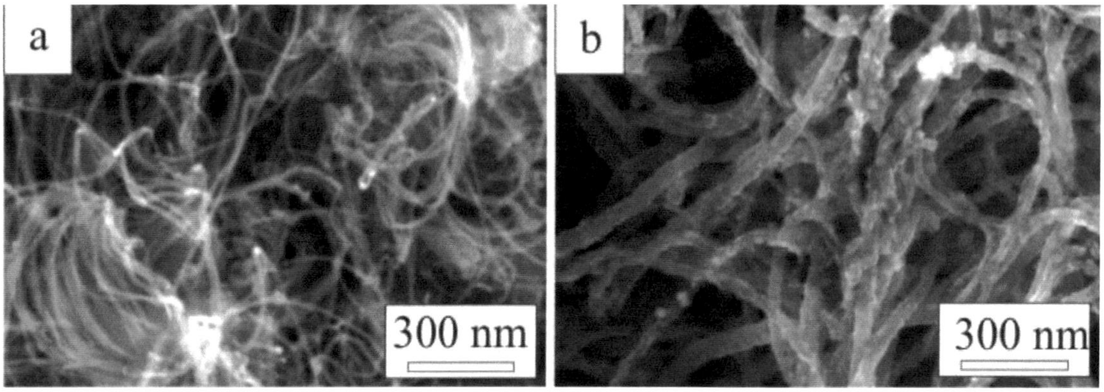

Figure 2. SEM images of clean CNT arrays (a) and NCc-15 (b).

The SEM image of NCc (fig. 2 b) show that the walls of CNT are uniformly coated by NiO nanoparticles with the average size of less than 10 nm. It was observed that with number of SILAR cycles increasing the total deposited layer thickness also increased, what can be proven by carbon nanotube diameter growth.

Scale photo and SEM image of electrode structure of the planar PsC manufactured according to the described technology are presented in Figure 3 and 4 respectively.

Figure 3. Scale image of planar PsC.

Figure 4. SEM image of planar PsC electrode.

The study of the SILAR layer thickness influence on the device capacity was performed using bulk samples, prepared according to described technique. For electrical test measurements the set of samples based on pure CNT arrays and CNT/NiO composites with different amount of SILAR cycles (5, 10, 15 and 20 cycles) was prepared. The charge-discharge characteristics measurements were made and subsequent analysis of the obtained curves was produce to determine the samples capacity. Figure 5a shows the capacity dependence on the SILAR cycles number. Figure 5b shows typical charge-discharge characteristics for samples based on pure CNTs (blue line) and CNT/NiO composite (black line.

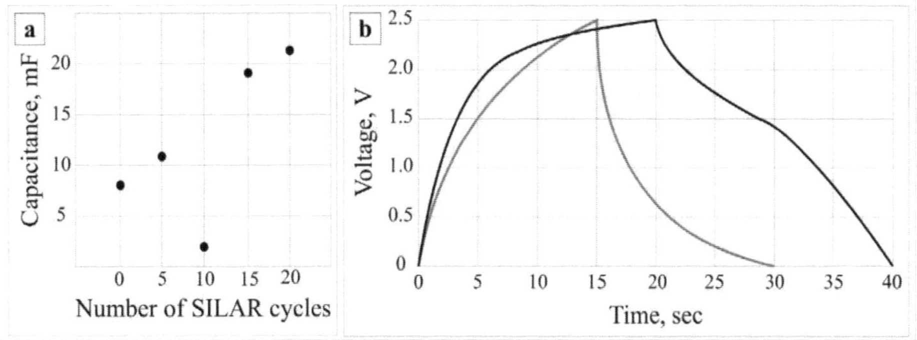

Figure 5. *"a"* – capacitance dependence on number of SILAR cycles and *"b"* - character of charge-discharge curves for clean CNT and CNT/NiO composite samples.

As can be seen in Figure 5a linear capacity dependence on SILAR cycles number was detected. Obviously, that point № 3 for NCc-10 is an experimental error. The greatest capacitance value was observed for NCc-20 and it exceeds the value for pure CNT at 2.5 times. Estimated value of the maximum nickel oxide layer thickness for NCc-20 was about 20 nm. It is necessary to conduct additional experiments with thicker layers and, most likely, the capacitance line reaches a maximum and then there will be a recession that is easily explained by gradually decreasing of the effective surface area of the electrode as a result of overgrowing of free space between the individual nanotubes. Figure 5b shows typical charge-discharge curves for pure carbon nanotubes and the NCc. In the first case (blue line), the curve has a "classic" character for EDLCs – rapid voltage lowering over time is observed. In the second case (black line), the clear inflection at a voltage value of 1.5 V on the curve is detected. This kink indicates the presence of pseudocapacitance, i.e. the appearance of an additional mechanism of accumulation and storage of energy.

4. Conclusions

We introduced the integral fabrication technology of planar PsC based on CNT arrays and nickel oxide thin layers deposition SILAR method. The influence of CNT's pre-treatment operations on the quality and morphology of the deposited layers was investigated.

The study of the PsC capacity dependence on the deposited nickel oxide layer thickness was produced. For these purposes the bulk experimental samples based on CNT's and oxide layers were made. Nickel oxide layer thickness was determined by the number of SILAR cycles. It was shown that with layers thickness increasing the device capacitance also increased. Capacitance value for the thickest oxide layer (20 SILAR cycles) was 2.5 times greater than that of devices based on pure CNT.

Further research and development should be directed towards to optimize the PC geometric dimensions, determine the oxide layers optimum thickness and to develop suitable the solid-state electrolyte.

Acknowledgments

This work was supported by the Ministry of education and science of the RF (№ 11.2551.2014/K).

References

[1] Conway B E 1991 *J. Electrochem. Soc.* **138** 6 pp 1539-1548
[2] Zhi M *et al* 2013 *Nanoscale* **5** pp 72-88
[3] Stoller M D *et al* 2011 *Energy Environ. Sci.* **4** pp 4685-4689
[4] Galperin V A *et al* 2014 *J. of Nano- and Microsystem technique* **6** 16
[5] Gavrilov S A *et al* 2010 *J. Semiconductor* **44** 13 pp 1649-11653
[6] Kitsyuk E P *et al* 2012 *Proc. Of SPIE – The Int. Soc. For Optic. Eng.* **8700** 87000A-2
[7] Conway B E 1999 *N.-Y.: Kluwer Academic Plenum Publishers*

Regimes of photon generation in Dynamical Casimir Effect under various resonance conditions

Levchenko A.V., Trifanov A.I.
ITMO University, 49 Kronverksky Avenue, Saint Petersburg, 197101, Russia
E-mail: alextrifanov@gmail.com

Abstract. We investigate spectral and statistical properties of the field generated in dynamical Casimir effect under various resonance conditions appeared from periodical motion of cavity boundary. In particular we are interested in single and two-mode squeezing effects which take place for each resonant frequency under the cases of bounded and unbounded cavity spectrum.

1. Introduction

Dynamical Casimir effect (DCE) is the relativistic quantum parametric process by which the accelerated motion of the cavity boundaries excites vacuum fluctuations [1]. This process is accompanied by the generation of photons and intermode interaction. In spite of the fact that it requires relativistic velocities [2] at present there are experimental realization of DCE in superconducting schemes [3].

The aim of the presented work is investigation of spectral and statistical properties of DCE radiation under various resonant conditions in one dimensional cavity. These conditions may appear due to periodical law of boundary motion, which we define following [4] as

$$q(t) = L\exp\left(\frac{q_0\cos\Omega t}{L}\right), \tag{1}$$

where L is the cavity length, Ω is modulating frequency and q_0 is initial boundary coordinate. Under this conditions effective Hamiltonian of the system has the form

$$H_{eff} = \sum_k \omega_k(t)\, a_k^+ a_k + i\sum_k \frac{\dot{q}(t)}{4q(t)}\left(a_k^{+2} - a_k^2\right)$$
$$+\frac{i}{2}\sum_{j\neq k}(-1)^{j+k}\frac{kj}{j^2-k^2}\left(\frac{k}{j}\right)^{1/2}\frac{\dot{q}(t)}{q(t)}\left(a_k^+a_j^+ + a_k^+a_j - a_ja_k - a_j^+a_k\right). \tag{2}$$

Here $\omega_k(t)$ is instantaneous frequency of k-th intracavity mode and a_k is its annihilation operator.

2. Resonance condition

In [4] it was only mentioned that the different regimes of photons generation and intermode interaction may occur and it depends on the choice of a particular resonant frequency of boundary motion. Here we investigate thoroughly three regimes of DCE field generation, which takes place when "mechanical" frequency Ω takes the values $\omega_k = \frac{\pi ck}{L}$ ($c = 1$ is speed of light) of stationary cavity for $k = 1, 2$ and 4. Besides that, we derived the general form of the Hamiltonian depenfing on parity of k.

2.1. Phase modulator

Using resonance condition $\Omega = \omega_1$ and rotating wave approximation (RWA) one can easily obtain resonance part of Eq. (2), namely:

$$H_{PM} = \frac{q_0\pi}{4L^2}\sum_k \sqrt{k(k+1)}(a^\dagger_{k+1}a_k + a^\dagger_k a_{k+1}). \tag{3}$$

This Hamiltonian has the form of that which describe phase modulation process. Under approximations of bounded spectrum corresponding spectral problem has an analytical solution. It is worthwhile to mention that there is no photon creation in this process. The ultimate resonator quality factor will not grow infinitely as a function $f(k) = \sqrt{k(k+1)}$, therefore we can assume approximately $f(k) \approx g(k) = \sqrt{(k - n_{min})(n_{max} - k)}$, where n_{min} and n_{max} are the minimum and maximum number of modes. In this case the problem can be solved analytically as in [5], where Hamiltonian of optical modulator was expressed in terms of SU(2) algebra generators

$$A_0 = \sum_{m=-S}^{S} m a^\dagger_m a_m, \quad A_+ = \sum_{m=-S}^{S-1} f(m) a_m a^\dagger_{m+1}, \quad A_- = \sum_{m=-S}^{S} f(m) a^\dagger_m a_{m+1}, \tag{4}$$

where $S = (n_{max} - n_{min} - 1)/2$. Then corresponding Hamiltonian has the form

$$H_{PM} = \frac{q_0\pi}{4L^2}(A_+ + A_-), \tag{5}$$

and my be diagonalized by the unitary transformation V

$$V H_{PM} V^\dagger = \frac{q_0\pi}{2L^2}A_0, \quad V = \exp\left[\frac{\pi}{4}(A_+ - A_-)\right]. \tag{6}$$

2.2. Single mode squeezing

Applying condition $\Omega = \omega_2$ together with RWA one can obtain the following Hamiltonian

$$H_2 = H_{SQ} + H'_{PM} = \frac{q_0\pi}{4L^2}\left(a_1^{\dagger 2} + a_1^2\right) + \frac{q_0\pi}{4L^2}\sum_k \sqrt{k(k+2)}\left(a^\dagger_{k+2}a_k + a_{k+2}a^\dagger_k\right), \tag{7}$$

which describes the process of photon creation in mode with $k = 1$ due to vacuum squeezing. We analyzed this process both theoretically and numerically. In particular the statistical properties of odd modes for single mode squeezing as a source of Casimir photons were investigated (since all even modes are in vacuum state).

Hamiltonian H_2 may be written in interaction picture of operator H'_{PM} which in fact is Hamiltonian of phase modulation process for odd modes. Using technique described in eq's (4)-(6) one can obtain time dependent operator H_2^{int} in the from

$$H_s = \frac{q_0\pi}{L^2}\left(\left(\sum_{\Delta p=-S}^{S} a_{\Delta p} R_{\Delta m(1)\Delta p}(t)\right)^2 + \left(\sum_{\Delta p=-S}^{S} a^\dagger_{\Delta p} R^*_{\Delta m(1)\Delta p}(t)\right)^2\right), \tag{8}$$

were function $R_{\Delta m(1)\Delta p}$ is expressed in terms of Wigner D-matrices.

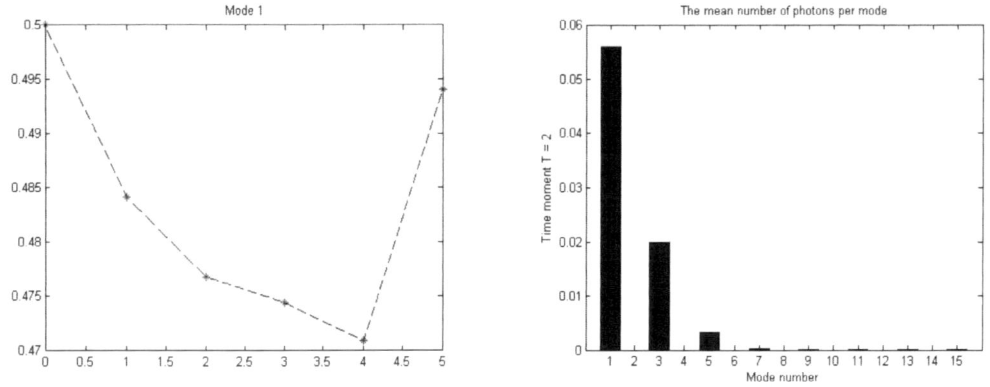

Figure 1. Photon generation in DCE; variance of quadrature for mode with number $m = 1$(left) and mean photon number per mode under the process described by Hamiltonian from Eq.(7) (right) at time $T = 2mcs$

2.3. *Two mode squeezing*

Resonance condition $\Omega = \omega_4$ with applied RWA gives the following form of effective Hamiltonian

$$H_4 = \frac{q_0\pi}{2L^2}\left(a_2^{\dagger 2} + a_2^2\right) + \frac{\sqrt{3}q_0\pi}{4L^2}\left(a_1^\dagger a_3^\dagger + a_1 a_3\right) + \frac{q_0\pi}{L^2}\sum_k \sqrt{k(k+4)}\left(a_{k+4}^\dagger a_k + a_{k+4}a_k^\dagger\right). \quad (9)$$

There are two processes in this case which insert into photon generation: single mode squeezing (the first term) and two-mode squeezing (the second term). Under assumption of weak interaction between modes with numbers k and $(k+4)$ spectral problem become solvable analytically since each term may be treated separately. Here the process of two-mode squeezing is of the most interest. Its Hamiltonian may be written in terms of SU(1, 1) algebra generators

$$K_+ = a_1^\dagger a_3^\dagger, \quad K_- = a_1 a_3, \quad K_0 = \frac{1}{2}(a_1^\dagger a_1 + a_3^\dagger a_3 + 1) \quad (10)$$

and diagonalized using coherent representations of this algebra.

2.4. *General case*

Resonance condition $\Omega = \frac{m\pi}{L} = \frac{2l\pi}{L}$ for m even gives the following Hamiltonian, which describes processes discussed above.

$$H'_m = \frac{q_0 l\pi}{4L^2}\left(a_l^{\dagger 2} + a_l^2\right) + \frac{q_0 m\pi}{4L^2}\sum_{k,j,k\neq j,k+j=m}\frac{kj}{j^2-k^2}\left(\frac{k}{j}\right)^{1/2}\left(a_j^\dagger a_k^\dagger + a_j a_k\right) - \frac{q_0 m\pi}{4L^2}\sum_k \sqrt{k(k+m)}\left(a_k^\dagger a_{k+m} + a_k a_{k+m}^\dagger\right). \quad (11)$$

it is worthwhile to notice that the first m modes undergo squeezing due to the DCE, while other modes become excited through the interaction mechanism similar to phase modulation. The mode with number $l = m/2$ undergoes single mode squeezing while others are being squeezed by pairs located simmetrically relative to "central" mode indexed by l. Similar situation takes place for resonance condition when m is odd. But in this case there is no process of single mode

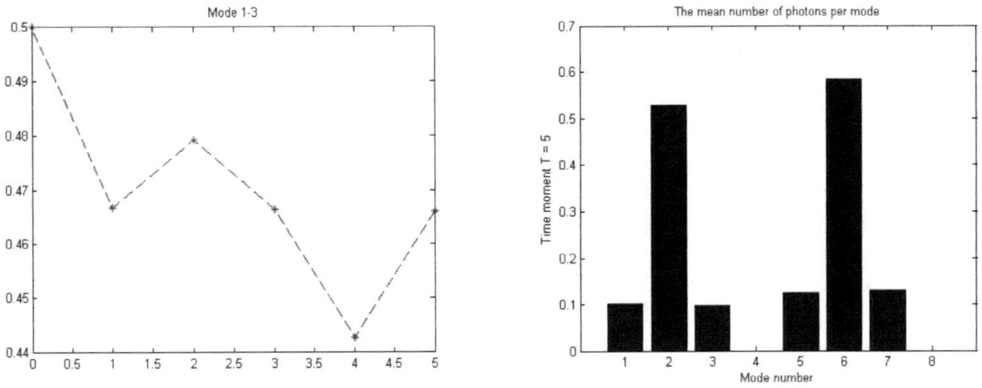

Figure 2. Photon generation in DCE; two mode variance of quadrature for modes with numbers $m = 1$ and 3 (left) and mean photon number per mode under the process described by Hamiltonian from Eq.(9) (left) at time $T = 5mcs$

squeezing

$$H''_m = -\frac{q_0 m\pi}{4L^2} \sum_{k,j,k\neq j,k+j=m} \frac{kj}{j^2-k^2}\left(\frac{k}{j}\right)^{1/2}\left(a_j^\dagger a_k^\dagger + a_j a_k\right) + \\ \frac{q_0 m\pi}{4L^2}\sum_k \sqrt{k(k+m)}\left(a_k^\dagger a_{k+m} + a_k a_{k+m}^\dagger\right). \quad (12)$$

2.5. Conclusion

Here we investigated the evolution of intracavity electromagnetic field in Dynamical Casimir Effect. Using effective Hamiltonian we show that there are three processes which takes place in the cavity: intermode interaction which is similar to phase modulation, single mode squeezing for the case of resonance with the mode indexed by even number and two mode squeezing which takes place for each resonance.

References

[1] Moore G 1970 *J. Math. Phys.* **11** 2679-2691.
[2] Dodonov V V 2010 *Physica Scripta* **82** 038105
[3] Wilson C M et all 2011 *Nature* **479** 376-379
[4] Law C K 1994 *Phys. Rev. A* **49** 433-537
[5] Miroshnichenko G P 2015 arXiv:1504.01632

Influence of ion irradiation on the resistive switching parameters of SiO_x-based thin-film structures

D S Korolev, A N Mikhaylov, A I Belov, V A Sergeev, I N Antonov, A P Kasatkin, O N Gorshkov and D I Tetelbaum

Lobachevsky State University of Nizhni Novgorod, 23/3 Gagarin prospect, Nizhni Novgorod, 603950, Russia

E-mail: dmkorolev@phys.unn.ru

Abstract. For the memristive $Au/SiO_x/TiN/Ti$ structures produced by magnetron sputtering and demonstrating reproducible bipolar resistive switching, the dependence of resistive states on the dose of irradiation with H^+ and Ne^+ ions with energy of 150 keV has been established. It is shown that, under proton irradiation, the resistive switching is not deteriorated up to the dose of $1{\cdot}10^{16}$ cm^{-2}, and, in the case of Ne^+ irradiation – up to the dose of $3.3{\cdot}10^{15}$ cm^{-2} (equivalent by the ionization losses). As the Ne^+ ions produce three orders of magnitude more elastic collisions than protons (for the same dose), the obtained results allow to predict the high radiation tolerance of the memristive structures to both ionization and displacement damage.

1. Introduction

One of the relevant problems of present electronics is the development of a new generation non-volatile memory (Resistive Random Access Memory – RRAM) based on using two stable states of a material (usually insulator): a high-resistance state (HRS) and a low-resistance state (LRS). Bipolar switching between these states is realized by applying voltage pulses of different polarity. Such a memory device is also called memristor [1] that can be implemented in a simple capacitor-like structure of the "metal-insulator-metal" (MIM) type [2]. Because of the simplicity of fabrication, scalability and low switching time, RRAM devices can replace in the nearest future the entire hierarchy of computer memory.

The development of new radiation-tolerant electronic components is also very important for the space and nuclear power applications. In view of the fact that irradiation leads to the loss of efficiency of traditional charge-storage Flash-memory, RRAM could be one of the main types of memory for the applications in radiation environments. However, the environmental test of radiation tolerance is quite complicated and often not available for researchers. We know only few works where HfO_2-based memristors were irradiated by high-energy protons (1-10 MeV) [3,4,5]. The authors of these studies have established that LRS, due to the formation of conductive channels (filaments) in oxide under the electroforming, shows high radiation hardness, and changes, observed in HRS, do not significantly affect the memory properties. However, the observed phenomena may differ for the memristive structures based on other oxide materials. Moreover, the data obtained in different experimental conditions cannot provide sufficient information about the mechanisms of radiation effects for the definite type of memristors.

Previously, we have shown that the tolerance of microstructures, which are located near the surface of solids, to the fast neutron irradiation can be examined by medium-energy ion irradiation [6]. The size of memristor active area in the MIM structure is limited by the thickness of insulator and electrodes and these regions are located at the depths of tens of nanometers. That is why we can successfully apply ion beams with medium energies to simulate the influence of high-energy particles, such as fast neutrons of cosmic protons.

At the present work, H^+ ions (150 keV) were used to simulate the impact of cosmic protons on the parameters of resistive switching (RS) in CMOS-compatible silicon oxide-based MIM-structures. To determine the contributions of both ionization and elastic losses, the MIM-structures were also irradiated by much heavier Ne^+ ions (150 keV).

2. Experimental

The capacitor-like MIM structure was deposited on CMOS-compatible TiN (25 nm) / Ti (25 nm) / SiO_2 (500 nm) / Si substrate. SiO_x film (40 nm thick) was deposited using RF-magnetron sputtering system MagSputt 3G-2 (Torr International) from fused silica in the argon-oxygen mixture (8% oxygen content) at substrate temperature of 200 °C. The top Au electrodes (40 nm) were deposited by DC-magnetron sputtering at 200 °C. The current-voltage (I-V) characteristics were measured by Agilent B1500A semiconductor device analyzer. The sign of bias on the device corresponds to the potential of top electrode (Au) relative to the grounded bottom electrode (TiN). After electroforming and switching to both HRS and LRS, the fabricated MIM structures were irradiated by H^+ and Ne^+ ions (150 keV) with doses in the range of $1\cdot10^{11} - 1\cdot10^{16}$ cm^{-2}.

3. Results and discussion

In the initial state, the MIM-structure shows typical characteristics of a capacitor with the silicon suboxide as dielectric layer. After electroforming, which was provided at negative bias of about 3-6 V, the current grows by several orders of magnitude. The structure begins to manifest reproducible bipolar resistive switching (RS) with the switching bias of 3-4 V and large number of switching cycles (more than 10^4) [7]. The typical RS hysteresis loops for the Au/SiO_x(40 nm)/TiN/Ti structures subjected to electroforming are given in figure 1.

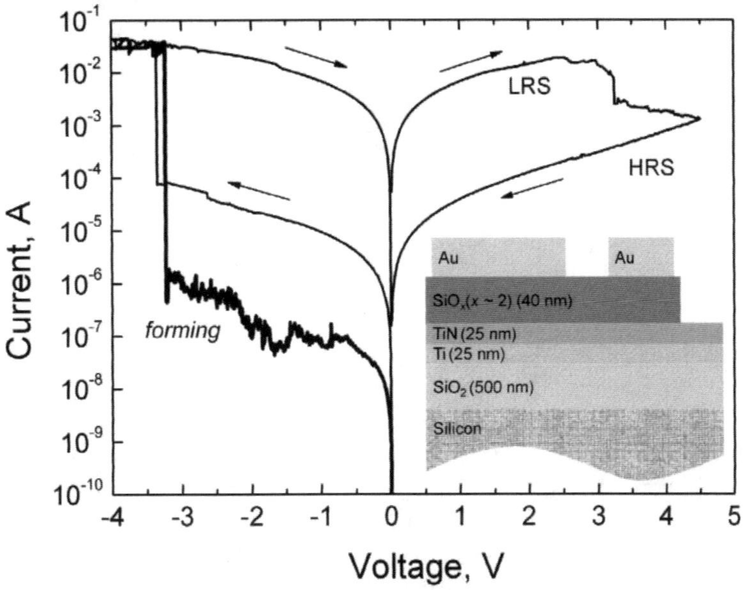

Figure 1. I-V characteristics of the Au/SiO_x/TiN structure. Schematic representation of the structure cross-section is shown on the inset.

To choose irradiation regimes for the present MIM-structure, the elastic and ionization losses under irradiation with 150 keV hydrogen ions and 10 MeV space protons, and also for 150 keV neon ions were calculated using the SRIM software [8] (figure 2). The figure shows that the elastic and ionization losses in SiO_x film under 10 MeV proton irradiation are almost 14 times lower than under irradiation with 150 keV H^+ ions. Elastic losses (a number of displacements/vacancies) for 150 keV proton irradiation are about three order of magnitude lower than for Ne^+ ions of the same energy.

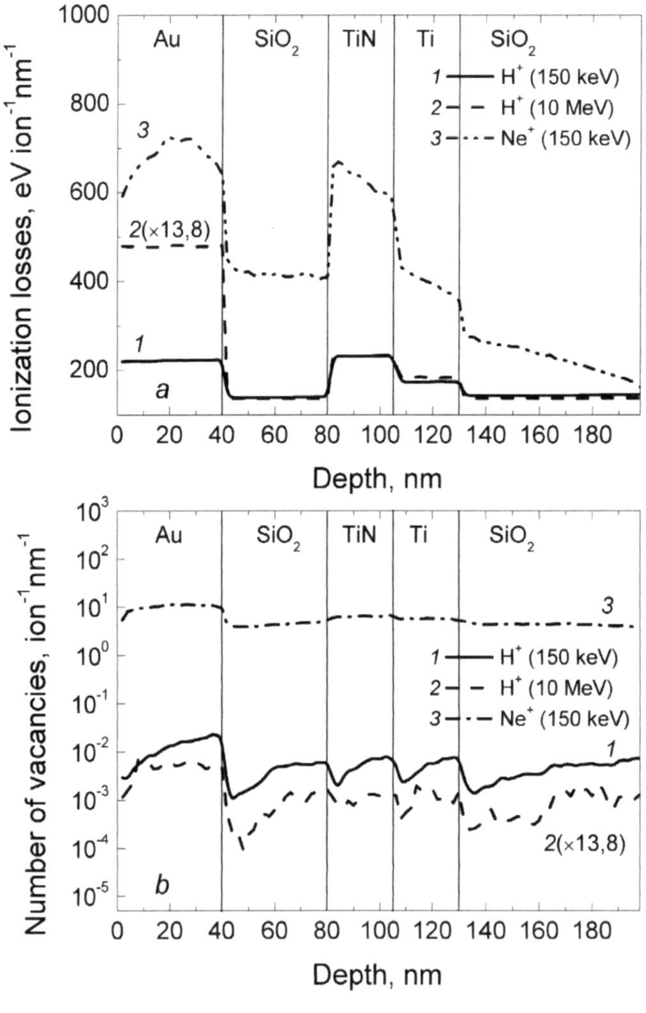

Figure 2. Depth distribution of ion energy ionization losses (*a*) and generated in elastic collision vacancies (*b*) calculated by SRIM [8].

Radiation tolerance of memristors may depend on the state in which they were under irradiation. Therefore, ion irradiation was performed for the structures which were before irradiation in each state – HRS and LRS. The values of the LRS and HRS currents measured at +0.5 V vs. the dose of H^+ and Ne^+ irradiation are given in figure 3. Experiments were carried in the following manner. Before the irradiation, the structures were switched into HRS or LRS with the subsequent current measurement at the reading voltage, after that they were irradiated to a certain dose and again subjected to switching and measurement (5 switching cycles) and so on (the accumulated doses are indicated in figure 3).

In the vast majority of cycles, there has been no significant change in currents for proton irradiation up to the maximal used dose ($1\cdot10^{16}$ cm^{-2}) independently of the structure state before irradiation. Since the RS in investigated structures is maintained after proton irradiation up to the dose of $1\cdot10^{16}$ cm^{-2} (figure 3), then one can conclude that the RS will be maintained under irradiation by cosmic protons with the doses much higher than maximal dose used in this work. In the case of Ne^+ irradiation, the situation is similar for currents in the HRS up to the maximal dose ($3.3\cdot10^{15}$ cm^{-2}). Note that for the used doses, the integrated values of energies spent on the ionization losses near the surface are equal for both types of used ions. At the same time, when the memristor is irradiated in HRS, there can be a spontaneous switching to LRS during irradiation with both types of ions, however the HRS is always reset after the subsequent switching cycle. The nature of spontaneous resistive switching under irradiation requires further investigation: it may be due to high energy ionization losses (i.e., charge accumulation) and possible elastic collisions of ions with atoms of oxide (formation of recoils and the corresponding radiation defects that is more significant for Ne^+ ions). In the case of Ne^+ irradiation, some noticeable degradation in the HRS (decrease in resistance) is observed for the highest dose, probably due to the accumulation of radiation defects in the entire volume of oxide material.

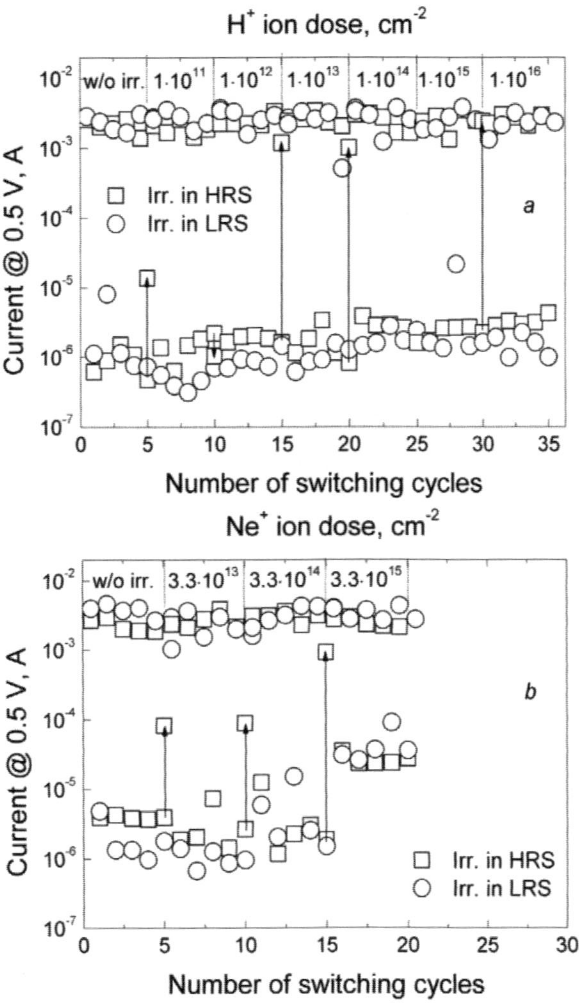

Figure 3. Currents measured in HRS and LRS of Au/SiO_x/TiN/Ti structures without irradiation and after irradiation with H^+ (*a*) and Ne^+ (*b*) vs. the number of switching cycles after each irradiation. Arrows show the increased currents measured after irradiation.

The nature of the observed reproducible bipolar RS in SiO_x-based thin-film memristive structures can be related to the formation and local oxidation of silicon conducting channels (filaments) in the bulk of SiO_x layer [9]. Such filaments produced by breaking Si-O bonds under high electric stress can not be affected by ion irradiation in LRS, whereas the local dielectric interlayer formed in the site of filament oxidation can be sensitive to both ionizing and defect-producing irradiation, which leads to some observed HRS changes.

4. Conclusions

In summary, the irradiation with medium-energy H^+ and Ne^+ ions has revealed high stability of resistive states of Au/SiO_x/TiN/Ti memristive structures. The obtained results allow to predict the high radiation tolerance of the SiO_x-based memristors to both ionizing irradiation and displacement damage.

The work has been supported by the Ministry of Education and Science of the Russian Federation (RFMEFI57514X0029). A.I.B. acknowledges support in the frame of the grant of the President of the Russian Federation (MK-3714.2015.2).

References

[1] Chua L 1971 *IEEE Transactions on Circuit Theory* **CT18** 507
[2] Strukov D, Snider G, Stewart D and Williams R 2008 *Nature* **453** 80
[3] Lee D, Lee J, Jung S, Kim S, Park J, Biju K P, Choe M, Lee T and Hwang H 2011 *IEEE Transactions on Nuclear Science* **58** 3317
[4] Butcher B, He X, Huang M, Wang Y, Liu Q, Lv H, Liu M and Wang W 2010 *Nanotechnology* **21** 475206
[5] Bi J S, Han Z S, Zhang E X, McCurdy M W, Reed R A, Schrimpf R D, Fleetwood D M, Alles M L, Weller R A, Linten D, Jurczak M and Fantini A 2013 *IEEE Transactions on Nuclear Science* **60** 4540
[6] Tetelbaum D I, Guseinov D V, Vasiliev V K, Mikhaylov A N, Belov A I, Korolev D S, Obolensky S V and Kachemtsev A N 2014 *Nucl. Instr. Meth. Phys. Res. B* **326** 41
[7] Mikhaylov A N, Belov A I, Guseinov D V, Korolev D S, Antonov I N, Efimovykh D V, Tikhov S V, Kasatkin A P, Gorshkov O N, Tetelbaum D I, Gryaznov E G and Yatmanov A P 2015 *Mat. Sci. Eng. B* **194** 48
[8] Ziegler J F, Ziegler M D and Biersack J P 2010 *Nucl. Instr. Meth. Phys. Res. B.* **268** 1818
[9] Mehonic A, Cueff S, Wojdak M, Hudziak S, Jambois O, Labbe C, Garrido B, Rizk R and Kenyon A 2012 *Journal of Applied Physics* **111** 074507

Specialized probes with nanowhisker structures for high resolution magnetic force microscopy

M.V. Zhukov[1,3], K.I. Belousov[1], A.M. Mozharov[2], I.S. Mukhin[1,2], A.O. Golubok[1,3]

[1] Department of material science and nanotechnology, ITMO University, 197101 Kronverkskiy pr., 49, Saint-Petersburg, Russian Federation
[2] Laboratory of renewable energy source, St Petersburg Academic University RAS, 194021 Khlopina, 8/3, Saint-Petersburg, Russian Federation
[3] Nanotechnology department, Institute for Analytical Instrumentation RAS, 198095 Ivan Chernykh, 31-33, Saint-Petersburg, Russian Federation

E-mail: cloudjyk@yandex.ru

Abstract. Creation and study of specialized nanowhisker probes with magnetic coating are performed for high-precision imaging of various objects by means of atomic force and magnetic force microscopy. Thin layers of Ni and Co are deposited on the surface of nanowhisker structures to perform visualization of magnetic fields on the sample surface, in particular, structure of pits on a hard disk drive (HDD). It is revealed that probes with nanowhisker structures covered with magnetic coating due to their high aspect ratio demonstrate a higher spatial resolution and contrast of magnetic fields visualization in comparison with standard magnetic probes.

1. Introduction

Scanning probe microscopy (SPM) is one of the main methods of high precision material investigation and modification with nanometer resolution. Advantages of SPM are ability to visualize geometric dimensions of surface features of various materials as well as to perform measurements of mechanical properties, the spatial distribution of adhesion forces, electric and magnetic fields on the surface of samples.

Magnetic force microscopy (MFM) should be also distinguish as one of promising technique in nanotechnology due to its ability to perform both highly accurate reading of magnetic fields distribution [1] and recording of magnetic information [2, 3]. The important role of nanoprobes for MFM resolution should be noted. Indeed, geometric parameters and the type of probe determine the sensitivity of this method, the resulting resolution and the contrast of obtained images.

The aim of this work is to create and study a new type of magnetic probes based on whisker structures, including the simulation of magnetic probes with different geometric shapes.

2. Experimental setup

Producing of nanowhisker structures on the summits of probes and control of their parameters were performed by means of scanning electron microscope CrossBeam Neon 40 (Carl Zeiss, Germany) [4, 5]. Study of nanowhisker probes possibilities were carried out on SPM complex Ntegra Aura (NT-MDT, Russia).

Deposition of magnetic materials (Ni, Co) on the top of probe pyramid surface was produced by electron sputtering with Auto 500 Edwards (BOC Edw., England).

3. Results and Discussions

Investigation of the probe with nanowhiskers (NW) in the air was carried out in tapping mode, increasing the stability of NW probes and reducing damaging effects of the probe to the sample. Study of micro- and nanostructures of test lattice TGQ01 with calibrated geometric parameters demonstrated a two-fold decreasing of broadening on vertical steps projection of the surface using NW probes in comparison with standard probes. Furthermore, imaging of nanoscale channels in electron resist PMMA on Si substrate revealed improvement of penetration ability using NW probes. The phenomenon of lateral resolution and penetration ability improvement in case of NW probes can be explained by the high aspect ratio of nanowhisker structures (length to diameter ratio) [6].

Distribution of magnetic fields on test specimens in the form of segments of a magnetic hard disk (HDD) was investigated with standard and specialized NW probes whose surfaces were coated with thin layers of magnetic material (Ni, Co). Magnetic fields study was performed with two-pass method. On the first pass information about topological features was obtained, while the second pass at a fixed height (10 nm) from the sample surface gave information about magnetic fields according to probe oscillation phase shift. The magnetic probe oscillated at the resonant frequency ($f_0 \approx 133$ kHz) during the distribution of magnetic fields measurement. The presence of magnetic field provided additional damping effect influenced on a phase shift between the driving force and the vibrations of probe. Our investigation showed that a NW probe with Ni or Co magnetic coatings gives opportunities to obtain better contrast of MFM phase in comparison with a standard probe (see Fig. 1, 2). In case of Co coating MFM images were more contrast. The thickness of magnetic coating was 50 nm. Increasing of magnetic thickness led to a deterioration of image quality due to reduction of probe sharpness, while decreasing resulted in a lack of sensitivity to magnetic fields.

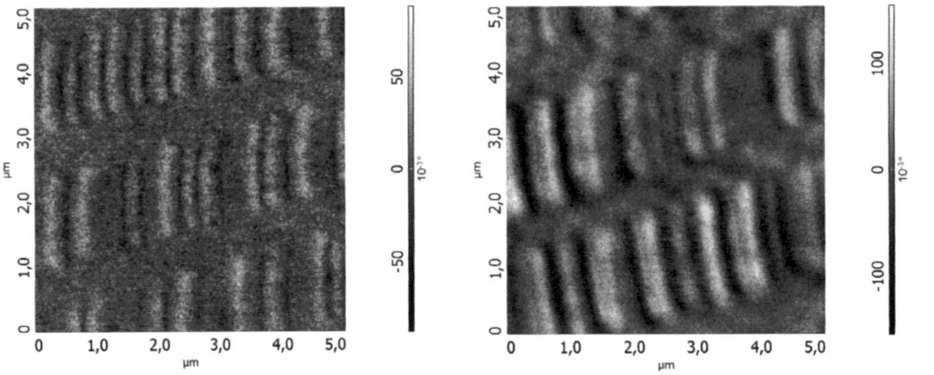

(a) (b)

Fig. 1. MFM images of HDD magnetic field obtained by standard probe (a) and NW probe (b) with magnetic 50 nm Ni coating.

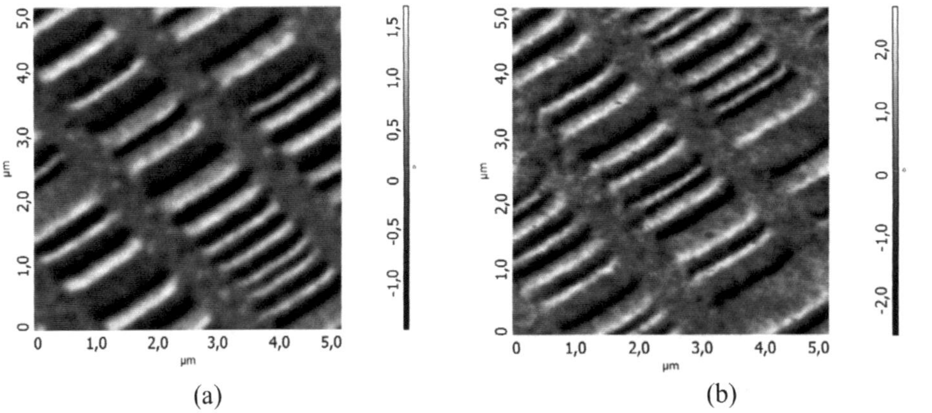

(a) (b)

Fig. 2. MFM images of HDD magnetic field obtained by standard probe (a) and NW probe (b) with magnetic 50 nm Co coating.

Journal of Physics: Conference Series **643** (2015) 012095 doi:10.1088/1742-6596/643/1/012095

Fig. 3 and 4 shows MFM images of individual pits obtained with standard and whisker magnetic probes covered with thin layer of Co. One can see that whisker probe is characterized with improvement of sensitivity, phase contrast and spatial resolution of MFM image.

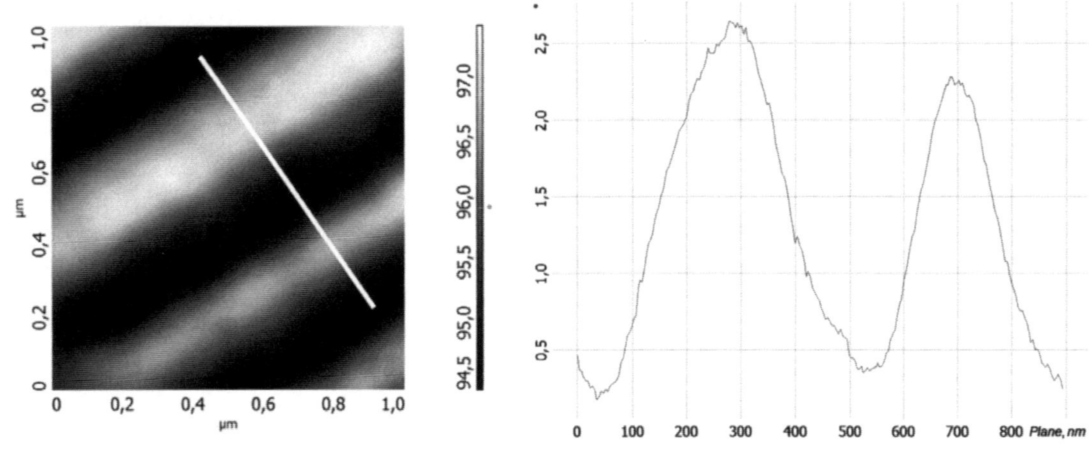

(a) (b)

Fig. 3. MFM image of individual pits (a) and cross-section (b) obtained with standard probe.

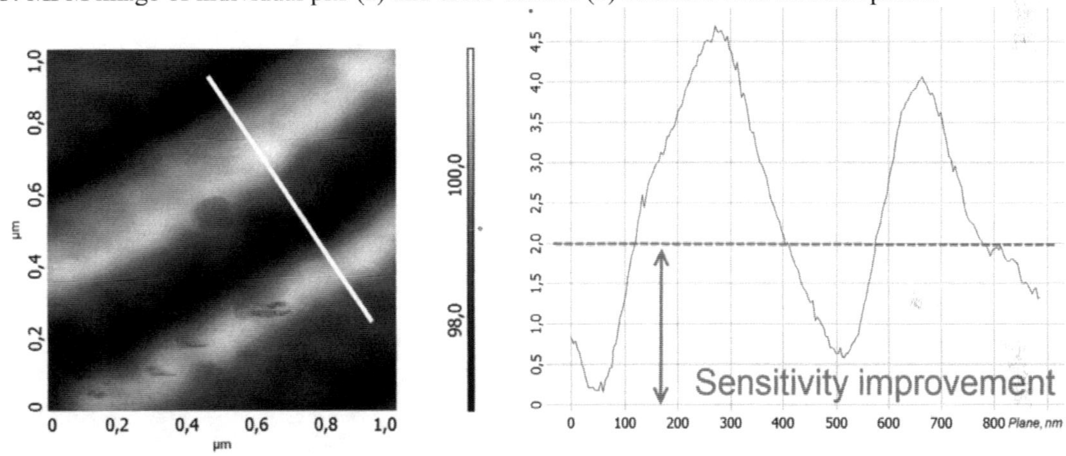

(a) (b)

Fig. 4. MFM image of individual pits (a) and cross-section (b) obtained with NW probe.

It should be also noticed that so-called "cross effect" linked with the influence of magnetic probe on sample magnetic field distribution was observed (Fig. 5). The phenomenon of "cross effect" is more significant in the case of NW probes compared to standard probes.

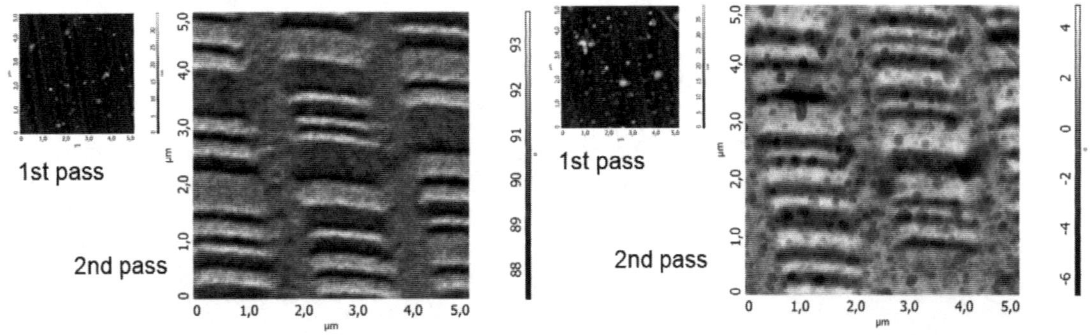

(a) (b)

Fig. 5. Observation of the "cross effect" with standard probe (a) and NW probe (b).

Numerical simulation of magnetic probes with geometry of standard probe and NW probe was carried out. Simulation was performed by means of COMSOL Multiphysics software based on finite element method for solving physics-based problems. A distribution of magnetic field was calculated by Maxwell's equations. Magnetized ferromagnetic layer of probes and HDD's bits were used as sources of magnetic field.

Field induced in a paramagnetic material of the probe, unmagnetized ferromagnetic material of HDD's surface and environment was calculated as:

$$\mathbf{B} = \mu_0 \mu_r \mathbf{H} \ ,$$

where **B** is the magnetic flux density (T), **H** is the magnetic field intensity (A/m), μ_0 is the permeability of vacuum ($4\pi \cdot 10^7$ H/m), μ_r is the relative permeability.

Resulting field in the bits and the magnetic part of the probe was performed with the following expression:

$$\mathbf{B} = \mu_0 \left(\mathbf{H} + \mathbf{M} \right),$$

where **M** is the magnetization vector (A/m).

Calculation of forces acting on the probe was obtained by integration of Maxwell stress tensor over the boundary of the probe. Maxwell stress tensor for only magnetic field can be written as:

$$\sigma_{ij} = \frac{1}{\mu_0} B_i B_j - \frac{1}{2\mu_0} B^2 \delta_{ij}$$

where B_i, B_j are the components of magnetic field, δ_{ij} is the Kronecker's delta.

During simulation magnetizations of ferromagnetic layer of the probe and HDD's bits were set as -300 emu/cm^3 and ± 600 emu/cm^3, respectively, where minus indicated the direction of the magnetization vector against the z-axis. The distribution of magnetization for standard probe and NW probe is shown in Fig. 6.

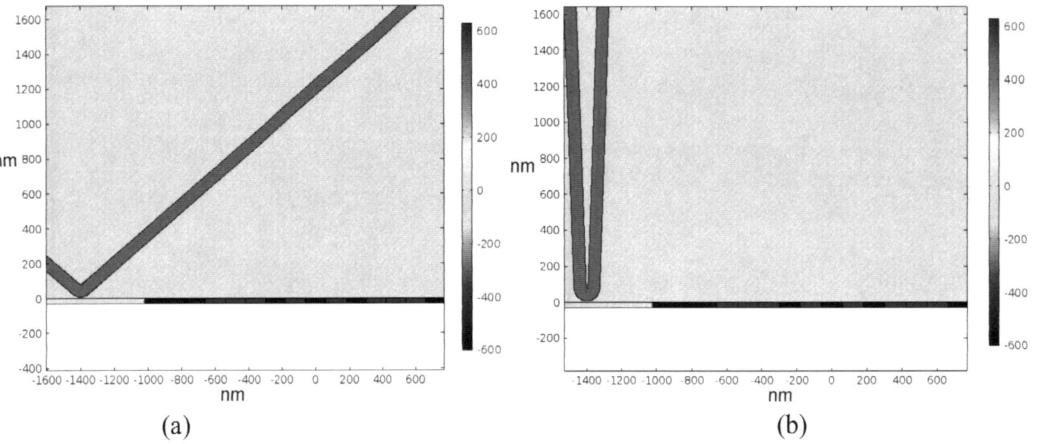

Fig. 6. The distribution of magnetization z component for Standard probe (a) and NW probe (b)

The relative permeability of paramagnetic material of the probe, unmagnetized ferromagnetic material of HDD's surface and the environment were set as 12, 4000 and 1, respectively. The thickness of magnetic layer of probes and HDD was set as 50 nm and 30 nm respectively. Width of bits was set as 120 nm. Calculation of forces acting on the probe was carried out at a distance of 8 and 12 nm from the HDD's surface. Force gradient was calculated at a distance of 10 nm from the HDD's surface by division of difference of force at 8 and 12 nm by 4 nm.

The simulation confirmed that NW magnetic probes are more sensitive in comparison with standard probe for individual bits and provides better MFM resolution (Fig. 7).

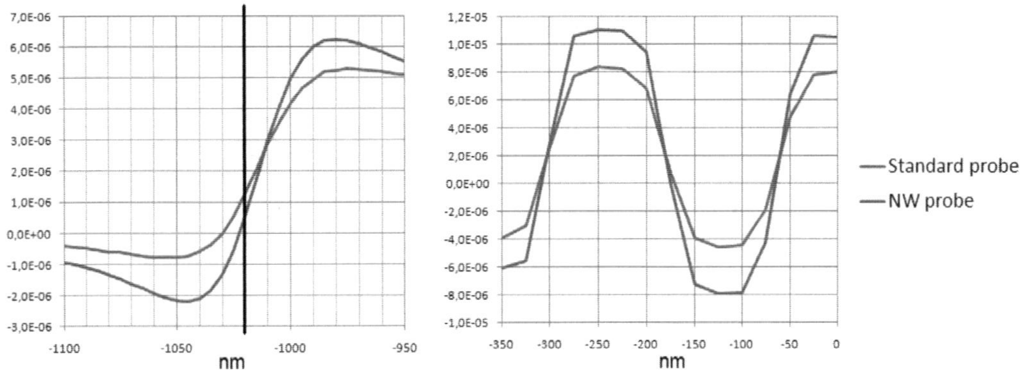

(a) (b)

Fig. 7. Force gradients for standard and NW probes: (a) near the beginning of bits (thick line); (b) for individual bits. Zero level indicates gradient above unmagnetized ferromagnetic material of HDD

4. Conclusions

Magnetic probes based on whisker structures were created and studied. The approbation of these probes was carried out on test samples with calibrated geometric parameters and magnetic disks. The simulation of MFM images was performed for magnetic probes with different geometry. A comparison of the results obtained by standard magnetic probes and specialized magnetic NW probes was carried out. Both theoretically and experimentally data showed that magnetic NW probes significantly improves the sensitivity, contrast and spatial resolution of magnetic force microscopy.

Work was carried out with the support of the leading universities of the Russian Federation (grant 074-U01), the Russian Foundation for Basic Research (14-02-31703 mol_a), and program U.M.N.I.K.

References

[1] M.R. Koblischka, U. Hartmann, T. Sulzbach 2003 Improvements of the lateral resolution of the MFM technique *Thin Solid Films* Vol. 428 pp. 93–97.

[2] Yu. P. Ivanov, A. I. Il'in, E. V. Pustovalov, L. A. Chebotkevich 2010 Influence of induced anisotropy on the processes of magnetization reversal of cobalt circular nanodots *Physics of the Solid State* Vol. 52 issue 8 pp 1694-1700.

[3] A.M. Alekseev, B.A. Gribkov, D.S. Nikitushkin, V.L. Mironov, S.N. Vdovichev and A.A. Fraerman 2008 Investigations of MFM tip induced remagnetization effects in CoPt ferromagnetic nanoparticles *Proceedings of MISM-2008* (Moscow, Russia) pp 260.

[4] V.V. Levichev, M.V. Zhukov, I.S. Mukhin, A.I. Denisyuk, A.O. Golubok 2013 On the operating stability of a scanning force microscope with a nanowhisker at the top of the probe *Technical Physics* Vol. 58(7) pp 1043-1047.

[5] I.S. Mukhin, I.V. Fadeev, M.V. Zhukov, V.G. Dubrovskii, A.O. Golubok 2015 Framed carbon nanostructures: Synthesis and applications in functional SPM tips *Ultramicroscopy* Vol. 148 pp 151–157.

[6] M.V. Zhukov, I.S. Mukhin, V.V. Levichev, A.O. Golubok 2015 Investigation of the contrast inversion effect on hydrophilic surfaces using Pt/C whisker probes in a scanning force microscope *Technical Physics Letters* Vol. 41(2) pp 149-152.

Effect of carbon doping on magnetic properties of Mn/Si interface

G S Grebenyuk[1], M V Gomounova[1], I I Pronin[1,2], B V Senkovskiy[3], D V Vyalikh[3]
[1] Ioffe Physical-Technical Institute, 26 Politekhnicheskaya, St Petersburg 194021, Russian Federation
[2] ITMO University, 197101 St. Petersburg, Russia
[3] Institut fur Festkorperphysik, Technische Universitat Dresden, Helmholtzstrasse 10, Dresden 01069, Germany

E-mail: georgijmail@gmail.com

High-resolution photoelectron spectroscopy with synchrotron radiation and magnetic linear dichroism in Mn 3*p* core-level photoemission has been used to study the effect of carbon-doping on the initial stages of Mn/Si interface formation and its ferromagnetic ordering. It is shown that the manganese deposition on Si(100)2x1 surface with and without carbon doping results in the formation of the interfacial MnSi silicide and the solid solution of silicon in manganese. The Mn film covered with segregated Si begins to grow on the samples surface at coverage of 9 Å of Mn. It is shown that in-plane ferromagnetic ordering of the interfaces arises after of 2-4 Å Mn in both systems, however, the remanent magnetization of Mn-C/Si is higher than that of Mn/Si.

1. Introduction

Formation of magnetic films on semiconductors surfaces are of great fundamental and technological interest because they are very promising for spintronic devices. Especially much attention was paid to the magnetic semiconductors and to the silicides of transition metals, in particularly manganese silicides, which can be integrated into the standard silicon-based technology [1-3]. Despite the fact, that there is no any bulk Mn-Si phases with a Curie temperature above 50 K, ferromagnetism near and above room temperature was observed in thin Mn/Si structures formed by vacuum deposition [4] and ion implantation [5-6] methods. The high-temperature ferromagnetism was found also in discrete manganese-silicon alloys created by delta-doping [7] and for Mn_xSi_{1-x} films formed by magnetron sputtering [8] and laser synthesis [9]. The mechanism of this interesting phenomenon is still discussed. Different explanations of high-temperature ferromagnetism have been proposed as a consequence of the reduced coordination near surfaces and interfaces of silicon or due to presence of structural defects of the silicon lattice which occur in the process of preparation of the alloy [10]. Another possible reason is the presence of carbon in the grown manganese silicide film, which can results in anisotropic modification of the local structure around the Mn sites [11]. It was found that the presence of carbon in Mn_5Si_3 films enhances its Curie temperature up to maximum of 350 K [12]. However, the thickness of formed films was about 100 nm, and the influence of carbon on the magnetic properties of ultrathin films was not revealed. The aim of the present work was to study the role of carbon doping in high-temperature ferromagnetism of Mn/Si interface. We have investigated the initial growth (up to 2 nm) of both pure and carbon-doped manganese films on silicon and for the first time studied their magnetic properties *in situ* using magnetic linear dichroism (MLD) in Mn 3*p* photoemission.

2. Experimental

The experiments were carried out in the Russian–German laboratory at the electron storage ring HZB BESSY II under ultra-high vacuum (UHV) conditions (10^{-10} mbar). The samples were prepared from phosphorus-doped (1 Ω/cm) silicon wafers misoriented by ~0.1° relative to the (100) plane. The Si surface was routinely cleaned by *ex situ* chemical processing followed by heating to 1450 K in UHV. This procedure provided the atomically clean reconstructed surface, which is characterized by the clear low energy electron diffraction pattern of the type Si(100)2×1. Manganese films were deposited on the silicon substrates from a thoroughly degassed source. Carbon doping of the films was carried by the use of C_3H_6 adsorption during Mn deposition. The pressure of gas was 10^{-8} mbar that provided the concentration of carbon of about 15%.

The phase composition and the electronic structure of the films were analyzed by high-resolution photoelectron spectroscopy with synchrotron radiation. The photon energy was varied from 110 to 600 eV, but most of the spectra were measured at 130 eV, an energy at which the surface sensitivity of the Si 2*p* spectra is maximum. The overall energy resolution of the analyzer and the monochromator was set to 100 meV full width at half maximum.

The magnetic properties of the interfaces were analyzed by magnetic linear dichroism in Mn *3p* photoemission. This effect consists in that the energy position and shape of the Mn 3*p* peak measured within a narrow solid angle depends on the direction of sample magnetization. The origin of the effect lies in the splitting of the spin-orbit doublet, $3p_{1/2}$ and $3p_{3/2}$, into sublevels with different magnetic quantum numbers, which is induced by core-level interaction with the magnetically polarized valence band [13-17]. The samples were magnetized along the two opposite directions lying in the plane of the sample surface by two Helmholtz coils placed inside a vacuum chamber. Current pulses passing through the coils generated a magnetic field of 1000 Oe to saturate the sample magnetization. The measurements of the Mn 3*p* spectra were performed in remanence at room temperature.

3. Results and discussion

The representative Si 2*p* and valence band spectra taken at increasing doses of Mn deposited on Si(100)2x1 with and without carbon doping are shown in Fig. 1.

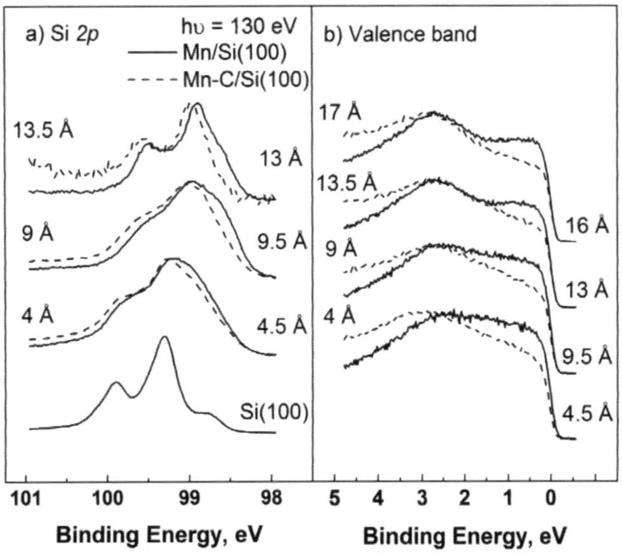

Figure 1. Si 2p (a) and valence band (b) spectra measured after the deposition of similar amounts of manganese on Si(100)2x1 surface with (dashed lines) and without carbon doping (solid lines).

The dynamics of change in these spectra for both systems is very similar. For better comparison the spectra are normalized to the maxima of the curves. It is seen that the line shape is quite sensitive to the Mn deposition that reflects a change in the state of silicon atoms in the analyzed surface layer and the formation of phases with the participation of Si, Mn and C atoms in the interface region.

More detailed information on the interfaces formation can be obtained by using the least-square fitting procedure, in which the spectra are decomposed into surface and bulk components. The results of such analysis for the undoped Mn/Si system are described in our recent paper [18]. Decomposition of Si 2*p* spectra measured for carbon-doped system is shown in Fig. 2. It is seen that after the deposition of 4 Å of manganese the spectrum consists of four modes. It is the mode B that corresponds to the bulk silicon atoms and three components (I, C and S) with binding energies equal to 99.55, 99.00 and 98.67 eV, respectively. These binding energies are close to the values obtained for Mn/Si interface corresponding to interfacial MnSi silicide, the film of a solid solution of silicon in manganese and the silicon segregated on the sample surface. Thus, in the case of Mn-C/Si interface I, C and S modes should be attributed to the $Mn_{1-x}C_xSi$ silicide, Mn-C-Si solid solution and the segregated silicon.

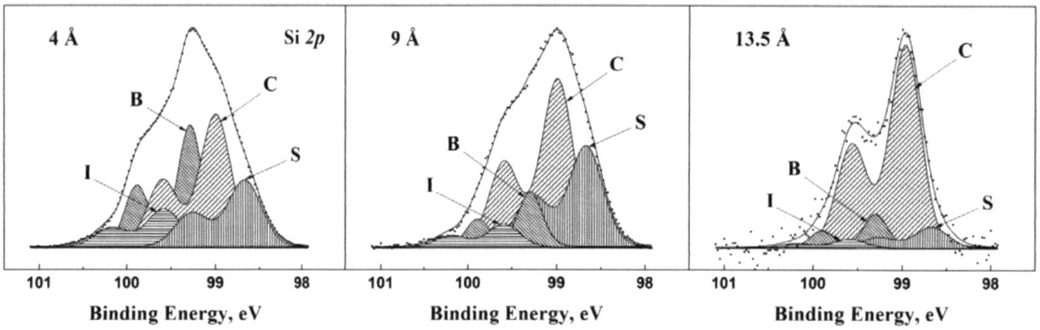

Figure 2. Decomposition of the Si 2p spectra from the Mn-C/Si sample with different amounts of Mn deposited at room temperature.

Based on the experimental dependences of intensity of all modes of Si 2*p* spectra on the coverage, we estimated the thickness of all layers forming the interface by assuming an exponential decay of the signal from each layer with its depth. The obtained data showed that at the early stage of manganese deposition the ultrathin layer (2.1 Å) of interfacial $Mn_{1-x}C_xSi$ silicide is rapidly formed on the silicon surface. Then Mn-C-Si solid solution begins to grow on this layer. After the deposition of 9 Å of Mn its growth is slowing down due to the beginning of the formation of $Mn_{1-x}C_x$ film. The manganese surface is covered with 0.2 monolayer (ML) of segregated silicon. Thus described model of the interface formation is very similar to the model that we obtained in the absence of carbon doping. The main difference between two systems is decrease of the amount of silicon segregated on the sample surface from 0.4 ML for Mn/Si system to 0.2 ML for carbon doped one.

Let us now consider the magnetic properties of the formed films. The Mn 3*p* spectra taken after the deposition of various amounts of manganese for two opposite directions of the magnetic field (M_{up} and M_{down}) for Mn-C/Si system are shown in the left part of Fig. 3. The spectra measured in the "up" direction are characterized by the picks slightly shifted to the higher binding energies in compare with I_{down} spectra.

To reveal the differences between the spectra we plotted the difference curves ($I_{up} - I_{down}$) that are shown in Fig. 3b. The difference curves demonstrate the maximum and the minimum corresponding to the sublevels of Mn 3*p* multiplet with different magnetic quantum numbers $m_j = \pm 3/2$. The MLD effect is commonly characterized by the magnetic asymmetry amplitude (A):

$$A = [S(E)_{max} - S(E)_{min}] \times 100\%, \quad (1)$$

$$\text{where } S(E) = [I_{up}(E) - I_{down}(E)]/[I_{up}(E) + I_{down}(E)] \quad (2)$$

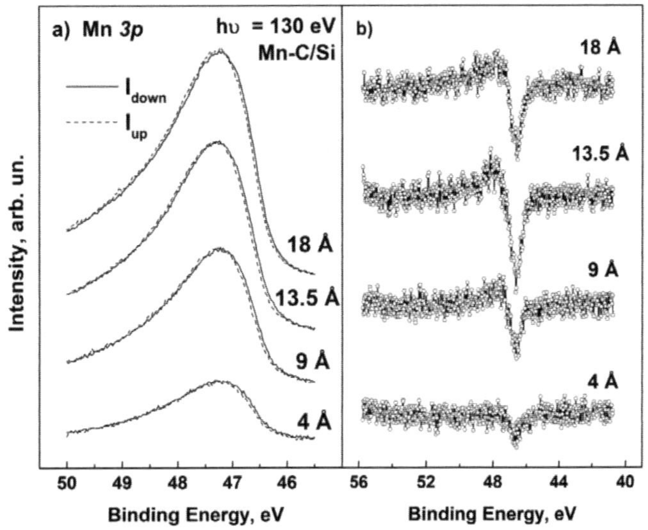

Figure 3. Mn 3*p* spectra $I_{up}(E)$ and $I_{down}(E)$ measured after the magnetization of the sample in two opposite directions in the surface plane (a) and the corresponding difference curves $I_{up} - I_{down}$ (b).

The parameter A is a direct measure of the sample remanent magnetization [16, 17]. Experimental data illustrating the dependence of A on the manganese coverage for both investigated systems are shown in Fig. 4.

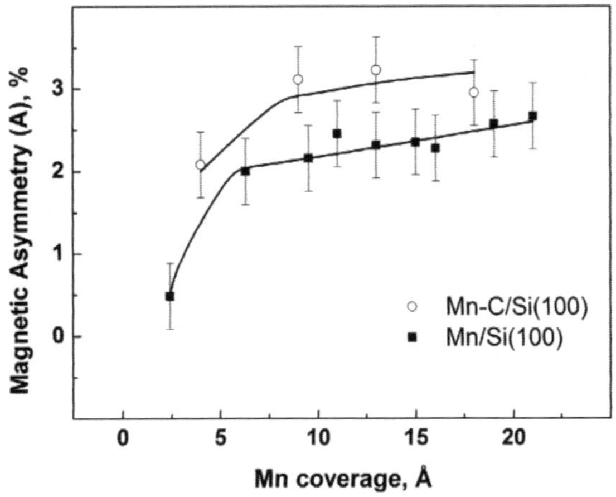

Figure 4. Magnetic asymmetry amplitude as a function of manganese coverage for Mn/Si and Mn-C/Si systems.

It is seen that the MLD effect appears in the range of 2-4 Å in both cases and enhances with increasing coverage. For example, at Mn coverage of 13 Å the value of magnetic asymmetry amplitude for undoped system becomes equal to 2.2% while for carbon-doped system it is approximately 30% higher than that for undoped one. Thereby our results demonstrate the influence of carbon on ferromagnetic ordering of ultrathin (up to 2 nm) manganese films on silicon surface.

4. Conclusion

Thus, the use of photoelectron spectroscopy with synchrotron radiation and MLD effect in the frame of a single experiment allowed us to reveal the specific features of both the formation of Mn/Si and Mn-C/Si interfaces and their ferromagnetic ordering. It is shown that the initial stages of growth of the carbon-doped and undoped manganese films are very similar. In both cases formation of interface silicide and Mn-Si solid solution precedes the growth of a metal film. The main difference is reduction of segregated silicon on the sample surface in carbon-doped system. The high-temperature ferromagnetic ordering arises in both systems at the coverage of 2-4 Å Mn. The magnetic asymmetry amplitude of the MLD effect is equal to 3 and 2.2% for carbon-doped and undoped films correspondingly that demonstrates essential influence of carbon on magnetic properties of the films.

Acknowledgments

The authors are grateful to the Helmholtz Zentrum Berlin for the possibility of using synchrotron radiation. G.S. Grebenyuk and I.I. Pronin acknowledge the support of the Russian Foundation for Basic Research (project no. 130200398).

References

[1] Magnano E, Bondino F, Cepek C, Parmigiani F and Mozzati M C 2010 *Appl. Phys. Lett.* **96** 152503
[2] Zou Z Q and Li. W C 2011 *Phys. Lett.* A **375** 849
[3] Azatyan S G, Utas O A, Denisov N V, Zotov A V and Saranin A A 2011 *Surf. Sci.* **605** 289
[4] Zhang F M, Liu X C, Gao J, Wu X S, Du Y W, Zhu H, Xiao J Q and Chen P 2004 *Appl. Phys. Lett.* **85** 786
[5] Bolduc M, Awo-Affouda C, Stollenwerk A, Huang M B, Ramos F G, Agnello G and LaBella V P 2005 *Phys. Rev.* B **71** 033302
[6] Zhou S, Potzger K, Zhang G, Mücklich A, Eichhorn F, Schell N, Grötzschel R, Schmidt B, Skorupa W, Helm M, Fassbender J and Geiger D 2007 *Phys. Rev.* B **75** 085203
[7] Chiu S H, Hsu H S and Huang J C A 2008 *J. Appl. Phys.* **103** 07D110
[8] Li T, Guo L, Liu C, Peng G, He B, Pan Z, Zhou Z, Xu S, Quan Z 2012 *Vacuum* **86** 1358
[9] Aronzon B A, Rylkov V V, Nikolaev S N, Tugushev V V, Caprara S, Podolskii V V, Lesnikov V P, Lashkul A, Laiho R, Gareev R R, Perov N S and Semisalova A S 2011 *Phys. Rev.* B **84** 075209
[10] Magnano E, Bondino F, Cepek C, Sangaletti L, Mozzati M C and Parmigiani F 2012 *J. Appl. Phys.* **111** 013917
[11] Surgers C, Gajdzik M, Fischer G, Lohneysen H V, Welter E and Attenkofer K 2003 *Phys. Rev.* B **68** 174423
[12] Gajdzik M, Surgers C, Kelemen M and Lohneysen H V 2000 *J. Appl. Phys.* **87** 6013
[13] Roth Ch, Hillebrecht F U, Rose H B and Kisker E 1993 *Phys. Rev. Lett.* **70** 3479
[14] Sirotti F and Rossi G 1994 *Phys. Rev.* B **49** 15682
[15] Bansmann J, Lu I, Meiwes-Broer K H, Schlatholter T and Braun J 1999 *Phys. Rev.* B **60** 13860
[16] Janke-Gilman N, Hochstasser M and Willis R F 2004 *Phys. Rev.* B **70** 184439
[17] Pronin I I, Gomoyunova M V, Malygin D E, Vyalikh D V, Dedkov Yu S and Molodtsov S L 2008 *J. Appl. Phys.* **104** 104914
[18] Varnakov S N, Gomoyunova M V, Grebenyuk G S, Zabluda V N, Ovchinnikov S G and Pronin I I 2014 *Physics of the Solid State* **56** 380

Model of quantum dot and resonant states for the Helmholtz resonator

A M Vorobiev[1] and I Yu Popov[1]

[1] ITMO University, Kronverkskiy, 49, St. Petersburg, 197101, Russia

E-mail: lenden31@yandex.ru

Abstract. The completeness of the system of resonant states for the operator extensions theory model of the Helmholtz resonator was proved. We deal with rectangular resonator with the Neumann boundary condition.

1. Introduction

Resonator with small boundary window is widely used model of a quantum dot weakly coupled with other parts of the nanosystem [1-8].The choice of the operator inside the dot depends on the physical motivation. The simplest case is when one use free Schrödinger operator (Laplacian). To describe the system effectively, it would be better to have a complete system of functions. . It is this property, that allows one to use separation of variables and Fourier series. Particularly, the Laplace operator for bounded domain has purely discrete spectrum and complete system of eigenfunctions. However, if one considers the Helmholtz resonator, i.e. the resonator coupled with the exterior through a small window, the situation changes. The eigenvalues and eigenstates transform, correspondingly, to resonances having non-zero imaginary part and to resonant states. The question about the resonant states system completeness was not earlier answered. In the present paper, we use a model of the Helmholtz resonator based on the theory of self-adjoint extensions of symmetric operators. Roughly speaking, in this model, small window is replaced by the point-like one. In the framework of the model, the resonant states can be found explicitly in the terms of the Green functions for the unperturbed problem. The main result is the proof of the completeness for a particular model problem of such type.

2. Model construction

Let us briefly describe the model. We start from the self-adjoint Neumann Laplacian, $-\Delta = -(\Delta^{in} \oplus \Delta^{ex})$ acting in the space $L_2(\Omega^{in} \oplus \Omega^{ex})$, $\Omega^{in} = [0;1]\times[0;1]$, $\Omega^{ex} = R^2 \setminus \Omega^{in}$. The Neumann Laplacian $-\Delta^{in}$ for Ω^{in} has the following eigenfunctions and eigenvalues:

$$\varphi_{n,m}(x,y) = 2\cos(\pi n x)\cos(\pi m y), \qquad (1)$$

$$\lambda^0_{n,m} = (\pi n)^2 + (\pi m)^2. \qquad (2)$$

To simulate coupling through small window with an exterior we use the model of point-like window [9-12]. We choose the window at $P = (x_1, y_1) = (0.5; 0)$ and construct a model by restriction-extension procedure [13]. Namely, we restrict the initial operator onto the set of smooth functions vanishing at the chosen boundary point (x_1, y_1). The closure of this restriction is symmetric non-self-adjoint operator with the deficiency indices (2,2). It has self-adjoint extensions, which are restrictions of the adjoint operator. The domain of the adjoint operator consists of elements (u^{in}, u^{ex}) having the following asymptotics near the point P :

$$u^{in,ex}(Q) = a^{in,ex} H_0^{(1)}(\mathrm{k}_0 |Q-P|) + b^{in,ex} + u_0^{in,ex}(\mathrm{Q}), \quad Q \to P \ ,$$

where $H_0^{(1)}(\mathrm{k}_0 |Q-P|)$ is the Hankel function, k_0 is some imaginary number (model parameter), $a^{in,ex}$, $b^{in,ex}$ are some constants, $u_0^{in,ex}(\mathrm{Q}) \to 0,\ Q \to P$. A self-adjoint extension is determined by a restriction – a correlation between the parameters. We choose the following condition, which gives us the model:

$$a^{in} = -a^{ex},\ b^{in} = b^{ex}.$$

Note that the initial operator has a system of eigenfunctions and eigenvalues $(\varphi_{n,m}(x,y),0)$ (see (1), (2)). As for the model self-adjoint operator, these eigenfunctions and eigenvalues convert into resonant states and quasi-eigenvalues, correspondingly. The initial system of eigenfunctions $\varphi_{n,m}(x,y)$ is complete in $L_2(\Omega^{in})$. The question is whether the perturbed system (resonant states) is also complete in $L_2(\Omega^{in})$.

3. Proof of completeness of the system of resonant states

We will consider perturbation of orthonormal system $\{ \varphi_{n,m}(x,y) \}$ using Green functions:

$$G^{in}(x,y,x_1,y_1,\lambda) = \sum_{n,m} \frac{\varphi_{n,m}(x,y) * \overline{\varphi_{n,m}(x_1,y_1)}}{\lambda_{n,m}^0 - \lambda} = \sum_{n:2,m} \frac{\varphi_{n,m}(x,y) * 2 * (-1)^{n/2}}{\lambda_{n,m}^0 - \lambda}$$

The expression for the Green function $G^{in}(x,y,x_1,y_1,\lambda)$is another, but it has no singularities, that will matter for us. Our model gives the main asymptotic term of the realistic problem for the Helmholtz resonator with a small (of radius ε) window. As for the realistic case, the asymptotic formula for resonances (quasi-bound state) is as follows (see [14])

$$\sqrt{\lambda_{s,t}} = \sqrt{\lambda_{s,t}^0} - \frac{\pi * \psi_0^2}{4 * \sqrt{\lambda_{s,t}^0}} * ln^{-1}\varepsilon - i * \frac{(\pi * \psi_0)^2}{8} * ln^{-2}\varepsilon..$$

Hence,

$$\lambda_{s,t}^0 - \lambda_{s,t} = 2 * \sqrt{\lambda_{s,t}^0} * \left(\sqrt{\lambda_{s,t}^0} - \sqrt{\lambda_{s,t}} \right) + o\left(\sqrt{\lambda_{s,t}^0} \right)$$

$$= \frac{\pi * \psi_0^2}{2} * ln^{-1}\varepsilon + i * \frac{\sqrt{\lambda_{s,t}^0} * (\pi * \psi_0)^2}{4} * ln^{-2}\varepsilon + o\left(\sqrt{\lambda_{s,t}^0} \right)$$

The perturbed functions $\{ \psi_{s,t}(x,y) \}$ are described by the following formula:

$$\psi_{s,t}(x,y) = c_{s,t} * (\varphi_{s,t}(x,y) + \alpha_{s,t} * G^{in}(x,y,x_1,y_1,\lambda_{s,t}))$$

We will use $G_{s,t}^{in}(x,y)$ for $G^{in}(x,y,x_1,y_1,\lambda_{s,t})$. $c_{s,t}$ is for normalization of ψ. $\alpha_{s,t}$ can be found from coupling condition for our operator extension:

$$\varphi_{s,t}(x,y) + \alpha_{s,t} * \left(G^{in}(x,y,x_1,y_1,\lambda_{s,t}) - G^{in}(x,y,x_1,y_1,k^2) \right) =$$
$$-\alpha_{s,t} * \left(G^{ex}(x,y,x_1,y_1,\lambda_{s,t}) - G^{ex}(x,y,x_1,y_1,k^2) \right) \text{ at } (x_1,y_1), \text{ where } k^2 < 0$$

So,

$$\alpha_{s,t} = \frac{-\varphi_{s,t}(x_1,y_1)}{G^{in}(x_1,y_1,x_1,y_1,\lambda_{s,t}) - G^{in}(x_1,y_1,x_1,y_1,k^2) + G^{ex}(x_1,y_1,x_1,y_1,\lambda_{s,t}) - G^{ex}(x_1,y_1,x_1,y_1,k^2)} =$$

$$\frac{-2*(-1)^{s/2}}{\sum_{n:2,m}\left(\frac{4}{(\pi n)^2+(\pi m)^2-\lambda_{s,t}}-\frac{4}{(\pi n)^2+(\pi m)^2-k^2}\right)+F(k)}$$ for $s \vdots 2$ and $\alpha_{s,t}=0$ for odd s. F(k)

appears from the Green function out of domain.
To
As mentioned earlier, $\|\psi\|$ should be 1, so $c_{s,t}=\frac{1}{\|\varphi_{s,t}+\alpha_{s,t}*G_{s,t}^{in}\|}$

Hence, $c_{s,t}{}^2=\frac{1}{\|\varphi_{s,t}+\alpha_{s,t}*G_{s,t}^{in}\|^2} \Rightarrow c_{s,t}=\frac{1}{\|\varphi_{s,t}^2+\alpha_{s,t}^2*(G_{s,t}^{in})^2+2*\varphi_{s,t}*\alpha_{s,t}*G_{s,t}^{in}\|}=$

$$=\frac{1}{\sqrt{1+\alpha_{s,t}^2*\left\|G_{s,t}^{in}\right\|^2+2*\alpha_{s,t}*\left(\varphi_{s,t}(x,y),G_{s,t}^{in}\right)}}=(*)$$

Note, that dot product $\left(\varphi_{s,t}(x,y),G_{s,t}^{in}\right)$ contains only one summand, where $n=s \wedge m=t$, if we sum by (n,m). So, we have $\frac{\|\varphi_{s,t}\|^2*2*(-1)^{s/2}}{\lambda_{s,t}^0-\lambda}$ as a product. Hence,

$$(*)=\frac{1}{\sqrt{1+\alpha_{s,t}^2*\left\|G_{s,t}^{in}\right\|^2+2*\alpha_{s,t}*\frac{2*(-1)^{s/2}}{\frac{\pi*\psi_0^2}{2}*ln^{-1}\varepsilon+i*\sqrt{\lambda_{s,t}^0}*\frac{(\pi*\psi_0)^2}{4}*ln^{-2}\varepsilon}}}$$

We use the theorem [15] :
Lemma. If set $\{\ \varphi_{s,t}(x,y)\ \}$ is complete orthonormal system of functions and $\sum_{n,m}\left\|\psi_{n,m}-\varphi_{n,m}\right\|^2<\infty$, then the set $\{\ \psi_{s,t}(x,y)\ \}$ forms a basis.

In our case this series is as follows:

$$\sum_{s,t}\left\|\varphi_{s,t}(x,y)*\left(c_{s,t}-1\right)+c_{s,t}*\alpha_{s,t}*G_{s,t}^{in}(x,y)\right\|^2$$

Convergence of this sum can be checked using integral criteria for summands. Consider difference of Green functions from α denominator:

$$\sum_{n:2,m}\left(\frac{4}{(\pi n)^2+(\pi m)^2-\lambda_{s,t}}-\frac{4}{(\pi n)^2+(\pi m)^2-k^2}\right)=\frac{4}{\pi^2}\sum_{n,m}\left(\frac{1}{4n^2+m^2-\frac{\lambda_{s,t}}{\pi^2}}-\frac{1}{4n^2+m^2-\frac{k^2}{\pi^2}}\right)$$

Note: we will use "~" in convergence sense below. With integral transition, we are interested in:

$$\int_0^\infty\int_0^\infty\frac{(\lambda_{s,t}-k^2)dndm}{\left(n^2+m^2-\lambda_{s,t}\right)*(n^2+m^2-k^2)}\sim\int_0^{\frac{\pi}{2}}d\varphi\int_0^\infty\frac{(\lambda_{s,t}-k^2)rdr}{\left(r^2-\lambda_{s,t}\right)*(r^2-k^2)}\sim-\ln\left(\frac{\lambda_{s,t}}{k^2}\right)$$

Hence, $\alpha_{s,t}\sim\frac{1}{\ln(s^2+t^2)}\sim\frac{1}{\ln(r)}$

We can note, that $\|G_{s,t}^{in}\| \leq \frac{c}{Im(\lambda_{s,t})}$ as a resolvent, so $\|G_{s,t}^{in}\| \leq \frac{c}{r}$

$$c_{s,t} - 1 \sim \alpha_{s,t}^2 * \|G_{s,t}^{in}\|^2 + 2 * \alpha_{s,t} * \frac{2*(-1)^{s/2}}{\frac{\pi*\psi_0^2}{2}*ln^{-1}\varepsilon + i*\sqrt{\lambda_{s,t}^0 * \frac{(\pi*\psi_0)^2}{4}*ln^{-2}\varepsilon}}$$ from the

Taylor series for $\frac{1}{\sqrt{1+x}}$ at 0. So, $c_{s,t} - 1 \sim (\frac{1}{\ln^2 r * r^2} + \frac{1}{\ln(r) * r})$

So, for convergence we have:

$$\begin{cases} \|G_{s,t}^{in}\| \sim \frac{1}{r} \\ c_{s,t} - 1 \sim \frac{1}{\ln^2 r \ * \ r^2} \\ \alpha_{s,t} \sim \frac{1}{\ln(r)} \end{cases}$$

Finally, using integral transition again, we can transform our series:

$$\int_0^\infty \int_0^\infty \left\|\varphi_{s,t}(x,y) * (c_{s,t} - 1) + c_{s,t} * \alpha_{s,t} * G_{s,t}^{in}(x,y)\right\|^2 ds\, dt \sim$$

$$\int_0^{\frac{\pi}{2}} d\varphi \int_0^\infty (\|\varphi\|^2 * \frac{1}{\ln^4 r * r^4} + \|G_{s,t}^{in}\|^2 * \frac{1 + o(r)}{\ln^2 r} + 2 * \|\varphi\| * (1 + o(r)) * \frac{\|G_{s,t}^{in}\|}{\ln(r) * \ln^2 r * r^2}) * rdr \sim$$

$$\int_0^\infty \|\varphi\|^2 * \frac{rdr}{\ln^4 r * r^4} + \int_0^\infty \|G_{s,t}^{in}\|^2 \frac{rdr}{\ln^2 r} + \int_0^\infty 2 * \|\varphi\| * \frac{\|G_{s,t}^{in}\| * rdr}{\ln^3 r * r^2} \sim$$

$$\int_0^\infty \frac{dr}{\ln^4 r * r^3} + \int_0^\infty \frac{dr}{\ln^2 r * r} + \int_0^\infty \frac{dr}{\ln^3 r * r^2}$$

All summands converge, so the set $\{ \psi_{s,t}(x,y) \}$ forms a basis in accordance with Lemma. The proof is complete.

Thus, the system of the resonant states (quasi-eigenfunctions) is complete in $L_2(\Omega^{in})$. The proof written above is valid for our explicitly solvable model

The work was partially financially supported by the Government of Russian Federation (grant 074-U01), by the Ministry of Science and Education of the Russian Federation (GOSZADANIE 2014/190, Project 14.Z50.31.0031), by grant MK-5001.2015.1 of the President of the Russian Federation.

References

[1] Delerue, C., Lannoo M. Nanostructures: Theory and Modelling. Berlin: Springer. 2004.
[2] Popov, I.Yu., Popova, S.L. Phys. Lett. A, 173 (1993), 484-488.
[3] Popov, I.Yu., Popova, S.L. Europhys. Lett. A, 24 (5) (1993), 373-377.
[4] Wulf, U., Krahlisch, M., Kučera, J., Richter, H., Höntschel, J. Nanosystems: Phys. Chem. Math. 4 (6) (2013), 800- 809
[5] Smrcka, L. Superlattices and Microstructures, 8 (1990), 221–224.
[6] Mil'nikov, G., Mori, N., Kamakura, Y. J. Comput. Electron, 10 (2011), 51–64.
[7] Tarel, G., Savona , V. Phys. Rev. B 81, 075305/1-8.
[8] Racec, P.N., Racec, R., Neidhardt, H. Phys. Rev. B., 79 (2009) pp. 155305/1--155305/14.
[9] Popov, I.Yu. ZAA, 32 (2013), 155-162.
[10] Popov, I.Yu. J. Math. Phys. 53 (2012), 063505/1-7. doi: 10.1063/1.4726176

[11] Popov, I.Yu. Math. USSR Sbornik, 71 (1), (1992), 209--234.
[12] Popov, I.Yu. J.Math.Phys. 41, (2000), 4391- 4405
[13] Pavlov, B.S. Uspekhi Mat. Nauk. 42 (6) (1987), 99-131.
[14] Gadyl'shin R R Russ. Math. Surv. 52 (1) (1997), 1-72
[15] Kato, T. Perturbation Theory for Linear Operators. Berlin: Springer. 1966.

Synthesis of luminescent ceramics from alumina nanopowder

V. Kortov, A. Kiryakov, D. Ananchenko and S. Zvonarev
Ural Federal University, Yekaterinburg, Russia

E-mail: Seni-i@yandex.ru

Abstract. The effect of the conditions of high-temperature synthesis in a reducing medium on the density, area and cathodoluminescence of ceramics made of compacted α-Al_2O_3 powder is the focus of this work. The dependence of the luminescence intensity on the temperature and duration of the synthesis of ceramics is presented. The optimal parameters for synthesis of luminescent ceramics from alumina nanopowder are defined.

1. Introduction

Alumina is a well-known functional material which has a wide range of practical applications. One of them is related to using oxygen-deficient single crystals and microcrystalline powders as dosimetric phosphors [1, 2]. High sensitivity of oxygen-deficient oxides to ionizing radiation is caused by vacancies in the anion sublattice. While the vacancies capture one or two electrons, they create F^+ and F-centers. Such centers are involved in the process of luminescence. The most effective method of formation of F and F^+ centers is growing single crystals from melting (Stepanov's method) in a highly reducing medium in the presence of carbon with a low partial pressure of oxygen. In this case an increasing temperature in crystal growing chamber leads to the formation of carbon monoxide (CO) which is a strong reducing agent and facilitates oxygen vacancy creation in the oxide.

Nowadays using nanopowders of various compounds as phosphors is proved to be promising. Ceramics with high mechanical durability are considered to be the most suitable materials for technological applications. Ceramics are useful for production of luminescent detectors and sensors of any shape and size. The existing methods allow us to obtain alumina nanopowders with a stoichiometric structure. To produce luminescent ceramics it is necessary to carry out their synthesis in reducing medium. Oxygen-deficient ceramics synthesis from alumina nanopowder has been little studied. A research of the effect of synthesis conditions on ceramic luminescent properties is required.

The aim of this paper is to study the processes of the ceramics synthesis from alumina nanopowders in a reducing medium and to define the influence of the synthesis conditions on its luminescence excited by means of a pulse electron beam.

2. Samples and methods

High purity α-Al_2O_3 powder (99.9%) with the particle size of 50-70 nm was used for the production of samples. The compacts were prepared by using cold uniaxial pressing in the hydraulic press Specac at a pressure of 1000 kgf/cm^2. The compacts 5 mm in diameter and 1.3 mm thick had the mass of 40 mg and the density of 1.56 g/cm^3. A high-temperature vacuum electric furnace was used to synthesize ceramics. The electric furnace allows the sample to be annealed at temperatures up to 1800°C and in vacuum of about 10^{-2} Pa. The synthesis of ceramics was carried out in the presence of graphite rods (20 g).

Pulse cathodoluminescence (PCL) of the ceramics was measured by an analyzer "CLAVI". Luminescence excitation was induced in the samples with an electron beam while they were irradiated in the air at room temperature. The duration of the beam was 2 ns, the maximum electrons energy was about 130 ± 10 keV, the current density was 60 A/cm^2. The spectral range of PCL measurements was 350-750 nm. The wavelength measurement inaccuracy was $\lambda = \delta \pm 0.75$ nm.

3. Results and discussion

3.1. The synthesis of ceramics under different conditions

The area of the samples synthesized at 900 – 1600°C for 60 minutes either in the presence of carbon or without it significantly changes. This effect can be explained by the ceramics sintering processes (Figure 1a). The sintering causes porosity decrease and nanoparticles agglomeration [3]. The area change of the samples sintered at above 1500°C without graphite is slowed down, which is caused by the sintering completion. In the samples annealed in the carbon presence at the temperatures over 1500°C the further area decrease leads to its evaporation.

It is known from the published data that high-temperature heating (more than 1500°C) in a vacuum causes thermal etching [4, 5]. It leads to decreasing mass of the sample. One can assume that in the presence of carbon a thermal etching is intensified due to the loss of oxygen in the oxide reduction. As a result, Al_2O_3 breaks down into lower oxides which are to be deleted during pumping. It means that the temperature of 1500°C is ultimate for the synthesis of oxygen-deficient alumina ceramics in a highly reducing medium.

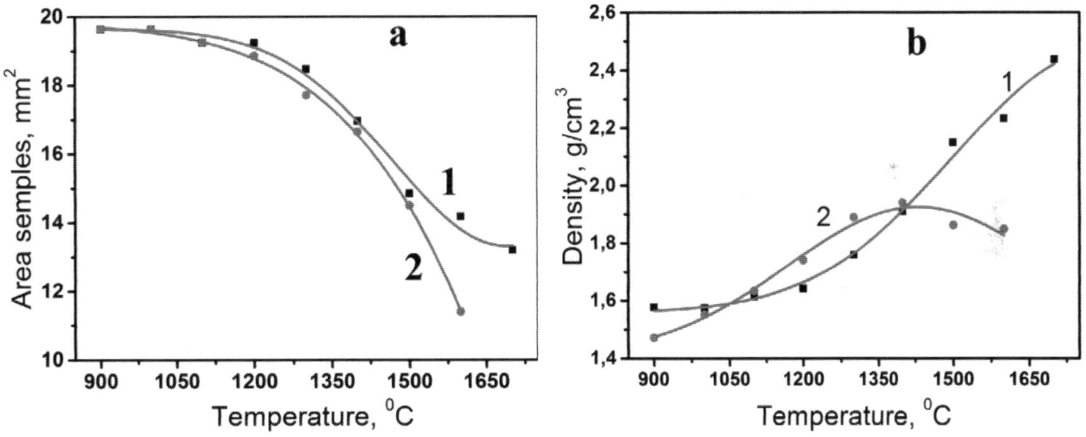

Figure 1. Dependences of the changes in the area (a) and density (b) of Al_2O_3ceramics on the synthesis temperatures: 1-without carbon, 2-with carbon.

To calculate the density of the samples, their mass, diameter and height were measured. Figure 1b shows that the temperature increase causes the density growth in the samples either annealed in vacuum or synthesized in the presence of carbon. In the latter case the density growth is replaced by the saturation and even by a decrease in its value in the temperature range of 1400-1500°C. This result correlates with the mentioned above temperature influence on the sample area, i.e. the result is caused by a loss of the sample mass during its intensive thermal etching and subsequent decomposition into lower aluminum oxides.

3.2. Pulse cathodoluminescence

Spectra of PCL were measured in order to define the ceramics synthesis conditions which lead to creation of highest concentration of oxygen vacancies. Such vacancies are luminescence centers. The PCL of ceramics synthesis in a vacuum in the presence of carbon were measured at different temperatures. It is known that F-centers created by oxygen vacancies are responsible for the emission band with the maximum at 420nm [5]. Observation of the changes in the intensity of this band at various temperatures of the ceramics synthesis shows relative changes of oxygen vacancies concentration. The results of the measurements of the PCL band intensity were obtained taking into account the change in the sample area after synthesis.

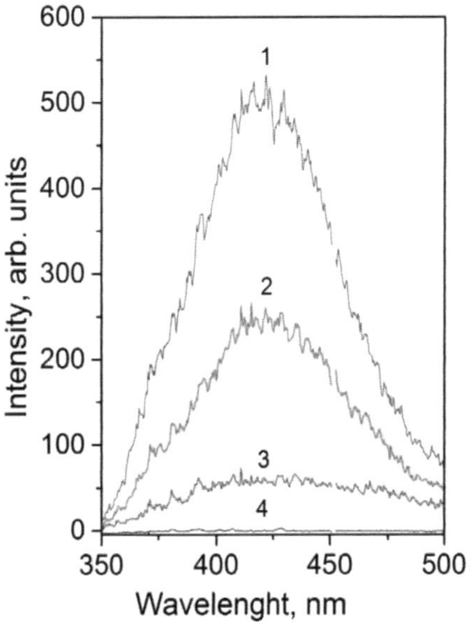

Figure 2. PCL spectra of ceramics synthesized in presence of carbon for 1 h at different temperatures: 1 – 1600°C, 2 – 1500°C, 3 – 1300°C; 4 – without carbon at T=1600°C.

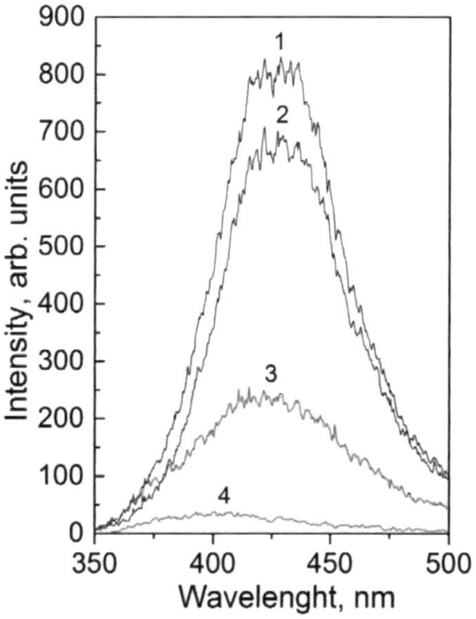

Figure 3. PCL spectra of ceramics synthesized at various durations of sintering: 1 – 300 min, 2 – 180 min, 3 – 60 min; 4 – without carbon for 300 min.

One can see that an increasing synthesis temperature leads to the intensity growth of the PCL emission band at 420 nm (Figure 2). The most significant PCL intensity growth occurs between 1300 and 1500°C. It should be noted that ceramics synthesis in vacuum without carbon is of low efficiency for creation of oxygen vacancies.

To define an effect of the annealing time on the PCL intensity, the period of isothermal heating was changed in the samples annealed in the presence of carbon at 1500°C (60 min, 180 min and 300 min.). The PCL spectra of these samples are shown in Figure 3. An increasing annealing time of alumina ceramics synthesis in a reducing medium also leads to luminescence intensity growth at the 420 nm band. The luminescence increase indicates anion defects and a growth of luminescence centers concentration in the oxygen sublattice of the oxide. A slight difference of the PCL intensities of the samples with a synthesis time of 180 and 300 min should be noted.

The experiments carried out allowed us to define the optimal parameters for synthesis of luminescent ceramics from alumina nanopowder, which give quite a high PCL yield with a relatively small loss of the sample mass. The temperature of 1500°C can be recommended for annealing for less than 3 hours.

4. Conclusion

1. Oxygen-deficient alumina ceramics have been synthesized under different temperatures in vacuum in the presence of carbon. It has been shown that the samples lose mass due to thermal etching at temperatures above 1500°C.
2. PCL spectra of ceramics synthesized at high-temperature in a strongly reducing environment have a band centered at 420 nm resulted from luminescence of F centers.

3. The increase in synthesis temperature up to 1500°C and annealing time up to 3 hours lead to the luminescence intensity growth in the samples without any substantial destruction of the samples due to thermal etching.
4. The obtained ceramics can be used to assess the possibilities of creation the luminescent detectors of ionizing radiation on their base and for other technical applications, in particular, in plasma display panels and optoelectronic devices.

5. Acknowledgments

The work was supported by the scholarship of the President of the Russian Federation for young scientists and graduate students engaged in advanced research and development in priority areas of modernizing the Russian economy number SP- 983.2015.2.

6. References

[1] Akselrod M S, Kortov V S, Kravetsky D J, Gotlib V I 1990 *Radiat. Prot. Dosim.* **32** 15
[2] Akselrod M S, Lucas A C, Polf J C, McKeever S W S 1998 *Radiat. Meas.* **29** 391
[3] Zvonarev S V, Kortov V S, Shtang T V, Ananchenko D V, Petrovykh K A 2015 *Appl. Radiat. Isot.* **95** 44
[4] Kaltaev Sh–o Kh, Sidelnikova S N, Dan'koYa A 2010 *Funct. Mater.* **17** 395
[5] Dan'koYa A, Rom A M, Sidelnikova S N 2007 *Funct. Mater.* **14** 460

On the spectrum discreteness of the quantum graph Hamiltonian with δ-coupling

M O Smolkina[1], I Yu Popov[1]
[1]ITMO University, Kronverkskiy, 49, St. Petersburg, 197101, Russia

E-mail: vega14@mail.ru, popov1955@gmail.com

Abstract. The condition on the potential ensuring the discreteness of the spectrum of infinite quantum graph Hamiltonian is considered. The corresponding necessary and sufficient condition was obtained by Molchanov 50 years ago. In the present paper, the analogous condition is obtained for a quantum graph. A quantum graph with infinite leads (edges) or/and infinite chains of vertices such that neighbor ones are connected by finite number of edges and with δ-type conditions at the graph vertices is suggested. The Molchanov-type theorem for the quantum graph Hamiltonian spectrum discreteness is proved.

1. Introduction

Many important properties of nanosystems are predetermined by the spectrum of its Hamiltonian. Particularly, spectrum discreteness shows that the bound states gives one full system in the state space of the corresponding system. The problem investigation starts from classical work of Weyl [1]. Early results are described in [2, 3]. The Hamiltonian spectrum discreteness condition (for the potential on axis or half-axis) was suggested by Molchanov in [15]. This condition becomes necessary and sufficient if we assume that the potential is bounded from below. Later, many authors obtained different sufficient conditions for the discreteness of the spectrum of the Sturm-Liouville or 1D Schrodinger operators[4]-[8]. Good review of results in the field is in [9]. Particularly, there are various necessary or sufficient conditions on locally integrable functions $q(x)$ in order that the Schrödinger operator $-\frac{d^2}{dx^2}+q(x)$ acting on $C_0^\infty(0,\infty)$ is bounded below and that its Friedrichs extension has only discrete spectrum. There are also some results for the case of operator valued potentials. In this paper the author reviews known conditions for both cases and presents results that unify and extend the previous work on the subject. Finally the result is applied to the Laplacian on non-compact complete Riemannian manifolds.

One-dimensional Hamiltonians are widely used in physical models. The most effective model of such type is a model of quantum graph (see, e.g., [10], [11]). It is a metric graph with the Schrodinger operator defined at the edges (according the physical motivation) and some coupling conditions at the graph vertices that specify the transition probabilities across the vertices (see, e.g., [12-14]). The problem of the spectrum discreteness is closely related with other classical problem starting from the Sturm theorem - the problem of zeroes of eigenfunctions. At present, it is intensively investigated (see, e.g., [16]). The Courant theorem with the problem of nodal set counting is not solved in general case yet.

The well-known Molchanov's condition for the potential on the real half-axis (or axis) that guarantees the discreteness of the spectrum of the corresponding one-dimensional Hamiltonian is represented in [17,18]. The main goal of the work is to obtain the similar condition for a special type quantum graph.

2. Molchanov-type theorem

Let Γ be a quantum graph, $V(\Gamma)$ is a set of it's vertices and $E(\Gamma)$ is a set of it's edges. Let also assume that the graph has finite number of infinite leads (edges) or/and infinite chains of vertices such that neighbor ones are connected by several (finite number) edges. We demand that the lengths of edges of these chains are bounded from below and above by some constants and attend that the periodicity or symmetry properties for this chain isn't needed. Moreover, we assume that after removing of these infinite chains, the graph contains finite number of edges (may be, some of them have infinite lengths).

We assume the following equation at the edges:

$$(H - \lambda I)u(x) = -u''(x) - \big(\lambda - q(x)\big)u(x) = 0. \qquad (1)$$

We deal with the Dirichlet boundary condition at the boundary vertices of the graph and δ-type coupling at the internal graph vertices:

$$\sum_{e \in E_v} \frac{du}{dx_e}(v) = \beta_v u(v),\ \beta_v > 0,\ x \in V(\Gamma), \qquad (20$$

where u is continuous on Γ, $V(\Gamma)$ is a set of vertices of Γ, E_v is a set of edges containing vertex v, $\frac{du}{dx_e}(v)$ is a derivative of the solution at the vertex v of the edge e in the outgoing direction from the vertex, β_v, $\beta_v > 0$, are some real positive numbers.

Theorem

Let $q(x)$ be bounded from below, i.e. there exists such constant $c > 0$, that for every x the following inequality takes place: $q(x) > -c$ and satisfy the following conditions on Γ:

Let us take arbitrary $w > 0$. Consider all paths L_a on Γ having lengths w ($|L_a| = w$), where a is the starting point of the path L_a. Let $q(x)$ be such function that (for any w)

$$\lim_{dist(a,v_0)\to\infty} \int_{L_a} q(x)dx = \infty\,. \qquad (3)$$

Here v_0 is some fixed vertex of Γ

Then, for every fixed λ there exists a set of β_v, such that any solution $u(x)$ of the problem (1) with the vertices conditions (2) has finite number of zeroes on Γ.

Sketch of the Proof

We will assume that some solution has infinite number of zeroes and will come to a contradiction. Two cases can be considered.

1. Let Γ has finite number of edges. Some of them can have infinite lengths. As we have δ-type conditions at the internal graph vertices and the Dirichlet condition at the boundary vertices, there are no accumulation points at the graph vertices.

If we have an infinite set of zeroes at an edge of finite length, then $u \equiv 0$ on the edge and we can throw off the edge and pose the Dirichlet condition at the end points of this removed edge. At every finite edge one can have finite number of zeroes and due to the finiteness of the number of edges, we have a contradiction. So if we have an infinite number of zeroes on Γ then there is an infinite edge with this property. For such edge we have the conventional Molchanov's condition for the potential on the real half-axis, i.e. this case simply reduced to the conventional theorem.

2. Let Γ has infinite leads (edges) and infinite chains of vertices such that neighbor ones are connected with finite number of edges.

If the infinite number of zeroes is on the part of the graph that has no chains of rings, the Theorem is proved as in point 1.

Let the infinite number of zeroes be at the infinite chains of vertices such that neighbor ones are connected with finite number of edges.

One can see that adding of a constant to the potential does not change the situation. Hence, it is sufficient to consider the case $q(x) \geq 0$. Let there exist $\lambda = \lambda_0 > 0$ such that there is a solution $u = u(x)$ of the equation (1) that has infinite number of zeroes at the chain.

Evidently, it is impossible for $\lambda = \lambda_0 < 0$ due to the sign of the second derivative of the solution: $u'' = -(\lambda - q(x))u$. Hence, if one has a root α then there are no other roots greater than α.

Let w be a such (small) positive number that $w < \frac{1}{\lambda_0+1}$. Let us choose N so large that for $dist(a, v_0) > N$, where $v_0 \in V(\Gamma)$, one has

$$\int_{L_a} q(x)dx > w(\lambda_0 + 1), \tag{4}$$

where L_a is a path on the graph with the starting point a and $|L_a| = w$. It is possible to choose such N due to the condition (3) for $q(x)$.

In accordance with our assumption that there is an infinite number of zeroes at the chain, there exists some path through the chain to infinity that contains an infinite number of zeroes. Below we will deal with the case of such path. One can enumerate zeroes at this path consequently. Hence, we can choose n for such path that $dist(\alpha_n, v_0) > N$, and, later, choose m, such that $m > n$ and $dist(\alpha_m, \alpha_n) > w$. Next, it is possible to suppose that $dist(\alpha_m, \alpha_n) = Pw$, where P is an integer (due to the conditions $q(x) \geq 0$, $\int_{L_a} q(t)dt > w(\lambda_0 + 1)$).

Let us rewrite equation (1) for $\lambda = \lambda_0$: $u'' = (q(x) - \lambda_0)u$. A difficulty is related with the graph vertices belonging to the path. Let us consider the case. Assume that vertex v belongs to the path. We multiply the both parts by u and integrate over the path from α_n to α_m . To come to a contradiction, we use integration by parts and the mean value theorem. By taking into account inequality (4) and the vertex condition (2), we show that there exists such set of the vertices parameters, β_v, that the following inequality takes place:

$$0 > \{1 - (\lambda_0 + 1)w\} \int_L \{u^2(t) + [u'(t)]^2\}dt,$$

which is impossible as $(\lambda_0 + 1)w < 1$. Hence, we arrive at a contradiction. The proof is complete.

One can use the obtained results for the description of many physical system for which the quantum graph model is appropriate approximation. It is interesting to generalize the result to the decorated quantum graph and to modifications of the Schrodinger operator, e.g., magnetic Schrodinger operator (Landau operator).

Acknowledgements

This work was partially financially supported by the Government of the Russian Federation (grant 074-U01), by Ministry of Science and Education of the Russian Federation (GOSZADANIE 2014/190, Projects No 14.Z50.31.0031 and No. 1.754.2014/K)), by grant MK-5001.2015.1 of the President of the Russian Federation. The authors thank Prof. P.Exner, Prof. K.Pankrashkin and other participants of "Pierre Duclos Workshop – MCQTN" (St Petersburg, 2014) for interesting discussion.

References

[1] Weyl H. 1908 Uber gewohnliche lineare Differentialgleichungen mitr singularenSteleen und ihre Eigenfunktionen. *Nachr. Acad. Wiss. Gottingen. Math. Phys.* KI, 37-64.
[2] Titchmarsh E. C. 1958 *Eigenfunction expansions* associated with second order differential equations, part 2 (Oxford,Oxford University Press (Clarendon Press)).
[3] Kats I. S. , Krein M. G. 1958 Criteria for the discreteness of the spectrum of a singular string. *Izv. Vyssh. Uchebn. Zaved. Mat.* no. 2, 136–153.
[4] Ahlbrandt C.D., Hinton D.B., Lewis R.T. 1981. Necessary and sufficient conditions for the discreteness of the spectrum of certain singular differential operators. *Canad. J. Math.* **33,** 229–246.
[5] Hinton D.B. 1979. Molchanov's discrete spectra criterion for a weighted operator. *Canad. Math. Bull.*, **22.** 425–431.
[6] Kwong M.K., Zettl A. 1981. Discreteness conditions for the spectrum of ordinary differential operators. *J. Differential Equations*, **40** 53–70.
[7] Oinarov R., Otelbaev M. 1988. A criterion for a general Sturm–Liouville operator to have a discrete spectrum. *Differential Equations*. **24**. 402–408.
[8] Curgus B., Read T.T. 2002. Discreteness of the spectrum of second-order differential operators and associated embedding theorems. *J. Differential Equations*, **184** (2), 628 –648.
[9] Bruning J. 1989 On Schrodinger operators with discrete spectrum. J. Funct. Anal. 85 (1) 117-150. DOI: 10.1016/0022-1236(89)90049-9.
[10] Kuchment P. 2002 Graph model of wave propagation in thin structures *Waves in Random Media* **12** (4), R1-R24.
[11] Lobanov I.S. Trifanov A.I. and Trifanova E.S. 2013 Genetic algorithm for constructing graphene nanoribbon with given electronic transport properties. *Nanosystems: Phys. Chem. Math.* **4** (4), 512-523.
[12] Exner P., Keating J. P., Kuchment P., Sunada T. and Teplyaev A. (eds.). 2008. *Analysis on Graphs and Its Applications.* Proc. Symp. Pure Math., 77 (Providence, RI: Amer. Math. Soc.)
[13] Popov I. Yu., Skorynina A. N., and Blinova I. V.. 2014 On the existence of point spectrum for branching strips quantum graph. *J. Math. Phys*. **55**, 033504, doi: 10.1063/1.4867604
[14] Popov I.Yu., Smirnov P.I. 2013 Spectral problem for branching chain quantum graph. *Phys. Lett.* A. **377**, 439-442.
[15] Molchanov A.M.. 1953 On conditions of spectrum discreteness for self-adjoint differential operators of second order. *Proc. Moscow Math. Soc.* **2**, 169-199.
[16] Band R., Berkolaiko G., Raz H., Smilansky U.. 2012 The number of nodal domains on quantum graphs as a stability index of graph partitions. *Commun. Math. Phys.* **311**, 815–838. DOI: 10.1007/s00220-011-1384-9
[17] Levitan B.M., Sargsyan I.S.. 1988 *Sturm-Liouville and Dirac operators*, (Moscow: Nauka).
[18] Eckhardt J., Kostenko A., Malamud M. , G. Teschl. 2014 One-dimensional Schrödinger operators with δ'-interactions on Cantor-type sets. *J. Differential Equations* **257** 415–449

Elastic moduli of alumina nanoceramics

J A Perevozchikova[1], E A Buntov and A F Zatsepin
Ural Federal University named after the first President of Russia Boris Yeltsin, 19 Mira Street, Yekaterinburg, 620002, Russia

E-mail: j-perevozchikova@mail.ru

Abstract. In this study we investigated the physico-mechanical properties of alumina nanoceramics, that is used as substrates in micro- and optoelectronics chips. The elastic characteristics (Young's modulus E, shear modulus G, Poisson's ratio ν, microhardness *H*) are determined by measuring the longitudinal and transverse wave velocities. The results suggest that the nanoceramics have unique properties that can be used in the future microelectronics to improve the stability of chips and to optimize the production technology.

1. Introduction
Currently, nanostructured materials that have special electrical, optical and mechanical properties represent considerable interest for micro- and optoelectronics. Nanoceramics studied in this work is used in microelectronics for chip substrates. For the stable operation of the device, the electrical properties must be combined with other physical and mechanical features, which can be obtained by measuring the sound velocity. In this work, we determined the elastic characteristics (Young's modulus E, shear modulus G, Poisson's ratio ν, microhardness *H*) of nanoceramics based on Al_2O_3.

2. Samples
In this study we investigated three samples of nanoceramics with alumina nanostructured phase received from plant of electronic instrumentation "Elion" (Zelenograd, Russia). For certification of the samples, the scanning electron microscopy and X-ray analysis were carried out. Investigation of the structure of the surface of the samples was performed by scanning electron microscope SIGMA VP (Carl Zeiss) production using secondary electron detector (In-lens) in high vacuum at an accelerating voltage of 5 keV. Pictures with a scanning electron microscope were obtained at different scales at several points samples cut. Images of the sample 2 and the sample 3 obtained with a scanning electron microscope are shown in Figures 1 and 2.

The particle size was determined by all the images using the analyzer of solid bodies microstructure fragments Clinker C7 by SIAMS company and the particle size of sample 1 was 200-400 nm, sample 2 had particle size of 1-3 micron, while the particle size of the sample 3 was 80-200 nm.

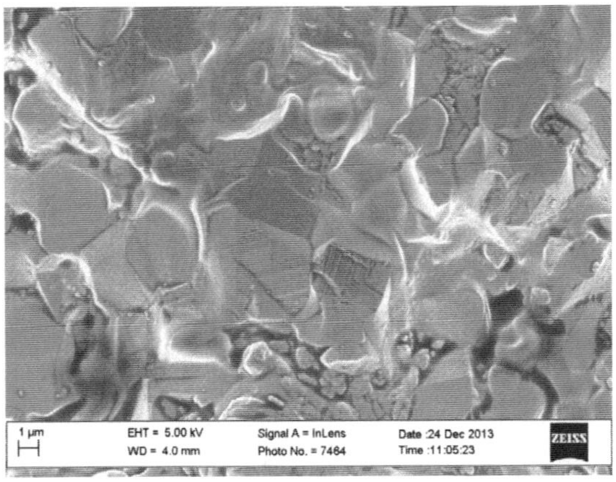

Figure 1. The surface of the sample 2 with a scale of 1 microns.

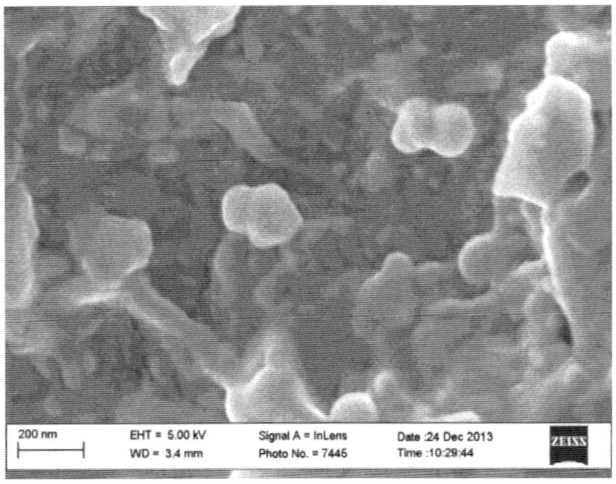

Figure 2. The surface of the sample 3 with a scale of 200 nm.

Another method to determine the particle size and phase composition performed was the X-ray analysis carried out on a diffractometer X'PERT PRO in the K_{α} copper radiation. The qualitative composition and size of coherent scattering regions (CSR) for samples was determined from the broadening of X-ray diffraction reflections. By the decoding results it was revealed that the main phase for all the samples was α-Al_2O_3 (corundum), and the average size of coherent scattering region for the sample 1 and sample 2 was more than 200, and CSR of the sample 3 was 100-150 nm.

The density of the samples was measured by the helium Pycnometer AccuPyc II 1340 with high accuracy: $\rho_1 = 3{,}7462 \pm 0{,}0006$ g/cm^3, $\rho_2 = 3{,}7563 \pm 0{,}0008$ g/cm^3, $\rho_3 = 3{,}7616 \pm 0{,}0006$ g/cm^3.

3. Methods

Determination of the propagation velocity of the longitudinal waves was performed on the device consisting of two piezoelectric transducers (5 MHz) and a sample between them. One of transducers was used as the signal source and the other was the receiver. The signal received by the second transducer was registered by a personal computer. Using a special program written in LabView, the value of the longitudinal ultrasonic velocity in the sample was calculated.

The size of nanoceramics sample was quite small, that is why a scheme was developed that allows single point measurements, in which the transverse wave reflected perpendicularly from the sample face in order to determine the velocity of shear waves. We used a specially shaped prism that conducts transverse waves. Aluminum was chosen as the material for the prism because its acoustic impedance is close to the acoustic impedance of the ceramic. The scheme of shear velocity measurements is shown in Figure 3.

The longitudinal ultrasonic wave *l* emitted in angle beam transducer *1*. On the border of the aluminum prism *2* it is transformed into a transverse wave *t*, which penetrates between the first α^{I} and second critical angles α^{II}. The shape of the aluminum prism *2* was calculated so that the transverse wave propagating through the prism goes through a thin contact layer *3* and the beam is perpendicular to the sample *4*. The silicate glue is used as the contact layer *3*. The signal from the pulse is received by ultrasonic measurement system. To determine the transverse velocity it is required to know the thickness of the sample and the time at which the ultrasonic wave has passed through it. Time is determined by the difference between the arrival times of the two pulses. The first impulse is a reflection from the near face of the sample. The second pulse is a reflection of the far face of the sample. The speed of transverse waves in this method is calculated as follows:

$$C_t = {}^{2D}/_{t}, \tag{1}$$

where D is specimen thickness, t is time between two pulses.

Measurements were performed using ultrasonic flaw detector PCUS-10 in pulse mode and the inclined transducer with 65° angle and 5 MHz frequency.

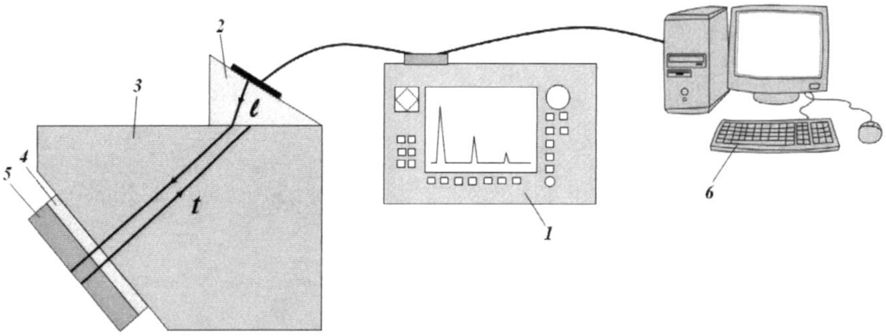

Figure 3. The scheme of shear velocity measuring is. Where *1* is ultrasonic flaw detector PCUS-10, *2* – tilted transducer, *3* – aluminum prism, *4* – contact layer, *5* – the sample, *6* – computer, *l* – longitudinal wave, *t* – shear wave.

4. Results

The results are shown in Table 1. Values of the elastic characteristics were calculated according to well known formulae (2)-(5) [1-5]:

- shear modulus:

$$G = \rho \cdot C_t^2, \tag{2}$$

- Poisson's ratio:

$$\nu = {(C_l^2 - 2 \cdot C_t^2)}\Big/{2 \cdot (C_l^2 - C_t^2)}, \tag{3}$$

- Young's modulus:

$$E = \rho \cdot C_t^2 \cdot {\left(3 - 4 \cdot \frac{C_t^2}{C_l^2}\right)}\Big/{\left(1 - \frac{C_t^2}{C_l^2}\right)}, \tag{4}$$

- microhardness:

$$H = {(1 - 2\nu)E}\Big/{6(1 + \nu)}, \tag{5}$$

where ρ is density of the sample, kg / m^3, C_l and C_t is velocity of longitudinal l and shear waves t, m/s.

Table 1. The properties of nanoceramics.

	Sample 1	Sample 2	Sample 3	Accuracy
Density ρ, kg/m^3	3746	3756	3762	± 0.8
Longitudinal waves velocity C_l, km/s	9951	9948	8728	± 3
Shear waves velocity C_t, km/s	4158	4030	4915	± 50
Young's modulus E, GPa	170	162	229	± 8
Shear modulus G, GPa	61	58	90	± 3
Poisson's ratio ν	0.39	0.4	0.28	± 0.01
Microhardness H, GPa	4.6	4	14	± 0.2

The values of the elastic parameters of the sample 1 and the sample 2 are equal within the experimental error. However, they are very different from those for the sample 3. The value of longitudinal wave velocity decreases with increasing samples density. Such dependence is anomalous, because the ultrasonic wave velocity usually higher for more dense samples [6]. In contrast, the dependence of shear wave velocity on the density is normal. The Young's and shear moduli, as well as the calculated microhardness of the sample 3 increase significantly, while the Poisson ratio decreases. This fact indicates that the sample is hard but more fragile than other samples. The data collection of X-ray diffraction and electron microscopy suggests that these unusual changes of the elastic properties may be associated with a size factor, i.e. with the presence of nanometer particles. Therefore, mechanical properties can be controlled by varying the particle size.

5. Conclusion

In this study we investigated the physico-mechanical properties of aluminum oxide nanoceramics. The findings suggest that the nanoceramics have unique properties that can be used in the future in microelectronics to improve the stability of chips and to optimize the production technology.

Acknowledgments

This work is supported by the Ministry of Education and Science of the Russian Federation (3.2015.2014/K).

References

[1] Mason U 1966 *Physical Acoustics. The methods and tools for ultrasound studies* (Moscow: Mir)

[2] Klyuev V V 2004 *Non-Destructive Testing: A Guide to 7 volumes* vol 3 (Moscow: Mashinostroenie)

[3] Rinkevich A B, Perov D V, Samoilovich M I and Klescheva S M 2010 Elastic properties and heat capacity of opal matrices and related 3D-nanocomposite materials *Phys. Sol. State* **52** 2570–76

[4] Marzouk S Y 2010 the acoustic properties of borosilicate glass affected by oxide of rare earth gadolinium *Physica B* **405** 3395-400

[5] Goljamina I P 1979 *Ultrasound. Small encyclopedia* (Moscow: Soviet Encyclopedia)

[6] Agranat B A, Dubrovin M N, Havskij N N and Jeskin G I 1987 *The fundamentals of Physics and the technology of ultrasound* (Moscow: Vysshaja shkola)

Hysteresis phenomenon of the field emission from carbon nanotube/polymer nanocomposite

S V Filippov[1,2], E O Popov[1,2], A G Kolosko[1,3] and P A Romanov[1]

[1]A.F. Ioffe Physico-Technical Institute of Russian Academy of Sciences, ul. Polytechnitscheskaya 26, St.-Petersburg, 194021, Russia
[2]St. Petersburg State Polytechnical University, ul. Polytechnitscheskaya 29, St.-Petersburg, 195251, Russia
[3]The Bonch-Bruevich Saint-Petersburg State University of Telecommunications, pr. Bolshevikov 22, St.-Petersburg, 193232, Russia

E-mail: s.filippov@mail.ioffe.ru

Abstract. Using the high voltage scanning method and the technique of multichannel recording and processing of field emission (FE) characteristics in real time mode we found out some subtle effects on current voltage characteristics (IVC) of the multi-tip field emitters. We observed the direct and reverse hysteresis simultaneously in the same field emission experiment. Dependence of the form of IVC hysteresis on time of high voltage scanning was observed.

1. Introduction

The field emission phenomenon is an inertialess process [1]. The basis of its study is the quantum-mechanical model of the tunnel effect, proposed by R. L. Fowler and Nordheim in 1928. This model describes well the field emission from the metal single-tip and multi-tip systems. To obtain reliable current-voltage characteristics of such systems was enough to measure the current in the range of several orders of magnitude. However in the past two decades with the advent of the huge variety of nanostructured materials and its nanocomposites, the results and interpretation of emission properties investigation became more complicated. This has led to a change in approach to the study of the emission properties of the objects, namely, repetition of the voltage-current characteristics of the passages with a cyclic increase and decrease in applied voltage. As a result, it began to appear a significant amount of work, indicating the presence of hysteresis in the obtained current-voltage characteristics.

For a single multi-walled carbon nanotube in the paper [2] was obtained IVC hysteresis, when at the falling branch of the voltage currents exceeding the current rising branches were observed.

There is an article about IVC hysteresis of emitter based on graphene sheets [3]. The effect of differences between branches of applied voltage is most noticeable on the FN plots. The study shows that a more rapid impact of high voltage leads to a broadening of the hysteresis loop. Hysteresis effects are explained by the balance between Joule heating and sorption processes.

In paper [4] the hysteresis of carbon nanotube fibers was investigated. The various field emission modes are identified in the forward and reverse sweeps and related to changes in the fiber morphology.

Oriented aluminum nitrides on a tungsten tip in [5] demonstrate hysteresis when the current on the falling branch less than on rising branch. The explanation is the charging of the insulating AlN layer films, when the electrons emitted into the vacuum are not compensated by electrons from source.

Discussion of the influence of adsorbed molecules on the appearance of hysteresis in an array of non-oriented CNT grown on a nickel substrate, is given in [6]. In this paper three types of adsorption were considered: internal adsorption, stimulated by the field and stimulated by the current. The dependence of the hysteresis of the vacuum level and the speed of the voltage was found. It is shown that the hysteresis decreases with increasing vacuum level.

There was a decrease in the emission current at the falling branch of the applied voltage for the array of carbon nanotubes coated with titanium [7] and tantalum. [8] It was shown that with increasing duration of the scan voltage, the hysteresis loop is expanded to lower currents. The special 'eight' shape of IVC was registered in [9] for CNTs.

The examples of works show that hysteresis appears in two basic ways: when the current of falling branch less then rising branch (so-called direct hysteresis, the most frequently observed in carbon nanotubes) and reverse hysteresis when the current with decreasing voltage become higher than the rising branch.

2. Exprimental details

We investigated thin film field emitter consisted of polymer / carbon nanotube composite. The technique for characterizing multi-tip field emitters was described in detail in our paper [10]. It includes the original program of multichannel recording and processing of data, written in LabView 2013. The power source is controlled by a computer FID Technology power supply. The rate of voltage change is set by the DAC board NI DAQ PCIe-6351.

For sample preparing we used two different types of nanotubes: MWCNT by Samsung (according to SEM length of tubes is ~ 10 μm and diameter is ~ 19 nm) and SWCNT (length of tubes exceeds 10 μm and diameter is ~ 2 nm). Firstly CNT powder was treated in an ultrasonic bath in o-xylene for 8 hours. Then this suspension was mixed with polystyrene solution and again was treated in the ultrasonic bath for 8 hours. Finally thin film of field emitter was deposited onto metallic substrate by spin coating technique.

Figure 1 and figure 2 shows SEM images of the created emitter surfaces: polystyrene with multi-walled nanotubes – "PS-MWCNT" and with single-walled nanotubes – "PS-SWCNT".

3. Results and discussion

We have previously shown that the shape of the scanning pulse voltage applied to the sample does not influence on the current-voltage dependence significantly [11]. In this study, we used only pulses with a triangular profile.

Research can be divided into two subgroups:

1) study the effect of the scanning voltage pulse duration on the IVC of different samples;

2) study the effect of the voltage pulse amplitude in the direction of the IVC hysteresis.

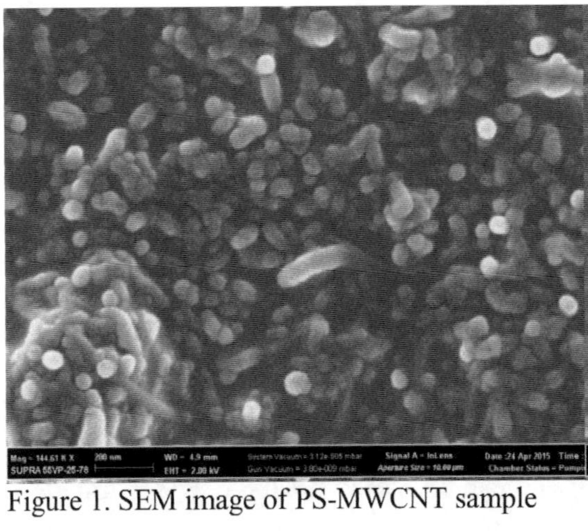

Figure 1. SEM image of PS-MWCNT sample

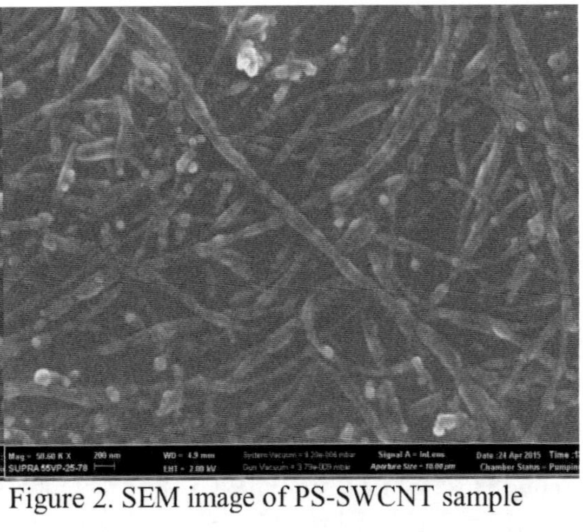

Figure 2. SEM image of PS-SWCNT sample

3.1. Effect of change in the scanning pulse voltage duration on the IVC

Figure 3 shows the shape of the scanning pulse voltage and the IVC of the PS-MWCNT sample. There is a hysteresis typical for many similar multipoint field cathodes hysteresis where rising branch passes over a falling branch (the so-called "direct" hysteresis).

Note that the repetition of this experiment leads to the almost complete repetition of the IVC in figure 3 (except for noise roughness).

Figure 4 shows the same IVC in the Fowler-Nordheim coordinates. There is the same hysteresis, but curve has also a characteristic crooked tail at low currents, which cannot be described by the known Fowler-Nordheim law for cold field emission.

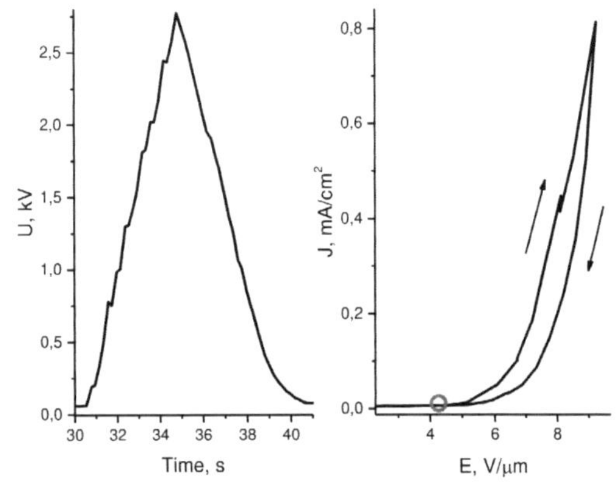

Figure 3. Example of a high voltage pulse and corresponding IVC for PS-MWCNT sample

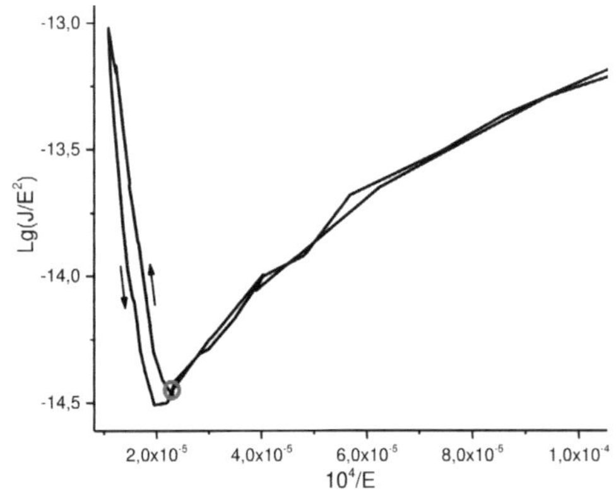

Figure 4. The IVC for PS-MWCNT sample in FN-coordinates

Figure 5 and figure 6 show the result of an experiment on the effect of the scanning pulse duration on the IVC of the sample PS-MWCNT. The graphs show that with increasing pulse duration the intersection point of the falling branch and rising branch appears in the "direct" hysteresis of IVC and gradually move to the high voltages.

Similar experiments with the PS-SWCNT sample showed that intersection point of the IVC branches for composite with single-wall carbon nanotubes appears at shorter pulses than for composite with multi-wall nanotubes (see figures 7, 8).

Note that the threshold voltage for PS-SWCNT is lower than for PS-MWCNT, which can be related with a smaller radius of the electron-emitting nanotips.

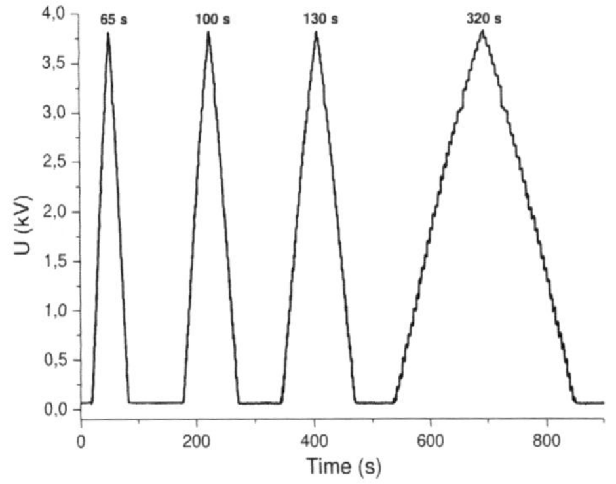

Figure 5. Sequence of slow pulses of voltage applied to PS-MWCNT sample with different pulse duration (see digits above)

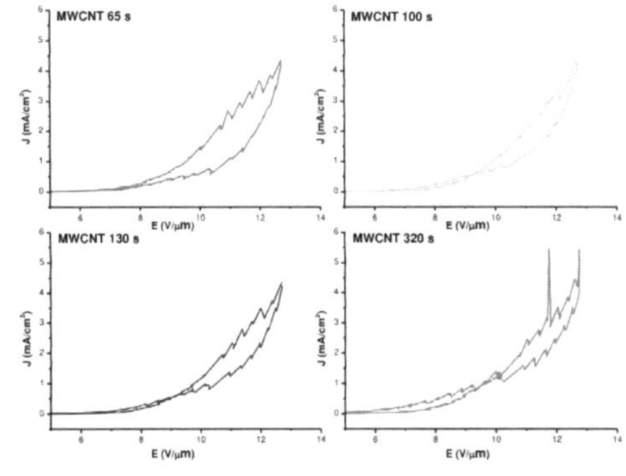

Figure 6. The "eight" shape hysteresis formation in IVC for PS-MWCNT sample at different pulse duration

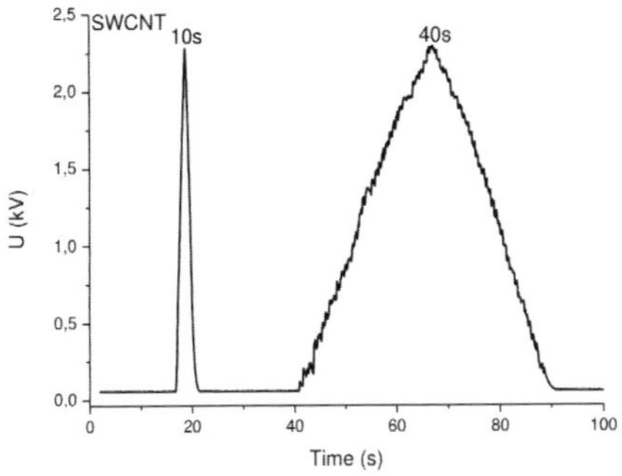

Figure 7. Sequence of slow pulses of voltage applied to PS-SWCNT sample with different pulse duration (see digits above)

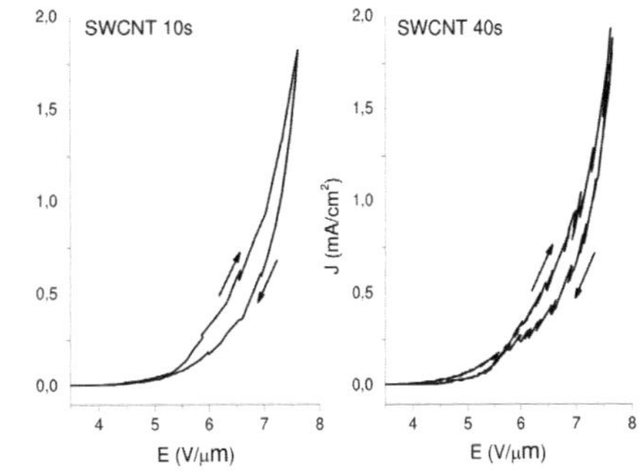

Figure 8. The "eight" shape hysteresis formation in IVC for PS-SWCNT sample at different pulse duration

3.2. Effect of change in the scanning pulse voltage amplitude on the IVC

In a second series of experiments we tried to find out the effect of change in the amplitude of the scanning pulse voltages, as well as their consecution on the shape of the IVC for PS-MWCNT sample.

Figure 9 shows three following each other voltage pulses with different amplitudes (the amplitude of each pulse is greater than the amplitude of the previous). Figure 10 shows the corresponding IVCs.

The graphs show that with increasing amplitude the IVC naturally increases and the difference between rising and falling branches of "direct" hysteresis increases too. Quite a different picture is observed with a gradual decrease in the amplitude of the pulse.

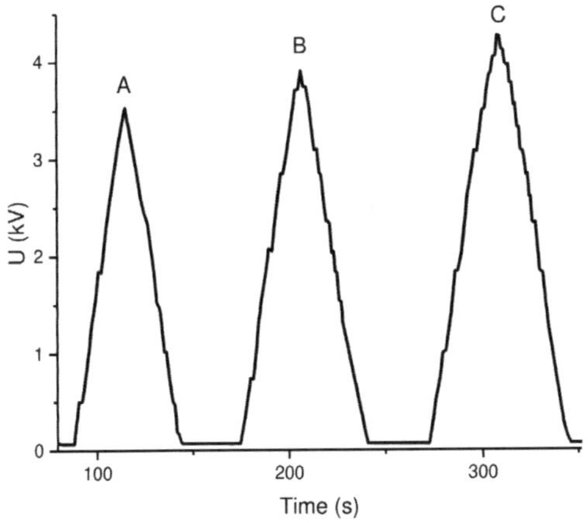

Figure 9. Sequence of slow pulses of voltage with different amplitudes applied to PS-MWCNT sample.

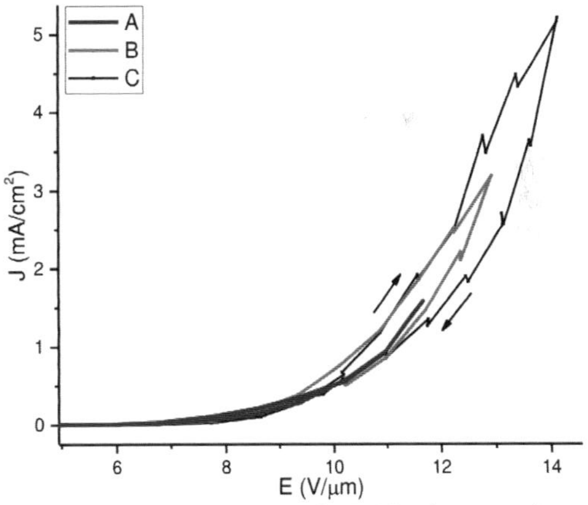

Figure 10. Change of IVC hysteresis with increasing pulse amplitude for PS-MWCNT sample.

Figure 11 shows the two following each other voltage pulses of different amplitudes (the second lower than the first) applied to the same PS-MWCNT sample. Figure 12 shows the corresponding IVCs. This experiment was repeated several times and each time with decreasing the pulse amplitude "direct" IVC hysteresis changed its direction turning to so-called "reverse" hysteresis, where a rising branch is held under the falling branch.

At the same time consistent decrease in the pulse amplitude retains the direction of the hysteresis - every pulse there is a "reverse" hysteresis (see figures 13, 14). Also, it was noted that the time delay between the pulses did not affect on the direction of IVC hysteresis.

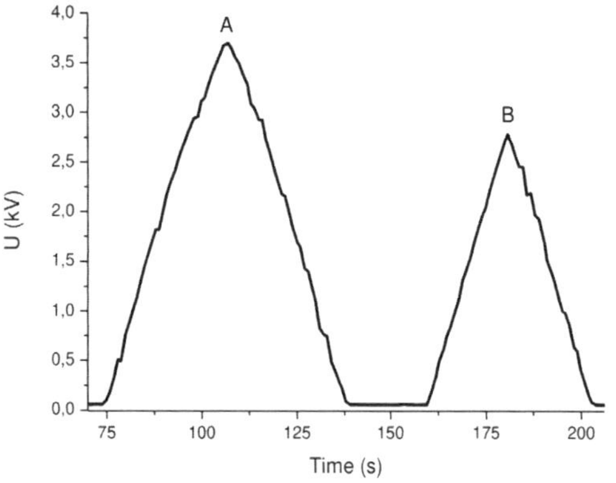

Figure 11. Two pulses of high voltage with different amplitude applied to PS-MCNT.

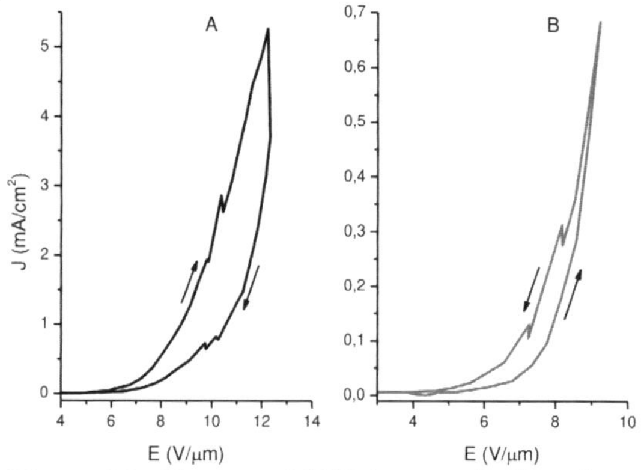

Figure 12. Change of IVC hysteresis direction after reducing of voltage pulse amplitude for PS-MCNT.

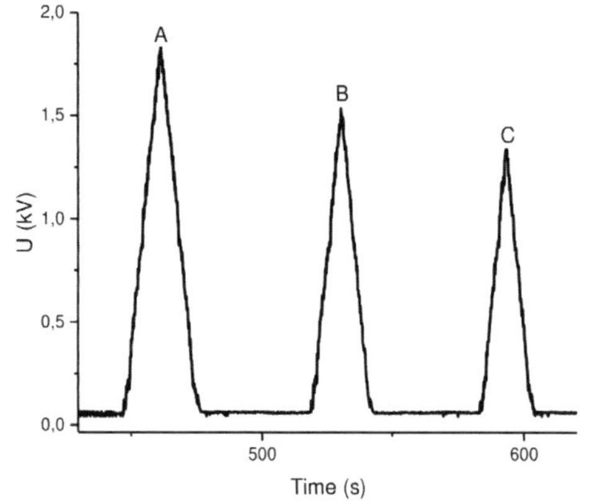

Figure 13. Three pulses of high voltage with different amplitudes applied to PS-SWCNT sample.

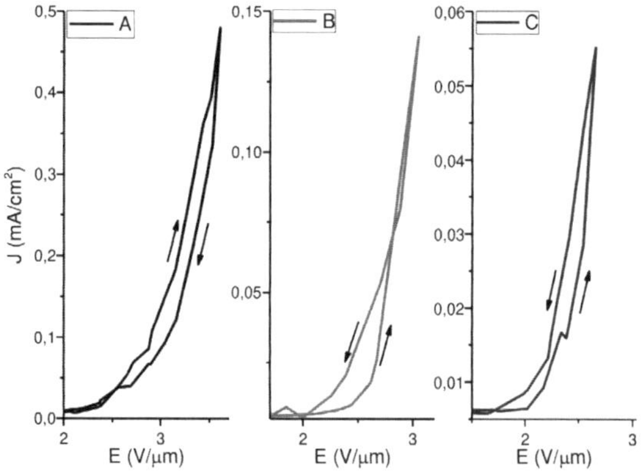

Figure 14. IVC hysteresis for three high voltage pulses with different amplitudes applied to PS-SWCNT sample.

Maintaining the pulse amplitude at a constant level leads to inevitable disappearance "reverse" hysteresis rise "direct." An example of this effect is shown in figure 15 and figure 16.

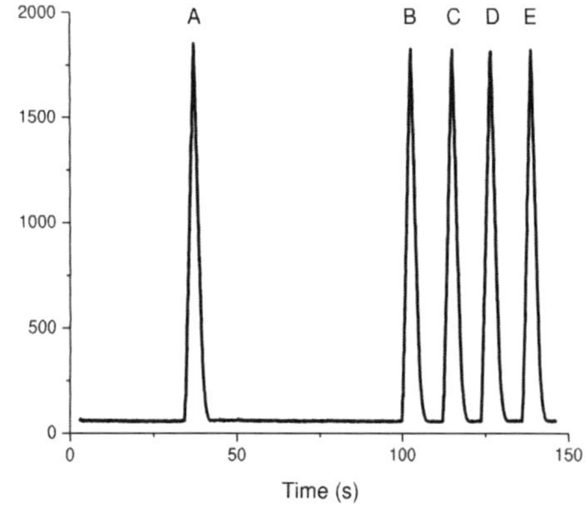

Figure 15. Two impulses of high applied voltage with different amplitude

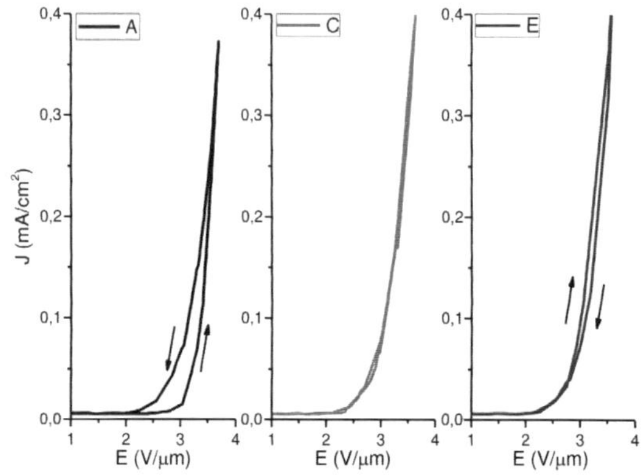

Figure 16. Change of IVC hysteresis direction after higher voltage applying PS-SWCNT

4. Conclusion

We managed to register the following effects connected with IVC behavior of nanocomposite field emitters:

1) In a mode of slow voltage change (an order of ten seconds on an impulse) the appreciable hysteresis is observed in the IVC.
2) The increasing in impulses duration leads to steepness change of a IVC falling branch and to emergence of an intersection point of falling and raising branches.
3) Using single-walled nanotubes in nanocomposite leads to earlier emergence of intersection point (at smaller duration an impulse).
4) Gradual growth of impulses amplitude keeps a direction of a hysteresis.
5) Amplitude reduction leads to emergence of a reverse hysteresis.
6) At maintenance of impulses amplitude the return hysteresis gradually disappears, turning to a straight line.
7) Delay time between impulses does not influence on the hysteresis direction, the important was only the previous voltage pulse amplitude.

Many works assume presence in nanocomposite materials a percolation effect associated with local charging of emission nanotips, separated from the cathode with the dielectric layer of polymer that can influence on emitting ability of a sample surface and to cause emergence of the IVC hysteresis.

Emergence of a point of intersection of growing and falling branches in a hysteresis can be associated with participation in emission process one more effect – thermionic emission.

Point 2 in the list resulted above confirms a hypothesis of presence of the thermionic emission contribution as at increase in duration of impulses nanotubes have time to heat up and the return branch IVC starts to cross a straight line.

Point 3 also confirms this hypothesis therefore as radius reduction nanotubes should lead to their faster warming up and consequently the thermionic emission contribution to a falling branch IVC becomes more powerful.

So, we assume that not only percolation effect can lead to IVC hysteresis emergence, but also thermionic emission is responsible for its appearance.

Acknowledgement

The work was carried out with the partial support of the Foundation for Assistance to Small Innovative Enterprises in Science and Technology (№0015113).

References

[1] Fursey G N, Field Emission in Vacuum Microelectronics, 2005 New York, *Springer* 205 p.
[2] Jin C H, Wang J Y, Wang M S, Su J, Peng L M 2005 *Carbon* **43** 1026.
[3] Chen J, Li J, Yang J, Yan X, Tay B-K and Xue Q 2011 *Appl. Phys. Lett.* **99** 173104
[4] Cahay M, Murray P T, Back T C, Fairchild S, Boeckl J, Bulmer J, Koziol K, Gruen G, Sparkes M, Orozco F and O'Neill W 2014 Appl. Phys. Lett. **105** 173107
[5] Zhang H Z, Wang R M, Zhu Y W 2004 *J. Appl. Phys.* **96** 264
[6] Li C, Fang G J, Yang X X, Liu N H, Liu Y P, Zhao X Z 2008 *J. Phys. D: Appl. Phys.* **41** 195401
[7] Zuo Y, Ren Y, Wang Z, Han X, Xi L 2013 *Organic Electronics* **14** 2306
[8] Wang Z, Zuo Y, Li Y, Han X, Guo X, Wang J, Cao B, Xi L, Xue D 2014 *Carbon* **73** 114.
[9] Okotrub A V, Kurenya A G, Gusel'nikov A V, Kudashov A G, Bulusheva L G, Berdinskii A S, Ivanova Y A, Ivanov D K, Strel'tsov E A, Fink D, Petrov A V and Belonogov E K 2009 *Nanotech. in Russia* 4 **9-10** 627
[10] Filippov S V, Popov E O, Kolosko A G, Romanov P A 2014 *J. of Phys.: Conf. Ser.* **572** 012026
[11] Cheryarina A S, Romanov P A, Filippov S V, Kolosko A G, Popov E O 2014 Proceedings of the youth scientific and practical conference. Science Week in SPbPU, **1** 85

Phase diagram calculation of $A^{III}B^{V}$ binary solutions of the eutectic type in the generalized lattice model

G A Panov and M A Zakharov

Department of Solid State Physics and Microelectronics, Novgorod State University, 173003, Novgorod the Great, Russia

E-mail: `Gennady.Panov@novsu.ru`

Abstract. The present work is devoted to the phase diagrams calculation of $A^{III}B^{V}$ systems within the framework of the generalized lattice model taking account of volume effects. The theoretically calculated phase diagram is compared with the corresponding experimental diagrams.

1. Introduction

Calculation of phase diagrams is one of the most interesting problem of physics of multicomponent systems. In particular, investigation of the thermodynamic properties and phase equilibria in $A^{III}B^{V}$ binary systems is of great importance for micro- and nanoelectronics. The aim of this work is a theoretical calculation of the phase diagram of $A^{III}B^{V}$ binary systems (In–Sb, In–As, Al–P, Al–Sb) within the generalized lattice model [1].

2. Model

It is well known, that the phase diagrams of the In–Sb, In–As, Al–P, Al–Sb systems are eutectic type phase diagrams with one intermediate phase of constant composition. The phase diagram of the eutectic type with intermediate phase of constant composition is schematically presented in Figure 1.

It is easy to see that the left and right parts of the initial phase diagram are similar to a conventional phase diagram of the eutectic type. The fundamental difference between both parts of the phase diagram in Figure 1 and the phase diagram of the eutectic type is that the mole fractions x of the B component vary in the range $0 \leq x \leq n/(m+n)$ for the left subsystem A–A_mB_n and in the range $n/(m+n) \leq x \leq 1$ for the right subsystem A_mB_n–B. As a consequence, the lengths of the segments that correspond in the concentration axis to the left and right parts of the diagram are not equal to unity, whereas the length of the concentration axis in any phase diagram should be equal to unity. These differences do not permit one to directly use the well-developed methods for calculating phase diagrams in the case of simulation of the components of phase diagrams for systems with intermediate phases. The method of calculation of phase diagrams of such type has been proposed in paper [2]. The method is based on the nonlinear transformations of the concentration axes and makes it possible to correctly reduce the calculation of the phase diagrams with several intermediate phases to a sequence of phase diagrams of individual subsystems. These cases were studied in detail in papers [2,3]. For

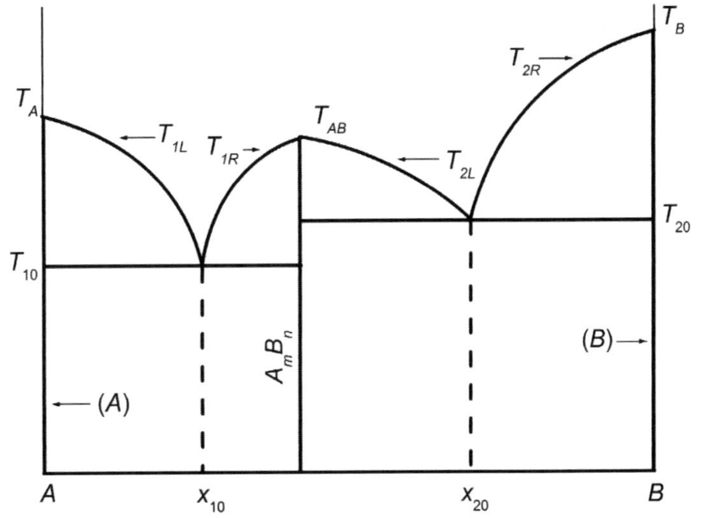

Figure 1. Phase diagram of the eutectic type with intermediate phase of constant composition in the absence of the mutual solubility of the components in the solid state.

this reason, we do not write out the explicit expressions for the nonlinear transformations of the concentration axes, which can be found in the papers mentioned above.

According to the basic principles of the generalized lattice model, the chemical potentials of a homogeneous binary solution (per mole of a material) can be represented in the following form [4]

$$\begin{aligned} \mu_1 &= \mu_{10} + RT\ln x + W\lambda \left(\frac{1-x}{x+\lambda(1-x)}\right)^2, \\ \mu_2 &= \mu_{20} + RT\ln(1-x) + W \left(\frac{x}{x+\lambda(1-x)}\right)^2, \end{aligned} \quad (1)$$

where μ_{i0} is the standard chemical potential of the ith component, R is the universal gas constant, T is the temperature of the system, x is the mole fraction of the first component, W is an analog of the energy of mixing in the generalized lattice model, $\lambda = \omega_2/\omega_1$, and ω_i is the characteristic atomic volume of the ith component.

Further let us consider the subsystem of an A–A_mB_n binary solution of eutectic type (see Figure 1). We can obtain all parameters of general lattice model for binary solutions of the eutectic type taking into account the chemical equilibrium in the three-phase system, i.e., equating the chemical potentials of the components in the solid and liquid phases.

Equations of the chemical equilibrium in the three-phase system form a closed system, whose solution allows us to find the concentration dependences of the liquidus branches and, hence, to construct the phase diagrams of binary system A–A_mB_n in the framework of the generalized lattice model.

It should be noted that the chemical equilibrium in the subsystem of an A_mB_n–B binary solution is considered in a similar way.

3. Phase diagram calculation

Using the technique developed in [2] calculation phase equilibria in the In–Sb, In–As, Al–P, Al–Sb systems was carried out (see, Figures 2,3). The parameters of the model are correspond to theoretical calculations of the phase diagrams was obtained (see, Table 1). It should be noted here we use the notation introduced work [2].

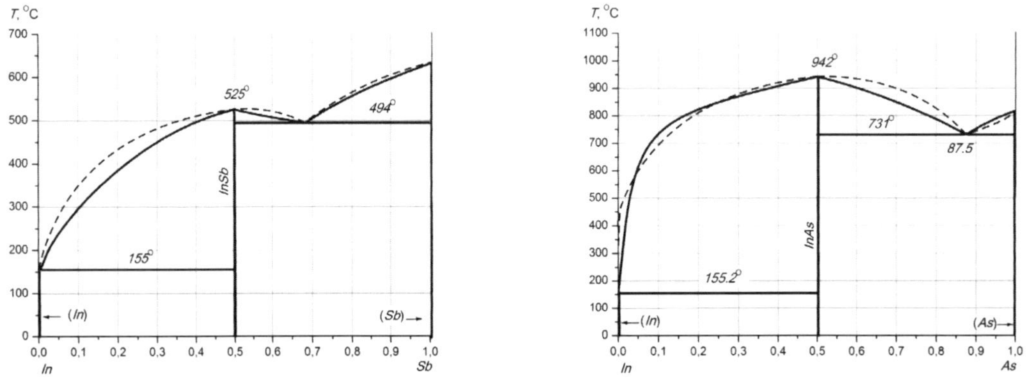

Figure 2. Phase diagrams of the the In–Sb, In–As systems. Dashed lines are the experimental data taken from [5]. Solid lines indicate the results obtained from theoretical calculations.

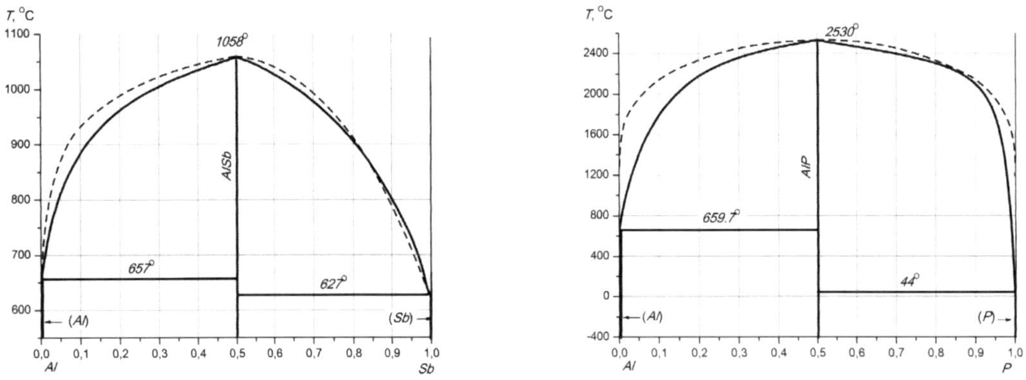

Figure 3. Phase diagrams of the the Al–Sb, Al–P systems. Dashed lines are the experimental data taken from [5]. Solid lines indicate the results obtained from theoretical calculations.

In addition it should be noted that explicit account of "volume" effects was carried by introducing "specific" the atomic volumes of components within the generalized lattice model.

It can been seen from Figures 2, 3 that the phase diagrams are consistent with the experimental data available in the literature [5].

Thus, comparison with experimental data shows that the theoretical approach can be used to adequately description and prediction of the phase equilibria in the $A^{III}B^{V}$ binary solutions.

Table 1. Parameters of the generalized lattice model.

A–B	x_0	T_0,°C	q_{AB}	λ	W/R, K
In–InSb	0.007	155.0	15.37	0.473	-1722
InSb–Sb	0.663	494.0	15.37	1.272	496
In–InAs	0.003	155.2	13.99	0.045	-439
InAs–As	0.875	731.0	13.99	1.516	-1569
Al–AlSb	0.004	657.0	23.53	0.178	-799
AlSb–Sb	0.992	627.0	23.53	2.256	-6016
Al–AlP	0.001	659.7	18.00	0.162	-4459
AlP–P	0.999	44.0	18.00	25.001	-4281

Acknowledgments. This study was supported by the Ministry of Education and Science of the Russian Federation (project 1755).

References

[1] Zakharov A Yu, Zakharov M A and Loginova O V 2004 *Int. J. Quant. Chem.* **100** 435
[2] Zakharov M A 2007 *Phys. Solid State* **49** 2312
[3] Panov G A and Zakharov M A 2013 *J. of Phys.: Conf. Ser.* **461** 012003
[4] Korzun E L and Terekhov S.V. 1987 *Zh. Fiz. Khim.* **61** 1186
[5] Lyakishev N P 1996 *Phase Diagrams of Binary Metallic Systems* (Moscow: Mashinostroenie)

The Structure of Metastable States in The Thomson Problem

A.N. Bondarenko[1], M.N. Karchevskiy[2] and L.A. Kozinkin[2]
[1] Sobolev Institute of Mathematics SB RAS, Novosibirsk, Russia
[2] Novosibirsk State University (NSU), Novosibirsk, Russia

E-mail: bondarenkoan1953@mail.ru, karchevskiy.michael@gmail.com, leon7archer@gmail.com

Abstract. A practical numerical method for the effective solution of the Thomson Problem is proposed. The developed iterative algorithm allows to conduct theoretical researches such as study of the number of asymptotic solutions depending on the particle count. Metastable states of charged point systems on a unit sphere were considered and the probability of falling into the basin of each state was obtained. Founded upon the algorithm and dual lattice representation approach, the framework for rapid construction of capped carbon nanotubes was created.

1. Introduction

What configuration of N unit point charges on a unit sphere minimizes the electrostatic potential energy of the system? Mathematically it is required to minimize the following energy functional for given N:

$$W(r_1, \dots, r_N) = \sum_{i<j} \frac{1}{|r_i - r_j|}, \tag{1}$$

where $r_i \in \mathbb{S}^2$ for $i = 1, \dots, N$.

The given problem was initially presented 1904 by Thomson and nowadays is one of the unsolved problems of the 20th Century. The development of numerical optimization methods for solving the Thomson Problem draws close attention of professionals due to its high complexity [1, 2, 3, 4, 5, 6, 7, 8, 9, 10, 11]. According to [2], the number of metastable configurations shows exponential asymptotic behavior with respect to N. Moreover, the region containing the global minimum is often small compared with those of other minima [4], which adds to the aforementioned difficulty of the Thomson problem. The Thomson problem has been tackled by several methods such as steepest descent [2], constrained global optimization (CGO) [3], generalized simulated annealing algorithm [5, 6], genetic algorithms [4], and variants of Monte Carlo [8, 11]. However, the lowest energy solutions were systematically catalogued only for $N < 400$ and for selected structures up to $N = 4000$ [8, 11]. Recent studies of the Thomson Problem revealed its significant application in the capped carbon nanotubes design [12, 13].

During the last decade a keen interest of researchers from various scientific fields has been caused by design and study of the properties of nanomaterials. Nanomaterials are generally materials containing structural elements, that do not exceed 100 nm along one dimension and thus develop entirely new properties.

Recently, increasing interest is caused by carbon nanotubes (CNTs) [14]. CNTs offer new opportunities for biological and medical applications: visualization of molecular, cellular and tissue structures, design of biosensors and its base electrodes, targeted delivery of various substances, radiation and photothermal therapy.

The most promising property of CNTs within biomedical applications area is its ability to penetrate various tissues of the body and transport large doses of agents, resulting in diagnostic and therapeutic effects [15].

2. Dynamic Monte Carlo simulation method

2.1. Methodology

We introduce a new statistical approach to the problem that allows to find not only ground state but also stable charges configurations and measure the degree of sustainability.

The proposed algorithm consists of 2 main stages:

2.1.1. Stage 1

(i) Let $m := 0$.
M sets each of N independent and uniformly distributed on a unit sphere $\xi_i \in U(\mathbb{S}^2)$ points are constructed. For random number generation puprose the Mersenne Twister PRNG is used.

(ii) Let $r_1, \ldots, r_N$ be radius vectors of on sphere distributed points, i.e. $|r_i| = 1$.

2.1.2. Stage 2

(i) Let $m := m + 1$.
The next set $r_1, \ldots, r_N$ of independent and uniformly distibuted on a unit sphere points is considered.

(ii) If $m > M$ then Stop.

(iii) For each pair $(i, j : i \neq j)$ the Coulomb force is modeled:

$$F_{ij} = \frac{(r_j - r_i)}{|r_j - r_i|^3}, \qquad \forall i, j : i \neq j. \tag{2}$$

(iv) For each point r_i the resultant is calculated:

$$F_i \equiv F(r_i) = \sum_{j, j \neq i} F_{ij}, \qquad \forall i. \tag{3}$$

(v) Now, let us displace the points on each iteration so that the potential energy of the system would decrease, i.e. let us find such $r'_1, \ldots, r'_N$, that

$$W(r'_1, \ldots, r'_N) < W(r_1, \ldots, r_N), \qquad W(r_1, \ldots, r_N) = \frac{1}{2} \sum_{i \neq j} \frac{1}{|r_i - r_j|} \tag{4}$$

It can be accomplished in the following steps:

(a) Let us choose k_1 from the condition:

$$F_{k_1} = \max_i |F_i| \tag{5}$$

(b) The k_1-th point is chosen and displaced according to the rule

$$r'_{k_1} = \gamma(r_{k_1} + \alpha F_{k_1}), \tag{6}$$

where α is specially selected factor responsible for the method's rate of convergence, γ is a normalizing factor that meets the terms $|r'_i| = 1$, r'_i is a new position of the i-th point.

(c) The transition from F_n to F'_n is carried out according to the rule

$$F_n{}' = F_n + \left(\frac{(r'_j - r_n)}{|r'_j - r_n|^3} - \frac{(r_j - r_n)}{|r_j - r_n|^3} \right), \qquad n \neq k_1, \quad F_{k_1}{}' = F(r'_{k1}). \tag{7}$$

(d) The shift to the next iteration is performed by the change of notation

$$r'_1, \ldots, r'_N \to r_1, \ldots, r_N \tag{8}$$

and return to beginning of the iteration process.

(e) The exit of the iteration process is carried out by the following condition:

$$|W(r_1, \ldots, r_N) - W(r'_1, \ldots, r'_N)| < \epsilon, \tag{9}$$

where ϵ is the energy accuracy coefficient.

2.2. Computational framework

Based on the proposed algorithm and three-dimensional visualization the computational software suite was created. Primarily aimed at the Thomson Problem solution, the program makes it possible to compute metastable configurations of point charges not only on a unit sphere but also on an arbitrary convex surface, therefore creating many significant applications. One of them is capped carbon nanotubes construction.

NanoDesigner is a multifunctional framework extension for the computer design of low energy fullerene and capped nanotube structures. This software provides an ideal tool to accompany the study of the finite carbon molecules using atomistic computer simulation.

The computational core of the framework was written in C++11 using OpenMP and Intel C++ Compiler. Moreover, the dynamic calculation precision contributes to the better rate of convergence (see figure 1).

The results' analysis and visualization modules were implemented by MATLAB routines.

3. Results and conclusions

The aim of this study was to investigate metastable configurations and the convergence probability of a number of initial point system to each metastable state. The exponential asymptotic behavior of the metastable states number mentioned in [2] is confirmed (see figure 2). As mentioned in [16], the global minimum will most likely be obtained comparing to any metastable state. However, our computational experiments show the opposite for some N – the specific cases are presented in tables 1, 2. They illustrate also the complexity of metastable states distribution.

Journal of Physics: Conference Series **643** (2015) 012103 doi:10.1088/1742-6596/643/1/012103

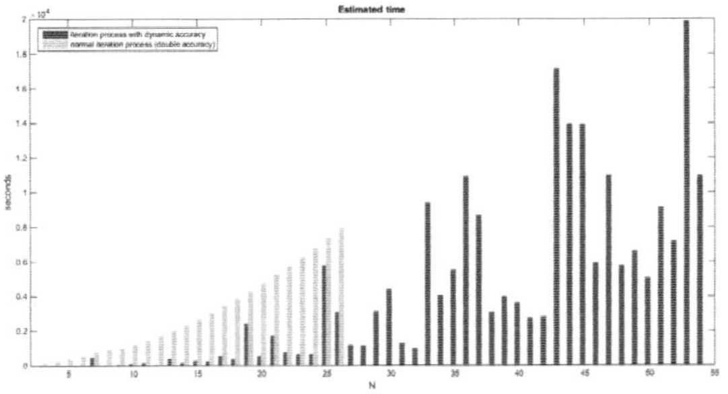

Figure 1. Comparison of the normal and dynamic iteration process (the computational precision grows near local extrema)

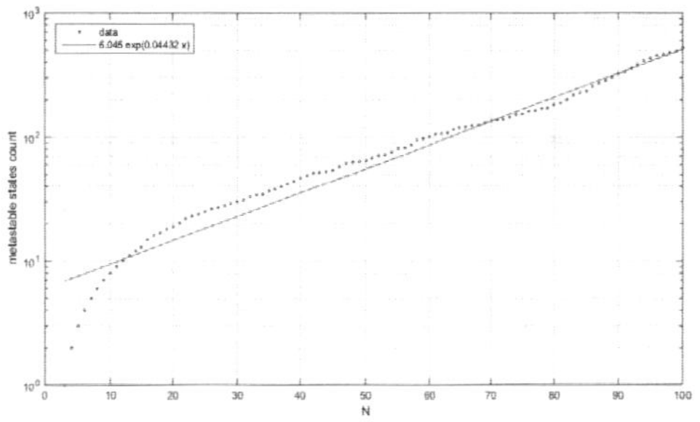

Figure 2. The metastable states number integral function and its exponential approximation

Table 1. Metastable states structure. It is important to note that already for $N = 37$ the global minimum is less likely to be obtained.

N	Stable state index	Occurence probability	Potential energy
16	1	0.75108	92.9125
16	2	0.24892	92.921
32	1	0.98558	412.2613
32	2	0.01442	412.469
37	1	0.24495	560.6213
37	2	0.75505	560.6295
46	1	0.15752	886.1687
46	2	0.67819	886.1716
46	3	0.15982	886.179
46	4	0.004477	886.2521

Despite the fact that numerical iterative methods aimed at the Thomson Problem solving have been intesively developing during last decade, the construction of the numerical algorithm that effectively enables us to find all solutions for $N > 500$ is a desirable goal for many applied researchers. Perhaps, our research provides another step in the right direction.

Table 2. Metastable states structure. The complexity grows rapidly as N increases.

N	Number of metastable states	Configurations structure
58	8	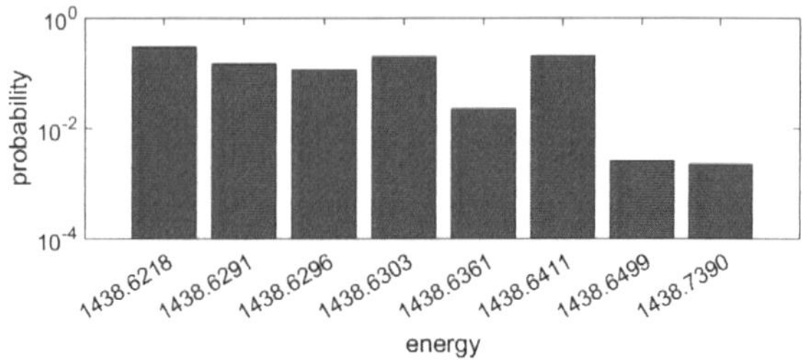
74	7	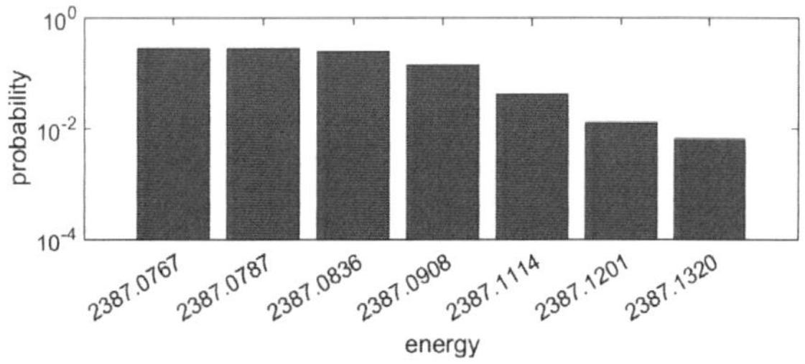
78	5	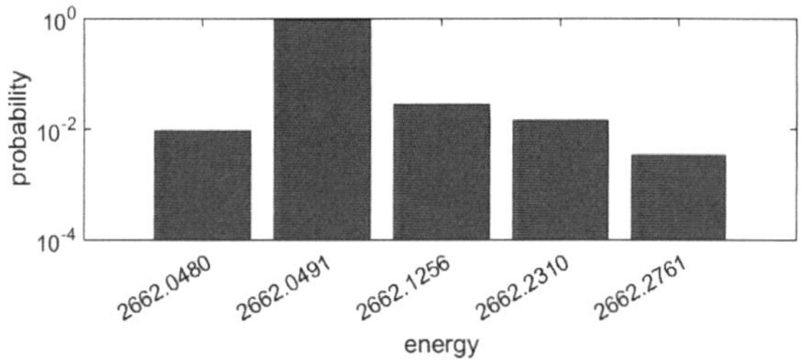
84	11	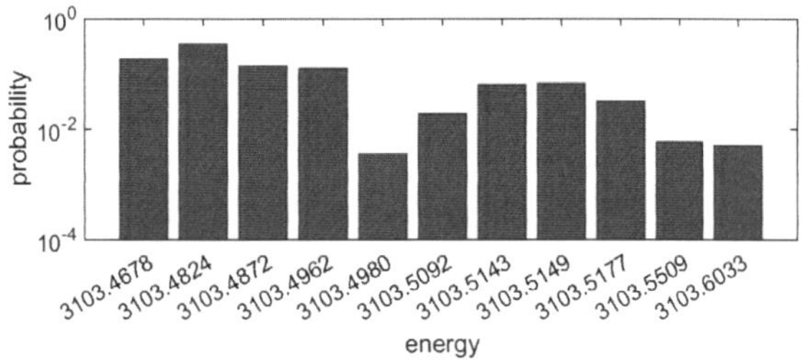
88	12	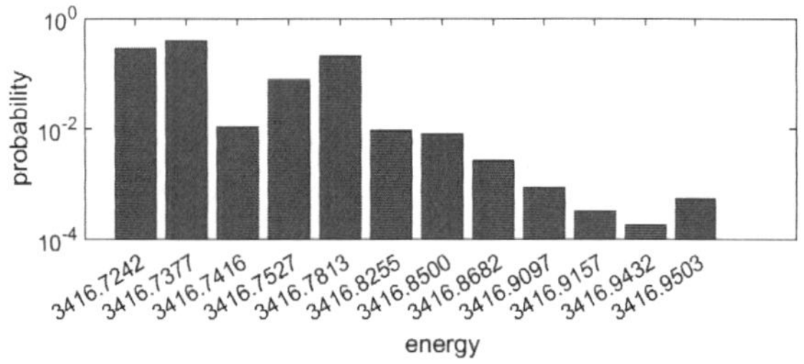
94	32	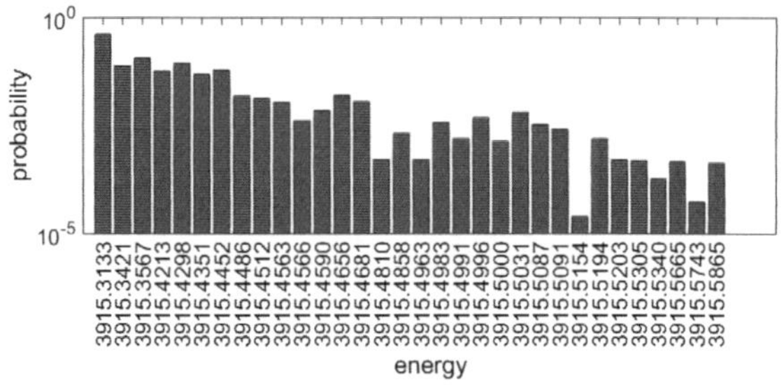

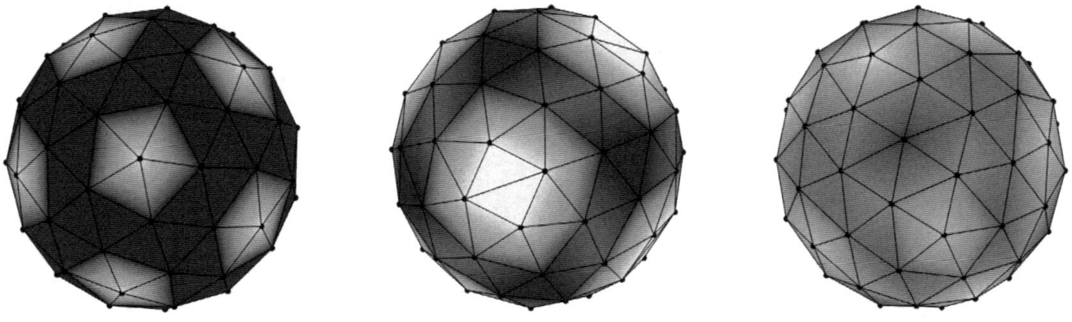

Figure 3. Stable point configurations for $N = 72$ (force absolute value distribution coloring)

Acknowledgments
The work was partially supported by RFFI foundation, grant N 14-01-00208.

References

[1] Altschuler E L and Pérez-Garrido A 2005 *Physical Review E* **71** 047703
[2] Erber T and Hockney G 1991 *Journal of Physics A: Mathematical and General* **24** L1369
[3] Altschuler E L, Williams T J, Ratner E R, Dowla F and Wooten F 1994 *Physical review letters* **72** 2671
[4] Morris J, Deaven D and Ho K 1996 *Physical Review B* **53** R1740
[5] Xiang Y, Sun D, Fan W and Gong X 1997 *Physics Letters A* **233** 216–220
[6] Xiang Y and Gong X 2000 *Physical Review E* **62** 4473
[7] Levin Y and Arenzon J J 2003 *EPL (Europhysics Letters)* **63** 415
[8] Wales D J and Ulker S 2006 *Physical Review B* **74** 212101
[9] Altschuler E L and Pérez-Garrido A 2006 *Physical Review E* **73** 036108
[10] Ji-Sheng C 2007 *Chinese Physics Letters* **24** 1825
[11] Wales D J, McKay H and Altschuler E L 2009 *Physical Review B* **79** 224115
[12] Robinson M, Suarez-Martinez I and Marks N 2013 *Physical Review B* **87** 155430
[13] Robinson M and Marks N 2014 *Computer Physics Communications* **185** 2519–2526
[14] Mubarak N, Abdullah E, Jayakumar N and Sahu J 2014 *Journal of Industrial and Engineering Chemistry* **20** 1186–1197
[15] Baeza A and Vallet-Regí M 2014 *Bio-and Bioinspired Nanomaterials* 23–48
[16] Pérez-Garrido A, Ortuno M, Cuevas E and Ruiz J 1996 *Journal of Physics A: Mathematical and General* **29** 1973

Electrophysical properties of phase change memory materials on the pseudo-binary line GeTe-Sb_2Te_3

A O Yakubov*, D Y Terekhov*, A A Sherchenkov*, S A Kozyuhhin, P I Lazarenko*, A V Babich* S P Timoshenkov*, D G Gromov* and A S Shuliatyev***
*Moscow State Institute of Electronic Technology (Technical University), 124498 Zelenograd, Russia
**Kurnakov Institute of General and Inorganic Chemistry, RAS, 119991 Moscow, Russia
E-mail: aa_sherchenkov@rambler.ru

Abstract. The temperature dependences of the resistivity and current-voltage characteristics of amorphous thin films on the basis of $GeSb_4Te_7$, $GeSb_2Te_4$, and $Ge_2Sb_2Te_5$ perspective for the phase change memory application were investigated. It was revealed that two-channel conduction mechanism with the transport of charge carriers by the localized states in the valence band tail and delocalized states of the valence band is characteristic feature of these materials.

1. Introduction

New generation of non-volatile memory is actively developed now. One of the most promising is phase change memory (PCM) due to the superior rate, durability and radiation resistance in comparison with the most widespread flash memory [1,2].

Two main types of PCM memory devices exist now - optical and electrical phase-change memories. The work of both types of devices is based on rapid reversible amorphous-to-crystalline phase-change processes, which can be initiated by laser or electrical pulses. PCM materials in amorphous and crystalline states have very different optical and electrical properties, and are attributed as "zero" and "unit" logical states. The electrical data storage is based on the large difference in resistivities of amorphous and crystalline phases.

Three compounds exist on this pseudo-binary line: $Ge_2Sb_2Te_5$ (GST225), $GeSb_2Te_4$ (GST124), and $GeSb_4Te_7$ (GST147). The most promising among them is considered to be GST225 due to the complex of properties.

Despite the successful commercial application of GST materials in optical storage devices, many fundamental and important practical issues related to the electrophysical properties of thin films, are not fully understood. For example, discussions of the possible charge carrier transport mechanisms in Ge-Sb-Te thin films are continued till now [4-7]. Understanding of the transport mechanisms is necessary for controlling the physical properties of programmable materials, and development of effective methods for improving and optimization of phase change memory technology. Therefore, the aim of the work was to study the influence of the composition variation of thin films along the pseudo-binary line GeTe-Sb_2Te_3 on the temperature dependences of the resistivity, current - voltage characteristics, and energy diagrams.

2. Experimental

The initial $GeSb_4Te_7$, $GeSb_2Te_4$ and $Ge_2Sb_2Te_5$ were synthesized with using of quenching technique originally described in [7] and later modified by us. Thin films were prepared by thermal evaporation in vacuum of the synthesized materials in the "explosive" mode. Residual pressure in the chamber was 10^3 Pa, substrate temperature did not exceed 50 °C. As-deposited films were amorphous, which was confirmed by X-ray diffraction (XRD).

Compositions of thin films were investigated by Rutherford backscattering (RBS) according to the procedure described in detail in (Ed = 1.0, Eα = 2.7 MeV, φ = 135 °). Overlapping of the peaks on the spectra was observed due to the neighbor position of antimony and tellurium in the Periodic table and their close atomic masses, which made impossible quantitative analysis. Therefore, ratios Ge/(Sb + Te) were determined, which for all investigated compounds were close to the theoretical values (Table 1).

Table 1. Results of RBS measurements

Composition	Ge/(Sb+Te)	
	Experimental	Theoretical
GST147	1/(10.7 ± 0.4)	1/11
GST124	1/(6.1 ± 0.2)	1/6
GST225	1/(3.4 ± 0.1)	1/3.5

For the investigation of electrophysical characteristics planar structures were prepared. For this purpose aluminum electrodes were fabricated on the oxidized silicon substrates with interelectrode distance of 7 mm. Amorphous chalcogenide thin film coated further by the protective SiO_x layer was deposited above the interelectrode gap. Geometrical parameters of electrodes were determined by optical microscope (Carl Zeiss Axiovert 40 MAC) and scanning electron microscope (Carl Zeiss NVision 40). The thicknesses of the investigated chalcogenide films were controlled by atomic-force microscope (AFM NT-MDT SolverPro) and ranged from 60 to 80 nm.

The electrical characteristics of thin films were studied on a special stand, which included two-electrode circuit based on the voltage control unit NI6008 and picoammeter KEITHLEY 6486. The temperature dependences of the resistivities of thin films were measured at the voltage of 0.7 V (E = 10^3 V/cm) in the temperature range from room temperature to 200 °C at a heating rate of 1 °C/min. Current-voltage characteristics (CVC) were investigated in the range of 0.1 – 10 V with the increment of 0.1 V, and from room temperature to 70 °C with the increment of 10 °C. The temperature gradient in the plane of the sample did not exceed 1 °C.

Investigations of the thermal properties of thin films were carried out on a differential scanning calorimeter (DSC-50 Shimadzu). Initial samples were thin films deposited on silicon wafers. Films were removed from the substrates by the sapphire scraper and pressed into aluminum crucibles. Measurements were carried out at heating rate of 10 °C/min in an inert atmosphere of nitrogen.

3. Results and discussion

Analysis of the temperature dependences of the resistivities (Figure 1) allowed to distinguish for all the investigated compounds three ranges with the similar temperature dependencies of the resistivities (designated as I, II, and III).

Range I (from room temperature to T = 120 °C) with the exponential temperature dependence.

Range II (from 100 to 140 °C) with a sharp drop in the resistivities of thin films on the several orders of magnitude.

DSC studies of $GeSb_4Te_7$, $GeSb_2Te_4$, and $Ge_2Sb_2Te_5$ thin films showed that in this temperature range, the appearance of the exothermic peak is observed, which is caused by phase transition from the amorphous to the crystalline state.

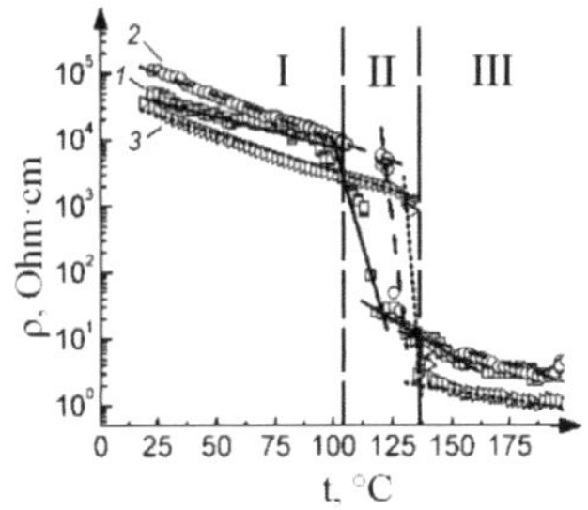

Figure 1. Temperature dependences of the resistivities of thin films: 1 - GST147; 2 - GST124; 3 - GST225.

It can be seen that the onset temperature of the exothermic peak is close to the beginning of the resistivity drop (T_x), defined as the intersection of approximating lines. Measurements by XRD, as well showed that the observed drop of the resistivity and exothermic peak corresponds to the phase transition from the amorphous phase to the rock salt cubic structure NaCl (fcc).

Range III. For temperatures $T > T_y$ (where T_y is the endset temperature of the resistivity drop) exponential temperature dependences of the resistivities are observed for GST thin films. However, the values of the resistivities are lower than in the range I, which is characteristic for chalcogenide films in the crystalline state [2].

The results of the analysis of the temperature dependence of the resistivity, DSC curves and literature data are presented in the Table 2, where T_x and T_y are the onset and endset temperatures of the phase transition, $\Delta T_n = T_x - T_y$, ρ_a and ρ_c are resistivities at 20 and 250 °C, respectively.

Table 2. Characteristics of thin film on the basis of compounds on the pseudobinary line Sb_2Te_3 - GeTe

Compounds	T_x - T_y, °C	ΔT_n, °C	T_x, °C		ρ_a, Ohm·cm	ρ_c, Ohm·cm	ρ_a/ρ_c
	ρ~T		DSC	[3]			
GST147	100 - 118	18	115	123	$4{,}9 \cdot 10^4$	2.6	$1.9 \cdot 10^4$
GST124	122 - 128	6	118	131	$1 \cdot 10^5$	3.8	$2.6 \cdot 10^4$
GST225	132 - 139	7	136	142	$2.8 \cdot 10^4$	1.0	$2.8 \cdot 10^4$

The data presented in Table 2 and Figure 1 shows that with the variation of $(GeTe)_n(Sb_2Te_3)_m$ from n:m = 1:2 (GST147) to 1:1 (GST124) and further to 2:1 (GST225), i.e. with the increasing of GeTe content, onset temperature of the phase transition increases, while the phase transition temperature range ΔT_n decreases. Slight difference between the onset crystallization temperatures determined by resistivity measurements, DSC data, and data presented in [3] can be explained by the difference in heating rates during measurements. Decrease of the crystallization temperature with the increase of Sb_2Te_3 while moving from GST225 to GST147, is due to the increase of the consentration of weaker chemical bonds Sb-Te (277,5 kJ / mol) compared with stronger chemical bonds Ge-Te (402 kJ/mol), which coincides with the data of [3].

CVC measurements showed that there are three characteristic regions with different current-voltage dependencies: for weak ($E < 10^3$ V/cm), middle (10^3 V/cm $< E < 10^4$ V/cm) and strong ($E > 10^4$ V/cm) electric fields. Figure 2 shows the current-voltage characteristics of GST225 thin film at different temperatures. The presence of different current-voltage characteristics is due to the dominating of different charge carrier transport mechanisms [7].

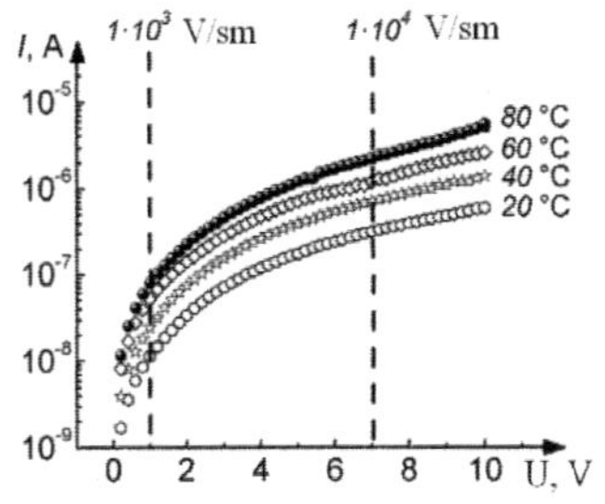

Figure 2. Current-voltage characteristics of the amorphous GST225 thin film in the temperature range from 20 to 80°C.

Amorphous thin films of the materials studied have nearly exponential temperature dependences of the resistivities in the temperature range from room temperature to 100°C (Figure 1). Such dependencies indicates on the activation nature of conductivity, which is characteristic for chalcogenide glassy semiconductors. In this case, the temperature dependence of the conductivity is described by the following expression:

$$\sigma = \sigma_0 \exp\left(\frac{E_F}{kT}\right), \quad (1)$$

where σ_0 is pre-exponential factor, E_a is activation energy of the conductivity, k is Boltzmann constant.

Analysis of the dependencies $\sigma = f(1/kT)$ with using of the expression (1) allowed to estimate activation energies of the conductivity, which are consistent with the literature data. Energy parameters for investigated thin films are shown in Table 3, where E_g is the width of the optical gap, E_0 is Urbach energy. Widths of the optical gap and Urbach energy are taken from, the error is ± 0.01 eV.

Table 3. Energy parameters of GST thin films

Material	E_g, eV	$E_g/2$, eV	E_a, eV	E_0, eV	σ_0, Ohm^{-1}·cm^{-1}
$GeSb_4Te_7$	0.50	0.25	0.22	0.09	0.1
$GeSb_2Te_4$	0.54	0.27	0.28	0.08	1.0
$Ge_2Sb_2Te_5$	0.61	0.30	0.29	0.13	2.2

When analyzing electrical conductivity, we used the Mott - Davis (MD) model of the energy bands. Characteristic features of this model are narrow tails of the localized states at the edges of the valence band and the conduction band, which are spreading in the mobility gap, and the boundary energies separating the localized states from delocalized in bands. The Fermi level in the MD model is fixed near the middle of the mobility gap by the localized states due to the defects. With using of the MD model, and taking into account experimental results, the energy band diagrams for amorphous GST thin films were designed (Figure 3).

Analysis of the diagrams revealed that the obtained values of the activation energies for conductivity are close to the middle of the mobility gap (Eg/2), except for GST124. For this composition activation energy is slightly higher $E_g/2$. Measurements of thermoelectric power carried out for investigated thin films, showed that GST has a negative value of thermoelectric power, indicating on the dominating p-type conductivity. The negative sign of thermoelectric power and the resulting value of the Seebeck coefficient (0.5 mV/K) for the amorphous GST225 thin film. Increase of the conductivity activation energy with the transition from GST147 to GST225 (i.e. in the direction from Sb_2Te_3 to GeTe on the pseudo-binary line) is due to the variation of the optical gap width and density of states in it. Decrease of the optical gap width is caused by the decrease of the conduction band bottom due to the broadening of the antibonding states band and increase of its population with the increase of the concentration of $SbTe_{3/2}$ structural units with weaker chemical bonds.

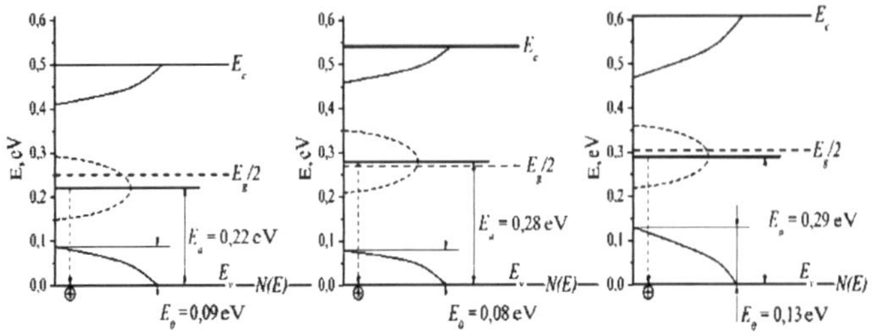

Figure 3. Energy diagrams for Ge-Sb-Te thin films: a – GST147; b - GST124; c - GST225

The spectral dependencies of the absorption coefficient for amorphous GST225, GST124 and GST147 thin films exist ranges with the exponential dependencies, which indicates on the presence of band tails in the mobility gap, and is caused by optical transitions with exponentially distributed states on the band tails. Maximum width of the band tail have GST225 thin films (E_0 = 0.13 eV). The values of E_0 for thin films of GST124 (0,08 eV) and GST147 (0,09 eV) are close. The obtained values of Urbach parameter indicates that thin films of GST124 and GST147 have less disordered structure in comparison with GST225.

As noted earlier, the activation energies of conductivities for all compositions are close to the middle of the mobility gap. This allows to suggest that significant contribution to the conductivity is caused by the transport of the charge carriers by delocalized states of the valence band. However, it is noted that for most disordered semiconductors with charge carrier's transport by the delocalized states pre-exponential factor σ_0 must be in the range from 10 to 10^3 $Ohm^{-1}\cdot cm^{-1}$. Obtained values of σ_0 for investigated GST materials are significantly less. Therefore, we used a two-channel conduction model proposed by Nagels. According to the model transport of the charge carriers by the localized states of the valence band tail and by the delocalized of state of the valence band can simultaneously contribute to the p-type conductivity of the chalcogenide glassy semiconductors:

$$\sigma = \sigma_0'\exp[-(E_F-E_V)/kT] + \sigma_{0h}\exp[-(E_F-E_0+W)/kT], \quad (2)$$

where the first term corresponds to the transport of the charge carriers by the delocalized of state, and second – to the hopping transport by the localized states of the valence band tail (W is the energy required for hopping). The possibility of using the two-channel model was demonstrated by Van der Plas, Buba and Nagels for the description of the previous generation of PCM materials on the basis of Ge_xTe_{1-x} and Si-Te-As-Ge (STAG). We used equation (2) for the description of the obtained dependences of the conductivity on 1/kT. For the modeling parameters σ_0, σ_{0h} and W were varied, obtained value of the conduction activation energy was used as E_f, while E_g and E_0 were obtained from the optical transmission spectra measurements (see Table 3).

Figure 4 present the temperature dependences of the conductivities, which show that the results of modeling and experimental data coincide.

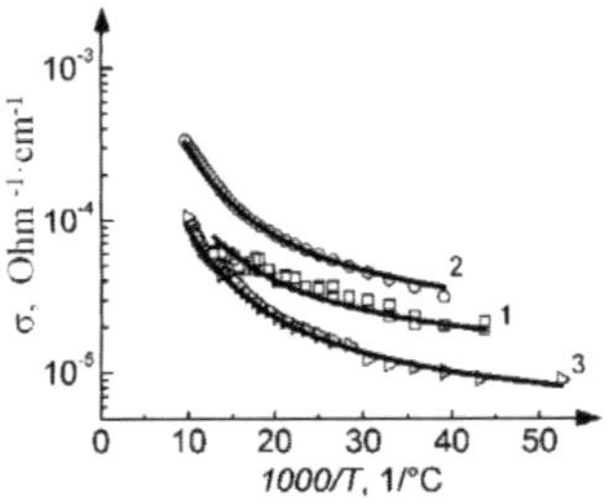

Figure 4. Dependences of the conductivities on 1000/T: 1 - GST147; 2 - GST124; 3 - GST225.
Markers denote the experimental data, continuous lines – results of modeling.

It should be noted, however, that the use of the equation (2) for fitting the temperature dependence of the conductivity does not allow strictly determine parameters σ_0, σ_{0h} and W. Only possible ranges of these parameters can be estimated, which are shown in Table 4.

Nagels et al gave following values for the pre-exponential factors: $\sigma_{0'}$ = 3000 $Ohm^{-1}\cdot cm^{-1}$ and σ_{0h} = 90 $Ohm^{-1}\cdot cm^{-1}$. Our results showed that the values of these parameters obtained with using of two-channel model (see. Table 4) differs from those of the work. However, it should be noted that the difference between estimated value of $\sigma_{0'}$ and σ_{0h} constitutes several orders of magnitude, which correlates with the data of. This result is also consistent with the MD model, where sharp drop of the carrier's mobility at the energy separating delocalized and localized states of the valence band is assumed. The difference in the values of pre-exponential factors can be explained by the peculiarities of the physico-chemical properties of GST materials. For example, the GST thin film has an extremely high rate, stability and reproducibility of switching from amorphous to crystalline states under the influence of electric pulses [1-3].

Table 4. Energy parameters for thin films on the basis of compounds on the pseudo-binary line Sb_2Te_3 – GeTe

Compound	Variable values			Fixed values			
	$\sigma_{0'}$, $Ohm^{-1}\cdot cm^{-1}$	σ_{0h}, $Ohm^{-1}\cdot cm^{-1}$	*W*, eV	E_g, eV	$E_g/2$, eV	E_f, eV	E_0, eV
GST147	0.08 ÷ 0.12	0.0008 ÷ 0.0012	0.005 ÷ 0.015	0.50	0.25	0.22	0.09
GST124	0.45 ÷ 0.60	0.0045 ÷ 0.0060	0.005 ÷ 0.015	0.54	0.27	0.28	0.08
GST225	2.0 ÷ 3.0	0.01 ÷ 0.02	0.005 ÷ 0.015	0.61	0.30	0.29	0.13

Thus, the experimental results obtained for GST thin films can be described by the two-channel conductivity model. However, additional measurements of the temperature dependences of thermoelectric power and estimation of the Hall mobility of charge carriers are necessary for strict determination of the model parameters.

This study was supported by Ministry of Education and Science of Russian Federation (project ID: RFMEFI57514X0096).

References

[1] Nicolescu, G. O'Connor, I. and Piguet, C. // Springer. – 2012. – P. 339-364.

[2] Wong, H.-S. P. Raoux, S. Kim, S.-B., Liang, J. and et al. // Proceedings of the IEEE. – 2010. – 98 (12). – P. 2201-2227.

[3] Yamada, N. Ohno, E. Nishiuchi, K. Akahira, N. Takao, M. // Journal of Applied Physics. – 1991. – 69. – P. 2849-2856.

[4] Nardone, M. Simon, M. Karpov, I. V. Karpov, V. G. // Journal of Applied Physics. – 2012. – Vol. 112. – P. 071101.

[5] Lazarenko, P. Sherchenkov, A. Kozyukhin, S. Shtern, M. Timoshenkov, S. Gromov, D. Redichev, E. // SPIE Proceedings. – 2014. – Vol. 9440. – P. 944006.

[6] Bogoslovsky, NA Tsendin, KD // Semiconductor Physics and Technology. - 2009 - v. 43 (10). - C. 3178-1382.

[7] Abrikosov, NH Danilov-Dobryakova, GT // Inorganic materials. 1965. - Vol.1. - №2. - S.204-208.

Problems of the experimental implementation of MTJ

L A Mazaletskiy[1,2], A S Rudy[2,1], O S Trushin[2,1], V V Naumov[2], A A Mironenko[2] and S V Vasilev[2]

[1] Department of Microelectronics and General Physics, Yaroslavl State University, Sovetskaya Street 14, Yaroslavl 150000, Russia
[2] Yaroslavl Branch of the Institute of Physics and Technology, Russian Academy of Sciences, Universitetskaya Street 21, Yaroslavl 150007, Russia

E-mail: boolvinkl@yandex.ru

Abstract. The results of experimental studies of MRAM technology based on standard magnetic tunneling junctions are presented. Basic steps of experimental fabrication of MRAM cell are considered. Experimental samples of MTJ with variable lateral sizes are fabricated. Current-voltage characteristics of the tunnel barriers are investigated. Main parameters of the tunnel barriers are estimated from comparison of the experimental data with the theory.

Magnetic Tunneling Junction (MTJ) is an important element of modern MRAM cell. It is believed that MRAM might be the next generation of computer memory. Therefore optimization of the technology for MTJ fabrication is of very important task. Achievement of high functional characteristics of MTJ requires intensive quality control at different stages of the technological route. In this work we discuss different aspects of quality control during this route. For this purpose we studied experimentally the technology of typical MTJ fabrication.

Standard spin-tunnel structure used to make MTJ consists of two magnetic layers separated by the layer of tunnel dielectric. This structure can be also used for magnetic field sensors [1]. MTJ fabrication involves complex technological route consisting of deposition of multilayer structure and subsequent MTJ pattern formation through lithography.

This work is devoted to the analysis of quality control methods at different stages of experimental fabrication of the MTJ samples. Two types of structures were considered: symmetrical Co(10 nm)/Al_2O_3(2 nm)/Co(30 nm)/SiO_2(200 nm)/Si(100) and non-symmetrical Ta(3 nm)/FeMn(15 nm)/NiFe(8 nm)/Al_2O_3(2 nm)/NiFe(10 nm)/Ta(10 nm)/SiO_2(200 nm)/Si(100). Multilayer structure was deposited by RF magnetron sputtering using industrial equipment Alcatel TETRA SCR-650. This machine is equipped with 4 magnetron sources, which allows depositing of multilayer structures in a single vacuum cycle. The magnetic layers were grown at external magnetic field of 100 Oe in the plane of substrate to promote easy axis formation.

Appropriate texture formation in the growing films is achieved through ion bombardment of the surface due to substrate bias potential. The layer of the tunnel dielectric was obtained by deposition of a thin Al layer (2 nm), which subsequently was oxidized in oxygen plasma. Low-energy ion bombardment in argon plasma was also used during antiferromagnetic (FeMn) growth to promote exchange bias effect in the underneath ferromagnetic (FeNi) layer and provide effective pinning of upper magnetic layer.

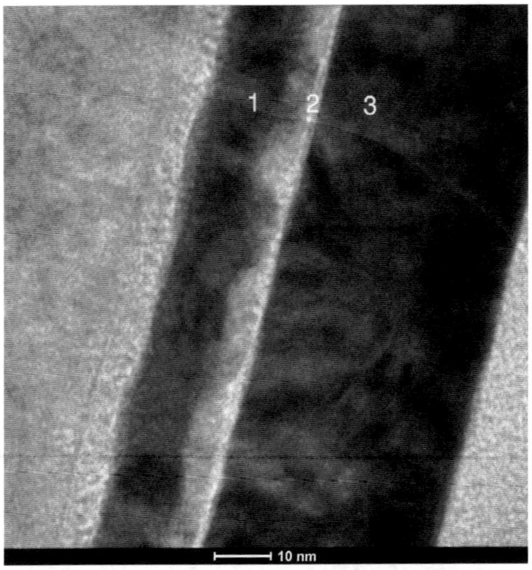

Figure 1. Cross-section of the multilayer spin-tunnel structure obtained using TEM. 1. FREE magnetic layer, 2. TUNNEL dielectric layer, 3. PINNED magnetic layer

It was found that changing the conditions of ion bombardment at this stage of the growth greatly affects pinning potential in the layer. The growth of buffer layer (at the bottom of the structure) and encapsulating layer (on the top of it) were done in separate vacuum cycles (before and after deposition of the core layers of spin-tunnel structure). We have already tested this kind of the technology before in studies of spin-valve structures [2].

The quality of multilayer structure has been studied using TEM. Cross-section of as deposited thin film structure is shown in figure 1. Bright color stripe in the center of the figure corresponds to the tunnel dielectric layer (Al_2O_3). Neighbor dark spaces correspond to magnetic and buffer layers. From this figure we can estimate the quality of tunnel dielectric layer (its thickness and homogeneity). Visual inspection does not find any major breaks in the dielectric layer. Another important characteristic, which could be obtained from the analysis of this figure, is roughness of the interface. It is well known that this parameter considerably affects magnetic characteristics of MTJ.

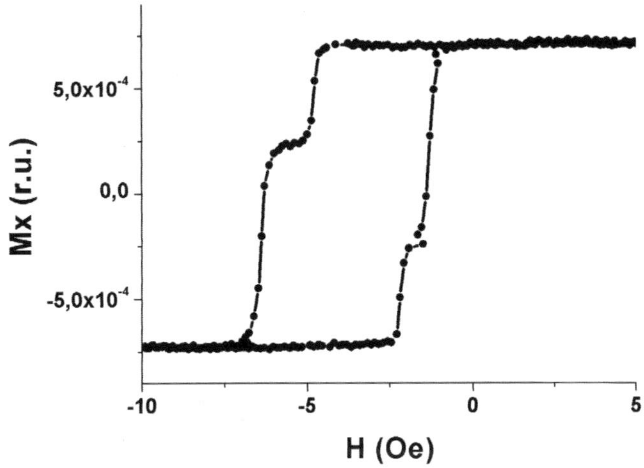

Figure 2. Hysteresis loop of non-symmetrical MTJ structure obtained by magnetooptic Kerr effect

Magnetic properties of the multilayer structure were measured on the next stage using magneto-optic Kerr effect. For this purpose home-made experimental set-up (measuring orthogonal Kerr effect) has been used. Hysteresis loop of one of our nonsymmetrical structures is shown on figure 2. Presence of characteristic steps on this curve shows that different magnetic layers switch separately. From other side we can see that loop is shifted to the area of large magnetic fields due to strong exchange bias effect.

Main problem in the formation of high quality MTJ is creating tunneling junction with appropriate transport characteristics. Pinholes and other defects could occur in a process of forming tunnel dielectric layer. Those defects deteriorate functional characteristics of the tunnel junction due to possible electric current short cuts. Usually, main source for those defects to appear is impurity inclusions (e.g. particles of dust). In case of high-class clean room facilities absent in the laboratory, one way to solve this problem (minimizing impurity inclusions) is the fabrication of MTJ with minimum lateral sizes.

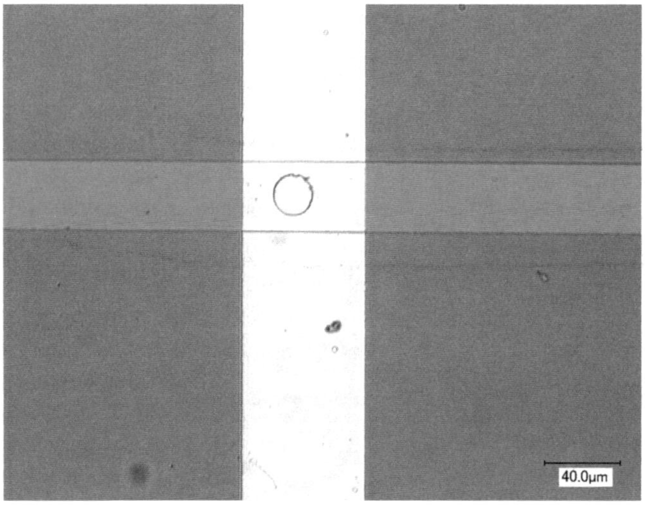

Figure 3. Optical microscope image of MRAM cell (Keyence VHX 2000)

On the next stage MTJ (in CPP geometry) with different shapes (rectangles, squares and circles) and lateral sizes (in the range from 3 μm to 45 μm) were fabricated using photolithography and ion beam etching methods. The whole technological route consisted of more than 15 operations [3]. Figure 3 shows optical microscope picture of sample MTJ structure at the final stage. As we can see circle-shaped tunnel junction is located in the center of two conducting wires crossing. Electrophysical properties of as-obtained structures were studied.

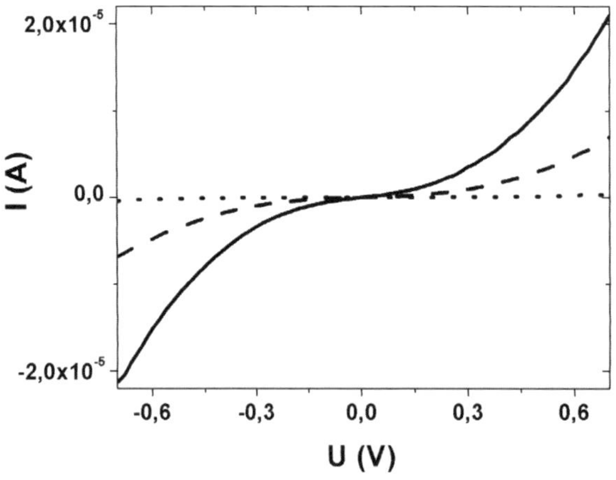

Figure 4. Voltage-Current characteristic of symmetrical MTJ structures with rectangular shape and different lateral sizes

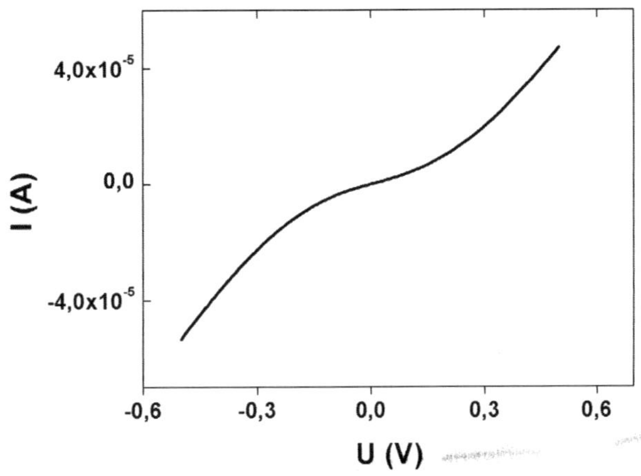

Figure 5. Voltage-Current characteristic of square-shaped non-symmetrical MTJ structure

Ratio of good MTJs in the experimental series was about 25%. In figure 4 typical voltage-current characteristics for symmetrical structures with rectangular shape MTJs (10x30 μm – bold line, 7x21 μm – dashed line, and 5x15 μm – dotted line) are shown. Similar curve but measured for non-symmetrical structure of square shape is shown in figure 5. We can see that I(V) dependence has similar nonlinear behavior as in Fig.4. Approximation of experimental curves using Simmons's theory [4] let us to estimate height of the tunnel barrier (2 eV) and its thickness (3.8 nm). The results obtained agree well with TEM results (from 3 to 5 nm) [5].

Measurements of magnetoresistance of non-symmetrical structures gave us result that didn't exceed 1.5%. Such low TMR values might be explained by degradation of magnetic films during fabrication process. Thus further work needed to improve the technology. In particular, magnetic annealing of MTJ sample could improve its performance.

In conclusion, we have studied typical MTJ fabrication process and developed methodology for quality control at different stages of the technological route. These results might be useful for optimization of MRAM technology.

Acknowledgments

This work has been done on the equipment of the Facilities Sharing Centre "Diagnostics of Micro- and Nanostructures" supported by grant of Russian Ministry of Education and Science (Agreement № 14.574.21.0099), program for fundamental research at the division of nanotechnology and informatics of RAS "Elements base for micro- nano-electronics and quantum computers, materials for micro- and nano-electronics, microchips techniques" project 1.11, and Russian Ministry of Education and Science state order 2014/258.

References

[1] Kasatkin S I, Meydan T, Murav'ev A M,Popadinec F F, Nikitin P I and Pudonin F A 2000 *Microsystems engeneering* **4** 23
[2] Trushin O S, Naumov V V, Barabanova N I and Bochkarev V F 2014 *Bulletin of the Russian Academy of Sciences. Physics* **78** 13
[3] Gallagher W J, Parkin S S P, et.al. 1997 *J. Appl. Phys.* **81** 3741-3746
[4] Simmons J G 1963 *J. Appl. Phys.* **34** 1793.
[5] Trushin O S, Naumov V V, Mironenko A A and Mazaletskiy L A 2014 *J. Integral* **77** 10-11

Design and long-term monitoring of DSC/CIGS tandem solar module

M F Vildanova, A B Nikolskaia, S S Kozlov and O I Shevaleevskiy

Solar Photovoltaic Laboratory, Institute of Biochemical Physics, Russian Academy of Sciences, Kosygin St. 4, 119334 Moscow, Russia

E-mail: mvildanova@sky.chph.ras.ru

Abstract. This paper describes the design and development of tandem dye-sensitized/Cu(In, Ga)Se (DSC/CIGS) PV modules. The tandem PV module comprised of the top DSC module and a bottom commercial 0,8 m^2 CIGS module. The top DSC module was made of 10 DSC mini-modules with the field size of 20 × 20 cm^2 each. Tandem DSC/CIGS PV modules were used for providing the long-term monitoring of energy yield and electrical parameters in comparison with standalone CIGS modules under outdoor conditions. The outdoor test facility, containing solar modules of both types and a measurement unit, was located on the roof of the Institute of Biochemical Physics in Moscow. The data obtained during monitoring within the 2014 year period has shown the advantages of the designed tandem DSC/CIGS PV-modules over the conventional CIGS modules, especially for cloudy weather and low-intensity irradiation conditions.

1. Introduction

In last decade dye-sensitized solar cells (DSCs) have attracted a considerable attention as a next generation low cost photovoltaic technology, which promises a possibility of mass production in the near future. Previously it was shown that amorphous silicon, as well as some other types of thin film PV cells, provides better light absorption capability and performance under non-direct irradiation in comparison with PV cells based on crystalline solids [1]. For this reason, the next generation PV cells based on nanocrystalline mesoscopic layers should be more effective under both direct and low-intensity irradiation conditions, especially in northern European countries and in Russia with a limited amount of sunny days. Long-term outdoor monitoring of PV cells performance, being performed for different types of solar modules, was generally focused on assessing the annual yield of a particular module under specific climatic conditions [2-4]. The monitoring data for Si-based modules, that show relatively high performance, are widely reported so far [5, 6]. In our previous publications we have shown the advantages of the designed tandem dye-sensitized solar cells (TDSCs) for improving the energy conversion efficiency and light absorption characteristics by combining two stacked DSCs sensitized with different dyes, absorbing light in the overlapped wavelength ranges [7, 8]. Later M. Graetzel and co-workers developed a new type of TDSC, where DSC cell was combined with a conventional Cu(Ga,In)Se_2 (CIGS) solid state cell with a total conversion efficiency exceeding 15% [9]. However, the reasonable advantage of DSC/CIGS tandem cell may rise not only from the improved performance characteristics, but also from the above mentioned possibility to utilize both direct and diffuse light, taking into consideration that the average content of scattered light in total irradiance during the year in Central European Region is about 55%.

2. Experimental

We have fabricated and studied DSCs with the different field sizes. The designed middle-size samples were used for the construction of tandem PV modules.

2.1. DSCs fabrication

Following the conventional technology [9], TiO_2 based pastes were fabricated from commercially available P25 powder, mixed with a solution of terpineol and ethyl cellulose in ethanol. Solar cells were fabricated on a fluorine-doped tin oxide (FTO) coated glass substrates, which were cut into the appropriate pieces, cleaned in detergent, ethanol and acetone. The electrodes were prepared from the pastes obtained using screen-printing method and sintered at 500°C for 30 min. The electrodes fabricated were dipped in a 0.3 mM of N719 (Sigma Aldrich) dye solution in a mixture of acetonitrile and tert-butyl alcohol (1:1) for 24 h. After sensitizer uptake the TiO_2 layers were washed in acetonitrile. Platinum-coated FTO glass substrates were used as counter electrodes. The counter and working electrodes were assembled into the sandwich-type cell and sealed with an ionomer film (Surlyn 1702). Commercially available electrolyte AN-50 Iodolyte (Solaronix) was introduced into the construction. Thus DSCs with the field size of 0,7 × 0,7 см^2 were fabricated. The I-V characteristics and energy conversion efficiencies of DSCs were measured under AM1.5 incident light illumination (1000 W/m^2) using Abet Technologies Solar Simulator and Semiconductor Characterization System 4200-SCS (Keithley, USA). The measurements of incident photon to current conversion efficiency (IPCE) were provided using QEX10 Solar Cell Quantum Efficiency Measurement System (PV Measurements, USA) in the wavelength range 300 – 1100 nm.

2.2. Design of DSC mini-modules and tandem DSC/CIGS modules

Following the technology, described in the previous paragraph, we have prepared DSC mini-modules with a field size of 20 × 20 cm^2, consisting of two electrically connected middle-size DSCs in aluminum frame. These devices were used for the construction of tandem DSC/CIGS PV-modules, in which 10 DSC-based top mini-modules were used for fabrication the top DSC module, combined with a bottom commercial 0,8 m^2 CIGS module (SunPower model AD-CIS-80, China). Two 0,8 m^2 tandem DSC/CIGS PV modules with optimized configuration of the top mini-modules, as well as two 0,8 m^2 conventional CIGS modules, were installed on a fixed rack facing South-East in a tilt angle of 40°.

2.3. The outdoor test facility

The outdoor test facility located on the roof of the Institute of Biochemical Physics in Moscow was equipped with two types of solar modules (tandem DSC/CIGS and conventional CIGS) and was connected to a measuring unit providing long-term monitoring of energy yield and electrical parameters of solar modules under study. The picture of installed PV test facility is given in figure 1.

Figure 1. Tandem CIGS/DSC (right) and CIGS (left) modules installed on the roof of the Institute of Biochemical Physics, Russian Academy of Sciences, Moscow, Russia.

3. Results and discussion

3.1. DSCs performance characteristics

First we have fabricated and investigated small laboratory DSC samples with the field size of 0.7 × 0.7 cm^2. IPCE spectrum for a DSC sample, given in figure 2(a), shows that the spectral area of light energy conversion for this type of PV cell ranges from 350 to 700 nm. Note that the appropriate spectral region for CIGS PV cell ranges to 1100 nm that opens a possibility for constructing efficient tandem DSC/CIGS PV-modules [9, 10].

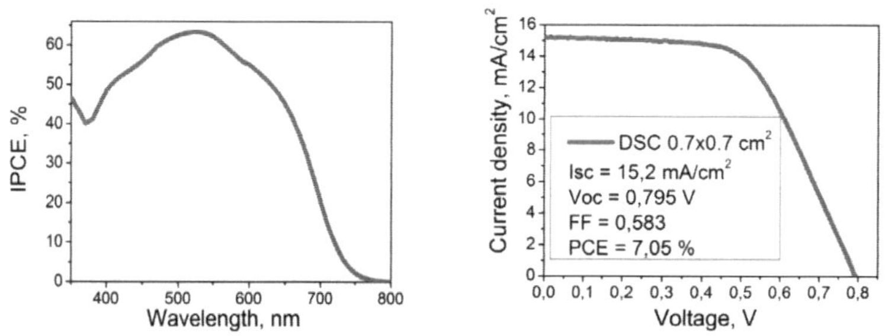

Figure 2. IPCE spectrum (a) and the I-V characteristic (b) of DSC PV cell (I_{SC} - short-circuit photocurrent, V_{OC} - open-circuit voltage, FF - fill factor, PCE – power conversion efficiency).

The photovoltaic performance and other characteristics of DSC PV cell are listed in figure 2(b). DSC with the field size of 0.7 × 0.7 cm^2 have demonstrated relatively high efficiency of 7,05%. The value of the Fill Factor is in a good agreement with that available in the literature [9].

At the following step we have fabricated DSCs with the field size of 20 × 20 cm^2, which were assembled in mini-modules. Ten DSC-based top mini-modules were arranged like a chessboard pattern for constructing 0,8 m^2 tandem DSC/CIGS PV module.

3.2. Monitoring of DSC/CIGS tandem module

The long-term monitoring of energy yield and electrical parameters of tandem DSC/CIGS PV modules in comparison with standalone CIGS modules was provided under outdoor conditions. The comparison of PCE for PV modules is given in figure 3.

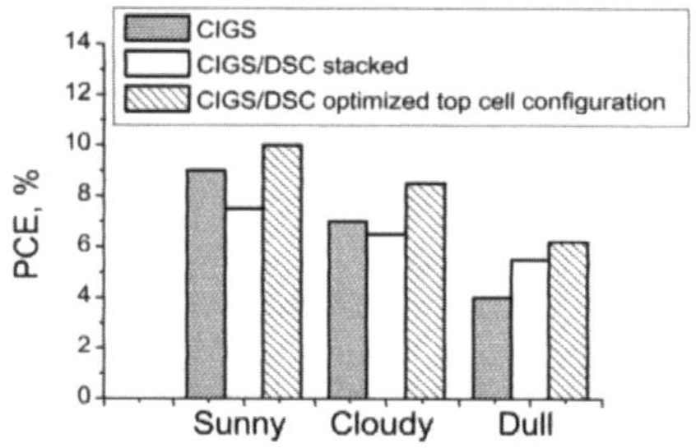

Figure 3. The average long-term monitoring PCE data for PV modules (conventional CIGS, CIGS stacked with DSCs and tandem CIGS/DSC modules with optimized top cell configuration).

CIGS module shows good performance at a sunny weather with PCE reaching 9%, however the efficiency sufficiently decreased under cloudy weather conditions. At the same time tandem DSC/CIGS module with optimized top cell configuration (a chess board pattern) have shown much higher potential abilities over the conventional CIGS modules, especially for cloudy weather and low-intensity irradiation conditions.

Figure 4 shows the daily energy yield of CIGS and tandem CIGS/DSC modules for the days with different weather conditions. Note that output power of CIGS module decreases to zero in the early morning and in late evening while tandem DSCs provided ~ 60% of power obtained at 12 a.m.

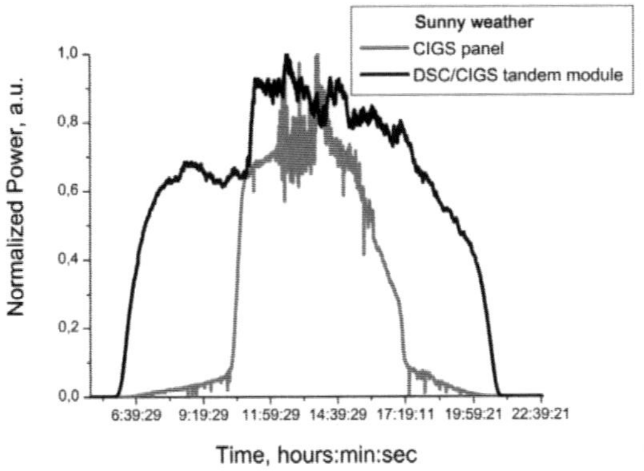

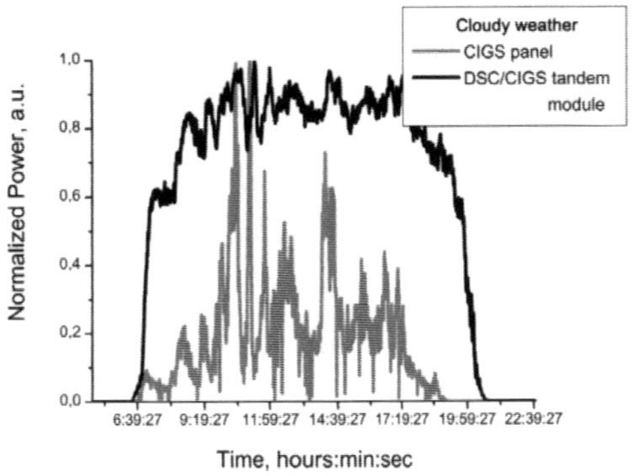

Figure 4. Power output for conventional CIGS and tandem CIGS/DSC modules under different weather conditions (left – sunny day, right - cloudy weather).

4. Conclusions

We have developed and fabricated tandem DSC/CIGS PV modules with optimized top cell configuration, in which the top DSC solar module covered about 50% area of the bottom CIGS solar module. To assemble the tandem PV module we have prepared DSC mini-modules with the field size of 20×20 cm^2 which were used for fabrication the top DSC module, combined with a bottom commercial 0,8 m^2 CIGS module. Tandem DSC/CIGS PV modules were used for providing the long-term monitoring of energy yield and electrical parameters in comparison with standalone CIGS modules under outdoor conditions. It was found that the developed tandem solar modules have shown a better photovoltaic performance in comparison with CIGS modules under cloudy weather and low-intensity irradiation conditions.

This work was partially supported by Russian Foundation for Basic Research (RFBR) grant №. 13-08-0140.

References

[1] Toyoda T, Sano T, Nakajima J, Doi S, Fukumoto S, Ito A, Tohyama T, Yoshida M, Kanagawa T, Motohiro T, Shiga T, Higuchi K, Tanaka H, Takeda Y, Fukano T, Katoh N, Takeichi A, Takechi K and Shiozawa M 2004 *J. Photochem. Photobiol. Chem.* **164** 203

[2] Kato N, Takeda Y, Higuchi K, Takeichi A, Sudo E, Tanaka H, Motohiro T, Sano T and Toyoda T 2009 *Sol. Energ. Mater. Sol. Cell.* **93** 893

[3] Jordan D C and Kurtz S R 2013 *Prog. Photovolt: Res. Appl.* **21** 12

[4] Van Sark W G J H M, Louwen A, de Waal A C, Elsinga B and Schropp R E I *Proc. of the 27th European Photovoltaic Solar Energy Conf. and Exhibition (Frankfurt, Germany, 25-28 September 2012)* pp 3247-9

[5] Kamei A, Yoshida S, Takakura H and Minemoto T 2014 *Renew. Energ.* **65** 78

[6] Kyritsis A, Tselepis S, Rikos E, Nikoletatos J and Halambalakis G *Proc. of the 28th European Photovoltaic Solar Energy Conf. and Exhibition (Paris, France, 30 September – 04 October 2013)* pp 3222-7

[7] Chevaleevski O, Larina L and Lim K S *Proc. of the 3rd World Conf. on Photovoltaic Energy Conversion (Osaka, Japan, 18-18 May 2003)* vol. 1 pp 23-26

[8] Shevaleevskiy O 2008 *Pure Appl. Chem.* **80** 2192

[9] Liska P, Thampi K R, Graetzel M, Bremaud D, Rudmann D, Upadhyaya H M and Tiwari A N 2006 *Appl. Phys. Lett.* **88** 203103

[10] Sacco A, Rolle L, Scaltrito L, Tresso E and Pirri C F 2013 *Appl. Energ.* **102** 1295

Equation of state for simulation of nanosecond laser ablation aluminium in water and air.

R. Davydov, V. Antonov, N. Kalinin

Peter the Great St.Petersburg Polytechnic University, Saint Petersburg, Russia

E-mail: davydovroman@outlook.com

Abstract. To analyze the physical processes at high energy densities, when laser is used, an adequate description the thermodynamic property of matter over a broad region of states including the normal conditions and plasma at high pressures and temperatures is required. For describing the thermodynamic properties of metals in nanoparticles production using laser ablation a semi-empirical equation of state model is proposed. To verify this model, an equation of state of aluminum was constructed. Using this equation was calculated ablation depths and crater profile for aluminum and compared with experimental data. Received results are in a good match with experiment.

1. Introduction

There is a growing interest in the nanofabrication of materials and their applications in various fields of life and technology, such as electronics, energy generation, health care and storage. A great deal of progress in this field has relied on the use of lasers. Production of nanoparticles can be done in several ways; one of them is laser ablation [1]. Evidence shows that this method is superior to other methods. Indeed, laser ablation in liquids, which consists of the pulverization of a solid target in liquid environment, gives a unique opportunity to solve the toxicity problems. In contrast to chemical nanofabrication methods, laser ablation can be performed in a clean, well-controlled environment, such as deionized water, giving rise to the production of ultrapure nanomaterial. The use of these particles decreases toxicity risks, which is especially important in vivo bio sensing and imaging applications [2].

To analyze the physical processes at high energy densities, when laser is used, an adequate description of the thermodynamic properties of matter over a broad region of states including the normal conditions and plasma at high pressures and temperatures is required. For that a semi-empirical equation of state model is proposed, based on Thomas–Fermi model with quantum, exchange and oscillation corrections.

2. Equation of state for metal

The thermodynamic properties of metal are described by the Helmholtz free energy, which is considered as a sum of three components, describing the thermal contributions of electrons (F_e) and atoms (F_a) and the elastic part of the interaction at T = 0 K (F_c):

$$F = F_e + F_a + F_c$$

The first component is calculated within the framework of the Thomas–Fermi model with quantum, exchange and oscillation corrections [3]. The advantage of this model is easy expansion on mixtures and materials in the liquid medium, but it needed some corrections at normal conditions and very high pressures. For that, two other components of the equation of state are calculated by empirical way.

The second component is given by formula:

$$F_a = 3RTln\left(\frac{\alpha(V_0/V)^{2/3}exp\left(\left(\gamma_0-\frac{2}{3}\right)\frac{B^2+D^2}{B}arctg\frac{Bln(V_0/V)}{B^2+D(ln(V_0/V)+D)}\right)}{T^2}+\frac{\beta(V_0/V)^{2/3}}{T}\right)$$

Here γ_0 is the value of Gruneisen coefficient under normal conditions and coefficients A, B, C and D are determined comparing with available thermodynamic data for metals on high pressures.

The third component is given by interpolation formula [4]:

$$F_c = a_0V_{0c}\ln(V_{0c}/V) - 3V_{0c}\sum_{i=1}^{3}\frac{a_i}{i}\left((V_{0c}/V)^{-\frac{i}{3}}-1\right) + 3V_{0c}\sum_{i=1}^{2}\frac{b_i}{i}\left((V_{0c}/V)^{\frac{i}{3}}-1\right)$$

The value of coefficient $b_1$and b_2 are determined by expressions:

$$b_1 = -\left[Z^2\frac{3}{10}\left(\frac{4\pi}{3}\right)^{\frac{1}{3}} + Z^{\frac{4}{3}}\frac{11}{36}\left(\frac{3}{\pi}\right)^{\frac{1}{3}}\right]a_BE_H(Am_uV_{0c})^{-\frac{4}{3}}$$

$$b_2 = Z^{\frac{5}{3}}\frac{1}{5}(3\pi^2)^{\frac{2}{3}}a_B^2E_H(Am_uV_{0c})^{-\frac{5}{3}}$$

Here V_{0c} is the specific volume at P = 0 and T = 0 K, E_H is the Hartree energy, a_B is the Bohr atomic radius, m_u is the atomic mass unit (amu), A is the atomic mass (in amu), Z is the atomic number of an element.

To determine the coefficients a_i we must solve the problem of minimization of the root-mean-square deviation of pressure at points $V_n = \frac{V_{0c}}{10n}, n = 1, \ldots, 100$ from the results of calculation by the Thomas–Fermi model with corrections subject to the conditions for the pressure, bulk modulus and its first and second derivative with respect to pressure at $V = V_{0c}$:

$$P(V_{0c}) = 0$$

$$B_c(V_{0c}) = -\frac{VdP_c}{dV} = B_0$$

$$\dot{B}_c(V_{0c}) = -\frac{dB_c}{dP_c} = \dot{B}_0$$

$$\ddot{B}_c(V_{0c}) = -\frac{d^2B_y}{dP_y} = \ddot{B}_0$$

The problem of conditional minimization is solved with the introduction of Lagrange factors [5]. The values of the parameters V_{0c}, B_0, $\dot{B}_0$, $\ddot{B}_0$ are fitted by iterations so as to satisfy under normal conditions the tabular value of specific volume V_0 and the values of isentropic compression modulus and its pressure derivative determined by the data of dynamic measurements.

The received coefficients for aluminium:

$$V_{0c} = 0.36, a_0 = 6940.1, a_1 = -4772.4, a_2 = 1265.4, a_3 = 50.1, b_1 = -5061.1, b_2 = 1597.1, \beta = 0.001, D = 0.4, B = 0.4, \gamma_0 = 1.95, \alpha = 0.2, V_0 = 0.3687$$

3. Mathematical model of laser ablation

In this work, the nanosecond laser ablation rate of aluminum is investigated from low fluence to high fluence in air and water. For nanosecond pulses the plasma induced by laser ablation of metal targets can be described by the hydrodynamic equations [6]:

$$\frac{\partial}{\partial t}\begin{pmatrix}\rho_1\\ \rho_2\\ \rho u\\ \rho v\\ E+0.5\rho(u^2+v^2)\end{pmatrix}+\frac{\partial}{\partial r}\begin{pmatrix}\rho_1 u\\ \rho_2 u\\ \rho u^2+P\\ \rho uv\\ u(E+0.5\rho(u^2+v^2)+P)+q_r+Q_r\end{pmatrix}$$
$$+\frac{\partial}{\partial z}\begin{pmatrix}\rho_1 v\\ \rho_2 v\\ \rho uv\\ \rho v^2+P\\ v(E+0.5\rho(u^2+v^2)+P)+q_z+Q_z-I\end{pmatrix}$$
$$+\frac{1}{r}\begin{pmatrix}\rho_{env} u\\ \mu_i \rho_i u\\ \rho u^2\\ \rho uv\\ u(E+0.5\rho(u^2+v^2)+P)+q_r+Q_r\end{pmatrix}=0$$

where ρ_1 and ρ_2 are the densities of the metal and air (or water), ρ is the total density defined as $\rho = \rho_1 + \rho_2$, u and v are the velocities in r and z direction, P the pressure, E the volumetric internal energy, I the net flux in laser radiation in the z direction, $q_r$and q_z the heat flux of thermal conduction in r and z direction, and Q_r and Q_z the radiative heat flux in r and z direction.

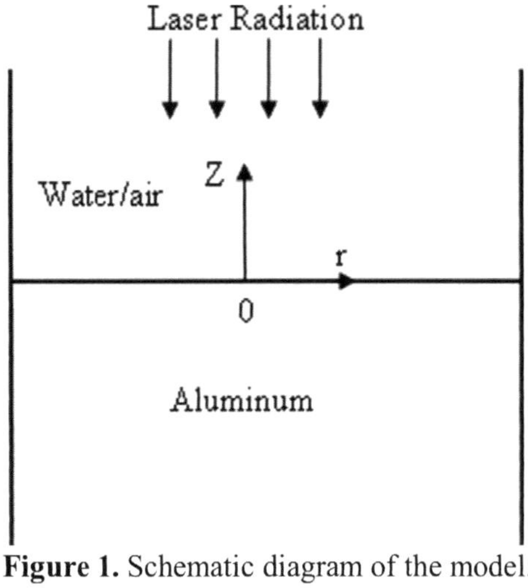

Figure 1. Schematic diagram of the model.

To solve the hydrodynamic equations, appropriate equations of state must be employed. For the aluminum targets we use equation of state constructed in previous chapter, for air and water we use equations described in articles [7, 8].

4. Results

Using the model were received the dependence of ablation depth on the laser pulse duration and also the profile of crater after a single laser shot:

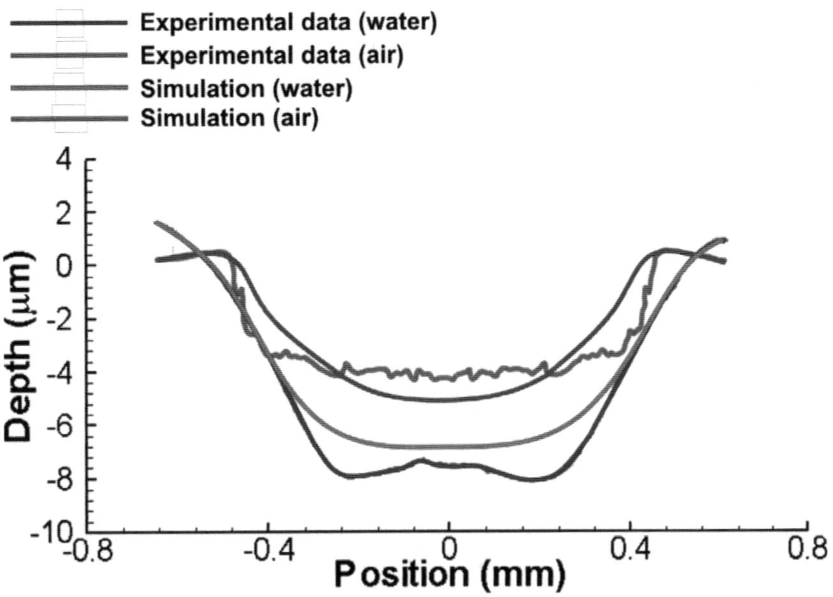

Figure 2. Crater profile after a single shot laser ablation (laser pulse duration 6 ns, wavelength 1064 nm, beam diameter 1 mm, laser fluence 24 J/cm^2, single shot laser ablation).

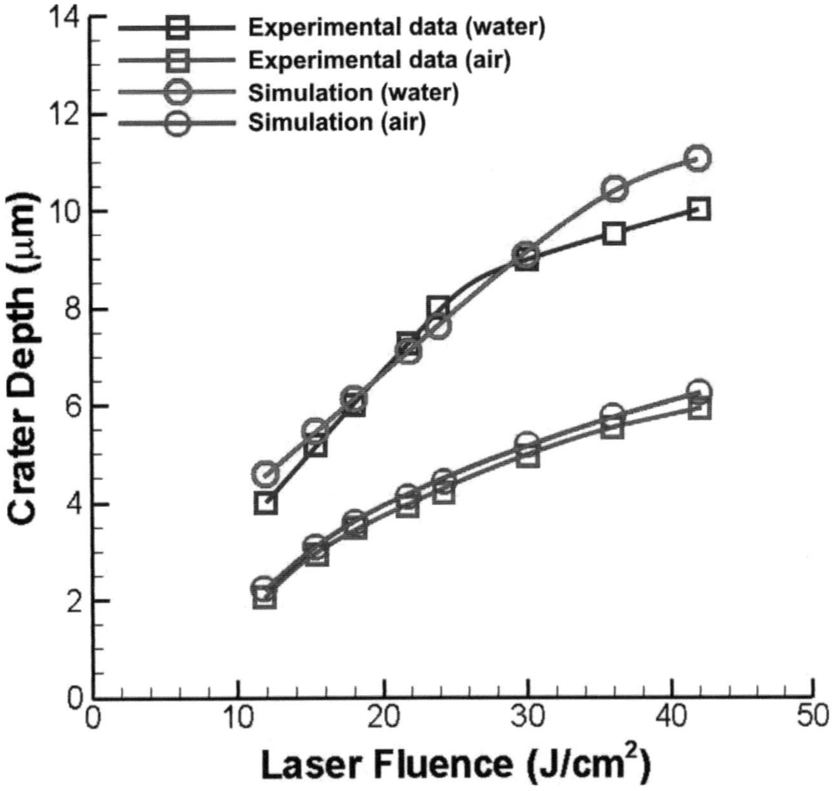

Figure 3. Crater depth under different laser fluences (laser pulse duration 6 ns, wavelength 1064 nm, beam diameter 1 mm, single shot laser ablation).

The ablation depths predicted agreed well with the experiments both in air and water. Also the simulated crater profile is relatively close to the experimentally (on the bottom and on the edges too). The described equation of state can be modified for other metals and used for improving accuracy simulation laser ablation.

References

[1] Makarov G N 2013 *Phys. Usp.* **56**, 643–682
[2] Mulvaney P, Linnert T, Henglein A 1991 *J. Phys. Chem.* **95** 7843-7846
[3] Nikiforov A F, Novikov V G, Uvarov V B 2005 *Quantum-Statistical Models of Hot Dense Matter and Methods for Computation Opacity and Equation of State* (Basel: Birkhauser Verlag) pp 285-294
[4] Khishchenko K V 2004 *Tech. Phys. Lett.* **30** 829-831
[5] Bushman A V, Fortov V E, Lomonosov I V 1991 *High-Pressure Equations of State*: *Theory and Applications* Proc. International School of Physics (Amsterdam: North-Holland) p 249.
[6] Wu B, Shin Y C 2007 *J. Appl. Phys.* **101** 103514
[7] Jeffery C A, Austin P H 1999 *J. Chem. Phys.* **110** 484-496
[8] Lemmon E W, Jacobsen R T, Penoncello S G, Friend D G 2000 *J. Phys. Chem. Ref. Data.* **29** 331-385

Features of nuclear magnetic resonance signals registration in weak magnetic fields for express - control of biological solutions and liquid medium by nuclear magnetic spectroscopy method.

A. Karseev, V. Vologdin, V. Davydov
Peter the Great Saint – Petersburg Polytechnical University, Saint Petersburg. Russia.
E-mail: antonkarseev@gmail.com

Abstract. This paper presents a method of nuclear magnetic spectroscopy in weak magnetic fields, as well as the peculiarities that arise when experimental studies conducting of biological fluids and liquid medium. The proposed method allows to measure of the condensed matter relaxation constants in weak magnetic fields by the method of nuclear magnetic resonance for express control of their condition at the place of study. The accuracy smaller than 1%. The results of experimental studies of various fluids and their mixtures are presented.

Keywords: Nuclear magnetic resonance, nuclear magnetic spectrometer, a biological solution, liquid medium, the magnetic field, times the longitudinal T_1 and transverse T_2 relaxation, the signal-to-noise ratio.

1. Introduction

The one of conditions for obtaining reliable results in the condensed matter research is the availability of reliable information about the state of the medium (for example, the presence of impurities, possible structural changes that have occurred for various reasons, etc.). This is especially important in the case of using the medium after a long period of storage, transportation, change of conteiner or other influences that can change its state. [1 - 5].

Chemical analysis of such solutions is difficult and time consuming task, because a complete analysis needs much expensive chemical reagents [6, 7].

Currently one of the most precise analytical techniques for studies of condensed matter is a nuclear magnetic (NMR) spectroscopy [2, 3, 5, 8]. Method of NMR spectroscopy is successfully applied to inorganic chemistry, solid-state physics, as well as for the study of biological objects, including farmacutico [1, 6, 7]. Unlike infrared spectroscopy (IR), NMR spectroscopy allows scientists to obtain more complex information about molecular structure and the chemical composition of the studied compounds. High resolution of NMR allows using it for the study of intermolecular and intramolecular dynamics in liquid media. That allows us to determine the rate constants of chemical reactions and relaxation [1 - 5, 8, 9].

NMR spectroscopy received the most practical application for protons (^1H, ^2H), and nuclei of carbon - 13 (^{13}C), fluorine - 19 (^{19}F), phosphorus - 31 (^{31}P), lithium - 7 (^7Li), sodium – 23 (^{23}Na), boron - 10

(^{10}B), nitrogen - 15 (^{15}N) and etc. The only condition for the NMR spectroscopy application is the presence in substance of nuclei with non-zero magnetic moment, because for the study the sample is placed in a strong magnetic field B_0 with a high homogeneity degree (no less than 10^{-3}). Currently, most of the produced NMR spectrometers based on the detection signal of the NMR pulse techniques (spin echo Hannah and Carr - Purcell, etc.). In pulse NMR spectrometers, a magnetic system with at least 7 T B_0 is used (operating frequency not lower than 300 MHz). In high-resolution NMR spectrometers the magnitude of the magnetic field changes, in this case, the fields with induction of not lower than 11 T are used. In addition, for studies of substances with complex structure, especially liquid media (for example, biological fluids), BRUKER has developed an NMR spectrometer with a superconducting magnet with an operating frequency of 1.82 GHz.

Most laboratories are not equipped with a spectrometer or analyzer for full research of environments, as these devices require special maintenance and such studies are expensive. This problem occurs not only on research, but also in environmental monitoring sampling in the "manual" (biological and liquid media), especially in hard to reach places of waterbodies [1 - 3, 5].

In this situation, methods of express-controls that allow scientists to determine the state of the medium at the site of sampling are most important. But recently the methods of express-monitoring were required that the study should not change the chemical composition and physical structure of the taken substance. This is necessary to obtain confirmation of identified abnormalities in the study of the substance with high resolution devices in stationary laboratories (for example, when the customs control or quality assurance of products), as well as to establish of possible causes this deviation. Using the method of nuclear magnetic spectroscopy, which satisfies this requirement, is one of the solutions.

But as the NMR spectrometer should be easy to carry and self-powered (rechargeable batteries), the NMR signal needs to be registered in a weak magnetic field ($B_0 \leq 0.1$ T - operating frequency is above 4.2 MHz). Because basically the weight and size of the NMR spectrometer are determined by the design of the magnetic system and service its devices, the registration of the spectrum of the investigated medium using pulse techniques in this case is impossible [1- 3]. All this creates a lot of features in the study.

2. Features of the study of biological fluids and liquids in low magnetic fields.

The main parameters of condensed matter, which can instantly set changes in its aggregate state (temperature increase, the appearance in it of other substances dissolved or not), are the times of the longitudinal T_1 and transverse T_2 relaxation [6 - 9].

The longitudinal relaxation time T_1 is determined by the interaction of the spin magnetic moments of nuclei with their surroundings (hence its second name is spin-lattice) and characterizes the substance containing the nucleus [6 - 9].

Transverse relaxation time T_2 is determined by the interaction of the spins (on each spin effects field created by neighboring nuclei being different for different spins matter), so its second name is spin-spin. It describes the time during which the broken phase matching of nuclei precessional rotation, ie, decreases the nuclear magnetization vector transverse component of the material [6, 7].

Knowing the temperature of the investigated medium it is possible to compare the measured values of relaxation times T_1 and T_2 with a base and then conclude its state [4, 5, 8 - 12].

The measurement of two relaxation constants is needed because in water, alcohol and dairy liquids are repeatedly embodiments with various mixtures of hazardous substances, in which T_2 is as the initial medium at the most a temperature range [1 - 3, 5 - 7]. Simultaneous measurement of T_1 and T_2 eliminates the error in determining the state of investigated medium.

Another feature of the study is that the registration of the NMR signal is supplied from a small volume of the test medium. Therefore, recording the NMR signal may not be possible with all of nuclei with magnetic moments. The main condition for registration of the NMR signal is the sensitivity of the substance to a method of NMR should be not below 10. This is because rapid control is carried out in a wide range of ambient temperature. When the temperature is more than 283 K the magnetic field changes. At signal-to-noise ratio is less than 1.3 it is quite difficult to configure the spectrometer device reception of the NMR signal on the resonance for measurements in automatic mode is. Therefore, the detection scheme is the NMR signal should have a high sensitivity and high gain that is hard.

Measurement of the relaxation constants T_1 and T_2 of biological fluids and liquids are made using modulation techniques for registration of the NMR signal. Therefore, in contrast to the stationary spectrometers, other than adjusting the frequency of reception of the NMR signal f_{NMR} on resonance, it is necessary to provide automatic adjustment of frequency f_m and amplitude H_m field modulation B_0 and generation level (H_1 field) of registration scheme of the NMR signal to the maximum signal-to-noise ratio. Otherwise, the measurement of T_1 and T_2 in automatic mode will not be possible.

Compact nuclear magnetic spectrometer and the experimental results.

On the basis of previous studies, we have developed a design of compact NMR spectrometer. Its structural scheme is presented in Fig. 1.

In this design of the NMR spectrometer the distance between the poles of the magnets 1 is 12 mm, magnetic induction B_0 = 62 mT, the heterogeneity of $0.5*10^{-3}$ cm^{-1}, the frequency of reception of the NMR signal for the protons f_{NMR} = 2614 kHz.

Registration of the NMR signal in such a weak magnetic field B_0 was carried out using the developed design of the low variance generator (autodin) 9, connected to the reception coil 6, which allowed us to record the NMR signal in the above-described magnetic system, from the volume of 0.2 ml and less. It was ensured that the low variance generator circuit is made on the basis of modern JFET with drain detection of the NMR signal and its further amplification. Such a scheme provides stability to operate in a wide frequency range with good rejection of common mode noise.

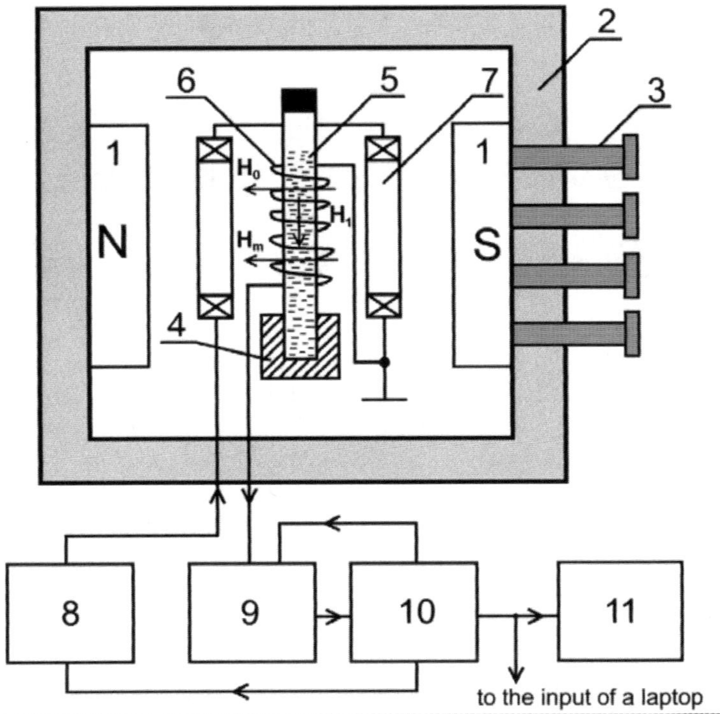

Figure 1. Structural diagram of NMR spectrometer: 1 - permanent magnet; 2 - neutral; 3 - adjustment screws for the orientation of the poles of the magnet; 4 - device for fixing the container; 5 - the container with the test environment: 6 - coil

reception of the NMR signal; 7 - modulation coil; 8 - RF generator; 9 - generator low variance (autodin); 10 - device processing and control; 11 - oscilloscope.

Fig. 2.a presents the recorded NMR signal at the output autodin. The NMR signal is registered in wide bandwidth (10 kHz) with modulation frequency of the magnetic field f_{mod} = 50 Hz.

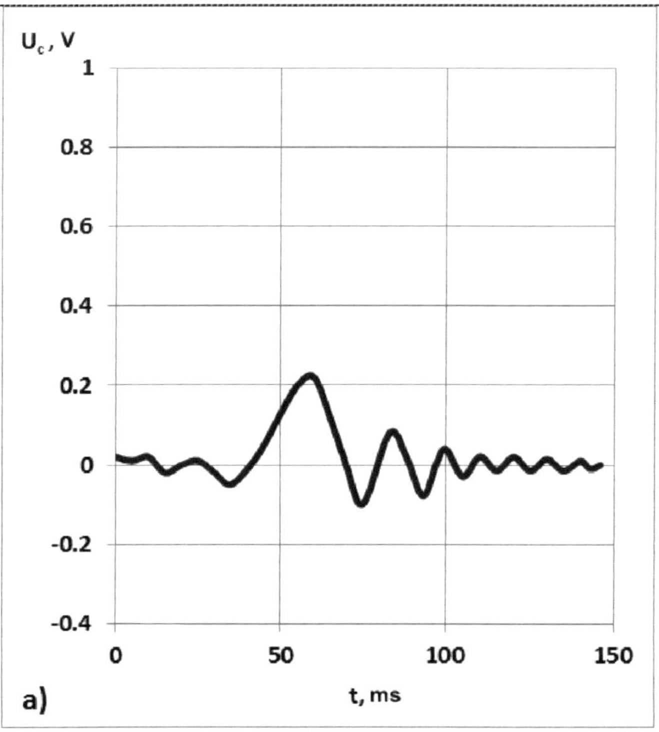

Figure 2.a Recorded signal from NMR acetic acid on the autodin output.

Determination of relaxation constants T_1 and T_2 using detectable NMR signals is discussed in detail in [1 - 3, 8]. Registered NMR signal (Fig. 2.a) despite the signal-to-noise ratio ≈ 4 does not allow measurement of T_1 and T_2 with an accuracy of 0.5 % [1 - 5]. Therefore, in the processing device and based on the STM32 microcontroller was implemented consistent scheme of accumulation of the NMR signal. The scheme is set to 100 and 1000 savings. This is due to the fact that the NMR signal can be registered in such fields not only protons, but also in the nuclei of fluorine - 19 (^{19}F), phosphorus - 31 (^{31}R), lithium - 7 (^{7}Li), sodium - 23 (^{23}Na), boron - 10 (^{10}B). But the attitude they SNR will be lower due to lower sensitivity to NMR method than that of protons.
Fig. 2.b presents the NMR signal at the output of the accumulation circuit. The line shape of the NMR signal obtained after accumulation allows the measurement of T_1 and T_2 with an accuracy of 0.5 %.

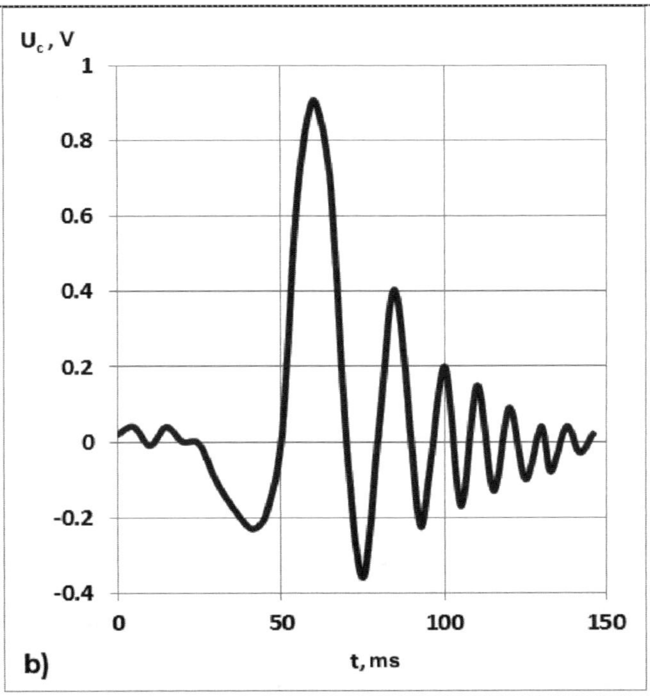

Figure 2.b Measured NMR signal from the acetic acid after 100 savings.

In Fig. 3 as an example of the detected NMR signal from tap water at T = 283 K without the use of circuits of accumulation is presented.

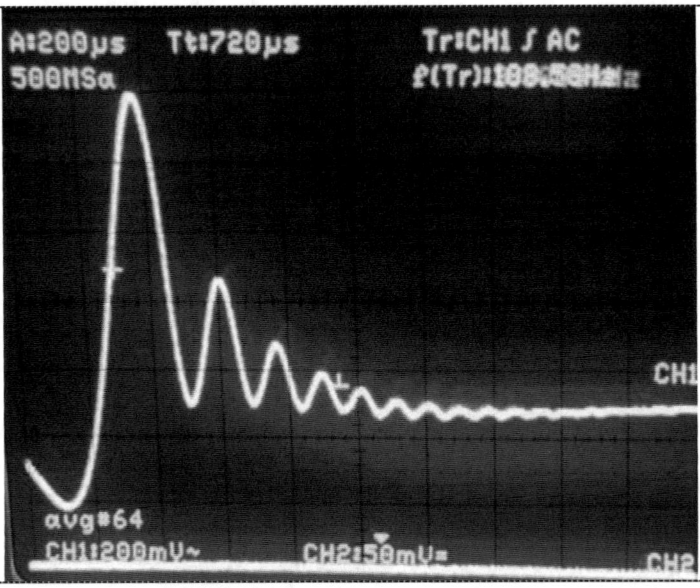

Figure 3. View of the NMR signal from tap water.

Modulation method of the NMR signal registration has a limitation on the application that is associated with the transversal relaxation time T_2 of condensed matter [2, 3, 5]. In environmental monitoring, especially when working with biological fluids T_2 may be above the limit of measurement

[6, 7]. In this case, the registration of the NMR signal will be impeded and the measurement impossible. Therefore, the developed NMR spectrometer has a breakpoint, allowing submitting the recorded NMR signal to the input of a personal computer. According to the observed NMR signal on the computer monitor it is possible to estimate the transversal relaxation time T_2. That will allow us to know the presence of impurities in a biological solution.

4. Conclusion

The peculiarities of the NMR signal reception in weak magnetic fields in the design of compact NMR spectrometer allowed us to perform measurements of the relaxation constants at ambient temperatures from 276 to 313 K with an error of < 1% in the following ranges: T_1 – from 1 ms to 20 s, T_2 – from 0.5 ms to 2.0 s. The measuring time of the two relaxation constants does not exceed 10 seconds. In addition, it is possible to tune of the NMR frequency for the different types of nucleus, which allows establishing their presence in the medium and their relative concentration. This feature fundamentally distinguishes the developed spectrometer by existing small NMR relaxometry for measuring T_2 of only one nuclei type [1-3, 5].

References

[1] Davydov V V, Cheremiskina A V, Velichko E N and Karseev A.Yu. 2014 Journal of Physics: Conference Series volume **541(2014)** 012006 (www:dx.dio.org/10/1088/1742-6596/541/1/012006) – proceeding the 1st International School and Conference «Saint – Petersburg OPEN 2014» on Optoelectronics, Photonics, Engineering and Nanostructures.
[2] Davydov V V, Dudkin V I and Karseev A Yu 2014 Measurement Techniques **57** 912.
[3] Davydov V V, Velichko E N, Dudkin V I and Karseev A Yu 2015 Instruments and Experimental Techniques **58** 62.
[4] Davydov V V, Dudkin V I, Karseev A Yu 2015 Technical Physics Letters **41** 355.
[5] Davydov V V, Velichko E N and Karseev AYu 2015 Scientific and Technical Journal of Information Technologies, Mechanics and Optics **15** 115.
[6] Lopez E, Ortiz W and Quintana I M 1998 Chemical Physics Letters **287** 429.
[7] Chen B, Ivanov I, Klein M L and Parrinello M 2003 Phys. Rev. Lett. **91(21)** 2155.
[8] Davydov V V 1997 J. of Physics B: Atomic, Molecular and Optical Physics **30** 3993.
[9] Davydov V V, Dudkin V I and Karseev A Yu 2015 Technical Physics **60** 456.

Excitation model of pacemaker cardiomyocytes of cardiac conduction system

M Grigoriev and L Babich

Tomsk Polytechnic University, Tomsk, Russia

E-mail: Mishatpu@mail.ru

Abstract. Myocardium includes typical and atypical cardiomyocytes – pacemakers, which form the cardiac conduction system. Excitation from the atrioventricular node in normal conditions is possible only in one direction. Retrograde direction of pulses is impossible. The most important prerequisite for the work of cardiomyocytes is the anatomical integrity of the conduction system. Changes in contractile force of the cardiomyocytes, which appear periodically, are due to two mechanisms of self-regulation – heterometric and homeometric. Graphic course of the excitation pulse propagation along the heart muscle more accurately reveals the understanding of the arrhythmia mechanism. These models have the ability to visualize the essence of excitation dynamics. However, they do not have the proper forecasting function for result estimation. Integrative mathematical model enables further investigation of general laws of the myocardium active behavior, allows for determination of the violation mechanism of electrical and contractile function of cardiomyocytes. Currently, there is no full understanding of the topography of pacemakers and ionic mechanisms. There is a need for the development of direction of mathematical modeling and comparative studies of the electrophysiological arrangement of cells of atrioventricular connection and ventricular conduction system.

1. Introduction

The most important features of the myocardium are automatism, excitability, conductivity, contractility and refractivity. Automatism of the heart lies in the ability of rhythmic contractions of myocytes under the influence of the pulses, occurring in the heart.

The myocardium includes typical and atypical cardiomyocytes – pacemakers, which form the cardiac conduction system. The latter provides automatism of cardiac contractions, as well as the coordination of the contractile function of myocardium of atria and ventricles. The first sinoatrial node of the system is a major center of automatism, i.e., the first-order pacemaker. Excitation is transmitted from the sinoatrial node to the cells. It reaches the atrioventricular node (second-order node) going along special conducting bundles from the atria. Atrioventricular node can generate pulses itself. Excitation from the atrioventricular node in normal conditions is possible only in one direction. Retrograde direction of pulses is impossible. Third-order node provides rhythmic automatism of the heart action. It is located in the bundle of His and in the Purkinje fibers. Automatism centers, which are located in the conduction system of the ventricles, are called third-order pacemakers. The frequency of myocardium activity is generally defined by sinoatrial node. It governs all underlying conductive system formations, and it "pushes" its rhythm upon them [1, 2].

Excitability of myocardium occurs under the influence of various stimuli: chemical, electrical, thermal and others that determine the ability to go into a state of excitement. The basis for this phenomenon is the negative electric potential which occurs in the initially excited area. Membrane of working cardiomyocytes in the excited tissue is polarized. It has a positive charge outside, and negative charge inside. This condition is caused by different concentrations of Na + and K + at both sides of the membrane. Different permeability of the cell membrane for these ions plays certain role. No Na + ions penetrate the heart cells membrane at rest. Potassium ions penetrate only partially. The diffusion of K + ions emerging from the cell increases the positive charge on the surface. The inner side then becomes negatively charged. The cell receives sodium ions under the influence of any stimulus. At this time, negative charge occurs on the cardiomyocyte membrane surface and potential reversal is provided. The amplitude of the action potential of cardiomyocytes is about 100 mV or more. This potential in turn depolarizes membranes of adjacent cells. As a result, they develop their own action potentials, that is, the spread of excitation along other cells of the myocardium takes place.

The conductivity of the myocardium is connected with the fact that the excitation waves go along its fibers at different speeds. Excitation along the atria propagates at a speed of about 0.8-1.0 m/s, along the ventricles at a speed of 0.8-0.9 m/s. At the same time it propagates along the special heart tissue at a speed of 2.0-4.2 m/s. Contractility of cardiomyocytes is due to the peculiarities of their structure. First atrial myocytes and then the papillary muscles with subendocardial layer of the ventricles are contracted. Then, the contraction process covers the inner layer of ventricles and thereby provides blood flow from the ventricular cavity to the aorta and pulmonary trunk. All changes to the contractile force of the cardiomyocytes, which appear periodically, are due to two mechanisms of self-regulation – heterometric and homeometric.

Refractoriness of cardiomyocytes is characterized by a decrease in tissue excitability throughout its activity. There are absolute and relative refractory periods. In case of electrical stimulation in the absolute refractory period the heart does not respond to the impact by stimulus and contraction. In addition, refractoriness continues as long as systole. In case of relative refractory period excitability of cardiomyocytes gradually returns to its initial level. However, in this case response of cardiomyocytes to the contraction can be above the threshold value. The relative refractory period is observed in atrial and ventricular diastole. The phase of the relative refractory is followed by the period of hyperexcitability. Moreover, it coincides with diastolic relaxation. It is characterized by the response of cardiomyocytes expressed by excitation flash even to the very small force pulses [3].

2. Review

To date, there is number of software and hardware systems, allowing observation of the reconstruction of the anatomical structure of the atria and visualization of the excitation dynamics: 1) CARTO - BiosenseWebster (USA); 2) EnSite- Endocardial Solutions (USA); 3) Bhotok-3D - research and production association (Tomsk); 4) Elkart-II navigator – Medical Industrial Company Electropulse (Tomsk).

These models have the ability to visualize the essence of excitation dynamics. However, they do not have the proper forecasting function for result estimation. Modern researches of excitation dynamics modeling, are aimed at scientific study of the heart as an object of study, or at modeling of the active medium property with its separate characteristic effects. In clinical practice, these techniques are not widely used due to their complexity also due to the high cost of individual model creation. Moreover, it is impossible to use reference solutions due to their uniqueness [4].

From the perspective of mechanics heart muscle is a complex multi-level system. Its behavior is predetermined by the two components. Passive, which is connected with the elastic and viscous properties of cardiomyocytes, defined as a composite material. Also active, which allows cardiomyocytes to perform their function of vascular blood flow provision. In this case mechanical activity of muscle takes place at the myocytes level and comprises contraction process. The contraction regulation consists of activation of actin filaments. It is caused and defined by changes in the content of cytosolic calcium. In turn, it is regulated by electric activation of the cell. At the same time an action potential develops, which depolarizes the cell membrane. The action potential arising in the myocardium to an external electrical pulse is conditioned by the participation of a number of transmembrane ion currents, including sodium, calcium and potassium currents. This whole cascade of processes is very complex and very significantly complicated by the presence of a number of feedbacks. Mechanical conditions of contraction, as well as the process of cross-bridges cycling can have significant impact on calcium and electrical activation of cells [5].

Thus, cardiomyocytes are very complex contractile “machine”. The principles of its arrangement and behavior can be predicted only with the help of complex mathematical modeling. Currently, there are certain developments and different mathematical models which describe myocardium/cardiomyocytes subsystems with varying level of detail. The models of passive mechanical properties of cardiomyocytes, models of their active mechanical behavior, as well as the model of the electrical activity of myocytes are described. However, there is still a significant lack of integrative models that describe the existing subsystems taking into account feed-forward and feedback between them. Integrative mathematical model allows for not only deep exploration of the general laws of the myocardium active behavior, but also for determination of the mechanism of violation of electric and contractile function of cardiomyocytes which accompany various cardiac pathologies, as well as for development and introduction of new methods for correction of various disorders. The possibilities of integrative methods for clinical practice are demonstrated by the analysis of the mechanism and methods of correction of arrhythmias that occur in case of calcium overload of cardiomyocytes, which take places in many diseases of the heart. This model helps to predict the role of mechanical and electrical feedback in the pathogenesis of the heart rhythm disturbance [6].

Developed comprehensive integrative model of cardiomyocytes provides the possibility of building one-, two- and three-dimensional models of the myocardium, in which electrical excitation and mechanical interaction between segments are matched with the development of action and stress potentials in the cells, considering feed-forward and feedback on the intracellular and tissue levels [7].

Modern muscle biomechanics accumulated contraction conditions that modulate the contractile activity of cardiomyocytes. Shifts and slopes of normalized isometric curves plotted at the contraction at different concentrations of constant calcium activation were changed in direct dependence on the degree of initial stretching of the studied cardiomyocyte preparation. The results are obtained repeatedly by many authors in the cardiomyocytes of different animals. It is noted that for the cardiomyocyte (unlike skeletal) contraction at constant activation is an artificial experimental mode, which requires its skinning (demembranization). More natural preparation is experiments with a response to an electrical pulse. There is a process of calcium transition in the cytosol of cardiomyocytes. This is rapid increase in the concentration of calcium, which is replaced by its decline.

Experimental conditions revealed effects indicating that the initial mechanical conditions of contraction subsequently affect its further course. The effect of short cyclic deformation occurs when in the course of isometric cycle myocyte undergoes drastic change in length (shortening or elongation) by a certain percentage, and then after a few milliseconds it is returned to the original length. Thereafter, the process of isometric contraction development occurs at this length differently than in cases where there is no deformation. In this case there is inactivation of contraction due to the fact that after the deformation the force level becomes lower and contraction-relaxation cycle ends faster. Inactivation occurs faster if it is deformed later during the cycle. This effect is especially pronounced in the deformation in the relaxation phase. The most important effect demonstrating the influence of mechanical impact on the development of contraction/relaxation cycle is the load depending relaxation. The essence of this phenomenon consists in the fact that the isotonic cycle ends much faster than isometric one. There is direct dependence on the load: the lower the load, the greater the difference between the duration of isometric and isotonic cycle. This is due to the paradoxical properties of isotonic cycles. The lower the load, the higher the speed of myocytes stretching under this load in the stage of relaxation. There are other features depending on various modifications of mechanical conditions, in addition to these cyclic deformations and load dependence [8].

The explanation of such phenomena lies in the assumption about modulating effect of mechanical conditions on the processes of calcium activation of contractile proteins. They affect the affinity of troponin C to calcium as calcium and troponin complex (CaTpC) play a key role in the regulation of cardiomyocyte mechanical activity. The formation of CaTpC complexes is accompanied by conformation of tropomyosin filament within the functional group. The latter consists of a fragment of the actin filaments (7 monomers), tropomyosin filaments which cover the center of the myosin head attachment to actin, as well as of troponin, which is located on tropomyosin.

The tropomyosin conformation reveals binding sites on the actin filament. This creates the necessary conditions for the formation of cross-bridge. The ross-bridge undergoes several specific stages of existence that are associated with the utilization of the ATP energy. The most important from the point of view of mechanical function are: force generating turn, which is a source of active stress and movement of actin towards myosin filaments (i.e. shortening); disconnecting of myosin head from actin, i.e. bridge detachment.

There are feedbacks between the mechanical conditions and calcium activation in cardiomyocytes. Mechanical and electrical feedbacks are explained by the fact that they are a consequence of ion currents through mechanic-sensitive channels. Studies which included recording of free intracellular calcium and the action potential (AP) during contraction, revealed the dependence of changes of all parameters of AP duration as a result of mechanical conditions (active shortening, deformation, load change) which occur simultaneously with changes in the shape and duration of calcium transition. This consistency conceptualizes the influence of mechanical conditions on the excitation of the cells. Their electrical activity takes place indirectly through a mechanical modulation of the intracellular calcium kinetics. It is very difficult to define the role of different mechanisms responsible for mechanical-calcium and mechanical-electrical feedback in complex and multi-level systems such as myocardium. This is revealed by using mathematical modeling [9, 10].

Models of ionic currents responsible for the formation of AP in cardiomyocytes were actively developed by many researchers. Together with Denis Noble models there are such known models as Luo&Rudy and its various modifications, model of Winslow group, a number of continuum models of macro-level processes with respect to the mechanical interaction or conduction of electric excitation in segments of the cardiac tissue, models of electromechanical coupling in multicellular segment. Despite the presence of a large number of models, description of the activation of cardiomyocytes is extremely simplified in all macroscopic models. The intracellular mechanisms that are responsible for feedback of electrical and mechanical activity are not considered at all.

The term "mechanical heterogeneity" means differences of mechanical characteristics of various interacting myocardial segments. These may be individual cardiomyocytes, sarcomeres or large sectors of the heart chamber walls. To some extent, the arranged heterogeneity is the property of normal myocardium. Most cardiac pathologies are accompanied by destruction of heterogeneous structure with increasing dispersion of mechanical properties of the ventricle wall. Mechanical heterogeneity is present at the molecular level. It is manifested in the change of ratios of the various isoforms of myosin (V1 and V3) in cardiomyocytes depending on their location. Under normal conditions, the ratio V1/V3 gradually increases in a direction from endocardial to epicardial layer.

Conditions of mechanical contraction of cells in a heterogeneous myocardial system directly depend on the mechanical properties of other cardiomyocytes interacting with it. Circuit of mechanical-calcium and mechanical-electrical feedbacks affect the course of the cell development (stress and/or shortening). This, in turn, affects the mechanical condition for contraction of neighbor cells. The resulting feedback circuit causes changes in mechanical behavior. Then mechanical conditions (stress and deformation fields) created by the interaction of cardiomyocytes are changed again. Mechanical conditions of contraction of each element of the system depending on dynamic changes in the properties of the other elements are modified dynamically in the process of continuous interaction. Process of continuous adjustment of the system elements to each other complicates the prediction of the behavior of each component and the system as a whole. As a consequence, the study of the phenomenon of mechanical heterogeneity requires more simplified experimental and theoretical models that provide the opportunity to identify and analyze the essential properties of the mechanical interaction of the elements of a heterogeneous system.

Methods of muscle duplets represent a pair muscle fibers or cardiomyocytes connected in series or in parallel and having mechanical interaction with each other. There are 6 duplet configurations implemented: 1) serial and parallel natural duplet (2 living muscles connected either in series or in parallel); 2) serial and parallel virtual duplet (a pair of virtual muscles or cardiomyocytes connected in series or in parallel); 3) serial and parallel hybrid duplets (living muscle in real time interacts with a virtual partner (living muscle attached to the levers of installation, the virtual object is simulated by computer). In various variants of virtual and hybrid duplets virtual elements were represented by different versions of mathematical models using models of active mechanical behavior of the myocardium and using versions of models of electromechanical coupling Yekaterinburg-Oxford (E0-2003 and E0-2006). One of the kinds of muscle duplets is a virtual serial duplet, which is based on the model E0-2006. It is used for theoretical studies of the influence of mechanical conditions in the occurrence of arrhythmia in case of overload of cardiomyocyte with calcium ions.

According to the principle of minimal sufficiency, each following extension allows for simulation of other new phenomena, and further clarification of the role of the additional parameters included in the model of the process and implementation of phenomena.

Mathematical models of the mechanical activity of the myocardium fully reproduce all the basic relationship of cardiomyocytes contraction-relaxation cycle in isotonic, isometric and physiological mode. It helped to find a mechanism that provides effect of mechanical conditions of contraction on the course of the calcium activation of myocytes. The basis of this effect is the cooperative interaction of regulatory substances, contractile proteins, as well as the dependence of the probability of formation of bridges on the available length of the sarcomere. In case of severe myocardial hypertrophy reduction in load dependence to complete disappearance is conditioned by the slowdown in the calcium absorption in the sarcoplasmic reticulum (SR). This is due to increased blocking of calcium ATPase in SR. The effect of the calcium pump speed in SR on load dependence is ambiguous. The load dependence declines as a result of lowering of the pump function be means of strengthening of its blockade, and also be means of its rising due to the increase in the speed of Ca2+ transfer by calcium ATPase molecules in SR. Myocardium viscosity affects the mechanical activity of myocardium. The viscosity of the end areas of the modules of the myocardium causes mismatch between the temporary increase in isometric contraction and shortening of the sarcomeres. The viscosity of the deep areas of preparations significantly modulates relaxation. It has an impact on the speed of this process. Combination of model of mechanical activity and electrical activity of "Noble-98" provides a new circuit of mechanic-electrical feedback. It arises from the interaction of Na-Ca of exchange current in myocardium and also due to the cooperative action of cross-bridges on the reciprocity of troponin C to calcium ions. The model allows for prediction of the ability of the feedback to perform function of trigger for the occurrence of acute heart failure in the event of calcium overload of cardiomyocytes. Ways of correction of this pathology are possible within this model [11, 12].

3. Conclusion

Analysis of the contemporary literature has shown that lateral areas of atrioventricular connection are studied fragmentarily; majority of further investigations are carried out on its central area, atrioventricular node, and limited mainly to the questions of excitation propagation; its pacemaker properties are studied to lesser extent. Question of the pacemaker arrangement of atrioventricular connection seems to be urgent. Idea of automatic activity of cells in the atrioventricular cusps is formed only on the basis of work performed on the heart of mammals. Processes underlying the diastolic depolarization of the subsidiary pacemakers cells in the atria are much less studied. The problem of accurate localization of atrial pacemakers requires new experimental studies; there is a need for revision of their data on topography. Obviously, these data are important not only theoretically but also clinically as pacemaker cells, in particular those in atrioventricular cusps represent a potential source of formation of ectopic pulses that can cause arrhythmia of supraventricular origin. Comparative aspects of this problem are still not sufficiently developed. The sufficient analysis of the formation of the atrioventricular connection in the course of vertebrate phylogeny was not conducted too. Pacemaker properties of cells of the atrioventricular heart area of fish, reptiles and birds are almost unexplored. There is no information about the ontogenetic changes of pacemaker function of atrioventricular connection [4, 13, 14].

Thus, currently there is no complete understanding of the topography of pacemakers and ionic mechanisms underlying functional properties of the cells constituting this very important area of the heart. There is a need for the development of direction of mathematical modeling and comparative studies of the electrophysiological arrangement of cells of atrioventricular connection and ventricular conduction system.

4. Acknowledgment

The research was financially supported by the Federal Targeted Programme "Research and Development in Priority Fields of S&T Complex of Russia in 2014-2020", the Agreement No. 14.578.21.0032 dated 05.06.2014 "Development of the experimental sample of a hardware-software complex for noninvasive recording of heart micropotentials in a wide frequency band without filtering and averaging in real time to early detect the symptoms of sudden cardiac death", the unique identifier of the contract: RFMEF157814X0032.

5. References

[1] Nikitina L.V., Kopylova G.V., Shchepkin D.V., Katsnelson L.B. Study of interaction of contracting and regulatory proteins of rabbit myocardium by method of artificial mobile systems // Biochemistry, 2008. T. 73, N 2. - P. 219-227.

[2] Bradd A.D., Al Abed A., Guo T., Lovell N.H., Dokos S. Study of cardiac pacemaker excitation using generic ionic models and realistic cell distribution // Conf. Proc. IEEE Eng. Med. Biol. Soc. 2012. P.195-198.

[3] Kopylova G.V., Katsnelson L.B., Ovsyannikov D.A., Bershitskiy S.Yu., Nikitina L.V. Application of in vitro method of mobile systems for study of calcium-mechanical link in skeletal and heart muscles // Biophysics, 2006. T. 51, N 5. - P. 781-785.

[4] Andreyev S.Yu. Modelling of excitation dynamics of atria in the tasks of heart rhythm restoration: thesis work…Candidate of Science (Engineering). Tomsk, 2006.

[5] Solovyeva O.E. Study of electrical and mechanical phenomena in myocardium by means of mathematical models: thesis work…Candidate of Science (Physics and Mathematics). Pushchino, 2006.

[6] Herrera-Valdez M.A., Lega J. Reduced models for the pacemaker dynamics of cardiac cells // J. Theor. Biol. 2011. N 7; 270(1). P. 164-176.

[7] Artebrant R., Tveito A., Lines G.T. A method for analyzing the stability of the resting state for a model of pacemaker cells surrounded by stable cells // Math. Biosci. Eng. 2010. N 7(3). P. 505-526.

[8] Shapkin A.S. Morphofunctional changes of heart conducting system in case of chronic cardiac failure: thesis work…Candidate of Medical Science, Kharkov, 2010.

[9] Roshchevskiy M.P., Bartusevich Ye.V., Gulyayeva A.S., Roshchevskaya I.M. Left ventricle of heart of ungulate animals: morphofunctional characteristics and orientation of muscle fibers // Report of Academy of Sciences – 2011. – T. 437, N 3. – P. 416-418.

[10] Viswanathan P.C., Coles J.A. Jr., Sharma V., Sigg D.C. Recreating an artificial biological pacemaker: insights from a theoretical model // Heart Rhythm. 2006. N 3(7). P. 824-831.

[11] Fuchs F. Length-dependent Ca2+ activation in cardiac muscle: some remaining questions / Fuchs F., D.A. Martyn //J Muscle Res Cell Motil, 2005. Vol. 26, N 4-5. - P. 199-212.

[12] Kurata Y., Hisatome I., Matsuda H., Shibamoto T. Dynamical mechanisms of pacemaker generation in IK1-downregulated human ventricular myocytes: insights from bifurcation analyses of a mathematical model // Biophys J. 2005. N 89(4). P.2865-2887.
[13] Maltsev V.A., Lakatta E.G. Synergism of coupled subsarcolemmal Ca2+ clocks and sarcolemmal voltage clocks confers robust and flexible pacemaker function in a novel pacemaker cell model // Am. J. Physiol. Heart Circ. Physiol. 2009. N 296(3). P. H594-615.
[14] Verkerk A.O., van Borren M.M., Peters R.J. et al. Single cells isolated from human sinoatrial node: action potentials and numerical reconstruction of pacemaker current // Conf. Proc. IEEE Eng. Med. Biol. Soc. 2007. N2007. P.904-907.

Two-dimensional porous anodic alumina for optoelectronics and photocatalytic application

L.S. Khoroshko

Belarusian State University of Informatics and Radioelectronics
P. Browki Str., 6 Minsk 220013 Belarus

Lyudmila.s.st@gmail.com

Abstract. Fabrication of porous anodic alumina film structures using anodizing, sol-gel synthesis and photolithography is reported. The structures receive interest as planar waveguides due to strong photoluminescence of the embedded trivalent lanthanides. Mesoporous structures comprising sol-gel derived titania in porous anodic alumina play a role of effective catalyst for water purification.

1. Introduction

Microstructured film pattern with different configuration in accordance with the photomasks on the porous anodic alumina (PAA) thin films (less than 5 μm) can be produced by the photolithography. Microstructures manufacturing requires the protective oxide metal mask formation on the surface which is technologically complicated and rather expensive operation. It is possible to use xerogels films as a metal mask alternative to planarization surface. The xerogels can be produced by the sol-gel method or dip-coating followed by the heat treatment [1]. For example, titanium oxide, silica, alumina, yttrium-aluminum composites and other xerogels may be used as masking materials. The selective etching of lithographic masks and windows helps to create a structure with the etched groove configuration like a waveguide. Xerogels doped with different luminescense ions expands the range of possible application.

There are operations which are implemented for the microstructured films manufacturing:

- higly ordered PAA fabrication;
- xerogel deposition to obtain the planar surface;
- photomask deposition and photolithography to form a pattern according to the photomask;
- chemical etching throught the lithographic mask windows using selective etchants;
- removal of the photoresist.

This paper presents the results of the morphological analysis of microstructure of PAA formed using electrochemical anodization, sol-gel synthesis, photolithography and chemical etching and discussed the route of manufacture of such structures and their perspectives for optoelectronics and photocatalytic application.

2. Experimental part

The following operations were perfomed to produce the microstructured PAA:

I. Synthesis of the PAA with the required morphology in various modes of the anodizing process. Two-stage anodic oxidation process was used for increasing regularity of oxide film in various electrolytes: orthophosphoric acid with a concentration of 9.8%, oxalic acid with a concentration of 3.6% and a mixture of sulfuric and oxalic acids.

II. The xerogel formation on the surface of anodic oxide: sols were deposited by spinning with intermediate drying of each layer in air at 100-200 ° C, thus 8-10 layers were deposited to full closure of the pores mouthes surface planarization.

III. Photomask deposition and photolithography.

IV. Anisotropic etching throught the photolithographic mask windows: the choice of etchant and the process parameters were done. The etchants based on the solution of chromic anhydride and phosphoric acid, ammonium fluoride solution in hydrochloric acid, solution of hydrofluoric acid were investigated.

Table 1 illustrates of the steps of synthesis the described microstructured films.

Table 1. The sequence of operations forming a microstructured porous anodic alumina film

№	Operation and its graphical representation	
1	Higly ordered PAA fabrication	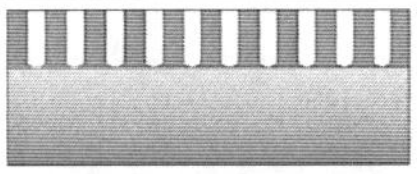
2	Xerogel deposition for the: – surface planarization – filling the mouth of the pores;	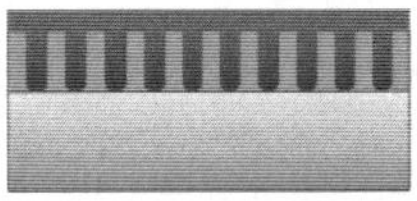
3	Photomask deposition	
4	Photolithography	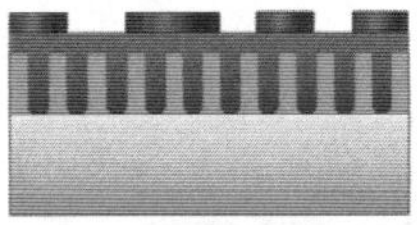
5	Anisotropic etching through the mask windows	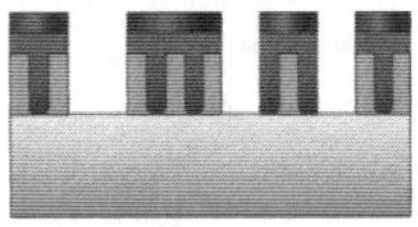
6	Photomask removing	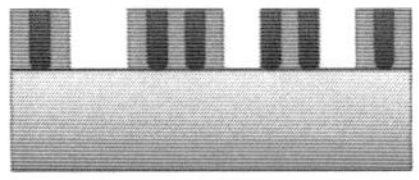

2.1. PAA and xerogel fabrication

PAA films was prepared by electrochemical anodization of aluminum foil, the process occurs in two stages. Anodic oxide obtained on the first stage was selectively vented in the solution containing chromic anhydride and phosphoric acid at a temperature of 60-70 ° C. The second anodizing stage took place in the conditions identical to the first. There are the prefered modes of the anodization process: for phosphate electrolyte the temperature was maintained at 10-12 ° C at a voltage of 100 V; for oxalic electrolyte 12-15 ° C the temperature was kept at 30 V and for combined electrolyte the temperature was maintained 12-15 ° C at a voltage of 20 V. Anodic films were finally dried in air at 100 ° C for one hour.

Titanium dioxide xerogel was deposited onto the PAA by sol-gel method using titanium ethoxide $Ti(OC_2H_5)_4$ as a titania precursor in ethylene glycol and citric acid; pH value of the sol equal to 1.0 was adjusted with concentrated nitric acid [2]. Sols were deposited on a substrate by spinning (rate

2700 rev / min, the duration of the operation for 30 seconds), each layer was dried in air at 100-200 ° C for 10-15 min, and then the next layer was formed. The structures xerogel/PAA were finally annealed for 30 minutes at a temperature of 400-500 ° C.

Protective mask with a pattern of grooves was formed on the alumina surface by the contact photolithography using the positive photoresist S 1813. The photoresist mask was removed in an organic solvent after the aluminum anodization in the open areas had made. Selective etching of alumina in the photolithography mask windows was carried out for 3 to 6 minutes depending on the type and temperature of the etchant.

The resulting structures were investigated by scanning electron microscopy (SEM).

3. Results and discuttion

In the phosphoric acid under the above conditions were obtained the porous oxide with a sufficiently large (90-100 nm) pore diameter, which hampers film xerogel formation because the pores were not filled after deposition more than 10 xerogel layers. Anisotropic etching of the structure through a mask of the photoresist leads to formation the grooves walls with the many defects due to he etchant flowing under the photomask. A partial penetration of photoresist into the pore channels complicates its removal during subsequent photolithography (Figure 1).

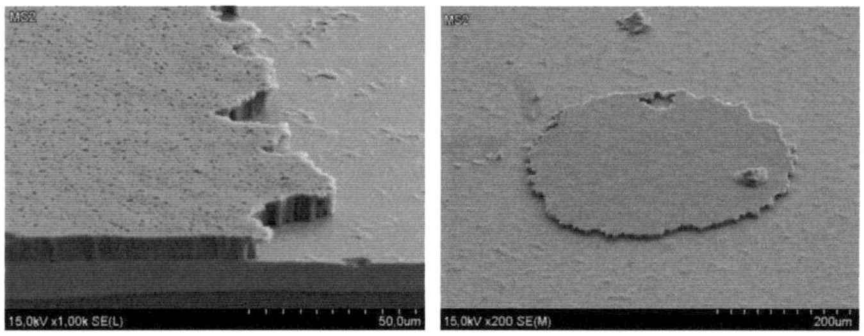

Figure 1. Titania xerogel on the PAA formed in phosphoric electrolyte after the photolithography, anisotropic etching and photomask removing

Anodic film formed in the various electrolytes may have pore diameter in the range of 30-50 nm (oxalate electrolyte) and 20-30 nm (combined electrolyte) that allows planarized oxide surface by the formation 8 layers of titania xerogel. Figure 2 shows SEM images of the PAA films obtained in the combined electrolyte (10-15% sulfuric acid with the addition of 2-5% oxalic acid) after photolithography and etching in a solution containing chromic anhydride and phosphoric acid. But etchant wicking under photoresist the mask is still observed on the PAA with smaller pore diameter.

Figure 2. Titania xerogel/PAA structures after the photolithography, anisotropic etching and photomask removing

For the alumina film with smaller pore diameters a flat vertical wall of etched groove may be obtained. The use of selective etching agents containing fluoride ions also gives good result as compared to chromium containing etchants (SEM images are presented in Figure 3).

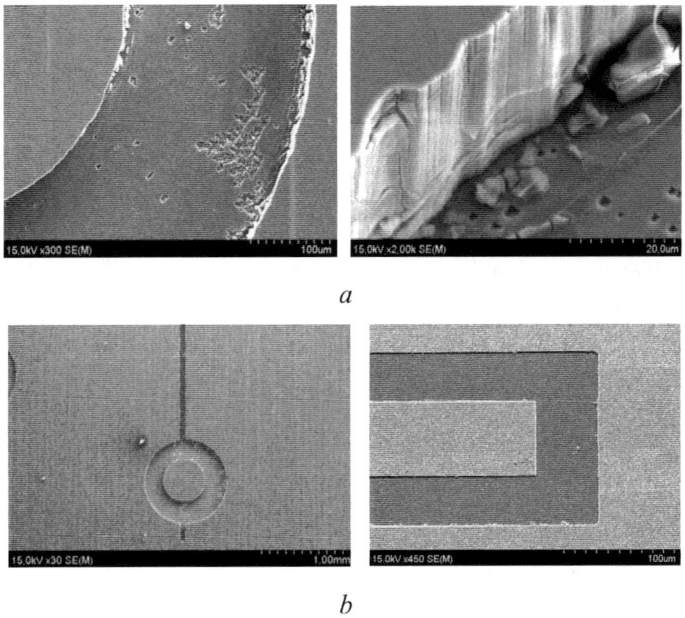

Figure 3. Titania xerogel/PAA structures formed in the combined electrolytes after the photolithography, anisotropic etching in the fluoride ions containing (*a*) and chromium ions containing (*b*) etches and photomask removing

Thus for an effective planarization PAA surface by the titania xerogel the PAA with diameter of the pores less than 50 nm are preferable. Vertical walls of the etching grooves with fewer defects obtained after the etching through a photomask in the ammonium fluoride solution in hydrochloric acid compared to the etching solution of hydrofluoric acid etchant or chromic anhydride and phosphoric acid etchant.

4. Microstructured films for the photocatalytic water purification

Titania in the anatase phase on the PAA structures with the longituginal etching grooves, in our view, are promising for environmental photocatalysis due to the developed surface which increases the contact plane between the titania and contaminated solution [3, 4].

Figure 4 shows SEM images of the structures containing more layers of titanium oxide. The xerogel is formed on the surface of the sample and in the etching grooves. Figure 5 presents the results of the photocatalysis with the microstructured PAA/titania. Photocatalytic degradation of the Rhodamine C was performed in double-walled glass reactor transmitting UV light at wavelength above 320 nm. Titania xerogel/PAA samples were placed in the reactor containing dye aqueous solution with the initial concentration of 0.04 mmol/l and were illuminated by the light for 2 h. The concentration of the dye was *in situ* monitored by recording absorption spectra of the dye solution at $\lambda = 551\text{-}554$ nm every 20 minutes.

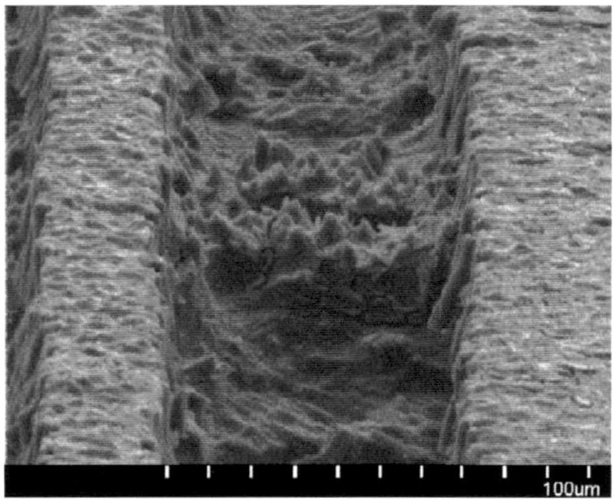

Figure 4. SEM image of microstructured PAA film with titania xerogel

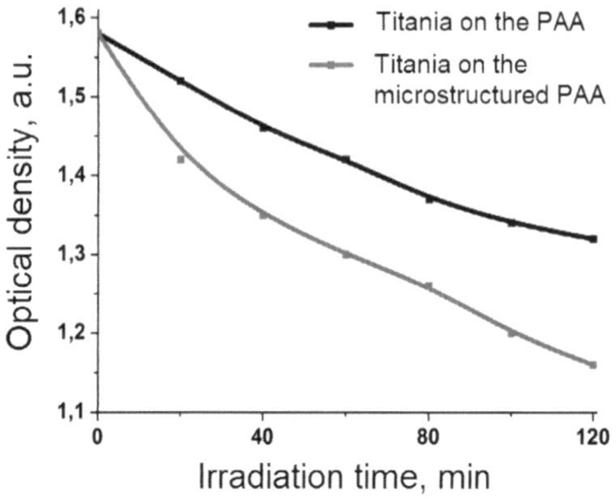

Figure 5. Photodegradation of Rhodamine C in aqueous solution under UV irradiation

The spatial structure of mesoporous titania on the interface plays a crucial role in photocatalysis with the irradiation, adsorption and desorption of the dye and its derivatives. Figure 5 shows that the patterning of PAA substrate leads to a marked increase of photocatalytic activity of TiO_2 xerogel/PAA sample that can be associated with more surface area.

5. Conclusion

The morphology of the microstructured PAA films formed by electrochemical anodization, sol-gel technology, photolithography and chemical etching without the use of protective metal mask were investigated by scanning electron microscopy. The developed elements can be further explored as a planar waveguide structures and resonators (in the form of a microcylinders) for use in planar optoelectronics. The possibility of doping xerogels by lanthanides with the luminescence spectrum in the optical range gives such microstructured films interesting for their distribution in the electromagnetic radiation and the formation of resonant modes. The titania xerogel/ PAA structures containing etched grooves is also of interest for further research on the photocatalytic water treatment.

Acknowledgments

The author wish to thank Prof. N.V. Gaponenko for stimulating discussion. The technical help of T.I. Orekhovskaya, M.V. Meledina, E.N. Krutko, M.V. Rudenko is gratefully acknowledged.

References

[1] N.V. Gaponenko 2001 *Synth. Met.* **124** 125
[2] O.P. Linnik, M.A. Zhukovskiy, G.N. Starukh, N.P. Smirnova, N.V. Gaponenko, A.M. Asharif, L.S. Khoroshko, V.E. Borisenko 2014 *J. Appl. Spectrosc.* **81** 901
[3] M. Zhukovskiy, O. Linnik, N. Smirnova, L. Khoroshko, A. Asharif, N. Gaponenko 2013 *Proceedings of the International conference «Nanomeeting-2013»* ed. by V.E. Borisenko, S.V. Gaponenko, V.S. Gurin, C.H. Kam (World Scientific: Singapore) pp 506-508
[4] L. Khoroshko, I. Nikolaenko, A. Asharif, N. Gaponenko, E. Krutko, A. Kulak 2013 *Proceedings of the International conference «Nanomeeting-2013»* ed. by V.E. Borisenko, S.V. Gaponenko, V.S. Gurin, C.H. Kam (World Scientific: Singapore) pp 503-505

Mathematical modeling of sedimentation process of nanoparticles in gradient medium

S I Ezhenkova, S A Chivilikhin

ITMO University, Dept. of Higher Mathematics,
Kronverkskiy pr. 49, St.Petersburg 197101, Russian Federation

E-mail: sveta.ejenkova@yandex.ru

Abstract. Mathematical model describing the motion of a light ray in the medium with a varying index of refraction formed by particles settling in a liquid has been built. We have received size distribution of particles settling in a liquid; calculated the light ray's trajectory in the medium; investigated the dependence of the light ray's trajectory on the initial particles concentration; received the solution of the equation of convective diffusion for nanoparticles.

Introduction

The development of nanotechnology has led to creating materials with unique optical properties that are not found in natural environment. Currently research is being carried out to study physical properties of such materials.

New areas of these investigations are connected with phenomena of light propagation in optically inhomogeneous media. In such media the refractive index changes continuously, so a gradient of refractive index serves as a characteristic of the medium. Such phenomena in gradient optics having been studied for a long time and the first educational experiments arising more than a century ago, this area of optics can be attributed to the modern area due to the fact that it is being developed intensively nowadays.

The statement of the problem

Let's consider the gradient medium formed by particles settling in a liquid. To calculate the trajectory of light ray the Fermat principle is used. Sedimentation process of nanoparticles is investigated in detail in the works [1, 2], diffusion is investigated in the work [3].

The purpose of the work is to construct a mathematical model describing the sedimentation process of nanoparticles in the liquid and to calculate the trajectory of ray's motion.

The model

In the experiment a glassful with liquid which contains particles of different sizes is used. Heavy particles settle to the bottom, and light particles stay at the surface, and thus they are distributed over the depth and create a gradient medium. A light ray goes through the medium and its trajectory can be observed.

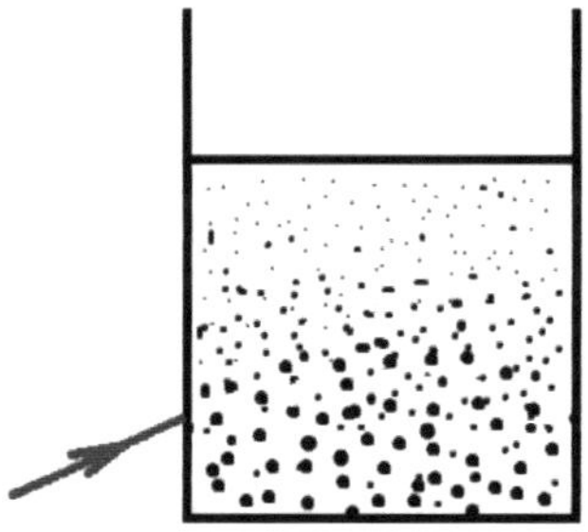

Fig.1. The object of modeling.

The calculation of the motion of the light ray's trajectory

We have investigated the time of ray's propagation through a small section of the route, which has a finite size:

$$dt = \frac{dl}{v} = \frac{n}{c} dl$$

where dt – the time of propagation; dl – part of the way; v – speed of propagation.

Using the formula for calculating particle's trajectory the time of light propagation through the given section has been found:

$$t_{1,2} = \int_{x_1}^{x_2} \frac{n(z)}{c} \sqrt{1 + \left(\frac{dz}{dx}\right)^2} dx, \qquad (1)$$

In geometrical optics the approach using Fermat's principle is the most commonly used, i.e. the stationary condition of the integral should be followed:

$$\int_{P_1}^{P_2} n(x, y, z) dz$$

where P_1 and P_2 – the fixed points in space, located on the ray's path; n – the refractive index; dS – element of the trajectory.

According to Fermat's principle a ray has to make its path with zero integral variation, and the function realizing the extremum of the integral has to satisfy the Euler equation:

$$\frac{d}{dx} \frac{\partial F}{\partial z'} = \frac{\partial F}{\partial z}$$

Rewrite this equation, according to (1):

$$\frac{d}{dx}\left(\frac{n}{c} \frac{z'}{\sqrt{1 + z'^2}}\right) = \frac{1}{c} \frac{dn}{dz} \sqrt{1 + z'^2}, \qquad (2)$$

where c – the speed of light.

Let's write refractive index of the medium taking into account the perturbation:

$$n = n_0 (1 + \hat{n}(z)), \qquad (3)$$

where $\hat{n}(z)$ – the relative deviation of the refractive index from the constant value.

Let's take for simplifying the relatively small inclination of the ray and small perturbations of the refractive index:

$$\begin{cases} z'^2 << 1 \\ \hat{n}(z) << 1 \end{cases}$$

$$z' = \frac{d\hat{n}(z)}{dz}$$

The boundary conditions are:

$$z\big|_{x=0} = 0$$

$$z'\big|_{x=0} = z'_0$$

We rewrite (2) considering (3), integrating the obtained equation (ranging from 0 to z) with into account boundary conditions and receiving integral form of the equation of the light ray's trajectory:

$$x(z) = \int_0^z \frac{dz'}{\sqrt{z'^2_0 + 2(\hat{n}(z') - \hat{n}(0))}} \qquad (4)$$

where $\hat{n}(z')$ – the relative deviation of the refractive index.

The trajectory chart of the light ray indicates the ray's path and its deviation from more dense part of the medium:

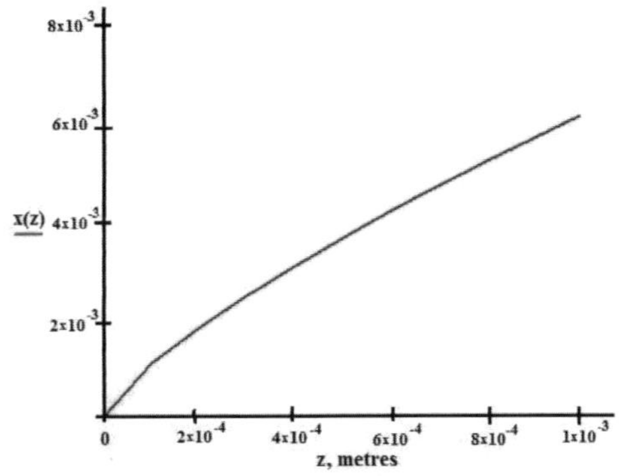

Fig.2. The trajectory of the light ray.

The calculation of the sedimentation process of particles in a liquid.

Let's consider a particle that settles at the constant speed and turns out to be at the depth z in the time t.

The attractive force (mg), the friction force (F_{fr}) and the buoyancy force (F_a) act on the particle.

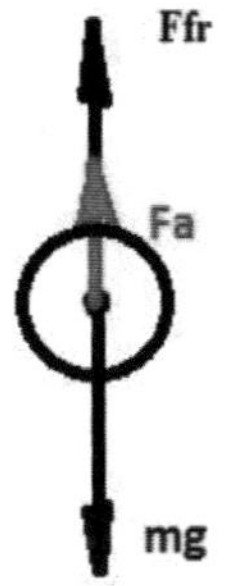

Fig.3. The particle and the forces applied to it.

The condition of equilibrium of forces applied to the settling particles is:

$$6\pi\mu R v = \frac{4}{3}\pi R^3(\rho - \rho_p)g,$$

where $v = \gamma R^2$ – velocity of the particle's settling.
This implies that:

$$\gamma = \frac{2}{9}\frac{(\rho - \rho_p)g}{\mu}$$

Let's consider $f(R,z,t)$ – the density of function of particle size distribution in the time t at the depth z. The equation for its calculation is written as:

$$\frac{df}{dt} + v\frac{df}{dz} = 0$$

The initial condition is:

$$f\big|_{t=0} = f_0(R,z)\Theta(z)$$

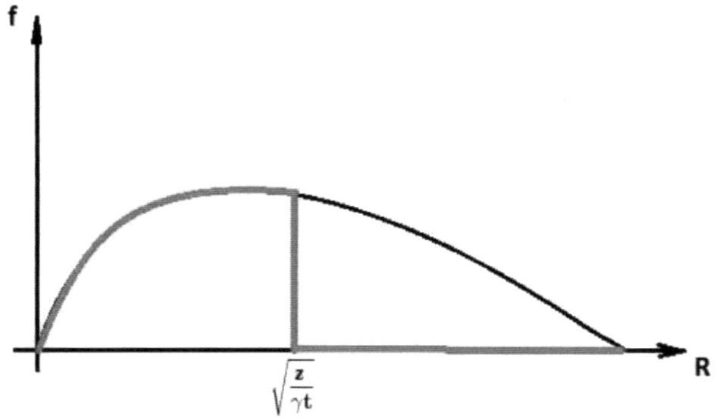

Fig.4. The function of particle size distribution.

Also we calculated the radius of particle, which was at the depth z in the time t:

$$R_z(z,t) = \sqrt{\frac{z}{\gamma t}}$$

The specific volume (volume of particles per unit volume):

$$V_p(z,t) = \frac{4}{3}\alpha^2 n\pi\left(\frac{24}{\alpha^4} - \frac{24e^{-\alpha R}}{\alpha^4} - \frac{24R\ e^{-\alpha R}}{\alpha^4} - \frac{12R^2 e^{-\alpha R}}{\alpha^3} - \frac{4R^3 e^{-\alpha R}}{\alpha^2} - \frac{R^4 e^{-\alpha R}}{\alpha}\right)$$

$$\hat{n}(z) = (\frac{n_p}{n_0} - 1)V_p(z,t)$$

where n_0 – refractive index at the initial time, n_p – refractive index medium formed settling particles.
Next we consider settling particles taking into account the Brownian diffusion.
Let us write the refractive indices of the medium at the initial time and at the moment when the particle has reached a depth of z by the time t:

$$n_0 = \int_0^\infty f(R)dR$$

$$n = \int_0^\infty f(R,z,t)dR = \int_0^\infty f_0(R)\Theta\left(\sqrt{\frac{z}{\gamma t}} - R\right)dR = \int_0^{\sqrt{\frac{z}{\gamma t}}} f_0(R)dR$$

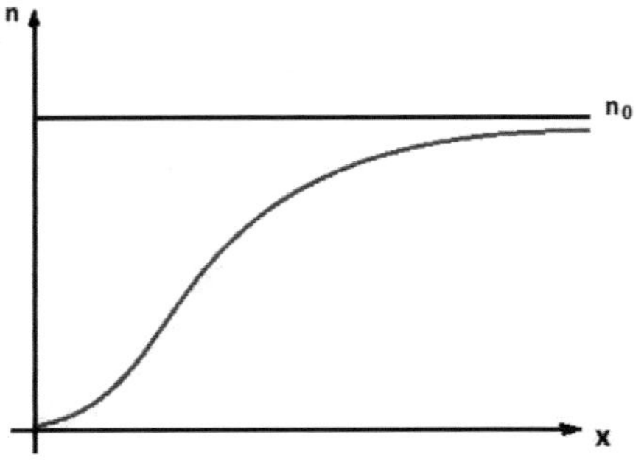

Fig.5. The refractive index distribution.

The equation for the density of function of the particle size distribution is:

$$\frac{\partial f}{\partial z} + v(R)\frac{\partial}{\partial z}(v(R)f) = 0$$

$$\frac{\partial f}{\partial t} + \frac{\partial j}{\partial z} = 0$$

where j – the particle flux density:

$$j = v(R)f - D\frac{\partial f}{\partial z}, \qquad (5)$$

Taking advantage of (4) we obtain the equation of convective diffusion:

$$\frac{df}{dt} + v(R)\frac{\partial f}{\partial z} = D(R)\frac{\partial^2 f}{\partial z^2}, \qquad (6)$$

The boundary conditions are

$$f\big|_{t=0} = f_0(R,z)\Theta(z)$$

$$j\big|_{x=0} = 0$$

$$f\big|_{x\to\infty} = f_0(R)$$

Then, using the image of the Laplace for f we have had the solution of equation (6):

$$f(z,t) = f_0 - 2vf_0\int_0^t\left(\frac{1}{\sqrt{\pi t}}e^{-\frac{1}{4D}\left(v^2\sqrt{t'}+\frac{z^2}{t'}\right)} - \frac{v}{\sqrt{D}}e^{\frac{zv}{2D}}Erfc\left(\frac{z}{2\sqrt{Dt'}} + \frac{v\sqrt{t'}}{2\sqrt{D}}\right)\right)dt', \qquad (7)$$

where v – velocity of sedimentation, f_0 – the cumulative distribution function at the initial time, D – diffusion, z – coordinate, t – time of sedimentation.

To sum up, analytical study was carried out. The graphs showing the change in the particle distribution function that depends on changing the sedimentation time were plotted:

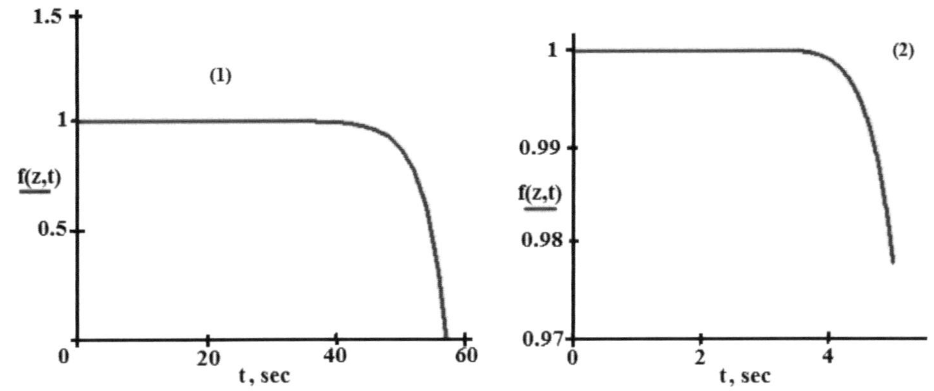

Fig.6. The density of the particle distribution function at large (1) and small times (2).

Acknowledgments
This work was financially supported by the Government of Russian Federation, Grant 074-U01.

References
[1] Suvankar Gangulya, Suman Chakrabortyb Sedimentation of nanoparticles in nanoscale colloidal suspensions. *Physics Letters A* **375** (2011) 2394–2399.

[2] B. Demeler, Tich-Lam Nguyen, G. E. Gorbet, V. Schirf, E. H. Brookes, P. Mulvaney, A. O. El-Ballouli, J. Pan, O. M. Bakr, A. K. Demeler, B. I. Hernandez Uribe, N. Bhattarai, R. L. Whetten Characterization of Size, Anisotropy, and Density Heterogeneity of Nanoparticles by Sedimentation Velocity. *Anal. Chem.* 2014, **86**, 7688−7695.

[3] Eun Chul Cho, Qiang Zhang, Younan Xia The effect of sedimentation and diffusion on cellular uptake of gold nanoparticles. DOI: 10.1038/NNANO.2011.58.

Charge transport in gap structures based on amorphous Al_2O_3

T M Borisova [1], R A Castro [1], S D Khanin [1]
[1]Department of Physical Electronics, Herzen State Pedagogical University of Russia, Moika Emb., Saint-Petersburg 191186, Russia
Tel: 8(985)266-63-74, e-mail: leonova_tatyana@list.ru

Abstract. The charge transport in amorphous layers of oxide of aluminum Al_2O_3 in the temperature range T=223-273K is studied. It was found the existence of hopping conduction mechanism in studied layers. The values of dispersion parameter s and activation energy E_A are determined.

1. Introduction

Aluminium oxide possesses a number of interesting physical and chemical properties - high melting point, heat conductivity and the dielectric parameters, increased radiation stability, high electrical specifications, wide band-gap, minor external impact resistance, shows no chemical reaction at contacted materials, thus causes its wide application as an active element in modern microtronics and electrooptics. During the recent years films of aluminium oxide have been analyzed as a dielectric layer in capacitor structures. Thanks to high dielectric capacitance amorphous films of Al_2O_3 are considered as one of alternatives of replacement of silicon oxide in the process of formation of superfine gate dielectric layers in technology of metal-oxide-semiconductor structure [1].

Bulk conduction of amorphous aluminium oxide layers was investigated by authors [2], [3]. The existence of several conduction mechanisms was discovered: from migration of weakly bound metal ions to hopping charge transport of over localized states near the Fermi level [2]. It is found that conduction of Al_2O_3 in the temperature range 223 K < T < 373 K is predominantly based on electron jump and is adequately described by equation [4]:

$$\sigma_1(\omega) = \frac{\pi^2}{96} N^2(E_F)kTa\omega r_0^4 e^2 , \tag{1}$$

where e – is electron charge; k – Boltzmann constant; N_F – density of states near the Fermi level; $a = 1/\alpha$ – localization radius; α – decay constant of the localized charge carrier wavefunction $\psi \sim e^{-\alpha r}$; ν_{ph} – phonon frequency.

The aim of this work was establishment of regularities of processes of charge transport in gap structures based on amorphous Al_2O_3, obtained by method atomic layer deposition (ALD).

2. Experimental details

The aluminium oxide layers in gap structures Al-Al_2O_3-Al were obtained by method ALD on p-type silicon Si (100) surface-supported substrate with its resistivity of ρ = 10 Ohm • cm. Drawing layers was carried out at a temperature of 300 K in the atmosphere of trimethylaluminum Al

$(CH_3)_3$ (TMA) with use of solution of nitric acid (HNO_3) and water as reagents. Aluminum contact points in area of 0.24 mm^2 were applied by thermal evaporation method. The thickness of oxide layers was 100 nm.

Measurement of surface conductivity was conducted in frequency range $10^{-1}...10^{6}$ hz and in the temperature range 223 K...273 K with spectrometer "Concept 81" (Novocontrol Technologies GmbH). System consist of frequency impedance analyzer ALPHA ANB, flow cell, temperature control system Novocool Cryosystem, Dewar flack containing vapor and feed control of nitrogen gas, and also automatic data collecting system with computer interface. The temperature stabilized with accuracy ≈ 0.5 ^{0}C, experimental uncertainty error exceed 1.0 %. The complex conductivity σ^* was calculated from the complex impedance Z^* by equation:

$$\sigma^* = \sigma' - i\sigma'' = \frac{-i}{\omega Z^*(\omega)}\frac{S}{d}. \quad (2)$$

3. Results and discussion

There is frequency dependence of the electrical conductivity σ' in different temperature range in Figure 1. The figure shows that the conductivity of investigated sample depend on temperature and increases with frequency increase of external variable field. There are two regions of σ' frequency dependence: low and medium frequency. There was discovered exponential dependence of electrical conductivity from frequency in the region of low frequency according to the law $\sigma' \sim \omega^s$. Exponent $s < 1$ steadily decreases with temperature in the range of (0.75 ± 0.01) to (0.50 ± 0.01). The temperature dependence of σ' has exponential character (figure 2), with activation energy E_a = (0.21 ± 0.01) eV. It is well-known that the conductivity of band type is generally frequency-independent up to $10^{10} - 10^{11}$ hz. Discovered dependencies reveal that charge transport occurs by electron jump between localized states in band-gap. The probability of jumps and their number increases with the increase of temperature and frequency, simultaneously delocalization of carriers within state groups occurs that is accompanied by decrease of coefficient s. Probably the role of localized states execute generally fluctuations of the substance density in layers of Al_2O_3 due to its disordered structure [5]. Increasing filling of such states with the rise of temperature leads not only to delocalization of the carriers but also to increase in conductivity.

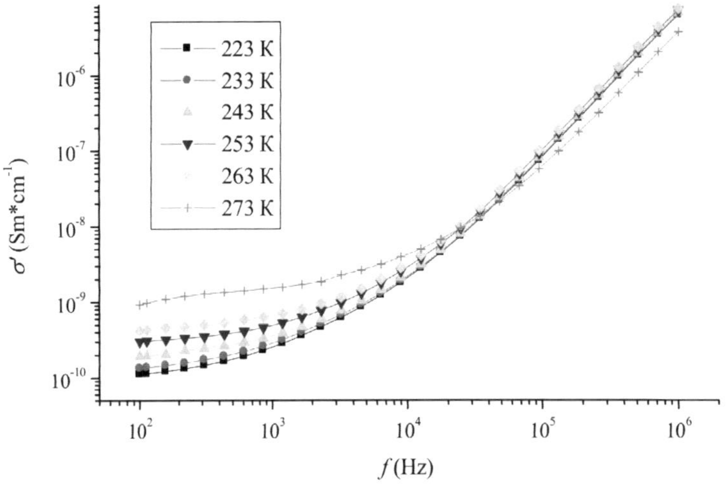

Figure 1. Frequency dependence of the electrical conductivity in gap structures on the basis of aluminium oxide.

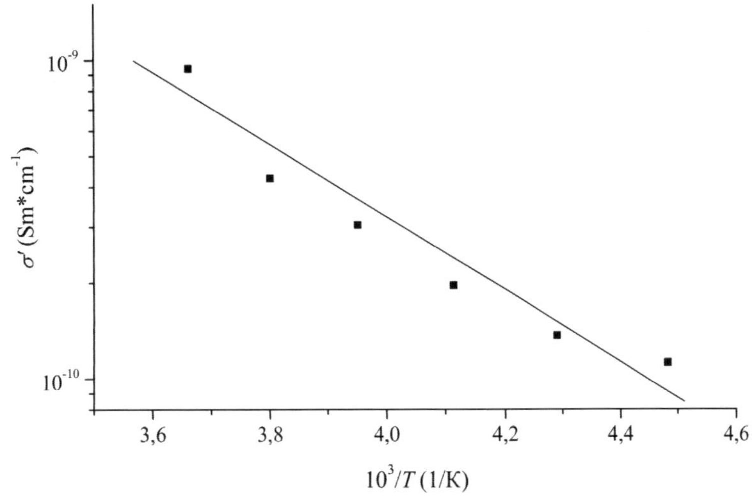

Figure 2. Temperature dependence of real conductivity in gap structures on the basis of aluminium oxide.

By the shape of I-V characteristic (Figure 3) that was measured at room temperature it can be concluded that in these films the conductivity mechanism is monopolar (electronic) [6]. The process of charge transport in amorphous aluminium oxide is considered on the basis of monopolar injection when charges of one type are introduced in dielectric material, in this context this is electronic monopolar conduction. It is assumed that charges are introduced from the side of aluminium electrode.

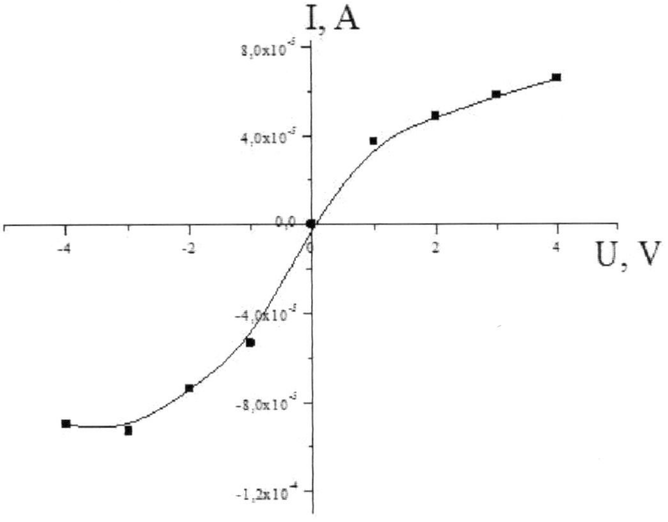

Figure 3. I-V characteristic of planar structures on the basis of amorphous aluminium oxide.

The samples of two types were investigated for discovery of reagent efficiency when amorphous aluminium oxide layers are synthesizing: No1 were used trimethylaluminium $Al(CH_3)_3$ (TMA) and water vapor, No2 were used TMA and nitric acid, sample thickness was 100 nm.

The use of nitric acid solution as the reagent leads to increase of surface conductivity value (figure 4), that probably related to structural feature of amorphous aluminium oxide. The structure of examples (type No1) is characterized by its short-range order that is the same with arrangement of atoms in crystal phase α-Al_2O_3 (hexagonal structure) [7]. In examples of type No2 short-range order depends on arrangement of atoms in mixture phase γ-Al_2O_3 (triclinic structure) and small quantity of pseudoboehmite (γ-$AlOOH$≈10%) (triclinic structure) [8], in the other words the used reagent determines the parameters of structure.

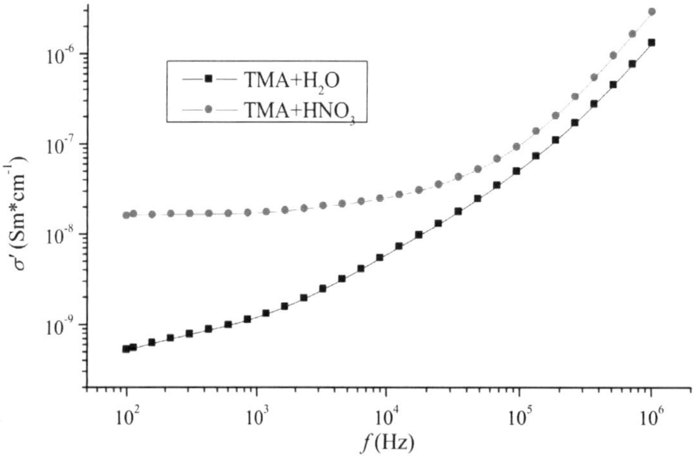

Figure 4. Frequency dependence of real conductivity in structures on the basis of Al_2O_3, that was received with the use off different reagents.

Modification of structure conductivity with use of various reagents correlates with structural mechanisms of polarization by imposition to structures of ac potential using frequency as well as temperature factors. The frequency dependence of electrical module M'' at room temperature is shown in Figure 5. The value of the most probable relaxation time τ_{max} changes with the change of reagent.

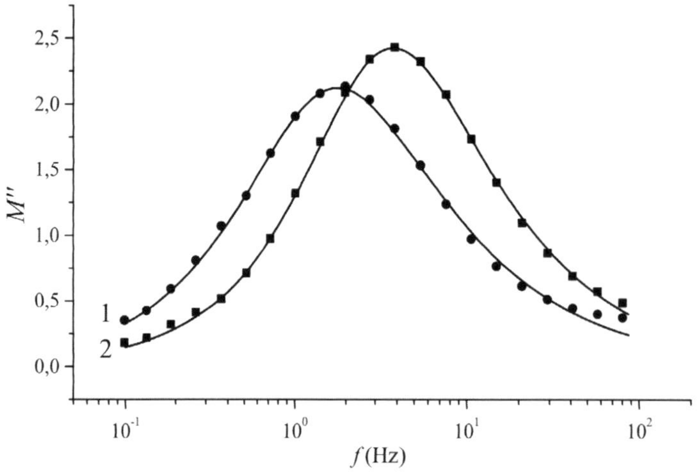

Figure 5. Frequency dependence of electrical module in structures on the basis of Al_2O_3 that were received with the use of various reagents.

4. Conclusions

Thus, it was found that layers conductivity of Al_2O_3 in the investigated frequency range is described with power law where the exponent of power is temperature-dependent. The charge transport in the samples under investigation occurs by jump mechanism in the band-gap via the localized states, and it is thermally activated process with activation energy $E_a = (0.21 \pm 0.01)$ eV. The use of nitric acid solution as reagent in synthesis of Al_2O_3 layers leads to increase of surface conductivity value that is connected with structural features of amorphous aluminium oxide.

The reported study was partially supported by RFBR, research project No. 14-32-50672.

5. References

[1] Wilk G D Wallace R M Anthony J M 2001 High-k gate dielectrics: Current status and materials properties considerations *J. Appl. Phys.* **89** p 5243

[2] Borisova T M Castro R A 2013 Mechanism of charge transport in Si/Al_2O_3/Al structures *Journal of Physics: Conference Series* **461** p 012017

[3] Novikov Yu N Vishnyakov A V Gritsenko V A Nasyrov K A 2010 Charge transport in aluminium oxide: phonon – assisted trap ionization *Izvestia: Herzen University Journal of Humanities and Sciences* **122** pp 46 - 52

[4] Pollak M 1971 Frequency dependence of conductivity in amorphous solids *Phil. Mag.* **23** pp 519 - 542

[5] Mott N F 1974 Electronic Processes in Non-Crystalline Materials (Mott N F Davis E Moscow Mir Publ **1**) p 472

[6] Novikov Yu N Gritsenko V A Nasyrov K A 2009 The Manyfon mechanism of trapionization in Al_2O_3: experiment and numerical modeling *Technic. Phys. Lett.* **89** 10 pp 599-602

[7] Choong - Ki Lee Eunae Cho Hyo-Sug Lee Kwang Soo Seol Seungwu Han 2007 Comparative study of electronic structures and dielectric properties of alumina polymorphs by first-principles methods *Physical review.* **76** p 245110 pp 1 - 7

[8] Yakovleva N M Anicai L Yakovlev A N Dima L Khanina E Ya Chupakhina E A 2003 Structure and Properties of Anodic Aluminum Oxide Films Produced in *HNO_3 Solutions Inorganic Materials* **39** 1 pp 50 - 56

Formation of metallic nanostructures on the surface of ion-exchange glass by focused electron beam

F E Komissarenko[1], **M V Zhukov**[1,3], **I S Mukhin**[1,2], **A O Golubok**[1,3] **and A I Sidorov**[1]

[1] ITMO University, 49 Kronverksky pr., 197101 St. Petersburg, Russian Federation
[2] St. Petersburg Academic University – Nanotechnology Research and Education Centre RAS, 8/3 ul. Khlopina, 195220 St. Petersburg, Russian Federation
[3] Nanotechnology Department, Institute for Analytical Instrumentation RAS, 31-33 ul. Ivana Chernykh, 198095 Saint-Petersburg, Russian Federation

E-mail: 1576406@gmail.com

Abstract. This paper presents a new method for formation of metallic nanostructures on the surface of ion-exchange glass. The method is based on the interaction of a focused electron beam with ions in ion-exchange glass. In experiments nanostructures with different shapes were obtained, depending on the electrons irradiation conditions.

1. Introduction

Currently, metal nanostructures of different configurations are widely used in photonics and of interest for optical measurements, detection of molecules and device fabrication. Nanoparticles and thin films of silver or gold show plasmon resonances in the visible spectral range [1] and they are used in different photonic devices such as plasmon waveguides, concentrators of surface plasmon waves, microresonators and others. Metal nanoparticles and films produced on the glass substrate amplify the light signal of the studied objects [2] and could be used in sensors, based on surface enhanced Raman spectroscopy [3]. Existing methods of obtaining nanoparticles include chemical synthesis [4], laser ablation [5], vacuum deposition [6] and thermal processing of metal-ion-containing glass in hydrogen atmosphere. [3,7]. Also, formation of metal nanoparticles in a subsurface layer of metal-ion-containing glass could be carried out by exposing them with an electron beam [8,9].

2. Experiment

This paper presents a method for creating metal structures with nanoscale sizes on the surface of ion-exchange glasses. The method is based on the interaction of a focused electron beam with ions in glass and gives opportunity to produce metal structures of various shapes. In the experiments we used Soda-lime glass with the composition of SiO_2-Na_2O-MgO-Al_2O_3, where Ag^+ ions were introduced in subsurface layer by ion-exchange mechanism by partly replacement of sodium ions with silver ions. Aluminum film with thickness of 90 nm was deposited on the surface of prepared glass for charges drain when glass is irradiated by electrons. Deposition of aluminum film was performed by electron beam evaporator Edwards Auto 500. Focused electron beam exposure was carried out with using a scanning electron microscope (SEM) Carl Zeiss CrossBeam Neon 40 with Raith Elphy Plus lithography attachment. On the sample surface pattern with an array of disks with a diameter of 200 nm and period of 600 nm was exposed. In experiments we used different exposure parameters: beam current and dose. Acceleration voltage in all experiments was 25 kV. After irradiation, the aluminum layer was removed with a ten percent solution of KOH. Next, the surface of glass was studied with a scanning electron microscope and with an atomic force microscope (AFM) NT-MDT Ntegra Aura. Metal structures were formed in the irradiated areas on the glass surface, and shapes of these structures were circles (figure 1), triangles (figure 2), rectangles or polygons (figure 3) depending on the exposure regime. In our experiments we used three different beam currents: 12, 215, 707 pA and four irradiation doses: 20, 40, 60 and 80 mC/cm^2. Triangle structures appeared in case of 12 pA beam current and 20 mC/cm^2 irradiation dose. Expositions with higher doses and 12 pA beam current led to formation of circle structures with diameter of 400 nanometers. Using beam current with 215 pA caused the emergence of polygon structures regardless of irradiation dose. In case of beam current with 707 pA any structures appearance was not detected.

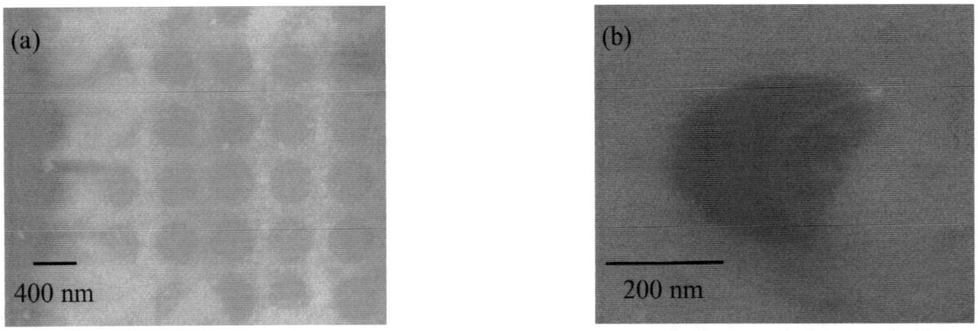

Figure 1. SEM images of array of circle structures (a) and single circle with wrapped edge (b).

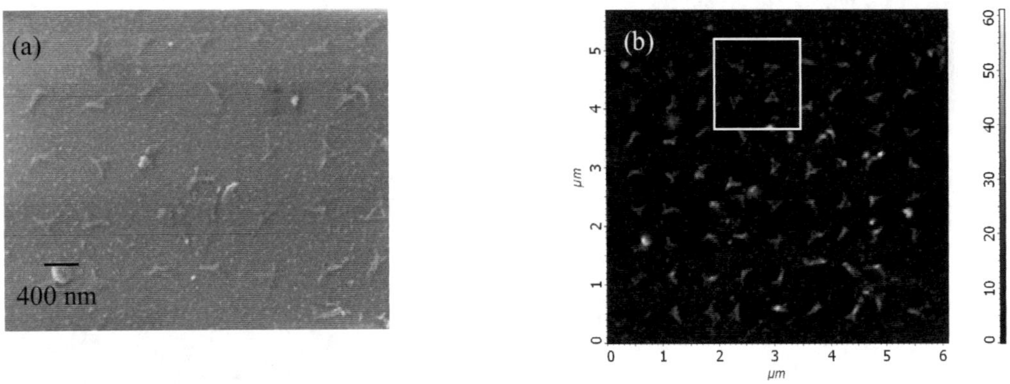

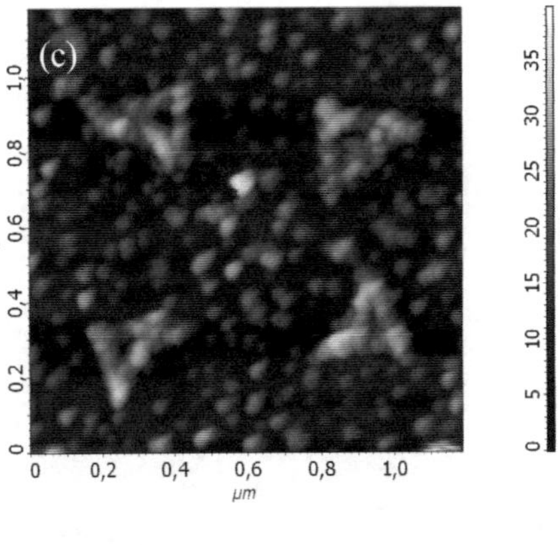

Figure 2. SEM (a) and AFM (b, c) images of array of triangles structures, image (c) presents area of image (b).

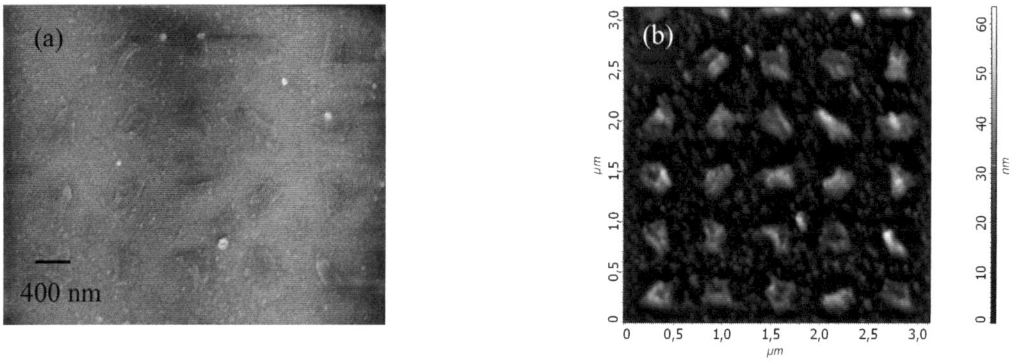

Figure 3. SEM (a) and AFM (b) images of array of polygon structures.

The study of nanostructures carried out by AFM showed that the height of structures was in a range of 30-40 nm (figure 4). It was also found that obtained nanostructures were inhomogeneous and formed from the multiple clusters. In addition, it was found that formed structures in all cases are surrounded by separated metal clusters with smaller diameters laying in the range of 25-60 nm. Creation of metallic structures with proposed method is reproducible.

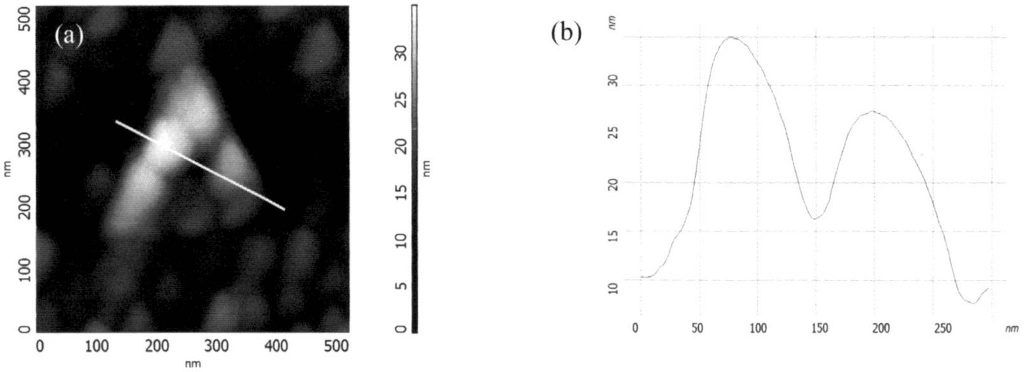

Figure 4. AFM image of single triangle structure (a) and its cross section (b).

3. Discussion

Interaction between electron beam and glass with metal film was considered with the use of Monte-Carlo method. In case of 25 kV acceleration voltage of electron beam and 90 nm aluminum film on the glass substrate electrons could penetrate into the glass on distance exceeded 6 microns (figure 5).

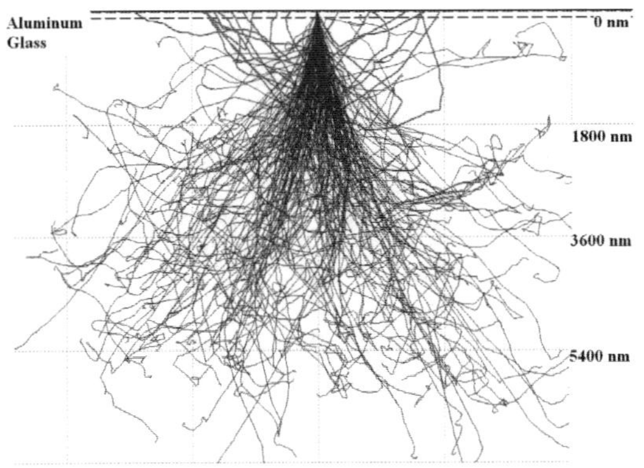

Figure 5. Result of Monte-Carlo simulation for glass substrate with 90 nm aluminum film and electron energy 25 keV

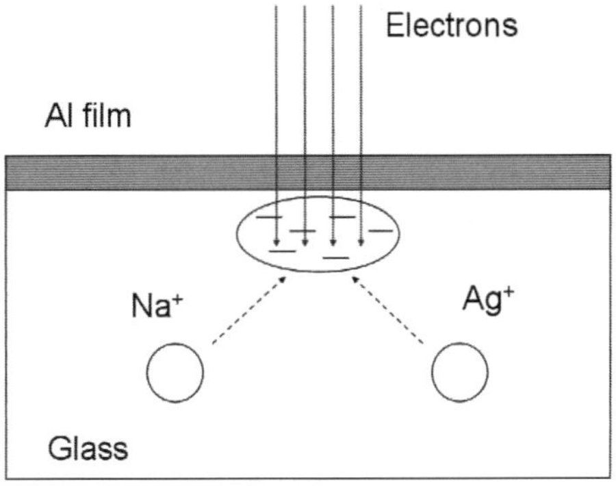

Figure 6. Scheme of cluster formation under electron beam.

Formation of metallic structures on the glass surface is explained as follows. Under electron beam irradiation negatively charged region is formed near the glass surface due to the accumulation of thermalized electrons. Positively charged silver and sodium ions migrate to that area (figure 6), reduce to neutral atoms after interaction with electrons and emerge on the glass surface, forming clusters, and combining in structures at locations where the current density of the electron beam was maximized [10]. In addition, electrons from incident electron beam could ionize neural atoms of silver and sodium, preventing formation of clusters. However, appearance of metal structures and separated clusters between them indicates that process of field migration and combining into clusters more efficient than process of neural atoms ionization by electron beam. Because of ion-exchange, concentration of silver ions exceeds concentration of sodium ions in subsurface layer, so appeared structures composed mainly from silver. Also, formed structures may consists aluminum ions, due to dissolution of the upper metal film under incident electron beam and following field migration of aluminum ions to the negatively charged region.

4. Summary

A new method of the formation of metallic nanostructures on the surface of ion-exchange glasses was presented. Structure formation occurs due to the interaction of a focused electron beam and free metal ions in glass. Shape of the structures can be modified with exposure parameters. This method of metallic nanostructures producing can be used in the fabrication of plasmon devices on glass, because it has the advantages of electron beam lithography (high resolution) while being productive due to small number of process steps.

Work was carried out with the support of the leading universities of the Russian Federation (grant 074-U01) and the Russian Foundation for Basic Research (14-02-31703 mol_a).

References

[1] Kreibig U and Vollmer M 1995 *Optical properties of metal clusters* (Berlin: Springer-Verlag) p 532
[2] Zayats A V, Smolyaninov I I and Maradudin A A 2005 *Phys. Rep.* **408** 131
[3] Zhurikhina V, Brunkov P, Melehin V, Kaplas T, Svirko Y, Rutckaia V and Lipovskii A 2012 *Nanoscale Res. Lett.* **7** 676
[4] Dykman L A, Bogatyrev V A, S. Yu. Shchegolev S Yu, and Khlebtsov N G 2008 *Gold Nanoparticles: Synthesis, Properties and Biomedical Applications* (Moscow, Nauka)
[5] Amoruso S, Ausanio G, Bruzzese R, Vitiello M and Wang X 2005 *Phys. Rev. B* **71** 033406
[6] Stoll S L, Gillan E G and Barron A R 1996 *Chem. Vap.Deposition* **2** 182
[7] Reduto I, Chervinskii S, Matikainen A, Baklanov A, Kamenskii A and Lipovskii A 2014 *J.*

Phys.: Conf. Ser. **541** 012073
[8] Sidorov A I, Nashchekin A V, Nevedomskiy V N, Usov O A and Podsvirov O A 2011 *Int. J. Nanosci.* **10** 1265
[9] Ignat'ev A I, Nashchekin A V, Nevedomskii V M, Podsvirov O A, Sidorov A I, Solov'ev A P and Usov O A 2011 *Tech. Phys.* **56** 662
[10] Brunov V S, Podsvirov O A, Sidorov A I and Churaev D V 2014 *Technical Physics* **59** 1215–19

Study of micro- and nanomechanical oscillators based on crystalline W and amorphous C whiskers

S.Yu. Lukashenko[1,3], V. V. Lysak[1], I.D. Sapozhnikov[3], I.S. Mukhin[1,2], A.O.Golubok[1,3]

1 Department of material science and nanotechnology, ITMO University, 197101 Kronverkskiy pr., 49, Saint-Petersburg, Russian Federation

2 Laboratory of renewable energy source, St Petersburg Academic University RAS, 194021 Khlopina, 8/3, Saint-Petersburg, Russian Federation

3 Nanotechnology department, Institute for Analytical Instrumentation RAS, 198095 Ivan Chernykh, 31-33, Saint-Petersburg, Russian Federation

E-mail: lukashenko13@mail.ru

Abstract. Nanomechanical oscillators (NMO) on the basis of crystalline W (tungsten) microwhiskers and amorphous C (carbon) nanowhiskers localized on the top of W tips are created and studied. Trajectories of single and coupled resonant oscillations are visualized, resonant frequencies and quality factor of NMO are determined. Trajectories of coupled resonant oscillations of NMO are simulated. Calculation results are qualitatively in agreement with the experimental data.

1. Introduction

Rapid development of nanotechnology methods in recent years makes nanomechanical oscillators (NMO) with small mass, high resonant frequency and high quality factor relatively accessible for both scientific research and practical applications. On the one hand, interest in NMO has purely scientific motivation given the possibility to research a fundamental physical problem concerning transition of a macroscopic object to the quantum state [1]. On the other hand, it is possible, using NMO, to conduct extremely-precise measurements of inertial mass, including determination of mass of single molecule or even atom [2], that being very attractive from the application-oriented viewpoint. Most studies in the field of NMO are usually dedicated to the research of oscillations of single micro-and nanoplates, nanobeams or nanowhiskers (NW) which are fixed on one or both ends. In this study we have researched oscillations of single amorphous C nanowhsikers (CNW) and crystalline W microwhiskers (WMW), and oscillations of more complex nanostructures which consist of CNW fixed on the top of crystalline WMW. These configurations may be considered as systems of coupled oscillators. Generally speaking, their behavior should differ from the behavior of single oscillators. To the best of our knowledge, such system of coupled micro- and nanooscillators has not been studied experimentally before.

2.Structures

For the convenience of conducting research CNW are localized on the top of W needles produced from polycrystalline W wire with base diameter 150 μm with the use of method of electrochemical etching in 5% solution KOH. By changing parameters of the process of electrochemical etching, the blunt (figure 1c) and sharp (figure 1b) W needles are created. Growth of CNW on the top of W tip was conducted both on the blunt and sharp needles (Figure 1b, 1c) directly in the chamber of SEM using

Journal of Physics: Conference Series **643** (2015) 012114 doi:10.1088/1742-6596/643/1/012114

the focused electron beam [3]. Depending on the parameters of growth it was possible to create CNW of diameter (50-100) nm and length (0,5-3) μm. Structure of CNW and WNW is studied using the method of electron diffraction in the transmission electron microscope (TEM) (Figure 1e). It was established that CNW are in amorphous state, and the tip of WMW are in crystalline state. Upper part of a needle might have different geometrical dimensions depending on the etching modes and in fact is a WMW with cross section of around 4 μm and length of (400-600) μm with tip radius around 30-200 nm (figure 1d).

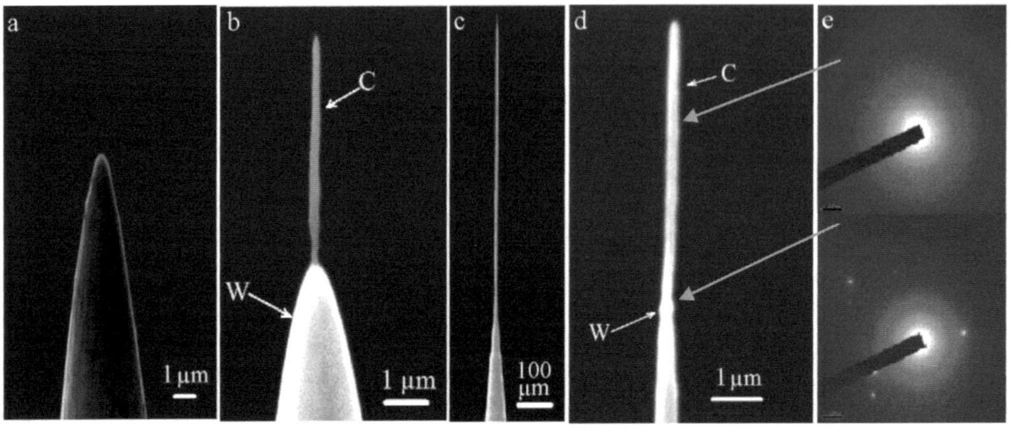

Figure 1. SEM images of etched tungsten needles and carbon whiskers that were grown on top of W tips: a) Blunt W tip, b) carbon nanowhiskers (CNW) on blunt tip c) wolfram microwhisker (WMW), d) CNW on WMW, e) diffraction image of CNW and WMW from TEM.

3. Experimental setup

Mechanical properties of NMO are analyzed using two methods are piezotransducer (Figure 2) and AC field (Figure 3). In the first case, the arrow-headed W wire is fixed in electromechanical transducer in the shape of piezoceramic tube which is located in the working chamber of SEM. In order to excite oscillations of NMO, the oscillations of electromechanical transducer powered by HF generator with frequency from 0 to 100 MHz is used.

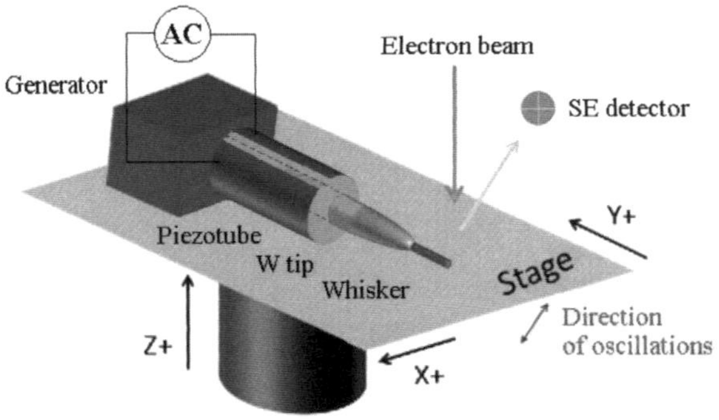

Figure 2. Setup for excitation the oscillations by force driving frequency with piezotransducer.

In the scheme, additional metal tip powered by alternating electric voltage from HF generator is applied, to create a capacitor with approach CNW.
Resonant frequencies of oscillations of NMO are determined in both methods through visualization of maximal oscillation amplitudes with the use of SEM detector. Resonant oscillations of hard W and complex NMO are analyzed in different frequency regime.

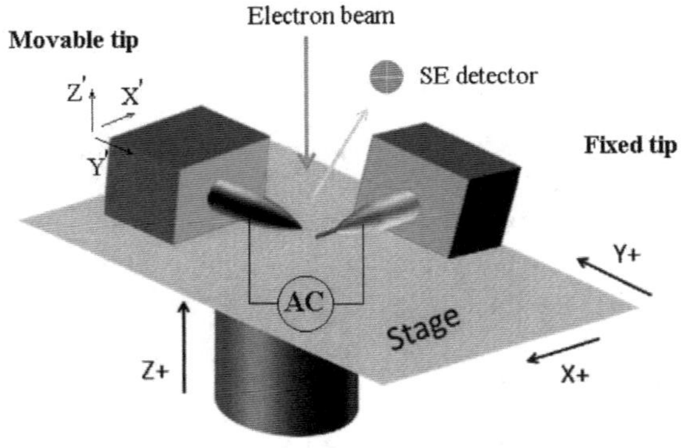

Figure 3. Setup for excitation the oscillations by AC field.

4. Results and Discussions

a. Mechanical properties of microresonators

Resonant oscillations of WMW are observed in three resonant modes with frequencies of around 30 kHz, 90 kHz, and 200 kHz, respectively (figure 4a, 4b, 4c) using piezotransducer scheme presented on Fig.2. We have also managed to excite and visualize rotational mode of oscillations of WMW with frequency of around 100 kHz (figure 4d). Quality factor of fundamental resonance of WMW was around 150 (figure 4f), and for resonance mode at 3.5 MHz was around 50 (figure 4g).

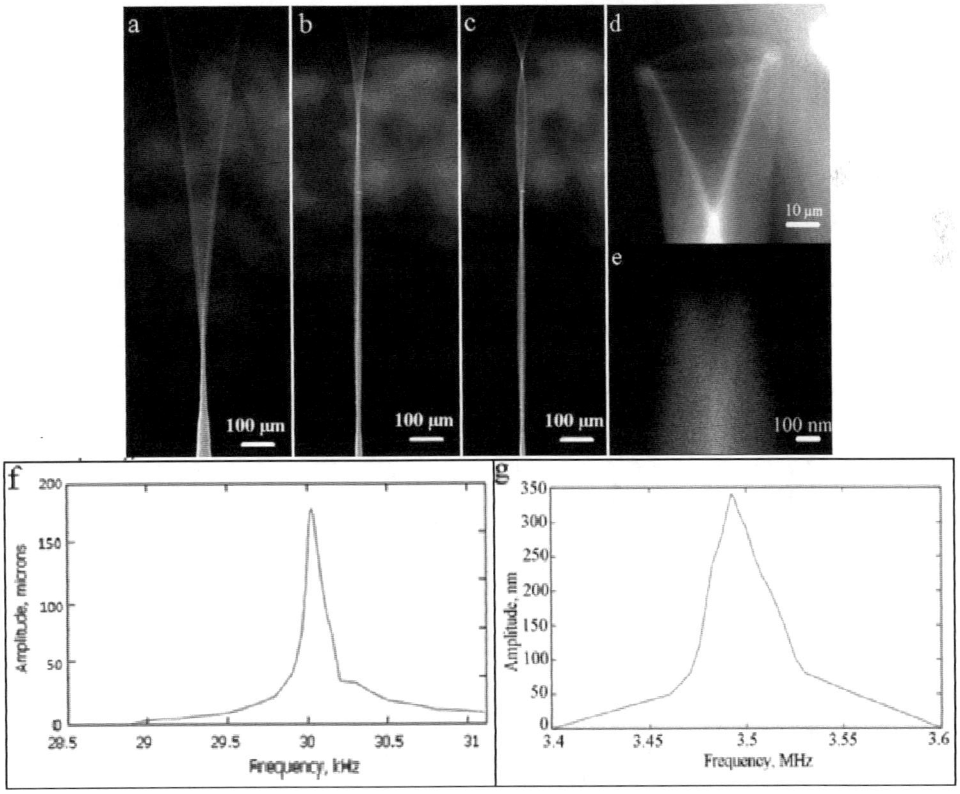

Figure 4. SEM images of resonant oscillations of WMW. a) Fundamental resonance of WMW, b) second resonant mode of WMW, c) third resonant mode of WMW, d) torsion resonant mode of WMW, e) high-frequency resonant mode of WMW at 3.5 MHz, f) amplitude-frequency characteristics (AFC) of first resonance of WMW at 30 kHz, g) AFC of resonance at 3.5 MHz.

b. Mechanical properties of nanoresonators

Frequencies of oscillations of first resonant modes of CNW are within (3 – 10) MHz depending on their geometrical dimensions for both methods AC field (figure 5b, 5c) and piezotransducer (Fig. 6a, and 6c). Quality factor of CNW on the sharp tip is around 80 and 40 on the blunt tip. (figure 6b, 6d).

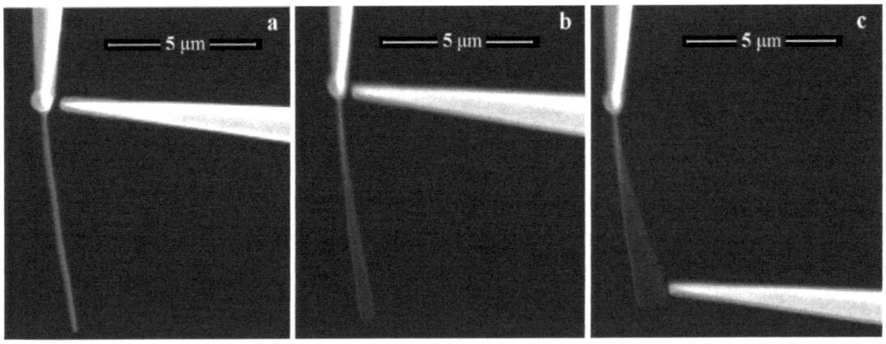

Figure 5. SEM images of visualization the resonance of carbon nanowhisker (CNW) excitated by alternating field. a) no oscillations, b) resonance of CNW at 1.41 MHz, c) resonance of CNW at 1.41 MHz.

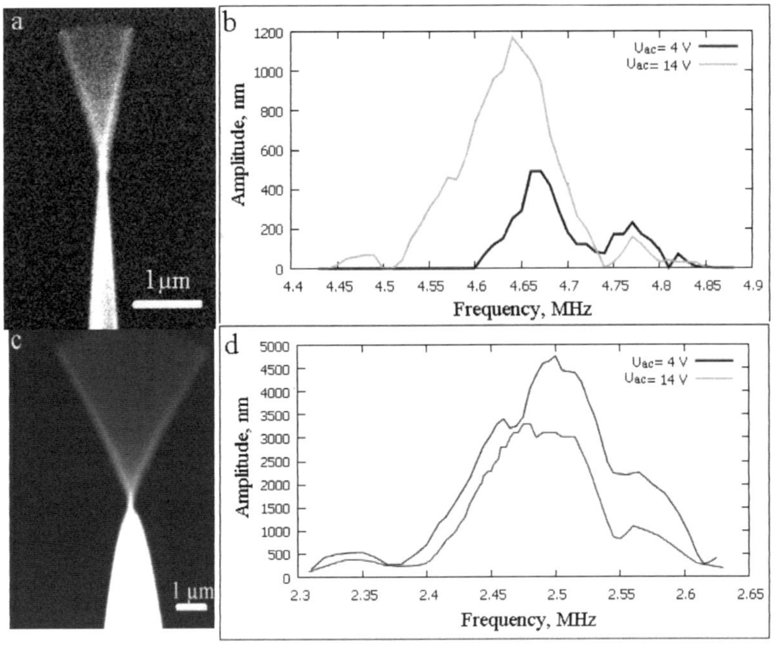

Figure 6. SEM images of first fundamental resonances (FR) of carbon nanowhiskers (CNW), excited by piezotube and its amplitude-frequency characteristics (AFCs). a) FR of CNW on sharp tip, b) AFC of CNW on sharp tip, c) FR of CNW on blunt tip, d) AFC of CNW on blunt tip.

Coupled oscillations (figure 7) of NMO in the form of CNW grown on the top of WMW are also visualized. In this case on the frequencies near the frequency of natural resonance of CNW (figure 7b) the coupled resonance is observed as simultaneously increasing the amplitude of oscillations of both the top of WMW and the top of CNW (figure 7a, 7c).

When the frequency is less than the fundamental resonance frequency was observed "in-phase" coupled oscillations. At higher frequencies were observed resonant oscillations with typical "anti-phase" configuration. As far as the frequency is increased several such coupled resonances is observed. It should be noted that, on a trajectory of oscillations of CNW, "bundle" with zero amplitude of oscillations is observed. Location of that "bundle" changed with the change of frequency of oscillations.

Simulation of trajectory of oscillations was conducted in COMSOL Multiphysics, and it is qualitatively corresponding with the results of experiment (Figure 7d-f).

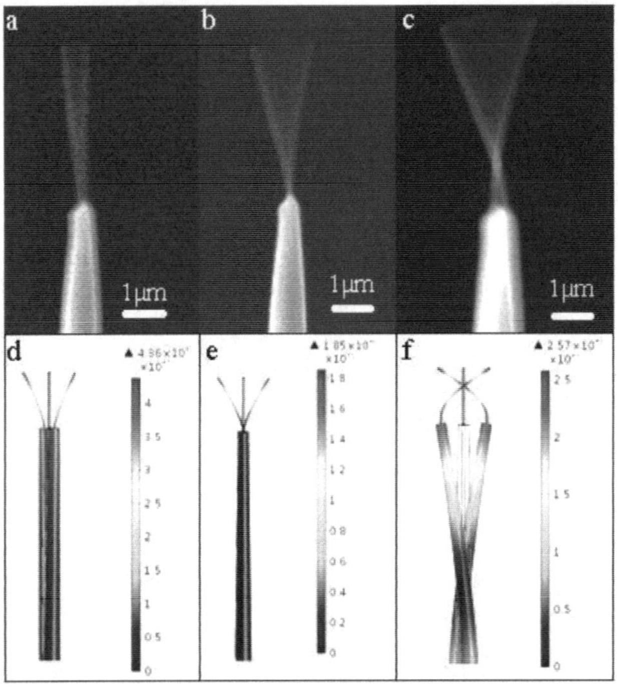

Figure 7. Resonant modes of coupled oscillators near fundamental resonant mode of carbon nanowhisker and results of simulation in COMSOL Multiphysics.

Before the determining the resonance peaks of the coupled system, we determined resonances of CNW and WMW in isolated cases. First, we determined the resonance of CNW at the blunt W needle. Then determined resonances of sharp W needle, and then under the same conditions was created CNW with the same dimensions. In the frequency diagram the peaks relevant to isolated cases was attended (blue and cyan peaks on figure 8). In addition, there were new peaks (red peaks on figure 8). Interestingly, new peaks of coupled system are equidistant from peaks of the sharp W needle.

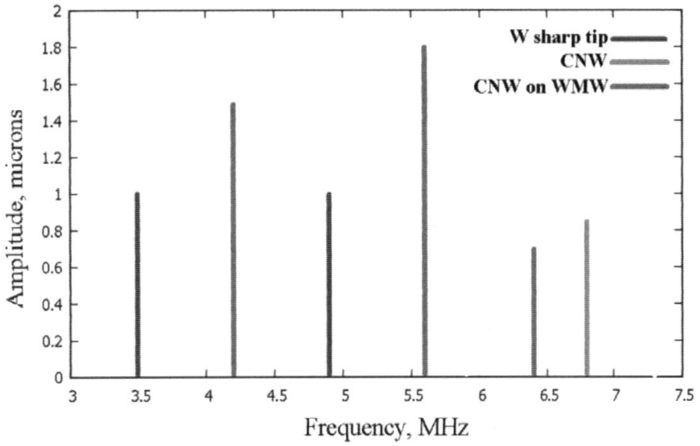

Figure 8. Resonance peaks of high-frequency resonant modes of tungsten sharp tip, fundamental resonance of carbon nanowhiskers (CNW) and new resonances of coupled system of CNW on the top of tungsten microwhisker (WMW).

Also, a sharp region was found where the instability of two configurations of oscillations. In addition, it was found jumps of oscillation amplitude, as well as a sharp amplitude decreasing. Such features of oscillatory systems we associate with nonlinear properties of coupled oscillators.

4. Conclusions

Mechanical micro- and nanooscillators in the form of crystalline and amorphous whiskers localized on the top of metal needles are created and studied. Resonant frequencies and quality factor of single and coupled micro- and nanooscillators is measured. Revealed peculiarities in the trajectory of coupled resonant oscillations of nanooscillators on the basis of nanowhiskers qualitatively correspond to the results of simulation. Also, we found regimes, in which there were bifurcation, jumps and decreasing of amplitude of oscillations. These features have sharp arias in frequency diagram, that interesting to use coupled oscillators as precise sensors, as well as a base to study nonlinear properties of oscillators.
Work carried out with the support of the leading universities of the Russian Federation (grant 074-U01), the Russian Foundation for Basic Research (14-02-31703 mol_a).

5. References

[1] O'Connell A Hofheinz M 2010 *Nature* vol 464 pp 697-703
[2] Jensen K Kwanpyo K 2008 *Nature Nanotech.* vol 3 pp 533-537
[3] Mukhin I Fadeev I Zhukov M Dubrovskii V Golubok A 2015 *Ultramicroscopy* vol 148 pp 151-157

Modeling, synthesis and study of highly efficient solar cells based on III-nitride nanowire arrays grown on Si substrates

A M Mozharov[1], A D Bolshakov[1], D A Kudryashov[1], N V Kryzhanovskaya[1], G E Cirlin[1], I S Mukhin[1,2], J C Harmand[3] and M Tchernysheva[3]

[1] St.Petersburg Academic University, Khlopina 8/3, 194021, St. Petersburg, Russia
[2] ITMO University, Kronverkskij 49, 197101, St. Petersburg, Russia
[3] CNRS-LPN, Route de Nozay, Marcoussis, 91460 France

E-mail: alex000090@gmail.com

Abstract. In this letter we investigate photovoltaic properties of GaN nanowires (NWs) – Si substrate heterostructure obtained by molecular beam epitaxy (MBE). Antireflection properties of the NW array were studied theoretically and experimentally to show an order of magnitude enhancement in antireflection comparing to the pure Si surface (2.5% vs. 33.8%). In order to determine optimal morphology and doping levels of the structure with maximum possible efficiency we simulated it's properties using a finite difference method. The carried out simulation showed that a maximum efficiency should be 20%.

1. Introduction

Highly efficient solar cells development and continuous search of new ways for photovoltaic efficiency improvement are topics of great interest for last few decades. Vertically aligned arrays of 1D nanostructures (NWs) due to their unusual properties are potential building blocks for optoelectronic devices, in particular solar cells [1, 2]. Arrays of III-nitride NWs (GaN, AlGaN, InGaN) grown by means of molecular beam epitaxy on crystalline Si substrates are known to possess high crystalline quality and low optical reflectance over the whole visible spectral range. Integrating of A3B5 and Si semiconductor technologies is one of the main priorities in modern photovoltaic researches and A3B5 NWs might be the key to overcome existing difficulties in this field [1].

Schematic image of a tandem solar cell consisting of GaN NWs array on top of emitter of the upper junction is presented on figure 1 a). In the simplest case it may only comprise of GaN NWs grown on Si substrate.

GaN NWs array obtained using AlN mask layer serves as a wide band gap window of the solar cell, has good antireflection properties and provides efficient direct electron transport to contacts (by means of ITO coating for example). It is worth noting that GaN possesses band structure parameters (electron affinity, band gap) good to achieve high quality n-n contact between GaN NW and wide band gap semiconductor used in upper junction of tandem cell.

2. Growth and postprocessing

Prior to the growth process a standard cleaning procedure of a Si substrate has been performed with a final treatment in HF to remove the oxide layer. The substrate was then degased in the growth chamber at 800 °C (figure 2). After that a thin 4 nm AlN layer was deposited on the substrate and then exposed under N2 flux for 1 minute to form porous material working as a mask for NWs formation.

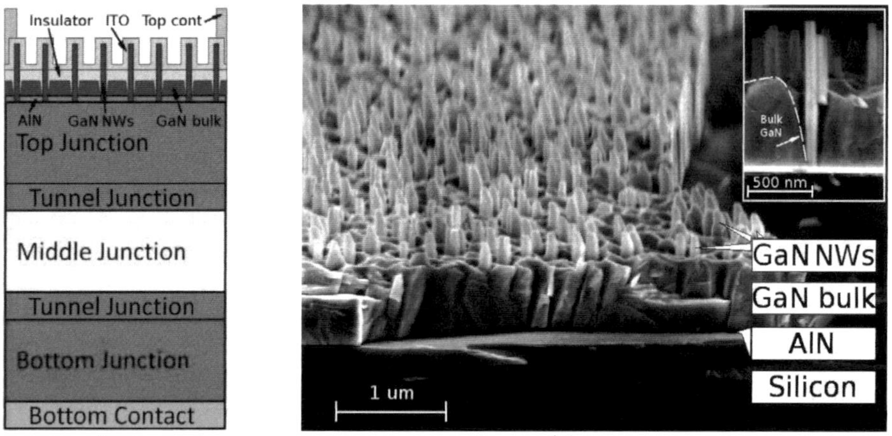

a) b)

Figure 1. a) Schematic image of a tandem solar cell consisting of GaN NWs array on top of emitter of the upper junction; b) SEM image of GaN NWs array on Si.

This thin porous layer works as a wide band gap window to confine charge carriers of both signs photogenerated in Si substrate and also hinders charge transport from Si into unwanted 3D GaN layer (saturated with dislocations) that is grown on top of mask during NWs formation (see figure 1 b). In this technique high crystalline quality GaN NWs providing good carrier transport with high carrier life time grow directly on Si substrate and have no mechanical contact with AlN layer as well as with 3D GaN layer (see inset on figure 1 b).

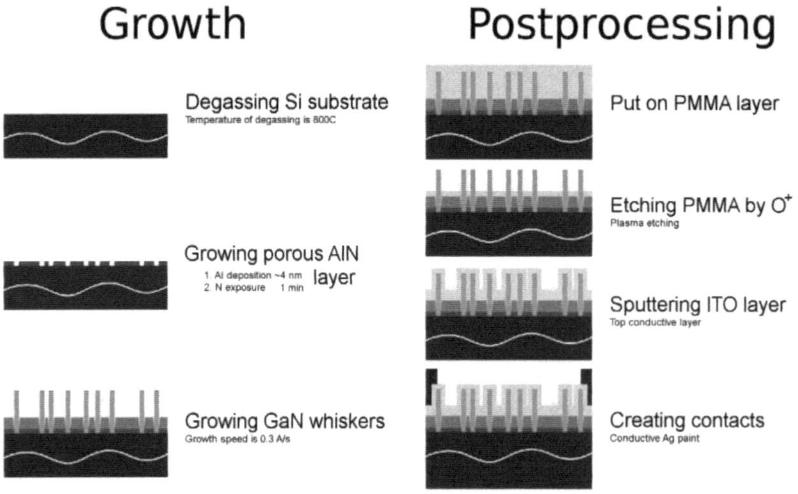

Figure 2. Illustration of sequences for growth and postprocessing procedures of GaN NWs – Si substrate collar cell fabrication.

Array of slightly unintentionally doped n-type GaN NWs on n-type Si substrate was grown using nitrogen plasma-assisted MBE technology with a Ga flux corresponding to the growth rate of 0.3 A/s for GaAs. For nitrogen beam, 2.1 sscm flux was used, with 520 W power providing N-reach growth conditions necessary for the NWs formation. Growth rate of the unwanted 3D GaN layer was two times slower. Typical length of the NWs was measured to be 1 μm, while the 3D layer thickness was

0.5 μm respectively. Mean NWs diameter was found to be 80 nm, surface density - 10^9 cm^{-2}. Elongated geometry of the wires enables us to process a top contact to them isolated with a dielectric from the unwanted GaN layer.

In order to decrease serial resistance of the cell we needed to exclude saturated with dislocation 3D GaN layer from the circuit. To this end we deposited a thin dielectric PMMA resist layer using spinning method on top of GaN layer (figure 2). This PMMA layer was then removed from NWs' top facets by means of oxygen plasma etching to provide photogenerated carriers transport to the top contact. The top contact was formed with ITO layer covering the entire cell upper surface that was deposited by the magnetron sputtering. A silver net was then made on the ITO surface. Silver paste was also used to derive bottom contact on Si.

3. Experimental results

Band diagram of the solar cell based on n-Si substrate – n-GaN NWs junction is presented on figure 3 a). GaN band gap is 3.2eV, Si band gap is – 1.1eV. At the Si-GaN heterojunction an electron transparent tunnel barrier and potential barrier for holes are obtained. Photogenerated in Si electrons travel through NWs to be collected at the face ITO contact. Collection of holes is implemented via back contact with Si.

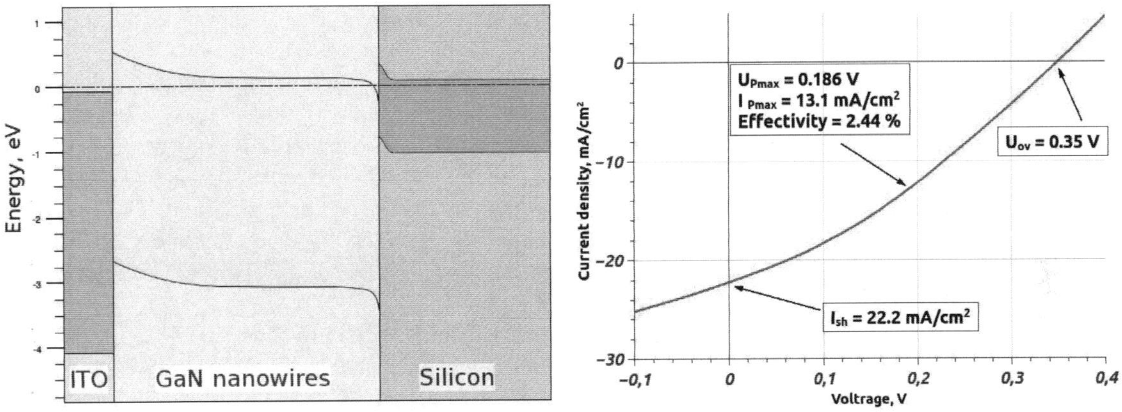

a) b)

Figure 3. a) Band diagram of the n-GaN NWs – n-Si substrate solar cell; b) I(V) characteristic of the n-GaN NWs – n-Si substrate solar cell under AM1.5 illumination.

Experimental current-voltage characteristic of the obtained cell under AM1.5 illumination (1000 W/m2) at a cell temperature of 25 °C is presented on figure 3 b). Open circuit voltage was found to be 0.35 V and corresponds to the potential barrier height for electrons at the Si-GaN heterointerface. Fill factor value was 0,31, which is more likely caused by high value of the ITO face contact serial resistance. Efficiency of the n-n solar cell was measured to be 2.4%. In case of tandem solar cell based on the heterojunction between n-type upper emitter and n-type GaN NWs potential barrier for electrons occurs at the heterointerface which leads to degradation of cell properties. Therefore, an optimization of doping level of both junction materials needs to be performed in order to lower the barrier height.

Dense array of self-assembled NWs having subwavelength diameters is known to have good antireflection properties [3]. Experimental data on antireflection characteristics of GaN NWs grown via molecular beam epitaxy on Si substrate is shown on figure 4. Oscillations of the reflectivity are caused by destructive interference of the incident light in the unwanted 3D GaN layer. Average value of the structure reflectivity was found to be 2.6%.

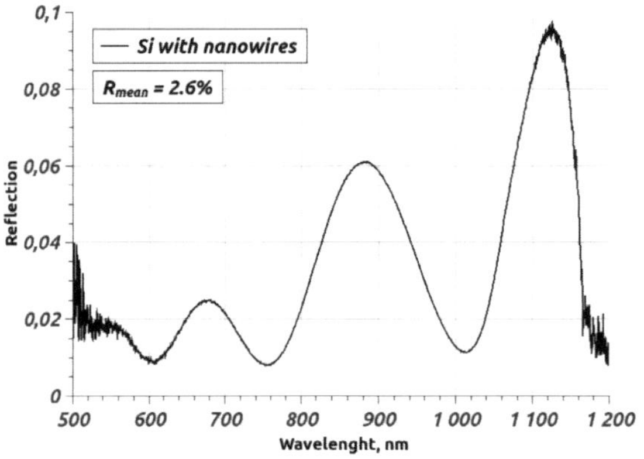

Figure 4. Antireflection characteristics of GaN NWs grown on Si substrate.

4. Modelling

In our study we have carried out modeling of a simple solar cell formed by n-type GaN NWs and p-type Si substrate. In such a photovoltaic device photogeneration of electron-hole pairs takes place in volume of Si and separation of the carriers occurs at the heterojunction. Band diagram of the cell is presented on figure 5 a).

In our simulations we have performed independent modeling of semiconductor and antireflection properties of the structure. The main aim of simulations was to obtain optimal NWs morphology, doping levels of both wires and Si substrate to determine maximum theoretical efficiency of the cell.

Antireflection properties modeling has been done via CST Microwave software package. In our simulations we used a geometrical model of hexagonally ordered array of cylindrical NWs on Si substrate (figure 5 b). Experimental data over dependencies of GaN 3D layer thickness on NWs average length and known special features of ITO sputtering on sidewalls of NWs were taken into account in calculations.

Simulations of the antireflection properties included twofold consecutive variation of the NWs surface density, radius and length. The following optimal geometrical parameters were obtained: lateral distance between the wires – 365 nm, radius – 60 nm, length – 1375 nm. Corresponding reflectivity value was 2.5%. For the sake of comparison spectral reflectivity of the pure Si is presented on figure 6 (its mean value equals 33.8%).

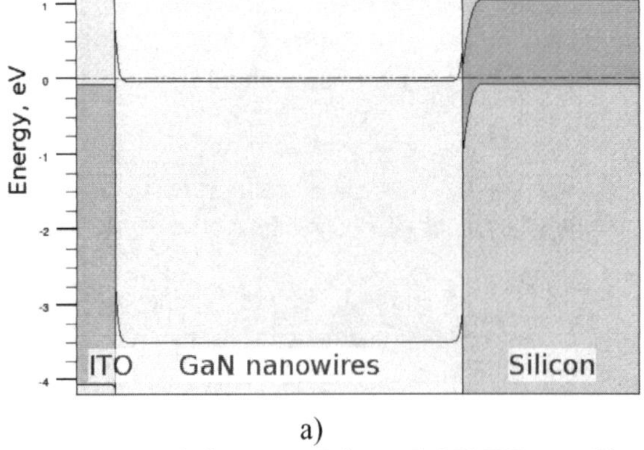

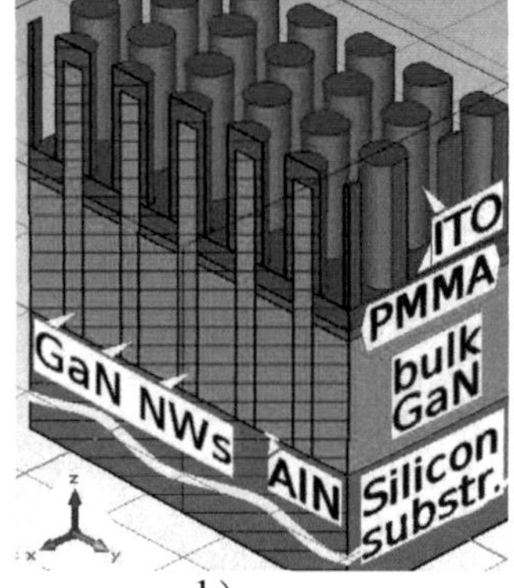

Figure 5. a) Band diagram of the p-GaN NWs – n-Si substrate solar cell; b) 3D model of cell.

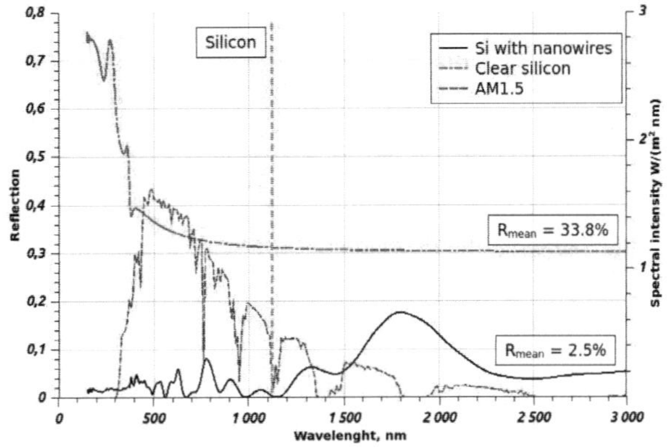

Figure 6. Modeling results of antireflection characteristics of GaN NWs on Si substrate.

Semiconductor properties of the structure were modeled using Newton method with Fermi statistics. We supposed that NWs have ideal crystal structure with negligible carriers capture at side walls that allowed us to reduce modeling to one dimensional problem. Instead of NWs we introduced GaN layer having thickness equal to an effective minimum distance that carrier needs to overcome between Si substrate and ITO layer (it consists of GaN 3D layer thickness + dielectric thickness, see figure 5 b) multiplied by ratio of substrate surface area to area covered with NWs. In this approximation sum $h_{bulk\ GaN} + h_{PMMA}$ equals 730 nm and area ratio – 10.2.

$$H_{GaN}^{eff} = \left(h_{bulkGaN} + h_{PMMA}\right)\cdot\frac{S_{surf}}{S_{GaNnw}}$$

The results of modeling are shown on figure 7. On figure 7 a) you find graphical dependencies of cell efficiencies on substrate doping level at different but fixed levels of wires doping. These curves have strongly marked maxima at substrate acceptors concentration around $1.3\cdot10^{18}$ cm^{-3} and maximum efficiency reaches 20%. Further increase of acceptors concentration leads to fall of both depletion region and mobility of carriers in Si which results in decrement of the cell efficiency.

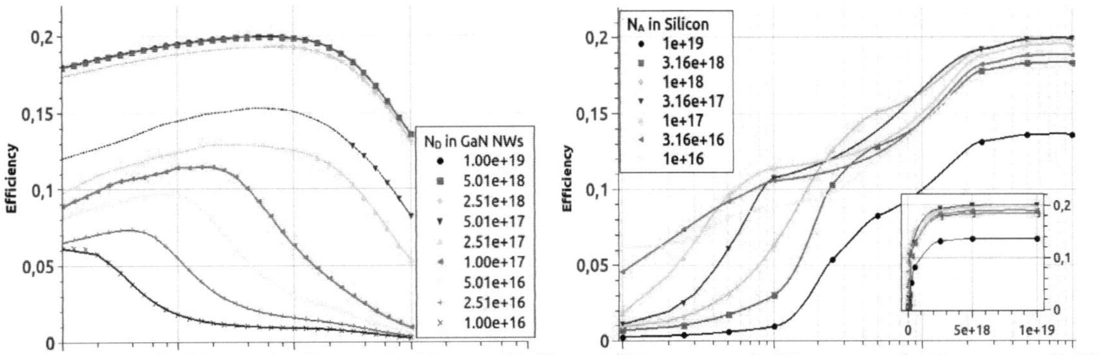

a) b)

Figure 7. a) Dependencies of cell efficiencies on substrate doping level at fixed levels of NWs doping; b) dependencies of cell efficiencies on NWs doping level at fixed levels of substrate doping.

On figure 7 b) you find graphical dependencies of cell efficiencies on NWs doping level at different but fixed levels of substrate doping. As can see at donors concentration above $4 \cdot 10^{18}$ cm^{-3} the curves demonstrate saturation and further increase of wires conductivity doesn't affect the cell efficiency. Modern MBE technology allows us to reach the obtained doping levels.

5. Summary

In this work we study solar cell based on GaN nanowires array grown epitaxially on Si substrate which appears to be the simple case of a tandem cell with an array of GaN nanowires on emitter of the cell's top junction. Nanowires in this case play a role of a top wide band gap window and provide great antireflection properties.

A simple solar cell consisting of GaN nanowires on top of n-type Si substrate has been obtained using MBE technology. We have shown that the structure has energy conversion efficiency of 2.4% and a high open circuit voltage of 0.35 V which can be explained via low doping level. To increase efficiency of the tandem cell a potential barrier on the boundary between GaN nanowires and top junction emitter should be decreased.

We have also performed properties simulations of the solar cell based on the junction between array of n-type nanowires and p-type Si substrate. It's been demonstrated theoretically that reflectivity of such a structure is only 2.5% and photovoltaic efficiency can reach 20%.

Work carried out with the support by Department for Nanotechnologies and Information Technologies of the Russian Academy of Sciences, President grant (14.W01.15.7736-MK), the Russian Foundation for Basic Research (15-02-06839 A and 13-02-12031 ofi), of the leading universities of the Russian Federation (grant 074-U01).

References

[1] Y B Tang, Z H Chen, H S Song, C S Lee, H T Cong, H M Cheng, W J Zhang, I Bello and S T Lee 2008 *Nano Letters* **8-12** 4191

[2] G E Cirlin, A D Bouravleuv, I P Soshnikov, Yu B Samsonenko, V G Dubrovskii, E M Arakcheeva, E M Tanklevskaya and P Werner 2010 *Nanoscale Research Letters* **5-2** 360

[3] L Hu and G Chen 2007 *Nano Letters* **7-11** 3249

Research of materials for porous matrices in sol-gel systems based on silicon dioxide and metallic oxides

E.V. Maraeva[1], A.A. Bobkov[1], A.I. Maximov[1], V.A. Moshnikov[1,2], S.S. Nalimova[1]

[1] Saint Petersburg Electrotechnical University "LETI", ul. Prof. Popova, 5, 197376, Saint Petersburg, Russia

[2]Peter the Great St.Petersburg Polytechnic University, 29, Polytechnicheskaya st., 195251, St.Petersburg, Russia

E-mail: *jenvmar@mail.ru*

Abstract. In this study silicon dioxide – stannic oxide and silicon dioxide – zinc nanomaterials oxide were obtained through sol-gel technology. The results of nitrogen thermal desorption measurements, atomic force microscopy measurements and particle sizes measurements are discussed.

1. 1. Introduction

Currently, the “quantum dots – poromeric inert matrices” hybrid systems have become broadly used. The advantage of quantum dots in porous inert matrices is stability of sizes; in systems with quantum dots the density of radiating centers can be significantly higher than that in polycrystalline materials. Porous matrices must be optically transparent when photoluminescent particles are introduced into them. To develop electroluminescent structures, it is necessary to form highly-conductive wide-gap layers. In this respect the materials of sol-gel systems based on silicon dioxide and metallic oxides are most promising.

The purpose of this study is researching different stages of sol-gel synthesis of materials in silicon dioxide – stannic oxide and silicon dioxide – zinc oxide systems [1-4].

2. Experiment

In this study an inorganic salt $Zn(NO_3)_2 \cdot 6H_2O$ as the precursor of zinc oxide, $SnCl_2 \cdot 2H_2O$ as the precursor of stannic dioxide and tetraethoxysilane as the precursor of silicon dioxide were selected. To control properties of the synthesized nanomaterial surface, the nitrogen thermal desorption and atomic force microscopy were used.

AFM experiments were performed using NTEGRA-Therma nanolaboratory (NT-MDT, Zelenograd, Russia). Commercial etched silicon tips NSG 01 (NT-MDT, Zelenograd, Russia) with typical resonance frequency of 150 kHz were used as AFM probes. Specific surface area measurements were made using Sorbi № 4.1 (CJSC «META», Novosibirsk, Russia) that realizes physical adsorption of noble gas by the sample to be studied [5].

3. Results and discussion

In the course of this study the samples of xerogels in the "silicon dioxide – stannic oxide" and "silicon dioxide – zinc oxide" systems were obtained, and specific influence of sol-gel synthesis conditions on the specific surface area and morphology of samples surface was investigated. As an example figure 1 presents AFM images of 85% SiO_2 – 15% SnO_2 layer obtained with various centrifuge speeds.

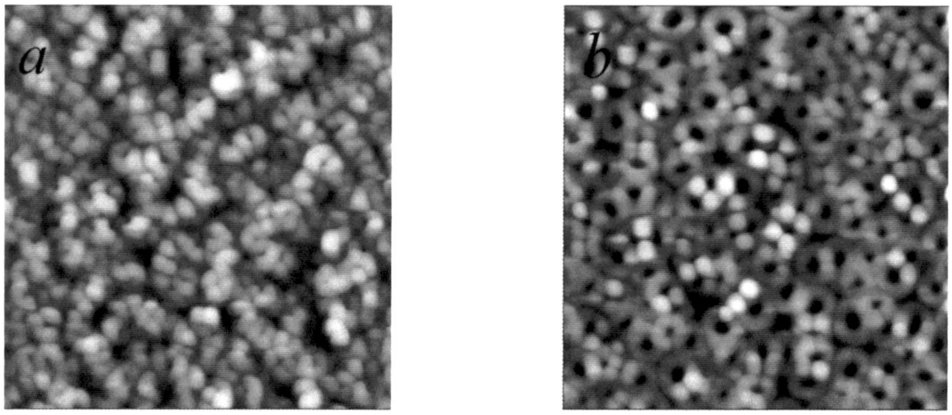

Figure 1. AFM images of 85% SiO_2 – 15% SnO_2 layers obtained with various centrifuge speeds: a – 5000 rpm , b – 1000 rpm (scan size area is 5 * 5 μm)

As it can be seen from figure 1 the reducing of centrifuge speed leads to formation of porous structure with a large pore size. This structure may be promising for deposition of the quantum dots. So by varying the centrifuge speed we can change the porous matrix and the sizes of deposited quantum dots.

Analysis of adsorption-desorption, capillary condensation and study of the pore size distribution were carried out on powders. Figure 2 presents the full adsorption/desorption isotherm for the sample of 85% SiO_2 – 15% SnO_2 powder. A relative error of the method is 6 %. The features of method are discussed in [6]. The results are shown in table 1.

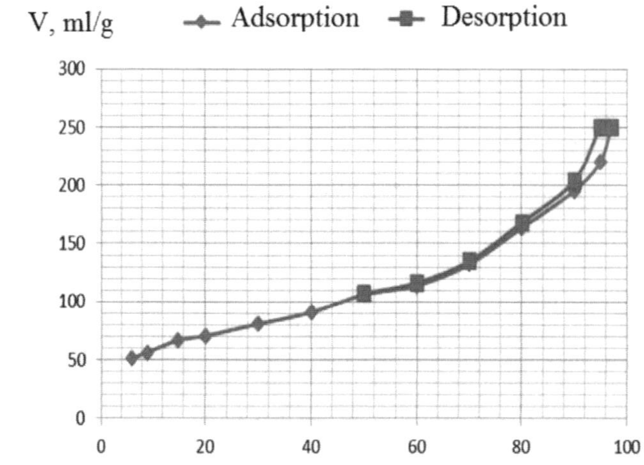

Figure 2. Full adsorption/desorption isotherm for the sample of 85% SiO_2 – 15% SnO_2 powder

Table 1. Sizes of pores of the 85% SiO_2 – 15% SnO_2 powder taken by full adsorption isotherm

D_i, nm	V_i/V_{total}, %
2.5	7
3	17
4	30
7	29
15	17

It was determined that the lion's share of pores in SiO_2 – SnO_2 materials has average dimensions of 3 nm, and 15 nm. Such systems can be perspective for injection of lead chalcogenide and cadmium chalcogenide quantum dots.

In case of silicon dioxide – zinc oxide system, the dependence of the heating temperature on layer structure was analyzed. Figure 3 presents AFM images of the surface of 80%ZnO-20%SiO_2 nanocomposites obtained at different annealing temperatures: a - 300°C, b - 400°C, c- 500°C,d - 600°C.

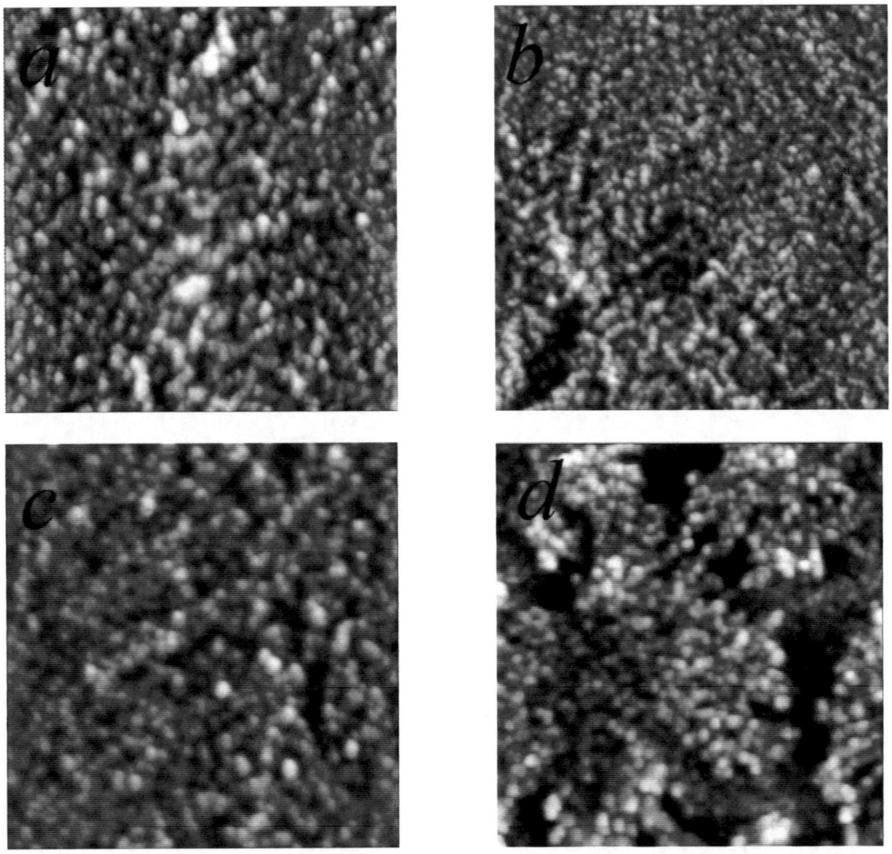

Figure 3. AFM images of 80%ZnO-20%SiO_2 layers obtained at different annealing temperatures: a - 300°C, b - 400°C, c- 500°C, d - 600°C (scan size area is 10 * 10μm)

As it can be seen from figure 3, the annealing temperature directly affects the distribution and size of the particles formed. Increasing of the annealing temperature leads to the reduction of particle sizes. The most uniform distribution of the particles is observed for the sample annealed at 500°C.

Conclusions

The experimental results show that by varying the synthesis parameters (such as centrifuge speed, annealing temperature) we can change the porous matrix and the sizes of deposited nanoparticles, for example quantum dots. The obtained results are perspective for optimizing the conditions of production of xerogels with required parameters.

Acknowledgments

The reported study was supported by by Russian Ministry of Education state task № 16.2112.2014 / K.

References

[1] Maximov A I, Moshnikov V A, Tairov Yu M and Shilova O A 2008 *Basics of sol-gel technology of nanocomposites* (SPb: Elmor)
[2] Gracheva I E, Moshnikov V A, Karpova S S, Maraeva E V 2011 *Journal of Physics: Conference Series* **291** 012017
[3] Gracheva E I, Moshnikov V A, Maraeva E V, Karpova S S, Alexandrova O A, Alekseyev N I, Kuznetsov V V, Olchowik G, Semenov K N, Startseva A V, Sitnikov A V, Olchowik J M 2012 *Journal of Non-Crystalline Solids* **358** 433
[4] Moshnikov V A, Gracheva I E, Kuznezov V V, Maximov A I, Karpova S S and Ponomareva A A 2010 *Journal of Non-Crystalline Solids* **356** 2020
[5] Lenshin A S, Maraeva E V 2011 *Izvestiya of Saint-Petersburg Electrotechnical University* **6** 9
[6] Belorus A O, Maraeva E V, Spivak Yu M, Moshnikov V A 2015 *Journal of Physics: Conference Series* **586** 012017

The study of metal sulphide nanomaterials obtained by chemical bath deposition and hot-injection technique

E.V. Maraeva[1], O.A. Alexandrova[1], N.A. Forostyanaya[2], V.S. Levitskiy[1], D.S. Mazing[1], L.N. Maskaeva[2], V.Ph. Markov[2], V.A. Moshnikov[1,3], A.A. Shupta[1], Yu.M. Spivak[1], S.S. Tulenin[2],

[1] Saint Petersburg Electrotechnical University "LETI", ul. Prof. Popova, 5, 197376, Saint Petersburg, Russia

[2] Ural Federal University, ul. Mira, 19, 620002, Ekaterinburg, Russia

[3]Peter the Great St.Petersburg Polytechnic University, 29, Polytechnicheskaya st., 195251, St.Petersburg, Russia

E-mail: *jenvmar@mail.ru*

Abstract. In this study lead sulphide – cadmium sulphide based layers were obtained through chemical deposition of water solutions and cadmium sulphide quantum dots were formed through hot-injection technique. The article discusses the results of surface investigations with the use of atomic force microscopy, Raman spectroscopy and photoluminescence measurements.

1. Introduction

Currently, the technology of chemical bath deposition of films of metallic sulphides [1-4] is especially attractive as far as such method of synthesis is relatively simple from technological point of view and effective at the same time; in comparison with vacuum thermal evaporation [5-8] it does not require expensive equipment. Microcrystalline layers of PbS – CdS solid solutions are traditionally used in infrared (IR) photosensitive devices and IR LEDs, operating at room temperature in 1 – 5 μm region. Increasing of cadmium concentration in a solid solution leads to increasing of bandgap energy. In nanoparticles with diameter reducing quantum-dimensional effects take place. An effective energy gap between discrete levels increases with nanoparticle's diameter reducing. Lead chalcogenides, as narrow-bandgap structures, are perspective materials as far as they can operate in IR region as well as in visible region depending on their diameters.

In this study the aspects of obtaining nanoparticles of lead chalcogenides and solid solutions via two various methods (chemical bath deposition, hot-injection technique) are researched, and its characterization with the use of such methods as atomic force microscopy (AFM), Raman spectroscopy and photoluminescence measurements is discussed.

1. Experiment

In this study lead sulphide – cadmium sulphide based layers were obtained through chemical bath deposition of water solutions using the following precursors: lead acetate $Pb(CH_3COO)_2$; sodium citrate Na_3Cit; cadmium chloride $CdCl_2$; thiourea $(NH_2)_2CS$. The pH level was adjusted by adding water solution of ammonia NH_4OH. The volumes of solutions and volume of water added into reactor were calculated according to law of mass action depending on expected content of cadmium sulphide in the solid solution. Deposition was made on the vitroceramic-based substrate within 2 hours at temperature of 80°C. Ethylenediaminetetraacetic acid (ADTA) was used for oxide phases etching.

Cadmium sulfide nanocrystals were synthesized using hot-injection technique in non-polar medium of octadecene (ODE). Cadmium precursor solution was prepared by dissolving of cadmium oxide in mixture ODE and oleic acid in tree neck flask under nitrogen flow which resulted in formation of cadmium oleate solution. The second precursor solution which consisted of sulfur powder dissolved in ODE was quickly injected in the cadmium precursor solution under intensive stirring using glass syringe. This resulted in rapid nucleation of CdS nuclei with subsequent growth. Thus prepared nanocrystals has hydrophobic surface due to layer of stabilizing oleic acid.

AFM experiments were performed using NTEGRA-Therma nanolaboratory (NT-MDT, Zelenograd, Russia). Commercial etched silicon tips NSG 01 (NT-MDT, Zelenograd, Russia) with typical resonance frequency of 150 kHz were used as AFM probes.

Absorption spectra were acquired on a PE-5400UV UV–vis spectrophotometer (LLC "Ekohim"). Photoluminescence measurements were carried out on equipment based on fast scanning spectrometer Ocean Optics USB-4000 using laser excitation on λ = 405 nm. The Raman spectroscopy was carried out with spectrometer LabRam HR800.

2. Results and discussion

As a result, a set of samples of solid solutions with different contents of cadmium sulphide was obtained, and properties of their surface were examined by analyzing the substrate area. The layers obtained were 100 to 500 nm thick depending on the original mix proportion. Regardless the proportion in original mix of solutions the most homogeneous film was formed in the center of substrate.

Figure 1 presents AFM images of lateral forces of obtained layers. The investigation of lead sulphide discovered that some crystallites consist of a number of nanolayers (figure 1a). By analysing of various scan areas it was determined that each layer thickness is about 40 nm.

According to [7, 8], when lead chalcogenides interact with oxygen the forming of oxygen phases takes place. In case of lead sulphide these phases are lead sulphate $PbSO_4$, oxysulphate $PbO{\cdot}PbSO_4$. Its formation takes place even in normal conditions. For more detailed surface investigation the samples were etched in ethylenediaminetetraacetic acid (ADTA). Figures 1b,c present AFM images of layers being obtained from water solutions with cadmium (b) and without adding cadmium (c) after etching.

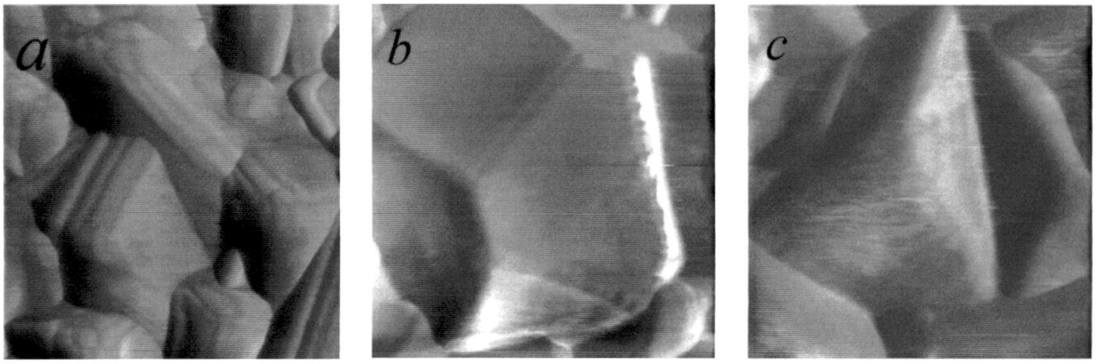

Figure 1. AFM image of lateral forces of PbS layer before etching (a), layers being obtained with cadmium (b) and without adding cadmium (c) after etching (scan size area is 1* 1μm)

It was determined that if cadmium was added into solution, crystallites with hexagonal structure were formed. In case of undiluted lead sulphide the crystallites had cubic or pyramidic form.

The Raman spectra were obtained in the backscattering geometry at room temperature with the spectrometer LabRamHR800 combined with a confocal microscope (manufactured by Jobin-YvonHoriba). The second harmonic of Nd:YAG-laser (wavelength 532 nm) served as the excitation source. The laser beam was focused to a spot with ~1-2 μm diameter on the sample surface. Typical power density was lower 5 kW/cm^2 to avoid the influence of laser irradiation on the structure of the studied objects.

For Raman spectroscopy the films obtained with the use of solution with various concentration proportions of $Pb(CH_3COO)_2$ and $CdCl_2$ in original mix were chosen. The concentration proportions of $Pb(CH_3COO)_2$ and $CdCl_2$ in initial solution are reflected in Table 1.

Table 1. Concentration proportions of $Pb(CH_3COO)_2$ and $CdCl_2$ in initial solutions

Sample	$[CdCl_2]/[Pb(CH_3COO)_2]$
1	2.7
2	2
3	1
4	0.5
5	0

Figure 2 shows the Raman spectra of Samples 1, 2 and 3 standardised to the maximum intensity (a) and the spectra of Samples 1, 2 and 3 in the first order oscillation area of LO phonon (b).

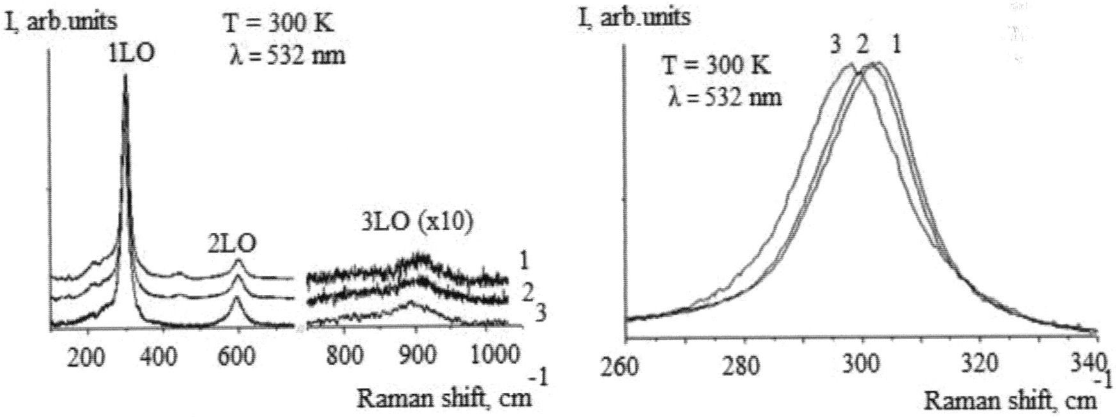

Figure 2. Raman spectra: 1 – sample 1; 2 – sample 2; 3 – sample 3

The Raman spectra of the samples are presented by broad bands with the maxima near 300 cm^{-1}, 600 cm^{-1} and 900 cm^{-1} (figure 2a). These bands are connected to multi-overtones with the longitudinal optical (LO) phonon mode CdS [9, 10]. As it was shown in [11-13], the most informative way to describe the features of the material's structural specifications is to analyse the position of the maximum of the first order LO phonon.

As it can be seen from the figure 2b, the positions of the band's maxima are different. Table 2 compares the positions of the maxima of 1LO phonon for these samples with single-crystal sample taken from the paper [14].

Table 2. Raman shifts for 1LO-phonons in CdS

Sample	Raman reflex maximum, cm^{-1}
Single crystal [11]	305
1	303
2	301
3	298

From the data presented in Table 2 it can be seen that the smallest shift is in the Raman spectra for Sample 1. An increased shift in the position of the maximum 1LO phonon is seen for samples containing Pb. From the data presented in Table 2, it is clear that in case of increase in the lead content of the resulting film there is an increase in the 1LO band shift. Apparently, this indicates an increase in the structural deformation of the CdS lattice due to the Pb atoms embedded there in.

Figure 3a shows the Raman spectra of samples 4 and 5, figure 3b shows the Raman spectra of Sample 4 before and after laser annealing at the power density ~4 times greater than used to obtain the Raman spectra. The same figure provides the spectrum of Sample 1 as an example (CdS).

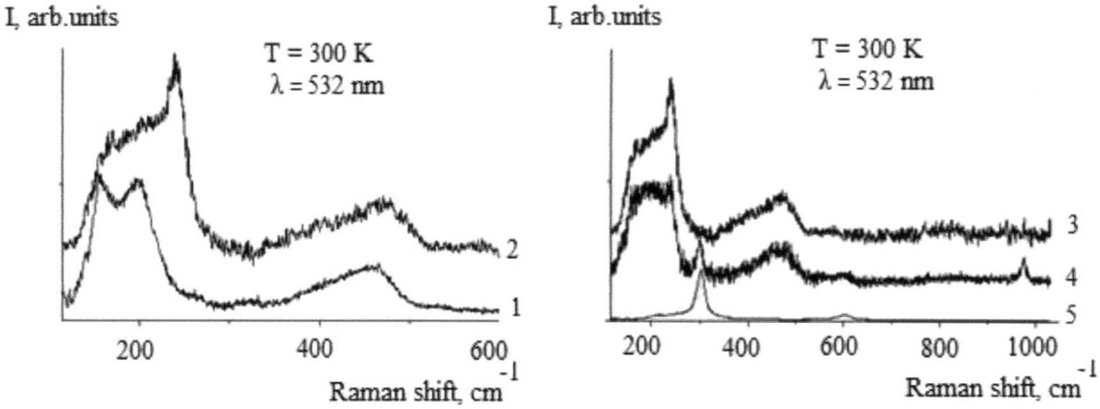

Figure 3. Raman spectra: 1 – sample 5; 2 – sample 4; 3 – sample 4 before laser heating; 4 – sample 4 after laser heating; 5 – sample 1

As it can be seen from the figure 3a, the Raman spectrum of Sample 5 is represented by three bands with the maximum positions near 150 cm^{-1}, 200 cm^{-1} и 440 cm^{-1}. This spectrum is very similar to the spectrum $Pb^{34}S$ from [15]. The Raman spectrum of Sample 4 is significantly different from the spectra of all samples reviewed in this study. As it can be seen from the figure, in this spectrum there is an intense line with the position of the maximum near 250 cm^{-1}. However, the rest of the spectrum is close to the PbS spectrum. In the CdS spectra the maximum position of 1LO phonon is ~300 cm^{-1}, so the band shift at ~50 cm^{-1} can be explained only by formation the solid solution $Pb_yCd_{1-y}S$.

As it can be seen from the figure 3b, the additional bands appeared after laser annealing in the spectra of Sample 3 with peaks close to 300 cm^{-1}, 600 cm^{-1} and 970 cm^{-1}. As mentioned earlier, the bands with maxima near 300 cm^{-1} и 600 cm^{-1} are typical of the first and second order of CdS LO phonon. Some researchers attribute the appearance of a band with a maximum near 970 cm^{-1} to the formation of the PbS oxide phases [16]. It can also be noted that the intensity of the band with ~ 250 cm^{-1} maximum position has notably decreased after annealing, but the position of band's maximum remained unchanged. The constant position of the band maximum and the reduced intensity indicate the quantitative reduction of $Pb_yCd_{1-y}S$ phase, while the phase composition of this compound is preserved. Apparently, the annealing results in selection of the CdS phase from $Pb_yCd_{1-y}S$ compound, as well as formation of PbS oxide phases.

Absorption and photoluminescence spectra of CdS nanocrystals prepared by means of hot injection method are presented in figure 4. Absorption band is blue shifted relative to what is expected for bulk CdS which indicates that nanocrystals are in the quantum confinement regime. As can be seen from spectra good monodispersity may be achieved using this approach as full width at half maximum parameter of photoluminescence peak is about 30 nm.

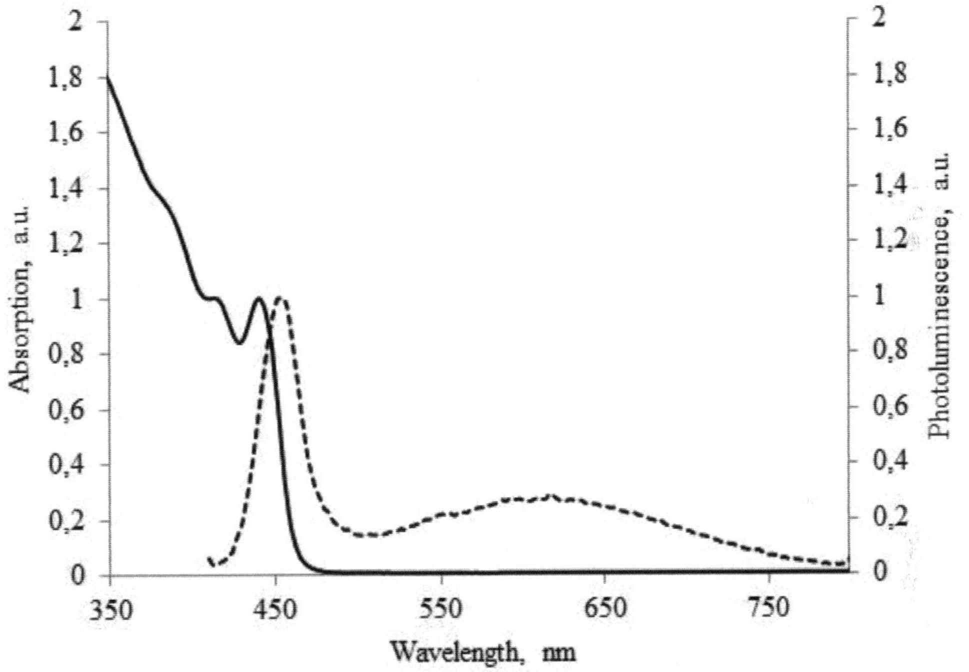

Figure 4. Absorption (full line) and photoluminescence spectra of colloidal CdS nanocrystals

Photoluminescence spectrum consists of two bands. The first one corresponds to interband transitions and reflects nanoparticle size distribution. The second one, of lower energy, is much broader and is associated with luminescence via trap states present within bandgap due to insufficient passivation of dangling bonds on nanocrystal surface. The mean size of nanocrystals was estimated to be approximately 4.8 nm. This assessment was made using equation from the work [17] which describes experimental correlation between nanocrystal size and wavelength of the first excitonic peak in absorption spectrum.

Conclusions

The paper discusses the aspects of obtaining lead sulphide – cadmium sulphide layers and cadmium sulphide quantum dots. The surface layers morphology are studied with the use of atomic force microscopy. AFM measurements after oxygen phases etching discovered that in case of adding cadmium into solution crystallites with hexagonal structure are formed. In case of undiluted lead sulphide the crystallites have cubic or pyramidic form. AFM data has a good correlation with the results of Raman spectroscopy.

Cadmium sulfide nanocrystals are synthesized using hot-injection technique and its absorption and photoluminescence spectra are studied. The results discussed are perspective for using in optoelectronics and for formation biological and medicinal markers for cellular structure studing.

Acknowledgments

The reported study was supported by grant of Russian Science Fund (project №14-15-00324) (a part of the study involving AFM investigation, quantum dots obtaining) and also by program 211 of Russian Federation Government № 02. A 03.21.0006 (a part of the study involving layers obtaining with the use of chemical bath deposition).

References

[1] Markov V, Maskayeva L, Ivanov P 2006 *Hydrochemical deposition of metallic sulphide films: simulation and experiment* (Ekaterinburg)
[2] Markov V F, Maskaeva L N, Kitaev G A 2000 *Inorganic Materials* **36** 1194
[3] Markov V F, Maskaeva L N 2010 *Russian Journal of Physical Chemistry* A. **84** 1288
[4] Katysheva A S, Markov V F, Maskaeva L N 2013 *Russian Journal of Inorganic Chemistry* **58** 833
[5] Chesnokova D B, Moshnikov V A, Gamarts A E, Maraeva E V, Aleksandrova O A, Kuznetsov V V 2010 *Journal of Non-Crystalline Solids* **356** 2010
[6] Moshnikov V A, Gamarts A E, Chesnokova D B, Maraeva E V 2011 *Inorganic Materials.* **47** 18
[7] Maraeva E V, Moshnikov V A, Tairov Y M 2013 *Semiconductors* **47** 1422
[8] Golubchenko N V, Moshnikov V A, Chesnokova D B 2006 *Glass Physics And Chemistry* **3** 337
[9] Leite R C C, Porto S P S 1966 *Phys. Rev.* **17** 10
[10] Chuu D S, Dai C M, Hsieh W F, Tsai C T 1991 *J. Appl. Phys.* **69** 8402
[11] Oladeji I O, Chow L, Liu J R, Chu W K., Bustamante A N P, Fredricksen C, Schulte A F. 2000 *Thin Solid Films* **359** 154
[12] Balandin A, Wang K L, Kouklin N, Bandyopadhyay S 2000 *Appl. Phys. Lett.* **76** 137
[13] Prabhu R R, Khadar M A 2008 *Bull. Mater. Sci.* **31** 511
[14] De Mello O., Hernandez L, Zeleya-Angel O, Lozada-Morales R, Becerril M, Vasco E 1994 *Appl. Phys. Lett.* **65** 1278.
[15] Sherwina R, Clarka R J H., Lauckb R, Cardonab M 2015 *Solid State Communications* **134** 565
[16] Pérez R G, Téllez G H, Rosas U P, Torres A. M, Tecorralco J H, Lima L C and Moreno O P 2013 *Journal of Materials Science and Engineering* **3** 1
[17] Yu W W, Qu L H, Guo W Z and Peng X G 2003 *Chem. Mater.* **15** 2854

A theory of spectral properties of disordered metal-semiconductor nanocomposites

N A Olekhno[1,2], **Y M Beltukov**[2,3] **and D A Parshin**[1]

[1]Peter the Great St.Petersburg Polytechnic University, 195251 Saint Petersburg, Russia
[2]Ioffe Institute, 194021 Saint Petersburg, Russia
[3]Université Montpellier II, CNRS, Montpellier 34095, France

E-mail: [1]`olekhnon@gmail.com`

Abstract. Random LC-networks are widely used to describe optical properties of disordered metal-dielectric nanocomposites. In such networks inductive bonds correspond to metal regions, while capacitive bonds represent dielectric. We show that parallel LC-circuits better describe metallic regions where values of L and C depend on a network lattice parameter and plasma frequency of metal. Spectral properties of disordered metal-dielectric and metal-heavily doped semiconductor composites are studied in a framework of this generalized LC-model. We derive analytical expressions for dependence of eigenfrequencies of metal-dielectric composites on permittivity of dielectric and relation between eigenfrequencies of metal-semiconductor composites before and after metal-dielectric transition in semiconductor. Obtained results are illustrated by numerical simulations.

1. Introduction

Disordered composites consisting of nanosize grains of noble metals placed into a dielectric matrix attract considerable attention because of their interesting optical properties associated with plasmon resonances, including second harmonic generation, enhanced optical absorption and surface-enhanced Raman scattering [1].

Discretized Maxwell equations in quasistatic approximation of metal regions with Drude-permittivity $\varepsilon_m(\omega) = 1 - \omega_p^2/\omega^2$ correspond to a cubic lattice of parallel LC-circuits with $L_m = 4\pi c^2/(a\omega_p^2)$ and $C_m = a/(4\pi)$, where ω_p is a plasma frequency of metal, a is a lattice parameter of discretization network and c is the speed of light. In the following we consider the units where $c = 1$. Dielectric regions correspond to a cubic lattice of capacitors $C_d = a\varepsilon_d/(4\pi)$ if permittivity ε_d is approximately constant in optical region. Thus, continuous heterogeneous medium is replaced by a discrete network with LC- and C-bonds. In literature the approximation when LC-bonds are replaced to L-bonds is extensively studied [1, 2, 3]. However, it is valid only for a low frequencies $\omega \ll \omega_p$ and the properties of the general case remain unclear.

Resonances in such networks can be studied in terms of generalized eigenvalue problem [4]

$$\sum_j K_{ij}\varphi_j - \omega^2 \sum_j C_{ij}\varphi_j = 0, \qquad (1)$$

where indices i and j enumerate sites of a network. Eigenvalues ω^2 are squared resonant frequencies of a system and eigenvectors φ describe distribution of potential at the sites of network for eigenmodes. K is a matrix of inverse inductances between sites of network: $K_{ij} = 1/L_{ij}$ if sites i and j are directly connected by inductance L_{ij}, otherwise $K_{ij} = 0$. Diagonal elements are defined by a *sum rule* $K_{ii} = -\sum_{j\neq i} K_{ij}$. C is a matrix of capacitances: C_{ij} is equal to capacitance of capacitor connecting sites i and j. If sites i and j are not connected with a capacitor, $C_{ij} = 0$. Diagonal elements $C_{ii} = -\sum_{j\neq i} C_{ij}$. These properties directly follow from Kirchhoff's rules for a network. Matrices K and C are symmetric, sparse and of size $N \times N$, where N is a number of sites in a network. Resonant spectrum of network is given by characteristic equation

$$|K - \omega^2 C| = 0. \tag{2}$$

2. Resonances in metal-dielectric nanocomposites

Resonant spectrum of LC–C–network, corresponding to metal-dielectric composite, depends on permittivity of a dielectric ε_d, since elements of matrix C involve capacitances $\varepsilon_d a/4\pi$. In this section we will show that initial generalized eigenvalue problem (2) can be transformed to a form, which does not depend on ε_d.

The total matrix of capacitances C is a sum $C = C_d + C_m$, where C_d is a matrix of capacitances of dielectric and C_m is a matrix of capacitances of metal. Matrices K, C_m and C_d have the following structure

$$K = \frac{a\omega_p^2}{4\pi} K^*, \quad C_d = \frac{\varepsilon_d a}{4\pi} C_d^*, \quad C_m = \frac{a}{4\pi} C_m^*, \tag{3}$$

where matrices with the asterisk (*) are dimensionless matrices with unit non-zero elements. These matrices are adjacency matrices with the additional sum rule for the diagonal elements: $K_{ii}^* = -\sum_{j\neq i} K_{ij}^*$ etc. Therefore, from Eq. (2) we obtain

$$\left|K^* - \frac{\omega^2}{\omega_p^2}(\varepsilon_d C_d^* + C_m^*)\right| = 0. \tag{4}$$

Matrices K^* and C_m^* both represent the same structure of metallic bonds, so $C_m^* = K^*$ and Eq. (4) can be rewritten as

$$\left|\left(1 + (\varepsilon_d - 1)\frac{\omega^2}{\omega_p^2}\right)K^* - \varepsilon_d \frac{\omega^2}{\omega_p^2}(C_d^* + K^*)\right| = 0. \tag{5}$$

It is a generalized eigenvalue problem

$$|K^* - \mu(C_d^* + K^*)| = 0 \tag{6}$$

with eigenvalues

$$\mu = \frac{\varepsilon_d \omega^2/\omega_p^2}{1 + (\varepsilon_d - 1)\omega^2/\omega_p^2}. \tag{7}$$

Eigenvalues μ do not depend on permittivity of dielectric ε_d and can be computed only for one value of ε_d. Therefore each eigenfrequency ω can be expressed in the permittivity ε_d and eigenvalue μ

$$\omega = \frac{\omega_p}{\sqrt{\varepsilon_d(1-\mu)/\mu + 1}}. \tag{8}$$

For example, surface plasmon in spherical granule with frequency $\omega_p/\sqrt{3}$ has $\mu = 1/3$ and Eq. (8) gives the well-known result $\omega = \omega_p/\sqrt{2\varepsilon_d + 1}$. Another example is volume plasmon in metal: it has frequency ω_p, which does not depend on the permittivity ε_d.

The spectral density of resonances in two-dimensional LC–C–networks for different values of ε_d is shown in Figure 1. We can see that the number of peaks of the spectral density does not depend on ε_d (Figures 1a and 1b). Both spectra are coincide up to statistical fluctuations in terms of spectral variable μ (Figure 1c).

Matrices K^* and C^*_d are both negative semidefinite because they are diagonally dominant matrices with negative diagonal entries due to sum rule. We can rewrite Eq. (6) in the following form

$$\left|K^* - \frac{\mu}{1-\mu} C^*_d\right| = 0. \tag{9}$$

The generalized eigenvalue problem with negative semidefinite matrices has non-negative eigenvalues $\mu/(1-\mu) \geq 0$ so $0 \leq \mu \leq 1$. Therefore, all spectra have no resonant frequencies higher than ω_p, which is clearly seen in Figure 1.

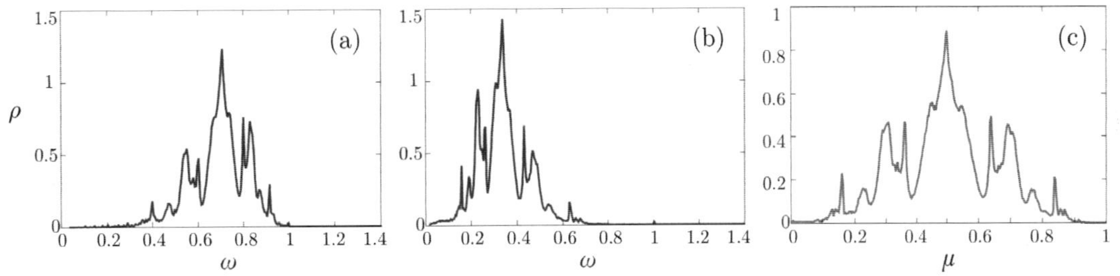

Figure 1. Spectral density of resonances $\rho(\omega)$ for random LC–C–networks with concentration of parallel circuits $p_{\mathrm{LC}} = 0.1$, $\omega_p = 1$ and permittivity of dielectric $\varepsilon_d = 1$ (a) and $\varepsilon_d = 5$ (b). In picture (c) both spectra are shown in terms of spectral variable μ and coincide up to statistical fluctuations. The results were calculated numerically for ensemble of 10^5 networks of size 20×20 sites with periodic boundary conditions in both directions.

3. Resonances in metal-semiconductor nanocomposites

Composite of metal and highly doped semiconductor can be considered as a mixture of two Drude-metals [5, 6] with different plasma frequencies ω_{p1} and ω_{p2}. Usually ω_{p1} lies in ultraviolet or visible range, while ω_{p2} lies in THz or infrared range. Such a system corresponds to a network with two types of LC – bonds with different resonant frequencies.

In this case matrices K and C in Eq. (2) are a sum of matrices for two metals $K = K_1 + K_2$, $C = C_{m1} + C_{m2}$ where

$$K_1 = \frac{a\omega_{p1}^2}{4\pi} K_1^*, \quad K_2 = \frac{a\omega_{p2}^2}{4\pi} K_2^*, \quad C_{m1} = \frac{a}{4\pi} C^*_{m1}, \quad C_{m1} = \frac{a}{4\pi} C^*_{m1}. \tag{10}$$

Thus, we come to the equation

$$|(\omega_{p1}^2 K_1^* + \omega_{p2}^2 K_2^*) - \omega^2 (C^*_{m1} + C^*_{m2})| = 0. \tag{11}$$

Matrices K_1^*, C^*_{m1} and K_2^*, C^*_{m2} are the same so

$$|(\omega_{p1}^2 - \omega^2) K_1^* + (\omega_{p2}^2 - \omega^2) K_2^*| = 0. \tag{12}$$

Finally, we come to a generalized eigenvalue problem

$$|K_1^* - \xi (K_1^* + K_2^*)| = 0 \tag{13}$$

with eigenvalues

$$\xi = \frac{1 - \omega^2/\omega_{p2}^2}{1 - \omega_{p1}^2/\omega_{p2}^2}. \tag{14}$$

One can note that the same expression for ξ can be obtained if permittivity of dielectric ε_d in Eq. (7) is replaced by the Drude-permittivity $\varepsilon(\omega) = 1 - \omega_{p2}^2/\omega^2$. Resonant spectrum in terms of spectral variable ξ is a function of concentration p and does not depend on values of ω_{p1} and ω_{p2}. One can note that the structure of problem (13) is equivalent to those of (6) so $0 \leq \xi \leq 1$, and all resonant frequencies of such networks take values in the range $\omega_{p2} \leq \omega \leq \omega_{p1}$. It is in agreement with results of numerical simulations shown in Figure 2. Physical interpretation is straightforward in the Drude-model: both permittivities are negative for $\omega < \omega_{p2}$ and there are no resonances in the composite; in region $\omega > \omega_{p1}$ both permittivities are positive and composite is transparent for incident radiation.

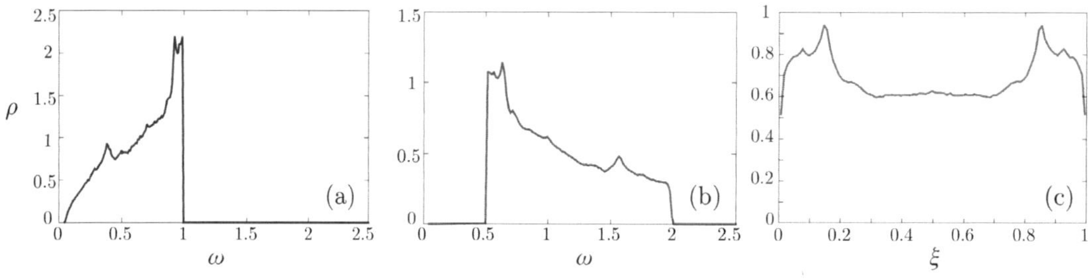

Figure 2. Spectral density of resonances $\rho(\omega)$ for random LC–C–networks (a) and LC–LC–networks (b), corresponding to metal-dielectric and metal-metal nanocomposites. In both cases $p = 0.5$ and networks are percolating. Picture (c) shows both spectums in terms of spectral variables μ and ξ respectively. The results were calculated numerically for ensemble of 10^5 networks of size 20×20 sites with periodic boundary conditions in both directions.

4. Mapping between different models

It is interesting to note that resonant spectrum of LC–C–model of metal-dielectric composite in terms of spectral variable μ as well as spectrum of LC–LC–model of metal-semiconductor composite in terms of variable ξ are coincide with spectrum of L–C–networks in terms of variable

$$\lambda = \frac{(\omega/\omega_p)^2}{1 + (\omega/\omega_p)^2}. \tag{15}$$

Indeed, if every bond of network is a single inductance or a single capacitor, problem (2) can be reduced to

$$|K^* - \lambda(K^* + C^*)| = 0, \tag{16}$$

where $C = C_d$ is a matrix of capacitances of dielectric. This formulation is directly maps with (6) because corresponding right and left matrices in these eigenvalue problems are equivalent. Detailed study of spectral properties of random L–C–networks is given in paper [3]. Thus, fundamental results of work [3], as well as results for networks with fluctuating entries [4] could be directly applied to a study of spectral properties of LC–C– and LC–LC–networks in terms of spectral variables μ and ξ.

Using mapping between problems (6) and (13) one can obtain the expression which connects the eigenfrequencies of a metal-semiconductor composite when semiconductor acts as dielectric and after it exhibits metal-dielectric transition (for example, after heating of an initial composite)

$$\omega = \omega_{p2}\sqrt{1 - \mu(1 - \omega_{p1}^2/\omega_{p2}^2)}, \tag{17}$$

where ω are a frequencies of eigenmodes of metal-semiconductor composite with semiconductor acting as metal and μ is a spectrum of composite with semiconductor acting as dielectric.

5. Conclusion

In present article we analytically derived expressions which describe dependence of resonant spectrum of disordered metal-dielectric and metal-semiconductor composites on matrix permittivity as well as transformation of resonant spectrum in the case of metal-dielectric transition in one of the components. These equations have a common property: both of them do not include information about geometry of the system in explicit way. All information about geometry manifests itself in initial resonant spectrum of the composite, i.e. spectrum, corresponding to initial matrix permittivity. Obtained formula also leads to correct expressions for some well-known cases.

Mapping between several types of network problems have been established. It takes place when special spectral variables are introduced. This makes possible a straightforward generalization of some properties of well-studied random L–C–networks to LC–C– and LC–LC–networks.

6. Acknowledgements

We are very grateful to Prof. V. I. Kozub for many fruitful discussions. NAO and YMB thank Dynasty Foundation for financial support.

References

[1] Sarychev A K, Shalaev V M 2000 *Physics Reports* **335** 275
[2] Clerc J P, Giraud G, Luck J M and Robin Th 1996 *J. Phys. A* **29** 4781
[3] Jonckheere Th and Luck J M 1998 *J. Phys. A* **31** 3687
[4] Olekhno N A, Beltukov Y M and Parshin D A 2014 *J. Phys. Conf. Ser.* **572** 012037
[5] Naik G V and Boltasseva A 2010 *Phys. Status Solidi RRL* **4** 295
[6] Martin van Exter, Grischkowsky D 1990, *Phys. Rev. B* **41**, 12140

THz radiation under non-central propagation of electrons through periodical channel

A A Ponomarenko[1], K V Lekomtsev[2], A A Tishchenko[1] and M N Strikhanov[1]

[1] National Research Nuclear University MEPhI, Moscow, Russia
[2] High Energy Accelerator Research Organization (KEK), Tsukuba, Japan

E-mail: AAPonomarenko@mephi.ru, tishchenko@mephi.ru

Abstract. Radiation from non-central electrons moving through channel with variable radius in the THz region is investigated using Particle In Cell (PIC) solver of the Computer Simulation Technology (CST) software package. Characteristics of radiation arising for non-central and central electrons propagation in the channel are compared, both for Smith-Purcell and Cherenkov kinds of radiation. It is demonstrated that the radiation is more intensive for the non-central propagation of electrons.

1. Introduction

There is a variety of possible schemes for generation of electromagnetic radiation in the THz (terahertz) frequency range. Powerful THz sources can be based on polarization radiation, when radiation occurs due to polarization of material by electrons Coulomb field. Using the channel in matter allows not to damage the target. Mechanism of Cherenkov radiation is mainly used for generation of powerful radiation [1], including THz spectral range. In our recent paper [2] we investigated source of THz radiation based on electrons moving through the channel with variable radius (see Fig.1), when radiation arises due to Cherenkov (CR) and Smith-Purcell (SPR) mechanisms. It was shown that SPR can be used as powerful source of THz radiation. Along with that CR and SPR permit to design comparatively compact THz radiation source.

In this work we consider electrons propagating at distance y from axis z of corrugated dielectric channel, see Fig.1. Theoretical calculations for the non-central propagation of electrons gives ambiguous expressions which are not that simple to analyse, and numerical simulation is of great use for preliminary simulation. In order to create powerful THz source in future experiment at LUCX facility in KEK in Japan [3,4], it is of use to investigate radiation under non-central propagation. For this purpose we shall perform numerical simulation using Particle In Cell (PIC) solver of Computer Simulation Technology (CST) software package.

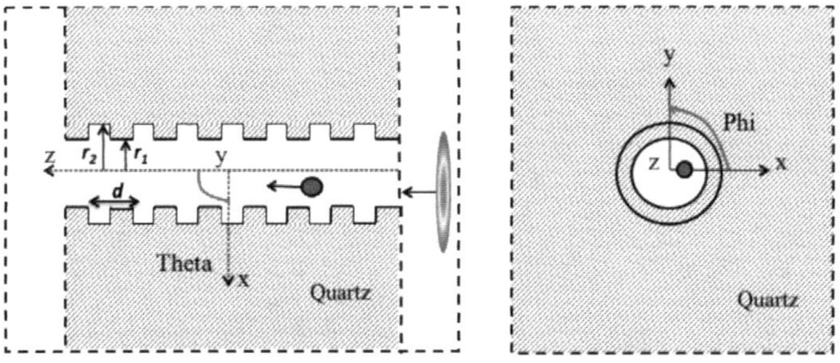

Figure 1. Scheme of THz source.

2. CST simulation of non-central propagation

In this section simulations of radiation emitted from non-central propagation through periodical channel in quartz will be discussed. Particle In Cell (PIC) solver of Computer Simulation Technology (CST) Particle Studio software calculates the development of the fields generated by a charged particle at discrete time samples. The fields are estimated at discrete points whereas the particles are tracked in continuous phase space. Values of the electric field are calculated at each discretization point in the domain, and the electric field values at the outer borders of the calculation domain are extrapolated into the wave zone [5] (which we can call also far-field zone), using the method described in [6].
Parameters of electron bunch and corrugated channel in material are shown in Table 1. Parameters in Table 1 were chosen to correspond to the experiment at LUCX. Internal radius r_1 was chosen to be larger in order to accommodate transversely larger bunch, propagating with an offset through the corrugated channel. Groove period and depth are chosen so that (-1) order of SPR corresponding to 300GHz is observed at the angle Theta = 89.94 deg.

Parameter	Value
Bunch Lorentz-factor (γ)	16
Bunch charge	0.1 nC
Longitudinal charge distr. $\sigma_{long} = l_b$	0.1mm
Transverse charge distr. $\sigma_{transv} = r_b$	0.05 mm
Frequency	300GHz
Wavelength	1mm
Material	Quartz
Dielectric permittivity	$\varepsilon = 3.75$
Internal radius, r_1	1mm
Groove depth, $a = r_2 - r_1$	0.3mm
Number of periods	15
Groove width, l	0.75 mm
Grating period, d	1.5 mm

Table 1. Simulation parameters.

In order to compare CST simulation with future analitical formulas or experimental dates it is of great interest to ivestigate the radiation distribution in far-field zone. To assume far-field zone, the propagation distance in free space between the target and observation point must satisfy the condition $L \gg \gamma^2\lambda$, L must be much larger than the formation length. Fig. 2 is plotted using data from Table 1.

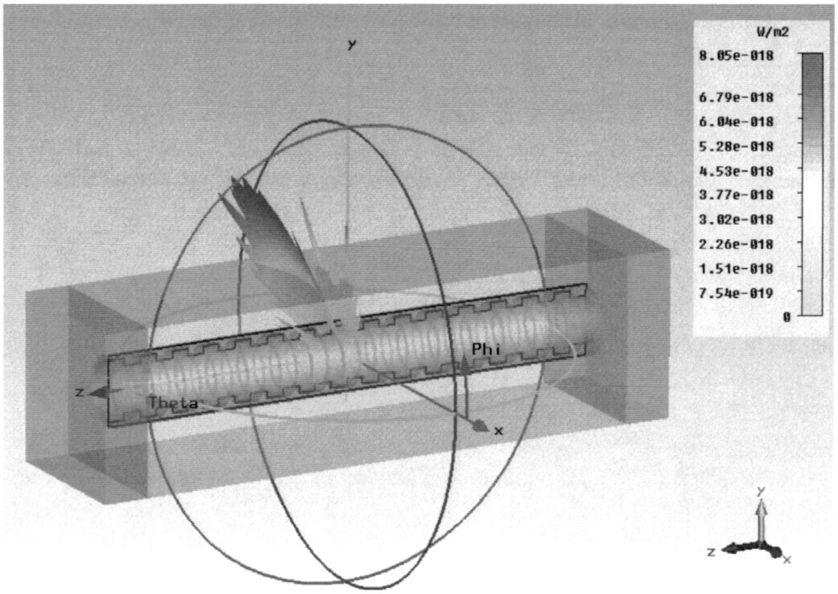

Figure 2. Radiation in the wave zone for the bunch position $y = 0.5\ mm$.

From Fig. 2 one may conclude that for $y = 0.5\ mm$ the radiation directivity is rather high, which is important from experimental point of view.

Now let us investigate and compare intensities for noncentral and central propagation of electrons for different bunch position (distances between bunch and axis y).

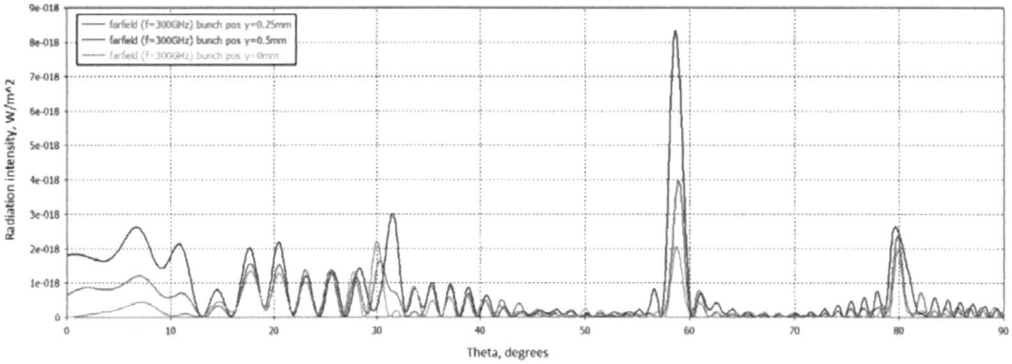

Figure 3. Dependence intensity on theta for different bunch positions.

As one can see from Fig. 3, the spectral-angular dependence on theta has several maxima, which correspond to SPR peaks for different m, $m = \frac{d}{\lambda}\left(\left(\beta\sqrt{\varepsilon}\right)^{-1} - \cos\theta\right)$ – SPR dispersion relation and the Cherenkov radiation peak for $m = 0$. Another interesting fact: if we increase the distance between bunch and axis, the radiation intensity also changes, but dependence intensity on bunch position y is not linear. The spectral-angular distribution leads to the maximum value for $y = 0.5\ mm$, which corresponds to CR, and its contribution is more pronounced, that that of SPR. It is important to note that situation with comparison SPR and CR is not simple, and intensity of radiation strongly depends on bunch and target parameters [7].

Now we shall move on to the investigation of CR radiation in case of non-central propagation, because for simulation parameters from Table 1 CR is more intensive then SPR. Fig. 4 demonstrates the polar plots of the CR for the five values of the bunch position.

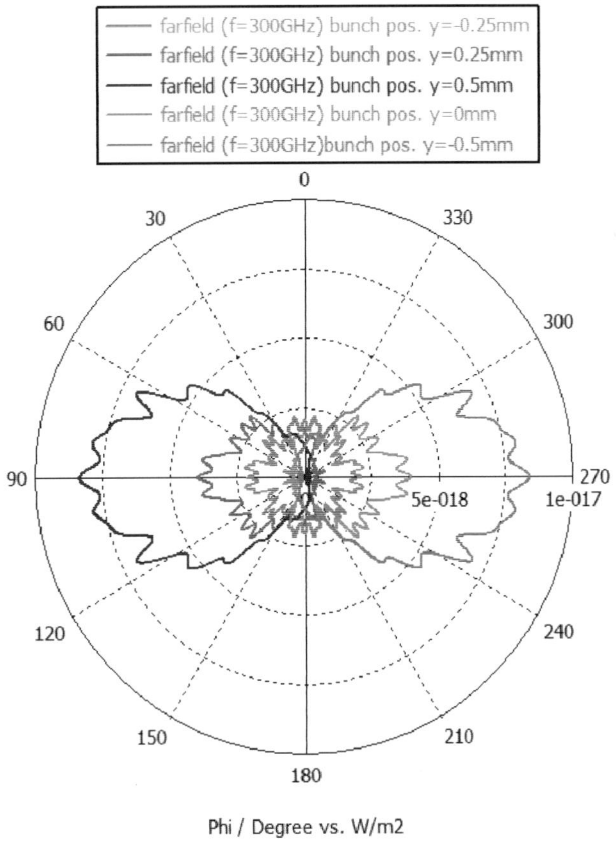

Figure 4. Polar plots for different bunch position.

We can see that the radiation is more intensive for the greater values of bunch position y. On the one hand, investigation of non-central propagation of electrons through corrugated channel can be useful for describing future experiment in LUCX facility in KEK. On the other hand, the problem of calculation of radiation is not that simple, and numerical optimization can be of great use. It means non-central propagation added one extra parameter, which should be optimized.

3. Conclusion

Non-central propagation of bunch of electrons through the channel shows two interesting features. First, without obtaining analytical expressions and corresponding optimization Cherenkov radiation can be more intensive as a THz source. Along with that the contribution of Smith-Purcell radiation seems to be comparable with CR. Radiation from non-central electrons moving through channel is proved to be more intensive, which is important from experimental point of view.

ACKNOWLEDGEMENTS

The work was partially supported by the Ministry of Education and Science of the Russian Federation, grant № 3.1110.2014/K.

References

[1] Zrelov V P Vavilov-Cherenkov Radiation and its Applications in High Energy Physics, Atomizdat, 1968 (in Russian).

[2] Ponomarenko A A, Ryazanov M I, Strikhanov M N and Tishchenko A A Terahertz radiation from electrons moving through a waveguide with variable radius, based on Smith- Purcell and Cherenkov mechanisms 2013 Nucl. Instrum. and Meth. B **309** 223.
[3] Fukuda M, Araki S, Deshpande A, Higashi Y, Honda Y, Sakaue K, Sasao N, Taniguchi T, Terunuma N and Urakawa J, Upgrade of the accelerator for the laser Undulator compact X-ray source (LUCX) 2011 Nucl. Instrum. and Meth. **637** 67.
[4] Aryshev A, Araki S, Fukuda M, Karataev P, Konkov A, Naumenko G, Potylitsyn A, Sakaue K, Sukhikh L, Terunuma N, Verigin D and Urakawa J, Observation of the stimulated coherent diffraction radiation in an open resonator at LUCX facility 2014 Nucl. Instrum. and Meth. A **763** 424.
[5] Potylitsyn A P, Ryazanov M I, Strikhanov M N, Tishchenko A A Diffraction Radiation from Relativistic Particles, Springer-Verlag; 2011.
[6] K.S. Yee, D. Ingham and K. Shlager, Time-Domain Extrapolation to the far field based on FDTD calculations, 1991 IEEE Trans. of Ant. and Prop. 39, 410.
[7] Ponomarenko A A, Lekomtsev K V, Tishchenko A A, Strikhanov M N and Urakawa J, CST simulation of THz radiation from a channel with periodically variable radius, Nucl. Instr. and Meth. B (2015) in print (doi:10.1016/j.nimb.2015.03.004).

Spectral correlations of phase modulated entangled photons

Sheremetyev V.O., Trifanov A.I.
ITMO University, Kronverksky, 49, Saint- Petersburg, 197101, Russia
E-mail: vlad.sheremetiev@gmail.com, alextrifanov@gmail.com

Abstract. We investigate spectral properties of entangled photons in EPR-pair, which interacts with phase modulators. In three mode approximation we got the exact analytical expressions for their common state evolution and calculated probabilities of coincidence between detectors clicks.

1. Introduction

Phase modulation (PM) process of quantized radiation have being actively investigated in last decades due to the development of optical communication systems. Quantum transmission lines require the possibility to control and shape wave packets at the single-photon level. Classical theory of modulation does not fit for this problem since it operates with infinity number of photons. A new theory of electromagnetic field modulation process was needed.

The foundation in this area was formed by works [1], [2], [3] and [4], where different approaches were suggested. It was shown in [5] that PM allows the complete control of single-photon wave packets. Nonclassical correlations between spectral modes of modulated wave packets is now at the most interest. Nonlocal spectral properties of entangled photon pairs were investigated in [6, 7]. In [8] the protocol of joint spectrum tunable control was suggested. Experimental study of group velocity dispersion effect on an entangled two-photon wave packet was reported in [9]. Statistical properties of field quantum state interacted with PM were studied in detail in [4]

Here using effective Hamiltonian method we investigate analytically modulation process of entangled photons in the system similar to that was studied in [2]. Restricted by three mode approximation we analyze probabilities of single and double clicks of detectors, monitoring single photon wave packets at each of three choosing modes. In our scheme polarization entangled photons (with horizontal and vertical states of polarizations) are generated due to spontaneous down-conversion process (see fig.(1)) and transmitted to the detector part with optical waveguides in use. Passing through these waveguides photons interacts with PM's PM_1 and PM_2 setting in each channel. Just after this interaction the photons are detected. The aim of this work is to obtain analytical formulas for joint probabilities of detectors clicks and calculate conditional states of modes in one channel using the measurement data from another.

2. The model of PM

According to [4] Hamiltonian of phase modulation process with phase φ may be written as

$$H = \omega_{opt} N + \Omega A_0 + \frac{2\gamma}{3}\left(e^{-i\varphi}e^{-i\tilde{\Omega}t}A_+ + e^{i\varphi}e^{i\tilde{\Omega}t}A_-\right), \tag{1}$$

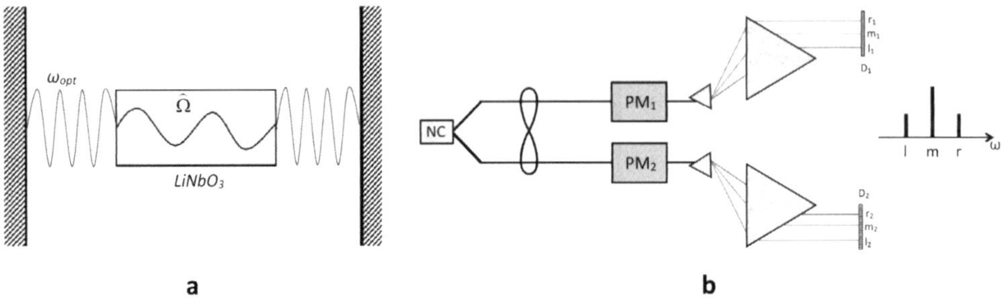

Figure 1. a Scheme of optical phase modulation process in the cavity; **b** Nonlocal phase modulation of polarization-entangled photons pair generated by nonlinear crystal (NC). Each photon is transmitted by optical waveguide and interacts with phase modulator. After that the probability of detectors clicks coincidence is monitored

where ω_{opt} is the frequency of modulated cavity field, $\tilde{\Omega}$ is the modulation frequency, Ω is frequency difference between adjacent optical modes and γ is mode interaction parameter. The operators A_+, A_-, A_0 and N are defined as

$$A_+ = \sum_{m=-1}^{0} f(m)\, a_m a_{m+1}^\dagger, \quad A_- = \sum_{m=-1}^{0} f(m)\, a_m^\dagger a_{m+1}, \tag{2}$$

$$A_0 = \sum_{m=-1}^{1} m a_m^\dagger a_m, \quad N = \sum_{m=-1}^{1} a_m^\dagger a_m. \tag{3}$$

Here $m \in \{-1, 0, 1\}$ is the index of mode relative to the center the cavity spectrum (with mode number $\tilde{m}$), $f(m)$ isoverlap function of the interacted modes. For our three mode approximation we assumed $f(m) = \sqrt{(2+m)(1-m)}$, which allows to A_+, A_-, A_0 to be generators of $SU(2)$ algebra:

$$[A_0, A_\pm] = A_\pm, \quad [A_+, A_-] = 2A_0. \tag{4}$$

Also it is worthwhile to note that N commute with all these generators, which follows form the fact that spectrum of N is degenerate. Then evolution operator corresponded to Hamiltonian (1) may has the form

$$U(t) = e^{\left(-i\left(\tilde{\Omega}t+\varphi\right)(\tilde{m}N+A_0)\right)} e^{-iQt} e^{i\varphi(\tilde{m}N+A_0)}, \tag{5}$$

where Q is the operator of "quasienergy"

$$Q = \omega\left(\tilde{m}N + A_0\right) + \frac{2\gamma}{3}\left(A_+ + A_-\right), \tag{6}$$

and $\omega = \Omega - \tilde{\Omega}$. Operator Q may be diagonalized by unitary transformation U_0

$$U_0 Q U_0^\dagger = 2\Gamma A_0 + \omega\tilde{m}N, \tag{7}$$

$$U_0 = exp\left(-\frac{\pi}{4}\left(A^\dagger - A^-\right)\right), \tag{8}$$

where we use resonance condition $\omega = 0$ and $\Gamma = \left|\frac{2\gamma}{3}\right|$.

3. Phase modulation of entangled state

Here it is assumed EPR state of photons polarization at the input of the modulation scheme. Denoting horizontal and vertical state of polarization by H and V correspondingly this state may by written as

$$|\Psi_0\rangle = \frac{1}{\sqrt{2}}[|010\rangle_{1H}|000\rangle_{1V}|000\rangle_{2H}|010\rangle_{2V} + |000\rangle_{1H}|010\rangle_{1V}|010\rangle_{2H}|000\rangle_{2V}], \quad (9)$$

where $|ijk\rangle_{n,\alpha} = |i\rangle_l |j\rangle_m |k\rangle_r$, $i, j, k \in \{0, 1\}$, l, m, and r are labels of modes corresponded to indexes $-1(l)$, $0(m)$ and $1(r)$, $n \in \{1, 2\}$ is number of the channel and $\alpha \in \{H, V\}$ is state of polarization.

State of the photon pair after interaction with PM's is taken by the formula

$$|\Psi_{PM}\rangle = U_2(t)\,U_1(t)\,|\Psi_0\rangle = \frac{1}{\sqrt{2}}[|000\rangle_{2H}|\Phi\rangle_{2V}|\Phi\rangle_{1H}|000\rangle_{1V} + |\Phi\rangle_{2H}|000\rangle_{2V}|000\rangle_{1H}|\widetilde{\Phi}\rangle_{1V}]. \quad (10)$$

Here $U_n(t)$ is evolution operator for n-th phase modulator and

$$|\Phi\rangle_{n,\alpha} = U(t)|010\rangle_{n,\alpha} = \left(ie^{-i\tilde{\Omega}t}e^{i\varphi_n}\sin(2\Gamma t)\right)|100\rangle_{n,\alpha} + \left(\sqrt{2}e^{-2i\tilde{\Omega}t}\cos(2\Gamma t)\right)|010\rangle_{n,\alpha} + \left(ie^{-3i\tilde{\Omega}t}e^{-i\varphi_n}\sin(2\Gamma t)\right)|001\rangle_{n,\alpha}. \quad (11)$$

Thus probabilities of detection photon in certain mode are given by

$$P_l = P_r = |\sin(2\Gamma t)|^2, \quad P_m = 2|\cos(2\Gamma t)|^2. \quad (12)$$

4. Joint probabilities of detection events

In Table 1 we present correlations between detectors clicks for each pair of modes in two channels. Namely, probability to get the clicks from detectors D_x and D_y, where $x \in \{l_1, m_1, r_1\}$ and $y \in \{l_2, m_2, r_2\}$ may be calculates as

$$P_{x,y} = \left\langle a_x^\dagger a_x a_y^\dagger a_y \right\rangle = \langle\Psi_{PM}|\, a_x^\dagger a_x a_y^\dagger a_y\, |\Psi_{PM}\rangle. \quad (13)$$

For instance, if detector D_{l_1} clicks in the first channel, one can obtain probability distribution to get detector click in the second channel. This distribution showed in the first column of Table 1. These probabilities as a functions of time is presented on fig. 2.

	l_1	m_1	r_1
l_2	$\frac{1}{4}\sin^4(2\Gamma t)$	$\frac{1}{2}\sin^2(2\Gamma t)\cos^2(2\Gamma t)$	$\frac{1}{4}\sin^4(2\Gamma t)$
m_2	$\frac{1}{2}\sin^2(2\Gamma t)\cos^2(2\Gamma t)$	$\cos^4(2\Gamma t)$	$\frac{1}{2}\sin^2(2\Gamma t)\cos^2(2\Gamma t)$
r_2	$\frac{1}{4}\sin^4(2\Gamma t)$	$\frac{1}{2}\sin^2(2\Gamma t)\cos^2(2\Gamma t)$	$\frac{1}{4}\sin^4(2\Gamma t)$

Table 1. Matrix of joint probabilities for D_x and D_y detectors clicks

In particular, one can find from here that conditional probabilities to detect photon in sideband modes r and l are equal which allow to determine if another part of EPR-pair is detected in central mode or in one of the sideband modes.

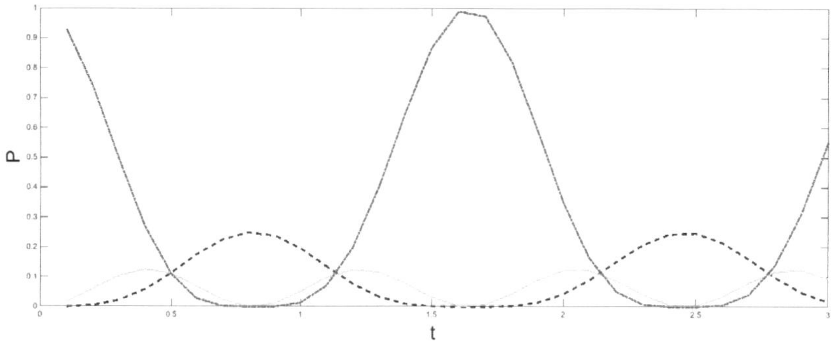

Figure 2. Joint probabilities as functions of time: solid green line is for $P_{m,l} = P_{m,r} = P_{l,m} = P_{r,m}$, dashed blue line is for $P_{l,l} = P_{r,l} = P_{l,r} = P_{r,r}$ and dashed-pointed red line is for $P_{m,m}$

5. Conclusion

In the present work we investigated analytically process of phase modulation of entangled photons. Using effective Hamiltonian approach and three mode approximation we studied conditional state of modes and spectral properties of one photon after detection its counterpart in another channel. We found that the scheme under consideration in some cases does not allow separation between sidebands. There are correlations between spectra of two entangled photons, but using detection result from one channel one can't separate statistical which mode was detected at another channel. Also we found that proposed measurement scheme drops out phase of modulator and phase difference between two modulators. Spectral correlations of two entangled photons may likely used for quantum key distribution protocol, but it requires more improved measurement scheme.

Acknowledgments

This work was partially financially supported by the Government of the Russian Federation (grant 074-U01), by the Ministry of Education and Science of the Russian Federation (GOSZADANIE 2014/190, Projects No 14.Z50.31.0031 and No. 1.754.2014/K)), by grant MK-5001.2015.1 of the President of the Russian Federation, by ITMO University from centralized resources (grant 415829).

References

[1] Pegg D T and Barnett S M 1989 *Phys. Rev. A* **39** 1665-1675
[2] Kolchin P, et. al. 2008 *Phys. Rev. Lett.* **101**, 103601
[3] Franson J D 1992 *Phys. Rev. A* **45**, 5
[4] Miroshnichenko G P 2015 Algebraic theory of an electro-optic modulator arXiv:1504.01632
[5] Specht H P 2009 *NATURE PHOTONICS Letters* **3**
[6] Harris S E 2008 *Phys. Rev A* **78** 021807(R)
[7] Sensarn S 2009 *Phys. Rev. Letters* **103** 163601
[8] Valencia A 2007 *Phys. Rev. Letters* **99** 243601
[9] Valencia A 2002 *Phys. Rev. Letters* **88** 183601
[10] Bellini M 2003 *Phys. Rev. Letters* **90** 043602
[11] Khan I A 2006 *Phys. Rev. A* **73** 031801(R)

The characteristics of semiconductor-to-metal transition in VO_2 of different morphology

Petukhova Yu.V., Osmolowskaya O.M., Osmolowsky M.G.

Saint Petersburg State University, Institute of chemistry, 198504, Russia, Saint Petersburg, Petrodvorets, Universitetsky pr. 26

Yulia.petukh.ova@yandex.ru

Abstract. Vanadium dioxide nanoparticles (NPs) of different morphologies were obtained using the hydrothermal technique. Their shape and size were studied by SEM, XRD and SSA estimation. The functional properties, structural changes and thermal behavior of samples obtained were investigated to clarify the SMPT peculiarities. It is shown that the introduction of a doping element changes a mechanism of the nanoparticles growth and so that the SMPT becomes less expressed. The detailed study of SMPT in undoped VO_2 showed the steps of the transition process. The testing of VO_2 coating on glass for «smart» windows was successfully performed.

1. Introduction

Bulk vanadium dioxide undergoes a semiconductor-to-metal phase transition (SMPT) at 340 K resulting in an abrupt change of magnetic, electrical and optical properties. Due to this fact, VO_2 becomes a perspective material for optical switchers and IR sensors [1], «smart» covers for windows [2], electrical switchers and elements of energy-independent memory [3]. What is more, vanadium dioxide has a rutile structure so that the different ions are able to intercalate easily into the lattice. Therefore, this material can be used as an electrode material for Li-ions batteries [4].

Nowadays powders of vanadium dioxide are mainly synthesized in hydrothermal route because well-crystallized nanoparticles with different morphologies can be obtained by variation of the reaction conditions [5 - 7].

It was already discussed that the dimension of the material is a determinant factor for expression of the SMPT characteristics. [8] The VO_2 thin films show weak change in a magnetic susceptibility after transition so it is needed to be verified if this tendency of size reduction influence on the SMPT will repeat for nanoparticles.

It is known that application of VO_2 depends on characteristics of the transition which in its turn is effected by VO_2 morphology. As we have shown before, the temperature of the SMPT can be lowered by doping [9]. But there is one more way to regulate morphology of the product obtained. The introduction of a chelating agent such as etidronic acid leads to change in a growth mechanism and thereby influences on particles size and shape.

So the hydrothermal synthesis of undoped and Fe and Cr-doped VO_2 particles in presence and in absence of the chelating agent are reported. The dependence of their physico-chemical and functional characteristics on oxide morphology is shown.

2. Experimental

VO_2 powders were synthesized from $H_2C_2O_4$ and V_2O_5. The yellow suspension in distilled water was mixed on a magnetic stirrer at a constant temperature 60°C. Over this, the solution became homogeneous and colored dark blue. On this step solutions of the doping ions and the chelating agent were added if necessary. The solution was transferred into a 500 ml teflon-lined autoclave. After hydrothermal synthesis the product was obtained as a black precipitate which was transferred to centrifuge tubes of 50 ml and washed several times with distilled water. Then the product was placed in an oven and dried at 80°C for at least 12 hours.

The morphology of powders obtained was analyzed by XRD, SEM and SSA estimation. X-ray powder diffraction was performed on the Bruker "D2 Phaser" X-ray diffractometer with CuKα radiation. Particles form and size were characterized by scanning electron microscope Hitachi S3400N. Specific surface area was determined by the Brunauer-Emmett-Teller metod on a surface area analyzer Micromeritics ASAP-2020MP. The temperature dependence of conductivity was determined using an electrical impedance spectroscopy (EIS) in a cell under normal pressure, in the temperature range of -160 to 165 °C with step of 20°C. The Novocontrol Concept 40 impedance analyzer was used in the frequency range of 10 to 10^7 Hz. The magnetization was measured on the vibrating magnetometer Lake Shore 7410 in the auxiliary magnetic field of 15 kOe and the temperature range of -196 to 127°C.The detailed investigation of SMPT in undoped VO_2 was conducted by DSC and XRD at temperatures ranges 7 – 400 K. The IR transmittance of VO_2 coating on glass at temperature range 279 – 333 K was studied by IR spectroscopy.

3. Results and discussion

3.1. The regulation of NPs morphology by variation of synthesis conditions

The results of XRD showed that the powders obtained contain no other phase of any vanadium oxides except VO_2.

By analysis of SEM images it is shown that the particles shape changes from ribbons and microplates obtained before with large amounts of dopant [9] to carambola-like and sponge-like spheres with the chelating agent (Figure 1) and smooth hollow spheres with one (Figure 2). The tendency of nanoparticle formation both for Fe and Cr-doped samples is mostly the same.

Figure 1. SEM image of Fe-doped vanadium dioxide without etidronic acid: carambola-like and sponge-like spheres morphology.

Figure 2. SEM image of Fe-doped vanadium dioxide with etidronic acid: smooth hollow spheres morphology.

An undoped VO_2 synthesized without the chelating agent represents rods with 80 nm thick and 1.5 μm length (Figure 3).

Figure 3. SEM image of undoped vanadium dioxide: rods morphology.

The results of particles size calculated by SEM correspond to values of specific surface area.

3.2. The dependence of SMPT characteristics on the morphology of VO_2

From the magnetic and electrical properties of doped vanadium dioxide it is found that the SMPT become less expressed and the temperature of the transition decreases to 200 K with the chelating agent and 150 K without one compared to the bulk material. There is no difference appeared between samples with varied amount of doping element which means that the characteristics of SMPT depend on morphology of the sample but not the quantity of dopant. The shape of doped samples makes them perspective for Li-ion batteries, because spheres are supposed to effectively interact with a matrix substance and distribute in it more evenly.

The detailed study of SMPT in undoped VO_2 has shown the abrupt change in magnetic characteristics near 280 K and the crystalline structure modification near room temperature. The thermal effects of transition and decrease in IR-transmittance were observed at about 310 K. It means that the transition consists of several steps appearing on different levels of structural organization. Thereby the SMPT takes place in temperature range which is convenient for "smart" cover application. The decrease in IR transmittance of VO_2 coating on glass above the room temperature was shown.

All conclusions about the transition are summarized in a Table 1. The SMPT peculiarities revealed require more detailed investigation.

Table 1. The dependence of SMPT characteristics on morphology of vanadium dioxide.

	Presence of the chelating agent	Morphology	Particle size, μm	T_{SMPT}, K	Expressiveness of SMPT	Application
Fe/Cr-doped VO_2	No	Carambola-like and sponge-like spheres	2–5	150	weak	Li-ion batteries
Fe/Cr-doped VO_2	Yes	Smooth hollow spheres	2–5	200	weak	Li-ion batteries
Undoped VO_2	No	Rods	0.08×1.50	280	strong	"Smart" covers, termosensors

Acknowledgements

Scientific researches were performed at the Centre of X-ray Diffraction Studies, Centre for Geo-Environmental Research and Modelling (GEOMODEL), Innovate Technologies of Composite Nanomaterials, Center of Thermal Analysis and Calorimetry and Center of Chemical Analysis and Materials Research of St. Petersburg State University.

References

[1] Chen X. and Lv Q. Theoretical analysis on the switching speed and driving power dissipation for optical switch based on VO2 thin film 2013 *Optik – Int. J. for Light and Electron Optics* **124** 579–82

[2] Huang Z., Chen C., Lv C. and Chen S. Tungsten-doped vanadium dioxide thin films on borosilicate glass for smart window application 2013 *J. of Alloys and Compounds* **564** 158–61

[3] Velichko A., Pergament A., Putrolaynen V., Berezina O. and Stefanovich G. Effect of memory electrical switching in metal/vanadium oxide/silicon structures with VO_2 films obtained by the sol–gel method 2015 *Materials Science in Semiconductor Processing* **29** 315–20

[4] Nethravathi C., Viswanath B., Michael J. and Rajamath M. Hydrothermal synthesis of a monoclinic VO2 nanotube–graphene hybrid for use as cathode material in lithium ion batteries 2012 *Carbon* **50** 4839–46

[5] Yina H., Yua K., Zhanga Z. and Zhua Z. Morphology-control of VO2 (B) nanostructures in hydrothermal synthesis and their field emission properties 2011 *Appl. Surface Science* **257** 8840–45

[6] Zhang Y., Chen C., Wu W., Niu F., Liu X., Zhong Y., Cao Y., Liu X. and Huang C. Facile hydrothermal synthesis of vanadium oxides nanobelts by ethanol reduction of peroxovanadium complexes 2013 *Ceramics Int.* **39** 129–41

[7] Li R. and Liu C.-Y. VO2(B) nanospheres: Hydrothermal synthesis and electrochemical properties 2010 *Materials Research Bulletin* **45** 688–92

[8] Osmolovskaya O.M., Murin I.V., Smirnov V.M. and Osmolovsky M.G. Synthesis of vanadium dioxide thin films and nanopowders: a brief review 2014 *Reviews on advanced material science* **36** 70–74

[9] Petukhova Yu.V., Osmolowskaya O.M., Fedorova A.V. and Osmolowsky M.G. Magnetic behavior of doped VO2 nanoparticles 2014 *Bulletin of the Russian Academy of Science: Physics* **78** 325–27

Modelling of field emitter array characteristics in diode system

K A Nikiforov and M F Sayfullin

Saint-Petersburg State University, 7-9, Universitetskaya nab., St. Petersburg, 199034, Russia

E-mail: knikiforov@cc.spbu.ru

Abstract. In terms of a mathematical model, this paper studies features of the current characteristics of a thin-film matrix field emission array cathode with vertical blade-shaped emitters. Microscopical distribution of electric field strength was calculated and its dependence on geometric parameters of the structures was defined.

1. Introduction

This study considers a thin-film matrix field emission cathode with vertical blade-shaped emitters [1, 2]. Such vertical blade-like field emission matrices are suitable for many applications: an emitter with circular blade-like margin was regarded as a prospective element of a vacuum nanotriode [3], a part of electron guns [4], with consideration of different production technologies [5], [6]. As such, modeling of such structures if of current interest [7−17].

Efficiency of a matrix field emission cathode is defined by the value of emission current density. The current density grows with increase of the electric field strength and area of emitting surface. However, increase of the latter via curvature radius of the emitter top does not necessarily lead to increase in emission current, as the field strength decreases in more blunt emitter margin. The goal of this study is thus defining the dependence between the emission current and the curvature radius of each emitter in the array in order to increase the efficiency of a cathode.

2. Mathematical model

In order to achieve the given goal, it is necessary to solve the problem of finding out the distribution of electrostatic potential in a bi-electrode system comprising of a cathode matrix and an anode. The surface of the matrix and the emitters were modeled as an ideal conductor with no adsorbed insulators. Metal emitters shaped as hollow round cylinders with several nanometers thick walls and sharp top margins are situated in the nodes of square matrix with a step of 5 μm on a flat substrate of highly doped silicon (Figure 1). Distance between the emitters is 2.3 μm, emitter hight is 0.2 μm, curvature radius of the cylinders is varied between 1 and 20 nm (the technology of their production allows to set those values independently).

The influence of the spatial charge was neglected. We used 3d Laplace equation for electrostatic potential φ with boundary conditions corresponding to the Dirichlet conditions of the equipotential surfaces of the cathode Γ_1, anode Γ_2 and the spherical surface Γ_3 around the computation area that models the behavior of the potential at the point at infinity.

$$\Delta\varphi = 0, \tag{1}$$

$$\varphi|_{\Gamma_1} = 0, \varphi|_{\Gamma_2} = V, \varphi|_{\Gamma_3} = 0. \tag{2}$$

The boundary problem (1), (2) can only be solved numerically. But for a singular emitter an analytical method for calculation of an electrostatic potential of diode system also exists [18]. This analytical method is not applicable to a matrix cathode, because it has mutual influence between different emitters that is absent at the open boundary of the cathode but influences homogeneity of emission throughout the array.

In this study an approximate solution for the boundary problem (1), (2) is found by the method of finite elements in form of a linear combination of quadratic basis functions using Comsol Multiphysics. Discretization parameters of adaptive tetrahedral finite-element lattice are chosen in a way that for a singular emitter there is a correspondence between numerical solution and the result of analytical computation by the method of pairing equations with reducing the boundary problem to a second-order Fredholm integral equation with quadrature right-hand side and the core represented explicitly [18].

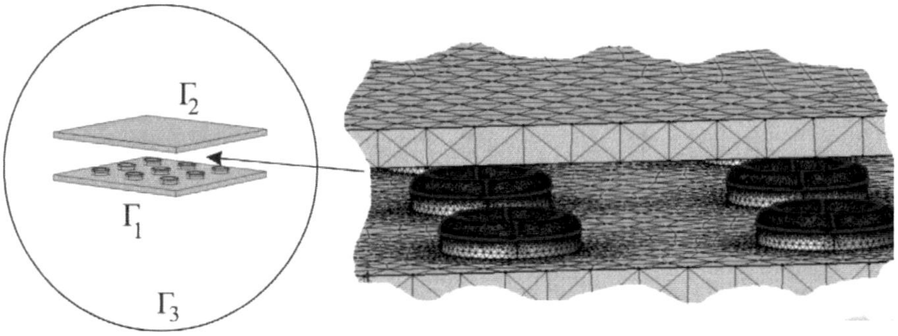

Figure 1. Computation area and depiction of a fragment of a matrix cathode discretized via finite element method.

3. Results

Software implementation of the model in Comsol Multiphysics with terminal-based input of geometrical parameters (that allowed to conduct computational experiments) takes in account the specifics of emission systems in which the computational area includes the boundaries of an emitter with highly curved surface and small size. This leads to a significant variance of characteristic measurements in the same geometric configuration, while the exponential dependence of current density on field strength requires particularly strict precision of defining the boundary conditions on the emitter. The solution has a rapidly changing gradient in the emission area (i.e. the blade margin), and thus the finite-element lattice grew more dense in the neighborhood of emitter blade so that the rate of convergence of the solution towards the exact solution would not significantly decrease and the number of unknown variables (i.e. dimensions of the finite-element linear algebraic system) would not grow.

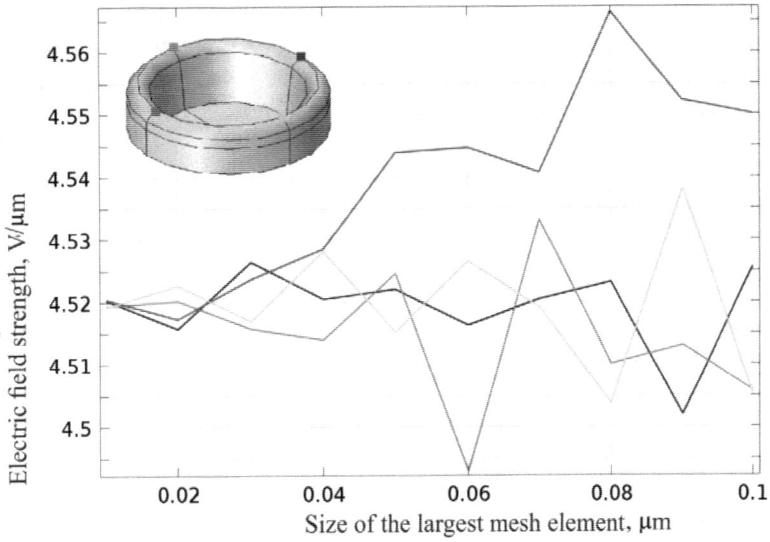

Figure 2. Field strength at the margin of one emitter in four symmetrically defined points (marked by dots colors of which correspond to colors of the plots) relative the size of the largest mesh element.

Figure 2 shows th calculated values for field strength in certain points on emitter margin with different levels of discretization. One can see that there is a definite upper bound for the lattice element size below which the symmetrical properties of the solution are conserved and there's reasonable convergence of the finite-element solutions in embedded lattices.

The current density was calculated by the Fowler-Nordheim equation (which does not influence the qualitative conclusions) and integrating of the current density for obtaining total current was performed numerically. As a result we've constructed a dependence between current characteristics and curvature radius of the emitter margin shown in figure 3.

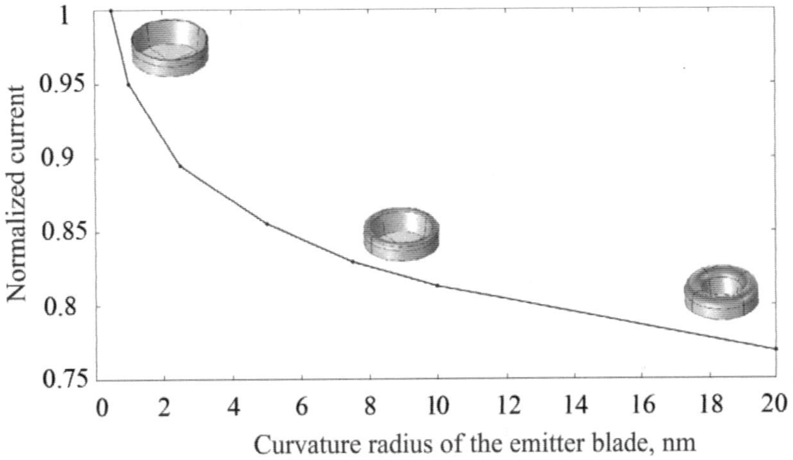

Figure 3. Influence of the curvature radius of the emitter blade on the emission current (on normalized values).

From figure 3 it is easy to see that in given range of the values of curvature radius of emitter margin its influence on the emission current is monotonous: the sharper the margin is, the higher the field strength and the emission current density, even though the emission area might decrease.

4. Conclusion

This paper studies the microscopic distribution of an electric field along the surface of a cathode matrix with cylindrical blade emitters in a flat diode configuration. The current characteristics and their influence on the geometric parameters were also assessed.

Acknowledgments

Financial support has been obtained from RFBR (13-01-00150) and partially from Saint-Petersburg State University (9.38.673.2013). Research was carried out using computational resources provided by Resource Center "Computer Center of SPbU" (http://cc.spbu.ru/en) and using experimental equipment of the Interdisciplinary Resource Center for Nanotechnology of St. Petersburg State University.

[1] Sayfullin M F and Nikiforov K A 2014 Studying fringe field effect of a field emitter array *Journal of Physics: Conference Series* **541** 012020

[2] Nikiforov K A, Antonova L I, Egorov N V, Trofimov V V, Makeev V V and Ogurtsov O F 2012 Non-gated field emission array as low-energy electron source: Experiment and simulation *RuPAC 2012 Contributions to the Proc. of 23 Russian Particle Accelerator Conf. (St. Petersburg)*, (Cern: JACoW) 218–20

[3] Hsu D and Grey H 1999 *Appl. Phys. Lett.* **75** 2497–9

[4] Khursheed A. 2013 *Ultramicroscopy* **128** 10–23

[5] Hsu D and Grey H 1995 Thin-film edge field emitter device and method of manufacture therefore *Pat. USA* 5382185

[6] Cheng H C, Wang C C, Ku T K and Hsieh L J 1995 *Jpn. J. Appl. Phys.* **35** 308–12

[7] Nikiforov K A and Egorov N V 2012 Program complex for vacuum nanoelectronics finite element simulations *RuPAC 2012 Contributions to the Proc. of 23 Russian Particle Accelerator Conf. (St. Petersburg)*, (Cern: JACoW) 409–11

[8] Sayfullin M and Nikiforov K 2014 Fringe field effect study of a field emitter array *2014 Tenth International Vacuum Electron Sources Conference, IVESC (Saint Petersburg)* (Piscataway: IEEE) 6892072

[9] Nikiforov K and Andrievskiy D 2014 Comparison of injection algorithms for electron beam simulation by particle-in-cell method *2014 20th International Workshop on Beam Dynamics and Optimization, BDO (Saint Petersburg)* (Piscataway: IEEE) 6890057

[10] Nikiforov K 2014 Modelling of emission processes in Matlab *International Conference on Computer Technologies in Physical and Engineering Applications, ICCTPEA (Saint Petersburg)* (Piscataway: IEEE) 6893317

[11] Gallyamov Z R and Nikiforov K A 2014 Model of electron transport in cell of a thin-film vacuum nanotriode *Journal of Physics: Conference Series* **541** 012035

[12] Nikiforov K A and Zartdinov A N 2014 Studying field emission characteristics of point and wedge–shaped surface defects *Journal of Physics: Conference Series* **541** 012009

[13] Nikiforov K 2015 Mathematical model and software complex for computer simulation of field emission electron sources *AIP Conf. Proc.* **1648** 450014

[14] Nikiforov K A and Egorov N V 2015 Simulation of specimen structure in atom probe tomography and field electron microscopy *2015 International Conference on Mechanics – Seventh Polyakhov's Reading (Saint Petersburg)* (Piscataway: IEEE) 7106762

[15] Nikiforov K A, Egorov N V and Shen C-C 2009 Surface reconstruction of a field electron emitter *Journal of Surface Investigation: X-Ray, Synchrotron and Neutron Techniques* **3** 833-9

Journal of Physics: Conference Series **643** (2015) 012122 doi:10.1088/1742-6596/643/1/012122

[16] Nikiforov K and Krasnova A 2014 Model of field electron emitter surface structure *2014 Tenth International Vacuum Electron Sources Conference, IVESC (Saint Petersburg)* (Piscataway: IEEE) 6892018
[17] Plšek, J, Hrubý P, Nikiforov K and Knor Z 2005 Properties of physisorbed water layers on gold revealed in a FEM study *Applied Surface Science* **252** 1553–60
[18] Vinogradova E M, Egorov N V, Klimakov A A 2010 Calculation of the electrostatic potential of the diode system based on a sharp-edged field cathode *Technical Physics* **55** 591−4

Modelling of static mode characteristics of a thin-film vacuum nanotriode

K A Nikiforov and Z R Gallyamov
Saint-Petersburg State University, 7-9, Universitetskaya nab., St. Petersburg, 199034, Russia
E-mail: knikiforov@cc.spbu.ru

Abstract. Mathematical modelling and computer simulation is performed for analysis of nanotriode characteristics in static mode. It is shown nanotriode has differential parameters, which significantly (by orders) differ from those characteristics of usual vacuum lamps, which, together with miniaturization significantly expands the possibilities of their use.

1. Introduction
The goal of this paper is to model characteristics of thin-film vacuum nanotriode (shown in figure 1) [1]. Thin-film vacuum nanotriode with cone-shaped emitters is considered. Each emitter has height 200 nm with 1 nm radius of surface curvature on the top. Current-voltage characteristics was calculated for single cylindrical triode cell (655 nm in diameter) in [1]. It was shown the value of the emission current comes to microamperes, when the voltage is less than 20 V applied on the anode-emitter gap 500 nm. Anode-cathode voltage 16-20 V can generate more than 0.5 mkA emission current flown to the anode when gate-cathode voltage is less than 14 V and gate diameter is 150 nm [1].

To achieve the goal of this paper it is necessary to investigate the behaviour of differential parameters when changing the geometric dimensions of the triode structures operating in static mode. Electron transport processes are described using electrostatic approximation and hydrodynamic flow function for finite element solution [2–4].

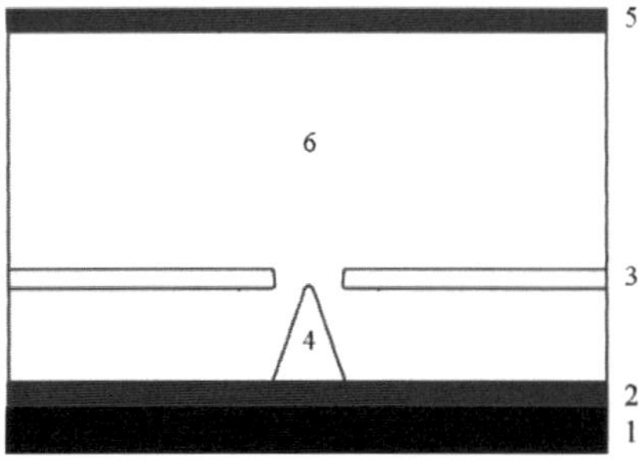

Figure 1. Scheme of the thin film vacuum field emission nanotriode cell: 1 – substrate, 2 – metal layer of cathode, 3 – metal gate electrode, 4 – metal emitter, 5 – metal anode, 6 – vacuum.

2. Mathematical model
The main characteristics of the vacuum triode in static mode are the following differential parameters. Slope S of anode volt-ampere characteristic (which is the equivalent conductivity

of the alternating current (AC)):

$$S = \frac{\partial I_a}{\partial U_g}|_{U_a=const},$$

where I_a – anode current, U_g – gate voltage, U_a – anode voltage.

Internal resistance AC R_i:

$$R_i = \frac{\partial U_a}{\partial I_a}|_{U_g=const}.$$

The gain of the triode to the anode current μ. It shows how many times stronger the gate potential affects the cathode current than the anode potential. The gain depends on the distribution of space charge:

$$\mu = SR_i = \frac{\partial I_a}{\partial U_g} \cdot \frac{\partial U_a}{\partial I_a} = -\frac{\partial U_a}{\partial U_g}|_{I_a=const}.$$

The permeability of the gate D. It characterizes the penetration of the electric field through the gate of the anode to the cathode. Permeability is numerically equal to the ratio of charges induced on the cathode by the anode and by the gate, in the absence of current in the triode. It follows that the permeability depends on the geometry of the electrodes. The permeability of the gate D is expressed through the gain μ from the equation $\mu \approx 1/D$ (when $U_g < 0$) .

Modelling of processes described above is a complex task, was carried out in the software complex developed in Matlab and Matlab PDE Toolbox [2–4]. Following is description of important aspects of calculation procedure with domain triangulation.

The solution has a rapidly changing gradient in the emission area (on top of the cathode), thus the finite-element mesh has to become tighter in the neighbourhood of the emitter tip to avoid reduction of convergence rate towards the exact solution and increase of the number of variables (i.e. the dimension of the finite-element system).

With adaptable mesh it is natural to employ an error indicator, that would include the residual norm of the equation and fluctuations of the gradient of finite-element solution, as they are connected with one of the basic values given for this problem – that is, with the electric field strength. This indicator is implemented in *pdejmps* function which is called by *adaptmesh* function aimed for the adaptive solution of the problem. For making the mesh more fine a way of splitting finite elements was chosen, such that they are split along their longest side (*rmethod=longest*), i.e. a triangle (based on the value of error indicator) is replaced with two smaller triangles by splitting its longest side in half.

There are other parameters of *adaptmesh* that can be modified for educational or research purposes, including through the GUI: *maxt* – the max total number of triangles (*maxt=inf* for no limit); *tripick* – choice of the triangles for splitting by relative error value (*pdeadgsc*), where the triangles with error value exceeding certain scaled constant are split, or by a conditional quality criterion (*pdeadworst*), in which a triangle is split when its error value is greater than some fraction (given by a coefficient between 0 and 1) of the worst error value among all the triangles; *ngen* – max number of iterations (*ngen=inf* for no limit).

In general case, if inital mesh is not explicitly given, *adaptmesh* calls *initmesh* for initialisation of the mesh using default *initmesh* values for all the parameters. The main parameters of *initmesh* are speed of increase of the triangle size (*hdgrad*, with values between 1 and 2), defining the level of non-uniformity of initial mesh, and the max allowed length of a triangle side (*hmax*). For being able to change those parameters *initmesh* is explicitly called during the initialisation of the mesh and the triangulation therefore obtained is passed on to *adaptmesh* for checking. Parameters can also be passed through GUI.

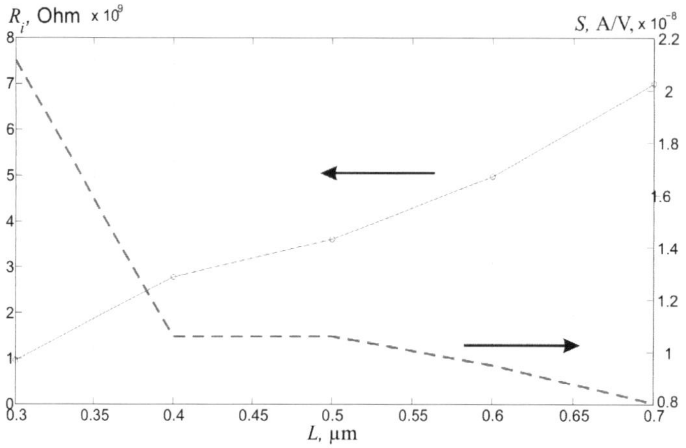

Figure 2. Internal resistance and slope of anode volt-ampere characteristic vs anode-gate distance.

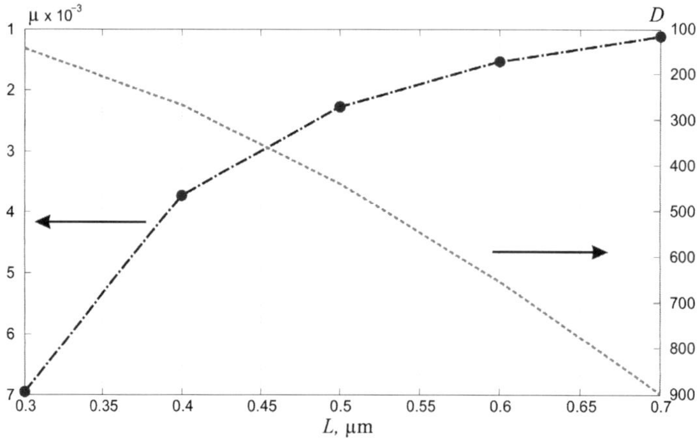

Figure 3. Triode gain and gate permeability vs anode-gate distance.

3. Conclusion

This paper presents a mathematical and computer model we've designed for solving problems of modern vacuum-nanoelectronics devices with complex sub-micron geometries with condition of large variance in electric field. The approach is based on a current function that is analogous to one used in hydrodynamics. For its analysis we use finite-element method over a non-uniform mesh and algorithms implemented in Matlab and Matlab PDE Toolbox. Results are also applicable to field emission devices simulation [4–13]. Future work is characterisation of proposed algorithm's complexity function by informational sensitivity [14–16] measurements.

Acknowledments

Financial support has been obtained from RFBR (13-01-00150) and partially from Saint-Petersburg State University (9.38.673.2013). Research was carried out using computational resources provided by Resource Center "Computer Center of SPbU" (http://cc.spbu.ru/en).

References

[1] Gallyamov Z R and Nikiforov K A 2014 Model of electron transport in cell of a thin-film vacuum nanotriode *Journal of Physics: Conference Series* **541** 012035

[2] Nikiforov K 2015 Mathematical model and software complex for computer simulation of field emission electron sources *AIP Conf. Proc.* **1648** 450014

[3] Nikiforov K A and Egorov N V 2012 Program complex for vacuum nanoelectronics finite element simulations *RuPAC 2012 Contributions to the Proc. of 23 Russian Particle Accelerator Conf. (St. Petersburg)*, (Cern: JACoW) 40911

[4] Nikiforov K 2014 Modelling of emission processes in Matlab *International Conference on Computer Technologies in Physical and Engineering Applications, ICCTPEA (Saint Petersburg)* (Piscataway: IEEE) 6893317

[5] Nikiforov K A, Egorov N V and Shen C-C 2009 Surface reconstruction of a field electron emitter *Journal of Surface Investigation: X-Ray, Synchrotron and Neutron Techniques* **3** 833-9

[6] Nikiforov K and Krasnova A 2014 Model of field electron emitter surface structure *2014 Tenth International Vacuum Electron Sources Conference, IVESC (Saint Petersburg)* (Piscataway: IEEE) 6892018

[7] Plek, J, Hrub P, Nikiforov K and Knor Z 2005 Properties of physisorbed water layers on gold revealed in a FEM study *Applied Surface Science* **252** 155360

[8] Nikiforov K A, Antonova L I, Egorov N V, Trofimov V V, Makeev V V and Ogurtsov O F 2012 Non-gated field emission array as low-energy electron source: Experiment and simulation *RuPAC 2012 Contributions to the Proc. of 23 Russian Particle Accelerator Conf. (St. Petersburg)*, (Cern: JACoW) 21820

[9] Sayfullin M and Nikiforov K 2014 Fringe field effect study of a field emitter array 2014 *2014 Tenth International Vacuum Electron Sources Conference, IVESC (Saint Petersburg)* (Piscataway: IEEE) 6892072

[10] Nikiforov K and Andrievskiy D 2014 Comparison of injection algorithms for electron beam simulation by particle-in-cell method *2014 20th International Workshop on Beam Dynamics and Optimization, BDO (Saint Petersburg)* (Piscataway: IEEE) 6890057

[11] Sayfullin M F and Nikiforov K A 2014 Studying fringe field effect of a field emitter array *Journal of Physics: Conference Series* **541** 012020

[12] Nikiforov K A and Zartdinov A N 2014 Studying field emission characteristics of point and wedgeshaped surface defects *Journal of Physics: Conference Series* **541** 012009

[13] Nikiforov K A and Egorov N V 2015 Simulation of specimen structure in atom probe tomography and field electron microscopy *2015 International Conference on Mechanics Seventh Polyakhov's Reading (Saint Petersburg)* (Piscataway: IEEE) 7106762

[14] Kiktenko A A, Lunkovskiy M N and Nikiforov K A 2014 Confidence complexity of computer algorithms *2014 2nd International Conference on Emission Electronics, ICEE (Saint Petersburg)* (Piscataway: IEEE) 6893971

[15] Kiktenko A, Lunkovskiy M and Nikiforov K 2014 Studying Informational Sensitivity of Computer Algorithms *Advances in Intelligent Systems and Computing* **285** 42131

[16] Kiktenko A, Lunkovskiy M and Nikiforov K 2015 Informational sensitivity as algorithms complexity function characteristic *AIP Conf. Proc.* **1648** 450015

Theoretical investigation of sedimentation process for nanoparticles statistical ensemble

Ksenia Mukhina, Sergey Chivilikhin

ITMO University, 49 Kronverkskiy pr., Saint-Petersburg, 197101, Russia

E-mail: mukhinaks@gmail.com, sergey.chivilikhin@gmail.com

Abstract. In this paper the sedimentation process of nanoparticles which have distribution in sizes was studied. The mathematical model under consideration gives a rise of velocity which resembles Rayleigh-Taylor instability. The numerical solution is consistent with results predicted by the model. The maximum value of concentration, where instability does not occur, was found.

1. Introduction

Sedimentation is the tendency for particles in suspension to settle out of the fluid in which they are entrained and come to rest against a barrier. Sedimentation of particles in a fluid environment is a familiar phenomenon in nature and has important technical applications. In 1950 J. Taylor [1] first paid attention to the problem of instability of liquid surface. But his studies have not dealt with some relevant factors as viscosity and surface tension. Bellman and Pennington [2] developed the problem by adding these factors to mathematical model. The instability occurs in situations between two fluids of different densities where the lighter fluid is pushing the heavier. The Rayleigh-Taylor instability (or RT instability) can be observed in a lot of situations in astrophysics, for instance supernova's explosion, in physics of plasma, also in electro- and magneto- hydrodynamics and nanohydrodynamics [3].

This process in suspension was studied rather well [4-5]. Evolution in time of border of glycerol and suspension was studied by C. Völtz, W. Pesch and I. Rehberg [6]. Experimental data showed existence of instability whish was similar to the RT instability. In Glowinski, Pan and Joseph's work [7] there was created two-dimensional simulation of this phenomenon and the three-dimensional simulation was made by Guda, Bukharina and Mucha [8]. Systems were bounded by walls. In Saveliev and Rozanov's article [9] there was considered the mathematical model of gas with particles of dust with fixed size. There were a number of similarities between results and RT instabilities. The investigations of the dynamics of interfaces between regions in the fluid with different densities might shed some light on the general understanding of interface dynamics in different branches of science.

The major objective of this study was to investigate the process of sedimentation of statistical ensemble of nanoparticles which had some distribution of size. In this paper the term 'nanoparticles' refers to nano-sized balls which don't interact with each other. Statistical ensemble can be defined as probability distribution for the state of the system. And emergence of velocity's instability is expected as result.

2. Methods

The sedimentation process of nanoparticles was modeled in a gravity field (see Figure 1). System is unlimited in horizontal direction and consists of gas and dust. Upper half-space contains mix of gas and dust, the lower half-space contains gas only. Particles of dust have some distribution in sizes with density function f_0. Origin point is located on the border of dusted area. Z-axis is directed to the ground. The friction force is $F = 6\pi\mu R\gamma R^2$.

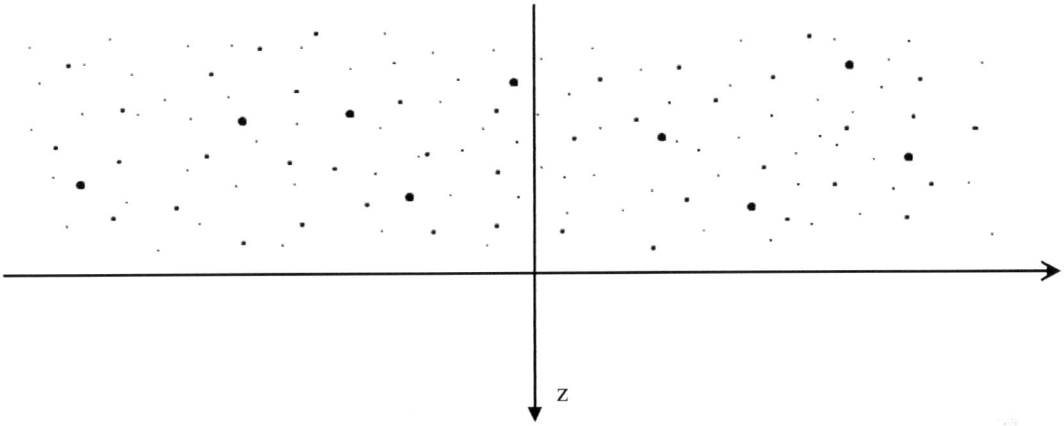

Figure 1. The scheme of the problem. Z-axis is directed as a gravity force.

In case of moving gas, dust gets gas velocity immediately. In vertical direction velocity contains velocity of gas and velocity of dust in zeroth-order approximation. Kinetic equation keeps same form as form one-dimension case.

$$\frac{\partial f}{\partial t}+\frac{\partial}{\partial x}(v_x f)+\frac{\partial}{\partial y}(v_y f)+\frac{\partial}{\partial z}(v_z f)=0$$

Probability density function will be found as sum of density function in zeroth-order approximation and perturbation function which is considering rather small.

$$f = f^{(0)} + \hat{f}$$

$$\frac{\partial f^{(0)}}{\partial t}+\frac{\partial \hat{f}}{\partial t}+\frac{\partial}{\partial x}\left[\hat{v}_x(f^{(0)}+\hat{f})\right]+\frac{\partial}{\partial y}\left[\hat{v}_y(f^{(0)}+\hat{f})\right]+\frac{\partial}{\partial z}\left[(\hat{v}_z+\gamma R^2)(f^{(0)}+\hat{f})\right]=0$$

Dusted gas is considered as continuous medium, thus we can use Navier-Stokes equations for describing our model.

$$\begin{cases} \rho\dfrac{\partial \hat{v}_z}{\partial t} = \mu(\dfrac{\partial^2 \hat{v}_z}{\partial x^2}+\dfrac{\partial^2 \hat{v}_z}{\partial y^2}+\dfrac{\partial^2 \hat{v}_z}{\partial z^2}) - \dfrac{\partial \hat{p}}{\partial z} + \displaystyle\int_0^\infty 6\pi\mu R\gamma R^2 \hat{f} dR \\ \rho\dfrac{\partial \hat{v}_x}{\partial t} = \mu(\dfrac{\partial^2 \hat{v}_z}{\partial x^2}+\dfrac{\partial^2 \hat{v}_z}{\partial y^2}+\dfrac{\partial^2 \hat{v}_z}{\partial z^2}) - \dfrac{\partial \hat{p}}{\partial x} \\ \rho\dfrac{\partial \hat{v}_y}{\partial t} = \mu(\dfrac{\partial^2 \hat{v}_z}{\partial x^2}+\dfrac{\partial^2 \hat{v}_z}{\partial y^2}+\dfrac{\partial^2 \hat{v}_z}{\partial z^2}) - \dfrac{\partial \hat{p}}{\partial y} \\ \dfrac{\partial \hat{v}_x}{\partial x}+\dfrac{\partial \hat{v}_y}{\partial y}+\dfrac{\partial \hat{v}_z}{\partial z}=0 \end{cases}$$

Using Navier–Stokes equations and kinetic equation, the system of equations describing the given model was completed:

$$\begin{cases} \frac{\partial \hat{f}}{\partial t} + \gamma R^2 \frac{\partial \hat{f}}{\partial z} + \hat{v}_z \frac{\partial f^{(0)}}{\partial z} = 0 \\ \rho \frac{\partial \hat{v}_z}{\partial t} = \mu\left(\frac{\partial^2 \hat{v}_z}{\partial z^2} + \frac{\partial^2 \hat{v}_z}{\partial y^2} + \frac{\partial^2 \hat{v}_z}{\partial x^2}\right) - \frac{\partial \hat{p}}{\partial z} + \int_0^\infty 6\pi\mu R\gamma R^2 \hat{f} dR \\ \frac{\partial^2 \hat{p}}{\partial x^2} + \frac{\partial^2 \hat{p}}{\partial y^2} + \frac{\partial^2 \hat{p}}{\partial z^2} = \int_0^\infty 6\pi\mu R\gamma R^2 \frac{\partial \hat{f}_{\hat{k}}}{\partial z} dR \end{cases}$$

Solving this system we get mathematical expression for Z-component of velocity.

$$\frac{\partial \hat{v}_{zk}}{\partial t} = \frac{\mu}{\rho}\left(\frac{\partial^2 \hat{v}_{zk}}{\partial z^2} - k^2 \hat{v}_{zk}\right) + k\frac{e^{kz}}{2\rho}\int_z^\infty e^{-kz'} J(z')dz' + k\frac{e^{-kz}}{2\rho}\int_0^z e^{kz'} J(z')dz'$$

The concentration value can be found by using this numerical solution:

$$V_{j,n+1} = V_{j,n} + \nu\Delta t\left(\frac{V_{j+1,n} + V_{j-1,n} - 2V_{j,n}}{(\Delta z)^2} - k^2 V_{j,n}\right) + \frac{k\Delta t}{2} e^{k\cdot j\Delta z} \sum_{j'=j}^{\infty} e^{-k\cdot j'\Delta z} \frac{J(j')}{\rho}\Delta z + \frac{k\Delta t}{2} e^{-k\cdot j\Delta z} \sum_{j'=0}^{j} e^{k\cdot j'\Delta z} \frac{J(j')}{\rho}\Delta z$$

3. Results and Discussion

If we assume that integral summands are equal to zero, we get the system which has some similarity with system which describes diffusion problem. It means that with growth of viscosity the velocity of particles will be decline (Fig. 2).

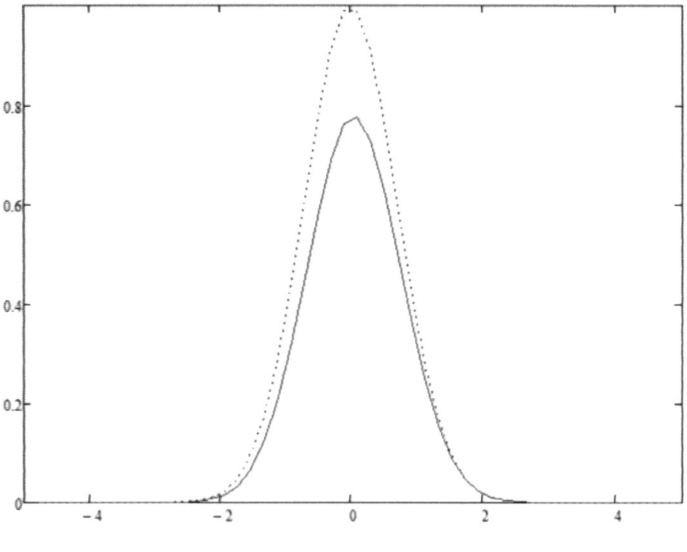

Figure 2. The dependencies of Fourier transform of velocity's Z-component from coordinate in first moment of time (dotted line) and after n iterations

Then let's the probability density function in first moment of time be $f_0 = n_0 \alpha^2 R e^{-\alpha R}$, where n_0 is the particles of dust concentration. The charts below show velocity's behaviour with different concentration of particles. It is clear that after certain level of concentration was reached, velocity has a sudden rise which can be explained by the instability similar to RT instability. The minimum value of concentration, where instability is appearing, is $n_0 = 3 \cdot 10^{10}\ m^{-3}$.

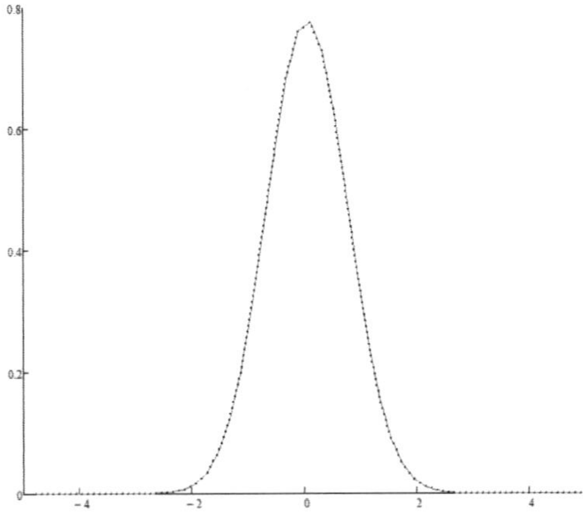

Figure 3. Comparison Fourier transform of velocity's Z-component dependence from coordinate with and without perturbation; $\nu = 10^{-6}; \gamma = 2.173 \cdot 10^6; n_0 = 10^{10}$

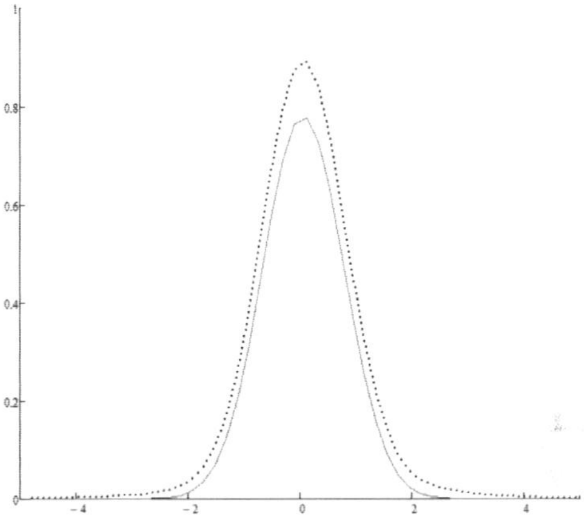

Figure 4. Comparison Fourier transform of velocity's Z-component dependence from coordinate with and without perturbation; $\nu = 10^{-6}; \gamma = 2.173 \cdot 10^6; n_0 = 10^{16}$

Thus, there was discovered a phenomenon which is close to process taking place during the Rayleigh-Taylor instability and the predicted rise of velocity was found. However, for better understanding of particles' behaviour the further investigations are required.

Acknowledgments
This work was financially supported by the Government of Russian Federation, Grant 074-U01.

References
[1] Taylor G.I. 1950 The instability of liquid surfaces when accelerated in a direction perpendicular to their planes *Proc. Roy. Soc. A.* **201** p 192–196
[2] Bellman R., Pennington R. 1954 Effects of surface tension and viscosity on Taylor instability *Quart. Appl. Math.* **12** p 151–162.
[3] Chandrasekhar S. 1981 Hydrodynamic and Hydromagnetic Stability NY.: Dover **652**
[4] Sharp D.H. 1984 An overview of Rayleigh–Taylor instability *Physica D.* **12** p 3–18
[5] Suvankar Ganguly and Suman Chakraborty 2011 Sedimentation of nanoparticles in nanoscale colloidal suspensions *Physics Letters A* **375** p 2394
[6] Kadau K 2004 et. al. Nanohydrodynamics Simulations: An Atomic View of the Rayleigh-Taylor Instability *Proc. Nat. Acad. Sci.* **101 (16)** p 5851–55
[7] Völtz C, Pesch W and I Rehberg 2001 Rayleigh-Taylor instability in a sedimenting suspension *Phys. Rev. E.* **65** p 1–7.
[8] Glowinski R, Pan T W and Joseph D D 2001 *J. Fluid Mech.* **434** p 23–37
[9] R.S. Saveliev, N.N. Rozanov, G.B. Sochilin, S.A. Chivilikhin Rayleigh-taylor instability of a dusty gas (in Russian) *Scientific and Technical Journal of Information Technologies, Mechanics and Optics* **3(73)** p 18-21

Modes of silver nanoisland film growth on the surface of ion-exchanged glass

A Kamenskii[1], I Reduto[2], J Kobert[2] and A Lipovskii[1,2]

[1]Peter the Great Saint-Petersburg Polytechnic University, 29 Polytechnicheskaya, St. Petersburg, 195251 Russia
[2] St. Petersburg Academic University, 8/3 Khlopina, St. Petersburg, 194021 Russia

E-mail: www.kamen@gmail.com

Abstract. The behavior of silver nanoisland film growth on the surface of ion-exchanged glasses processed in the air and in hydrogen atmosphere is compared. The films grown in air degrade with the increase of processing time because of oxidizing. The growth of the film in hydrogen atmosphere tends to saturate with time due to the decrease in the flow of atomic silver towards the glass surface. This is because of the formation and growth of silver nanoparticles in the bulk of the glass, these nanoparticles being the preferable sink for silver atoms.

1. Introduction

Last decades metal island films (MIF) are of interest for photonics and plasmonics because of high local electric field formed in the vicinity of metal nanoislands irradiated with a light capable of excitation of surface plasmon resonance (SPR) [1]. This enhancement of the electric field provides high efficiency of Raman scattering in the agent deposited onto MIF [2], their good catalytic abilities [3] as well as high optical nonlinearity of transparent media embedded with metal nanoparticles [4]. Recently a technique to grow MIF, which is based on out-diffusion of metal from glass containing reducible ions under thermal processing of the glass in hydrogen, has been proposed [5]. This paper is aimed at the study of MIF growth dynamics on the surface of silver ion-exchanged glass under their thermal processing in the air and in hydrogen.

2. Experimental

In the experiments we grow silver nanoisland film (SNF) on the surface of ion-exchanged glass in the course of out-diffusion of neutral silver from the glass matrix enriched with silver ions in silver-sodium ion exchange processing as described elsewhere [6,7]. Soda-lime glass slides were ion-exchanged in the bath containing 5 wt.% of $AgNO_3$ and 95 wt.% of $NaNO_3$ at 325°C for 20 min. We annealed the ion-exchanged slides in the air and in hydrogen at temperatures varied in the range of 78-300°C during 10-320 min. The behaviour of SPR in the grown films was characterized using optical absorption spectroscopy, for this we used differential optical spectra obtained via subtraction of the spectra measured after removal of the grown film from the glass surface from the spectra measured after preparing the samples. This allowed us obtaining optical absorption spectra (optical density, OD) of the film itself without the influence of silver nanoparticles which grow in the bulk of the glass [8]. Topography of the silver nanoislands and their concentration were studied with atomic force microscopy (AFM). We also characterized the SNF grown on the surface of the ion-exchanged glasses stored at room temperature in air for a long time.

3. Results and discussion

Thermal processing of the glasses in the air resulted in the growth of SNF, which was evidenced by the formation of optical absorption peak corresponding to the SPR in the silver nanoislands, which is definitely because of water vapors in-diffusion [9], silver reduction and out diffusion. One can see that the increase in the SPR absorption is followed by its degradation at higher temperatures, see Figure 1, where the height of the SPR peak is plotted as the function of the annealing temperature.

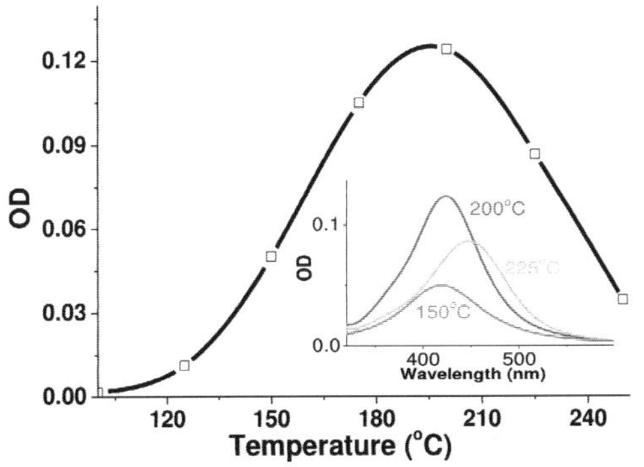

Figure 1. The height of the SPR absorption peak vs processing temperature. The samples are annealed for 10 min in the air. Inset: typical absorption spectra of the grown SNF.

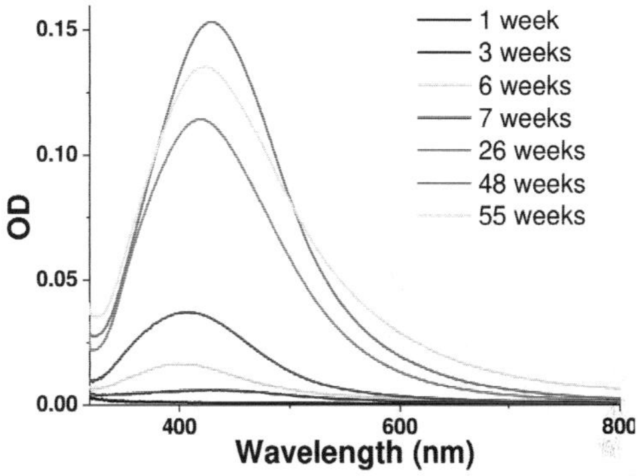

Figure 2. The absorption spectra of the ion-exchanged glass stored in the air at room temperature for 3-55 weeks.

AFM image of silver nanoislands grown in such anneal (Figure 3) show that longer processing in the air results in decreasing of both surface concentration of the nanoislands and their average size. This allows concluding that the most probable reason of the degradation is the oxidation/sulfidation of silver nanoparticles and back diffusion of silver into the glass matrix.

It is essential that the ion-exchanged glasses stored in the air at room temperature have also demonstrated the growth of SNF (Figure 2). In this case, optical absorption peak is wider comparatively to thermally-processed samples. Also the dispersion in the SPR peaks height is much wider, most probably this is because of defects of the glass surface. Despite this, the peak magnitude saturates with time. This allows concluding that only silver ions reduced in a very thin subsurface layer of the ion-exchanged glass make input in the film growth. AFM image of a SNF formed after storage in air of the ion-exchanged glass slide is presented in Figure 4.

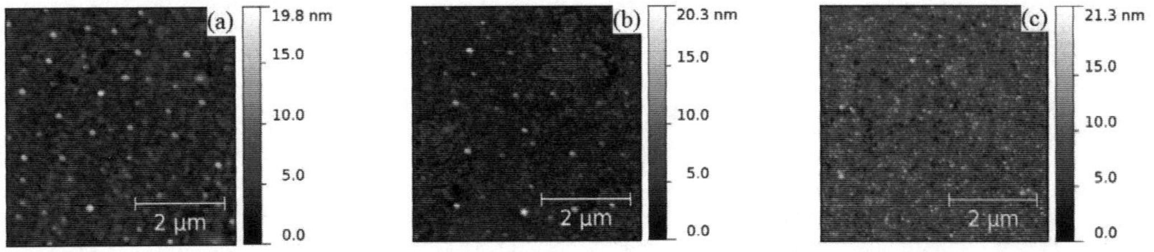

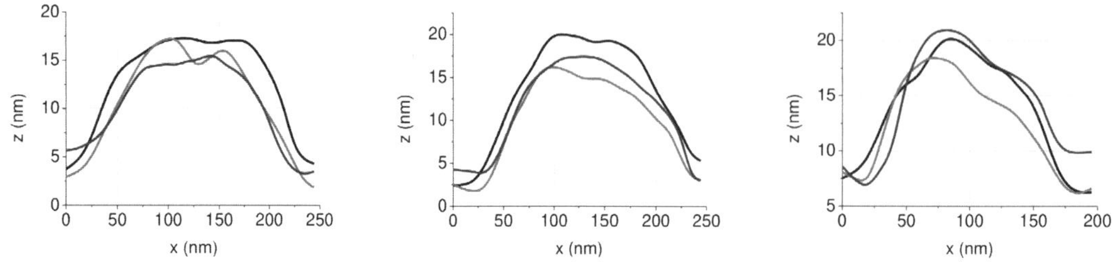

Figure 3. AFM images [3] and corresponding profiles for three nanoislands of MIF formed after 10 min processing of ion-exchanged glass in the air at 150°C (a), 200°C (b) and 250°C (c).

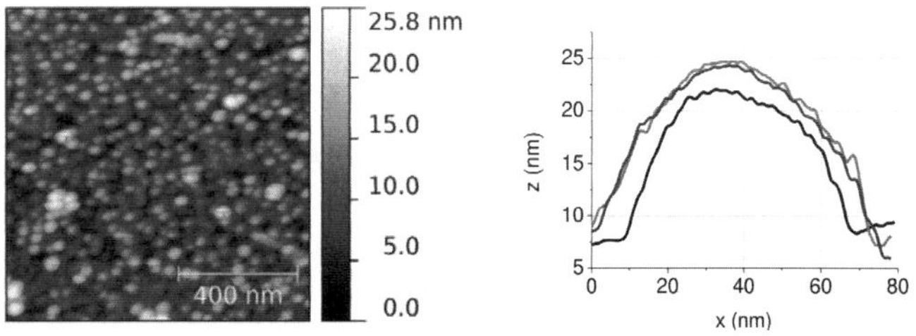

Figure 4. AFM image and corresponding profiles for three nanoislands of MIF formed after storing of ion-exchanged glass in air atmosphere at room temperature for 55 weeks.

Contrary, SNF formed under thermal processing of the glass slides in hydrogen are highly stable at room temperature and do not demonstrate any change after long storage in air. This allows concluding about the absence of silver ions and atoms in a subsurface layer of the glasses processed in hydrogen. Annealing of the glass samples in hydrogen resulted in a monotonous growth of the SPR absorption peak with the increase of the processing temperature (Figure 5). However at higher temperatures temporal dependence of the peak magnitude shows saturation (Figure 6) because of the nanoparticles growing in the bulk of the glass, which behave as powerful sink for silver atoms. Measured with AFM lateral size of silver nanoislands grown in hydrogen varies from 20 to 400 nm depending on the mode of hydrogen anneal.

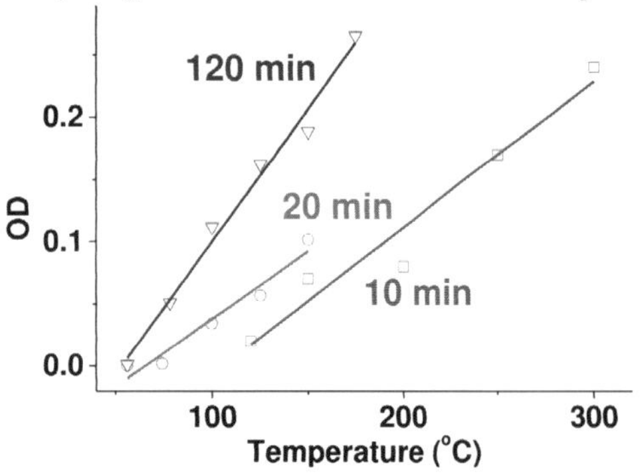

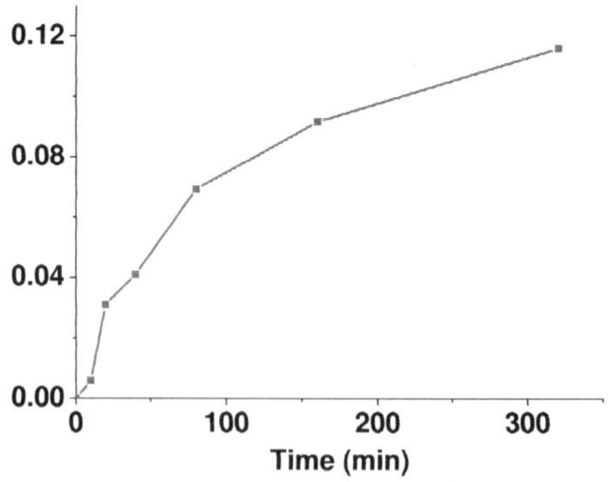

Figure 5. The dependence of the height of the SPR absorption peak on the temperature of annealing the samples in hydrogen for different durations of the annealing.

Figure 6. The dependence of the height of the SPR absorption peak on the duration of annealing the samples in hydrogen at 100°C.

Annealing of ion-exchanged glass slides in hydrogen at lower temperatures has also allowed us precise formation of SNF, AFM images of which are presented in Figure 7. Here one can see that formed nanoislands are only slightly less in size than the nanoislands formed at higher temperature, however their concentration is essentially less. This evidences the preference of surface diffusion over the diffusion from the bulk of the nanoparticles in the nanoislands growth [10].

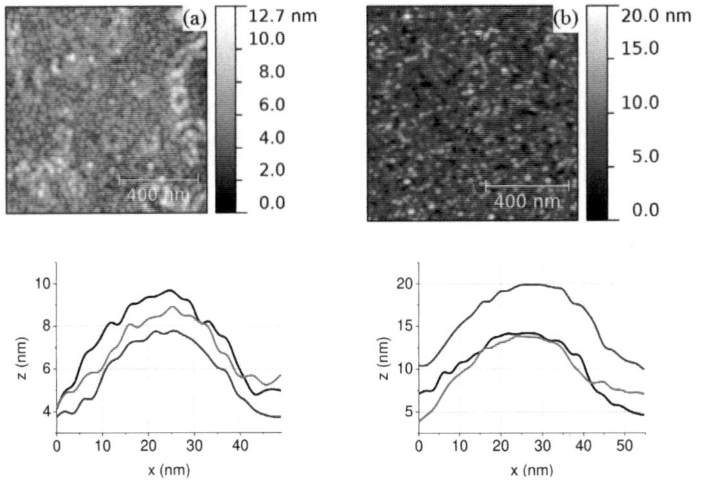

Figure 7. AFM images and corresponding profiles for three nanoislands of MIF formed after 120 minute processing of ion-exchanged glass in hydrogen atmosphere at 78°C (a) and 175°C (b).

4. Conclusion

The behavior of silver nanoisland film growth on the surface of ion-exchanged glasses processed in the air and in hydrogen atmosphere is compared. The SNF grown in hydrogen demonstrate linear growth of the SPR absorption peak amplitude with the temperature increase. This indicates the augmentation of silver concentration on the glass surface during the process. In contrast, the films grown in air degrade with the increase of processing temperature that is most probably because of oxidizing of the film. Films grown in H_2 atmosphere are highly stable in the air because of the absence of silver ions and atoms in the subsurface layer of the processed glasses.

Acknowledgments
The study has been supported by Russian Ministry for Education and Science project # 16.1233.2014/K. The SNF characterization has partly been performed using the equipment of the Joint Research Centre «Material science and characterization in advanced technology» (Ioffe Institute, St. Petersburg, Russia), authors thank Alexander Baklanov for the participation in AFM measurements.

References

[1] Chen Y and Ming H 2012 *Photon. Sens.* **2** 37–49
[2] Lee S J, Guan Z Q, Xu H X and Moskovits M 2007 *J. Phys. Chem.* C **111** 17985–88
[3] Hochleitner G, Steinmair M, Lugstein A, Roediger P, Wanzenboeck H D, Bertagnolli E 2011 *Nanotechnology* **22** 015302
[4] Bigot J Y, Halte V, Merle J C, Daunois A 2000 *Chem. Phys.* **251** 181–203
[5] Zhurikhina V V, Brunkov P N, Melehin V G, Kaplas T, Svirko Yu, Rutckaia V V and Lipovskii A A 2012 *Nanoscale Res. Lett.* **7** 676
[6] Chervinskii S, Matikainen A, Dergachev A, Lipovskii A and Honkanen S 2014 *Nanoscale Res. Lett.* **9** 398
[7] Reduto I, Chervinskii S, Matikainen A, Baklanov A, Kamenskii A and Lipovskii A 2014 *J. Phys.: Conf. Ser.* **541** 012073
[8] Redkov A V, Lipovskii A A and Zhurikhina V V 2013 *J. Non-Cryst. Sol.* **376** 152–7
[9] Kaganovskii Yu, Mogilko E, Lipovskii A A and Rosenbluh M 2007 *J. Phys.: Conf. Ser.* **61** 508–12
[10] Redkov A, Chervinskii S, Baklanov A, Reduto I, Zhurikhina V and Lipovskii A 2014 *Nanoscale Res. Lett.* **9** 606

Investigation of the dielectric properties of composites based on silicon dioxide with carbon nanotubes

A A Goshev, M K Eseev, S N Kapustin, A V Malkov, C S Osokin

Northern Arctic Federal University named after M.V. Lomonosov, Severnaya Dvina Emb. 17, 163002, Arkhangelsk, Russia

E-mail: m.eseev@narfu.ru

Abstract. The aim of this work is an experimental study of the dielectric properties of nanocomposites with a matrix of silica reinforced with multi-walled CNTs at different concentrations and to determine the role of the CNT in the scattering and absorption of electromagnetic radiation in the range of extra-long and medium radio waves (25 Hz-1 MHz). The study found that in addition to the standard behavior of the electrical characteristics of nanocomposites with increasing concentration there was an optimum concentration of carbon nanotubes at which conductivity and absorption of electromagnetic radiation in the material greatly increased. This can be used to enhance the functionality of nanocomposites.

1. Introduction

Recently, active research of composite materials is being conducted which include nanostructures, in particular, carbon nanotubes (CNTs) as filler. Due to the large surface area of the CNT, as well as significant differences of the physical properties of nanoobjects from macrostate material properties of the nanocomposites its properties are not additive characteristics of each phase and can be radically different from those of each of its components [1]. In the review [2] data is provided on the study of the electrical properties of nanocomposites with CNT. Particularly urgent task is to study the strongly nonlinear dependence of properties of the composite on the concentration of the filler, which is necessary for the selection of the optimum functional properties of materials (strength, thermal, dielectric and other characteristics). In the literature [3] data is presented on the influence of impurities on the dielectric properties of CNTs of different composites at microwave frequencies associated with the search for promising materials for "stealth"-technologies [3]. This paper [4, 5] investigates the microwave characteristics at a controlled modification of carbon nanotubes and the nanocomposite materials based on them.

2. Obtaining nanocomposite material

Our composite consist of SiO2 (matrix). It is reinforced with multi-walled CNTs at different concentrations – 0%, 0.05%, 0.1%, 0.5% (by weight of silica + water), with an aspect ratio of $10\text{-}10^2$ (see Figure 1), obtained by CVD-method.

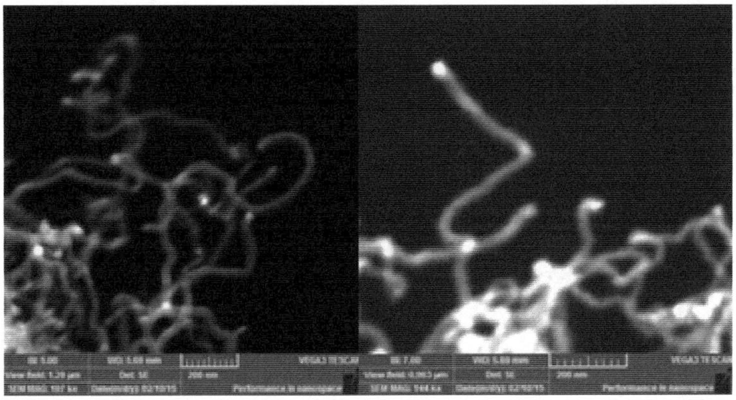

Figure 1. CNT is under the electron microscope Tescan VEGA 3Sem

Silicon were obtained from silicate glue (see Figure 2). The chemical reaction is shown below.

$$Na_2SiO_3 + 2HCl \rightarrow 2NaCl + H_2SiO_3,$$

$$H_2SiO_3 \rightarrow H_2O\uparrow + SiO_2.$$

Figure 2. The Silica particles and the silicon composite.

3. The frequency dependence of the nanocomposite electrical properties

Electrical characteristics of the samples were studied by dielectric relaxation spectroscopy method [2], according to which the sample is placed between the plates of the capacitor (area of S=16 cm^2 and the distance between the plates d=3.3mm). The sample was exposed to an alternating electric field with a frequency varying in the range of 25Hz -1MHz. The measured values were electrical capacitance *C*, the quality factor *Q*, the resistance R.. Study of electrical properties of the composites in electric fields were produced using immittance meter E7-20.

Microwave properties of the composite are described by the frequency electrical field *f* dependence of the complex permittivity:

$$\varepsilon = \varepsilon' - i\varepsilon'', \tag{1}$$

$$\varepsilon'(f) = \frac{\left(C(f) - C_0\right)d}{\varepsilon_0 S}, \quad \varepsilon''(f) = \frac{d}{\varepsilon_0 S}\left(\frac{C(f)}{Q(f)} - \frac{C_0}{Q_0}\right), \tag{2}$$

$$\sigma(f) = 2\pi f \varepsilon''(f)\varepsilon_0. \tag{3}$$

Where the real and imaginary parts characterize the dielectric polarization and losses in the composite respectively. Where C, C_0 - capacitance of the cell with the composite and empty cell respectively and Q, Q_0 – quality factor of the filled and empty cells.

4. Results of experiment

According to the table it can be argued, that there is an optimum concentration at which maximum absorption and scattering of electromagnetic waves at low frequencies occurs. It is known that the change of CNTs concentration leads to a nonlinear modification of the nanomaterials functional properties. In [6] this dependence is noted for the mechanical properties. The dependence of the electrical properties (1), (2) of the silicon dioxide powder on concentration of carbon nanotubes at different frequencies in Table 1.

Table 1. The dependence of the electrical properties at different frequencies.

Composition, CNT, n%	Frequencies of electrical field			
	50 Hz		500 Hz	
	$\varepsilon',\cdot 10^4$	$\varepsilon'',\cdot 10^4$	$\varepsilon',\cdot 10^4$	$\varepsilon'',\cdot 10^4$
Natural, no CNT	59.4	283	0.95	23.8
0.01	53.6	31.5	18.4	13
0.05	326.3	2331	22.8	243
0.1	41.9	72.3	13.3	12.3
0.5	8.8	8	2.3	2.4

Based on the data (see Figure 3) it can be concluded that there is a sharp increase in the absorption properties of the sample at a concentration ~ 0.05% at low frequencies. There is also strongly marked maximum in dispersion of electromagnetic waves for all concentrations in the range of ~ 10kHz.

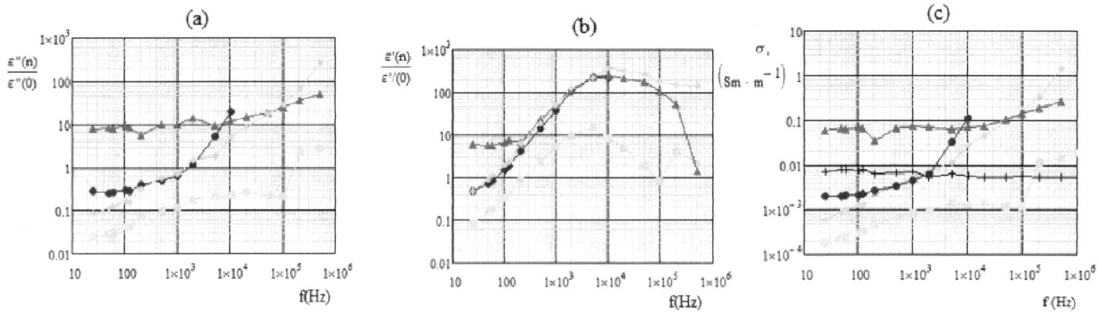

Figure 3. Depenence of the electrical properties (the imaginary (a), and real (b) part of the permittivity (2), conductivity (c), (3)) on the frequency of the electromagnetic field at different concentrations of CNTs: 0.5% 0.1% 0.05% 0.01% 0%.

The behavior of ε ' is associated with a decrease of electromagnetic field change rate compared to the typical recharge time (relaxation time) of effective capacitor formed of adjacent conductive particles. At high frequencies the capacitor does not have time to recharge, and the value of the dielectric permittivity tends to the same value of the composite permittivity with zero concentration of CNTs.

Dependence of the imaginary part of permittivity and conductivity on the frequency also shows a clear concentrational nonlinearity. When n = 0,5% apparently occurs the transition into the zone of agglomerate conductivity where there is deterioration in the conductivity of the sample. With increase of the electromagnetic field frequency the conductivity is caused by the bias current and increasing for all samples.

5. Model of agglomerates

These nonlinear dependence suggests the restructuring of a complex system fine powder SiO_2 + multiwalled CNTs + H_2O (free and bound). With the increase of concentration structures are formed (agglomerates). Dielectric properties of samples support this hypothesis. Their surface area is less than the summary surface area of their components (see Figure 4a,4b). The Figure 4c shows an equivalent circuit ща electrical rechargeable capacitor nanosystem. The reason of the formation of this structures is the great difference between silicon and CNTs surface energy.

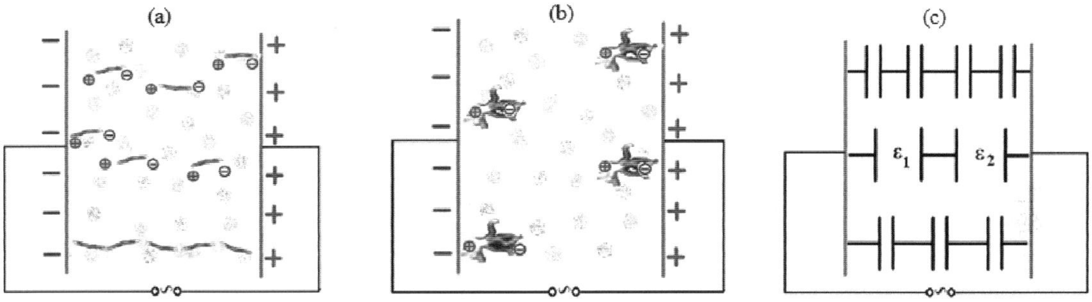

Figure 4. Model representation of the CNT polarization in an alternating electromagnetic field and percolation conductivity (a), the formation of agglomerates with increasing concentration of CNTs (b), rechargeable capacitor system (c).

6. Conclusions

We formulate the main conclusions and results.

- The obtained results allow to judge about the essentially nonlinear dependence of dielectric properties (conductivity and permittivity) on the CNT concentration in this frequency range. There is an optimum concentration of CNTs (~ 0.05%) at which the absorption and dispersion of electromagnetic waves increase sharply. At this concentration even distribution of CNTs by volume of the sample is achieved.
- There is an opportunity to create a nanomaterial with desired dielectric functional properties ε', ε''.
- Such features of dielectric properties of the sample possibly relate to the formation of agglomerates, which surface is less than the total surface of the constituent nanotubes.
- Verification of the hypothesis of agglomerates formation with the increase of CNTs concentration is possible by means of positron annihilation microscopy [7, 8, 9], which is in contrast to the AFM and electron microscopy allows to investigate internal structure of the sample.

Acknowledgements

This work was supported by the project the Ministry of Education and Science of the Russian Federation №3635 "Investigation of the properties nanocomposites with controlled modification of the structure reinforced with carbon nanotubes".

The research was done using the equipment of Core Facility Center "Arktika" of Northern (Arctic) Federal University under financial support of Ministry of Education and Science of the Russian Federation (project RFMEFI59414X0004).

References

[1] Rakov E G 2013 *Russ. Chem. Rev.* **82** 538
[2] Eletskii A V, Knizhnik A A, Potapkin B V, Kenny J M 2015 *Phys. Usp.* **58** (3) in print

[3] Muradyan V E, Sokolov E A, Babenko S D, Moravsky A P 2010 *Tech. Phys.* **55** 242
[4] Usanov D A, Skripal' A V, Romanov A V 2011 *Tech. Phys.* **56** 102
[5] Usanov D A, Skripal' A V, Romanov A V 2014 *Tech. Phys.* **59** 873
[6] Kablov E N, Kondrashov S V, Yurkov G Yu *Nanotechnologies in Russia* 2013 **8** 163
[7] Ahmanova E V, Eseev M K, Kobets A G, Meshkov I N, Rudakov A Y, Sidorin A A and Yakovenko S L 2012 *Physics of Particles and Nuclei Letters* **9** 373
[8] Sidorin A A, Meshkov I, Ahmanova E, Eseev M, Kobets A, Lokhmatov V, Pavlov V, Rudakov A and Yakovenko S 2013 *Mater. Sci. Forum* **733** 291
[9] Sidorin A A, Meshkov I, Ahmanova E, Eseev M, Kobets A, Lokhmatov V, Pavlov V, Rudakov A and Yakovenko S 2013 *Mater. Sci. Forum* **733** 322

Application of GaN for photoelectrolysis of water

M V Puzyk[1], A S Usikov[2,3], S Yu Kurin[4], A M Puzyk[1], A D Fomichev[2], I A Ermakov[2], D S Kovalev[2], B P Papchenko[2], H Helava[3], and Yu N Makarov[3,4]

[1]Herzen University, Nab. r. Moyki 48, St. Petersburg 194186, Russia
[2]University ITMO, Kronverkskiy pr. 49, St. Petersburg 197101, Russia
[3]Nitride Crystals Inc., 181 E Industry Court, Suite B, Deer Park, NY 11729, USA
[4]Nitride Crystals Group, Ltd., pr. Engel'sa 27, St. Petersburg 194156, Russia

E-mail: puzyk@mail.ru

Abstract. GaN layers of n-type and p-type conductivity grown by HVPE on sapphire substrates were used as working electrodes for water electrolysis, photoelectrolysis and hydrogen gas generation. Specifically the water splitting process is discussed. Corrosion of the GaN materials is also considered. The hydrogen production rate under 365-nm UV LED irradiation of the GaN and external bias was 0.3 ml/(cm^2*h) for an n-GaN photoanode (n~8×10^{16} cm^{-3}) in 1M Na_2SO_4 electrolyte and 1.2 ml/(cm^2*h) for an n-GaN photoanode (n~1×10^{17} cm^{-3}) in 1M KOH electrolyte.

1. Introduction

At present, several basic technologies are used for the production of hydrogen: the conversion of methane, technologies for converting biomass into valuable fuel, and the decomposition of water. Semiconductor material immersed in an aqueous solution of electrolyte allows decomposition of water into molecular hydrogen and oxygen by solar irradiation of its surface. The required energy to split water molecule is generated by sunlight absorption in the semiconductor. This method of H2 generation is a simple, low-energy consumption, carbon-free technology not demanding high temperatures and pressure. Another key benefit of this process is the high purity of the output hydrogen gas stream, an important requirement for its use in fuel cells.

The solar water splitting process may be spontaneous under illumination if the electrochemical redox potentials of the oxygen evolution reaction (OER) and the hydrogen evolution reaction (HER) in an electrolyte are bracketed by the energy gap of a semiconductor [1]. This is fulfilled for GaN and InGaN materials [2]. A generic diagram of photoelectrochemical water splitting using n-type GaN material and an alkaline electrolyte is given in figure 1. The absorption of solar photons in a semiconductor material forms excited electron–hole pairs that are split by a specific band-bending at the GaN electrode (anode)/electrolyte interface. The band-bending is due to the establishment of an equilibrium between the n-GaN material and the electrolyte. Excited electrons coming to the cathode through an external circuit take place in HER that gives ions of OH^- (anion) and molecules of H_2. The anions come to the GaN-anode to participate with exited holes leaving the semiconductor in OER that gives molecules O_2 and H_2O.

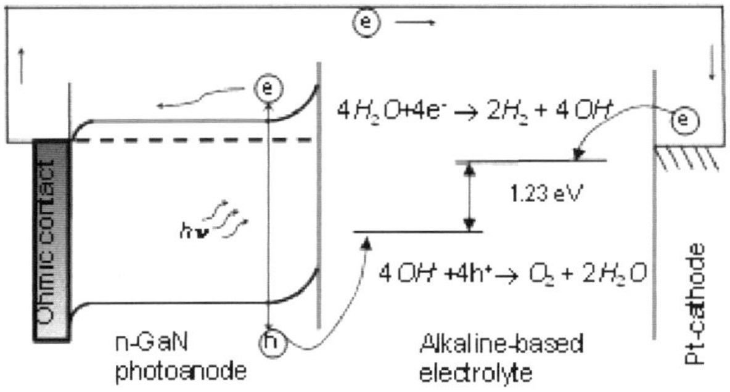

Figure 1. Schematic diagram showing photoelectrochemical water splitting using p-type GaN as a working electrode (photocathode).

In this work we report on first results of electrolysis and photoectrolysis of water using different electrolytes and n-type and p-type GaN layers as working electrodes (anodes or cathodes). GaN layers and multilayer structures were grown on sapphire substrates by chloride hydride vapor phase epitaxy.

2. Experimental

5-7 μm thick GaN layers and AlGaN/GaN p-n structures were grown by chloride hydride vapor phase epitaxy (HVPE) on c-plane 2-inch sapphire substrates. The growth procedure included in-situ sapphire substrate treatment and an AlN/AlGaN (x~ 0.6) buffer layer deposition followed by GaN or AlGaN/GaN multilayer p-n structure growth. Details of the structure growth and characterization can be found elsewhere [3].

The energetics of hydrogen reduction in water solutions of various pH-electrolytes using GaN-based working electrodes were evaluated under conditions of external electrical power and solar irradiation. Sulfuric acid (pH=1), sodium sulfate (pH=7) and potassium hydroxide (pH=14) were used as the electrolyte.

The sapphire substrate orientation made it possible to cut a 2-inch wafer into six sectors (plates). These sectors were used for contact deposition on selected areas of the plates and electrode preparation. As contacts, Ti/Ni/Au or Ni/Au thin films were evaporated on select areas of the p- or n-type GaN layer, respectively. Wire welding completed the electrode fabrication. The contact pads and contact wire were protected by chemical-resistant varnish (wax).

3. Results and discussion

First, the over potential on GaN working electrodes needed for H_2 generation was measured using different aqua electrolytes and an external power source. The aqueous contact area of the GaN electrode was about 2.5-3 cm^2. A Pt plate (4 cm^2) was used as the counter-electrode. The reference electrode was made of Ag/AgCl/KCl (potassium-chloride-saturated silver-chloride electrode). The potential of hydrogen reduction at the GaN electrode (Na_2SO_4-based electrolyte, pH=7) was -0.80 V that is -0.40 V lower (over potential) than that at the Pt electrode. The over potentials measured on GaN electrodes in other electrolytes were -0.4 V (H_2SO_4, pH=1) and -1.20 V (KOH, pH=14). The over potentials could be due to a potential barrier for carriers on the GaN/electrolyte interface, higher specific resistivity and worse current spreading on the GaN electrode. Optimization of the GaN electrode contact configuration is required.

Then, the effectiveness of water electrolysis with GaN working electrodes to produce one milliliter of hydrogen was studied. Figure 2 shows the I-V (V-I) characteristics and energy consumption in relation to the type of electrolyte and the electrolyte concentration. Graduated tube was used to collect generated hydrogen. I-V characteristics showed linear dependencies with various slopes that are in line with different resistivity of GaN electrodes and the electrolyte.

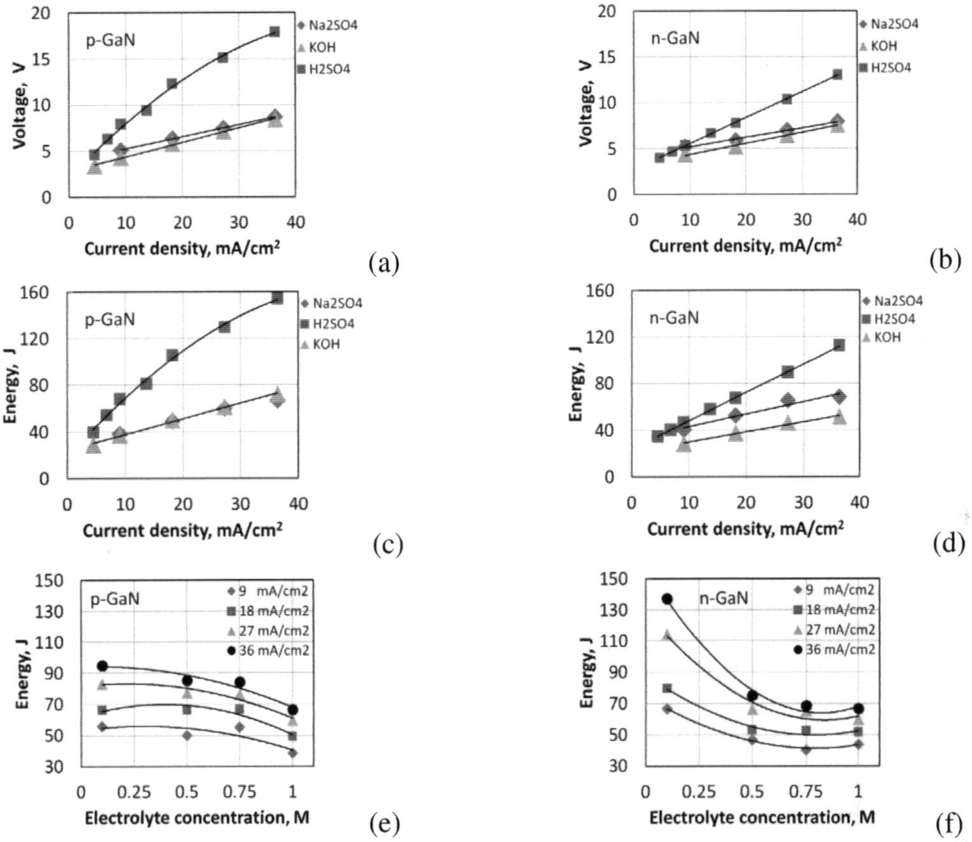

Figure 2. I-V characteristics (a, b) and energetics of a 1 ml H_2 generation (c, d, e, f) using n-GaN and p-GaN working electrodes (as cathode), different 1.0 M-aqua electrolytes (a, b, c, d) and different concentration of Na_2SO_4-based electrolytes (c, d).

Energy consumption in Joules (J) was evaluated based on I-V characteristic and a time needed to generate 1 ml of hydrogen. The water electrolysis took place with lower energy consumption for n-GaN and p-GaN electrodes in the alkaline electrolyte (pH=14). Opposite to that, the process is more energy-consuming for n-GaN electrodes in sodium sulfate (pH =7) electrolyte at lower electrolyte concentration and higher currents (figure 2 c, d). The energy consumption showed sublinear dependencies that demonstrate a band bending effect at the GaN electrode surface that creates an energy barrier in n-GaN to be overcome by electrons in order to participate in the redox reactions in the electrolyte (figure 2 e, f). It seems that the polarity of the external bias (+2.5 - +5 V on the GaN cathode to begin H_2 generation) was a driving force to form an additional barrier for electrons at the interface the n-GaN electrode/electrolyte and to reduce the barrier at the p-GaN electrode surface. At high concentration of electrolyte and high current the band bending effectively becomes lower and overall energy consumption for the process on the n-GaN electrode is somewhat lower than at the p-GaN electrode due to higher carrier mobility and better current spreading in the n-GaN material.

Experiments on water photoelectrolysis were performed under irradiation by 365-nm UV LED and with external power source. Optical power of the UV LED was of 16 mW that corresponds to a concentration factor of ~20x in comparison to the AM1.5-Global standard spectrum of solar radiation in the UV spectral range. Hydrogen generation was observed on Pt cathode using n-GaN as the photoanode. It started at low potential values depending on pH of the electrolyte. n-GaN layers were found to be electrochemically inactive (no corrosion) at potentials from -1.5 V to +1.5 V in all aqueous electrolytes applied in this work. Oxygen generation on the p-GaN anode began at potential more than +2 V (hydrogen was observed on Pt-counter electrode in this case) and accompanied by the p-GaN electrode corrosion. Under UV LED irradiation and with external bias of + 2.5- +3.0 V photocurrent (I_{photo}) was in a range of 3-7 mA/cm^2 and sufficient for hydrogen evolution.

The hydrogen production rate was as high as 0.3 ml/(cm^2*h) for the n-GaN photoanode (n~8×10^{16} cm^{-3}) in 1M Na_2SO_4 electrolyte (UV LED irradiation and +2.5- +3.0 V of external bias). Under the similar conditions using an other n-GaN photoanode (n~1×10^{17} cm^{-3}) in 1M KOH electrolyte, the hydrogen production rate was as high as 1.2 ml/(cm^2*h). These values compare to those from GaN with NiO catalyst (no external bias, Xe lamp irradiation, 0.37 ml/(cm^2*h)) [4], n-$In_{0.18}Ga_{0.82}N$ photoanode (with external bias, Xe lamp irradiation, 1.44 ml/(cm^2*h), and p-n $GaInP_2$/GaAs double electrode (no bias, 5 suns, 1.1 ml/(cm^2*h).

The experiments have shown that the doping and crystalline qualities of the GaN electrodes need to be optimized. A surface passivation and (or) catalyst application are required to increase the hydrogen production rate and diminish the GaN electrode corrosion. Utilizing GaN/InGaN structures as working electrodes may improve hydrogen evolution by effectively absorbing a larger portion of the solar spectrum.

4. Conclusions

In this paper we report on comparative studies of electrolysis and photoectrolysis of water using different electrolytes and n-type and p-type HVPE-grown GaN layers as working electrodes (anodes or cathodes). GaN layers and multilayer structures were grown on sapphire substrates by chloride hydride vapor phase epitaxy.

Potential of hydrogen reduction at the GaN electrode (Na_2SO_4-based electrolyte, pH=7) was -0.80 V that is -0.40 V lower (over potential) than that at the Pt electrode. The over potentials measured on GaN electrode in other electrolytes were -0.4 V (H_2SO_4, pH=1) and -1.20 V (KOH, pH=14). The lowest energy of hydrogen generation was observed in 0.5 M-1.0 KOH electrolyte.

Corrosion of the GaN working electrode used as a cathode was not observed. Obvious corrosion was observed on GaN-anode electrodes.

Hydrogen generation at n-GaN cathode started at low external bias values depending on pH of the electrolyte. The hydrogen production rate was as high as 0.3 ml/(cm^2*h) for n-GaN photoanode (n~8×10^{16} cm^{-3}) in 1M Na_2SO_4 electrolyte (UV LED irradiation and +2.5- +3.0 V of external bias) and 1.2 ml/(cm^2*h) n-GaN photoanode (n~1×10^{17} cm^{-3}) in 1M KOH electrolyte.

Acknowledgments
Work at University ITMO was supported by the Ministry of Education and Science of Russian Federation (grant agreement 14.575.21.0054, unique identifier of research activities is RFMEFI57514X0054).

References

[1] Pinaud B A, Benck J D, Seitz L C et al. 2013 *Energy Environ. Sci.* **6** 1983-2002
[2] Fujii K, Ohkawa K 2006 *Phys. Status Solidi C* **3** 2270–2273
[3] Kurin S, Antipov A, Barash, I, Roenkov A et al. 2014 *Phys. Status Solidi C* **11** 813-816
[4] Ohkawa K, Ohara W, Uchida D, Deura M 2013 *Jpn. J. Appl. Phys.* **52** 08JH04
[5] Aryal K, Pantha B, Dahal R, Li J, Lin J, Jiang H 2011 *APS March Meeting (Dallas)*
[6] http://www.hydrogen.energy.gov/annual_review14_report.html

Nanomaterial disordering in AlGaN/GaN UV LED structures

E I Shabunina[1], M E Levinshtein[1], M M Kulagina[1], S Yu Kurin[2], A E Chernyakov[3], V N Petrov[1], V V Ratnikov[1], I N Smirnova[1], S I Troshkov[1], N M Shmidt[1], A S Usikov[4,5], H Helava[5], and Yu N Makarov[2,5]

[1]Ioffe Physical Technical Institute, Polytekhnicheskaya 26, St. Petersburg 194021, Russia
[2]Nitride Crystals Group Ltd., 27 Engels av., St Petersburg 194156, Russia
[3]Submicron Heterostructures for Microelectronics Research and Engineering Center RAS, Polytekhnicheskaya 26, St. Petersburg 194021, Russia
[4]University ITMO, Kronverkskiy pr. 49, St. Petersburg 197101, Russia
[5]Nitride Crystals Inc., 181E Industry Ct., Ste. B, Deer Park, NY 11729, USA

E-mail: Natalia.Shmidt@mail.ioffe.ru

Abstract. Multifractal analysis was applied to characterize quantitatively nanostructural disordering in HVPE-grown AlGaN/GaN UV LED structures. A higher level of leakage currents shunting the active region of LEDs by an extended defect system is correlated with higher values of multifractal parameters (MFs). As a result, the concentration of injected carriers participating in radiative recombination in the active region is reduced. MFs and the conductivity of quasi-ohmic shunts localized in an extended defect system are higher in AlGaN/GaN structures than in InGaN/GaN structures. It is one of the reasons behind the low external quantum efficiency of AlGaN/GaN UV LEDs.

1. Introduction

AlGaN quantum well (QW) based ultraviolet (UV) light emitting diodes (LEDs) are expected to replace conventional UV lamps based on mercury vapor [1, 2]. Compact UV LEDs are environmentally friendly and are of interest in various applications such as air, water and surface sterilization and decontamination, bio-agent detection, UV curing, etc. The UV LED emission can be selected in a wide and continuous range of wavelengths as opposed to the fixed emission lines of the conventional UV lamps. Another prospective application of (Al)GaN/GaN (and InGaN/GaN) p-n structures is photoelectrochemical water splitting for H_2 generation. [3, 4].

During the last decade much efforts has been focused on developing high power and high efficiency AlGaN-based UV LEDs emitting at wavelength shorter than 365 nm [1, 2]. The major problems in UV AlGaN/GaN LEDs development are found to be low external quantum efficiency (about 2%) and short life-time (about 2000 hours) [5]. Development of high brightness InGaN-based blue LEDs has demonstrated that structural quality is one of the important factors to get higher external efficiency and longer life-time of LED lamps [6]. It has been reported for GaN layers and InGaN/GaN blue LEDs that the structural quality is not merely in a strict dependence on total dislocation density but also is connected with the peculiarities of nanomaterial disordering (nano-structural arrangement (NA)) [7, 8]. It was found that changes of nanomaterial disordering caused by different angles of tilt and twist of the mosaic structure results in the existence of an extended defect system. This system contains different types of dislocations, V-defect among them, and domain boundaries enriched by dislocations. These features are apparent in the surface morphology. This allows examining the surface with the help of atomic force microscopy for quantitative characterization of nanomaterial disordering on the basis of multifractal analysis [7]. A correlation of nanomaterial disordering with multifractal parameters determined from the surface morphology examination of GaN layers and InGaN/GaN blue LEDs [7, 8] has been observed.

Quantitative estimation of nanomaterial disordering in UV AlGaN/GaN LED structures has not been assessed and its connection with the LED parameters has not been studied previously. In this work we apply a multifractal analysis for quantitative characterization of nanomaterial disordering in AlGaN/GaN UV LEDs with relation to their electrical and structural properties.

2. Experimental

The AlGaN-based UV LED structures were grown by hydride vapor phase epitaxy HVPE on c-plane 2-inch sapphire substrates. The growth procedure included in-situ sapphire substrate treatment and an AlN/AlGaN (x~ 0.6) buffer layer deposition followed by AlGaN/GaN multilayer p-n structure growth. Details of the structure growth and characterization can be found elsewhere [9]. The structures grown were processed to form mesas and deposit contacts. Fabricated chips had planar dimensions of 850 μm x 850 μm with light extraction through the surface. Electroluminescence (EL) peak emission was usually observed at 350-360 nm.

Three types of UV AlGaN/GaN LED structures having different nanomaterial disordering were grown by changing the buffer layer growth conditions. Quantitative estimation of nanomaterial disordering in the structures grown was performed by a technique certified earlier for InGaN/GaN blue LEDs [7]. The technique is based on the fact that surface morphology implies nanomaterial disordering of the material. The different tilt and twist angles of domains in the GaN-based mosaic structure result in co-existence of numerous forms of nano-structural arrangement (NA) that influences the surface morphology. Multifractal analysis of digital multitude obtained by atomic force microscopy (AFM) corresponding to the surface morphology of UV AlGaN/GaN LED structures makes it possible to evaluate two parameters (degree of disorder - Δ_p and level of self-organization – D) for quantitate characterization of the nanomaterial disordering. The AFM morphology of AlGaN/GaN UV LED structures having different multi-fractal parameters is given in figure 1. An AFM image of 450-460 nm InGaN/GaN blue LED structures with internal quantum efficiency of 45% is given for comparison.

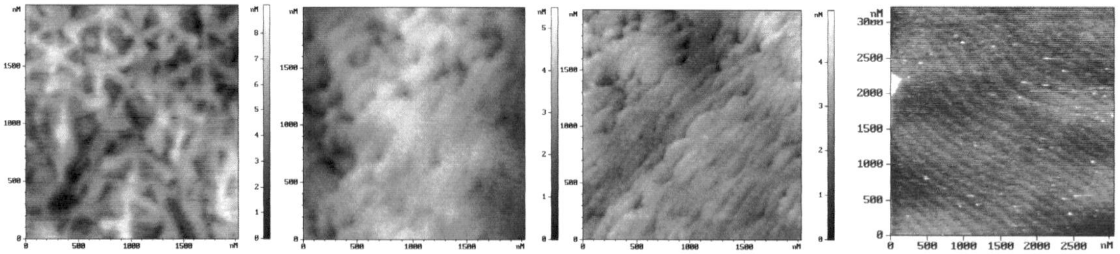

(a) (b) (c) (d)

Figure 1. AFM views of surface morphology of AlGaN/GaN (a, b, c 2 μm × 2 μm scan) and InGaN/GaN (d: 3 μm × 3 μm scan) LED structures with different nanomaterial disordering (multifractal parameters Δ_p and D): (a) Δ_p = 0.390 and D=1.84; (b) Δ_p = 0.370 and D=1.78; (c) Δ_p = 0.348 and D=1.62; (d) Δ_p = 0.370 and D=1.78.

I-V characteristics, EL spectra and–dependencies of the spectral current- noise density (S_I) on the forward current density (j) of UV AlGaN/GaN LEDs obtained from the LED structures having different nanomaterial disordering were studied. Low frequency noise analysis was also applied in the frequency range from 1 Hz to 10 kHz and the power spectra were measured at a current density of 10^{-3} – 10 A/cm^2 in all studied LEDs.

3. Results and Discussion

Increasing parameters of nanomaterial disorder are accompanied by an increase of direct and reverse currents at low voltage (U < 3 V) (figure 2).

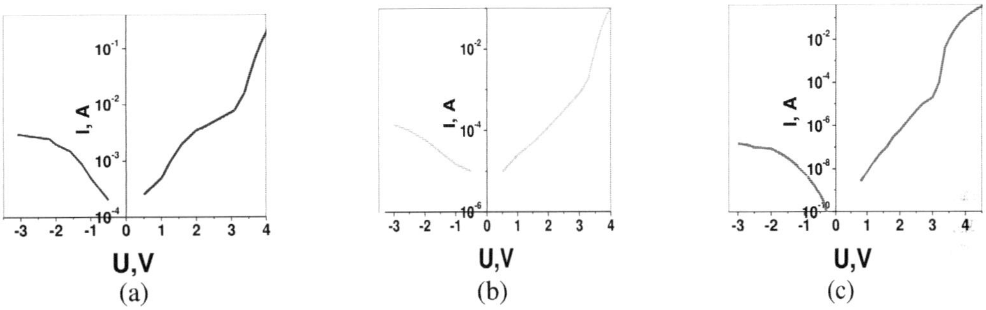

Figure 2. I-V characteristics of AlGaN/GaN UV LEDs fabricated from LED structures with different nanomaterial disordering (multifractal parameters Δ_p and D): (a) Δ_p = 0.390 and D=1.84; (b) Δ_p = 0.370 and D=1.78; (c) Δ_p = 0.348 and D=1.62. Respective AFM images are given in Figure 1.

Moreover, I-V characteristics of all AlGaN/GaN LEDs demonstrated shunts in the p-n junction area whose conductivity becomes higher with increase of Δp and D values. A similar correlation has been reported in InGaN/GaN LEDs [8] and it was proposed that the shunts localize at an extended defect system. This suggestion is in a good agreement with data obtained by other methods [2, 10-12]. A comparison of AFM data of high-defect and defect-free areas has proved that leakage paths (shunts) are observed only in open-core threading dislocations [2]. By using a defect-sensitive etch technique, it was demonstrated that AlGaN-based UV LEDs having different densities of the open-core dislocations are also differ in values of leakage paths (shunts) [2, 10]. Higher density of leakage paths corresponds to higher density of this type of dislocation. The open-core dislocations known as nano-pipes or V-defects are 2-50-nm tubes with the (10-10) facets penetrating the epitaxial layer [11]. The nano-pipes were observed by means of AFM, conductive AFM, transmission electron microscopy (TEM), and even optical microscopy with the help of defect-sensitive etching [11, 12]. The density of this type of dislocation is two orders of magnitude lower than the total dislocation density. I-V characteristics at low voltage of AlGaN-based UV LEDs determine the cumulative level of leakage path (shunt) conductivity of the extended defect system. The interrelation of the open-core dislocation conductivity with nanomaterial disordering is thought to be due to the worse nano-structural arrangement that accompanies the enrichment of growing layers by indium or gallium metals. It has been reported indium or gallium segregation results in V-defects formation [13].

The results obtained in this work demonstrate the values of disordering and shunt conductivity in AlGaN/GaN structures are higher than that in InGaN/GaN structures with high external quantum efficiency, but similar to InGaN/GaN structures with low EQE values (less than 5%). Thus, as in the InGaN/GaN LED structures, multifractal parameters together with the shunt current at voltages less than 2 V can be used for quantitative estimation of nanomaterial disordering in AlGaN/GaN LED structures. It has been also demonstrated that high values of conductivity of quasi-ohmic shunts localized in the extended defect system leaded to decreasing of EQE [14]. This is correct for both

InGaN/GaN and AlGaN/GaN LEDs [8, 14]. Moreover, carrier leakage through the shunts was revealed in bulk GaN having low dislocation density of 10^6 cm^{-2} [15].

Thus, the higher nanomaterial disordering and higher values of conductivity of quasi-ohmic shunts in AlGaN/GaN LEDs than in InGaN/GaN LEDs is one of the reasons behind the low external quantum efficiency values. In addition, the correlation between nanomaterial disordering in AlGaN/GaN LEDs and the conductivity of the quasi-ohmic shunts can be applied for selection of LED lamps from different manufacturers by structural quality of the LED structure.

Figure 3 shows that the dependency of spectral noise density on current density has a near monotonic tendency S_I (j)~ j at a current density less than 10^0 A/cm^2 and is similar to that in InGaN/GaN structures, which is, to a large extent, related to processes localized in an extended defect system shunting the p-n-junction.

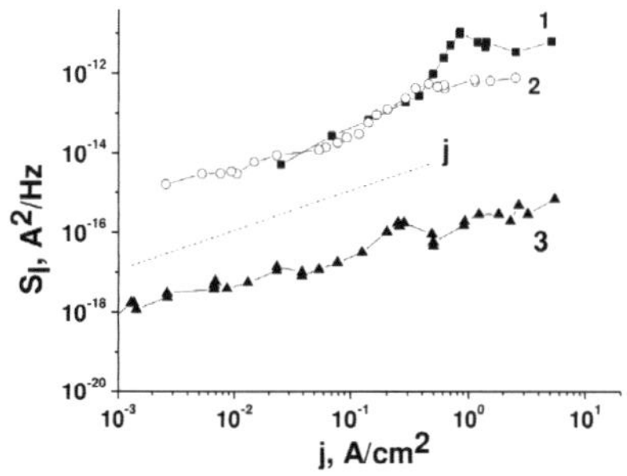

Figure 3. The spectral noise density (S_I) in relation with forward current density (j). Curves 1, 2, 3 corresponds to I-V curves (a), (b), and (c) in figure 1, respectively.

4. Conclusions

Nanomaterial disordering of AlGaN/GaN UV LED structures grown by HVPE on sapphire substrates in comparison with InGaN/GaN LED structures has been studied. A correlation between nanomaterial disordering and conductivity of the quasi-ohmic shunts localized in the extended defect system and shunting the p-n-junction of LEDs was revealed. The shunt conductivity became larger with the value of nanomaterial disordering that can be represented by mulifractal parameters. It was demonstrated that the disordering in AlGaN/GaN LED structures is higher than in InGaN/GaN ones. As a result, a concentration of injected carriers bypassing radiative recombination in the active region of AlGaN/GaN LEDs is much higher than that in InGaN/GaN LEDs and it is one of the reasons behind the low external quantum efficiency of AlGaN/GaN UV LEDs.

Acknowledgments

Work at University ITMO was supported by the Ministry of Education and Science of Russian Federation within the grant agreement 14.575.21.0054 (unique identifier of research activities is RFMEFI57514X0054).

References

[1] Shatalov M, Sun W, Lunev A, Hu X et al. 2012 *Appl. Phys. Exp.* **5** 082101
[2] Moseley M W, Allerman A A, Wierer J J et al. 2015 *Journal of Applied Physics* **117** 095301
[3] Ohkawa K, Ohara W, Uchida D, Deura M 2013 *Jpn. J. Appl. Phys.* **52** 08JH04
[4] Walter M G, Warren E L, McKone J R et al. 2010 *Chem. Rev.* **110**, 6446–6473
[5] http://www.nichia.co.jp/
[6] http://www.cree.com/
[7] Shmidt N M et al. 2001 *Nanotechnology* **12** 471
[8] Shmidt N M et al. 2001 *Nanotechnology* **12** 471

[9] Kurin S, Antipov A, Barash I, Roenkov A et al. 2014 *Phys. Status Solidi C* **11** 813-816
[10] Moseley M, Allerman A, Crawford M, Wierer J J et al. 2014 *J. Appl. Phys.* **116** 053104
[11] Hong S, Yao T, Kim B, Yoon S, and Kim T 2000 *Appl. Phys. Lett.* **77** 82
[12] Lee S W, Oh D C, Goto H, Ha J S et al. 2006 *Appl. Phys. Lett.* **89** 132117
[13] Lee L C, Zhao D G, Jang D S, Li L, Wu L L, Chan P et al. 2012 *Appl. Phys. Lett.* **101** 252110
[14] Pinos A et al. 2011 *Journal of Applied Physics* **109** 103108
[15] Scajev P et al. 2012 *Journal of Applied Physics* **111** 023702

Laser cutting of GaN/Al_2O_3 structures

O S Medvedev[1], O Khait[2], S Yu Kurin[3], A S Usikov[4,5], B P Papchenko[4], H Helava[5], and Yu N Makarov[3,5]

[1]St. Petersburg State University, Universitetskaya nab. 7-9, St. Petersburg 199034, Russia
[2]Multitech Ltd., Rizhsky pr. 26, St. Petersburg 190103, Russia
[3]Nitride Crystals Group Ltd., pr. Engel'sa 27, St. Petersburg 194156, Russia
[4]University ITMO, Kronverkskiy pr. 49, St. Petersburg 197101, Russia
[5]Nitride Crystals Inc., 181 E Industry Court, Suite B, Deer Park, NY 11729, USA

E-mail: o.s.medvedev@spbu.ru

Abstract. We report on intra-volume laser scribing process developed for GaN layers grown on single-side and double-side c-plane sapphire structures by hydride vapor phase epitaxy. Dies of 10 mm × 10 mm and 15 mm × 15 mm were cut from 2-inch wafers using a 1.064-μm Nd:YAG pulsed laser. The impact of laser cutting on the mechanical properties of GaN/Al_2O_3 structures is discussed.

1. Introduction

Gallium nitride (GaN) and related III-N materials are widely used for applications in optoelectronic, high power and microwave devices. A post-growth processing is an essential part to manufacture semiconductor devices. One of the main processes at a final step is dicing of GaN-based epitaxial structures on a common substrate into small-size individual chips while keeping crack-free, clean, smooth chips surface. When processing the sapphire substrates usually used to grow III-N LED structures, the chip separation using a diamond scriber and breaking is near common standard. However, there is a strong demand for improved throughput and yield ratios of the processing using a laser micro machining technology. The laser processing has rapidly become more prevalent because it reduces chance of damage to process material due to shock or handling, reduces waste handling costs, provides one-step alternative to chemical etching process, allows removal of one type of material without damage to the underlayers.

Laser micro machining technology implies a variety of processes including hole drilling, ablation, milling, and cutting (scribing, dicing). Lasers with various parameters of emitted light (pulse energy, shape and width, wavelength) combined with precise positioning (stepper) system are used for this technique. Typically a UV laser is used for dicing or ablation (laser lift-off) of GaN based structures grown on sapphire (Al_2O_3) substrates having optically polished back side (double-side polished (DSP) sapphire) [1-4]. However single-side polished sapphire (SSP) substrates, that are cheaper than DSP ones, are also widely used in the III-N growth technology.

In this work we apply a 1.064-μm Nd:YAG laser to cut chips of 1-3 cm^2 from 2-inch GaN/sapphire wafer. These large-size chips (or plates) were used to fabricate GaN working electrodes for water splitting experiments [5]. We report on intra-volume laser scribing process developed for GaN/sapphire structures grown on SSP or DSP sapphire substrates.

2. Experimental

GaN layers of a 2-10 μm thick were grown by chloride hydride vapor phase epitaxy (HVPE) on cplane 2-inch SSP and DSP sapphire substrates. Thickness of SSP and DSP substrates was 430 μm and 330 μm, respectively. The growth procedure included in-situ sapphire substrate treatment and an AlGaN buffer layer deposition followed by GaN layers growth. Details of the structure growth and characterization can be found elsewhere [6].

The precise laser machining system used in this work is schematically given in Figure 1. It consists of diode-pumped solid state pulsed Nd:YAG laser emitting in near-infrared spectral range (1064 nm) with repetition rate up to 10 kHz and average power up to 1 W, a stepper system to position precisely a substrate under the laser beam (not shown in Figure 1), a dichroic mirror, a recording unit comprising a microscope with a camera. The stepper has two horizontal orthogonal coordinates with 2.5 μm precision, one vertical coordinate with positioning precision 1 mm and rotation stage for orientation of a sample in horizontal plane. The recording unit is connected with a computer to monitor the material processing with 2-μm resolution in real time.

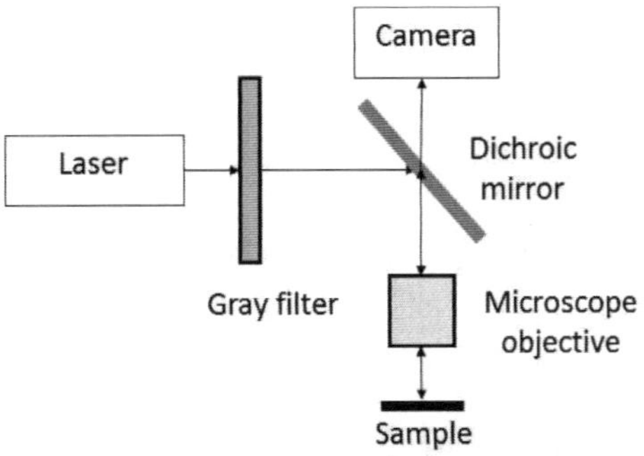

Figure 1. Schematic of the precise laser machining system.

The laser machining system may implement a traditional method of the sample surface ablation or a method of intra-volume laser scribing depending on the laser radiation parameters (wavelength, pulse duration, optical power density in the pulse, duty cycle, and beam divergence). The system has also a capability to process a wafer from both front and back sides adjusting a quality of the dice edges.

In this work we applied intra-volume laser scribing to fabricate 10 mm × 10mm or 15 mm × 15 mm chips from 2-inch GaN/sapphire wafer. This scribing technique is associated with a plasma breakdown in transparent dielectric media under irradiation by a short pulse of high power density laser beam [7]. Under a very strong electric field of the laser wave, which is of hundreds of MV/cm, valence electrons are accelerated to energy sufficient for impact ionization of host atoms in the crystal lattice. As a consequence of the atoms ionization, electron avalanche is quickly developed reaching a critical density. Breakdown in transparent media is connected with rapid increasing of electron density to a critical level needed for strong absorption of the laser radiation by emergent plasma. As a result, the media is collapsed in local areas under blast waves in the plasma that results in the rupture of the material lattice leading to cracks and cavities formation.

GaN layers grown on a 350 μm thick DSP sapphire substrates were processed from the back side of the substrate to fabricate 10 mm × 10 mm and 15 mm × 15 mm chips. Short pulses of the focused laser beam with adjusted duration and power density created micro-cavities and micro-cracks inside the sapphire substrates. A spacial set of these cavities fabricated in the substrate promoted to cut (to break) the wafer into chips.

The same approach was used to process GaN layers grown on a 430 μm thick SSP sapphire substrates but the laser beam was focused into sapphire substrate through the GaN layer. Special attention was paid to avoid destroying of the GaN layer by adjusting the laser beam spot. The fabricated chips were studied by means if scanning electron microscopy (SEM).

3. Results and discussion

GaN layers grown on a 350 μm thick DSP sapphire substrates were processed by laser scribing from the back side of the substrate. Figure 2 shows cross section of GaN layer grown on DSP sapphire substrate. During the scribing process, the laser beam was firstly focused in a point positioned at distance of 310 μm from the substrate back side. The short impulse of the laser irradiation (5 ns, 2 mJ) creates a carrot-like or ellipsoid-like cavity (~15 μm (height) × ~6 μm (width)) in the focused point of the substrate. The cavity pulls out across the substrate. After that, the machining system stepper shifts the wafer in other position and new cavity creates. By this way a row of cavities can be formed as shown in Figure 2a.

An average distance between the cavities is 20 μm. Each cavity is surrounding by micro cracks that form a destroyed area between the cavities shown in Figure 2b. The micro cracks are spread in-plane the substrates (along the cavities row). To form another row of cavities, the laser beam was focused in a point positioned at a distance of 210 μm from the substrate back side. Two rows of cavity with destroyed area between them are clearly seen in Figure 2a. These two rows of cavities formed in two orthogonal directions in the GaN/sapphire wafer found to be sufficient to break the wafer easily into rectangular dies of 10 mm × 10 mm and 15 mm × 15 mm.

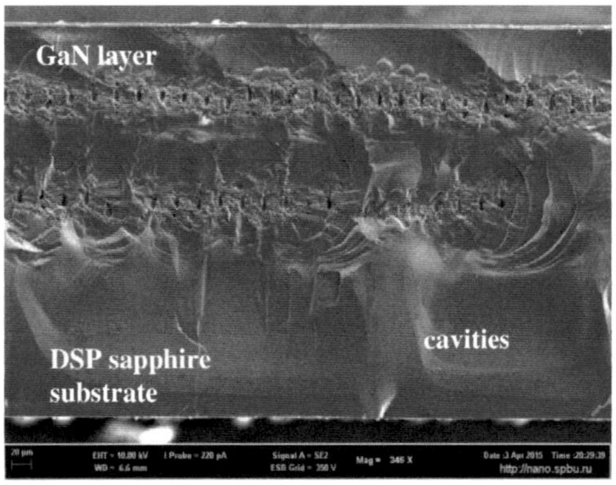

(a)

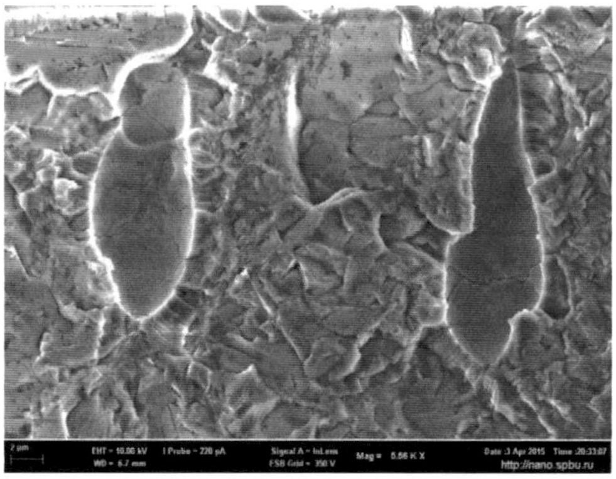

(b)

Figure 2. Scanning micrographs of cross section view of a 3.6 μm thick GaN layer grown on DSP sapphire substrate after the laser scribing from the backside of the substrate (a) and cavities under high magnification (b). In-plane cracks surround the cavities. A spacial set of cavities resulted in a net of cracks that promote to break the structure after the laser scribing.

Damaged area of dice edge found to be a 25 μm wide without any ablation. The GaN layer surface pollution (contamination) by products of the scribing process was not observed. Cracks in the GaN layer attributed to the scribing process were not observed too.

The main advantage of the intra-volume laser scribing is the process requiers low-power laser. Depending on the laser impulse duration and the laser beam focusing the radiation power density can exceed the threshold level of the rupture of the material. It gives selectivity of the scribing process and avoids destroying of transparent layers of the structure by laser radiation passing through.

GaN layers grown on a 430 μm thick SSP sapphire substrates were processed by laser scribing through the GaN layer. Figure 3 shows cross section of GaN layer grown on SSP sapphire substrate. Three cavity rows were formed by the intra-volume laser scribing process on a distance of 100 μm, 280 μm, and 370 μm from the layer surface. The cavity dimensions are 10 μm (height) × 3 μm (width). The SSP sapphire substrate is thicker than SSP substrate. Our experiments had shown that three rows of cavities formed in two orthogonal directions in the GaN/SSP sapphire wafer are sufficient to break easily the wafer into rectangular dies of 15 mm × 15 mm.

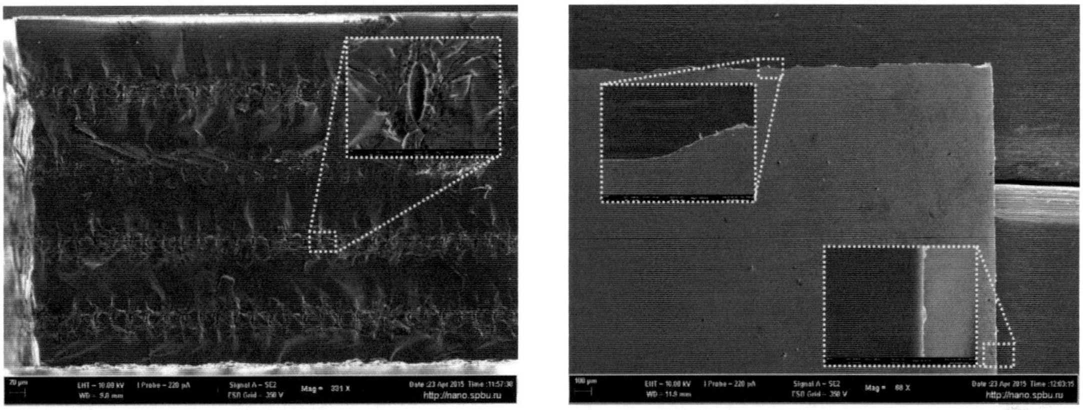

(a) (b)

Figure 3. Scanning micrographs of cross section view of a 3.6 μm thick GaN layer grown on SSP sapphire substrate after laser scribing through the GaN layer (a, the insert shows the ellipsoid-like cavity at higher magnification) and a plain- view of the same die (b, inserts show specific of the die edges at higher magnification.) The GaN layer appears as a thin light area on the top of the Figure 2a.

The plain-view of the die edges shown in Figure 3b. Two orthogonal edges have different sharpness that can be associated with two crystallographic directions (axes) on the c-plane sapphire surface: along *a*-axis and along *m*-axis. Typically 2-inch sapphire has a flat along *a*-axis. The laser cribbing in our experiments was performed in a direction parallel and perpendicular to the flat. It looks that rougher edge of the die in Figure 2b having fluctuation up to 25 μm is along the *m*-axis of sapphire. The other die edge has the sharpness of 2-3 μm. The similar results were observed for dies cutting from GaN/DSP sapphire wafers. More experiments are required to optimize the die edge sharpness.

4. Conclusions

An intra-volume laser scribing process with 1,064-μm Nd:YAG laser was applied for dicing of 3-10 μm thick GaN layers grown by HVPE on SSP and DSP c-plane sapphire substrates. The laser scribing process is associated with a plasma breakdown in transparent dielectric media under irradiation by a short pulse of high power density laser beam.

GaN layers grown on a 430 μm thick SSP sapphire substrates were processed from the front side (from the GaN layer surface) to fabricate 15 mm × 15 mm dies. GaN layers grown on a 350 μm thick DSP sapphire substrates were processed from the back side of the substrate to fabricate 10 mm × 10 mm and 15 mm × 15 mm chips. Surface damaged area of a dice edge found to be a 25 μm wide without any ablation in both set of dies. A surface pollution (contamination) of the GaN layers by products of the scribing process was not observed. Cracks in the GaN layers attributed to the scribing process were not observed also.

Acknowledgments

Work at University ITMO was supported by the Ministry of Education and Science of Russian Federation (grant agreement 14.575.21.0054, unique identifier of research activities is RFMEFI57514X0054). SEM images were obtained using equipment of Interdisciplinary Resource Centre for Nanotechnology of St. Petersburg State University.

Journal of Physics: Conference Series **643** (2015) 012129 doi:10.1088/1742-6596/643/1/012129

References

[1] Gradoboev Y G, Kazaryan M A, Mokrushin Y M and Shakin O V 2012 *AIP Conf. Proc.* **1464** 721
[2] Savriama G, Jarry V, Barreau L, Boulmer-Leborgne C and Semmar N 2012 *AIP Conf. Proc.* **1464** 169
[3] Tamhankar A and Patel R 2011 *J. Laser Appl.* **23** 032001
[4] Dai Y, Xu G, Cui J and Bai F 2010 *Proc. SPIE* **7590** 75900O – 8
[5] Ohkawa K, Ohara W, Uchida D and Deura M 2013 *Jpn. J. Appl. Phys.* **52** 2–4
[6] Kurin S, Antipov A, Barash I, Roenkov A et al. 2014 *Phys. Status Solidi* C **11** 813–816 [7] As-Chun Tien et al. 1999 *Phys. Rev. Letters* **82** 3883–3886

Journal of Physics: Conference Series **643** (2015) 012130 doi:10.1088/1742-6596/643/1/012130

Force sensitive resistor on the base of metal-carbon structures

A Tikhomirov[1] and Y Izotov[2] and N Sobolev[3]
[1] Petrozavodsk State University, Department of Physics- Engineering, Researcher, Lenin Avenue 33, Russia.
[2,3] Petrozavodsk State University, Department of Physics- Engineering, Undergraduate student, Lenin Avenue 33, Russia.

E-mail: izotov93@yandex.ru

Abstract. The aim of the work was to explore the opportunity of creation of force sensitive resistor on the base of metal-carbon structures in size hundreds of nanometers.

1. Introduction

Methods of obtainment of nanostructures on the basis of electric discharge have significant perspectives [1]. The electric discharge in this application has some advantages such as energy efficiency, quality of the process [2, 3].

Transfer to the submicrosecond range of pulse in the electropulse discharge has a significant impact on the size of obtained powder structures [4] and results in their decrease.

The use of non-conducting liquid in the electropulse discharge has a great importance for extracting the structures obtained in it [5].

Force sensors are a very perspective direction at present. Researches of the sizes of nanoparticles metal-carbon for ensuring the best tensor effect were conducted. The aim of this work is epy elaboration of a sensor with the acceptable properties for the use measurement of power loading.

2. Experiment

Metal-carbon structures were obtained on basis of electropulse discharge in liquid. Solid switch has been applied in this work to get short high-voltage pulses. Presently the area of IGBT-devices is developing very fast [6] and characteristics of abovementioned semiconductor devices allow to bring together transistors consequentially and parallel in the integrated.

The experimental set-up consists of the pulse generator and the power module. The basis of pulse generator is a program microcontroller pic16f877 of Microchip with the control's clock speed of 20 MHz. The work frequency of generator is 1 kHz with the duty of pulse about 2 us. The power module is shown in figure 1.

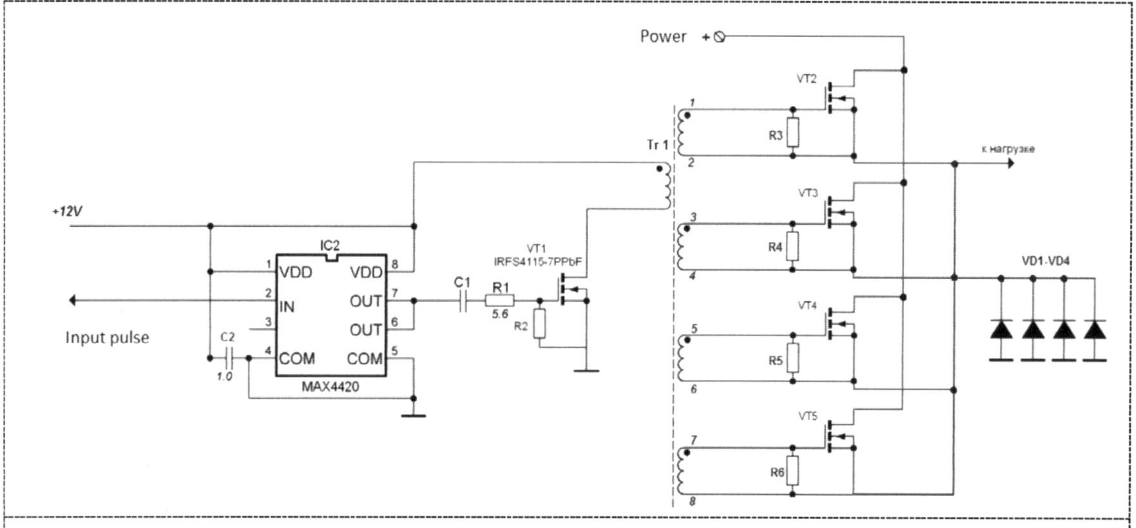

Figure 1. Power module

The signal from microcontroller goes to six ampere fast driver MAX 4420 (time of front is 25 ns on the load 2500 pF).

Taking into account low-powered outputs of microcontroller which are not able to charge gate capacity of powerful transistor VT1 (IRFS4115-7PPbF), controlling the gates of power IGBT-switch VT2-VT5. Considering the fact that to ignite the electro-erosion discharge it is not required to use high voltage, but necessary to provide discharge current the IGBT-switch is connected parallel.

The capacitor C1 is used to remove the constant component, resistor R1 – to limit current of gate charging VT1 (maximum of output current is six amperes), R2 – to discharge gate after ending electro-pulse.

Transformer Tr.1 is used to the galvanic separation between pulse generator and power module. Primary and secondary transformer coils Tr1 contain five turns each. Resistors R3-R6 are used for gate discharge when transistors close.

3. Results and Discussion

The electropulse method as aforesaid was used to obtain metal-carbon powder. Samples of molybdenum-carbides powder were obtained in erosion electropulse discharge. It find out that electric resistance of samples are depend from the time of electropulse processing. More after, electric resistance of samples is depending from mechanic force.

The samples were investigated by using scanning electronic microscope Hitachi SU1510 (Figure 2).

Figure 2. Metal-carbon structures

On the basis of samples of molybdenum-carbon powders the force sensors are created. In figure 3 the structural scheme of a force sensor is presented.

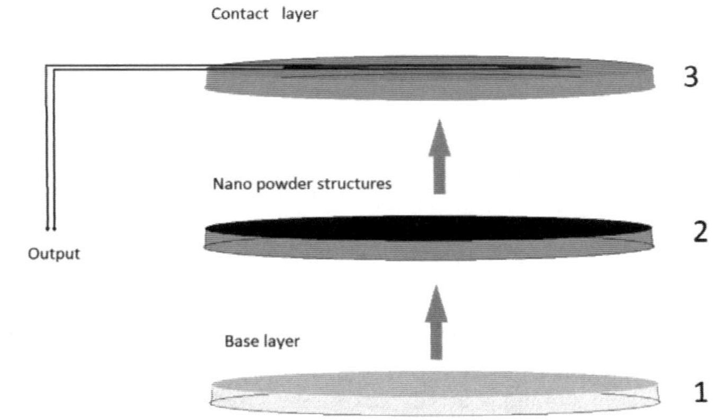

Figure 3. Structural scheme of a sensor

Where (1) is the lower layer deposited molybdenum-carbon structures on top (2) which contacts the of output (3).

The average size of obtained structures is about 100 um. In the result of the conducted research the dependence of electric resistant on the value of the external force has been found out.

4. Force Curve

For interpretational convenience, the data is plotted on logarithmic scales. This force is influenced by the substrate and overlay thickness and flexibility, size and shape of the actuator, and spacer-adhesive thickness (the size of the internal air gap between membranes). A typical resistance vs. force curve is shown in figure 4.

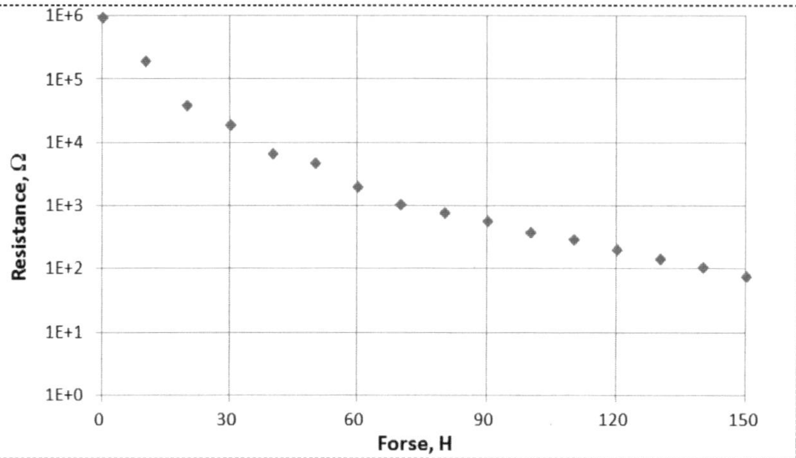

Figure 4. A typical sensor resistance vs. force

After switching on, the resistance will decrease close to exponentially dependence from force. The dependence of the electric resistance repeats again with each new cycle of applying the force.

5. Circuit FSR Voltage Divider

To convert force into voltage, force sensitive resistor sensor is attached to the measuring impedance (Figure 5) and the output voltage is described by equation:

$$V_{out} = \frac{V_{+} \times R}{R + R_{FSR}}$$

If R_{FSR} and R are swapped, the output swing will increase with increasing force.

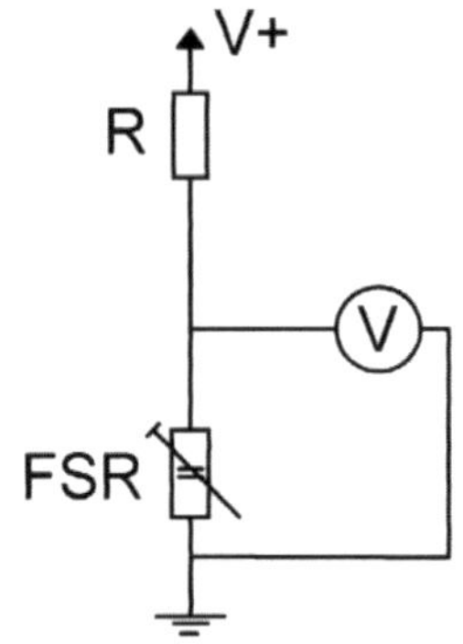

Figure 5. Circuit and equation voltage divider

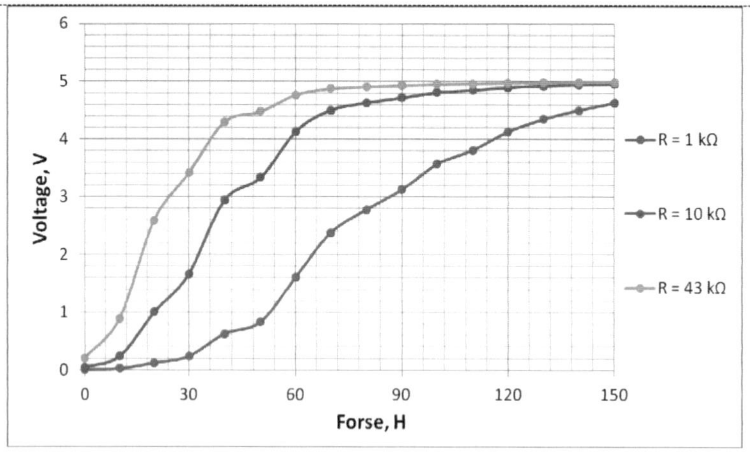

Figure 6. Force-to-voltage conversion in FSR

The measuring resistor, R, is chosen to maximize the desired force sensitivity range and to limit current.

In that scheme, the output voltage will decreases with increasing force (Figure 6). In figure 6 dependence of three different obtained sensors are shown.

6. Conclusion

The structure of molybdenum-carbon powders is homogenous enough, at the same time the size of carbon structures significantly exceeds the size of molybdenum structures. Molybdenum particles have a spherical form and a size of about some hundreds nanometers. The force sensitive resistor has been obtained.

The work is done by the support of Program of strategic development of Petrozavodsk State University from 2012 to 2016, Ministry of Education of Russian Federation, contract № 3.757.2014/K.

References:

[1] L. V. Kogitov, I.V. Zaporotskova, V.V. Kozlov Perspecive nanomaterial on the basis of carbon, Science journal of Volgograd state university. 10 series № 4 – 2009, p. 63-85

[2] Abu Zaid OA (1997) On the effect of electro-discharge machining parameters on the fatigue life of AISI D6 tool steel. J Mater Process Technol 68 (1): 27-32

[3] Ho KH, Newman ST (2003) State of the art electrical discharge machining (EDM). Int J Mach Tool Manufac 43:1287–1300

[4] Ekmekci B, Elkoca O, Erden A (2005) A comparative study on the surface integrity of plastic mould steel due to electric discharge machining. Metall Mater Trans 36B:117–124

[5] Puertas I, Luis CJ (2003) A study on machining parameters opti- mization of electrical discharge machining. J Mater Process Technol 143–144:521–526

[6] Krishna Mohana Rao G, Satyanarayana S, Praveen M (2008) Influ- ence of machining parameters on electric discharge machining of Maraging steels—an experimental investigation. Proceedings of the World Congress on Engineering. Vol. II, London, UK